Selection

Selection
The Mechanism of Evolution

Second Edition

BY

Graham Bell

OXFORD
UNIVERSITY PRESS

Great Clarendon Street, Oxford OX2 6DP

Oxford University Press is a department of the University of Oxford.
It furthers the University's objective of excellence in research, scholarship,
and education by publishing worldwide in

Oxford New York

Auckland Cape Town Dar es Salaam Hong Kong Karachi
Kuala Lumpur Madrid Melbourne Mexico City Nairobi
New Delhi Shanghai Taipei Toronto

With offices in

Argentina Austria Brazil Chile Czech Republic France Greece
Guatemala Hungary Italy Japan Poland Portugal Singapore
South Korea Switzerland Thailand Turkey Ukraine Vietnam

Oxford is a registered trade mark of Oxford University Press
in the UK and in certain other countries

Published in the United States
by Oxford University Press Inc., New York

This edition first published in paperback 2008
First edition © Chapman Hall 1997

British Library Cataloguing in Publication Data
Data available

Library of Congress Cataloging in Publication Data
Data available

Typeset by Newgen Imaging Systems (P) Ltd., Chennai, India
Printed in Great Britain
on acid-free paper by
Antony Rowe, Chippenham, Wiltshire

ISBN 978–0–19–856972–5 (Hbk.) 978–0–19–856973–2 (Pbk.)

10 9 8 7 6 5 4 3 2 1

For Susan

Preface to the second edition

The advances made by experimental evolution in the last decade provided the main reason for revising *Selection*. In the course of revising it, however, the text changed so much that what has emerged is for the most part a new book. The first edition was written from a conviction that selection is a subject unto itself, the basis of a distinct kind of science that requires a separate treatment. To make it accessible to undergraduates, it was organized into sections with declarative headings, written as continuous prose with all the academic furniture relegated to boxes, eschewed mathematical and quantitative arguments, and was twinned with an undergraduate version. The conviction remains, but the format has been abandoned. The second edition is organized in a more conventional manner, gives a much more quantitative treatment, and even makes modest use of 'the awesome power of population genetics' (B. Charlesworth), at least where this yields simple rules and verifiable predictions. I hope that it will provide a useful guide to what we know about the mechanism of evolution.

Acknowledgements

I wrote most of the text while on sabbatical leave, and thank McGill University for giving me the opportunity to renew my scholarship in the field. I owe the same debt to my hosts at Université du Québec à Montréal, and in particular to Jean-Francois Giroux and Luc-Alain Giraldeau for generously providing me with space and facilities. My main intellectual debt is to my students and colleagues, whose work over the last decade is the basis for much of this book: Rowan Barrett, Austin Burt, Nick Colegrave, Sinéad Collins, Shaun Goho, Oliver Kaltz, Rees Kassen, Vassiliki Koufopanou, Craig MacLean, Gabriel Perron, Sébastian Renaut, Taica Replansky, Kathy Tallon, Gabriel Yedid and Clifford Zeyl. Austin Burt read through the entire manuscript in proof and made many valuable susggestions.

For many are called, but few are chosen. Matthew 22:14

Contents

The Second Science

The ancient dream of philosophy is that all knowledge should be governed by a few general principles. This was abandoned in the Enlightenment, when general philosophy (dealing with questions that cannot be decisively answered) took one path, while natural philosophy (dealing with questions that can be answered decisively) took another. It was still possible to hope that all issues within natural philosophy could be brought under the same head and answered in terms of the same set of principles. This hope still flourishes today in some circles, but in fact expired in the nineteenth century with the advent of economics and evolutionary biology. The conventional science that everyone is familiar with from high school shows how the properties of physical things are determined by the shapes and velocities of atoms and molecules. It provides the mechanism for physics, chemistry, cosmology, geology, and large parts of biology, including most of physiology and genetics. It fails to provide any understanding of how living organisms become adapted to their conditions of life; this is the province of the second science, the science of evolution. Adaptation is just as mechanistic as particle physics or biochemistry, and its principles can be established and verified in the same way, by observation and experiment. Its mechanism is profoundly different from physicochemical mechanism, however, and nothing that applies in the one case has any validity or force in the other. It will often be very useful to understand physical principles in order to understand the context of evolutionary change; and it will often be very useful to understand evolutionary principles in order to understand any aspect of the physics or chemistry of organisms. Otherwise there is no conversation between these two disciplines: no knowledge of physical principles, no matter how profound or detailed, can lead to any understanding of evolution, and vice versa. There is a fence that separates fundamentally different scientific modes of inquiry that does not coincide with conventional disciplinary boundaries. On one side of this fence lie physics and chemistry, physiology and genetics, ecosystem ecology and neurobiology. On the other side lie evolutionary biology, much of community ecology and animal behaviour, economics and parts of history, whose distinctive feature is the operation of selection on variable populations, and which cannot be understood in terms of nor reduced to the principles of physics and chemistry. To some it may be a cause of regret that the natural world cannot be understood in terms of a single set of rules and laws, but at least, so far as we know, the whole of the natural world can be understood in terms of just two general systems, and no more. This book is an account of one of these, in the context of biological systems. It is an account of the Second Science.

CHAPTER 1

Simple selection

The main purpose of evolutionary biology is to provide a rational explanation for the extraordinarily complex and intricate organization of living things. To explain means to identify a mechanism that causes evolution, and to demonstrate the consequences of its operation. These consequences are then the general laws of evolution, of which any given system or organism is a particular outcome. The very complexity that stimulates investigation, however, for long obscured the possibility of a mechanical interpretation, and the richness of the subject matter continues to encourage a piecemeal and anecdotal description of nature, and to inhibit the development of a strong central research programme in experimental evolution. Nevertheless, to explain complexity it is first necessary to study simplicity; although the products of evolution are exceedingly complicated, the processes of evolution can be seen most clearly and manipulated most easily when these complications are stripped away. One way of asking, what is the most fundamental principle of evolution? is instead to ask, what is the simplest system capable of evolving?

We can imagine organisms that lack fins or photosynthesis or any of a thousand other attributes, but we cannot imagine organisms that are unable to reproduce, because they would lack both ancestors and descendants. The simplest type of organism would be one that reproduced, but did nothing else. There is, of course, no cellular organism of this kind, because cell growth and division are supported by multifarious processes of metabolism. Underlying the reproduction of cells and individuals, however, is the replication of the nucleic acid molecules that encode this metabolism and ensure its hereditary transmission. The simplest evolving system is therefore a self-replicating nucleic acid freed from its dependence on cellular machinery.

1.1 RNA viruses are the simplest self-replicators

Qβ is a virus that infects bacteria. It uses RNA, rather than DNA, as its genetic material. In cellular organisms, RNA is used only as messages to decipher the DNA code, and there is no machinery for replicating RNA. The Qβ RNA encodes a number of proteins, including one that specifically catalyses the replication of Qβ RNA—the Qβ replicase. The host cell provides the rest of the apparatus for producing the protein, and a supply of raw material from which new Qβ genomes can be constructed. The virus can thus be thought of as a small wormlike creature with an unusually simple morphology that is completely specified by the sequence of nucleotides in a single RNA molecule (Figure 1.1). Simple though it is, it can be simplified further. The viral genome encodes several kinds of protein that are used to transmit itself from one host cell to another, but not while replicating within the host cell. Once inside the cell, all that is needed is the replicase, together with a supply of nucleotides got from the host. Because the replicase can be isolated and purified from infected bacteria, these simple requirements can be provided in a culture tube. A solution of replicase and nucleotides provides a chemically defined environment in which Qβ RNA will replicate itself as though it were inside a bacterial cell.

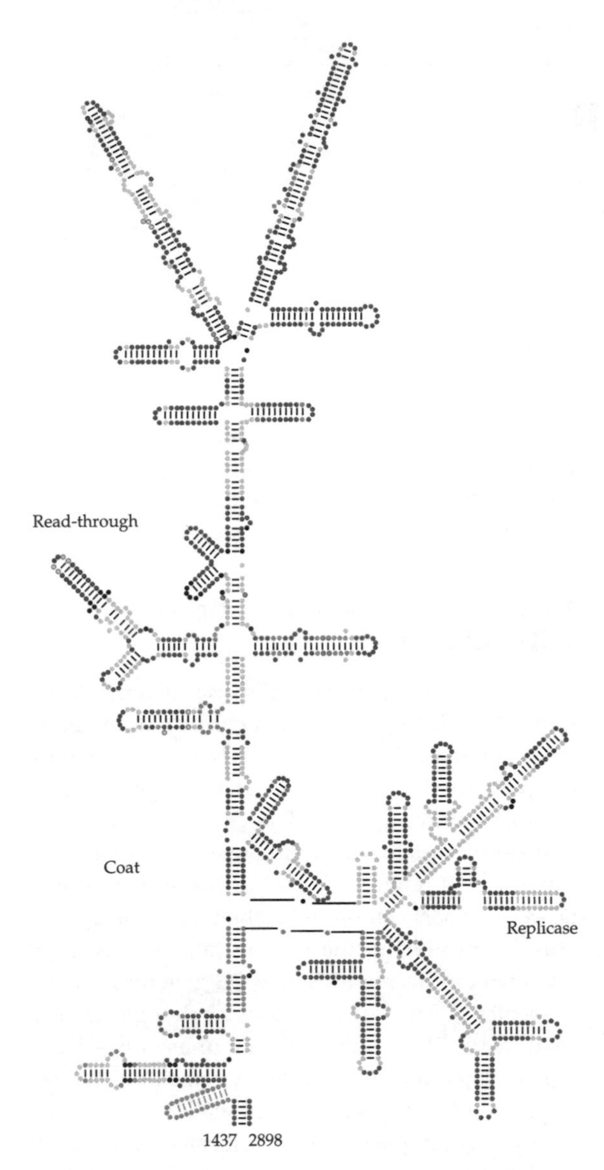

Figure 1.1 RNA virus structure. The coliphage Qβ comprises just four genes: two encode proteins responsible for attachment, penetration, and lysis, one is a coat protein, and the fourth is the replicase. The second maturation gene is expressed by occasionally reading through the stop codon delimiting the coat protein gene. Magnified, the organism consists of a complex bundle of loops formed from a single strand of RNA. Magnified still further, the RNA strand is defined by a linear sequence of the four nucleotides *A*denine, *U*racil, *C*ytosine and *G*uanine. This sequence completely determines the properties of the organism—including self-replication—in a particular environment, such as the interior of a bacterial cell. The four genes of the phage are embodied by a sequence of somewhat more than 4000 nucleotides. This is a proposed secondary structure for phage Qβ. From Zuker (2003), Figure 47.

1.2 Exponential growth can be maintained by serial transfer

When the culture medium is inoculated with a small population of Qβ, each RNA molecule immediately proceeds to produce copies of itself; the copies themselves produce copies, and the

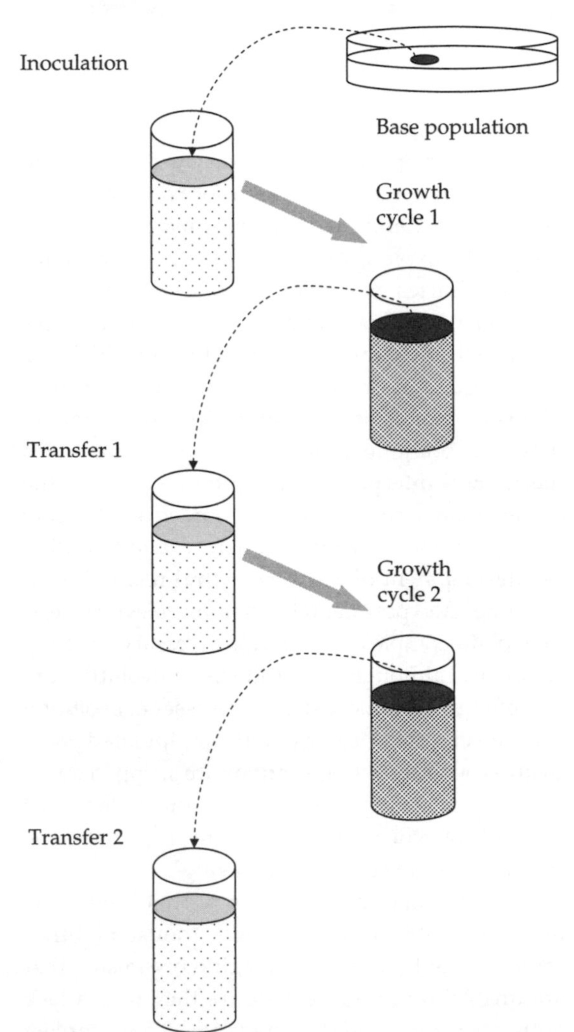

Figure 1.2 Serial transfer. To grow an organism in batch culture, a number of individuals are inoculated into a volume of growth medium. They will continue to reproduce until the nutrients in the medium are exhausted and growth ceases. However, a sample of the population can be withdrawn at any time and transferred to a fresh batch of medium. So long as this process is repeated, the population will continue to reproduce. The rate of growth of the population is set by the frequency of transfer and the volume of inoculum.

copies of copies do likewise, until the population is hundreds or thousands of times as large as the original inoculum. However, this process cannot continue for very long. Because each copy of one of the original molecules is itself copied, the number of molecules increases exponentially. The finite supply of nucleotides, the building blocks of RNA molecules, soon begins to become exhausted, and after a time no further growth is possible. Growth can be renewed only when a fresh supply of raw materials is provided. This is most easily done by using a small amount of the grown culture to inoculate a fresh culture tube. When this too has become depleted, a small amount is withdrawn and used to inoculate a third tube, and so on, until it is the patience of the investigator that has become exhausted. This is the technique of *serial transfer* (Figure 1.2). It is a simple and effective way of keeping a population in a state of perpetual growth.

1.3 Replication is always imprecise

As long as any raw materials remain, each RNA molecule produces copies of itself, through the mediation of the replicase. However, the replicase is not a perfect machine, and many of these copies will not be perfect simulacra of the parent molecule. In some cases, the wrong nucleotide will have been added at one or more positions along the growing RNA molecule; in others, extra nucleotides may have been inserted, or resident nucleotides deleted—more drastically, severely abnormal molecules may be produced by the failure to copy large parts of the RNA sequence. Errors of this sort are quite often made during RNA replication, because the replicase has no ability to review its work and correct any errors that it has made. It has no ability to engineer molecules, either: errors are made at random, or, rather, without respect to their present or future utility. It is therefore not possible to create a large clone of RNA molecules, all of which are exactly alike; instead, a large population descending from a single founding molecule will consist of numerous variations on a theme, each of which is a sequence that differs to a greater or lesser extent from the original sequence (Figure 1.3). Populations of self-replicating RNA molecules are markedly diverse, because of the low fidelity of the RNA replicase, but the same principle applies to the replication of DNA, although this is a more rigorously controlled process, and therefore to clones of DNA-based organisms. Indeed, it is a very general principle that applies to any copying process. Information cannot be transmitted without loss: therefore, no message can be copied perfectly with certainty. Self-replication is always to some degree imprecise, and variation is a property of self-replicating systems that does not in itself require any special explanation.

Each variant sequence is itself copied, although of course imprecise copies are themselves copied imprecisely. A growing population of $Q\beta$ does not, therefore, consist merely of a diversity of sequences; it consists of a diversity of lineages, each propagating the altered sequence of its founder.

1.4 Imprecise replication leads to differential growth

The technique of serial transfer is based on a very simple ecological principle: a population that grows exponentially rapidly exhausts the resources that support its growth. In diverse populations, this principle has an evolutionary corollary. Each variant lineage will tend to increase exponentially (Figure 1.4), but different lineages may increase at different rates. After any given interval of time, some lineages will have become more abundant than others, even if all were initially equal in numbers. Lineages that are increasing at different rates will therefore diverge in frequency, some becoming more common and others more rare. This can occur very quickly, because different exponential processes diverge exponentially. The diversity of the growing population is thus not merely a passive consequence of the infidelity of self-replication, but also reflects variation in the rates of self-replication among the variant lineages. The composition of the population will change through time as more rapidly replicating lineages become increasingly prevalent.

1.5 Selection acts directly on rates of replication

The serial transfer of self-replicating RNA molecules was the basis of one of the most remarkable

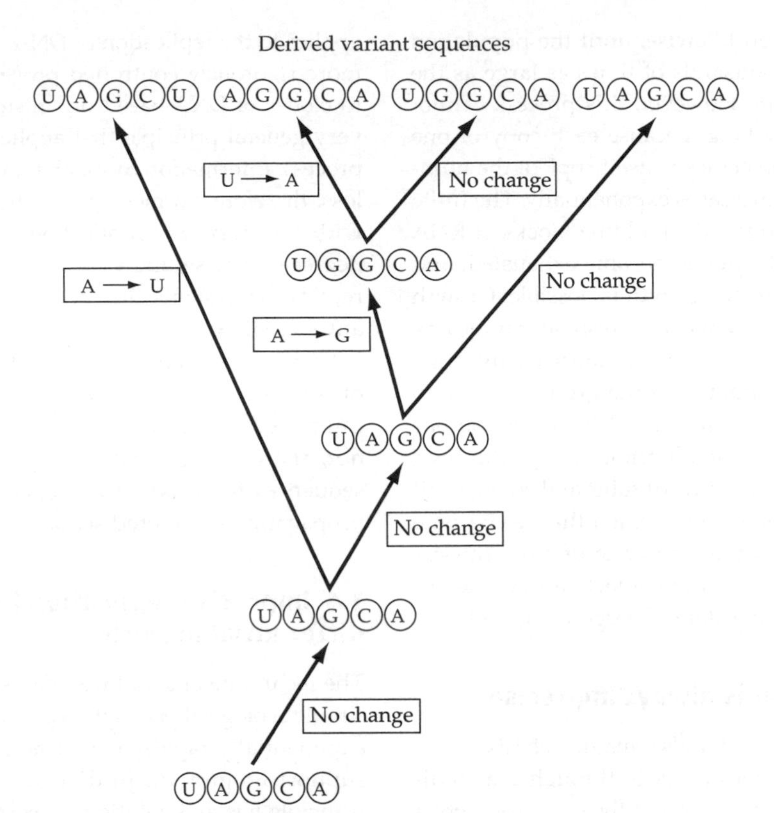

Figure 1.3 Diversification through imprecise replication. The relationship between ancestors and descendants can be represented as a family tree, also called a tree of descent or a phylogenetic tree. This diagram is the most direct representation of descent, as the copying of a sequence of nucleotides. The lines connect sequences that are copies, the arrow pointing from the ancestral to the descendant sequence. A line is thus a succession of copies. Any line, or any set of lines, descending from the same ancestral sequence is a lineage. When a sequence is copied incorrectly, the altered or mutated sequence is the ancestor of a distinct lineage. However, a descendant copy of this altered sequence may itself be copied incorrectly, giving rise to a variant that differs in two positions from the original ancestral sequence. This alteration of alterations is responsible for the branching, treelike ramification of lineages through time.

experiments in modern biology. It was performed by a group of scientists in Sam Spiegelmann's laboratory at the University of Illinois in 1967. They designed their experiment around the following question:

> What will happen to the RNA molecules if the only demand made on them is the Biblical injunction, *Multiply*, with the biological proviso that they do so as rapidly as possible? (Spiegelman 1971)

The elementary considerations that I have outlined are enough to provide an answer. The initial inoculum of viral RNA encounters a new and strange world, an unusually benign world in which it does not have to deal with the usual problems of parasitizing complex and hostile bacterial cells. Being provided with replicase and nucleotides, it begins to increase exponentially in numbers. This increase is soon checked by the finite supply of these resources. The tendency to increase in numbers while resources are in short supply creates competition, because not all can prosper: there will arise, as one might put it, a struggle for existence. But the growing population is necessarily diverse, and not all variants will be equally able to replicate themselves. Those that replicate more rapidly will increase in frequency, replacing their competitors: this process can be referred to as the selection of the more rapidly replicating types.

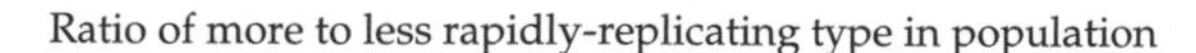

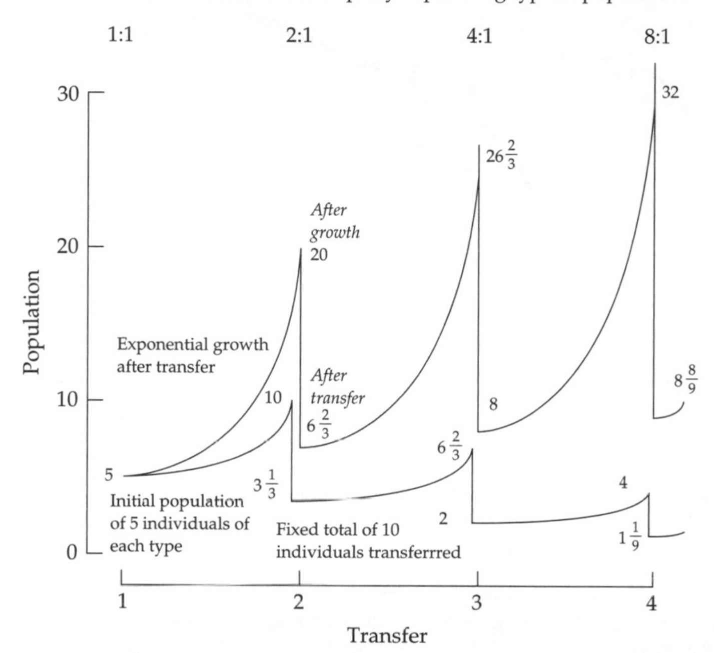

Figure 1.4 Exponential processes diverge exponentially. A population of 10 individuals initially comprises two equally frequent types, one of which produces twice as many offspring as the other. After three cycles of reproduction and density regulation, the more fecund type is $2^3 = 8$ times as frequent as the other.

Because each type tends to reproduce itself, selection will involve the replacement of some lineages by others, or in other words will cause a permanent change in the genetic composition of the population, which constitutes modification through descent, or evolution. The experiment will result in the evolution of RNA molecules that are better adapted to the novel conditions of growth furnished by culture tubes: better able, that is to say, to replicate themselves at high rates in this novel environment.

This chain of reasoning was first forged by Darwin and Wallace more than 100 years before the Qβ experiments. However, it predicts the leading result of these experiments very clearly: the emergence and establishment of variants with much greater rates of self-replication . This happens quite quickly. After about 70 transfers, amounting to about 300 rounds of RNA replication, the population consists of types able to replicate in their new environment about 15 times as fast as the original virus (Figure 1.5). The early Qβ experiments were reviewed by Spiegelman (1971; Kacian *et al.* 1972), who also gives a wealth of background information about the system. There are subsequent reviews by Orgel (1979), Eigen (1983), and Biebricher (1983). Biebricher and Orgel (1973) reported similar experiments with the RNA polymerase of *E. coli.*

In the simplified world of the culture tube, many of the attributes that are essential to the success of Qβ as a parasite are no longer necessary. The virus no longer needs to be able to encode proteins for packaging its genome or for lysing the host cell, because the experimenter transfers it from tube to tube. It no longer even needs to encode a replicase, because that too is provided for it. All the elaborate metabolic machinery deployed by living organisms has been eliminated, and we are left with their essence alone, a naked replicator with unlimited opportunity to replicate itself. It is not really very surprising that the experiment succeeds in illustrating the Darwinian thesis. The more important point is not that the rate of self-replication is an attribute that can be selected, but rather that the rate of self-replication is the only attribute that can be selected directly. There is no natural tendency that favours strength or beauty or cunning, or any other quality, as a quality in its own right. There is a natural tendency that favours high rates of self-replication, because more efficient replicators will tend to increase in frequency without the intervention of

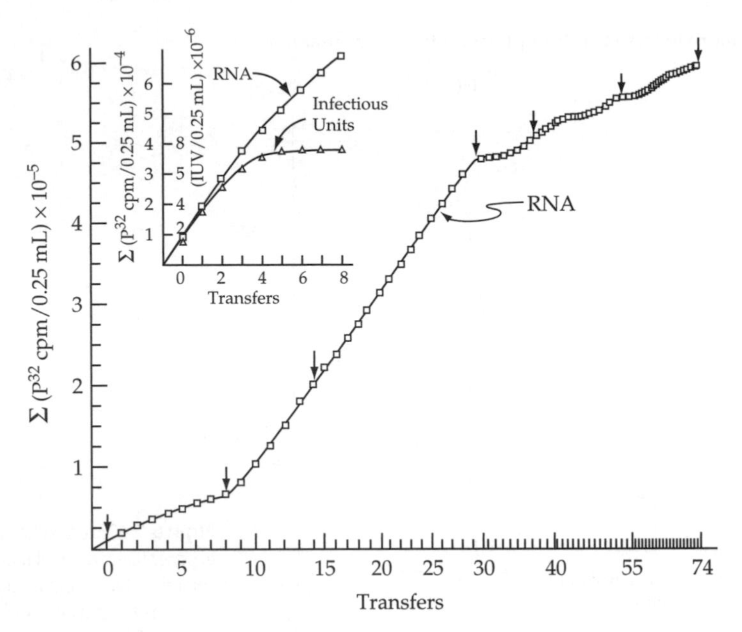

Figure 1.5 Evolution of Qβ through serial transfer. This is the original Qβ experiment, from Mills *et al.* (1967) by permission of PNAS. The y-axis is the cumulative incorporation of labelled radionucleotide into RNA and the x-axis the number of serial transfers. The cultures were initially allowed to grow for 20 minutes; this was reduced to 15 minutes after transfer 13, to 10 minutes after transfer 29, for 7 minutes after transfer 52, and thereafter for 5 minutes only. The changes in the length of the growth period, and the logarithmic scaling of the time axis, make it difficult to infer changes in the rate of increase of the population of RNA molecules, but there is obviously a sharp advance between transfers 8 and 9. The inset diagram shows that the descendants of Qβ had completely lost their ability to infect *E. coli* after only four or five transfers.

any external agency. So far as we know, this is the only natural tendency of this kind. The process of selection acts directly on rates of self-replication. It does not act directly on any other attribute or characteristic whatsoever. The direct response to selection is therefore always an increase in the rate of replication in given circumstances.

1.6 Selection may act indirectly on other characters

The ability to grow, or to grow more efficiently, in a given environment cannot exist in itself, without any external reference: it must be caused by morphological or physiological or behavioural attributes of some sort. These attributes can also be selected; indeed, we are usually far more concerned with them than we are with replication rate itself. They are, however, selected only indirectly; selected, that is, through their connection with altered rates of self-replication. Qβ has little in the way of morphology, physiology, and behaviour, especially when cultured in tubes; it is defined by a particular sequence of nucleotides, and it is changes in this sequence that cause changes in rates of replication, and that are selected indirectly as a consequence. One change that is easy to demonstrate, and whose effect on rates of replication is easy to appreciate, is that Qβ gets smaller. The intact virus with which the experiment is originally inoculated is a chain of 3300 nucleotides, encoding the proteins necessary for functioning as a transmissible intracellular parasite. After 70 transfers, the evolved variant, with its much greater rate of replication, was only 550 nucleotides in length. The reason is very simple: other things being equal, a smaller molecule will be replicated more rapidly than a larger one. In the benign environment of the culture tube, more than 80% of the viral genome is unnecessary, and variants which lack the unnecessary sequences are favoured by selection by virtue of their greater rates of replication. What eventually remains is not a random fragment—random pieces of Qβ RNA are unable to replicate—but rather a minimal sequence that supports efficient replication. The experimenters, however, made no attempt deliberately to select small molecules. Rather, they contrived a situation in which novel types with higher rates of replication would evolve, and small size evolved as a side effect of this procedure because of its correlation, in these circumstances, with greater rates of replication. All attributes of organisms evolve in this way, as an indirect consequence of selection for greater rates of increase. Their evolution thus represents an indirect or correlated response to selection.

1.7 The indirect response to selection is often antagonistic

In its natural state, Qβ cannot jettison 80% of its genome, because it would then be unable to infect bacteria. The intact virus is not very well adapted to growing in culture tubes; equally, the much shorter RNAs that evolve in culture tubes are not very well adapted to growing in bacteria. Indeed, the ability to infect bacteria is almost completely lost after 4 or 5 transfers (Figure 1.5). This is a type of correlated response: the loss of an ability that was formerly present, as the indirect consequence of selection for higher rates of increase in a novel environment. It can be distinguished as an antagonistic response, to recognize how selection that has enhanced performance in one environment has had the effect of reducing performance in another environment. Selection will cause progress, but it is a constrained and restricted sort of progress, with advance in one direction being associated with regress in others. Perfection is unattainable.

1.8 Evolution typically involves a sequence of alterations

After 70 or so transfers, Qβ has become fairly well adapted to the comfortable environment of the culture tube. What if we now make it uncomfortable? The difference between a bacterium and a culture tube is so great that it is difficult to interpret the evolutionary changes that take place, except in a very broad sense (the overall reduction in size, for example). Comfort is a general state of affairs. However, once the population has approached an evolutionary equilibrium, with further change taking place only very slowly, it can be made uncomfortable in some specific way, by introducing a defined source of stress. Any further evolution can then be attributed to that stress, and if the biochemistry and physiology of the stress and the response to it are known, then the evolutionary sequence of events that builds adaptation can be analysed in great detail. RNA molecules can be made uncomfortable in a number of ways: one is to add to the culture medium a small amount of ethidium bromide (EtBr), a substance that binds to RNA molecules and makes it difficult for them to replicate themselves. A very low concentration of EtBr is at first sufficient to inhibit RNA replication almost completely, but Saffhill *et al.* (1970) were able to evolve resistance in Qβ. The base population was the V-2 variant, produced by selecting V-1, the variant evolving in the original experiment by Mills *et al.* (1967), for growth from extremely dilute inocula. This was grown in batch culture by serial transfer in the presence of a low concentration (2 μg/mL) of EtBr until, after 8 or 9 transfers, it grew about as fast in these conditions as the original V-2 did in the absence of EtBr. The concentration was then increased to 4 μg/mL, and by repeating this procedure a population capable of growing at 50 μg/mL EtBr was eventually selected, after 108 transfers (Figure 1.6).

The physiological basis of this adaptation is straightforward: the evolved molecules bind EtBr less strongly. For this reason, their rate of replication exceeds that of the original type whenever the toxin is present in appreciable quantities. However, the reverse is also true: in the original culture medium, with no ethidium bromide, the resistant variant grows less rapidly than the original type. Kramer *et al.* (1974) selected for EtBr resistance in MDV-1, a 221-nucleotide molecule evolving from much shorter sequences. Very small inocula—to maximize the rate of selection—were grown in 15 μg/mL for 25 transfers. The evolved strain grew much better than MDV-1 when the concentration of EtBr exceeded about 2 μg/mL, but the original MDV-1 grows faster than the evolved strain when

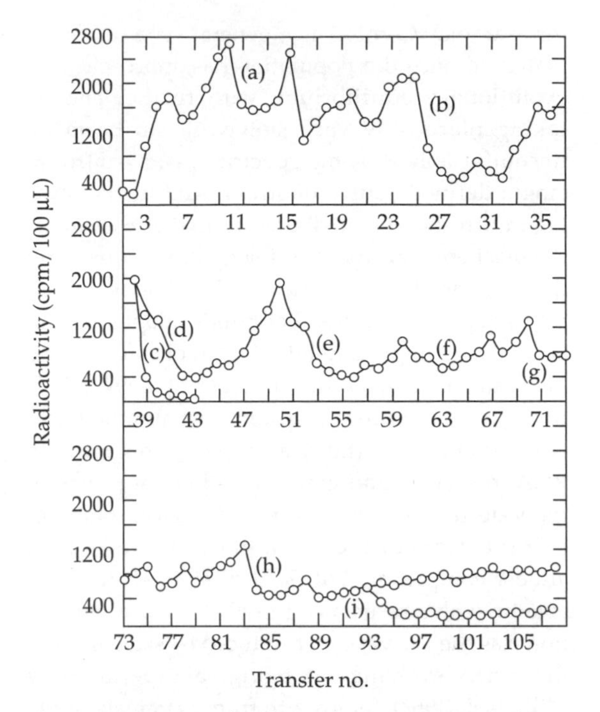

Figure 1.6 Experimental evolution of resistance to EtBr in phage Qβ. The y-axis is a rate of replication. Heavy lines and letters mark transfers when the concentration of ethidium bromide was increased. From Saffhill *et al.* (1970).

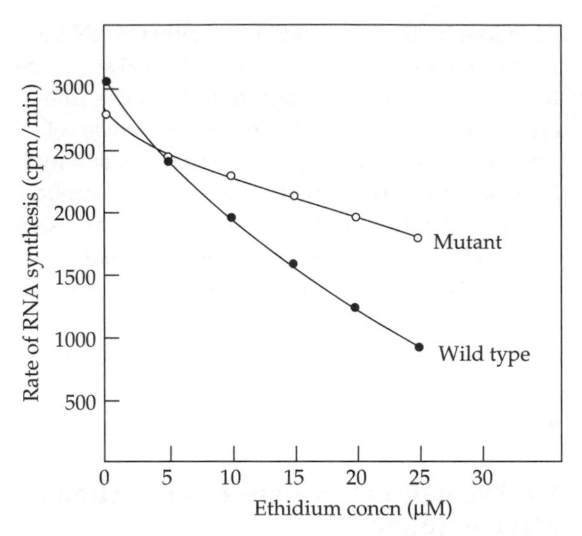

Figure 1.7 Cost of adaptation during evolution of EtBr resistance in phage Qβ. From Kramer *et al.* (1974), Figure 3.

there is no EtBr in the medium (Figure 1.7). This is another example of an antagonistic correlated response, showing how advance and regress may be coupled even when alternative environments differ only by a single defined chemical feature.

The simplicity of the system makes it possible to analyse the genetic basis of this physiological change. In one experiment, for example, resistance to EtBr was caused by altering three nucleotides at different positions along the RNA molecule (Figure 1.8). Note that the genetic change is rather slight; from a structural point of view, the molecule has hardly changed at all. The functional change, on the other hand, is pronounced. Indeed, it is as pronounced as it can possibly be—the functional difference is life or death. There is no simple scaling between genotype and phenotype: very different genotypes may behave in very similar ways, whilst a trivial genetic change may have the most profound phenotypic consequences.

From a physiological or genetic point of view, this is as deep as any analysis can be pursued: nucleotide sequence is the ultimate character, from which all other attributes are eventually derived. From an evolutionary point of view, however, the fundamental issue is the manner in which these alterations occur in time. There are two possibilities. The first is that the original population was so diverse that it contains all possible sequences, including the sequence that happens to be successful when EtBr is present in the growth medium. Evolution through selection is then simply a matter of sorting this initial variation, until the best-adapted sequence has replaced all others. The second possibility is that the original population includes only a small fraction, perhaps an extremely small fraction, of all possible sequences, so that the well-adapted sequence that eventually predominates evolves through the successive replacement of sequences by superior variants that have arisen during the course of the experiment. The evolution of resistance to EtBr occurs sequentially, with the three altered nucleotides being substituted one at a time, so that the well-adapted sequence that eventually evolves is built up in a step-like manner. It is easy to appreciate that this will almost always be the case. The number of possible combinations of

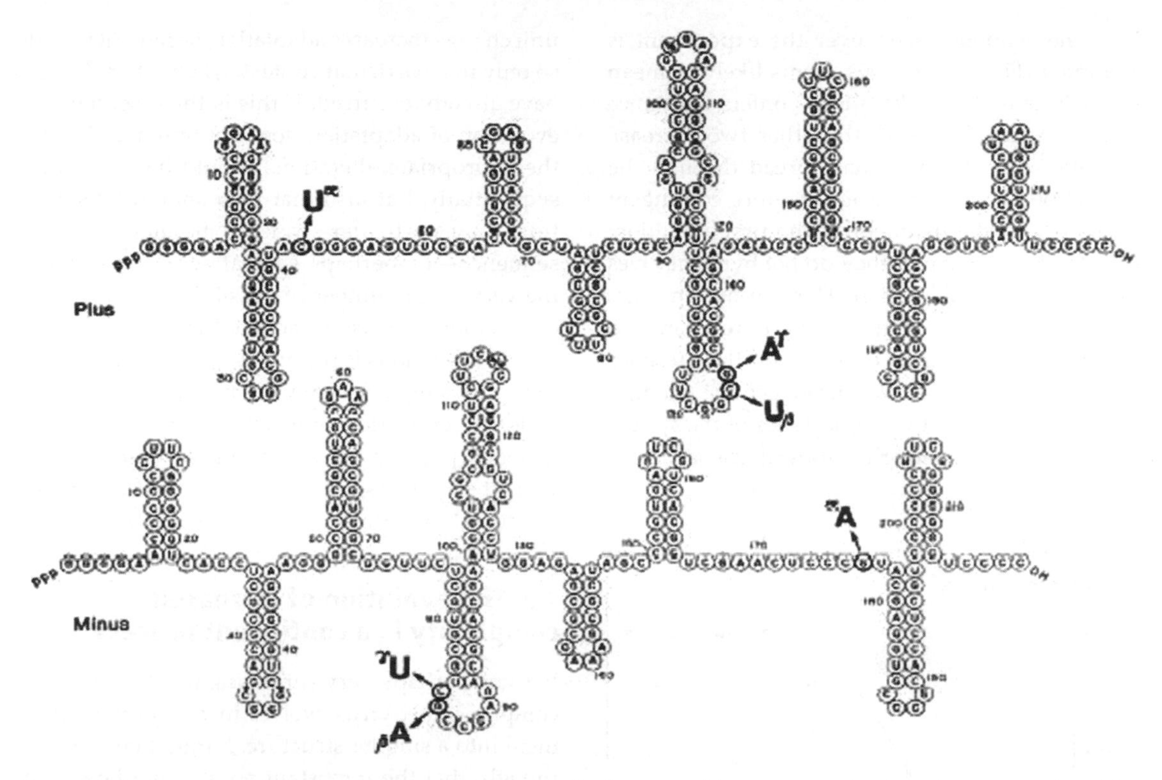

Figure 1.8 Resistance to ethidium bromide in phage Qβ is encoded by three nucleotide substitutions. The figure shows the position of the three alterations on the complementary strands. From Kramer *et al.* (1974), Figure 6.

nucleotides is so large that it is inconceivable that any but the tiniest fraction of them will be present in the original population. The evolution of any but the simplest modifications will generally involve the sequential substitution of several or many slight alterations, leading eventually to a state that was not originally present.

The variant with three alterations is resistant to EtBr. Whether or not any or all of the three variants with a single alteration show any resistance has not been ascertained directly, but at least one of them must do so: otherwise, sequential change would be impossible. Similarly, once this variant has become established in the population, at least one of the two variants with two alterations must show increased resistance; and the evolved variant that is the final outcome of selection must be more resistant than its predecessors. Resistance can evolve only if the initial, susceptible genotype is connected with the evolved, resistant genotype through an unbroken series of intermediate states, each of which confers an advantage. Whether or not any particular attribute can evolve depends on the connectance of the system; on whether or not there is a route leading to a genotype specifying that attribute, along which a process of sequential substitution can lead the population, step by step, each successive alteration representing an improvement. The 'steps' are, generally speaking, unit alterations in nucleotide sequence. This need not be taken quite literally. If two alterations are necessary to create a better-adapted sequence, while neither confers any advantage by itself, the probability that both should occur in the same genome will be small, but such sequences may nevertheless arise quite often in large populations. However, sequences with several predefined alterations will be vanishingly rare, even in the largest populations. Evolution can leap only very small gaps.

The three alterations conferring resistance not only occur one after another, they also occur in the same sequence whenever the experiment is attempted (Figure 1.9). This seems likely to mean that only one of the alterations confers resistance when it occurs by itself; the other two increase resistance once the first has spread through the population. Their spread is therefore contingent on the prior establishment of the first alteration; until this has occurred, they do not by themselves cause increased adaptation. They do not, in other words, have independent effects on resistance to EtBr; rather, their effect depends on the presence of another alteration. Whenever genetic changes interact in this way, the connectance of the system is reduced. It is not enough to show that a particular unit change increases adaptation, because it may do so only in a particular context, when other changes have already occurred. If this is the case, then the evolution of adaptation does not only require that the appropriate alterations should be substituted sequentially, but also that they should be substituted in a particular sequence, because only this sequence—or perhaps several sequences, out of the very large number of possibilities—involves a continuous increase in adaptedness. Evolution in these populations is presumably more complicated than this simple scheme, because the low fidelity of RNA replication will rapidly create a heterogeneous population containing several hundred variant sequences; the dynamics of the system are discussed in more detail by Davis (1991).

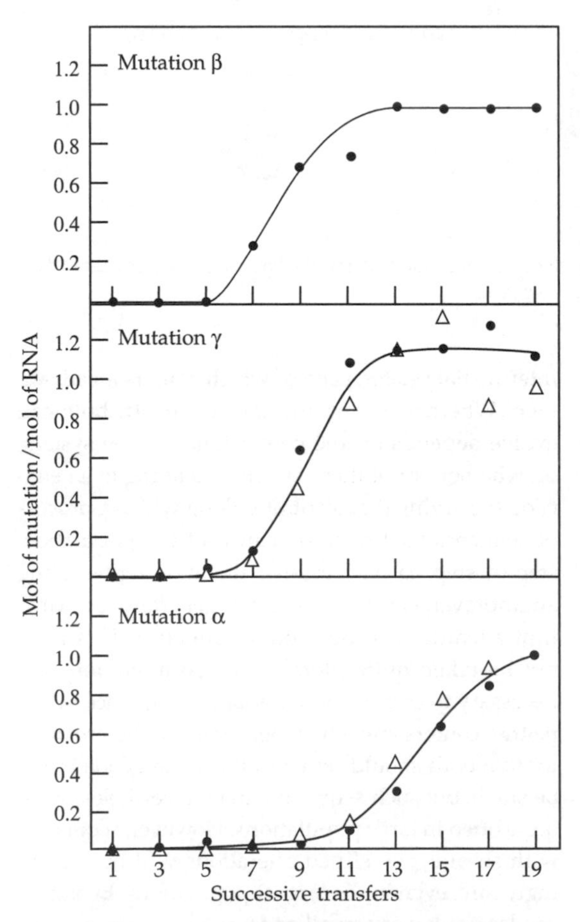

Figure 1.9 Sequential substitution of three mutations causing resistance to ethidium bromide in phage Qβ. From Kramer *et al.* (1974), Figure 8.

1.9 The evolution of increased complexity is a contingent process

It is not, perhaps, very surprising that the relatively complex intact virus evolves in a benign environment into a simpler structure. It might be objected, though, that the important point is not how complex things can become simpler, but how complexity can evolve in the first place. This is much more difficult to study. Nevertheless, it has been found that self-replicating RNA molecules will appear in the culture tubes, even if the cultures are not inoculated with Qβ RNA. It seems that they evolve from very short RNA sequences that form spontaneously in solutions of single nucleotides, or else are present in minute quantities as contaminants, and that have some very rudimentary ability to replicate themselves in the presence of the replicase. The evolution of these molecules *de novo* was first reported by Sumper and Luce (1975) and Biebricher *et al.* (1981a, 1981b), and has since been confirmed by Biebricher and Luce (1993), although Hill and Blumenthal (1983) and Chetverin *et al.* (1991) have argued that they, or similar ancestral molecules, are present as contaminants in the original reaction mixture. The novel sequences are themselves quite short, typically between 25 and 50 nucleotides in length, and indeed include the smallest and simplest self-replicators known (Figure 1.10): they resemble the defective viral genomes often found

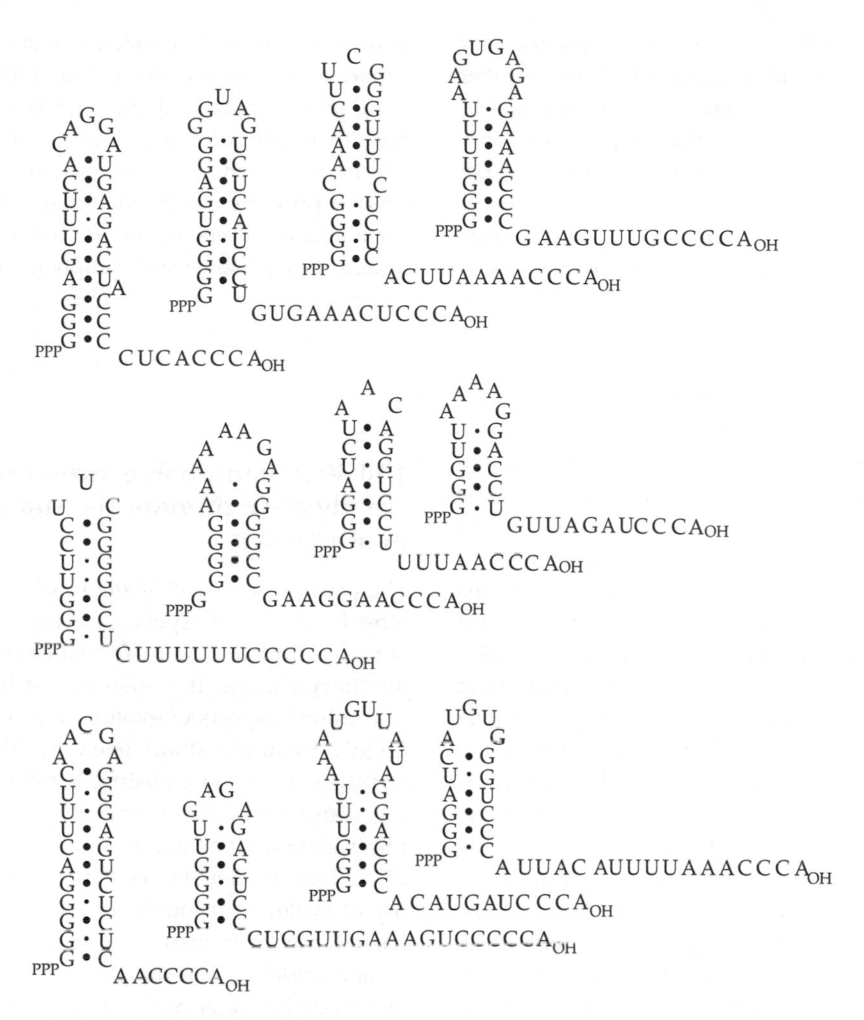

Figure 1.10 RNA molecules evolving de novo in medium containing Qβ replicase but lacking template. From Biebricher and Luce (1993), Figure 4.

late in viral infections, being replicated by the replicase supplied by intact virus. Some remarkable experiments were reported by Bauer *et al.* (1989) and McCaskill and Bauer (1993). They stabilized nucleotide triphosphate solutions in very long, narrow capillary tubes, and observed waves of evolving RNA spreading in both directions from nucleation sites representing a single initial template molecule of RNA. When no template was supplied, similar wavefronts were detected, although they appeared only after a much longer initial lag.

These spontaneously evolving RNAs are very variable; as a general rule, every experiment yields a different result, although all are efficient self-replicators. I presume that in any given chemically defined environment there is some sequence superior to all others that would eliminate all its competitors if it ever evolved. There is no sign that it ever does evolve, however. Selection does not seem to hack out a straight path to an optimal solution, giving the same result in every case. It seems instead to drive historically unique processes of evolutionary change, because of the contingent nature of sequential substitution in poorly connected systems. At the same time, evolution is not completely unpredictable: the novel RNA molecules often share a common feature (such as an unpaired 3′ terminus), showing that the number of major evolutionary themes may be limited.

Populations that are evolving in similar environments come to differ because evolution through selection has no foresight: it does not provide any way in which a population can map out the best route towards a predefined goal. There may be several, or many, states that represent a high degree of adaptation to particular circumstances, as there are many small RNAs that replicate well in the same standard conditions in culture tubes. A population of much less well adapted types cannot choose one of these states and proceed to evolve towards it. The route taken by the population will instead be *contingent*: it will depend on the fortuitous occurrence of variation on which selection can act, and the enormous range of possible variation means that different variants are likely to arise at different times in different populations. Each population is likely to evolve by a different route, and arrive at a different destination. Evolution is modification through descent: selection can do no more than build on the foundations laid by previous generations. The course of evolution may therefore be irreproducible, with populations diverging through time even though the environments in which they are being selected are the same. The diversity of self-reproducing RNAs evolving from simpler ancestors is an example of this pattern. All of them behave in much the same way, of course, being able to replicate fairly efficiently in culture tubes: it is the genetic basis of this adaptation that varies so widely.

A population may fail to become perfectly adapted because superior variants are poorly connected. The 550-nucleotide RNAs evolving by simplification from intact virus do not continue to evolve into the 100- or 200-nucleotide RNAs evolving from simpler predecessors, or at least do not do so in the short term, despite the fact that the shorter molecules have higher rates of replication. Evolution through selection may produce improvement: but it must produce continual improvement. Only those variants that possess an immediate advantage over their competitors can increase in frequency, and the route taken by the population must be traced out by a succession of variants, each superior to its predecessor. This blind and fumbling process may lead the population towards a state that cannot be improved upon by any feasible modification, even though superior designs exist. If progress can be made only by the demolition and wholesale reconstruction of the present design, then progress will simply not be made; the abrupt appearance of a novel creature superior to its ancestors is so unlikely that it can be neglected. Selection provides no blueprints, and no alternative engineer is available.

1.10 Very improbable structures rapidly arise through the cumulation of alterations

The process of *sequential substitution* is one of the most fundamental aspects of evolution, because it is responsible for the historical character of evolutionary change. It is also one of the least well understood aspects of evolution. It has been thoroughly misunderstood even by well-educated scientists, with the amusing, if rather embarrassing, consequence that every few years an eminent physicist or astronomer announces that they have discovered an elementary logical flaw in the theory of evolution through selection. The argument has often been framed in terms of the complexity of haemoglobin, the oxygen-carrying protein of vertebrate blood. A molecule of haemoglobin is a chain of 128 amino acids, and functions properly only if the appropriate amino acid is present at each place along the chain. If evolution depends on the selection of variants arising by chance, what is the probability that this sequence of amino-acids arose by chance, so that it could subsequently be selected? There are 20 different kinds of amino acids, and therefore 20^{128}, or about 10^{200}, sequences 200 amino acids in length. This is a very large number. If the surface of the earth were filled a metre deep with protein molecules 128 amino acids long; if all the molecules had different sequences; if every molecule changed its sequence at random every second, without the same sequence ever recurring; and if this process had been going on since the origin of the earth, nearly 5 billion years ago—then it is still exceedingly improbable that the haemoglobin sequence would yet have been generated. Selection is therefore powerless, because the appropriate variation will never arise. Evolution (it is concluded) requires a guiding hand, supplied in some versions by the inheritance of acquired characters, and in others by divine intervention.

To understand why this argument is fallacious, consider the design of a dartboard. Figure 1.11 shows the arrangement of the board, familiar to anyone

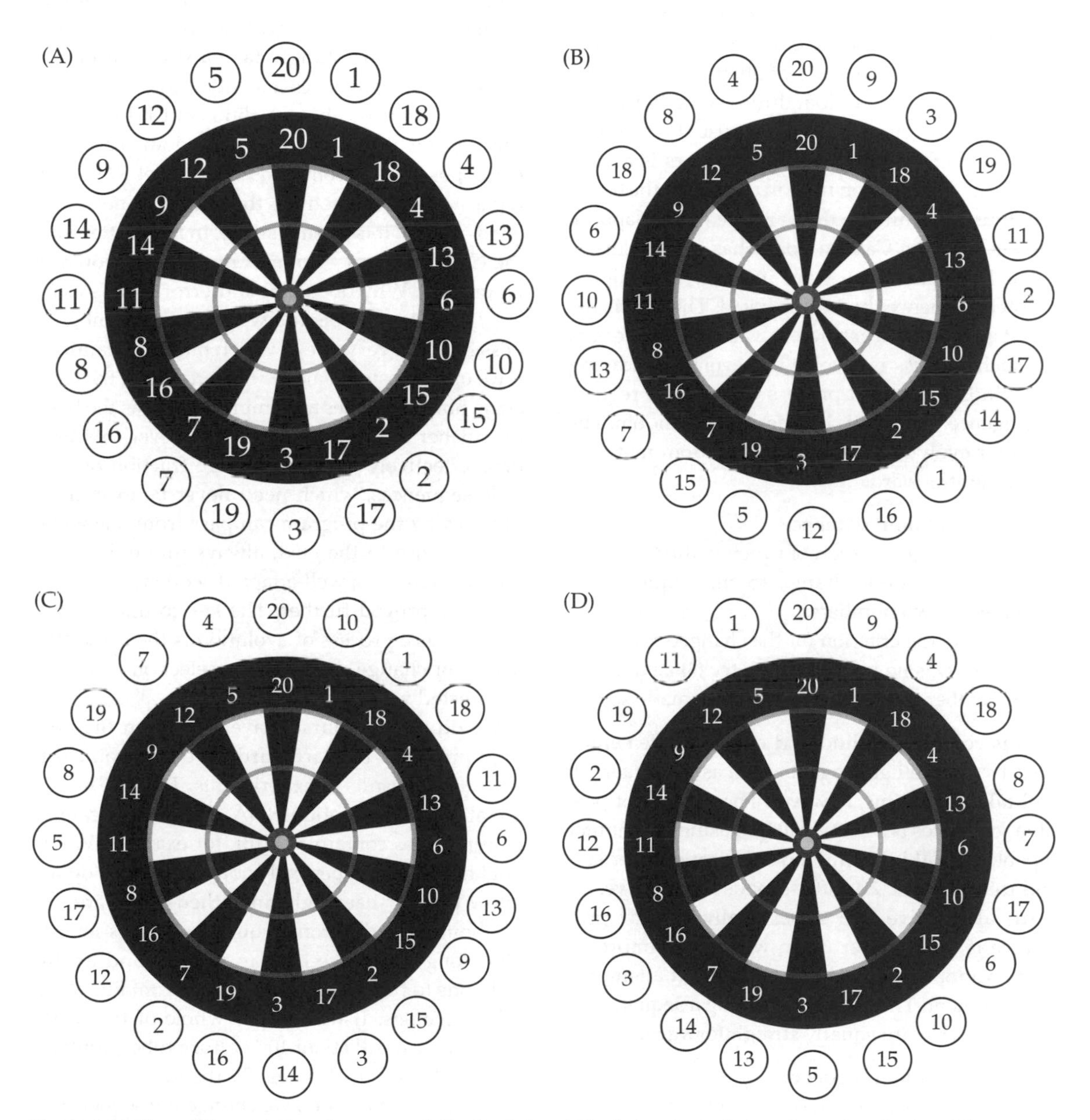

Figure 1.11 The evolution of dartboard sequences. A. The Gamlin board. The design criterion is the variance of overlapping triplets Vartrip = 3.702. B. An evolved board, with 20 located arbitrarily at the top (Vartrip = 0.4035). C. A second evolved board with a different sequence (Vartrip = 0.6725). D. The best evolved sequence observed so far (Vartrip = 0.1696).

whose life has not been entirely blameless. It was invented by Brian Gamlin, a Lancashire carpenter, in 1896 and consists of a circular sequence of numbers that has since spread throughout the world, like a particularly successful bacterial chromosome. It is clearly designed to reward skill: a clumsily thrown dart that misses the numbered sector it was aimed at and instead strikes one of the neighbouring sectors has an expected score of about 10 points, the expected score of a random throw, no matter which sector was the target. This is because the summed score of neighbouring triplets of sectors is almost invariant. The criterion used in designing the board was thus to minimize the variance of overlapping triplets, and the Gamlin board has Var(triplets) = 3.702. There are three ways in which this highly adapted sequence can be generated. The first is the way it actually was generated, by intelligent design. No such process is involved in evolution, leaving two kinds of natural process that do not require intelligence. The first is by guessing at random. The rules for evolving a sequence by random mutation are straightforward:

1. Make a **random** sequence.
2. Calculate the criterion of functionality.
3. Make a **random** change to the sequence by exchanging two numbers.
4. Calculate the criterion for the changed sequence.
5. If there was no improvement, try (3) again.
6. Remember the best score you have made so far.

This is completely blind and can be implemented even by unintelligent agents such as computers. It will take some time, however. The number of possible sequences is $20! = 2.4 \times 10^{18}$, so using a computer capable of making a million substitutions per second this will take 2.4×10^{12} seconds = 76 103.5 years.

An alternative and superficially similar procedure is trial and error, in which any improvement is adopted as the basis for the next attempt at improvement. The rules for evolving a sequence by trial and error are equally straightforward:

1. Make a **random** sequence.
2. Calculate the criterion of functionality.
3. Make a **random** change to the sequence by exchanging two numbers.
4. Calculate the criterion for the changed sequence.
5. If there was no improvement, try (3) again.
6. If there was an improvement, adopt the changed sequence as the new standard and then try (3) again.

This is also completely blind and can be implemented even by unintelligent agents such as computers. It is astonishingly fast: a simple program run on a laptop reaches a solution in a fraction of a second, that is, about a thousand billion (10^{12}) times faster than guessing at random. There are many possible outcomes (as with novel RNA self-replicators) and sometimes the solution is not a very good one, but it is often (usually, in fact) better than the Gamlin board—some examples are shown in Figure 1.12. Why does trial and error work so rapidly? The process begins with a huge number of possibilities, every one of which must be examined if we use guesswork alone. The process of trial and error, however, takes any improvement as the basis for further modification, and whenever this happens effectively discards a large fraction of possible sequences, which need never be examined. This is why the program can pass from one working sequence to the next, always improving, and rapidly arrive at a well-adapted sequence that cannot be improved further. The key to understanding the basic process of evolution is the *cumulative* nature of change produced by selection acting on undirected variation.

We do not, of course, have to rely on the analogy provided by dartboards; we can point directly to the results of experiments. Take the highly adapted 218-nucleotide RNA able to replicate in culture tubes containing EtBr, for example. We do not know that it is perfectly adapted, but we do not know that for haemoglobin, either. There are 4^{218}, or about 10^{128}, different sequences of this length. How long would we have to wait around for the right one to turn up? A culture tube contains about 10^{16} molecules. If we use the whole of the world ocean, we can fill about 10^{22} culture tubes, containing in total 10^{38} molecules. By the same ingenious biochemistry as before, we change the sequence of every molecule every second, never allowing any sequence to be repeated. We shall have tried out all the possible sequences, and found our adapted RNA, after about $10^{128}/10^{38} = 10^{90}$ seconds, or about 10^{72} years. This is about 10^{62} times as long as the universe has existed. Because highly adapted RNAs can, in fact, evolve within a few days or weeks, it is clear that their evolution is not a matter of waiting for something to turn up. No one has suggested that the normal rules of RNA replication are suspended for the benefit of these experiments, however, or that the experiments are the target of continual divine intervention. The paradox disappears when it is realized how evolution through selection proceeds in a stepwise fashion. Random alterations in the sequence of nucleotides occur from time to time, because the replicase is not a perfect copying machine. Most changes are deleterious, and quickly die out without giving rise to a lineage. Occasionally a change is advantageous, in the sense of permitting more rapid replication in the particular circumstances of an experiment. Any such variant quickly replaces the resident type. The whole population has now changed: and the altered sequence can serve as the basis for further improvement. In this way, it is possible very rapidly to evolve sequences which at first sight seem hopelessly improbable.

1.11 Competitors are an important part of the environment

When a culture tube is inoculated with RNA, all resources are at first present in excess, and growth is exponential, or nearly so. Under these conditions, variants with greater exponential rates of growth are selected, as I have described. The outcome of competition among a number of variants could therefore be predicted by growing each variant for a short time in a separate tube: the variant which grew fastest would be the one which replaced all others in a mixed population. If growth is continued, however, resources will begin to become scarce before very long. In particular, there will no longer be enough replicase to go round. The nature of competition now changes; it is no longer purely reproductive, but operates more directly between molecules for access to resources. Variants with high rates of replication will, of course, continue to be selected; but they must now achieve these high rates of replication by obtaining an adequate supply of replicase, and thereby directly or indirectly preventing other molecules from doing so. In practice, this means that the most successful sequences will be those which are best at seizing scarce replicase molecules whenever they encounter them. Now, there is no reason to suppose that sequences that can replicate fast when replicase is abundant will also be best able to scavenge for replicase when it is scarce. In fact, new variants do appear and spread when selection is prolonged to the point where the availability of replicase becomes limiting. These variants perform poorly in the early stages of growth, and their superiority becomes evident only when molecules begin to compete amongst one another so that resources cease to become available (for a time) to others, and thereby deny them the opportunity to replicate. The crucial feature of the environment is no longer the physical conditions of growth—the presence or absence of EtBr, for example—but rather the behaviour of the other variants in the population. Absolute measures of performance, such as the exponential rate of increase in pure culture, may then be unreliable predictors of evolutionary success or failure. What matters is not so much the ability to do well, but rather the ability to do better.

1.12 Evolution through selection is a property of self-replicators

The very simple nature of the $Q\beta$ system, and in particular the tight coupling between nucleotide sequence and behaviour, makes it possible to exemplify evolutionary principles in a particularly direct and straightforward manner. In other respects, choosing to open a book about selection with an account of degenerate viruses in culture tubes may seem rather eccentric. Evolution is the study of living complexity: are viruses (let alone RNA molecules a few hundred nucleotides in length) even alive? The question is important, not because it can be given any definite answer—'alive' is a category invented long before the nature of viruses was elucidated—but rather because it is not necessary to give it an answer at all. Whether or not they are regarded as living, viruses, and even their degenerate descendants in culture tubes, can evolve; and *evolvable* is a more fundamental category than *living*. Anything that replicates itself will do so imprecisely; some of the variants that appear will have altered rates of replication; those with higher rates of replication will be selected. Evolution through selection is therefore a property of all self-replicating things.

1.13 Selection can be used to engineer the structure of molecules

We are only now beginning to understand—and therefore to be able to exploit—the generality of this principle. Qβ RNA maintained in culture tubes is not, strictly speaking, an autonomous self-replicator, because the replicase, which normally it encodes, is supplied by the experimenter. However, RNA, like protein, is capable of acting as an enzyme; and in particular RNA enzymes, or ribozymes, can catalyse the breakage and reunion of other RNA sequences. It is perfectly conceivable that there exist RNA sequences that are able to catalyse their own synthesis, without the need for a protein replicase. No such molecule has yet been identified (although it is only a question of time, I think, before it is), but the evolutionary potential of self-catalytic RNA has begun to be studied. Joyce (1992) used an RNA isolated from the ciliate *Tetrahymena* that is able to cut a different substrate RNA molecule and attach part of the substrate to itself. This reaction normally requires the presence of magnesium ions. How might it be possible to create a molecule that would carry out the same reaction when calcium ions but not magnesium ions were present? It would be extremely difficult to design such a molecule; the relevant biochemistry is just not known in enough detail. However, it is possible to devise a system in which variants that can perform the reaction, however inefficiently, succeed in replicating, whereas those that cannot perform the reaction are unable to replicate. The system devised by Lehman and Joyce (1993a, b) uses a ribozyme from the ciliate *Tetrahymena*. This ribozyme binds a partially complementary substrate RNA molecule, cleaving it at

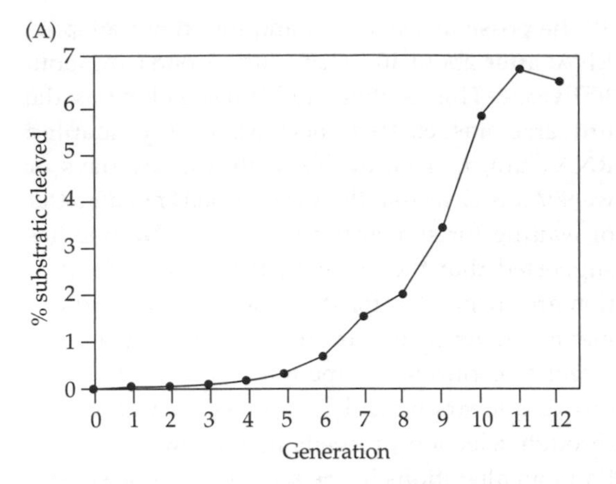

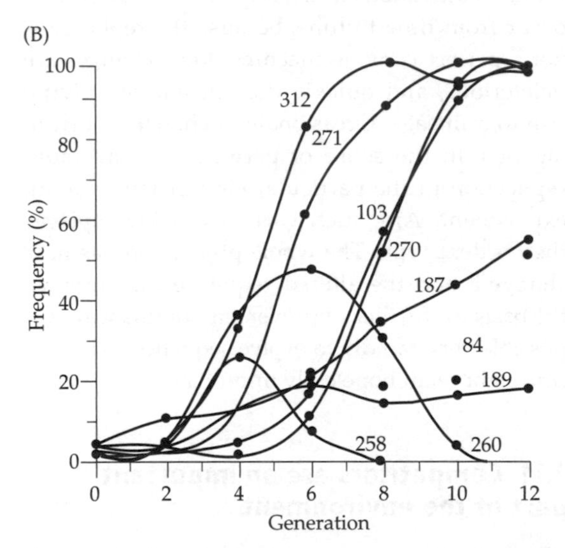

Figure 1.12 The evolution of ribozyme sequences. A. Phenotypic improvement over twelve cycles of selection. From Lehman and Joyce (1993a), Figure 3. B. Underlying changes in genotype frequencies. From Lehman and Joyce (1993a), Figure 6.

a particular site and bonding part of the cleaved product to the ribozyme itself. It normally requires magnesium or manganese ions to operate in this way. To evolve variants functional in the presence of calcium ions alone, it is necessary to amplify lineages that are more successful in catalysis under these conditions. Very briefly, a heterogeneous population of catalytic RNA molecules binds the substrate RNA. This complex is translated into cDNA by reverse transcriptase, using a primer corresponding to the bound substrate sequence. RNA is transcribed from this DNA in such a way that only the catalytic part of the molecule is synthesized, and then amplified by RNA polymerase in an error-prone fashion that recreates a large, highly variable population of catalytic RNA molecules. Because only those RNA molecules that bind the substrate are translated into DNA and subsequently amplified, the procedure is highly selective for successful catalytic activity. The results of engineering molecules in this way are described by Joyce (1992), Beaudry and Joyce (1992) and Lehman and Joyce (1993a, b). Figure 1.12A shows the increase in the catalytic ability of the ribozyme in the presence of calcium ions alone as the result of 12 cycles of selective amplification. This was attributable to a complex series of substitutions. Figure 1.12B shows how the frequency of some of the mutations that were detected changes through time; the numbers are their positions on the ribozyme molecule. Some spread through the population, to become fixed by the end of the experiment; others increase in frequency during the early stages of selection, but then decline and are eventually lost. Not surprisingly, genetic variance increased rapidly in the first few cycles, but later fell as the population converged on a few highly superior variants that shared mutations at seven sites. Many of the alterations were highly *epistatic*, their effect depending on their genetic context; some were even mutually exclusive, being effective only if the other did not occur. This ribozyme system, in short, undergoes the same process of sequential substitution as evolving Qβ RNA. After only a dozen cycles of selection and amplification, a new RNA capable of cleaving the substrate RNA in the presence of calcium ions alone with quite high efficiency has evolved. This is an interesting evolutionary system in its own right; it is also a remarkable illustration of how evolution through selection can produce functional structure with an ease and rapidity that a conventional engineering approach could not hope to emulate.

1.14 Self-replicating algorithms evolve in computers

Among the most illuminating and certainly the most bizarre of self-replicators are the digital organisms invented by Ray (1991). They are algorithms: strings of instructions written in the machine-code language of a computer. The basic algorithm, a list of 80 instructions, is designed to replicate itself—to place a copy of the same set of instructions into the memory of the computer. The process of replication is deliberately made somewhat imprecise, so that instructions are occasionally modified slightly as they are being copied. The algorithm is designed to replicate itself, and it is not designed to do anything else; that is to say, no inherent tendency to change in certain directions, or in any direction, is programmed into the list of instructions that defines each individual. The creatures proliferate within the central processing unit of a computer. The resources they require are memory space (for storing their code, and that of their descendants) and computing time (for executing their code). It is not too fanciful to imagine them as algorithmic parasites, utilizing the computer's operating system to achieve their own self-replication, somewhat as Qβ uses the protein-synthesis machinery of its bacterial host. The space and time that has been allotted to them is sooner or later insufficient to allow all the copies of the original sequence to replicate themselves. Some must then be discarded, by erasing them from the memory of the computer; those whose code has suffered the most errors during copying are the most likely to be lost in this way. The population is at this point a clone whose members reproduce and die in their allotted reserve within the computer. If they are allowed to do this for long enough, however, something happens: they begin to evolve (Figure 1.13).

The first sign that the system is something more than an eternally recurring set of computer instructions is that new creatures 80 instructions in length appear, usually differing from each other, and from their ancestor, by a minor change in a single instruction. In different runs, different variants arise and spread, often replacing the ancestral type. This was unexpected: the original sequence had been very carefully written as a minimal self-replicating algorithm, and yet several thousand different self-replicating variants were derived from it through slight modifications of one sort or another. Much more surprising, however, was the abrupt appearance of

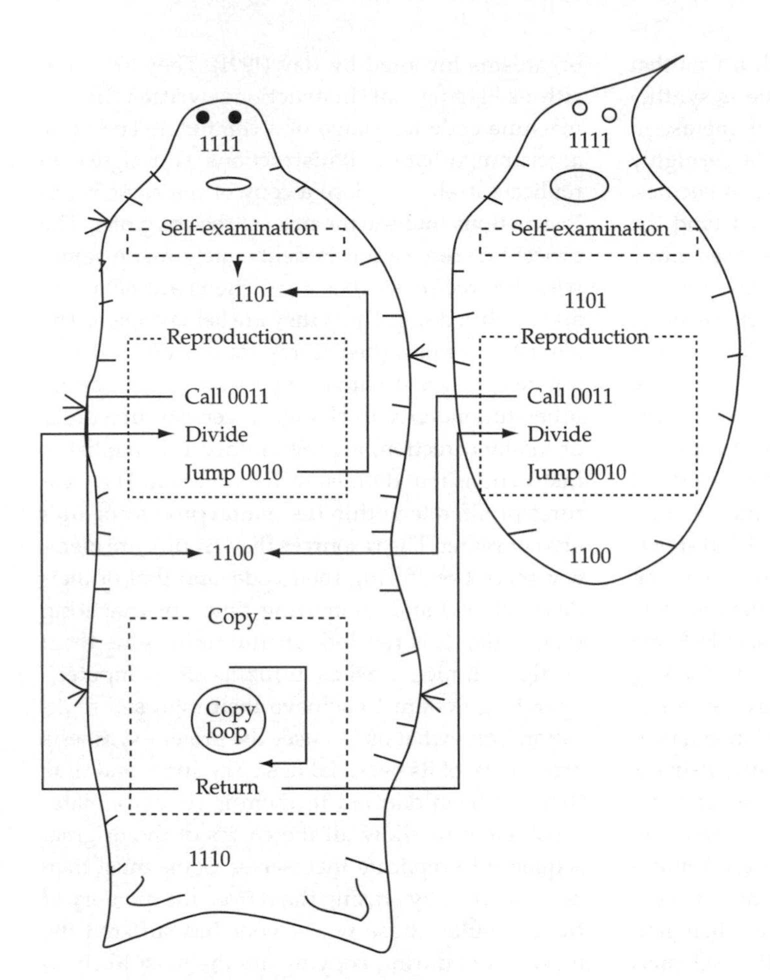

Figure 1.13 The evolution of algorithm sequences. At the left is a cartoon of the ancestral 80-instruction Tierran organism; to the right is an evolved 45-instruction parasite.

a completely different creature, in which a trivial change to a single instruction has a marked effect on its appearance, reducing its size from 80 to 45 instructions. This variant does not have the ability to replicate itself autonomously, because the code that directs the copying procedure has been lost. However, it is able to use the copying procedure of the intact algorithm, and is therefore maintained in the population provided that there are enough 80-instruction self-replicators to supply it with the missing instructions. It is, in short, a parasite that is able to utilize the replication machinery encoded by the intact host. The parallel with Qβ is astonishingly close: in both systems, defective variants arising through slight modifications that cause the deletion of a large part of the genome parasitize the ancestral sequence by utilizing its replication machinery—a replicase in the one case, a copying procedure in the other.

This does not by any means exhaust the evolutionary possibilities of the system, whose full extent, indeed, remains to be explored. Host variants 79 instructions in length appear that are resistant to the parasite; they are out-manoeuvred by modified parasites 51 instructions long. A peculiar creature that is 80 instructions long, but differs from the ancestral sequence by 20 or so alterations, has sometimes been observed; it is a hyperparasite able to force parasites to replace themselves with copies of itself. Even more remarkable are 61-instruction sequences that are unable to replicate when alone, but achieve a cooperative self-replication when associated in groups, occupying adjacent blocks of computer memory. They are eventually exploited by tiny creatures only 27 instructions long that insert themselves between these cooperating sequences and exploit the cooperative system of replication to their own advantage. This whole menagerie of different kinds of creature descends from the single self-replicating sequence with which the computer is originally inoculated. The process is completely transparent, because the complete sequence of any or all algorithms in the population can be ascertained at any time. Despite the complete lack of any specific directing principle, the sequential substitution of slight variants drives the evolution of a diversity of replicators of different kinds. It is perhaps the purest example of evolution through selection that can be imagined, and provides a striking demonstration that evolution through selection is a general property of self-replicators, irrespective of whether by any conventional standards they would be considered to be living.

1.15 Evolution through selection is governed by a set of general principles

The experiments that I have described exemplify a series of generalizations about the behaviour of systems of self-replicators:

- Heritable variation in the rate of replication causes evolution through selection.
- Heritable variation arises as random, or undirected, alterations of nucleotide sequence; it does not in itself direct the course of evolution.
- The rate of replication is the only attribute that is selected directly.
- Characters that cause changes in the rate of replication will be selected indirectly, and may evolve as a consequence.
- Adaptation caused by selection in given conditions is likely to be associated with loss of adaptation in other conditions.
- Evolution proceeds through the sequential substitution of superior variants, not exclusively by sorting pre-existing variation.
- A given state can evolve from a prior state only if the two states are connected by a continuous series of slight modifications, each of which is individually advantageous.
- Selection causes the modification of prior states of organization, but cannot abruptly give rise to wholly novel states; the course of evolution is an historically unique, contingent process conditioned by the fortuitous occurrence of particular variants.
- Selection tends to improve performance in given conditions, but does not always and may never optimize performance.
- Because selection is caused by differences in rates of replication, the outcome of selection will often depend on what kinds of competitor are present, rather than on the physical conditions of growth.

The way that these general laws, and others yet to be discussed, mould the evolution of living organisms will be modulated by the developmental, physiological, genetic, and ecological circumstances in which they operate. That is what the rest of this book is about. It is important to realize, however, that they are not themselves developmental, physiological, genetic, or ecological laws. Nor can they be described in terms of or reduced to developmental, physiological, genetic, or ecological principles. They are a separate set of laws that defines a separate discipline, based on the single irreducible quality of self-replicators of any kind: *evolution through selection*.

CHAPTER 2

The genetic and ecological context of selection

2.1 History, chance, and necessity

Variation and selection recur in every generation. There are therefore two theories of evolution: the first is that it is directed by variation, and the second that it is directed by selection.

2.1.1 Lamarckian evolution

The first is Lamarck's theory of the inheritance of acquired characteristics, published in the year of Darwin's birth (1809). During their lifetimes, individuals perceive the state of their environment and respond appropriately to it. This may involve some trivial change, such as the thickening of part of the cuticle or the enlargement of a muscle, or a more profound reorganization, perhaps through the formation of new body cavities as the result of a change in the pattern of circulation of fluids. This change is transmitted to their offspring. There is thus an inherently progressive principle in nature, causing a general increase in the level of organization. Selection may occur, but plays a subordinate role; adaptation to novel environments is caused primarily by the spontaneous appearance of appropriately directed variation. We can translate this into modern terms by saying that the environment elicits favourable mutations.

So far as we know, this theory is wrong. It is not wrong as a matter of principle; it is an internally consistent and intellectually satisfying theory of evolutionary change. It is wrong as a matter of fact. No mechanism that would act as a specific directing principle to produce appropriate genetic variation has yet been identified. During the early years of the study of bacterial genetics a number of experiments were devised to investigate this issue. They involved challenging bacteria with a new and hostile environment, and then tracing the origin of the adaptations that evolved. In all cases, it was found that the mutations that conferred adaptation, and which increased in frequency through selection in the new environment, occurred before the environment changed. These results have been very widely accepted as showing that adaptation occurs through the selection of pre-existing variation, and not through the elicitation of appropriate variation. There are some poorly understood processes in bacterial populations that have been interpreted as adaptive mutations arising in response to specific environmental stresses (see Chapter 4). When the conventional wisdom is challenged from time to time, however, it is usually the pastime of literary gents such as Samuel Butler or Arthur Koestler, who might hesitate to pronounce on the principles of inorganic chemistry or hydrodynamics, but regard evolution as a subject where everyone has the right to their own opinion. Together with the haemoglobin paradox, this is the main source of the periodic announcement of the demise of Darwinian orthodoxy.

2.1.2 The selection of undirected variation

The remaining possibility is Darwin's theory of evolution through selection. This does not require that mutations occur at random: on the contrary, it is well known that some genes mutate more frequently than others, and that different sites within a gene may have different rates of mutation. Nor does it assert that mutations are not induced by the environment: it is equally well known that many mutations are caused straightforwardly by physical agents such as ionizing radiation. The crucial point is that mutations are not *appropriately* induced by environmental factors. Suppose that we expose bacterial cultures, or populations of any organism, either to ultraviolet light or to EMS (ethyl methyl sulphonate, a mutagenic chemical). Both are toxic, and inhibit growth; both are also powerful mutagens, and in the appropriate doses will cause a range of mutations. It will not be found, however, that the mutations induced by ultraviolet light are specifically resistant to ultraviolet light, and not to EMS; nor that the mutations induced by EMS are specifically resistant to EMS, and not to ultraviolet light. Adaptation to either agent, or to any less dramatic environmental challenges, is through the selection of *undirected* variation. The direction (or lack of direction) taken by evolution is determined by the action of selection in each generation, and not by any inherent directional property of variation itself.

Because heritable variation in fitness is nearly universal, selection will operate in most populations at most times, driving evolutionary change. The outcome of evolution, however, will not depend on selection alone. Selection has no power to bring into being novel kinds of organism *ex nihilo*; it merely acts so as to modify the pre-existing state of the population. Evolution is therefore an historical process, whose outcome depends in part on initial conditions. Nor will selection determine precisely the outcome of evolution, even if its starting point be known, because each generation is formed through a process of sampling rather than through deliberate inerrant choice. Evolution is therefore a stochastic process, whose outcome depends in part on chance. History, chance, and necessity will always contribute, in different degrees, to any kind of evolutionary change.

2.1.3 Descent

The first component that we must invoke in order to explain the current state of a population is historical, the state of the previous generation from which it descends. Populations that start from different points will necessarily follow different trajectories even if they eventually reach the same end point. A more interesting possibility is that populations may reach different end points, whether or not they were initially identical. The reason for thinking that this may often be the case involves three general principles. First, most populations will include only a very small sample of possible combinations of alleles. Secondly, the effects of a gene may depend on the state of other loci elsewhere in the genome, so that any combination of genes has effects that cannot be predicted precisely from knowing the independent effects of each gene. Finally, evolution is a process of cumulation, in which each episode of selection modifies the prior composition of the population. Replicate selection lines may initially diverge because a different beneficial mutation arises and becomes fixed in each. To the extent that the effects of a gene depends on the state of other loci, the properties of new mutations arising subsequently in each line will depend on which mutation happens to have been substituted first in that line. The initial divergence of lines will thereby predispose selection in each to favour different mutations in the next period of time. The differences among lines will continue to cumulate, with each line evolving high fitness, or a similar character state, but with a unique genetic basis. This is the principle of contingency: lines diverge genetically because of the contingent nature of cumulative change under continued selection.

This is an important principle, because it colours our whole approach to evolution. If evolution is a regular and predictable process, then the history of life on Earth was more or less preordained from the beginning, and mankind, or something very like mankind, is the inevitable outcome of 3 billion years of continued selection. If evolution is a contingent process, Earth history is only one of an uncountable number of possible outcomes, and we are the unrepeatable product of a unique experiment. What applies to us applies, of course, with equal force to the crab under the rock or the rose in the hedge. Which of these two extreme views is more nearly correct cannot be established simply by documenting the course of evolution to date; the question can be decided only by comparing the history of life on different planets, or, more conveniently, by analysing the genetic basis of adaptation in replicate selection lines.

2.1.4 Selection

The second component is the systematic process of selection, which tends to cause an increase in the frequency of some kinds of individual and a decrease in others. It is possible to imagine a population that is perfectly adapted, in which offspring are always exact replicas of their parents, and which inhabits an environment that never changes. In these circumstances, evolution would come to a halt, and the population would propagate itself, generation after generation, without any kind of change. In practice, this is never the case. Even if the environment were to remain constant, the population would change. Genotypes inevitably suffer changes in structure, especially when being replicated. Genetic alterations, or mutations, are of various kinds, from a change in a single nucleotide of DNA to the duplication, deletion, or rearrangement of large sections of chromosomes. Supposing that perfect adaptation were attainable, therefore, it could not be permanently maintained, because of this internal process of genetic change. Even if mutation could be neglected, the environment would change. The physical environment is continually changing, on all time-scales: from hour to hour, from day to night, from season to season, from year to year. Organisms that live on different timescales—bacteria on a scale of hours, mammals on a scale of years—will be affected by different types of change, but none are insulated from change altogether. Perfect adaptation to any given environment could not be permanently maintained, because of this external process of environmental change.

Adaptedness is not maintained by default; there is no passive process ensuring that a high level of functional organization will continue to be preserved as long as no specific catastrophe occurs. Change is generally for the worse. A well-adapted population is fitted to its environment, in the colloquial sense of being appropriately formed, like a key to a lock. If either the population or the environment changes haphazardly, it is most unlikely that the fit between them will be improved. Both internal and external change therefore cause in every generation an increasing lack of fit between population and environment, which, if it were to continue unchecked, would break down adaptedness completely. The maintenance of adaptation therefore requires the perpetual operation of selection, acting in every generation to restore the fit of population to environment. There are two broad categories of natural selection. In the first place, it will be a conservative force that eliminates new mutations so as to preserve the current adaptedness of the population. This is called *purifying selection*. Secondly, it will be a progressive force that favours new mutations when the environment changes. This is called *directional selection*. These categories are not exclusive. Both internal and external change will occur in every generation, eliciting both purifying and directional selection in different degrees on different characters.

2.1.5 Drift

The third component is chance. The actual composition of the population will inevitably differ from what we expect on the basis of descent and selection, because the life of each individual is an historically unique succession of events whose eventual outcome is influenced by a multitude of factors. The next generation is formed in a stochastic, or probabilistic, fashion from the success or failure of many such lives. We may be able to predict its average properties with some assurance, but its composition will fluctuate to a greater or lesser extent through time in ways that we cannot predict or account for.

If it were possible to distinguish two variants which were in all other respects functionally equivalent, and which in all circumstances had equal fitness, then neither would tend to change in frequency under selection; but they would nevertheless fluctuate in frequency from generation to generation through sampling error. Sampling error is in this case the only process causing the composition of the population to change through time. It will continue to operate if the two variants, rather than being strictly equivalent, express an extremely small difference in fitness. In a perfect world, any difference in fitness, however small, would drive a gradual but perfectly regular directional change in the composition of the population. However, it is easy to appreciate that if the difference in fitness be sufficiently small, the change in composition caused by selection may be smaller than the change caused by sampling error. The expected outcome of selection is unaffected; but in any particular case, this expectation is unlikely to be realized. If sampling error could be neglected, then two populations identically constituted will change in precisely the same way when selection is applied, however slight the selection might be. But this is not the case. If selection is very slight, its effect may be small relative to sampling error, and replicate populations will then come to differ, despite being selected in the same way. Sampling error, acting at the level of the population, is analogous to mutation, acting at the level of the individual. It is impossible, in either case, to transmit information with perfect fidelity: mutation makes it impossible for a nucleotide sequence to be transmitted perfectly, whereas sampling error makes it equally impossible for a population distribution to be transmitted perfectly. Both are undirected processes of variation that reduce adaptation, albeit the variation created by mutation is required for selection to be effective.

The effect of sampling error will be most pronounced in small populations. Rolling dice provides a familiar analogy. If six fair dice are rolled, we expect one to show a six. However, we are not astonished if none do; the probability that this will happen is $(5/6)^6$, or about 1 in 3. If 6000 fair dice are rolled, we expect 1000 to show a six, but are not astonished if 990 or 1010 do so. If none do, we would dispute that they are fair, because the probability of this happening by chance is $(5/6)^{6000}$, a probability too small for my calculator to deal with or my credulity to accept. In large populations, large fluctuations caused by sampling error are very unlikely to happen, and selection proceeds nearly deterministically. In small populations, the frequency of an allele under weak selection will fluctuate appreciably because of sampling error. A slightly beneficial allele may be lost; a slightly deleterious allele may be fixed. Sampling error thus reduces the effectiveness of both directional and purifying selection, hindering adaptation to a new environment and the maintenance of adaptedness in an unchanging environment.

Roughly speaking, sampling error will be a serious hindrance to selection if $s < 1/N$, where N is the number of individuals in the population (§3.2). For most purposes, the effects of sampling error are negligible in populations which always consist of a few thousand individuals or more. It cannot be neglected in populations of a few hundred individuals or fewer. The expected outcome of selection is the same in populations of any size, but the reproducibility of evolution is less in small populations. Small populations with the same initial composition will diverge through time until their genetic composition becomes widely different, despite being exposed to the same agents of selection acting in the same manner. The sorting of pre-existing variation by selection would have caused them all to change at the same rate in the same direction; thus, genetic drift has hindered or prevented at least some of them from attaining the degree of adaptedness that we would expect, if selection were the only force operating.

The effects of history, chance, and necessity are almost inextricably mixed in most evolving lineages, but in laboratory experiments they can be cleanly separated. We begin with a series of populations which differ initially in their level of adaptation to some novel environment; these initial differences are clearly historical. Each population is divided into a number of replicates, each of which has the same ancestry and will be exposed to the same conditions of growth; any variation among them must be attributable to chance. The extent to which these populations resemble one another after being cultivated separately, despite their initial differences and subsequent chance divergence, must be caused by the systematic operation of selection. Travisano *et al.* (1995) showed how the overall outcome of selection can be partitioned between these three processes (Figure 2.1). They selected *E. coli* in maltose-limited chemostats, using lines that had previously been selected for 2000 generations in glucose-limited chemostats and that differed in their ability to utilize maltose. Fitness itself evolves in parallel in all lines, the effect of selection being paramount. Other characters, such as cell size, are indirectly affected by selection but are much more sensitive to chance and history. This might serve as a rough general guide to evolutionary change: fitness itself evolves predictably, but the genetic and physiological bases of adaptation are channelled by history and diverted by chance. It has been discussed at great length in the literature of evolutionary biology whether we can as a first approximation neglect one of these two processes, and this is a subtle and difficult question to which the answer is, no. Even in the very simplest systems, history, chance, and necessity all play their parts in adaptation, and unravelling this knot is the focus of experimental evolutionary biology.

2.2 The rate of genetic deterioration

It is very generally thought that almost all mutations are deleterious. This is usually justified by arguing that randomly adjusting or removing some component of a complex and highly integrated device is unlikely to improve its performance, following Fisher's well-known analogy of the precisely focused microscope. In fact the assertion that most mutations are deleterious conflates two assertions, one linking genetic to phenotypic change and the other linking phenotypic change

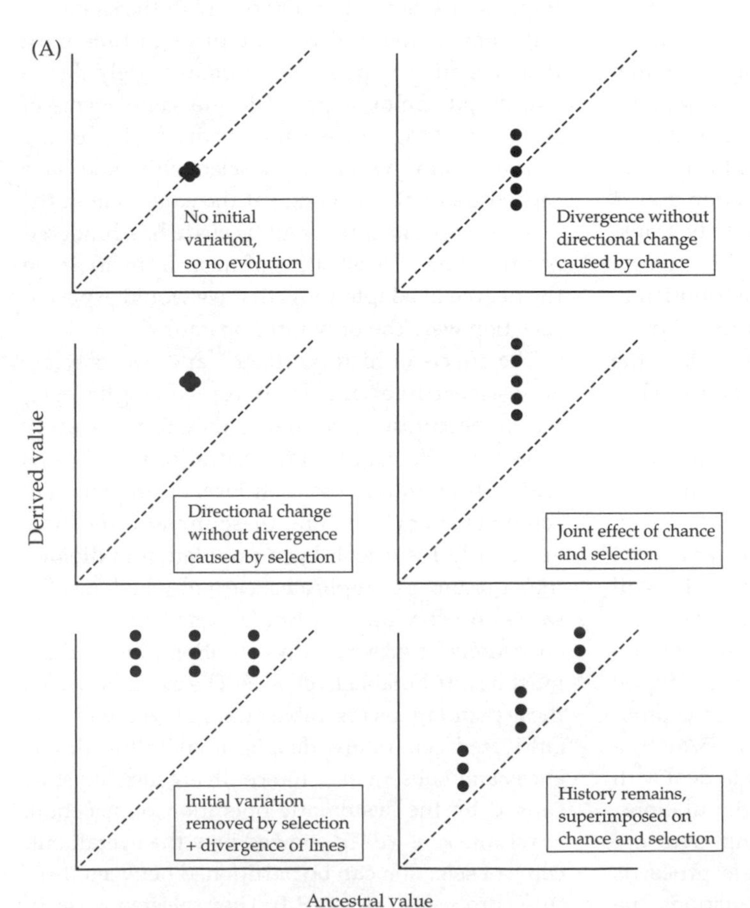

Figure 2.1 *(Continued)*

to fitness. The first is that mutations of larger scale are likely to have larger effects on character state. Replacing a single nucleotide often has little or no effect on protein structure and correspondingly little effect on metabolism or morphology, whereas a large insertion or deletion is likely to result in the production of a completely novel protein, or in failure to produce any protein, which is likely to cause a pronounced metabolic change or deficiency and thereby an altered phenotype. This is not always the case, there being many examples of single point mutations with large effects on fitness, but mutation accumulation experiments suggest as a general rule that fitness declines as the number of defective nucleotides increases. The second idea is that larger changes in character state are likely to cause larger reductions in fitness. More extensive injuries (for example, loss of a limb compared with loss of a digit) are more perilous, and the effects of injuries may be additive, as the removal of digits from frogs seems to be (McCarthy and Parris 2004).

The mechanical analogy refers more closely to the effect of altering the phenotype. Any protein, control system, morphological structure, or behaviour makes some distinctive contribution to the overall well-being of the organism, and the degree to which a given character state fulfils this contribution is its performance. Any alteration of character state is likely on balance to reduce performance, and the greater the alteration the greater the impairment it will cause. The consequences of

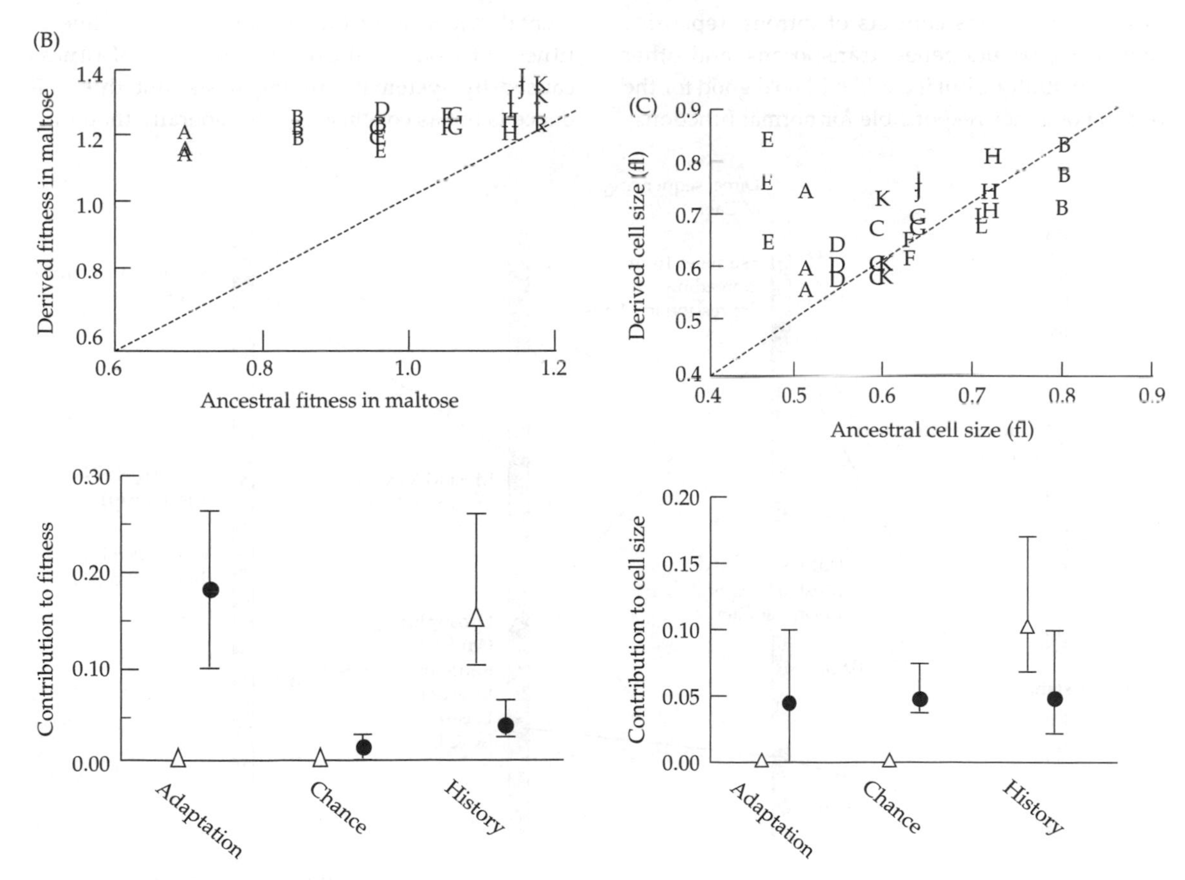

Figure 2.1 History, chance, and necessity. A. Possible outcomes of continued culture in relation to initial conditions, drift and selection. B. Response of fitness. The upper diagram shows the ancestral and derived populations, as in the schematic; the lower diagram shows estimates of the effects, as variance components. C. Response of cell size. From Travisano *et al.* (1995)

accidents of development, or of random injuries in adults, are seldom beneficial. The individual as a whole is seen as a well-adjusted machine whose performance, in most cases, is almost as good as it could possibly be. Genetic mutations, in turn, cause alterations of character state, and are therefore likely to be deleterious, and the greater the genetic effect the more severely deleterious it is likely to be. Gross chromosomal aberrations such as aneuploidy almost always have severely deleterious effects. The genome is viewed as a well-harmonized system of genes each with a specific, indispensable function in the economy of the body. This is, I think, the classical view of the genome, underlying analogies such as 'the Parliament of the genes' (Darlington 1958, Leigh 1971). It has not survived the discovery that much of the genome in most organisms consists of introns, repetitive sequences, pseudogenes, transposons, and other non-contributors, but it could still hold good for the residue of genes responsible for normal function.

Most newborn individuals are capable of surviving if they are protected from disease and starvation. If the entire seed output of a plant or the entire egg output of a fish is collected, most seeds and most eggs can with sufficient care be got to grow to adulthood. Mutation may be inevitable; most mutations may inevitably be deleterious; but nevertheless most individuals that are born can live. Most mutations, therefore, are not very deleterious. The assault on adaptedness is not carried out primarily by a storm of mutations that kill or maim, but rather by a steady drizzle of mutations with slight or inappreciable effects on health and vigour. Some classical estimates of the rate and effect of mutation are shown in Figure 2.2. They illustrate the doctrine that the bulk of mutations have slightly deleterious effects. The main experimental evidence for the doctrine is the decline in fitness of isolate cultures and the loss of fitness caused by systematic mutagenesis, but this evidence is not as conclusive as is generally thought.

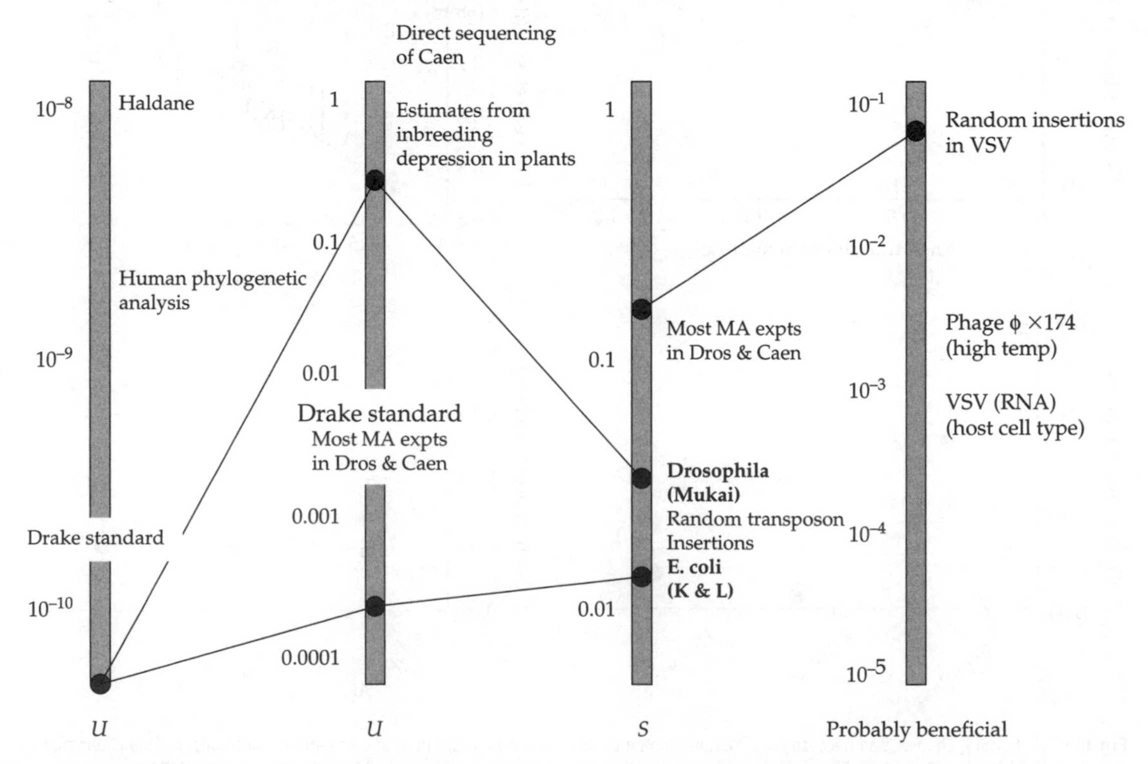

Figure 2.2 Some classical estimates of the rates and effects of mutations. Dros is *Drosophila*, Caen is *Caenorhabditis elegans*, MA is mutation accumulation; K & L refers to Kibota & Lynch (1996); Drake refers to Drake *et al.* (1998); Mukai refers to Mukai (1964).

2.2.1 Two scaled mutation rates

The *fundamental mutation rate* is set by the probability that a base will not be replicated faithfully, designated as u. The mutation rate per locus is then $u_L = Lu$ where L is the length of the gene in nucleotides. From what has been argued above, u is primarily the rate of slightly deleterious mutation, and is often regarded as being nearly constant over organisms, genes, and environments and therefore justifying considerable effort to measure precisely. It almost always affects evolutionary dynamics through its contribution as a component of two scaled mutation rates: the genomic rate of mutation and the mutation supply rate.

The *genomic rate of mutation* is the number of novel mutations U arising anywhere in the genome during a single replication. It is assumed that rates and probabilities are interconvertible, and that numbers are obtained by multiplying a probability by the number of trials, such that $U = Gu$, where G is the size of the coding genome in nucleotides. It is an important quantity because it determines the mean fitness of a population. It might seem obvious that mean fitness will be depressed further by deleterious mutations of larger effect, but this is not the case. Supposing that deleterious mutations of average effect s arise at a rate U per genome per generation, the mean fitness of the population is $\bar{w} = \exp(-U)$ regardless of s (Haldane 1937). This is because mutations of large effect are rapidly removed by selection, whereas those of less effect linger. G is a large number that is characteristic of a species and can be determined precisely, so we would expect that U is also characteristic of a species in most circumstances.

The amount of genetic variation created by mutation governs the rate at which adaptedness is degraded; at the same time, it equally governs the rate at which adaptedness can be maintained or restored by selection. It depends on the input of new mutations at rate u per generation, and their loss through drift at rate $1/N$. The ratio of these two quantities is $M = u/(1/N) = Nu$, the net *mutation supply rate* per generation. In a diploid population there are twice as many genes as individuals, and so the mutation supply rate $= 2Nu$. This is an important quantity because it determines the dynamic regime within which selection will operate. If $M \ll 1$ then the population will be genetically uniform except for brief periods of time during which a beneficial mutation is spreading; otherwise the population will be a complex and changing mixture of genotypes of varying fitness. N often varies over many orders of magnitude within a species and is often very difficult to estimate precisely, so we would expect that M is characteristic of a population but not necessarily of a species.

2.2.2 Rate of deleterious mutation

The classical methods for estimating the rate of mutation at a given locus involve counting mutants in replicate lines or at different stages of population growth. Only mutants that have a characteristic morphological or biochemical phenotype can be detected. Careful studies in seven model organisms gave an average estimate of $u = 2.3 \times 10^{-10}$ (Drake 1991), with no indication of any consistent difference between prokaryotes and eukaryotes, nor between unicellular and multicellular organisms. The same data gives an average estimate of $U = 5 \times 10^{-3}$.

There are several ways of estimating mutation rates indirectly (reviewed in Drake *et al.* 1998). Deleterious recessive mutations will persist in diploid populations at a frequency of $\sqrt{(u/s)}$, so the mutation rate can be estimated if the selection coefficient is known. Haldane (1932) used this method to estimate $u_L \approx 10^{-5}$ per generation for haemophilia. Alternatively, affected individuals born to unaffected parents for a dominant disorder must bear a novel mutation, and this yields similar estimates for u_L in the range 10^{-4}–10^{-6} per generation (Vogul and Motulsky 1997). If a typical gene consists of 2×10^3 bp, the estimates of u_L imply $u \approx 2 \times 10^{-7}$–$2 \times 10^{-9}$ per generation. These are presumably underestimates because only a fraction of mutations will cause disease.

In an obligately selfed diploid population deleterious mutations are quickly exposed to selection, and the extent of inbreeding depression will depend on the influx of new mutations. If selfed progeny have fitness w_s and randomly outcrossed progeny w_x then the relative fitness of the selfed progeny is $(w_s/w_x) = \exp(-HU)$ where $H = (½ - h)$ expresses the degree of dominance of new deleterious mutations (Charlesworth *et al.* 1990). Consequently the genomic rate of deleterious mutation can be estimated as $U = -2[\ln(w_s/w_x)]/(1 - 2h)$. Applying this procedure to two species of annual plant, Johnstone and Schoen (1995) found $U = 0.24$–0.87 for viability and seed production, implying $U = 0.1$–0.4 per haploid genome. Similar techniques have been developed for estimating U from comparisons of inbred and outbred populations (Deng and Lynch 1996). Estimates of U based on inbreeding depression seem to fall in the range 0.1–1 (Drake *et al.* 1998 Table 7).

There are also phylogenetic methods that involve counting the number of different bases in orthologous genes for two species whose last common ancestor can be dated. When it is reasonable to ignore selection, as for pseudogenes, this has yielded estimates of $u \approx 2 \times 10^{-8}$ per generation for the human genome (Kondrashov and Crow 1993, Nachman and Crowell 1997). There are about 30 cell divisions in the human germline, so these estimates imply $u \approx 10^{-10}$–10^{-11} per replication.

2.2.3 Decay of isolate lines in the absence of selection

Deleterious mutations cannot usually be studied directly because they may occur anywhere in the genome and most have no discrete phenotype. They can be accumulated blindly, however, by the serial transfer of random individuals: a selection experiment without selection. A single individual founds a population that is allowed to reproduce for a certain period, after which a single descendant is chosen to initiate the next cycle of growth. By the end of the growth cycle each individual will have inherited a certain number of deleterious mutations from its ancestors. Because these mutations arise independently in different lineages, individuals will usually bear mutations at different loci, and all of these mutations will be rare in the population as a whole. When a single individual is chosen at this point, however, all the mutations it bears are necessarily fixed in the population it founds. Artificial selection for characters associated with higher fitness is precluded by choosing the single founder of each cycle strictly at random. The isolate line therefore tends to accumulate an increasing burden of deleterious mutations, and the decline in fitness this causes can be used to estimate the mutation rate. A random individual will bear on average U new deleterious mutations. If the average selection coefficient associated with each mildly deleterious mutations is s, then the fitness of an individual bearing U mutations relative to its immediate ancestor will be $\ln w = U \ln (1 - s)$, assuming that fitness effects are multiplicative. Consequently, the population mean fitness at time t will be linear in U, $\ln w_t = Ut \ln (1 - s) \approx -Uts$. Because lineages accumulate mutations independently, replicate lines will diverge linearly through time as $\mathrm{Var}(\ln w_t) = Uts^2$. Both U and s can then be estimated from replicated mutation accumulation experiments. Let D_W be the change in $\ln w$ and D_V the change in $\mathrm{Var}(\ln w)$, from the slopes of the linear regressions. Then $U = D_W{}^2/D_V$ and $s = D_V/D_W$. These estimates are unbiased only if all mutations have equal negative effect on fitness, otherwise U is biased downward and s biased upward. For example, if mutational effects are exponentially distributed $U = 2D_W{}^2/D_V$ and $s = D_V/2D_W$. A third useful parameter is the mutational heritability, the mutational variance $V_M = Us^2$ standardized by the environmental variance, which expresses the quantity of novel variation that becomes available to selection per generation. Estimation procedures are discussed in detail by Keightley (1998). There is a useful short review by Bataillon (2000).

The first modern mutation-accumulation experiments were performed by Mukai (1964), who found a decline in fitness of roughly 1% per generation leading to an estimate of $U = 0.4$ per generation in *Drosophila*, corresponding to $U = 0.016$ per germline replication. They became very fashionable when Kondrashov (1988) persuaded population geneticists that the rate of deleterious mutation was a crucial population parameter, and many careful and extensive experiments have been reported in *Drosophila* and *Caenorhabditis* (e.g. Fry *et al.* 1999, Vassilieva and Lynch 1999). Most of these found much slower declines in fitness of roughly 0.1%

per generation, yielding much smaller estimates of U of about 0.05–0.005 per generation, corresponding to roughly 1×10^{-3} per germline replication. This is rather surprising, as the lethal mutation rate in *Drosphila* is well known to be about 0.01–0.05 per genome per generation (Crow and Simmons 1983). The discrepancy between the original estimates and later work provoked a great deal of discussion but may reflect nothing more than variation of u among species and strains (Baer *et al.* 2005). Mutation-accumulation experiments with microbes are somewhat more difficult to interpret because new deleterious mutations may be lost stochastically during culture growth. Taking this into account, Kibota and Lynch (1996) obtained $U \approx 2 \times 10^{-4}$ in *E. coli* (Figure 2.3). Experiments with yeast have also generally yielded low estimates of $U \approx 1 \times 10^{-4}$ per replication (e.g. Joseph and Hall 2004). All these results imply that u is roughly 10^{-11} per replication.

An emerging consensus that $U \ll 1$ was upset by sequencing and directly counting mutations in mutation accumulation lines of *Caenorhabditis*, which showed that $u = 2 \times 10^{-8}$ per generation (i.e. about 2×10^{-9} per germline replication) and thus $U \approx 2$ per generation (Denver *et al.* 2004). The most obvious source of this large discrepancy is the effect of mutations in laboratory conditions, as any beneficial mutations will mask the effect of those that are deleterious.

2.2.4 Mutation rate in other replicators

Other replicators have higher rates of mutation than chromosomal genes in cellular organisms. Mitochondrial genes have long been known to evolve faster than chromosomal genes in animals (Brown *et al.* 1979), and this seems to be attributable to a higher mutation rate, perhaps caused by oxidative stress or the lack of some nuclear DNA repair mechanisms. Sequencing of lines from a *Caenorhabditis* mutation-accumulation experiment gave $U = 2 \times 10^{-7}$ per nucleotide, two orders of magnitude greater than the chromosomal rate (Denver *et al.* 2000). I cannot find any direct estimates of chloroplast mutation rates, although phylogenetic analyses of land plants suggests a value of $U = 1–2 \times 10^{-9}$ per nucleotide per year, and presumably about the same per generation (Curtis and Clegg 1984, Kawai and Otsuka 2004). RNA genomes lack exonuclease proofreading and so have very high mutation rates. In the cases summarized by Drake *et al.* (1998), $u \approx 5 \times 10^{-4}$ and $U \approx 1-5$ per replication. Elena and Moya (1999) analysed several experiments with vesicular stomatitis virus (VSV), an RNA virus, and found $U = 1.2–2.4$. In short, the values for

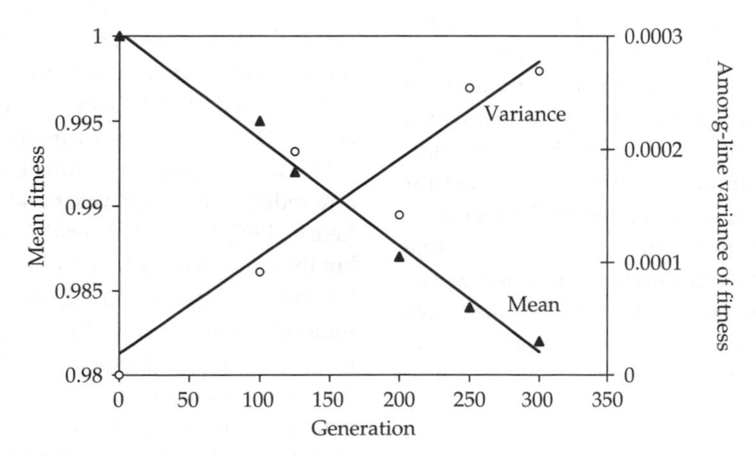

Figure 2.3 Effect of isolate culture on fitness in *E. coli*. From Kibota and Lynch (1996).

chromosomal genes may have to be modified when other kinds of replicators are studied.

2.2.5 The genomic mutation rate

A compendium of genome sizes can be found at the Database of Genome Sizes (*www.cbs.dtu.dk/databases/DOGS*). The smallest genomes of free-living microbes consist of about 10^6 nucleotides and the largest of about 10^7 nucleotides; *E. coli* and *Bacillus subtilis* both have about 4.2×10^6 nucleotides. The genomes of budding yeast *Saccharomyces* (1.2×10^7) and fission yeast (1.4×10^7) are about the same size as the largest bacteria. The unicellular green alga *Chlamydomonas*, the small plant *Arabidopsis*, and the small nematode *Caenorhabditis* all have rather compact genomes of about 10^8 nucleotides; the *Drosophila* genome is about twice as large. Other genomes are much larger: mouse and human genomes are both about 3.4×10^9 nucleotides, and the largest genomes yet measured belong to amoebas, lilies, and lungfish and exceed 10^{11} nucleotides. Much of these large genomes, however, consists of sequences in which mutations would not be deleterious. The human genome contains only about four times as many genes as yeast, suggesting that 5×10^7 is a reasonable upper limit to the number of functional nucleotides in multicellular organisms, and is close to the value actually observed in model organisms whose genetics is well known. These figures suggest that in bacteria $U \approx 2 \times 10^{-9} \times 5 \times 10^6 = 10^{-2}$ and in eukaryotic microbes $U \approx 0.02$–0.2. In multicellular organisms the evolutionarily relevant timescale for U is the generation, so it seems reasonable to multiply by the number of replications in the germline, which is about 10 in *Caenorhabditis*, about 25 in *Drosophila*, and about 30 in humans. This yields $U \approx 1$–5 for plants and animals, as suggested by Denver *et al.* (2004). Although there has been considerable debate about the genomic mutation rate in the past, these figures seem likely to be reliable. Their most important implication is that evolutionary processes will be different in bacteria ($U \ll 1$) and multicellular eukaryotes ($U \approx 1$) to the extent that they are governed by the genomic mutation rate, with eukaryotic microbes being intermediate.

2.2.6 The effect of mutations

The original experiments by Mukai gave estimates of $s = 0.03$ in homozygotes. Most subsequent experiments have suggested higher values of 0.1–0.2, although in *E. coli* the Kibota–Lynch experiment gave $s = 0.012$. The distribution of fitness effects in these experiments is unknown, however, and it is possible that a large class of neutral mutations occurs but cannot be detected. The distribution of fitness among mutation accumulation lines of an RNA virus had a long tail of subvital lines representing severely deleterious mutations (Elena and Moya 1999), but the number of mutations in each line is unknown. Random nucleotide substitutions in VSV caused an average fitness decline of 19%, although this was largely attributable to 40% of lethal mutations, so that $s = 0.138$ among viable mutants (Sanjuan *et al.* 2004). The distribution of effect is often well described by a compound distribution in which one component is a beta or similar distribution for the bulk of mutations of small effect ($s \approx 0.01$–0.03), a second comprises a few mutations of much larger effect ($s \approx 0.1$–0.2) scattered more or less uniformly, and the third is the mode at lethality.

The environment in which a new mutation is expressed may modify its effect on fitness, and there seems to be a general feeling that the effect should be more severe in more stressful environments. This should cause a faster drop in Us under stressful conditions, which was observed in a *Drosophila* experiment by Kondrashov and Houle (1994), while Vassilieva *et al.* (2000) found that s was much greater when mutation-accumulation lines were assayed at 12 °C instead of the temperature of 20 °C at which they were propagated.

Disrupting genes by transposon insertion usually reduces fitness in yeast and *E. coli* (Elena and Lenski 1997, Elena *et al.* 1998, Thatcher *et al.* 1998), but the effect is usually surprisingly slight and may be expressed in some conditions but not in others. Random insertion of Tn10 into *E. coli* reduced fitness by about 3% (Elena *et al.* 1998). About 80% of insertions reduced fitness, whereas the rest were neutral; none increased fitness. The distributions of fitness had a long but very low tail of about 5%

of subvital lines with fitness less than 0.95; if these are excluded the average effect falls to $s = 0.016$.

The systematic deletion of almost all genes in the yeast genome has made it possible to provide an exhaustive account of the effects of loss-of-function mutations (Winzeler *et al.* 1999, Giaever *et al.* 2002). As these represent the largest possible genetic effect it is reasonable to expect they will have large negative phenotypic effects. Surprisingly, only about 20% are lethal, whereas most of the rest have no obvious qualitative effect on growth. In most cases a barcode technique has been used to measure quantitative effects on growth: a mixture of all the deletion strains is propagated and its composition monitored by measuring the hybridization intensity of the unique DNA tags borne by the strains. This is a powerful way of identifying the genes responsible for maintaining normal activity under a defined stress. To estimate the average effect of a deletion, however, it is necessary to compare the strains with their ancestor, either in pure culture or in mixture.

I grew all 4846 viable haploid strains of the yeast single-deletion set in such a way that each strain could be compared with two replicate cultures of the ancestral undeleted strain growing on the same row on the same plate. The measure of growth was the optical density of the stirred culture after 48 hours of growth at 28 °C in rich medium (yeast peptone dextrose, or YPD). A standard growth score was then calculated as Y_{standard} = (strain − ancestor)/(ancestor − blank), using the average of uninoculated wells on each plate as a blank. A score of 0 indicates that the deletion has no effect, whereas a score of −1 indicates that it is lethal. The expected outcome on the Fisherian view is an exponential decline in frequency from a mode of severely impaired strains with large negative Y_{standard}. As we know from previous experiments that most strains are not very severely impaired, however, a plausible if theoretically unfounded alternative is a mode at small negative values of Y_{standard} with a tail of subvital strains extending towards lethality. Neither of these possibilities was realized. Instead, Y_{standard} was narrowly and almost symmetrically distributed around a mode near zero (Figure 2.4). A few strains (about 20) were nearly lethal and grew only very slowly; only a very few strains (about 10) had intermediate values. Excluding these cases, the phenotypic standard deviation of the means was $\sigma_P = 0.052$ and the genetic standard deviation $\sigma_G = 0.039$. This shows that genes fall into two sharply delineated categories, defined by the effect of deletion: *Fisherian genes*, whose deletion is lethal, and *non-Fisherian genes*, whose deletion is more or less inconsequential. Moreover, the overall effect of deletion was beneficial: Y_{standard} exceeded zero for about 2/3 of all loci (3489/4846), with a mean value of +0.0147, or +0.0195 excluding the few nearly lethal strains.

Fisherian genes tend to be concentrated in certain functional categories, especially those that include the underlying infrastructure of the cell, such as ribosome assembly, RNA metabolism, transcription, and the cell cycle. Genes for metabolism and transport are seldom essential because complete medium supplies most substrates at high concentration, and likewise genes for stress response or the sexual cycle are dispensable because they are not required in normal conditions of vegetative culture. A similar pattern is shown by Y_{standard}, so that the fraction of Fisherian genes in a functional category is correlated with the effect on growth of deleting non-Fisherian genes (Figure 2.5).

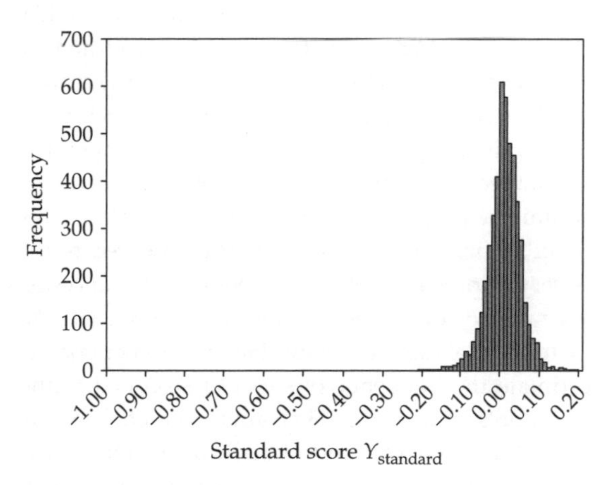

Figure 2.4 Effect of complete loss-of-function mutation on growth. This shows the distribution of growth relative to the ancestor in rich complete medium of the deletion of each of 4829 genes in yeast (lethal deletions are not shown). Note the almost complete absence of deletions with intermediate effects on growth.

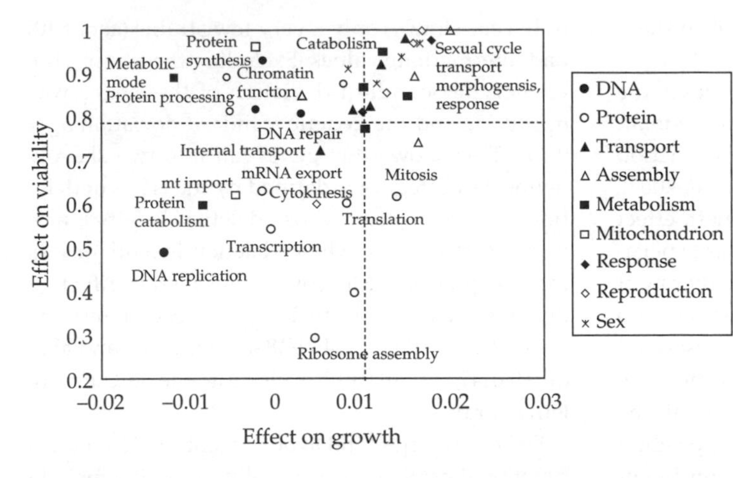

Figure 2.5 Effect of deleting genes in different functional categories. Effect on viability is probability that deleting gene in given category permits growth; effect on growth is reduction in standard score. Broken lines are means over all categories.

The two chief limitations of this result are that it applies only to yield in pure culture, and only to a rich artificial medium. The fitness of the ancestor in competition with an arbitrary mixture of deletion strains was evaluated by forming pools of 46–48 strains, in equal proportions, and then culturing these with a roughly equal quantity of the intact ancestor. The frequency of the ancestor was then measured at intervals of 3–4 generations for 35 generations (10 transfers) in each of 48 non-overlapping pools, representing about ½ of all viable deletions. Over all these 48 experiments the average frequency of the ancestor declined steadily through time, driven by a selection coefficient of −0.125 (Figure 2.6). Moreover, the ancestor declined in every experiment. This result does not clearly distinguish between two extreme possibilities: each deletion strain might possess about the same competitive advantage relative to the ancestor, or each pool might contain a single superior strain that replaced all other strains and the ancestor. The observed result is very improbable if strains superior to the ancestor are rare, however, whereas it is nearly certain if 20% or more of strains are superior to the ancestor. To estimate the frequency of superior strains directly, two pools comprising 93 strains were chosen and each strain competed individually against its ancestor for about 30 generations. Selection coefficients from these pairwise competitions had a mode at a small negative value and a right tail of positive values (Figure 2.7). The mean selection coefficient of the ancestor was −0.0177 (se 0.0078) and the ancestor was inferior in about 2/3 (64/97) of all cases. The outcome of competition is thus very similar to that expected from growth in pure culture. Moreover, the selection coefficient of deletants is positively correlated with standard growth (Pearson correlation coefficient (r = 0.42), with a regression coefficient (b = + 0.88) close to +1 and an intercept (c = 0.0045) close to zero. Some of the genes that were deleted were subsequently found to be non-functional. Deleting these should have no effect, but in fact gave $y_{standard} = 0 + 0.0350$. Hence, the slight elevation of growth shown by the deletion strains was caused by the genetic manipulation used to construct them, the insertion of a cassette containing a gene for antibiotic resistance. The true effect of deletion is (0.0195 – 0.0350) = –0.0155,

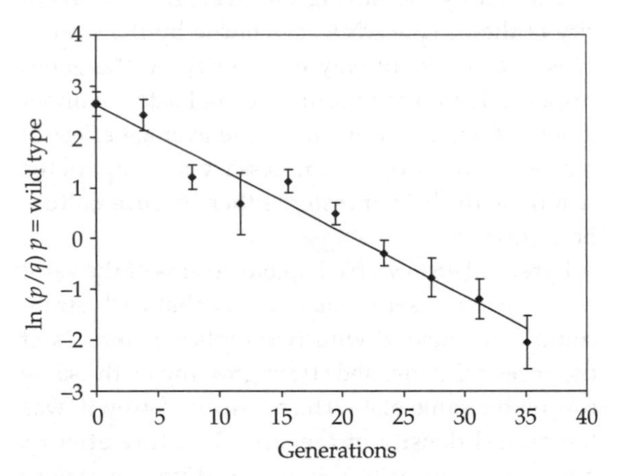

Figure 2.6 Loss of undeleted ancestor from pools of 48 deletion strains. Regression is $y = -0.125x + 2.64$, where the slope is the selection coefficient acting against the undeleted ancestor.

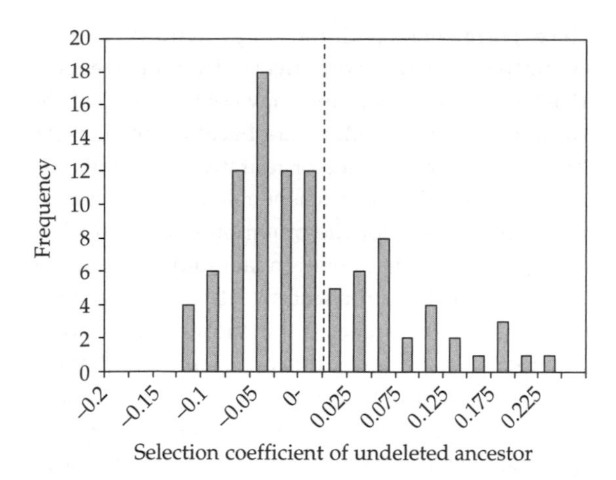

Figure 2.7 Fitness of undeleted ancestor in competition with a single deletion strain.

a value comparable to that obtained in E. coli and other organisms.

The almost universal opinion that almost all mutations are deleterious is therefore correct only for a minority of genes. For most genes, even complete loss-of function mutations are almost as likely to be beneficial as to be deleterious, with an average effect close to zero. This applies to growth in rich medium in laboratory conditions. There is strong evidence that mutation is generally deleterious in natural environments, because nucleotides that change amino acids are conserved relative to those that do not (§7.5). This demonstrates that there is some feature of natural environments that is not exhibited by laboratory conditions of growth, and also that estimates of the effect of mutations made in the laboratory cannot be assumed to hold in the field.

The unexpectedly mild effect of most deletions leads to a difficulty: if complete deletion has little or no effect on fitness, how is the integrity and normal function of non-essential genes preserved? One response is that deletions are not really deletions because much of the genome is duplicated. It is true that the probability that deletion is lethal is less for duplicated genes (about 10%) than for unique genes (about 15%). This does not apply in most cases, however, nor is it easy to understand how the integrity of both copies can be maintained.

Alternatively, deletions may not be quasi-neutral because small multiplicative effects on fitness are decisive. The effect of loss-of-function mutations on the population can be expressed in two ways:

- The mean number of mutations per haploid genome W: the fraction of the genome bearing loss-of-function mutations is then $m = W/G$.
- The number of individuals killed by genetic impairment K, resulting in the loss of a fraction $k = K/N$ of the population.

In a haplont, loss-of-function mutations are always expressed, and hence $W = G(u_L/s)$, and the fraction of zygotes that are killed by genetic impairment is

$$k \approx 1 - e^{-Ws} = 1 - e^{-U}.$$

This is the well-known result that the mean fitness of the population depends only on the genomic mutation rate. If $U \ll 1$ then $k \approx U$. If s is small then W can be large, even for modest mutation rates. For example, if $s = 0.01$ and $u_L = 10^{-6}$ then 1% of all loci will be loss-of-function, so that a genome of size 10^4 will bear 100 non-functional genes. Harrison *et al.* (2002) found about 200 disabled loci in a standard laboratory strain of yeast. Many of these may be pseudogenes that have become fixed, but 35 loci are disabled by a single stop codon and may represent the predicted loss-of-function mutations; more extensive sequencing of whole genomes or large parts of genomes will resolve the issue. The accumulation of non-functional genes will be checked if selection is more effective because of interactions with the environment or with the rest of the genome. These possibilities are discussed in later sections.

2.2.7 Beneficial mutations

The rate of deleterious mutation can be estimated without regard to the state of the population if almost all mutations are expected to be deleterious provided that the population is reasonably well adapted to the conditions of growth. The rate of beneficial mutation, on the other hand, is crucially dependent on the level of adaptation. If a population has been maintained in the same conditions for thousands of generations all or almost all the accessible beneficial mutations will already have been fixed. The transposition lines in which Elena *et al.* (1998) failed to find any beneficial mutations, for example, were derived from an ancestral line that had been propagated in the same environment for 10 000 generations. Nevertheless, mutations caused by transposition are occasionally beneficial. Insertion of Tn10 into experimental populations of *E. coli* generated beneficial mutations that spread to fixation (Chao *et al.* 1983). Insertion of the transposable element *Ty1* into yeast consistently reduced stationary density, but after competition genotypes lacking insertions were usually eliminated (Wilke and Adams 1992). These mutations were amplified by selection, so their initial frequency is unknown, but in novel conditions of growth beneficial mutations may not be as rare as is commonly supposed. When *E. coli* lines were cultured in maltose, a substrate they were not well adapted to, several beneficial insertions were found (Remold and Lenski 2001). Mutation-accumulation lines may decline only slowly because their load of deleterious mutations is balanced by beneficial mutations that are trapped at the same time, and Shaw *et al.* (2002) found that roughly half of all mutations in *Arabidopsis* are beneficial. This claim was vigorously rebutted (Keightley and Lynch 2003), but the effect of gene deletion in yeast shows that it is probably sound in laboratory conditions. Joseph and Hall (2004) found that about 5% of mutations in yeast mutation-accumulation lines were beneficial. The most precise estimate of the beneficial mutation rate at a nucleotide level is the study of an RNA virus by Sanjuan *et al.* (2004), who found 2/29 viable single-nucleotide mutations were beneficial. They used a chimaeric ancestral sequence that had not been cultured long under laboratory conditions and may well have been poorly adapted. Other virus experiments yield much lower estimates: the initial probability that a random mutation will increase fitness is about 0.0034 for phage φX174 adapting to high temperature and about 0.0009 for RNA VSV adapting to a new host cell type (see § 3.7.11).

In brief, a substantial fraction of random mutations are neutral or may actually enhance growth in laboratory conditions. There is no such thing as a fixed or even typical rate of beneficial mutation, because the potential supply of beneficial mutations depends on how closely the population is adapted to the experimental conditions of growth. When they occur they will dilute the effect of the more frequent deleterious mutations and may contribute to of the discrepancy in mutation rate between phenotypic and molecular analyses of mutation-accumulation experiments. The most reasonable estimate that we currently possess for the mutation rate per nucleotide per replication in benign laboratory conditions is $u \approx 2 \times 10^{-9}$ per replication. This is higher than most previous estimates, but in view of their known and likely biases it is not inconsistent with them. If we take the typical length of a gene as 2×10^3 nucleotides ($L = 1.2 \times 10^7$ for the 6000 genes of the yeast genome) then $u_L \approx 10^{-6}$ per replication, which allowing for the number of replications in the human germline is not greatly different from the estimate made by Haldane more than 70 years ago. Taking the average of the φX174 and VSV estimates suggests that the initial rate of beneficial mutation per nucleotide is about $(2 \times 10^{-3}) \times (2 \times 10^{-9}) = 4 \times 10^{-12}$ per replication, but the yeast data suggest that this may be an underestimate for cellular organisms in benign laboratory conditions.

2.2.8 The rate of accumulation of genetic variance in fitness

A serial transfer experiment that is originally inoculated with a single cell or individual will give rise to a large population after a few generations, from which we can extract a number of cells and measure their fitness. They will probably all have nearly the same fitness, because there has not been enough time for very many mutations to have occurred. If we continue to transfer the population for hundreds or thousands of generations, lineages will diverge through the stochastic accumulation of mutations, and when we repeat the assay we shall expect to find substantial genetic variation in fitness among cells. This mutational variance can be estimated from the among-line variance in mutation-accumulation experiments, where each line represents a single lineage. The mutational heritability of offspring production in *Caenorhabditis* estimated in this way increased by $V_M = 1\text{–}2 \times 10^{-3}$ per generation (Keightley and Caballero 1997). An alternative procedure is to isolate individuals from mass cultures and measure the rate of increase of variance through time (Goho and Bell 2000). Although natural selection will in this case act with full force, an experiment with *Chlamydomonas* yielded a similar estimate of 1–4 $\times$ 10^{-3} per generation (Figure 2.8). Thus, a level of genetic variance corresponding to the quantity of microenvironmental variation among cultures or developmental variation among individuals will be supplied by mutation within a few hundred generations. A mutation-accumulation experiment using *Daphnia* found that equilibrium levels of genetic variance for life-history characters would be approached within a few hundred generations in natural populations (Lynch *et al.* 1998). This slight incessant input of mutational novelty is the original source of all subsequent evolutionary change. Because most mutations are deleterious in field conditions, it will generally reduce the fit between population and environment, which must then be restored by selection. On the other hand, should the environment change it is the only process by which new favourable variants can appear and become available to selection.

2.2.9 The replication limit

This argument can be used to show that there is a limit to the rate of mutation that genomes can sustain. A replicator such as a self-replicating RNA molecule or a DNA gene is a sequence of nucleotides, and functions properly only if this sequence is preserved. Mutation causes changes in the sequence. Suppose that each molecule replicates R times before being destroyed by some thermodynamic accident. On average, at least one of these R new molecules must be a perfect copy of the parental sequence; if this is not so, then the number of copies of that sequence must inevitably decrease through time. Thus, if the precision of replication (the probability that no errors are made during replication) is Q, it must be the case that $Q > 1/R$ if adaptedness is to be stable. Suppose that the replicator is a sequence of G nucleotides, and that the error rate per nucleotide during replication is u. The probability that no errors are made is $Q = (1 - u)^G$, or about e^{-Gu}. The minimal requirement for stability is thus $u < (1/G) \ln R$. Replicators that are long sequences of nucleotides can maintain themselves only if the mutation rate is sufficiently low.

The existence of this replication limit places severe constraints on the design of self-replicating systems. Nucleic acids can replicate themselves without enzymes, but the error rates for non-enzymic replication are about 10^{-1}–10^{-2}, so no very sophisticated replicator can be maintained. RNA replication mediated by specific replicases has error rates of about 10^{-3}–10^{-4}, so the largest single self-replicating RNA molecule is about 10^4 nucleotides long. This is why the genomes of RNA viruses such as $Q\beta$ do not in fact much exceed a length of about 10^4 nucleotides. DNA replication has a much lower error rate, of about 10^{-9}–10^{-10} per nucleotide per replication, allowing the genomes of DNA-based organisms to range up to about 10^9 nucleotides. The genome of the bacterium *E. coli* is made up of about 3.8 $\times$ 10^6 nucleotides; the human genome is about 1000 times larger, with about 3.5 $\times$ 10^9 nucleotides, approaching the upper limit for stable replication, although the constraint applies to functional nucleotides only.

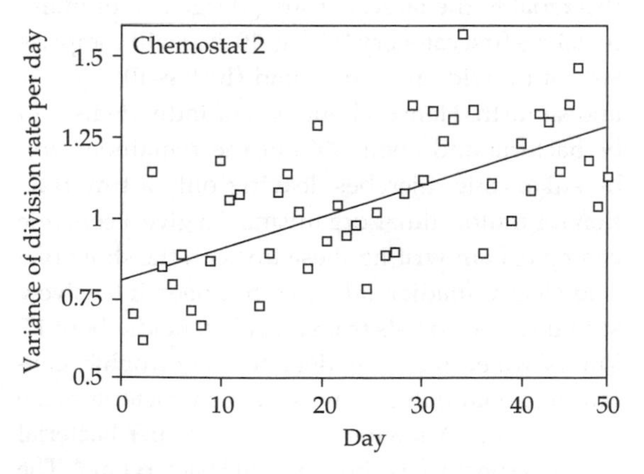

Figure 2.8 Rate of accumulation of variance of fitness despite countervailing natural selection in a chemostat population of *Chlamydomonas* (from Goho and Bell 2000). The regression is $y = 0.0096x + 0.81$, there being several generations per day.

2.2.10 The size spectrum

The rate of adaptation will be strongly influenced by the number of individuals of novel type that enter the population each generation. This is why theoretical accounts of selection or drift often begin, 'A diploid population of N individuals . . .'. It is by no means straightforward, however, to measure N, to develop reliable generalizations about population size, or even to identify to identify with confidence the set of individuals that constitutes a population.

When electronic particle counters came into use in the 1970s they soon produced a remarkable result: in ocean water the mass of particles in equal logarithmic size classes is independent of size (Sheldon *et al.* 1972). The width of logarithmic classes increases with size, so it is usual to express the rule for standardized data by dividing the biomass in each class by the width of the class. The outcome is a power law, $B = kW^{-y}$, relating biomass B to size (mass or weight) W through the exponent y, governing the shape of the relation, and the scaling factor k. When the non-normalized Sheldon distribution is flat $y = 1$, which I take to be the paradigmatic form of the size spectrum (Figure 2.9). This applies to living organisms as well as to suspended detritus: for example, Rodriguez and Mullin (1986) collected plankton of 10^{-7} g to 1 g in size from the surface waters of the North Pacific Gyre and obtained $y = 1.16$. This provides the rule that biomass is conserved on a logarithmic scale as the basis for a general theory of

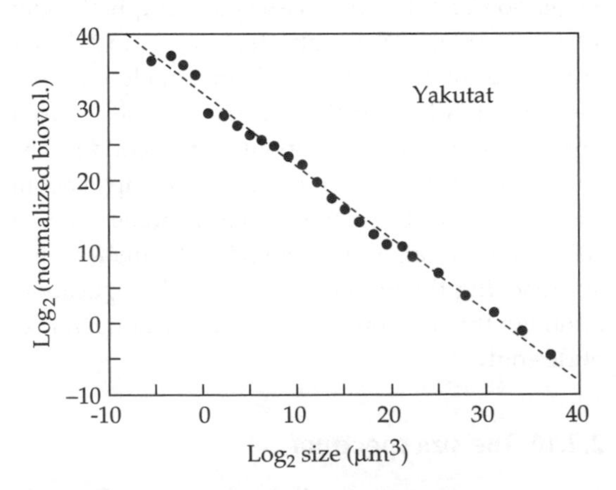

Figure 2.9 Size distribution of biomass in the ocean. This example is a sample from a seamount off the coast of Nova Scotia, from Quinones *et al.* (2003). The broken line is a slope of −1.

abundance, although many studies have documented exceptions to it.

The size-spectrum rule is a property of a broad class of distributions with long upper tails, such that extremely large events are occasionally observed. The best-known of these is the *Pareto distribution*, used by the nineteenth-century economist Alfred Pareto to characterize the frequency of people whose incomes x exceeded a certain amount: the cumulative Pareto distribution is $\mathrm{Prob}(W > x) = \kappa^{\psi}W^{-\psi}$, where κ is the scaling factor (usually the smallest observation) and ψ is an empirically determined constant. If this is expressed in terms of frequency in equal logarithmic categories the exponent is $-\psi + 1$; hence, a Pareto distribution with $\psi = 1$ is equivalent to a Sheldon size spectrum, yielding a normalized spectrum with $y = \psi$. The normalized spectrum refers to biomass; the underlying probability density distribution is the power law $\mathrm{Prob}(W = w) = W^{-\psi-1}$. That is, the power law for total abundance $A = kW^{-y}$ will have $y = 1 + \psi = 2$ for the Sheldon size spectrum. The relationship between the Pareto distribution and the power-law description of size spectra was pointed out by Vidondo *et al.* (1997).

The general feature of the Pareto distribution is the overwhelming predominance of small events. The probability density of the frequency of particles falling within the range W_0, W_1 is $\kappa^{\psi}(W_0^{-\psi} - W_1^{-\psi})$. If we order a planktonic community into equal log intervals of size from the smallest at 10^{-12} g (bacteria) to the largest at 10^{6} g (large fish or mammals) the first category (10^{-12} g–10^{-11} g) will comprise 90% of individuals, the second (10^{-11} g–10^{-10} g) 9%, and so forth. Hence, about 99% of individuals will be bacteria and about 99% of the remainder will be eukaryotic microbes, leaving only a tiny fraction for multicellular organisms. To give a concrete example, I am writing these lines on the shore of a cold clear Canadian lake that has been intensively studied by ecologists from McGill. It holds about 1.7 km^3 of water and is moderately oligotrophic, containing about 0.18 g C m^{-3} of which bacteria make up 0.01275 g. Allowing 4×10^{-14} g C per bacterial cell, this constitutes about 3×10^{11} bacteria m^{-3}. The lake is a moderately well-stirred vessel, with a residence time of $1 - 2y$, supporting a total population of planktonic bacteria of about 6×10^{20} cells. The comparable rough figures for other groups are 6×10^{18} eukaryotic microbes, 1×10^{17} ciliates and microscopic metazoans (mostly rotifers), 6×10^{14} macroscopic metazoans (mostly crustaceans), and 2.5×10^{7} fish. Calculations like this show that small organisms are extremely abundant. If the bacteria in the lake could be regarded as a single population, it would contain all possible combinations of three mutations, given a genome of 4×10^{6} nucleotides. This lake holds only a small fraction (about 10^{-5}) of the world's fresh water; and fresh water itself holds only a small fraction (about 10^{-3}) of aquatic organisms. The world population of bacteria in aquatic habitats is about 1.2×10^{29}, which in turn is exceeded by the number in soil (2.6×10^{29}) and in subterranean habitats ($4-6 \times 10^{30}$) according to Whitman *et al.* (1998). Disregarding subterranean habitats, whose contribution may be overestimated, the population of bacteria living on the surface of the Earth is about 4×10^{29} cells. Hence, using $y = 2$ in the power law for world abundance $A = kW^{-y}$ we have $k = 10^{5.6}$ (units g). With Pareto distribution of abundance, and taking all organisms 10^{-12}–10^{-10} g in mass to be bacteria, the world population of multicellular organisms in excess of 1 mg mass (1 mm length) is about 4×10^{20} individuals. Regardless of absolute density, Pareto distribution will ensure that bacteria will always be vastly more abundant than metazoans, with eukaryotic microbes occupying an intermediate place. Hence bacteria and metazoans are likely to occupy different dynamic regimes with respect to mutation supply rates and the evolutionary processes they influence.

2.2.11 Distribution of species abundance

For most evolutionary processes, however, the important number will be the abundance, or evolutionary extent, of the species, which I interpret as the number of individuals that will eventually bear a successful beneficial mutation. This is the *substitutional population*. Organisms of some given size do not constitute a single species, because lineages are separated by incomplete horizontal transmission in bacteria and by sexuality in eukaryotes. The average abundance of species is therefore governed not only by how overall abundance varies with size but also by how diversity varies with size. Given that overall abundance $J = kW^{-y_1}$ suppose that species richness $S = J^{y_2}$; then mean abundance will be $J/S = kW^{-y_1+y_1y_2}$. Siemann *et al.* (1996) examined about 10^5 individual insects collected by sweep-netting grassland and herbage and found $S = 1.05J^{0.51}$, which taking $y_1 = 2$ and $y_2 = 0.5$ leads to $J/S = kW^{-1}$.

Duarte *et al.* (1987) surveyed aquatic organisms from bacteria to fish, ranging from 10^{-12} g to 10^5 g in size, and found that maximum abundance varied with $W^{-0.95}$ (Figure 2.10). This is the abundance that can be achieved under ideal conditions as single-species cultures in the laboratory. It corresponds to a volume of living tissue amounting to 6×10^8 μm^3 mL^{-1}, or so that maximum abundance constitutes a fraction 0.0006 of the occupiable space, regardless of body size. In natural communities abundance is much lower yet follows the same allometric rule. Peters and Wassenberg (1983) surveyed species of animals ranging between 10^{-8} and 10^5 g in size and found that mean abundance varied with $W^{-0.98}$. The pelagic community of lakes occupies $1-50 \times 10^6$ μm^3 mL^{-1} depending on resource status (Peters 1986), so the upper limit of abundance in natural communities represents between 10^{-6} and 5×10^{-5} of the occupiable space. Thus, a very rough idea of the number of individuals N of a common species

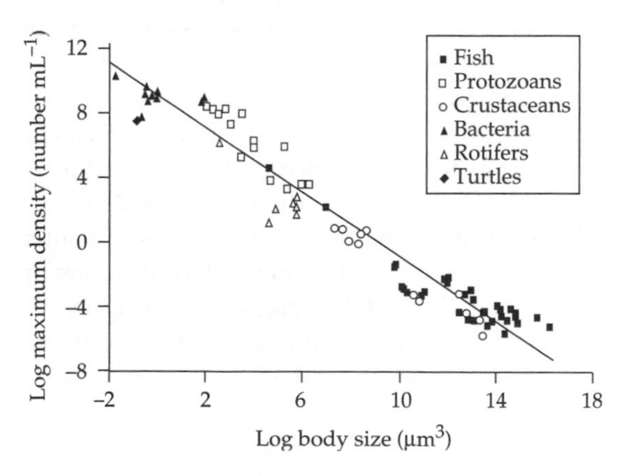

Figure 2.10 Relationship between maximum abundance and body size for aquatic organisms, from Duarte *et al.* (1987). The regression is $y = -0.95x + 8.53$.

of body mass W occupying a volume of water V is $N = 10^{-6}\ V/W$. For bacteria this suggests a maximum density of 10^6 cells mL^{-1}. In soil bacterial biomass constitutes about 1–3% of organic carbon and occupies about 10^{-5} of total volume (Sparling 1985 cited by Grundmann 2004). Dykhuizen (1998) estimated from rates of DNA re-annealing that about 20 000 common species of bacteria are found in 30 g of soil; supposing that 10 000 will be found in a 1 g sample, their realized density will be $(6 \times 10^8)/10^4 = 6 \times 10^4$ cells mL^{-1} for such species in soil and hence rather more than this (since there will be fewer species) in water. Most soil bacteria are found in pores, and Grundmann *et al.* (2001) estimated that *Nitrobacter*-like organisms occupied about 6×10^{-6} of soil porosity, which is comparable to the rule of thumb for aquatic systems. There is thus a broad general rule for the whole range of values found in nature that population size varies as the reciprocal of body size, or equivalently that the mass of a population is independent of the body size of its members.

Surveys that are taxonomically more restricted, and thus cover a narrower size range, usually report smaller exponents (see Cotgreave 1993). Damuth (1987) showed that the exponent for mammals was about -0.75, and argued that this reflected a principle of energy equivalence, as metabolic rate scales as $W^{0.75}$. For very narrow size ranges of an order of magnitude or less the relationship between size and abundance becomes triangular rather than linear, presumably because rare species may be of any size (Lawton 1989). We can then ask how abundance is distributed among organisms of similar size. For the extensive insect survey conducted by Siemann *et al.* (1996) the abundance of species within the same size class followed the power law $J_R = J_1R^{-z}$, where J_R is the abundance of the Rth most abundant species and z is an empirically determined parameter. This is the *Zipf distribution*, which relates the size of an event to its rank R, and was originally devised to describe how the frequency of usage of words decreased as a power function of their rank. It has a scale factor k equal to the size of the first-ranked event and a shape factor ζ representing the exponent of the power law. The Zipf is essentially a reformulation of the cumulative Pareto distribution, because the number of events exceeding a given size (10 species of the 100 surveyed equal or exceed 100 g) is equivalent to the size of the rth event (the tenth-largest species of the 100 surveyed weighs 100 g). Thus, the Zipf and Pareto distributions amount to plotting the same data with axes reversed, and consequently their parameters are related very simply as $\zeta = 1/\psi$. Hence the corresponding exponent of the power law for the abundance of the Rth ranked species $A_R = kR^{-z}$ will be $z = 1 + 1/\psi$, where ψ is the Pareto shape function, so that $z = 2$ for the Sheldon size spectrum for which $\psi = 1$. Hence the insect data are consistent with the general rule of size invariance.

Concordance with Pareto is exceptional, however. Log species abundance is (more or less) linear with rank, not with log rank. Two major attempts were made to identify the underlying distribution. Fisher *et al.* (1943) fitted the abundance of moths to a novel distribution, the log series, in which the abundance of species if increasing rank was given by αn, $\alpha n^2/2$, $\alpha n^3/3$, . . ., where α is a fitted parameter that expresses diversity independently of sample size. This distribution always gives a mode of singletons (species represented by a single individual), with a long upper tail. Preston (1962) objected that species of intermediate abundance were the most frequent in many communities, and proposed instead a truncated log-normal distribution, the rarer species lacking from most samples. A good point of entry into this very large literature is given by Rosenzweig (1995). It has recently been shown that both patterns are generated by neutral models in which all species are demographically equivalent: communities experiencing higher rates of immigration or speciation have many rare species and resemble the log-series, whereas lower rates produce a distribution resembling the truncated log-normal (Bell 2000) (Figure 2.11). Total abundance is taken to be fixed, a constraint consistent with the fixed total biomass of organisms of given size. The underlying distribution is a multinomial which yields analytical solutions for species abundance in terms of the parameter $J\nu$, where J is the fixed total community size and ν the rate of immigration or speciation, such that $J\nu$ is the number of individuals of novel type appearing per generation

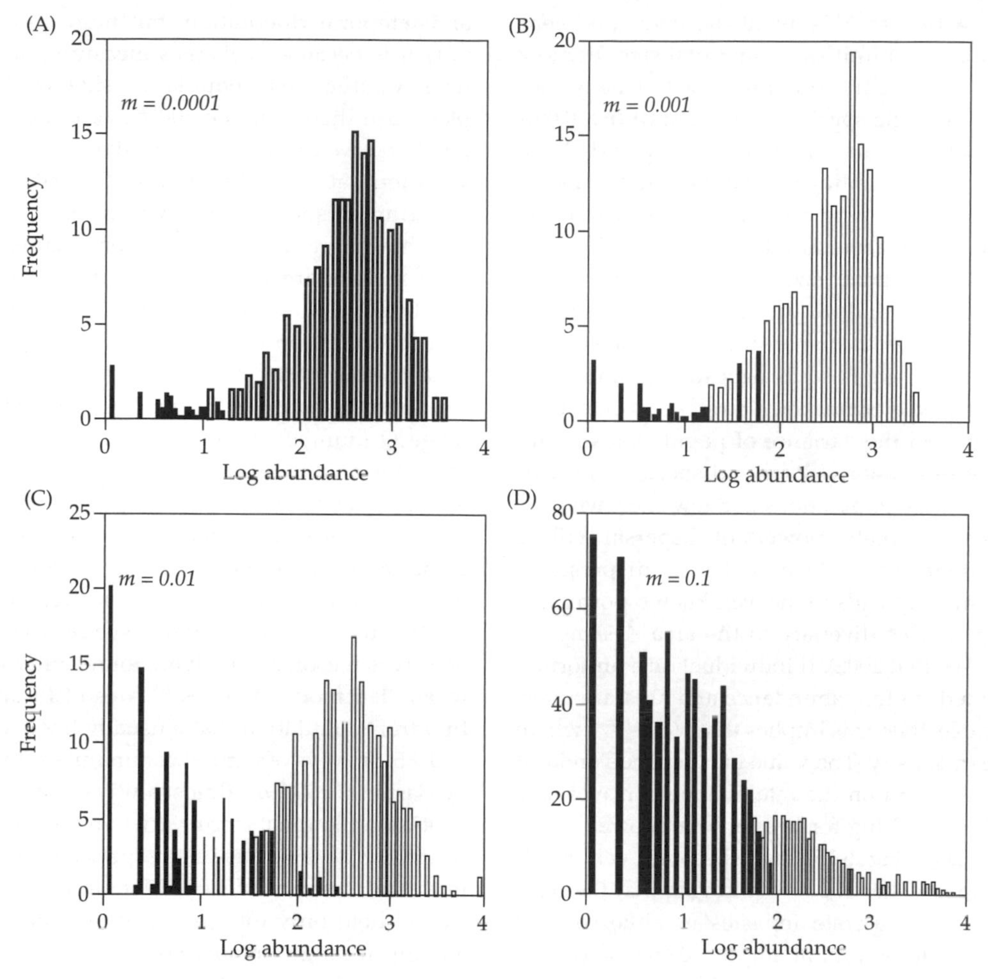

Figure 2.11 Distribution of abundance in neutral communities as a function of the immigration rate. Open bars represent long-resident lineages, solid bars recent immigrant lineages. Note that the mode of the overall distribution shifts to the left as the immigration rate *m* increases. From Bell (2000), Figure 2.

(Hubbell 2001). The neutral theory provides a simple mechanistic basis for the statistical form of the species abundance distribution and thus anchors it to the underlying ecological and evolutionary processes that govern abundance; it is moreover robust to modest competitive differences among species. In a neutral metacommunity the distribution of species abundance at a site will resemble a truncated log-normal, whereas in the community as a whole it will resemble the log-series (Hubbell 2001, p. 126). This has given us a general theory of abundance for ecologically equivalent species which are tacitly assumed to be of equal body size. Neither the neutral multinomial nor the log-series nor the truncated log-normal give rise to the power-law distribution of abundance characteristic of Pareto, however, except in the limit of very high immigration or speciation, or $\alpha \to 1$, which might hold for insects flying freely between adjacent fields.

The mean population size in the metacommunity will be $N = J/S$. For the entire community, or a random sample of individuals from it, $S = 1 - 2J\nu \ln 2\nu$ for $J \gg 1 \gg \nu$ (Hubbell 2001, p. 165). For high rates of appearance (i.e. $J\nu > 1$) S increases nearly linearly with J, so N is about the same (and very small) for communities of any total size. For low rates ($J\nu \ll 1$) small communities will usually contain only a single species, so $S = 1$ until $J\nu > 0.001$, after which it begins to rise, and consequently $N = J$. In most cases, sampling will be restricted to a local population occupying a particular site within the range of the metacommunity, and species diversity and mean population size will then depend on the size of the site and on the rate of local dispersal m between sites. If dispersal is unrestricted then sampling any site is equivalent to randomly sampling the whole community, but if $m < 1$ then both the mean and the variance of population size are sensitive to ν: poorly dispersed species will usually have large populations at a few sites, whereas species with greater powers of dispersal will be evenly distributed. Restricted local dispersal in neutral models leads to the well-known power law relating species diversity to the area A sampled, $S = cA^z$ (see Bell 2002). If individuals are uniformly distributed, so that abundance and area are interchangeable, then this implies that $N = d^{-z}J^{1-z}$ where d is mean density. The value of z depends on local dispersal m and on the rate of appearance of new types $J\nu$, increasing for higher $J\nu$ and lower m (see Hubbell 2001, Fig. 6.15). As $J\nu$ becomes very small and m approaches 1, $z \to 0$ implying that the community is depauperate and sites are alike, so $N \approx J$ for the single dominant species; conversely, high $J\nu$ and low m yields $z \to 1$, implying a diverse but spatially structured community in which the species present have similar and low abundance. For less extreme situations where m lies between 0.01 and 0.1, while $J\nu$ lies between 0.1 and 1, values of z from neutral models lie mostly between 0.15 and 0.5. Observed values are generally between 0.15 and 0.4 (Williamson 1988).

These arguments provide a framework for a general theory of abundance in communities of fixed size in terms of simple demographic processes that lead to a range of statistical generalizations that have broad support from empirical studies. Nevertheless, they do not directly provide estimates of population size N relevant to simple evolutionary theory. This is partly because of complications such as trophic structure, inbreeding, and temporal fluctuation, but more fundamentally it is because ecologists measure population density rather than population size. In the simplest case there may be 100 frogs in an isolated pond and we can be confident that $N \approx 100$ in the short term at least. If there are 10^7 *Bacillus subtilis* in a gram of soil, however, we still need to know how many grams the population occupies; likewise, if there are 10^9 *Keratella cochlearis* (a rotifer) in a lake we do not know whether this should be treated as one population or several. We could continue the allometric approach by using the *range–abundance relation*. Neutral models predict that the range Q (number of sites occupied) of a species will be related to its mean local density D (in sites where it occurs) as a power law $Q = kD^y$ (Bell 2003). The values of k and y depend on the rate of local dispersal m: for very low rates ($m < 0.001$) the relation is very loose, whereas for very high rates ($m > 0.5$) the curve saturates because species whose local density is reasonably high are found almost everywhere. For moderate rates between 0.01 and 0.1 I find from simulation that y usually lies between 1 and 1.5. Surveys of natural communities (reviewed by Gaston *et al.* 1997) almost always find a positive relation, although its form varies and the relation is often very loose because many species are abundant in a small fraction of sites. Moreover, the relation should hold only for species of the same size; at any given range, smaller species should be more abundant. If we write $N = QD$ then $Q = k(W_0/W)^z D^y$ where k is the range at unit abundance of a species of mass W relative to a reference species with mass W_0, and z is the exponent of the density–body mass relation. Substituting gives $N = ka^{1+y}W_0^{\ z}W^{zy}$. Re-plotting the data on common British birds in Gaston *et al.* (1998) gives $Q \approx 0.1D^{1.2}$, with range as the fraction of 13 census sites occupied and local density as territories ha^{-1}. Expressing range in km^2 (by multiplying by the area of Great Britain) and local density in individuals km^{-2} gives $k = 1220$ and $y = 1.2$. This predicts a total British population size for species with mass 0.1 kg of 2.2 million individuals, which is roughly correct as a maximum. Taking W_0 as 0.1 kg and using the general relationship between density and body mass in Peters (1983) yields $N = cW^{-1.2}$ with $c = 3 \times 10^6$. Thus, the total abundance of a bacterial species with mass 10^{-12}g would be $N = 3 \times 10^{20}$. If there are about 10^{10} distinct kinds of bacteria, as suggested by Curtis *et al.* (2006), then this predicts a total world bacterial population correct to within an order of magnitude. Unfortunately, arguing from birds to bacteria from a shaky statistical base means that this is little more than a guess.

2.2.12 Genetic variation and species abundance

A species comprising N individuals will experience the effects of genetic drift and inbreeding in a manner equivalent to an ideal population of N_e individuals with equal expected fecundity. The *effective population size* N_e then calibrates the rate at which heterozygosity is lost and thereby predicts the amount of diversity expected at equilibrium between drift and mutation. It is usually less than the census size because of unequal sex ratios, variation in fecundity, or fluctuations in population size (§2.3). For an outbreeding diploid sexual population the expected heterozygosity for neutral alleles is $H = 4N_e u_L/(1 + 4N_e u_L)$ (Crow and Kimura 1970). From surveys of genetic variation the effective size can then be inferred to be $N_e = [H/(1 - H)](1/4u_L)$, where u_L is the overall rate of neutral mutation for the locus or loci considered. The expected heterozygosity is calculated from allele frequencies assuming random mating; observed heterozygosity is usually less because of inbreeding. The hundreds of studies of allozyme variation reported in the 1970s provide a rich source of data. The observed heterozygosity of species in large surveys increases with the fraction of loci that are polymorphic, with almost all values from large-scale surveys lying between 0.01 and 0.2 (Figure 2.12). About one-fifth of mutations will yield an electrophoretically detectable phenotype; the fraction of these which is neutral is unknown, but an overall rate of neutral mutation per locus of 10^{-5} would imply that N_e lies between 250 and 6250 for most animals and plants. Both the mean and the range seem much too low, especially as most large-scale surveys will involve common species. One possibility is that observed heterozygosity is depressed by inbreeding, for example because most

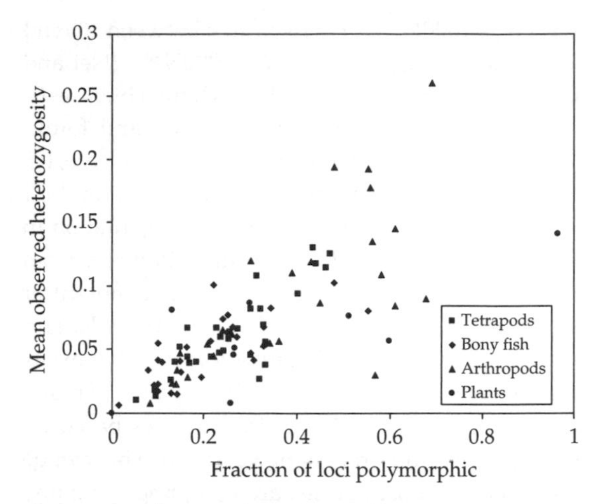

Figure 2.12 Allozyme polymorphism and heterozygosity. Data from Nevo *et al.* (1984).

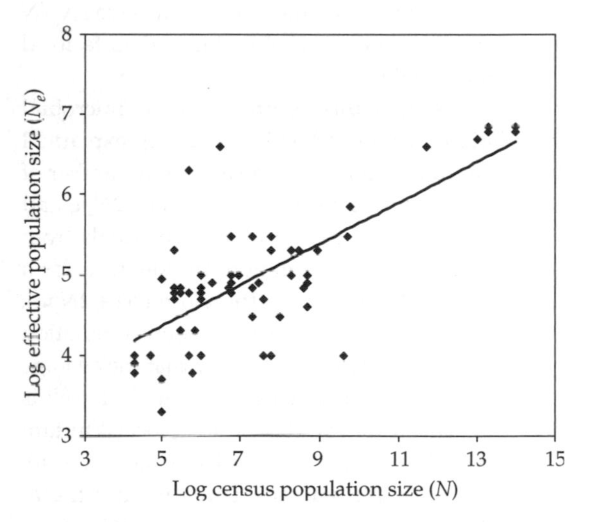

Figure 2.13 Relationship between census and effective population size. Data from Nei and Graur (1984).

individuals breed within relatively small, isolated groups. There is certainly little correlation between observed and expected heterozygosity (indicating that species vary in the level of inbreeding), but the range of expected heterozygosities is much larger, including extreme values near unity. This increases estimates of N_e, as inferred from neutral theory, but they remain much less than N. The relation between effective and census size is strongly allometric, with $N_e \approx 1000N^{0.25}$ for census size between 10^5 and 10^{14}, and consequently $N_e/N \approx 1000N^{-0.75}$ (Nei and Graur 1984; Figure 2.13). Frankham (1995) estimated N_e/N from demographic data and found values of about 0.6 when corrected for sex ratio, 0.5 when corrected for variation in fecundity, and 0.1 when corrected for both factors and fluctuation in size. It is not possible to determine allometry from the data he presents, but mean values do not differ strikingly between insects, birds, mammals, and plants, and consequently any allometry that exists must be weak. There is thus a discrepancy of many orders of magnitude between estimates based on demography and on genetic variation. This can be reconciled if most species are comprised of many small local populations, so that drift affects local frequencies quite quickly and global frequencies much more slowly. This would produce estimates of N_e that are much smaller than N, whereas N_e/N estimated from demographic data refers to local populations only.

The same difficulty is presented by microbial populations, where it has been lucidly explained by Maynard Smith (1991). The expected number of neutral alleles per locus in a haplont is $1 + 2N_e u_L$ and the probability that two random individuals from an asexual haploid population are identical for a given nucleotide sequence is therefore $1/(1 + 2N_e u_L)$, where as before u_L is the overall neutral mutation rate. Electrophoretic surveys show that the average number of detectable alleles per locus in *E. coli* is about 10, with wide variation among loci (Whittam *et al.* 1983). This suggests $N_e u_L = 4$ and hence $N_e < 10^7$ or 10^8 for plausible guesses at the neutral mutation rate. Sequence-based surveys usually turn up a few common clones, with a long tail of rare or unique genotypes. Under reasonable assumptions the molecular data suggests $N_e < 4 \times 10^8$ for bacteria, about as many as can be cultured in 1 mL of rich medium and many orders of magnitude lower than a plausible census size. There are two reasonable explanations for the discrepancy: that the environment varies in time or in space. Temporal variation will induce periodic selection that temporarily depresses variation and thereby reduces N_e (Levin 1981). If this acts locally then a lineage bearing a beneficial mutation will become fixed in the local population but will be unable to spread to other sites because it lacks the requisite alleles at other loci. The mutation itself will spread much more slowly through the global population by horizontal transfer. Hence, population structure offers a possible explanation of anomalously small estimates of N_e both in plants and animals and in microbes.

The relationship between the two scaled mutation rates U and M is illustrated in Figure 2.14, assuming that the bacterial estimate of $N_e u_L$ is correct. It shows that microbes and macrobes inhabit distinctively different adaptive regimes, microbes having $U < 1$ and $M > 1$ whereas macrobes have $U > 1$ and $M < 1$. I think that this dichotomy is likely to be reliable, at least for abundant species (there are doubtless many rare microbes with $M < 1$). Estimates of population size are so uncertain, however, that the diagram should be interpreted with caution.

2.3 The rate of environmental deterioration

2.3.1 Aggregation

If individuals are able to move about freely, so that each may interact with any other, a species comprises a single large population. This is seldom true, either because individuals are relatively small and sessile (like trees in a forest) or because they are aggregated (like forests in a landscape). The species then consists of many local populations, within which most interactions occur. Selection then takes place largely within these populations, and its effects are exported to the wider metapopulation by dispersal. A population thus corresponds intuitively to a cluster of individuals.

A customary way of expressing aggregation is to divide the occupiable space into quadrats and count the number of individuals in each. If individuals are distributed at random then the distribution of abundance among quadrats will be Poisson and the mean and variance of abundance will be equal. More generally, Taylor (1961) showed that the variance of abundance among samples of a species increases with the mean as $\mathrm{Var}(N) = aN^z$,

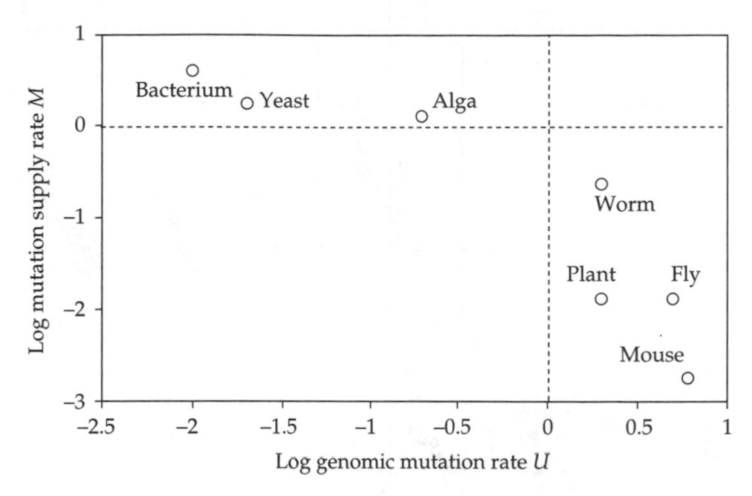

Figure 2.14 Possible relationship between the mutation supply rate and the genomic mutation rate for model organisms. Calculations assume $u = 2 \times 10^{-9}$. Census population size calculated from allometry with body size, pinned on mouse value at 0.025 kg of 2.5×10^8 from range–abundance relationship for birds. Effective population size is calculated from Nei and Graur (1984), corrected to value of 4 for bacteria. The relative locations of the taxa are probably reliable, but their positions on the *y*-axis are uncertain because of the debatable estimate of census size. The taxa as illustrated are abundant; rare types will fall below these.

where z expresses the degree of aggregation (Figure 2.15A). Taylor's power law is readily generated by neutral community models (Figure 2.15B). At high rates of local dispersal m individuals are more or less randomly distributed so that $z \to 1$ as $m \to 1$, whereas low rates of dispersal lead to more pronounced aggregation with $z \to 2$ as $m \to 0$. This is the consequence of drift in local communities: when dispersal is very low most sites will contain either no individuals or many individuals of a given species. Suppose that a fraction f_0 of sites contain no individuals while the rest contain N_{occ}, so that the variance is given approximately by $f_0(0 - N)^2 + (1 - f_0)(N_{occ} - N)^2$. It follows that $\mathrm{Var}(N) = [f_0/(1 - f_0)]N^2$, so that the log-log plot will have a slope of 2 and intercept $\log[f_0/(1 - f_0)]$. The average value of z among 444 species of birds was 2.1 (Taylor *et al.* 1983). This is consistent with neutrality, although a similar conclusion may follow from functional hypotheses: for examples, if individuals are able to disperse preferentially to sites with an excess of resources local abundance will follow an Ideal Free Distribution (Fretwell and Lucas 1970), which likewise leads to a power law with exponent 2 (Gillis *et al.* 1986). For present purposes the important point is that most species are rather strongly aggregated, so that most sites contain either many individuals or none. The number of individuals in occupied sites can be estimated from the intercept as $N_{occ} = N/(1 - f_0)$. The intercept is entirely dependent on sampling scale, however, since smaller quadrats will have higher f_0 and thus greater variance at given mean. If the quadrat size is smaller than the (unknown) grain of population structure then N_{occ} will be underestimated, and if they are larger N_{occ} will be overestimated. Without a range of quadrat sizes, therefore, it is not possible to infer N_{occ} from survey data. If f_0 is large, however, N_{occ} could be taken as a minimal estimate of local population size.

An alternative to quadrat sampling is to express density and aggregation in terms of the distance d between neighbouring individuals. This is proportional to body length, so that the allometric relation $d = aW^z$ has exponent $z = 1/3$ over a wide range of body size (Duarte *et al.* 1987). When d has units of μm the intercept is a = 10–20 for maximum densities in pure culture and 200–300 for natural populations of animals. Hence cells should be separated by about 200–300 μm on average in bacterial populations. Grundmann *et al.* (2001) estimated d = 375 μm between local clusters of

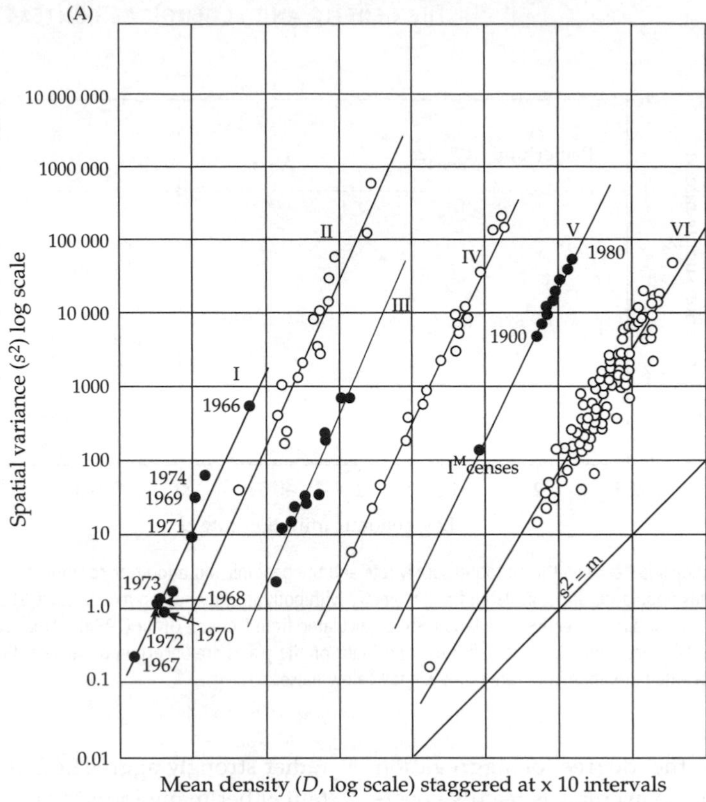

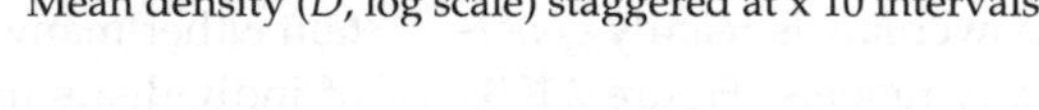

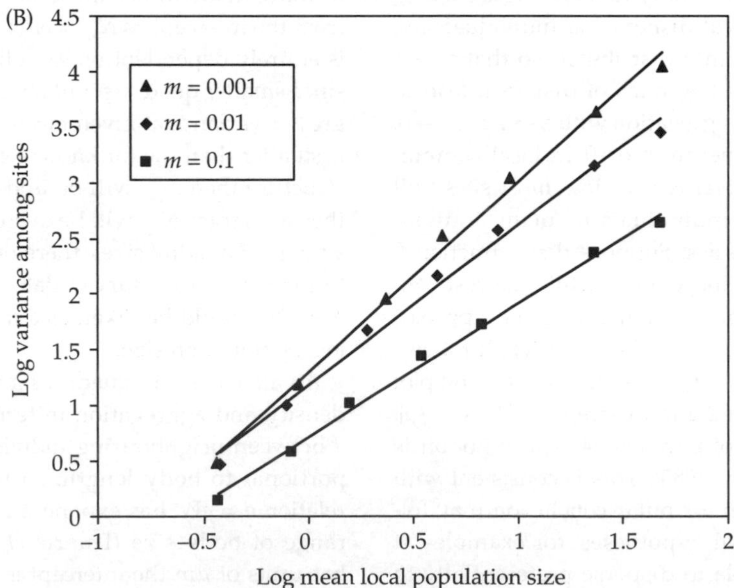

Figure 2.15 Aggregation. A. Taylor's power law fitted to various communities: I Garden dart moth, II haddock, III sea urchin, IV cattle tick, V human (USA), VI Hudson Bay fox. Note that slopes are between 1 and 2. From Taylor *et al.* (1978), Figure 6a. B. Power law description of aggregation in neutral communities. The variance of abundance decreases with local dispersal m, but its relation with the mean remains more or less the same. The slope of the central regression ($m = 0.01$) is 1.43.

Nitromonas-like organisms in soil and $d = 35\ \mu m$ within clusters.

2.3.2 The ecological population concept

Aggregation suggests the existence of local populations, but despite the importance of the concept there is no generally agreed operational definition of a population. Waples and Gaggiotti (2006) list some of the ways in which the concept has been interpreted, and suggest that they fall into two categories. The first is an ecological concept: individuals belong to the same population if they encounter one another and therefore compete directly for resources. The second is a genetic concept: individuals belong to the same population if they mate with one another or compete for mating partners. I shall describe the ecological concept in the context of estimating population size, and the genetic concept in the context of estimating the dispersal rate.

Local populations may take a number of forms: a cluster of individuals in space, a moving herd or shoal, or the group of neighbours whose home ranges intersect, for example. In the simplest case, the local population consists of the individuals living together in some site, where they compete with one another but not with individuals at other sites. Under the neutral model, for any given species the frequency of populations p_i falls off with the fraction of the total number of individuals of that species they comprise f_i as $p_i = k_1 \exp(-k_2 f_i)$ where k_1 is the fixed total abundance of individuals per site and k_2 depends on the local dispersal rate.

Neighbours may be more likely to interact with one another than with more distant individuals and so for some purposes could be considered to constitute a population. The home range of motile animals increases allometrically with body size, so that larger species have larger home ranges (Peters 1983). In lizards, for example, the area of the home range is about $0.1W^{1.0}$ km^2. This implies that the number of individuals within the home range area also increases with body size: for lizards, $N_{home} = 7.5W^{0.18}$. A lizard with body size 0.1 kg thus occupies a home range of about 10^4 m^2 within which it interacts with about 5 other individuals. The allometric parameters vary among groups according to motility and diet, and the allometry of home range over many orders of magnitude of size has not yet been described. Nevertheless, it seems likely that the exponent of the home range relation will remain positive. This is consistent with the very broadly applicable rule that the velocity of motile animals increases directly with body length, amounting to about 1 body length per second, which implies an exponent of about 0.5–1.0 depending on the dimensionality of the environment.

Large plants and animals often form discrete clusters to human senses, like frogs sitting on water lilies in a pond. The microbes that we generally encounter only in bulk, however, may also be spatially aggregated. Direct visualization of bacterial cells in thin soil sections shows that typical soil species such as *Bacillus subtilis* are usually found as very small colonies of 2–5 individuals (Nunan *et al.* 2002). *Nitrobacter*-like organisms (which may include several species capable of oxidizing nitrite) present at bulk densities of $1\text{–}3 \times 10^6\ g^{-1}$ occupy favourable microsites of about 50 μm diameter containing an average of 7 cells (Grundmann *et al.* 2001). Extrapolating the allometric relation for lizards suggests a home range for bacteria of about 100 μm^2, which in this case at least is accurate to within an order of magnitude. The 4×10^5 such sites in each mL of soil tend to be somewhat aggregated, forming larger clusters that may contain several hundred bacteria. Comparable figures can be derived for other kinds of bacteria. In short, the high bulk density of microbes in soil conceals a complex spatial structure in which local interactions may involve large numbers of very small groups of individuals. In aquatic environments interactions between suspended cells may be more global; but even here many individuals may belong to relatively small communities living on carcasses, detritus or small particles. The fraction of the population attached to particles varies greatly among studies, probably as a function of particle density, but it is usually between 10% and 50%. Moreover, these attached individuals tend to be more active metabolically than suspended individuals. Iriberri *et al.* (1987) found 3–6 cells per particle, a number comparable to the size of microcolonies in soil.

In short, natural populations are often strongly aggregated, even when this is not readily apparent, and even exceedingly numerous organisms may often live mostly in small groups. The outcome of selection in these groups is then exported to the wider community through dispersal.

2.3.3 Dispersal

A metapopulation of P local populations each with N individuals coheres through dispersal, the exchange of individuals among the local populations. The rate of emigration m is the probability that an individual leaves its natal site and settles and grows elsewhere; the rate of immigration n is the probability that an individual growing in a site was born in another site. Since immigration and emigration must balance, the immigration rate assuming random allocation of emigrants to sites is $n = m(1 - 1/P) \approx m$ for large P. Emigration is the underlying mechanism of dispersal, expressing the response of individuals to local conditions of growth, giving rise to immigration, which has more profound evolutionary consequences through altering the genetic composition of local populations.

The active dispersal of motile organisms is limited by their velocity v, which depends primarily on length and thus scales with body mass as $v = aW^z$ with $z \approx 1/3$ for swimming and flying (Bonner 1965, Peters 1983). This does little more than set an upper limit to movement. The home range represents the individual scale of adult individuals and includes the neighbours with whom an actively moving individual will interact directly. These may be rather few. In vertebrate ectotherms, for example, the area of the home range is $0.1W^{1.0}$ and average density is $64W^{-0.77}$ (Peters 1983) so that an individual lizard weighing 100 g will have $(0.1W^{1.0})(64W^{-0.77}) = 6.4W^{-0.23} \approx 11$ neighbours. The lineage scale includes both adult activity and the dispersal of propagules or offspring. Wright (1946) defined the local population in terms of the dispersal distance $\sigma^2 = t\sigma_1^2$, where σ_1^2 is squared daily movement and t is longevity. The neighbourhood size is then the number of individuals in circle of radius 2σ, i.e. $N_{neighbourhood} = 4\pi\sigma^2 D$ where D is density. The neighbourhood size defined in this way is equivalent to N_e (Hedrick 2000). The median dispersal distance of juvenile animals increases with size as $\sigma^2 = aW^z$, with $z \approx 0.67$ (Sutherland *et al.* 2000). The scaling parameter a is a function of vagility, with $a = 13$ for birds and $a = 2$ for mammals (units in kg and km). In birds, for example, $\sigma^2 = \sigma^2_{juv} + L\sigma^2_{adult}$ where σ^2_{juv} is dispersal in first year, σ^2_{adult} is dispersal per year as adult, and L is longevity (Barrowclough 1980). In well-studied small passerines such as the great tit the juveniles are responsible for most lifetime dispersal, and direct observation of marked individuals shows that $\sigma \approx 1$ km. Given a breeding territory size of 1 ha and thus an average density of 200 km^{22} in suitable habitat we have $N_{neighbourhood} \approx 2500$ individuals for a common species. More generally, using values for North American birds in Peters (1983) we can calculate for a species of average body size (55 g) that $N_{neighbourhood} = 4\pi(13W^{0.67})(6.2W^{-0.19}) = 250$ individuals for an average species.

Organisms that are unable to move are dispersed passively by currents of air and water or by other organisms. Even in motile organisms, individuals may travel passively much further than they could by their own efforts. In particular, the spores of microbes can be transported great distances, and one common theory of microbial community structure is that 'everything is everywhere' (see Fenchel and Finlay 2004, Finlay and Fenchel 2004). Any biologist who has visited a boreal bog and a tropical rain forest will have noticed that whereas the large animals and plants are utterly different the protists and micrometazoans that can be squeezed from a handful of moss are remarkably similar. There is thus a cosmopolitan fauna of small eukaryotes, less than a few hundred μm in length, which are very readily dispersed and thus form the same community in suitable sites anywhere in the world. This view has been challenged because microbial species richness increases with area or volume just as it does in large organisms (Bell *et al.* 2005b), but the observation is not inconsistent with cosmopolitan distribution: neutral theory predicts a power-law increase of diversity with community size even in very large communities with immigration, provided that immigrants are not frequent relative to offspring born to residents. Bohonak and Jenkins (2003) admit that freshwater invertebrates have the potential for rapid dispersal, but question whether realized rates are equally high. Eukaryotic microbes certainly colonize newly available habitats very rapidly, however. Maguire (1963) followed the colonization of sterile vials at varying distances from a pond and obtained estimates of $\sigma^2 = 1\text{–}1.5$ m day^{-1} for flagellates, amoebas, ciliates, unicellular algae, and micrometazoans such as rotifers and gastrotrichs. Even large multicellular animals > 1 mm in length are often effectively dispersed: for example, Louette and de Meester (2003) surveyed newly constructed artificial ponds that accumulated an average of 4 species of cladoceran within 15 months, equivalent to a specific immigration rate of about 0.01 day^{-1}. The underlying dispersal rate cannot be estimated because the size of the source population is unknown, but clearly it cannot be very low.

Bacteria are dispersed even more effectively than small eukaryotes and colonize any suitable site very soon after its appearance, which is why microbiologists work in flow hoods. Over the open sea, air contains about 10 viable bacteria m^{-3}; rural air has about 100 viable bacteria m^{-3}; and city air about 1000 viable bacteria m^{-3} (with much greater densities in enclosed spaces) (Bovallius *et al.* 1978). A large fraction of these are borne on fairly large particles, comparable in size with small eukaryotic microbes, and will be deposited on surfaces quite readily. The number of cells of a common bacterial species entering a site over a given period of time is therefore likely to be rather large. The resident population may also be large, however, so the relative importance of immigration is not as clear. Colony size provides a simple estimate of dispersal rate, since each colony of C cells has grown from a single dispersal event and $(C - 1)$ divisions without dispersal, so the dispersal rate is $1/(C - 1)$ (Tom Curtis, unpublished). In natural populations inhabiting soil or water colonies seem to be very small (see above), and local dispersal rates at least must then be rather high, of the order of 0.1 over the lifetime of the colony, implying a rate > 0.02 per generation.

Dispersal is least effective for large sessile organisms with dense non-motile propagules, such as most seed plants. In most cases the bulk of the propagules produced by a plant are necessarily deposited nearby, roughly speaking within a distance comparable with its height. They fall, therefore, within the individual distance of the parent, over which it integrates environmental conditions through its roots and leaves. The number of propagules P that are dispersed further typically declines exponentially with distance d as $P(d) = \text{constant} \times \exp(-zd)$, although other functions sometimes provide a better fit. Summing over all distances and normalizing this to unity gives the fraction falling within one unit of the parent as approximately $2z/(2 + z)$ for small z. Willson (1993) collated data for seed dispersal and found $z \approx 2$ on average for wind dispersal by unspecialized herbs, with a mode at < 1 m: most seeds will fall within 1 m of the parent and experience similar conditions of growth. Species with specialized structures that aid wind dispersal have $z \approx 1$, and only about 2/3 will fall within 1 m of the parent. Wind-dispersed tree seeds have $z \approx 0.05$, but the individual distance of trees is perhaps 10 m rather than 1 m, so $z \approx 0.5$ in comparable units and about 40% of seeds will fall within the individual distance of the parent. If the wind is blowing, however, this value drops to 10% for trees or specialized herbs as most seeds are deposited at some considerable distance downwind. Pollen dispersal is surprisingly like seed dispersal: most successful pollen fertilizes ovules only a short distance from the parental plant, and success falls off very sharply with distance because pollen grains condense on intervening structures and soon die (review in Ellstrand 1992).

2.3.4 The genetic population concept

Selection or drift acting independently in local populations creates spatial variation in gene frequency. This is effaced by dispersal in proportion to the number of immigrants received by a population relative to the number of offspring produced by residents. For neutral alleles, the variance of the frequencies of neutral alleles among local populations F_{ST} is related to the number of immigrants per generation Nm as $F_{ST} = 1/(1 + 4N_em)$, given a number of assumptions about metapopulation dynamics (Wright 1951). The parameter m is often termed a dispersal rate, although it is in fact an immigration rate that will be approximately equal to the emigration rate m as defined above. As F_{ST} can be measured rather easily, it serves as a convenient means of estimating N_em as $\frac{1}{4}(1 - F_{ST})/F_{ST}$, and hence m if N_e is known or can reasonably be guessed. Empirical estimates of F_{ST} from allozyme surveys of animals have been collated by Bohonak (1999) (Figure 2.16). They range from very small values (<0.01) to about 0.8; more than 90% are less than 0.5. The average of nearly 200 studies was 0.15 (sd = 0.19). We humans are average animals: estimates of F_{ST} for human populations from electrophoretic variation, blood groups, and other kinds of loci are 0.11–0.16 (Tishkoff and Verrelli 2003). A very detailed survey of human microsatellite variation (Rosenberg *et al.* 2002) gave smaller estimates of 0.035–0.05 among continental regions and 0.01–0.05 among populations within continents, perhaps because the high rate of mutation at microsatellite loci creates high levels of variation within populations. The distribution of F_{ST} among animals is modal at small values and the frequency declines exponentially, roughly as $0.35\exp(-5F_{ST})$. Taxa in which dispersal was judged to be more restricted tended to have larger values of F_{ST} or equivalent measures of genetic structure (Bohonak 1999). Briefly, F_{ST} in animals with feeble powers of dispersal (with benthic eggs and direct development, for example) usually lies between 0.1 and 0.5, whereas effective long-distance dispersal (by planktonic larvae, for example) is associated with values of less than 0.1, and sometimes less than 0.01. The average value of F_{ST} in almost 500 species of outcrossing plants was 0.15, the same as for animals (Hamrick and Godt 1996). The least differentiation (mean F_{ST} = 0.1) is found among wind-dispersed outcrossing species, and the most pronounced differentiation (mean F_{ST} = 0.5) for gravity-dispersed selfers. In both animals and plants, therefore, genetic differentiation among populations tends to be effaced by physical or genetic dispersal, and this provides a reasonable justification for using F_{ST} as an estimator of dispersal rate. For comparison, the average value of F_{ST} at allozyme loci in *E. coli* is 0.02 (Whittam *et al.* 1983).

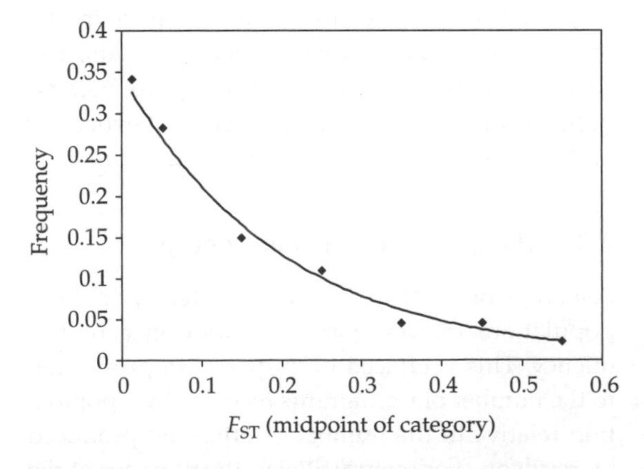

Figure 2.16 Genetic structure of populations. Data for various groups of animals from Bohonak (1999). The fitted curve is $y = 0.35\exp(-4.9x)$.

The number of immigrants is confidently estimated only in well-studied populations with marked individuals, such as small songbirds. A survey of the Wytham population of great tits (*Parus major*) showed that 689 individuals were the offspring of residents, 35 had immigrated from a few kilometres radius and 553 had immigrated from further afield (Verhulst *et al.* 1997). F_{ST} for this species, as for most birds, is rather small; Kvist *et al.* (1999) found little geographic structure across Europe, perhaps because of postglacial expansion, and calculated F_{ST} = 0.014 within regions. A value of 0.01 implies Nm = 25, which is about right for local dispersal but much too small for broader areas. Barrowclough (1980) used a dispersal distance of 0.8 km to calculate an expected F_{ST} of 0.009. In other species the observed local dispersal distances also gave reasonable (that is, very low) expected values for F_{ST}. An alternative is to use the allometric scaling rule for population density $N = 32W^{-1}$ (Peters and Wassenberg 1983) to compute the expected immigration rate. Taking $N_e = 0.67N$ (Nunney and Elam 1994), W = 0.02 kg and F_{ST} = 0.01 we have $m = (25 \times 0.02)/(0.67 \times 32) = 0.023$, which again is about right for local dispersal. It is difficult to have much confidence in these calculations, however, or in any use of F_{ST} to infer dispersal rates. Perhaps the strongest statement that can be made is that there must be frequent local dispersal if species generally consist of rather small spatially separated subpopulations, as suggested above. Even this cannot be taken completely at face value because of the many assumptions required for the estimate of $N_e m$ to be valid (Bossart and Prowell 1998, Whitlock and McCauley 1999). The most important is that alleles are neutral, and in particular are not under selection that varies in direction from site to site: uniform selection will reduce genetic variance among populations and thereby reduce F_{ST}, whereas local selection will inflate F_{ST}. It is also assumed, moreover, that populations are equal in size and each receives immigrants indifferently from all others, regardless of location. Despite these shortcomings, it is unlikely that estimates of F_{ST} are wildly erroneous unless the metapopulation has extreme characteristics such as very large Nm or Var(N) (Neigel 2002). A more troubling weakness is that any estimate of F_{ST} depends on prior identification of populations as the categories for the partition of variance. The value of $N_e m$ depends on the subdivision of the sample into P local populations because P affects both immigration rate and (much more strongly) local population size. Alleles will drift to fixation more rapidly in smaller populations, giving rise to larger values of F_{ST}. Consequently, a sampling scheme that pools individuals from several real local populations into a single nominal population will underestimate F_{ST} and thereby overestimate $N_e m$. In microbes, for example, clone and colony are not co-extensive: cells are not arranged in neat coherent groups like colonies on agar. In *Bacillus subtilis*, repeated sampling has shown that the clone is more extensive than the colony and ranges over a distance of 1 cm or more (Istock *et al.* 1992); similarly, in *Agrobacterium* there is no relation between spatial separation and genetic distance within a sphere of about 1 cm radius (Vogel *et al.* 2003). At this spatial scale, therefore, colonies overlap, through active movement or passive displacement. There are many reports of genetically identical isolates being obtained from sites hundreds of kilometres apart, but whether these are strictly clonal, sharing a recent common ancestor, has not yet been conclusively demonstrated. Furthermore, repeated sampling at small scales (< 1 mm) suggests that microcolonies may contain more than one clone: there may be several genotypes of *Nitrobacter*-like organisms within a soil sample 50-μm in diameter (Grundmann and Normand 2000), or among the nitrogen-fixing *Rhizobium* within a single nodule on a plant root (Souza *et al.* 2002).

2.3.5 Long-distance dispersal

The average or median dispersal distance governs the flow of poorly adapted immigrants from nearby sites and thereby acts like deleterious mutation in setting the level of mundane purifying selection. Dispersal distances are roughly exponentially distributed and long-distance migrants are therefore rare, but they may have a disproportionate effect on adaptation by colonizing vacant sites outside the general range of the species or introducing novel and potentially beneficial alleles into distant populations. Sutherland *et al.* (2000) show that maximum recorded dispersal distance in birds and mammals is a power function of the median distance, with exponent +1 and intercept about 10, so that the maximum distance is about 10 times the median distance for any given species. This is obviously sensitive to sampling effort, but provides a useful yardstick for evaluating the effect of long-distance dispersal on adaptation and genetic structure.

Clearly, our knowledge of dispersal is almost as limited as our knowledge of population size. As a very broad generalization, many organisms are aggregated in space, forming a large number of relatively small subpopulations among which there is frequent migration. This applies even to very numerous organisms occupying what are rather homogeneous environments to the human eye, such as the microbial communities of soil and water. In these circumstances, the ecological interactions underlying the operation of selection are mainly local.

2.3.6 Five theories of the environment

The distribution and abundance of organisms, the operation of local selection and the effect of dispersal depend on how the environment varies in space and time. Broadly speaking, there are five general accounts of environmental structure and its effect on adaptation. The first is the *uniform model*: individuals are distributed at random among identical and unchanging sites, so that distribution and abundance are determined by chance alone. This is, of course, a statistical ideal that is approached by microcosms and underlies a great deal of simple theory but is unlikely to hold for natural situations. The second is the *neutral model*, where restricted local dispersal will lead to a patchy distribution of species or genotypes, even in a uniform environment, as the outcome of chance and history combined. The third is the *ecological model*, which permits any amount of environmental variation in productivity, provided that this affects all individuals in the same way, without implying any degree of local adaptation. This leads to a patchy distribution as the outcome of chance, history and environmental variance. The fourth is the *evolutionary model*, which incorporates genotype–environment interaction reflecting the variation of relative fitness among sites, so that populations tend to become adapted to local conditions of growth. This leads to a patchy distribution of locally adapted types as the outcome of chance, history, environmental variance and selection. Finally, the fifth is the *co-evolutionary model*, which recognizes that the growth of a species or genotype at a site will change the characteristics of the site and thereby alter the relative fitness of competitors and enemies, which will in turn feed back on the focal type. This may lead to complex ecological and evolutionary dynamics as adaptation is continually frustrated and renewed. In this section I shall suggest some general propositions about environmental variation, leaving evolutionary and co-evolutionary considerations for later chapters.

2.3.7 Environmental variation in space

The current state of the environment experienced by an individual with respect to some factor or combination of factors can be represented as the sum of the regional average value E and the deviation from this value ϵ at the site where the individual is growing, $E = \mathrm{E} + \epsilon$. Conditions of growth change from generation to generation because offspring grow up at a different time and in a different place (reproduction without dispersal is growth). The regional average will change over time from E to E′, and the new site occupied by an offspring has deviation ϵ'. Thus

$$\Delta E = (\mathrm{E} - \mathrm{E}') = (\mathrm{E} + \epsilon) - (\mathrm{E}' + \epsilon') = (\mathrm{E} - \mathrm{E}') + (\epsilon - \epsilon').$$

This partitions the overall change into temporal and spatial components. The temporal component often has a long-term trend, for example as the region becomes warmer, and we may be interested in how populations become adapted to an environment that is changing consistently in a particular direction. The relevant environmental change will usually be $\Delta E = (\mathrm{E} - \mathrm{E}')$, whose sign and magnitude might both be important. The spatial component, on the other hand, has no trend (unless dispersal generally occurs in a particular direction, for example downstream), and we may interested in how conditions fluctuate from generation to generation as the result of dispersal, in which case the relevant change is $\Delta\epsilon = (\epsilon - \epsilon')$, whose sign is unimportant.

The spatial structure of the environment from the point of view of a dispersing propagule is expressed by the relationship of $\Delta\epsilon$ to dispersal distance d. Imagine travelling along a line away from some fixed point while recording the state of the environment. For a short distance conditions are likely to remain more or less the same, but sooner or later they will change more or less abruptly as one passes from one kind of site to another, for example from grassland to fen. Any plot of environmental similarity on distance along this line would trace a more or less horizontal line punctuated at intervals by steep drops. A second line from the same point would look similar, except that the drops would occur at different points; thus, the average of many lines will be a more or less smooth decline of similarity. The measure of similarity for any pair of sites is their squared difference, because we wish to express the magnitude but not the sign of the difference between them, which is an environmental variance. We expect that the variance will be greater for sites that are farther apart because the farther we walk the more likely we are to encounter different conditions of growth. The environmental variance for the parental site and the offspring site is $\mathrm{Var}(\epsilon) = \frac{1}{4}(\epsilon - \epsilon')^2$, so that $\ln \mathrm{Var}(\epsilon) = \ln \frac{1}{4} + 2 \ln \Delta\epsilon$, the sign of $\Delta\epsilon$ being lost. For all the pairwise combinations of sites from a random or representative set of sites within a region we can calculate both the distance between them and the value of Var(ϵ). This leads to the empirical but very robust conclusion that the increase of environmental variance is described by a power law: Var(ϵ) = Xd^z (Bell *et al.* 1993). At this point I should admit that the very extensive and sophisticated literature on geostatistics has not accepted this general rule. Nevertheless, I have conducted very detailed surveys of spatial variation, in collaboration with my colleagues Martin Lechowicz and Marcia Waterway, and am convinced that the power-law rule holds in almost all cases and represents a general and fundamental feature of environmental variation.

We carried out a detailed spatial survey of about 10 km² of old-growth forest at Mont St-Hilaire in southern Quebec, collecting soil samples from every hectare to build up a quantitative picture of how conditions of growth vary in an undisturbed natural environment. Figure 2.17 shows the spatial pattern of soil pH as an example, summarized by the variance plot, the regression of log

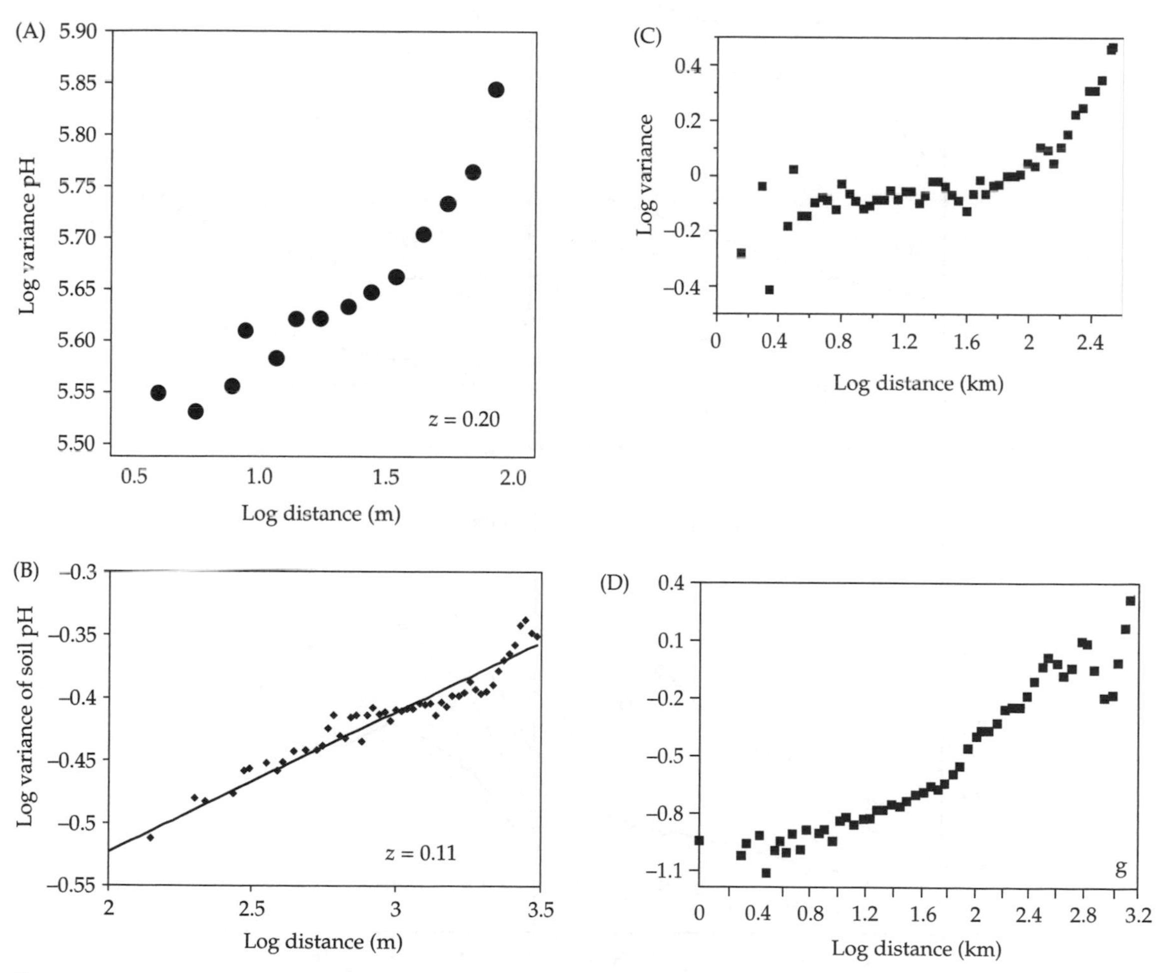

Figure 2.17 Spatial structure of the physical environment. Variance plots for soil pH at different spatial scales. A. Distances up to 0.1 km within 1 ha of old-growth forest (from Richard *et al.* 2000). B. Distances up to 4 km in the entire fragment of old-growth forest (G. Bell, M. Lechowicz, and M. Waterway, unpublished). C. Distances up to 300 km in Wisconsin, USA (Bell *et al.* 1993). D. Distances up to 10 000 km in North America (Bell 1992c).

variance on log distance. Environmental variance continues to increase indefinitely; there is no spatial scale at which the entire range of conditions has been encountered: the power law for pH holds at all scales from a single-hectare plot at Mont St-Hilaire (Richard *et al.* 2000) to the continent of North America (Bell *et al.* 1993). In short, the variance plot provides a description of spatial variation that is independent of the spatial extent of the region being surveyed. The exponent z is also independent of the units of measurement both of distance and of the environmental factor, and can therefore be used to compare the power laws for different kinds of factor on a common scale. It describes the variance of pairs of points in two-dimensional space. The expected distance between random points along a line of length l is $l/3$, and the exponent of the power law relating variance to the length of a line is therefore also z. The expected distance between random points within a circle of

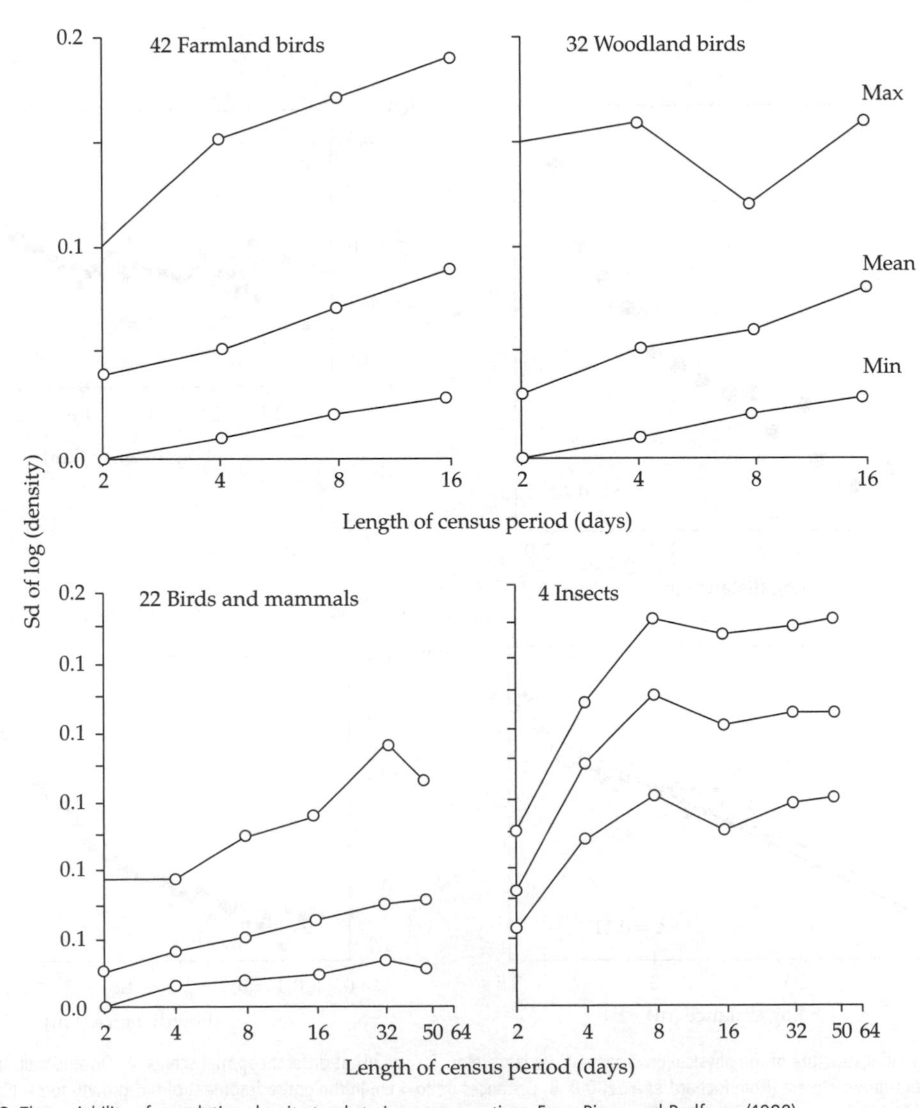

Figure 2.18 The variability of population density tends to increase over time. From Pimm and Redfearn (1988).

radius r is $2r/3$, and because its area increases with r^2 the exponent of the power law relating variance to area has exponent $z/2$.

The exponent z of the power law has a simple interpretation. If z is large then distant sites are much less similar than nearby sites and the region is coarse-grained. Conversely, if z is small then nearby sites are nearly as dissimilar as distant sites and the region is fine-grained. The power law thus expresses the 'patchiness' of the environment. Because it applies at all spatial scales, it shows that the environment is self-similar in the sense of being equally patchy at all scales. In other words, all organisms will encounter spatial variation, no matter what spatial scale they live at. Suppose that the spatial scale of growth is d_G: this would be the root extension of a plant, the mycelial extent of a fungus or the home range of a bird, for example. Likewise the spatial scale of reproduction is d_R: this is the dispersal distance of a seed or a fledgling. The difference in conditions experienced by an offspring $\Delta_R w$ relative to the range of conditions encountered by its parent $\Delta_G w$ is then $\ln(\Delta_R w/\Delta_G w) = (z/2)\ln(d_R/d_G)$. As a very broad generalization we would expect d_R and d_G to be positively correlated, because both will tend to be small for small or sedentary organisms and large for large or mobile organisms. If d_R/d_G is roughly constant over species then the degradation of fitness caused by dispersal, relative to the range of growth conditions experienced by the parent, will also be roughly constant.

The power laws for growth and relative growth that we have found all have small exponents of $z \approx 0.1$–0.2. If this result is confirmed by more extensive surveys of other kinds of organism and other kinds of site it would show that environmental variance is generally governed by weak power laws, that is, that the environment is generally rather fine-grained.

2.3.8 Environmental variation over time

Temporal variation can also be described by power laws (Bell *et al.* 1993). A physical factor will vary on all timescales, and at any given timescale will vary by some characteristic amount. If the amount of variation is independent of timescale then events that occur with different frequency will contribute equally to the overall variance, as in pure white noise. Natural time series, however, very commonly have the property that the contribution C made by a category of events is inversely proportional to its frequency f, such that $C(f) = f^{-\gamma}$ with $0 < \gamma < 2$ (Steele 1985, Halley 1996). The two extreme cases are white noise, which has $\gamma = 0$, and a random walk, or Brownian motion, which has $\gamma = 2$. Most situations will be intermediate, with $\gamma \approx 1$ (a reddened spectrum, or 'pink noise', from the optical analogy), and environmental variance will then increase approximately with the logarithm of the length of the time series, or with elapsed time t, $\text{Var}(E) = at^z$. The spectral and power-law approaches are equivalent, with $z = 1 - \gamma/2$. Both

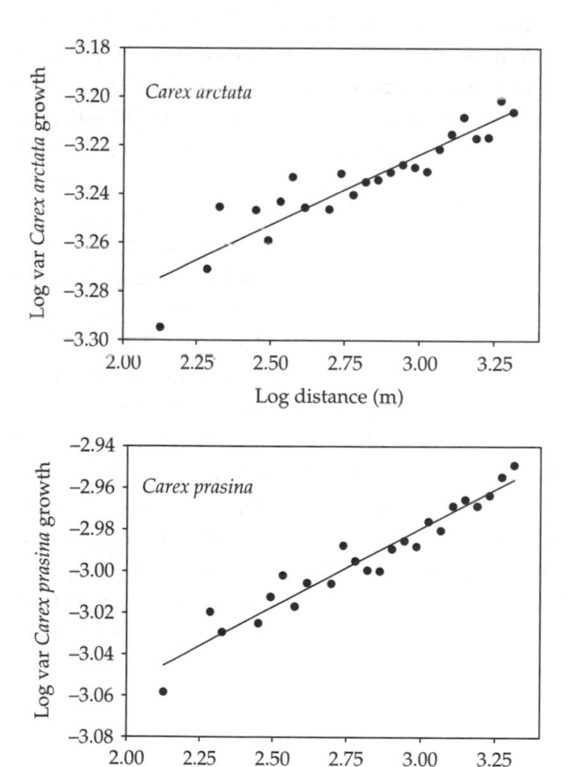

Figure 2.19 Explant bioassays of forest spatial structure. Power-law exponents are 0.084 for *C. arctata* and 0.118 for *C. prasina*. Unpublished data of G. Bell, M. Lechowicz, and M. Waterway.

formulations imply that variance will continue to increase indefinitely with the length of the time series, albeit at a decreasing rate, as it does for the extent of a spatial survey.

Physical variation often follows a power law very closely. For example, Koscielny-Bunde *et al.* (1998) analysed deviations of daily maximum temperature from their seasonal average values, estimated with very voluminous and exact data taken at weather stations around the world over a period of more than a century. They found that plots of ln variance on ln elapsed time are almost perfectly linear with $\gamma = 0.7$, $z = 0.65$, which they suggest constitutes a 'universal persistence law'. Pelletier (1997) found $\gamma \approx 0.5$ from ice-core samples for elapsed times between 1 month and 2000 years, with some evidence of larger values for longer and shorter time periods. This environmental variability could drive changes in the growth and abundance of populations. Pimm and Redfearn (1988) showed that ln abundance N of animals also follows a power law (Figure 2.18). Since the ln variance of ln N is equivalent to ln $[N(t+1)/N(t)]$ this implies that the variance of realized growth rate increases indefinitely over time. Inchausti and Halley (2002) used the extensive records held in the Global Population Dynamics Database (*http://cpbnts1.bio.ic.ac.uk/gpdd/*) to obtain a median value of $z = 0.36$ across more than 500 long-term data sets. The distribution of z showed a mode at small values of 0–0.3 and declined in a roughly exponential manner for larger values. The distribution of γ was roughly symmetrical around a mean of 1.0. Rather surprisingly, there were no very pronounced differences attributable to taxon, trophic level, habitat, or body size. Temporal variation in abundance has a natural tendency to cumulate, whereas spatial variation does not, and this might result in time being more coarse-grained than space. I have the impression that z is indeed greater for temporal than for spatial variation in growth and abundance, although I am not aware of any systematic comparison.

2.3.9 The biotic environment

Growth and absolute fitness w follow power laws that are qualitatively and quantitatively similar to those for physical factors. The growth of *Arabidopsis* phytometers in soil samples brought into the greenhouse follows the same rule as physical factors (Bell and Lechowicz 1991). We chose two resident species of sedge, one very common (*Carex arctata*) and the other rather rare (*Carex prasina*) in our survey area at Mont St-Hilaire, and grew them in soil samples from each hectare and found the variance plots shown in Figure 2.19, which again are linear over the scale of the survey. Trials with genetically uniform plants have shown that variance continues to decrease at scales smaller than 1 ha (Bell and Lechowicz 1991). These results came from explant trials, because implant trials that measure the growth of cloned plants at sites in the forest are much more arduous to perform. We transplanted cloned ramets of 15 species of sedge at 10 m intervals along three 1 km transects and found that survival followed a power law (Bell *et al.* 2000). Information on relative fitness is also scarce, but we found that the relative growth of *C. arctata* and *C. prasina* follows a power law with an exponent similar to either species considered separately.

This chapter has provided some basic quantitative information about the arena within which populations evolve. I now turn to the evolving population itself, in the simplest possible conditions of growth, before attempting to interpret populations in a state of nature.

CHAPTER 3

Natural selection in closed asexual populations

Few questions in biology, or in the whole of natural science, can be more intriguing than to view a population of organisms introduced into an unfamiliar environment and to ask, what are the mechanisms by which this population will become more closely adapted to the new conditions it has encountered? Although simple to state, it might be difficult to answer in a natural environment where the population is buffeted by innumerable conflicting forces. Even under more controlled conditions on the farm or in the laboratory, the course of adaptation will be shaped in part by the breeding system and the pattern of dispersal. The dynamics of evolving populations can be seen most plainly when these complexities are stripped away and the entire system, population and environment, is reduced to an isolated laboratory microcosm supplied with a single limiting resource and inoculated with a single kind of haploid asexual microbe. The reader may object that this is evolutionary biology with the interesting bits missed out. Even the simplest evolving system is by no means straightforward, however, and the struggle to understand the elementary mechanisms of adaptation is the necessary first step towards a general account of evolution.

3.1 Microcosmologia

3.1.1 Dallinger's experiment

Early in 1879 a curious device was delivered to a vicarage in Liverpool, England. It resembled a boiler: a large copper vessel with heating tubes, an elaborate thermostat, and three stoppered glass flasks immersed in the water (Figure 3.1). Once it had been set up the flasks were filled with 'putrefactive fluid' maintained at 60 °F (15 °C) and soon acquired a rich microbial community. Over the next 5 years the Reverend W.H. Dallinger slowly raised the temperature of the cultures and observed the response of three species of flagellates. At 73 °F most of them died, but after a few weeks they had recovered and after this progress was smooth until the next crisis, at 78 °F. In this way, in fits and starts, sometimes pausing for months on end when any further increase would be fatal, he eventually obtained cultures capable of surviving at 158 °F (70 °C). The reason for this extraordinary experiment was a desire to test the theories of Charles Darwin, with whom Dallinger was corresponding. It seemed hopeless to study domestic animals and plants, despite the changes that had manifestly been wrought by artificial selection, because of the great length of time that would be required (as was thought) to observe any appreciable change.

> Now in the Infusoria... the cycle of life is so relatively short, and the generations succeed each other so rapidly, while the successive progenies can be so easily observed, that if we can devise apparatus and conditions that will enable us to institute slow changes of environment, we should be able to observe critically how far changes in the organisms led to responsive adaptations and successive survival. (Dallinger 1887 *p.* 192).

With this insight, Dallinger became the founder of experimental microbial population biology, and succeeded in demonstrating that organisms can rapidly and repeatedly adapt to conditions far

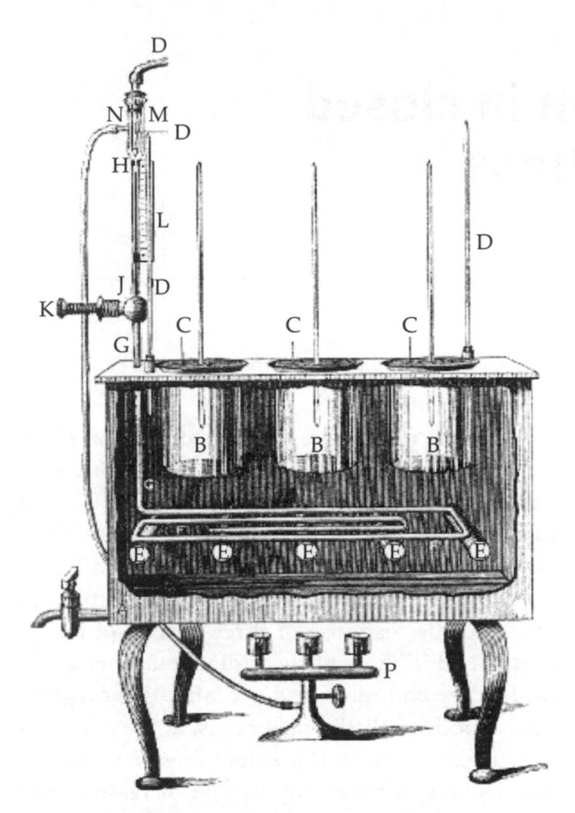

Figure 3.1 Dallinger's apparatus. From Dallinger (1887).

outside their initial range of tolerance, through an agency that could only be natural selection.

The experiment itself was brought to an abrupt conclusion: 'The accident happened, destroying the use of the instrument . . .'. We are not told the cause of the catastrophe, but we are not at all surprised that an apparatus designed to warm up vats of 'putrefactive fluid' in a Victorian vicarage should sooner or later suffer an accident. It is more difficult to explain why it should have had no successors. It provided the first hard experimental support for Darwinian evolution, and the experimental method was becoming firmly established in physiology, development and genetics. In evolutionary biology, however, experimental work was initiated by the *Drosophila* school in the 1920s, and no microbial experiments were reported until the development of the chemostat in the 1950s, by which time Dallinger's experiment had long been forgotten.

3.1.2 The laboratory microcosm

Microcosms have become simplified since Dallinger's time. The ideal microcosm is a simple, cheap vessel in which a culture can be kept in a state of continual growth with little effort and no risk of contamination by other organisms. In practice these objectives cannot all be realized at the same time, and one of three designs is commonly used: serial transfer in batch culture, continuous culture in a chemostat, and growth on solid medium in a Petri dish (Figure 3.2).

The simplest vessel is a glass vial supplied with sterile nutrient medium which is inoculated with a small volume of culture. The organisms adjust physiologically to the new conditions of growth and then proceed to reproduce. After a few cycles of reproduction the limiting substrate of the nutrient medium has been exhausted, growth slows or ceases altogether, and the organisms persist in a quiescent state before the experimenter transfers them to a fresh vial. Selection for rapid growth early in the history of the culture will dominate evolution in these conditions, although reducing the lag phase and increasing survival when starving may also be important. The basic design features of batch culture are the inoculum volume and the transfer interval. The inoculum volume is commonly set at 1% or less of the culture volume, allowing 6–7 doublings before starvation: very small inocula reduce the pace of adaptation because rare types such as novel mutations may be excluded, whereas very large inocula reduce the effect of selection per transfer. The optimal inoculum volume that maximizes the probability of fixation of a beneficial mutation is surprisingly large, at 13.5% of the culture volume (Wahl *et al.* 2002). Batch culture is a simple technique that is readily automated, and by using robotic transfer with microwell plates very highly replicated experiments involving $\approx 10^4$ cultures become feasible. It is also resistant to contamination, because transfers can be done inside a sterile flow hood. Its main drawbacks are that it is often inefficient, because much time is wasted as the grown cultures await transfer, and that the culture environment is continually changing.

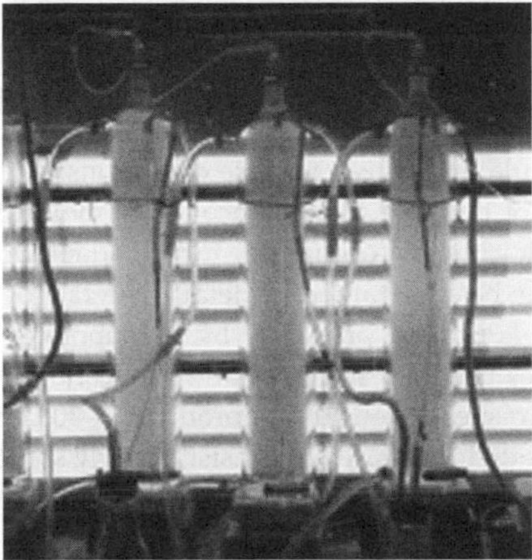

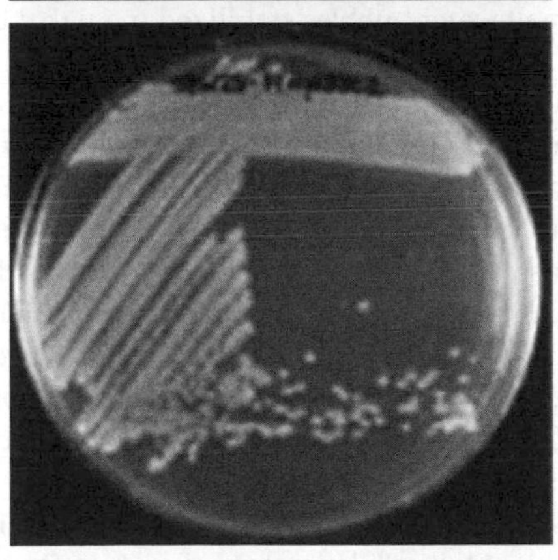

Figure 3.2 Microcosms. From top: batch culture, chemostat, Petri dish. Photo credits: *io.uwinnipeg.ca/~simmons*; *www.uregina.ca/biology/faculty/weger*; *www.cpb.bio.ic.ac.uk*.

A simple chemostat is a glass vial which is supplied with nutrient medium continuously from a reservoir rather than intermittently, the excess volume being removed at the same rate. After an initial period following inoculation the growth rate of the population is governed by the rate of inflow and the concentration of the limiting nutrient in the reservoir. The population dynamics of microbial populations in chemostats are very well characterized, and a basic review of the field from the standpoint of experimental evolution is given by Dykhuizen and Hartl (1983). Selection for sustained growth in famine conditions will dominate evolution in chemostats. The main design decision is the choice of a dilution rate, which is again a compromise between a high turnover that accelerates growth and a low turnover that minimizes stochastic loss. Chemostats are elegant devices that permit very exact experiments, but they are troublesome to maintain, cannot as yet be highly replicated, and are notoriously prone to contamination.

The Petri dish is usually used to isolate, screen, or store samples from liquid-culture experiments, but can be used as an experimental vessel in its own right. It is always used for serial transfer, spreading a new plate at intervals with a sample from the previous cycle. It differs from liquid culture in two main ways. First, interactions between immobilized cells will be local rather than global, so that taking up substrates or releasing metabolites will affect only neighbouring cells, whereas in liquid culture all are equally affected by all. Secondly, the great bulk of interactions take place between members of the same lineage, whereas in liquid culture members of all lineages encounter one another indifferently. The spatial relationships between lineages can be preserved if desired by replica plating, using a pad of velveteen or some similar material to print from the old to the new plate. The behaviour of individual genotypes can be directly visualized on solid medium, and contamination is immediately detected and easy to remove. The main drawbacks of the Petri dish are that growth is relatively slow and is more difficult to measure, since the spectrophotometer or particle counter used to measure cell density in liquid cultures cannot be used.

3.1.3 The inhabitants of the microcosm

The microcosm is a time machine, allowing evolutionary change to be observed within a period of time that is short relative to human lifespan. Adaptation occurs more rapidly in calendar time when lives are short and populations are large, so microcosm creatures are small and reproduce rapidly (Figure 3.3). Some are so small and simple, indeed, that they are not creatures at all in the eyes of most people. *Genetic algorithms* are strings of instructions able to replicate themselves within a computer. They can evolve to replicate more rapidly in given conditions, or to perform simple logical tasks when appropriately rewarded. The rate of adaptation is limited by the clock speed of the computer. Because the genotype of every individual can be stored and made available for inspection later, it is possible in principle to provide a complete accounting of adaptation; in practice the flood of data often overwhelms the capacity to understand it. *Ribozymes* are short RNA molecules that can evolve *in vitro* the ability to bind a target molecule or survive adverse conditions. Although it is not possible to sequence every molecule, the genetic changes occurring during adaptation can usually be ascertained in detail. The rate of adaptation is limited primarily by the laboratory procedures used for selection and replication. *Viruses* are parasites of cells that reproduce 10–100 times as fast as their hosts. They consist of fairly short RNA or DNA sequences in which the alterations responsible for adaptation can often be identified. *Vesicles* are small water bubbles stabilized in oil that bear a gene, translation machinery, and a substrate. They can be used for precise biochemical studies of adaptation, although they are the least frequently used system. *Bacteria* are the staple of experimental evolution, being the smallest self-sustaining cellular organisms. To a first approximation, all experimental work concerns the mammalian gut symbiont *E. coli*, although *Pseudomonas* and *Klebsiella* have also been used. *Eukaryotic microbes* such as *Saccharomyces* and *Chlamydomonas* are relatively long-lived and have rather seldom been used, although for some issues, such as the evolution of sex, they are indispensable. *Multicellular organisms* such as *Drosophila* and mice have been used mainly to investigate the consequences of artificial selection.

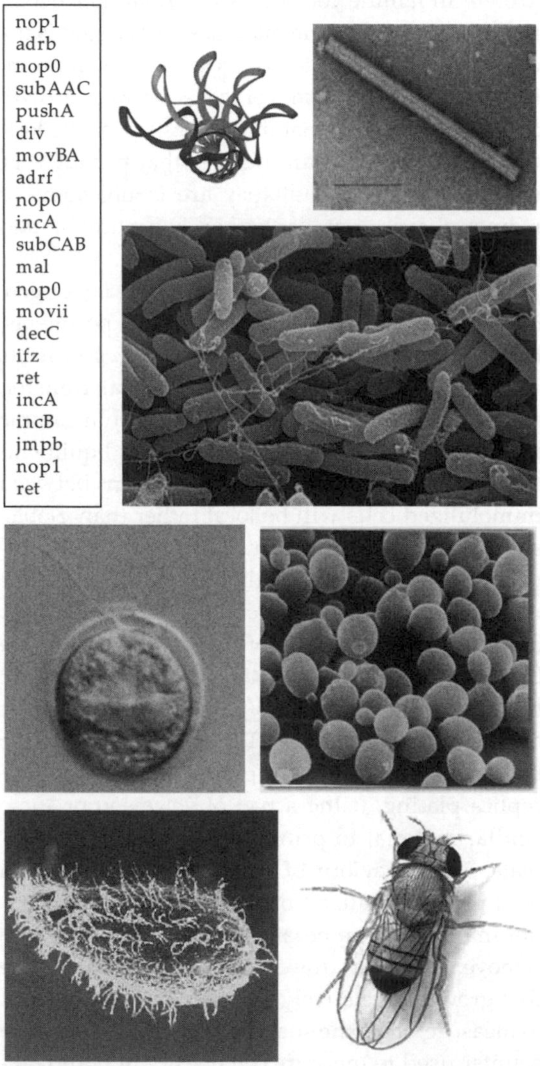

nop1
adrb
nop0
subAAC
pushA
div
movBA
adrf
nop0
incA
subCAB
mal
nop0
movii
decC
ifz
ret
incA
incB
jmpb
nop1
ret

Figure 3.3. Microcosm inhabitants. A Tierran algorithm; a ribozyme; a virus (tobacco mosaic virus); a bacterium (*E. coli*); a green alga (*Chlamydomonas*); a yeast (*Saccharomyces*); a ciliate (*Paramecium*); and a metazoan (*Drosophila*). Photo credits: *www.yale.edu/ocr*, *www.ncbi.nlm.nih.gov/ICTVdb*; *zebu.uoregon.edu/~imamura*; *rydberg.biology.colostate.edu*; *www.utoronto.ca/greenblattlab*; *www.life.uiuc.edu/hing*.

3.1.4 The selection experiment

The defining feature of experimental evolutionary biology is the selection experiment, a modern invention that differs in many respects from other types of experiment and is still little appreciated even within the field of evolution studies. Most experiments in biology concern individuals—how an individual lives, grows, reproduces, or responds to stimuli. Experimental design and statistical analysis have been developed to set up and interpret experiments of this kind. In some cases a population is used in place of an individual, for example in testing the effect of an herbicide, but the population is treated as set of essentially equivalent individuals so no new principle is involved. In selection experiments the entity that responds is neither the individual nor the population, but the lineage. Lineages are different from individuals and populations because they necessarily incorporate history. They tend to diverge because historical change often cumulates through time. They magnify rare events, so that a mutation occurring in only one individual in a million cannot necessarily be neglected. Consequently, procedures intended for other kinds of experiment should not be used uncritically to interpret selection experiments. For example, suppose that we set up 50 selection lines that experience some novel factor and find that after a few hundred generations one selection line has evolved a new phenotype whereas the remaining 49 have not changed. If the ancestral value of the phenotype is 1 and the new value 10, applying the usual logic tells us that the average change of +0.18 is small relative to its standard error of 1.26, so nothing interesting has happened. In practice the exceptional line may furnish decisive evidence of some hitherto unsubstantiated process. This difference between physiology and evolution arises from the weight that is given to rare events in the two disciplines: they are a nuisance in physiology whereas they are crucially important in evolution.

A selection experiment proceeds by configuring the chosen microcosm so as to provide the conditions necessary to test some hypothesis, introducing the organisms, and then allowing them to reproduce. Whether or not anything interesting happens depends on how much variation is present and how long the experiment is run. If there is no variation to begin with then it must be supplied by mutation, and the chance of a mutation occurring depends on how large the population is and how long we wait for. The time unit for evolutionary change is the generation time of the organisms, but the time unit for running the experiment is the year or two that we can devote to it. Within the reasonable constraints of social life, we would like our experiments to be as extensive as possible, where the extent of an experiment is the number of individuals available to be screened by selection: for an experiment involving a population of N individuals with generation time g over time t, extent = Nt/g. This will influence our choice of organism. Most research in molecular genetics and development concerns a very few well-understood organisms which serve as surrogates for the whole of nature. The intense focus on these model organisms, such as *E. coli*, yeast, *Drosophila*, *Caenorhabditis*, *Arabidopsis* and mouse, is justified on evolutionary grounds: the fundamental similarity of organisms following from their common descent implies that what is found in yeast will apply to most eukaryotes. They are not all equally suitable, however, for experimental research in evolutionary biology. A dedicated researcher might follow 10^4 mice for 10 years, at most 100 mouse generations, in an experiment with an extent 10^6. A colleague transferring 1% of a single vial of bacteria supporting 10^8 cells daily for 10 days would have performed an experiment of extent 10^7. This is the straightforward argument for using microbial models with very large population numbers and very short generation times.

Model systems, whether single species or communities, have never been widely used in ecology, I think because most ecologists do not believe that it is possible to extrapolate from a flask to a lake, that is, do not believe that there are interesting general ecological principles that apply to all communities regardless of scale. Evolutionary biology falls somewhere between molecular genetics and ecology. Laboratory experiments using model systems are reasonably well established in evolutionary biology, although I have the impression that they have always been less highly regarded than field studies. Many people would still doubt the relevance of bacteria in glass vials, but this scepticism again seems to concern the ecological realism of compressing the environment into a glass vial rather that the evolutionary realism of compressing the Pleistocene into a couple of weeks.

Early work in experimental evolution almost invariably involved artificial selection, and contemporary experiments on natural selection are exceedingly scarce. Fernandus Payne (1912) cultured *Drosophila ampelophila* in the dark for 69 generations in order to see whether it would evolve the characteristic adaptations of cave animals; he found no response in either morphological or behavioural characters because he did not select for long enough for neutral mutations to become fixed, although he can hardly be faulted for not anticipating this. Microbial experiments began when the chemostat was developed in the 1950s and used to measure mutation rates (Atwood *et al.* 1951a and 1951b, Novick and Szilard 1951). Two important schools of experimental microbial evolution developed from this. The first was led by Robert Mortlock, Patricia Clarke and others and flourished largely in England. It was primarily concerned with the evolution of biochemical pathways and specialized in selecting strains able to grow on exotic carbon sources. The work of this school is admirably summarized in an edited volume, *Microbes as Model Systems for Studying Evolution* (Mortlock 1984b) but publications dwindled thereafter as microbiologists turned away from metabolic studies to work on molecular genetics. The second school includes researchers such as Daniel Dykhuizen, Julian Adams, Lin Chao and many others, who mostly work in the USA and are primarily concerned with charting evolutionary mechanisms. This school spurred a considerable growth of interest in experimental evolution in the 1990s, when Rich Lenski founded the long-term *E. coli* lines that have served as the poster child of the experimental school and provided material for a small army of colleagues and postdocs. The accessibility of viral genomes led Jim Bull and others to lead the effort to understand evolutionary change at the level of sequence, following the earlier virus experiments of Sam Spiegelmann's Qβ school in the 1960s and 1970s. At the time of writing it is becoming feasible to monitor point mutations in evolving microbial populations, and the ability to follow the precise pathway of adaptation seems likely to be the next phase in the history of experimental evolution.

Francis Bacon famously distinguished between experiments of light and experiments of use. Selection experiments can be employed in either sense. The purpose of a selection experiment of light is to identify and evaluate the evolutionary processes that contribute to adaptation. Experiments of use can also be conducted. The most familiar employ artificial selection to modify economically important features of crop plants and livestock in some desirable direction, which will be discussed in Chapter 6. Natural selection in microcosms can also be used to produce strains that cannot otherwise be engineered in our current state of understanding. For example, lines that have become adapted to growth on exotic carbon substrates could be used as bioremediation agents to clean polluted sites. This potential remains almost wholly untapped.

3.1.5 Fitness and adaptedness

Evolution happens because of differences in fitness and results in adaptation. Adaptation is the ability to grow, if you like, whereas fitness is the ability to grow better. These are the two main ways of evaluating the outcome of selection experiments.

Adaptedness is measured by growth as pure culture. This is very rapid and reliable, and can readily be automated to score thousands of cultures. Growth or yield can be measured as number of individuals by techniques such as plating dilute suspensions, haemocytometry or by particle counter. In practice the most convenient method is usually to record optical density with a spectrophotometer. This measures the amount of light scattered by the culture, so the wavelength used is of little concern, although the size of the cells may affect the reading. The main limitation of the method is that the relationship between optical density and cell density is highly non-linear, and in the early stages of growth the readings are uninformative. Measuring growth on plates is more difficult. Colony area can be measured and recorded automatically, but colonies differ greatly in the number of cells per unit area. The culture can be washed off the plate into a vial and read with a spectrophotometer, although this is time-consuming, destructive, and imprecise. Measuring growth in pure culture has two main drawbacks. In the first place, most models of growth have at least two parameters, growth rate and yield, and it is not always easy to decide which is the more appropriate. In practice growth rate requires fitting a curve to a time series whereas yield is a single measurement, so yield is used more often. For assaying an experiment the usual advice is that the appropriate parameter is the yield achieved after a period of time equal to the transfer interval. This may be doubted, since rapid early growth might be favoured in batch culture as a pre-emptive strategy without much regard to the ability to continue growth at high density. Secondly, apparently trivial alterations in conditions of culture, such as the size or shape of the vessel, can affect growth and it is normally recommended to use identical conditions in assay and experiment, which is not always feasible. Even then, it will not always be easy to detect a difference in growth between two strains if growth is sensitive to slight differences in conditions. Whether or not the two strains differ, replicate measurements of the same strain will certainly differ, because of the slight differences between culture tubes, even under laboratory conditions. This makes it difficult to identify any real difference between the two strains if this difference is small in comparison with the difference between replicate cultures of the same strain. If the ratio of the real variation between cultures of the two strains to the total variation among cultures is R, then the number of cultures of each strain that we shall have to measure, in order to establish the difference between them beyond reasonable doubt, is roughly speaking proportional to $1/R$. It might easily be that the genetic variation of fitness amounts to only 1% of the environmental variation; then to be 95% sure of establishing that the 1% difference that we observe between the average fitnesses of the two strains would not arise by chance alone in more than 5% of experiments, we would have to measure fitness in more than 1000 cultures of each strain.

Fitness is measured as relative growth in mixed culture. This requires marking the two competing strains so that they are readily distinguished and counted. When two strains are mixed together and allowed to propagate for a number of generations, their frequencies will change through selection. We first mix the two strains in equal proportions, and spread a sample of the mixture on to plates. The initial frequencies of the two strains can be estimated by counting the number of colonies of each type, provided, of course, that they have been made distinguishable; this can be done, for example, by incorporating into one of the strains a gene that is not expressed during growth in culture tubes but that causes a visible phenotypic change when the cells are growing on plates using a different culture medium. A second sample is then used to inoculate a culture tube, and the mixed culture propagated by serial transfer as a selection line for a number of generations. This culture is then again plated out and the frequency of each strain estimated by counting colonies. If the two types differ in fitness by 1% ($s = 0.01$), selection will cause the frequency of the fitter strain to increase from 0.5 to 0.55 in about 40 generations. This difference in frequency can be measured easily and precisely by plating out the culture and counting 1000 or so colonies. Replicate selection lines must be run to eliminate the possibility that gene frequencies are changing at random, sometimes one strain increasing and sometimes the other, but the precision with which the frequencies can be estimated in each case means that a few replicate lines—of the order of 10—will be adequate. Weak selection is therefore readily detected by allowing selection to sort a mixed population so that a small, almost imperceptible difference in fitness gives rise to a substantial and easily-estimated difference in frequency.

In continuous culture a single strain i grows exponentially at rate r_i so that its abundance is given by $N_i(t) = N_i(0) \exp(r_i t)$. When a second strain j is also present and grows in the same way, the ratio of abundances will be

$$[N_i(t)/N_j(t)] = [N_i(0) \exp(r_i t)/N_j(0)\exp(r_j t)].$$

The ratio of abundances will be the same as the ratio of frequencies, $p = N_i/(N_i + N_j)$ and $q = 1 - p$. Hence $\ln(p_t/q_t) = \ln(p_0/q_0) + (r_i - r_j)t$. The difference in exponential growth rates $(r_i - r_j)$ is called a *selection coefficient* and is usually denoted by s_{ij} or simply s when it is understood to refer to selection of i in the presence of j. The selection coefficient can then therefore be estimated as the slope of the linear regression of $\ln(p_t/q_t)$ on time t. Calculated in this way, s has units of time. If the dilution rate of the chemostat (the fraction of its volume drawn off in unit time) is D then the average growth of the population at quasi-steady-state is likewise D. The generation time, or doubling time, G is then defined by $\exp(GD) = 2$, or $G = (\ln 2)/D$. The selection coefficient in units of generations is then $s = (\ln 2)\,[(r_i - r_j)/D]$. A fuller account of the estimation of dynamic parameters in chemostats is given by Dykhuizen and Hartl (1983). Similar conclusions apply to batch culture, where there is no continuous dilution but the population grows at average rate r. The selection coefficient in units of generation time is then $s = (\ln 2)\,[(r_i - r_j)/r]$, which if i is rare is approximately $s = (\ln 2)\,[(r_i - r_j)/r_j]$ as used by Lenski *et al.* (1991). In a population with discrete generations the basic recursion equation is $p' = p(1 + s)/[p(1 + s) + q]$, hence $(p'/q') = (p/q)(1 + s)$ and by extension

$$(p_t/q_t) = (p_0/q_0)(1 + s)^t \approx (p/q)\exp(st).$$

Then the slope of the regression of $\ln(p_t/q_t)$ on t is $\ln(1 + s)$, or approximately s if s is small. Despite the differences in detail between these systems we can analyse them in nearly the same way, especially when a type is rare or the difference in growth rate is small, basically because the principle that exponential processes diverge exponentially justifies the use of the ln frequency ratio as an estimator of the strength of selection. This method assumes that the genotypes grow independently of one another, that is to say that their growth rates r_i are constants. The same result should therefore be obtained if the genotypes are cultured separately in different vessels, being mixed and redistributed at longer or shorter intervals. Thus, growth in pure culture should yield the same estimate of the selection coefficient as competitive growth in mixture. When this is not true the competitive estimate is preferred, as reflecting more closely the actual difference in fitness, but it should be noted that the basic assumption of the estimation procedure has broken down.

Figure 3.4 is an example of this analysis from Dykhuizen and Dean (1980), using two strains of *E. coli* cultured in a chemostat. The open symbols represent the frequency ratio in replicate experiments in glucose-limited chemostats. The slope of the pooled data is +0.0045, with standard error 0.0078; thus, selection is so weak that it cannot be detected reliably. The filled circles represent similar experiments in lactose-limited chemostats. Here, the slope is −0.0334, with standard error 0.0023; in this environment, therefore, selection involving a difference in fitness of about 3% is readily detected. Selection coefficients as small as $s = 0.01$ can routinely be estimated by this procedure, and the actual limit of detection in longer-term experiments is substantially lower.

An evolved line could be tested against several different kinds of competitor. The obvious choice is the immediately preceding wild type, since the fitness difference measured thus is that responsible for the success or failure of the isolate. This competitor may not be available, however. The usual choice is the ultimate ancestor, kept in a non-evolving state by freezing or maintained in conditions where it is assumed not to change. This is satisfactory so long as it can be assumed that fitness differences are transitive, which may not always be the case (see §3.5.3). When the selection lines are exposed to some definite stress or deficiency not present in the environment of the ancestor, the competitor may be obtained from a control line maintained under ancestral conditions. The fitness difference measured will then correspond to the specific adaptation of the experimental line to the environmental factor under investigation. When neither

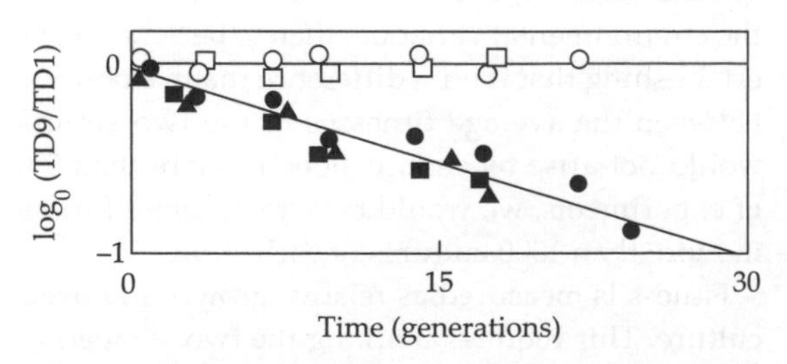

Figure 3.4 Estimating the selection coefficient from the frequency ratio.

control line nor ancestor can be used as competitor (because they are not appropriately marked, for example) then the ancestor and the evolved line could be competed separately against an arbitrary tester strain, the fitness difference between ancestor and evolved line being inferred from the difference between the two selection coefficients. This is not often used, I think because genotype-specific interactions might occur. This objection could be circumvented by competing both ancestor and evolved line against two or more arbitrary testers and statistically eliminating genotype-genotype interactions should they occur. In short, competition against ancestor is the default procedure, but control lines are preferable in some circumstances and (in my view) the use of arbitrary testers should be considered if they offer some substantial advantage. The most likely advantage is automated screening, since the principal drawback of competition trials is that they are very laborious.

3.1.6 Microcosm genealogy

The history of a lineage can be represented as a phylogenetic tree, linking ancestors and descendants. This may be evolution on the grand scale, depicting the ancestry of the Metazoa, or the origins of land plants. However, if we examine any part of such a tree in more detail, we see the same tree-like structure on a smaller scale (Figure 3.5). The first tree represents tetrapod vertebrates, where the earliest branching shown occurred about the middle of the Carboniferous (lower Pennsylvanian), about 300–320 million years ago. The second shows placental mammals, with most orders diverging in the Eocene or before, about 50–70 million years ago. The third is the family tree of one of these orders, the carnivores; the separation of canoid and feloid lineages is Eocene, but subsequent divergence was spread over about 25 million years, into the middle Oligocene. The fourth tree is for canids, the dogs, where the earliest branching shown occurred in the early Miocene, about 10 million years ago. The final tree shows the ancestry of dogs and wolves, *Canis*, which have diverged over the last 1 million years or less. Continuing to magnify branches in this way will lead eventually to a population of dogs at a scale of generations, in each of which there is an episode of variation followed by an episode of selection. This is the lowest genetic scale; it cannot be magnified any further. The swarming denizens of the microcosm are all connected to their ultimate ancestor, the founding cell, by a miniature phylogenetic tree of this sort.

The genealogy of the microcosm is strictly branching because its inhabitants are asexual, with no recombination or horizontal gene transfer. It can therefore be dissected into a number of lineages, or monophyletic groupings, of given depth. Any pair of cells descend from a single individual, their most recent common ancestor (mrca), which lived at some time in the past. A third cell might also descend from that individual, in which case it belongs to the same lineage, or not, in which case it belongs to a different lineage. If we go back only a single generation then each lineage comprises the progeny of a single parent, and there will be

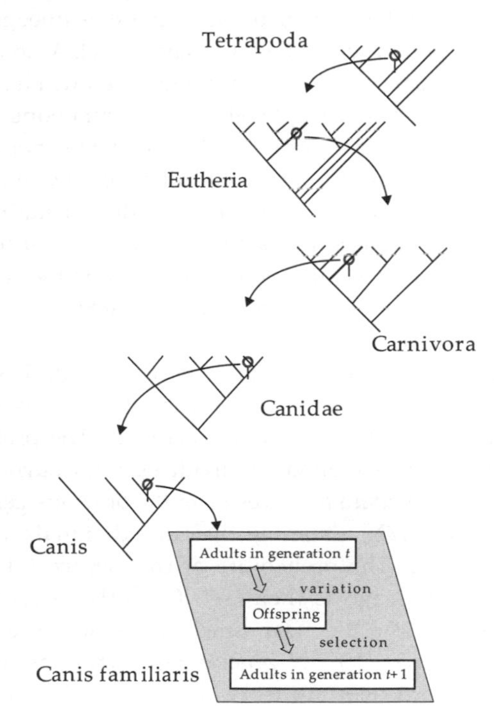

Figure 3.5 Phylogenetic trees at different genetic scales. Each of the trees here is a magnified view of the branch shown as a thick line in the preceding tree.

a large number of very small lineages. Reaching back further in time we shall be dealing with more remote ancestors which may have left many descendants, and there will be fewer but larger lineages. Eventually we shall find only a single lineage descending from the mrca of the population as a whole. This is the coalescent, or point of coalescence, of the current population. The universal mrca is not necessarily the ultimate ancestor, which may have given rise to many lineages all of which, save one, subsequently became extinct; the universal mrca is the founder of the single surviving lineage. This implies that if we look ahead in time the entire population at some point in the future will have descended from a single lineage in the current population. I shall call this the generating lineage (it is the same as the 'progenitor tail' of Rice 2002). If we project only a few generations the generating lineage is almost certain to be the most frequent lineage in the current population, or at least one of the most frequent. Further into the future, however, the generating lineage will necessarily be smaller because all other lineages contemporary with it have become extinct. At some very remote period the entire population will have descended from a single cell in the current population, their mrca. Now: if a cell is not a member of the generating lineage it will ultimately leave no descendants at all. Furthermore, as the generating lineage is reduced to a single cell, the mrca of the population of the distant future, all cells in the current population, save one, will ultimately leave no descendants.

The theory of the coalescent (Kingman 1982; review by Donnelly and Tavaré 1995) leads to estimates of the population genealogy. The probability that two random individuals in the current population share an ancestor in the previous generation is $1/N$, assuming that all individuals are equivalent. The probability of coalescence $t + 1$ generations ago is then $(1/N)(1 - 1/N)^t \approx (1/N) \exp(-t/N)$ and the mean time since coalescence is $\Sigma\ (t/N)\exp(-t/N) \approx N$ generations, corresponding to standard neutral theory (see below). This will be a very long time in microcosms whose inhabitants number in the billions. It is reduced by variation and selection. The coalescence time (expected time since the mrca) of a sample of n individuals is $(2N/\sigma^2)(1 - 1/n)$ where σ^2 is the variance of the number of surviving offspring: coalescence is accelerated if individuals vary. If this variance is genetic then selection will further reduce the coalescence time (Neuhauser and Krone 1997) to the forward time required for a beneficial mutation to arrive and sweep through the population.

The significance of a genealogical view is that almost all of the innumerable cells living in the microcosm at present will not contribute to adaptation because ultimately they will leave no descendants. The average properties of the current population are of little or no importance in the long term: only the generating lineage has an evolutionary future and adaptation will therefore be governed by extreme values and rare events.

3.2 Sorting: selection of pre-existing variation

In a more or less unchanging environment, purifying selection will preserve the adaptedness, and the existing genetic composition, of the population through time. If the environment fluctuates without any consistent trend, each episode of selection will be directional, but the genetic changes that occur in each generation will not cumulate through time, the effects of one episode being cancelled by the next. However, there may be a consistent secular trend persisting over many generations; or, there may be an abrupt long-lasting shift in the state of the environment; or, migrants may enter and permanently occupy an environment different from that of their parents. In these circumstances, successive episodes of selection will have the same direction: the same set of variants will be selected, generation after generation. The variation that was present in the population before the change will be *sorted* by selection, until, if the altered state of the environment lasts for long enough, the type that was originally common has been largely supplanted by another variant. Sorting is a particularly simple kind of selection with a predictable outcome; it occurs in many situations and is served by a large body of descriptive and predictive theory. Here I describe the main features of sorting in haploid asexual populations, before turning to evolutionary processes where novel variation appears during the course of selection.

3.2.1 Single episode of selection

At the limit of magnifying the phylogenetic tree we would be looking at a single generation in a single population: a cohort of newborn offspring that grow up, reproduce, and die. These offspring are not exact replicas of their parents; setting the complexities of sexuality aside, the imprecision of reproduction will introduce alterations of various kinds, causing them to vary to some extent. This new variation will be sorted by selection, with some variants surviving and reproducing more successfully than others. This coupling of an episode of variation with an episode of selection during a single generation is the unit event in evolution; evolutionary history on the grand scale consists of a long series of such events occurring successively through geological time.

In the simplest case, a population comprises two types with arbitrary frequencies p and $1 - p$, which produce w_1 and w_2 offspring respectively. At the beginning of the next generation the frequency of the first type will have changed to $p' = pw_1/(pw_1 + (1 - p)w_2)$, so that the change in the frequency of the first type is $\Delta p = sp(1 - p)/(1 + sp)$, where $s = (w_1 - w_2)/w_2$ is a selection coefficient as explained above. If selection is weak then a good approximation will be $\Delta p = sp(1 - p)$: the response is equal to the product of the strength of selection and the quantity of genetic variance. For a rare type, the factor by which its frequency is increased by selection in a single generation is roughly equal to s, so that differences in fitness of a few per cent will be difficult to detect. A sudden grave stress, on the other hand, may propel a hitherto obscure type to high frequency at a stroke. This can readily be observed by spreading a bacterial culture onto medium containing an antibiotic: the few surviving individuals from which the population subsequently descends bear mutations whose selection coefficient is nearly the maximum possible $s = 1$.

At the other extreme, the population might be a complex mixture of many types that cannot readily be distinguished yet differ with respect to some phenotypic trait symbolized as Z. The trait is liable to selection, in the sense that the mean value of this character among individuals that succeed in reproducing differs from the overall mean of the population by a quantity known as the *selection differential*, D_z. This drives a change in the mean from one generation to the next by a quantity known as the *response to selection*, R_z. If each selected parent transmits its phenotype perfectly to its offspring then $R_z = D_z$. In practice, this is unlikely to happen because the phenotype of an individual is not determined solely by its genotype but also by accident and environment. These non-genetic sources of variation will contribute to the superiority of the highest-ranked parents and to the inferiority of the lowest, but cannot be transmitted to their offspring. Consequently, the offspring of the most highly-ranked parents are likely to be somewhat inferior to their parents whereas the offspring of the lowest-ranked will be somewhat superior. This regression of the offspring towards the parental mean can be expressed as $Z_o = \text{constant} + b_{op}Z_p$ where Z_o and Z_p are the mean values of offspring and parent respectively. The regression coefficient b_{op} expresses the heritability of the trait Z, such that $b_{op} = 1$ indicates that offspring resemble perfectly their parents whereas $b_{op} = 0$ indicates that the offspring of different parents do not differ consistently. This is equivalent to the genetic variance of Z relative to its overall variance, $b_{op} = V_g/V$, which is the fraction of the selection differential that will be transmitted to the next generation. Hence $R_z = D_z V_g/V$, which is equivalent to the corresponding expression for two discrete types.

3.2.2 Sorting of a single type

Successive episodes of selection cause a sustained rise of the type with greater fitness. The change produced by selection in terms of the ln frequency ratio is $\ln(p_t/q_t) = \ln(p_0/q_0) + st$, so solving for t shows that the frequency of an arbitrary favoured type increases logistically from some initial value f_0 to any defined greater value f_t in $t = (1/s)\ln(F_t/F_0)$ generations, with $F_t = f_t/(1 - f_t)$ and $F_0 = f_0/(1 - f_0)$ (Figure 3.6). Consequently, it will pass

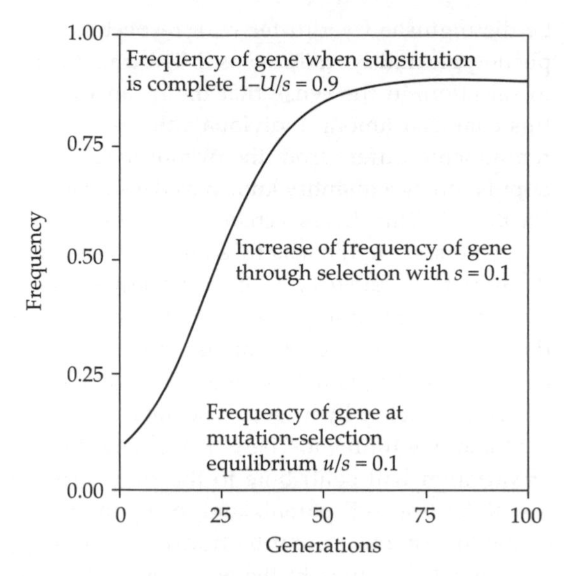

Figure 3.6 Sorting of a single type. The diagram shows the (unrealistic) situation in which $s = 0.1$ and $u = U = 0.01$. From its frequency at mutation-selection equilibrium of $u/s = 0.1$, the gene spreads rapidly through the population, attaining its final frequency of $1 - U/s = 0.9$ within 100 generations. In more realistic situations, the rates of selection and mutation will be much lower.

through equal log intervals of frequency in roughly equal times. Over the series..., e^{-10}, e^{-9}, ..., e^{-1}, $(1 - e^{-1})$, ..., $(1 - e^{-9})$, $(1 - e^{-10})$...the frequency of the favoured type will pass from any given term to the next in roughly the same length of time, about $1/s$ generations. This rule can be used to obtain a rough impression of the timescale of change through sorting. A type with an advantage of 1%, for example, will increase through 10 ln units of frequency, from 0.25% to 99.75%, in about 1000 generations, whereas a more strongly favoured type with an advantage of 10% would spread over the same interval in about 100 generations.

An interesting special case is a mutation arising in a single individual and spreading through a population until only a single individual of the inferior type remains, which will be complete after $(2/s)$ ln $(N - 1)$ generations, where N is the total number of individuals in the population. N does not appear in the expression given in the previous paragraph because it was assumed implicitly that the population was infinite in size. In finite populations, the expected time required for complete substitution increases with the logarithm of population size because larger populations include more ln intervals of frequency. The logarithmic scaling implies that this increase is modest: increasing N by a factor of 100 only doubles the substitution time.

3.2.3 Mixture of discrete types

More generally, the initial population will be a mixture of several types. Those which happen to be better adapted to the new environment will tend to increase in frequency, until the population has become dominated by one or a few types adapted to the current circumstances, and selection produces no further change. Selection sorts the initial population, favouring types with greater fitness, and thereby increasing the mean fitness of the population as a whole as its genetic variance is consumed. The population will eventually come to consist almost entirely of one of the types, initially rare, that have the greatest fitness in the changed conditions, so the mean fitness of the population when selection is complete cannot exceed the fitness of the fittest type present before selection. The end point of the sorting process is thus set by the range of types present in the initial population.

In a mixture, the best type is simultaneously in competition with several other types. As the better types increase in frequency the mean fitness of the population increases and the selection coefficient favouring the best type falls. In this way, above-average types impede the spread of the best type, and instead of a rapid, logistic rise to fixation, the best type spreads slowly and follows a more linear trajectory (Figure 3.7). This process of clonal interference may greatly extend the time to fixation. If the only two types present are a wild type with unit fitness and a rare type with fitness $1 + s$ then the time to fixation t_0 can be calculated straightforwardly. Now suppose that there are also n evenly spaced intermediate types, such that the ith type has fitness $1 + is/(n+1)$. If the best type and all the intermediate types initially have the same low frequency, simple simulations show that the time to fixation t_n is a linear function of n. The regression of the standardized time t_n/t_0 on n has a slope

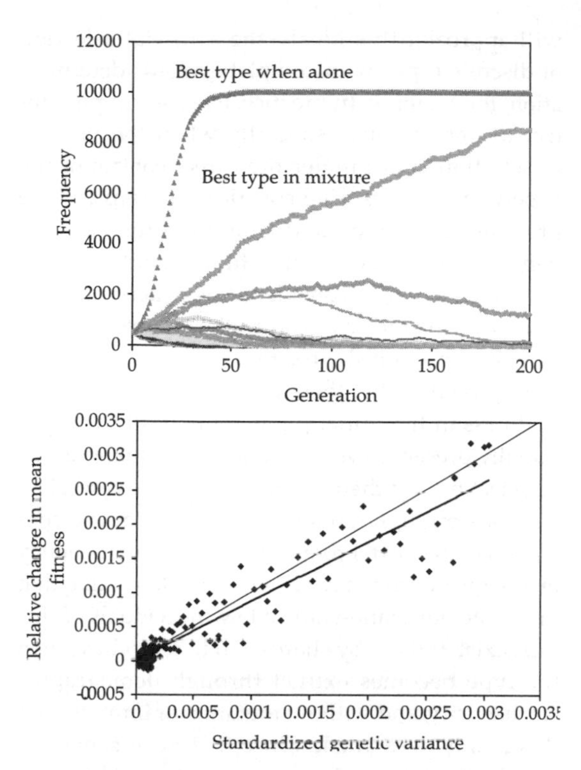

Figure 3.7 Sorting many types. In the upper diagram, the uppermost line shows the spread of a beneficial mutation with fitness 1.2 from an initial frequency of 5% to fixation in a population of 10 000 individuals when competing against a wild type with unit fitness. The other lines show the flux of frequencies when 20 genotypes with fitnesses equally spaced between 1 and 1.2 are all competing together. The lower diagram shows the relative change in mean fitness $(1/\overline{w})\Delta\overline{w}$ per generation as a function of the standardized genetic variance $V^2(w)$. The upper line is the line of equality predicted by the fundamental theorem. The lower line is the regression of the data, which fall very close to expectation.

of about ½, so that the effect of adding one more intermediate type is to extend the time to fixation by about one-half the time taken when only the best type is spreading. This is no more than a very rough rule of thumb, but correctly indicates that the number of competing inferior types may have a much greater effect on the time to fixation than the selection coefficient relative to the wild type. The most complex sorting experiments involve the 4000 or so single deletions of yeast genes (e.g. Giaever *et al.* 2002), but their long-term dynamics have not yet been described.

3.2.4 Fundamental theorem of natural selection

The dynamics of mixtures is complicated, because the rate of change of mean fitness is continually modulated by the increase in mean fitness and the decline in the variance of fitness. In the limit, however, a mixture of an infinite number of types is easier to analyse: this is primarily because each type is present at an infinitesimal frequency that is not affected by selection, so the genetic variance of fitness remains constant through time. This is the *infinitesimal model* that has been extensively used in theoretical population genetics because of its mathematical tractability. Populations described by this model would move smoothly towards the best-adapted state through the selection of a long succession of imperceptibly superior variants, the extreme expression of the classical Darwinian view that evolution will almost always proceed slowly and in very small steps. Although the model is obviously unrealistic, it makes it possible to develop a statistical mechanics of evolving populations. In sufficiently large and variable populations, we turn from enumerating a few discrete types to estimating overall parameters without any direct knowledge of the underlying distribution of frequencies. The most important parameters are the mean and variance of fitness, so that the relationship between them is the basis for a statistical description of sorting.

The rate at which sorting occurs depends on the magnitude of the variation among individuals in the initial population, and on the fidelity with which these differences are transmitted to descendants. If there is initially a large difference in fitness between alternative types, the rate of selection will be greater. If the differences among individuals are predominantly inherited, selection will be more efficacious in causing evolutionary change. The rate of sorting does not depend on the range, which may be strongly influenced by the fortuitous occurrence of a single extreme individual, but rather on the variance of the population. The greater the variance of fitness, the greater will be the change in mean fitness caused by a single episode of selection. This increase will be expressed in the following generation only if the differences in fitness between individuals are inherited; that is, the effect of selection on mean fitness in successive generations depends on the genetic variance of fitness. The greater the genetic variance of fitness, the more rapidly adaptation will occur.

In the simple situations that we are presently considering, we can phrase this conclusion even more tersely: the rate of evolution is equal to the genetic variance of fitness. If we write the mean fitness of the population at a given time as $\bar{w}$, and the standardized genetic variance of fitness as V^2_w, then:

$$(1/\bar{w})\, d\bar{w} = V^2_w$$

The standardized genetic variance of fitness is the genetic variance divided by the square of the mean, $V^2_w = \mathrm{Var}(w)/\bar{w}^2$. This conclusion was first reached by Sir Ronald Fisher, who called it the *fundamental theorem of natural selection*. This the most succinct way of summarizing the complex dynamics of a diverse population undergoing sorting, by relating the change in the mean to the quantity of variation (Figure 3.7). It is very generally applicable, being an algebraic identity that relates the unweighted mean of a set of numbers (mean fitness before selection) to their weighted mean, the weights being the numbers themselves (mean fitness after selection). Fisher regarded it as fundamental because it attaches an arrow to evolutionary change, pointing in the direction of ever-increasing adaptedness. Later commentators (there have been many) have expressed a range of opinions. Some have been unwilling to believe that average population density and productivity are much greater now than in the Mesozoic, and consequently relegate the principle to an indicator of short-term change following environmental disturbance. Fisher himself pointed out that the same pattern of weak selection is unlikely to continue over long periods of time and would likely have agreed. The fundamental theorem is discussed in more detail in §7.2.

3.2.5 Sorting in finite populations

In principle, sorting is a deterministic process leading to a predictable outcome, and some situations will approach this ideal—the artificial selection of discrete types with complete genetic determination, for example. In practice, even sorting is subject to uncertainty, especially when the process of selection is not under the direct control of the experimenter. Offspring constitute a sample of the preceding generation, and sampling error will inevitably lead to genetic drift in finite populations.

When sorting is considered as a deterministic process, selection will lead to the complete loss of diversity through the fixation of the best-adapted type, provided that there is some genetic variance of fitness in the initial population. If all the types initially present have the same fitness there can be no selection, but their frequencies will nevertheless fluctuate through random sampling from one generation to the next. From time to time the frequency of any given type may fall to such a low value that in a finite population only a few individuals of this type exist. If these by chance fail to reproduce, then the type becomes extinct through demographic stochasticity, and without mutation or immigration this loss is permanent. Since this must sooner or later be the fate of all types, except one, the eventual outcome of drift will be the fixation of a single type and the complete loss of diversity. Both selection and drift end with the exhaustion of diversity, but whereas the deterministic process of selection leads to the predictable fixation of the best type, the stochastic process of drift leads to the unpredictable fixation of the luckiest type. Luck depends entirely on initial frequency: if a given type has frequency p in the initial population, then the probability that it will be fixed is likewise p.

Because all types except one must eventually become extinct, all individuals in the population at some time in the future will descend from a single individual in the present population, their most recent common ancestor. Drift is much slower than selection, however. The number of individuals descending from a single ancestor cannot be much greater than the number of generations that have elapsed (Fisher 1930), and indeed the expected length of time required for the lineage descending from this individual to become fixed is $2N$ generations. Thus, the timescale of substitution through drift is proportional to N, whereas through selection it is proportional to ln N. Moreover, substitution time under drift is intrinsically variable, with a standard deviation of roughly N.

3.2.6 Drift and selection

Drift and selection are not alternatives. All populations are finite, and few are completely devoid of variation in fitness, so both drift and selection will almost always occur together. The most important effect of drift is that it makes the outcome of selection less predictable (Figure 3.8). The expected response to selection per generation in a finite population is $sp(1 - p)$, its deterministic value, but through sampling error it will deviate from this precise value, the variance of this deviation being $p(1 - p)/N$. Since the expected time before fixation depends on p, any increase of p resulting from drift will accelerate, whereas a decrease will retard fixation. Each generation offers an independent trial of

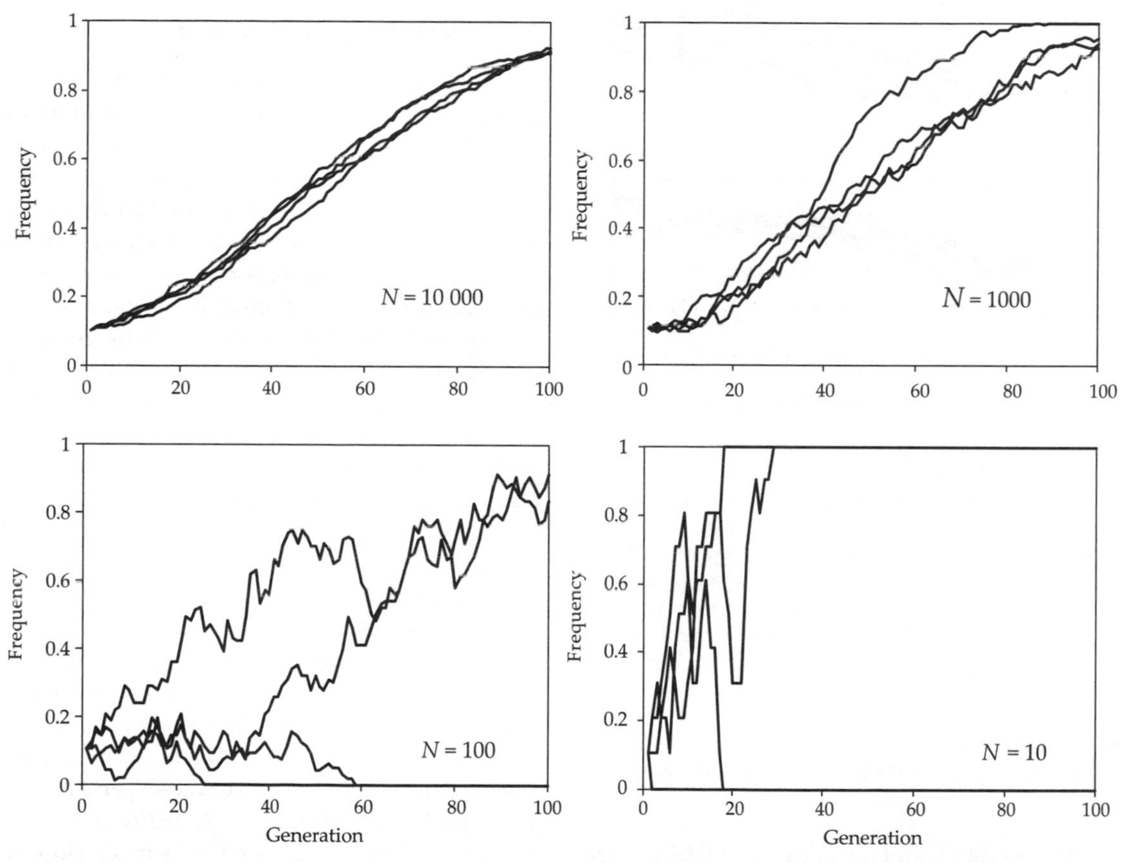

Figure 3.8 Selection and drift. An allele with a selective advantage of 5% is introduced into a population of N haploid individuals. In large populations of $N = 10\,000$ or more it increases in frequency smoothly through time, and all replicate populations behave in the same way. In smaller populations of $N = 1000$ its increase in frequency is noticeably somewhat erratic, and replicate populations follow somewhat different courses, primarily because they diverge slightly in the early generations of selection, when there are only 100 or so copies of the gene present. Nevertheless, the replicate populations follow similar courses, and the allele will approach fixation at about the same time in each of them. In small populations of $N = 100$ or so, only 10 copies of the gene are present initially; allele frequency fluctuates widely through time, and replicate populations diverge markedly. Two of the populations shown here eventually become fixed, although at different times, and the allele is lost from the other two. When population size $N = 10 < 1/s$ only a single copy of the allele is initially present, and whether the allele spreads at all is largely a matter of chance. In this case, it quickly became extinct in two populations and becomes abruptly fixed in the other two. It cannot be predicted with confidence how long it will take for the allele to be fixed, or even whether it will be fixed at all.

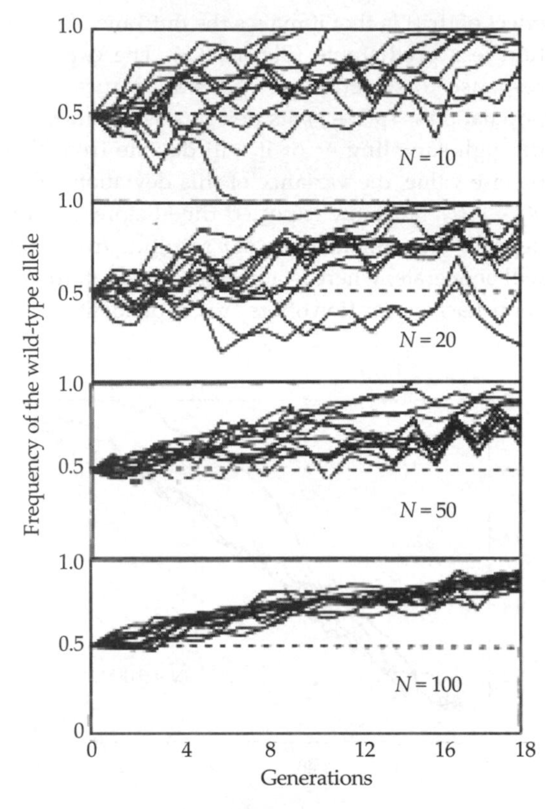

Figure 3.9 Selection and drift. This is an experimental demonstration of the principles illustrated in Figure 3.8, from populations of the flour beetle *Tribolium* studied by Rich *et al.* (1979). A mutation for body colour, *black*, and the wild-type allele are initially equally frequent. The organism is diploid, but heterozygotes can be distinguished, so in subsequent generations the frequency of black can be estimated directly by counting. Wild-type individuals are fitter, and in populations of $N = 100$ (since the organism is diploid, with 100 individuals there are 200 copies of the gene) the wild-type allele increases in frequency fairly regularly and predictably. In small populations of $N = 10$ the increase is much more erratic, the behaviour of a given population cannot be predicted in detail, and in one case the allele, despite its greater fitness, is lost from the population.

these possibilities, and the outcome will be a range of fixation times (Figure 3.9). How much uncertainty is introduced into the schedule of fixation depends on the magnitude of the deviation relative to the expected value of the response, which will be small if $s < 1/N$.

In populations of a few thousands of individuals or more, then, drift proceeds much more slowly than selection, and its effects are overwhelmed even by weak selection driven by differences in fitness of a few per cent. It is quite plausible, of course, that some types are so nearly equivalent that the difference in fitness between them is less than 1 part in 1000; what may be less plausible is that the environment should be so stable that such a minute difference would be preserved over the thousands of generations that must pass before variation is lost. This is Fisher's objection to the importance of drift as an agent of evolutionary change, and it retains much of its force today.

3.2.7 Fluctuating population size

If population size fluctuates, then the effect of genetic drift will vary from generation to generation. Suppose that for most of the time a population were too large for drift to matter, whereas from time to time it became so small that types were likely to be fixed or lost by chance. In the long run, the outcome of sorting would obviously be affected much more strongly by the rare periods of small size than by the much longer intervening periods of large size. This intuition can be made more quantitative by calculating the size of an unchanging population that would be equivalent in terms of genetic drift to a population whose size fluctuated over time. This is the fixed *effective population size* N_e of an ideal population for which the sampling variance of genotype frequencies is the same as that of a real population of size N. The effective population size is the harmonic mean of successive populations N_t over a given period of time T: $N_e = 1/[(1/\Sigma N_t)/T]$, where the summation is over time. The harmonic mean is always less than the arithmetic mean. For example, if the population fluctuates between a two sizes N_1 and N_2 with equal probability then $N_e = 2N_1N_2/(N_1 + N_2)$, and if $N_1 \gg N_2$ then $N_e \approx 2N_2$ and so is much closer to the lesser than to the greater value. This is particularly important in serial transfer experiments in which a small inoculum is allowed to expand by several orders of magnitude in each growth cycle; regardless of how large the population becomes, its effective size is only about twice the size of the inoculum. This implies that beneficial mutations are likely to be lost stochastically at transfer.

3.3 Purifying selection: maintaining adaptedness despite genetic deterioration

Sorting produces an optimally adapted population in which a single genotype has become fixed. This adaptation will be degraded by recurrent mutation. The reduction in fitness that mutations cause may be individually slight, but collectively substantial. It is easy to appreciate that a large number of slightly deleterious mutations may have collectively the same effect as a single crippling mutation. Adaptedness is therefore endangered by the accumulation of mutations with slight effects. The greater the mutation rate, the faster this accumulation will proceed, eventually to the point where adaptedness can no longer be maintained.

3.3.1 Mutation–drift equilibrium

Neutral alleles are introduced by mutation and lost by drift; their dynamics form the core theory of population genetics, and a full account is given in any textbook. Briefly, the rate of fixation of neutral alleles at equilibrium must be equal to the rate at which they appear, that is to the genomic rate of neutral mutation $U_{neutral}$. If the population fluctuates in size this will hold only as a long-term average, as fixation will accelerate in declining populations and slow down in expanding populations. The time between the appearance of successive mutations that will ultimately become fixed is thus $1/U_{neutral}$. The expected passage time required for each successful mutation to be fixed is $2N$ generations in a haploid asexual population, with a standard deviation of about N, or $4N_e$ generations in a sexual diplont, given a set of assumptions about population size and fecundity. This period corresponds to the date of the last common ancestor of all individuals in the present population with respect to a given gene or set of linked genes. The overall heterozygosity expected for neutral alleles at a locus in a sexual diplont is $4N_e u_{neutral}/(1 + 4N_e u_{neutral})$. These results appear many times in any discussion of evolutionary dynamics.

3.3.2 Mutation–selection equilibrium

Imagine a population that at some point becomes perfectly adapted to its environment. Every gene has attained the best possible state; no individual bears any deleterious mutations. This perfection cannot be maintained even for a single generation. Some progeny will receive one or more new mutations, so that the population includes, not only the class of individuals with no mutations, but also classes of individuals that have suffered different degrees of mutational damage. Each will transmit these mutations to its offspring, together with any new mutations that have occurred in its germline. The number of mutations borne by any lineage will tend inexorably to accumulate, eroding the adaptedness of the population. This process is countered by selection. Individuals that bear more mutations are less likely to reproduce successfully than those that bear fewer. The next generation will therefore be recruited predominantly from individuals in the current generation that bear fewer mutations than average, so that selection tends to reduce the average number of mutations per individual. In this way, the tendency for mutations to accumulate without limit is checked by the tendency for individuals that bear fewer mutations to produce more offspring. At some point the number of old mutations removed through selection will be equal to the number of new mutations appearing. The opposed tendencies of mutation and selection are then balanced: the population does not consist entirely of perfect, undamaged individuals (that is forbidden by mutation), nor of a few hulks riddled by mutations and scarcely able to survive (that is prevented by selection), but rather of a mixture of individuals, some with fewer mutations and some with more. This mixture represents an equilibrium state, a population in which the frequency distribution of mutations per genome is stable from generation to generation. In each new generation there is an episode of variation, in which new mutations arise, and an episode of selection, in which an equivalent number of mutations are removed through the death or sterility of individuals that bear them, restoring the fit between population and environment. At this point, the population is said to be in *mutation–selection equilibrium*.

The point of balance is set by the relative rates of mutation and selection. Suppose that each generation begins with an episode of mutation. The mutation rate of a given gene is u. This means that the frequency of the mutant after mutation (p') is related to its frequency before mutation (p) in this way: $p' = p(1 + u)$, assuming that mutation is rare. There is then an episode of selection. The rate of selection acting on the mutation is s, in the sense that the frequency of the mutant after selection (p'') is related to its frequency after mutation but before selection (p') in this way: $p'' = p'(1 - s)$ assuming that selection is weak. At equilibrium the rate of mutation u is balanced by the rate of selection s, and the frequency of the mutant in the population is $p = u/s$. In other words, selection cannot completely eliminate deleterious mutations from the population, but instead drives them down to a frequency u/s that is sustained by recurrent mutation (Figure 3.10). The mutation rate is usually taken to be the collective rate of all changes in the structure of a gene, so that the selection rate is an average taken over all the different kinds of change, which may have quite different effects on fitness. With $u_L = 10^{-6}$ mutant variants of a gene that depress fitness by 1–10% should occur in the population frequencies of the order of 10^{-3}–10^{-4}. This argument can be extended to the genome as a whole. The average number of deleterious mutations borne by each individual at mutation–selection equilibrium is U/s, where U is the genomic mutation rate, which is likely to be 1 or less for bacteria and 10–100 for animals and plants. These figures should be taken only as very crude rules of thumb. They serve to emphasize how the continuous operation of selection stems the continuous accumulation of mutational load.

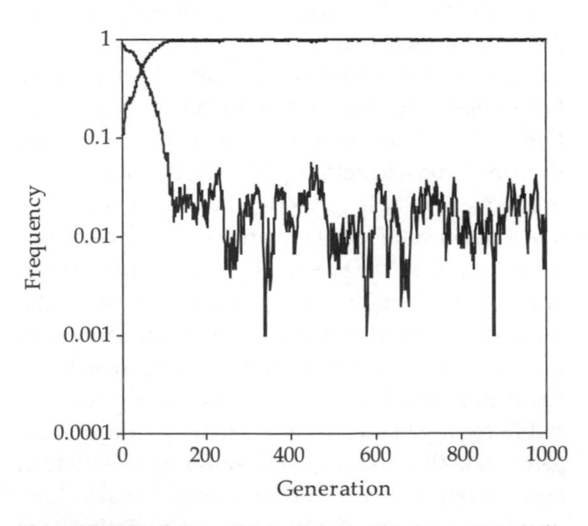

Figure 3.10 Mutation–selection balance. This represents a similar situation to that in Figure 3.8 with $N = 1000$ and $s = 0.05$, but a mutation rate of $u = 0.001$. The inferior allele is not completely lost because it is continually regenerated by mutation, so a fluctuating reservoir of rare variants persists and provides a perpetual source of variation.

3.3.3 Muller's ratchet

If the population is small then mutation and selection do not necessarily come to a permanent balance. A population can be divided into classes of individuals bearing different numbers of deleterious mutations. Each class has a certain frequency, and once the population has reached equilibrium these frequencies do not change consistently through time. However, they may, and in fact they must, fluctuate to some extent through sampling error. Even if the population is rather large, a class that occurs at low frequency may comprise very few individuals, and the number of individuals in this class will fluctuate wildly from one generation to the next. It is usually the extreme classes that will be the least frequent, including the class of individuals that bear no deleterious mutations. This class is therefore very vulnerable to sampling error, and very likely sooner or later to disappear entirely. Once lost, it cannot be regained. The least damaged individuals in the population, following the loss of the completely undamaged individuals, are those which bear a single mutation. In principle, this class could give rise to undamaged individuals by back-mutation; but the alteration of a non-functional gene to produce a functional one is much less likely than loss of function, so rates of back-mutation are very low, and long before the class of undamaged individuals would have been restored, the class with a single mutation has itself disappeared. The geneticist H.J. Muller was the first to realize that asexual populations are subject to a ratcheted process, in that the number of mutations borne by the least heavily loaded class of individuals can never decrease (Muller 1932). If this class is sufficiently small, as it is likely to be in small populations, the combination of mutation and sampling error will cause a cumulative and irreversible loss of adaptedness. There is a more extensive discussion of this process in §12.1.

Mutation–accumulation lines use Muller's ratchet in very small populations to trap mutations and thus cause a continual decline in fitness; the rate of decline provides an estimate of the rate of deleterious mutation (§2.2). In many cases this decline was very modest: in a large and careful experiment with *Caenorhabditis*, for example, little if any deterioration in offspring production could be detected after 60 generations of transferring single larvae (Keightley and Caballero 1997). Replicate lines diverged widely during this time, however, largely because a small minority of lines had very low productivity. This pattern would arise if deleterious mutations of large effect are occasionally trapped, whereas the bulk of mutations of small effect are balanced by beneficial or compensatory mutations trapped by the same procedure (see above). In other cases, however, severe and prolonged drops of fitness have been observed. The replication rate of isolate lines of the RNA vesicular stomatitis virus (VSV) consistently dropped about 1% per replication through a series of daily transfers to the point where the lines were barely 4% as fit as the ancestor (de la Pena *et al.* 2000). Declines as severe as this endanger the survival of the population (Zeyl *et al.* 2001). It has been claimed that Muller's ratchet is a serious threat to rare species, but this is not necessarily the case. Rather modest declines of U/s have been reported in most experiments, even with an effective population size $N_e = 1$ under self-fertilization, so that populations would be expected to persist for many hundreds of generations.

3.4 Directional selection: restoring adaptedness despite environmental deterioration

Evolution on short timescales occurs by simple sorting, because the variation on which selection acts is dominated by the variation initially present in the population, and the input of new variation can be ignored. On longer timescales this balance shifts, as novel mutation becomes an increasingly important source of variation. Over the course of hundreds or thousands of generations the balance reverses, and the variation initially present, however great, is of little consequence relative to the new variation arising during evolution.

In any given generation, of course, selection sorts the variation that happens to be present. However, when selection is continued for long periods of time, this sorting process has in every generation a new point of departure, as novel mutations modify the genotypes that have been established in the population as the result of previous generations of sorting. Long-term evolution is therefore a process of *cumulation*. There is an important difference between sorting and cumulation. Simple sorting is a repeatable process whose outcome can be predicted when the environment can be controlled and the composition of the initial population specified. This is not necessarily the case for cumulative change. We cannot in general predict which mutations will occur, nor the order in which they will occur. If we establish replicate selection lines, maintained in identical environments, they may each experience a different sequence of beneficial mutations, and therefore follow a different pathway to adaptation. The opportunistic nature of selection means that we cannot in general predict the route of cumulative change; we may not even be able to predict its eventual outcome.

Fortunately, there are two considerations that mitigate this pessimistic evaluation and prevent the study of adaptation from becoming merely a string of anecdotes. The first is that it is sometimes wrong: cumulative change may involve a repeatable succession of beneficial mutations that give rise to a predictable outcome. The second reason is that there may be general features of adaptation that do not depend on the particular genes involved. These include, for example, the expected effect of beneficial mutations, the number of mutations involved, and the general time course of adaptation. This section is largely about the predictable features of adaptation by a population to a new or changed environment. Rather than containing a stock of genetic variation available for sorting, the population has become fixed for the type best adapted to the previous conditions of growth, so that all adaptation is based on the selection of novel mutations. For the time being, the population remains haploid and asexual.

3.4.1 Probability that a beneficial mutation will be fixed

Some of the mutations appearing in each generation will be deleterious, and the lineages that bear them are impaired and will usually soon die out, contributing nothing to the future of the population. Others will be beneficial, and these are crucial because each is potentially the ancestor of a new and better-adapted lineage that will eventually constitute the dominant type in the population. When a beneficial mutation first begins to spread in a population the number of individuals initially bearing the mutation is necessarily very small, and the dynamics of the gene are at this point strongly influenced by sampling error. The fate of a new beneficial mutation is thus determined by the balance between the directional process of selection and the non-directional fluctuations caused by sampling error: the greater the rate of selection, the greater the probability that the mutation will continue to increase in frequency. It is therefore often the case that a mutation fails to become established in a population, even though it increases fitness. I shall designate the probability that a newly-arisen beneficial mutation will eventually become fixed as Φ, because it is one of the most important factors contributing to adaptation and has been the subject of much theoretical work. The classical approach developed by Haldane, Fisher, Wright, and Kimura has been summarized by Moran (1962 p. 104) and Crow and Kimura (1970 p. 418). The problem is similar to the 'Gambler's Ruin' in which a unit sum is wagered repeatedly until the gambler has won all the stakes available or lost his initial holding (Feller 1957). Φ is governed mainly by the selection coefficient s and the initial number of mutants (or initial capital of the gambler) m. We are interested chiefly in novel mutations ($m = 1$), and the exact value of Φ then depends on how population growth is modelled. Classical models assume that the number of offspring (or copies of an allele) left by an individual after one generation of reproduction follows a Poisson distribution with mean 1 for the wild type and c for the mutant, so that $s = \ln c = c - 1$ for c small. The branching-process model used by Haldane (1927) leads to the result $\Phi = 1 - e^{-(1+s)\Phi}$, with solution $\Phi = 2s - 2s^2/3 + 5s^3/9 - \ldots \approx 2s$ if s is small. Subsequent work with models in continuous time has shown that

$$\Phi \approx (1 - e^{-2sm})/(1 - e^{-2sN}) \; \{s \text{ small}\}$$
$$\approx 2s/(1 - e^{-2sN}) \; \{m = 1\}$$
$$\approx 2s \; \{N \text{ large}\},$$

where {} gives the conditions for the approximation. In the Gambler's Ruin scheme a random individual reproduces or dies in each time step, causing its lineage to increase or decrease by a single individual. On average the wild type adds 1 individual and the mutant lineage adds $1 + s$ individuals, so the probability that the mutant lineage increases is $(1 + s)/(2 + s)$ and $\Phi \approx (1 - e^{-sm})/(1 - e^{-sN})$ $\{s \text{ small}\} \approx s$ $\{m=1, N \text{ large}\}$. In microbial populations the obvious model for reproduction is binary fission resulting in two new individuals (both survive), one new individual (one survives and one dies) or none (both die). If the population is stable then the probability of survival is ½, and the probabilities of 2, 1 and 0 survivors are ¼, ½ and ¼. For a type whose survival is $w = 1 + s$ relative to the current wild type, the corresponding probabilities are $(1 - w/2)^2$, $2(w/2)(1 - w/2)$ and $(w/2)^2$, from which it follows that the probability that a lineage founded by a single individual of this type will become extinct is $(1 - s)^2/(1 + s)^2$, or equivalently that the probability of survival of a new beneficial mutation is $4s/(1 + s)^2$ (Gerrish and Lenski 1998). In short, reasonable population-dynamic models lead to the conclusion $\Phi = Cs$, where $C = 1$–4. The original Haldane result $C = 2$ will be close enough for most practical purposes. Thus, a mildly beneficial mutation with $s = 0.01$ will spread quite rapidly once it has become established; but it is unlikely ever to become established at all (Figure 3.11).

It is important to understand that the magnitude of the fluctuations caused by sampling error,

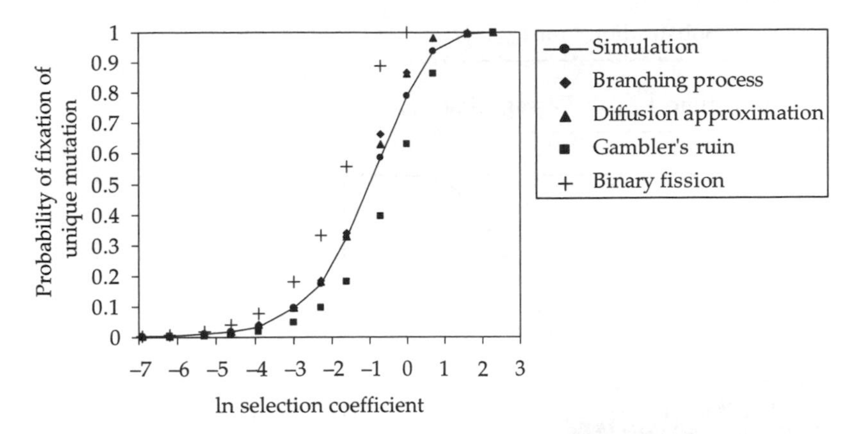

Figure 3.11 The probability of survival of a novel beneficial mutation. Various analytical models are shown relative to a simulation that represents a single mutant producing 1 + *s* offspring appearing in a population of 1000 individuals; the next generation is formed as random sample without replacement of these offspring.

and thus the probability that a novel mutation will by chance become extinct, does not depend primarily on its frequency in the population, but rather on the number of individuals bearing it. Once several hundred individuals bearing the mutation are present in the population, its dynamics are nearly deterministic. In a large population, genes will escape from sampling error and begin to spread deterministically at very low frequencies; in small populations, their dynamics may be largely stochastic even when they are quite common. If the population fluctuates in size a more exact expression for the probability of fixation is $\Phi = C(N_e/N)\, s$, and beneficial mutations are likely to be lost at population minima. In batch-culture experiments designed to study the spread of beneficial mutations, therefore, the size of the inoculum used at each transfer will affect the evolutionary dynamics. The probability of fixation in an experiment with a dilution ratio (ratio of inoculum to culture volume) D is $2sD(\ln D)^2$ and is therefore maximized by $D = 1/e^2 \approx 0.135$ (Wahl *et al.* 2002). Adaptation is not necessarily accelerated by enforcing greater expansion in the growth phase with a small inoculum: the optimal design is to transfer frequently while allowing only about three doublings between transfers.

The stochastic fate of novel mutations does not cause any qualitative change in the process of adaptation. Novel mutations are a random sample of possible mutations; a second round of randomization caused by sampling error does not randomize the sampling of mutations any more completely. However, there is a quantitative effect. A mildly beneficial mutation may have to occur dozens or hundreds of times before it evades sampling error and becomes sufficiently numerous to spread in a nearly deterministic fashion. Sampling error thereby slows down adaptation to new environments.

3.4.2 Periodic selection

The overall time required for a beneficial mutation to be fixed in the population can be divided into two phases (Figure 3.12). The first is the waiting time until the ancestral individual appears; this is not necessarily the first individual to bear a particular beneficial mutation, because a mutant lineage may die out soon after its appearance, but rather the founder of the successful lineage itself. The second phase is the passage time, until this lineage shall have expanded from a single individual so as to constitute the entire population, or in practice to exceed some given high frequency. The substitution of a beneficial mutation is readily detected by propagating a bacterial population that is isogenic except for a genetic marker that does

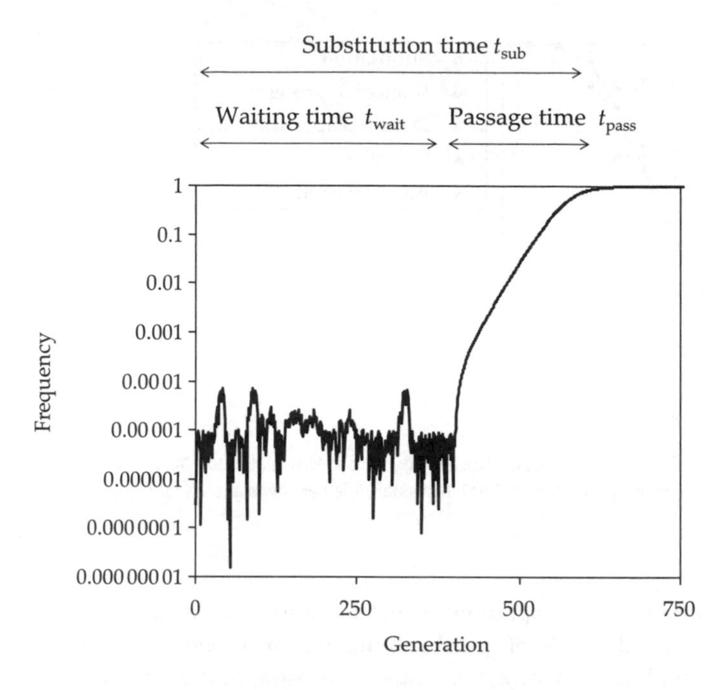

Figure 3.12 The waiting time and passage time for the substitution of a beneficial mutation. The time axis begins immediately after the substitution of the previous beneficial mutation. Note that the frequency axis is logarithmic.

not affect normal growth but leads to a visibly different kind of colony when cultured on appropriate selective medium. The two marker states are at first almost equally frequent, and remain so, with minor fluctuations, for several transfers. Within a few days or weeks, however, one will abruptly rise in frequency and quickly displace the other. These events mark the passage of a beneficial mutation that has occurred in an individual bearing one or the other marker state: the waiting time during which the two markers fluctuate neutrally in frequency, followed by the passage time during which one completely replaces the other because of its linkage to a beneficial mutation. The process is often called *periodic selection*, and was first observed in chemostat cultures of bacteria by Atwood *et al.* (1951a, b). (The term is unfortunate, as the process is stochastic rather than periodic, but deeply entrenched.) The original experiments followed the frequency of spontaneous resistance to T5, which is neutral in glucose-limited media in the absence of the phage and consequently increases slowly under mutation pressure. From time to time it abruptly drops because a beneficial mutation has occurred in a T5-sensitive cell whose descendants then replace all others. Fluctuations in the frequency of T5-resistant cells thereby chart the spread of successive beneficial mutations through the population. (Figure 3.13). Periodic selection is an extremely important aspect of bacterial evolution because it implies that one genotype—bearing not only the beneficial mutant allele but also one particular allele of every other gene—replaces all others, thereby removing almost all genetic variation from the population.

The waiting time depends on two factors. The first is simply the mutation supply rate, given by Nu. The second is the loss of a novel mutation through demographic stochasticity before it has become abundant, which will occur with probability *constant* × s, where we can take the constant as 2 to represent either offspring production or binary fission with strong selection. The overall rate at which successful beneficial mutations appear is the product of the mutation supply rate and the probability of stochastic extinction, so that the overall waiting time is $t_{wait} = 1/(2Nus)$.

The passage time depends only on the selection coefficient s, except that we must define a frequency above which the mutation is regarded as fixed. If

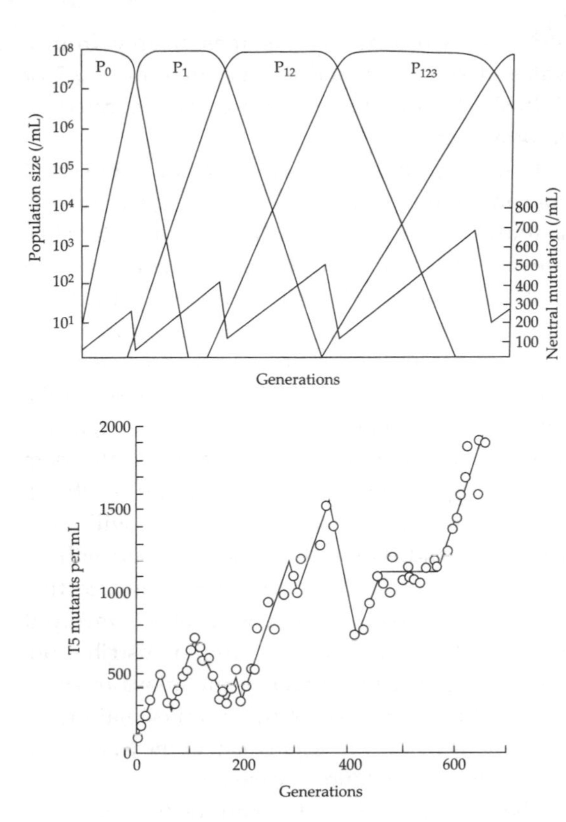

Figure 3.13 Periodic selection. Change in the frequency of T5-resistant mutants in a chemostat population of *E. coli*. From Dykhuizen and Hartl (1983), after Novick (1958).

the mutation was already present before the environment changed, it was a deleterious allele with a frequency $m = u/s$, determined by the balance between mutation and countervailing selection. The rate u is essentially the frequency with which the common type gives rise by mutation to this particular variant. After the change, it is favourably selected, and will continue to increase in frequency until the allele that was common before the change has been reduced to the frequency which is just maintained by recurrent mutation. At this point, the combined frequency of all the variant forms of the gene is $m' = U/\bar{s}$, say, where U is the frequency with which the type now common gives rise by mutation to any variant; $\bar{s}$ is the average rate of selection acting against these variants. The allele favoured by the change has therefore risen in frequency from m to $1 - m'$. In the simplest case, there would be only two alleles, favoured in different environments, and the allele favoured in the new environment would increase in frequency from m to $1 - m$. The time taken for this substitution is roughly $t_{pass} = 2(1/s)(-\ln m)$, provided that selection is fairly weak. For any arbitrary final frequency, we can write $t_{pass} = F/s$, where F is a function of the initial and final frequencies that does not depend on u or s (§3.2.2).

The total time expected to elapse before the first adaptive substitution is complete is then $t_{sub} = t_{wait} + t_{pass} = 1/(2Nus) + F/s$. The dynamics of adaptation will depend on the distribution of s. At any particular time, and especially after some pronounced environmental change, several or many mutants will have greater fitness than the current wild type, each having a characteristic value of s. The distribution of s among these candidate mutants will govern the subsequent distribution among successful mutants. If the mutation supply rate is very high ($M = Nu \gg 1$ for each mutant) then the best mutant will appear several times in each generation and is almost certain to be fixed, provided that its selection coefficient s_{max} is not too small ($s_{max} > 1/Nu$). The population moves in a single step from its currently maladapted state to the best-adapted state. If the mutation supply rate is small ($M \ll 1$ for each mutant) this will not necessarily be the case, because an inferior mutant may appear and spread before the best mutant has appeared. The population will then move towards the best-adapted state by a series of steps, and we would like to know how many steps are likely to be involved, and how large they are likely to be.

3.4.3 Fisher's geometrical analogy

Fisher (1930) likened the phenotype of an individual to a point in n-dimensional space, where each dimension represents the state of a character. The optimal phenotype is a second point lying some distance away. The effect of a shift in character state or states is that the phenotype as a whole moves closer to or further from the optimum; shifts that result in a closer approach to the optimum are beneficial. The essentials of the situation can be represented in two dimensions, such that the current population lies on the circumference of a circle of diameter d whose centre is the optimal state; this circle encloses all the points that constitute an increase in adaptedness, and the current lack of adaptedness is expressed by d. If the current state is displaced by mutation a distance r in any direction, adaptedness is increased if this displacement carries it within the circle, but worsened if it is moved outside the circle (Figure 3.14). If r is small relative to d, the chances of it being moved inside or outside the circle are about the same; in fact as r becomes very small, they approach equality. On the other hand, if r is large relative to d then the character state will almost certainly move outside the circle, regardless of the direction of change. In fact, for $r < d$ the probability of improvement is $\frac{1}{2}(1 - r/d)$. (If $r > d$ this probability is of course zero.) Thus, as r increases relative to d the probability that a change will increase adaptedness becomes steadily smaller. Moreover, as the number of dimensions (independent characters) increases the probability that a random shift will move the phenotype closer to the optimum declines with $\sqrt{n}$. We can take into account both effects by defining a standard magnitude of change $x = (r/d)\sqrt{n}$, in which case the probability that a shift of size x will move the phenotype closer to the optimum is $1 - N(x)$, where $N(x)$ is the standard cumulative normal distribution (Kimura

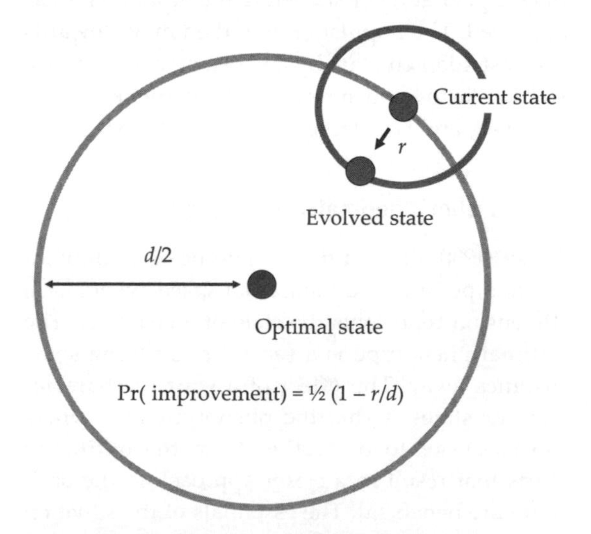

Figure 3.14 Fisher's geometrical analogy of adaptation.

1983). This probability declines steeply from a value of ½ at $x = 0$ and is very small for $x > 2$ or 3. In short, only small shifts are likely to cause an improvement.

Fisher's geometrical analogy can be converted into an evolutionary model by making two further assumptions: first, that changes in character state are caused by mutation, and secondly, that the population consists of a single type most of the time. We can then imagine that a beneficial mutation will occur from time to time, whereupon selection will move the population as a whole closer to the optimum, after which the process can be repeated. The point is that only mutations of very slight effect have an appreciable probability of being beneficial, so the first mutation to be substituted will be of this kind, and its effect on mean fitness will be small. Moreover, it has been widely inferred that the whole spectrum of mutational effects involved in adaptation will follow a similar distribution, with a great predominance of mutations of small effect. This is the basis of the infinitesimal model of adaptation, a vivid restatement of the traditional gradualism of evolutionary biology.

The infinitesimal model requires that the great majority of beneficial mutations that arise in a population should have very small effects on fitness. It does not follow, however, that this will hold for those mutations that become fixed. Kimura (1983) pointed out that mutations of small effect are likely to become extinct shortly after their appearance, so that the mutations most likely to be fixed first will be of moderate effect whose rarity will be compensated by their greater probability of survival. Consequently, the distribution of effects for the first mutation to be fixed will be modal, with a peak at intermediate values. The expected standard effect of this first substitution is $(4/3)\sqrt{(2/\pi)} = 1.06$, with variance $(3/2) - [(4/3)\sqrt{(2/\pi)}]^2$ (Kimura 1983). This corresponds to an average displacement towards the optimum of $r/d = x/\sqrt{n}$, which depends on the dimensionality of the organism. The net movement towards the optimum will decrease with n because the number of possible directions that could be taken increases, but it is difficult to make a convincing guess at the likely value of n, even to an order of magnitude. In principle, all the characters expressed by an organism could be collapsed statistically into n independent dimensions. This value of n is unknown but might be quite large; a simplistic one gene-one character theory would suggest $n \approx 10\,000$ for multicellular organisms. The relevant value of n, however, is the number of these dimensions undergoing selection in a given environment. Pursuing the one gene–one character notion, we can ask how many genes will be responsible for restoring adaptation when a well-adapted population is moved away from the optimum by a change in the conditions of growth. A guess that this will be more than 10 but fewer than 100 suggests that that the first step in adaptation is likely to take the population 10–30% of the distance back towards the optimum.

Fisher's analogy has been extraordinarily influential. It gave simple and precise mathematical shape to the gradualistic concept of a population moving slowly and smoothly through an isotropic phenotype space towards a single optimal state. Nevertheless, it incorporates two assumptions about fitness that are not always realistic.

In the first place, it is assumed that the maladapted population can persist indefinitely. This is not necessarily the case. If a new pathogen appears, a preferred prey species disappears, a spawning area dries up or oxygen is depleted, the maximal rate of increase of the population may become negative, and it will sooner or later become extinct. Mutations of small effect that give some protection against disease or oxygen lack may occur and even spread, but being of small effect are unlikely to prevent the extinction of the population. If conditions have deteriorated so badly that the population is inviable, only mutations of large effect that restore it to the viable zone can produce permanent adaptation. Once they have become established, however, further mutations of small effect will fine-tune adaptation within the viable zone. Thus, severe environmental perturbation may enforce the successive substitution of mutations of decreasing effect.

Secondly, Fisher's model assumes that the characters concerned are not only independent and uncorrelated, but also that they have independent effects on fitness. If X and Y are two characters with independent effects on fitness and optimal states x_{opt} and y_{opt}, then the effects on fitness of character states x and y can be given as $z_1 = [(x - x_{opt})/x_{opt}]^2$ and $z_2 = [(y - y_{opt})/y_{opt}]^2$, the overall effect being $z = z_1 + z_2$. This generates a fitness surface that is a segment of a sphere, with the consequence that movement from any point to any point closer to the joint optimum is always a movement from lower to higher fitness. Although this is a simple and tractable concept, it may not always represent the process of adaptation correctly because uncorrelated characters may not have independent effects on fitness. For example, the performance of a structure made up of several components often depends both on its overall size and on the proportion of its parts, as in the different parts of a jointed limb or the pistils and stamens of a flower. The effects of size and proportion could then be given as $z_1 = [(x + y - z_{1,opt})/z_{1,opt}]^2$ and $z_2 = [(x - y - z_{2,opt})/z_{2,opt}]^2$, with overall effect $z = z_1 + z_2$. In this case the fitness surface is anisotropic and movement to a point closer to the optimum does not necessarily cause an increase in fitness. The optimum is instead approached by a ridge: movement along this ridge in the direction of the optimum leads to greater fitness, whereas movement in any other direction is likely to lead to lower fitness, even for points closer to the optimum. Thus, if fitness depends on the interaction between characters the effect on fitness of mutations that have the same magnitude of effect on phenotype will depend on their direction.

For a simple geometrical analogy representing selection on interacting characters, consider a population on a fitness ridge at some distance from the optimum. A maladapted population is likely to be on or near the crest of the ridge, because this is the most likely destination of displacements that increase fitness. The optimum itself is never a point in space, since changes of character state of less than a certain magnitude will have inappreciable effects on fitness. The optimum is thus the nearly flat summit of a fitness ridge, within which movement is unconstrained. Movement from the current state to the neutral zone is limited (in two dimensions) to a triangle whose base is the diameter v of the neutral zone. For a population at a distance d from the edge of the neutral zone the angle θ at the apex of the triangle is given by $\tan(\theta/2) = 1/(1 + 2R)$

where $R = d/v$, and the probability that a random mutation lies within this triangle increases as the neutral zone is approached, with a limit of ½ at the boundary. In this simple case, adaptation is likely to involve a succession of mutations of decreasing effect but increasing frequency (Figure 3.15).

3.4.4 The variable-mutation model

Phenotypic models framed in terms of fitness are difficult to relate either to the changes in proteins that underlie fitness, or to the changes in genes that underlie both of these. Genes and proteins do not evolve smoothly and continuously, but rather in discrete steps each involving a change from one nucleotide or amino acid to another. Focusing on the effect of a mutation, rather than of a phenotypic change, naturally suggests a discrete genetic model rather than a continuous phenotypic model.

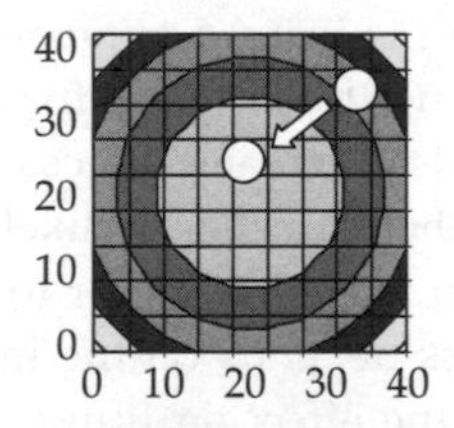

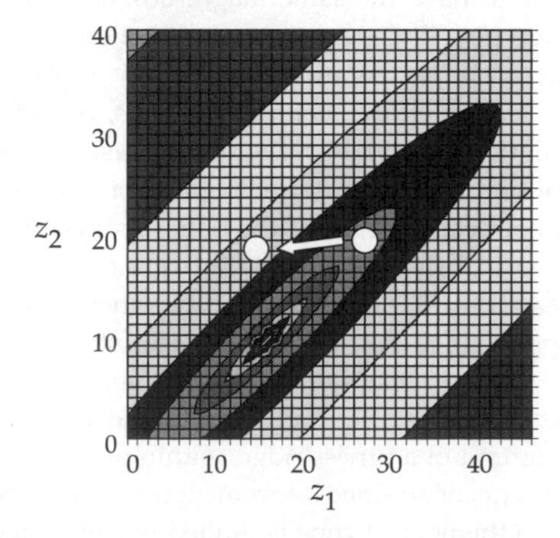

Figure 3.15 Fisherian and non-Fisherian space. In Fisherian space, any closer approach to the optimum increases fitness. (small upper diagram). If characters interact in their effects on fitness then a phenotypic change that moves the population closer to the optimum may reduce fitness (lower diagram).

What is the distribution of effect among beneficial mutations? The infinitesimal model does not preclude the existence of beneficial mutations of large effect, but it does require that they be very rare. This is equivalent to assuming that the rate of mutation declines with effect. This might be because they are inherently unlikely, or because a beneficial mutation will have less effect when it occurs in a genome that is already very fit. Suppose that the effect of beneficial mutations be divided into categories representing equal increments Δ of s, so that the fitness of individuals bearing mutations belonging to the ith category is $w_i = 1 + i\Delta$, whereas the rate of appearance of mutations in the ith category is some declining function of i. What is then the expected effect of the first mutation to be fixed? A reasonable analogy might be provided by competitors in a race, each at the starting blocks in their lane. They will start at different times and run at different speeds, but the winner will cross the finishing line first because their overall time will be the least. This suggests a principle of least time: the category that is most likely to provide the first beneficial mutation to be fixed is that with minimal t_{sub}. The passage time always becomes less as s increases, whereas the waiting time may either increase (because mutations of larger effect are less frequent) or decrease (because a mutation of large effect is more likely to survive). Thus, a mutation of intermediate effect is most likely to be fixed first, provided that mutation rate falls off steeply enough with effect. Using the expressions for t_{pass} and t_{wait} given above, the minimum time to fixation is associated with the selection coefficient category i^* given by: $du/di = -(u/i^*)(1 + t_{pass}/t_{wait})$, subject to a slightly more complicated condition that this minimum should exist: $(d^2u/di^2)|_{s=s^*} < (u/i^2)[2 + (1 + t_{pass}/t_{wait})(1 + 2t_{pass}/t_{wait})]$.

To get any further we need to know how mutation rate and effect are related. The simplest possibility is that mutation rate declines exponentially with effect, such that $u_i = u_0 \exp(-ki)$. There is some experimental justification for this assumption (Kassen and Bataillon 2006). If all categories have the same fitness ($\Delta = 0$), and a population initially consisting of the wild type alone is allowed to propagate itself for a few generations, then the frequency of any given category under mutation and drift alone likewise falls off exponentially with slope k. One of these categories will sooner or later become fixed, with probability equal to its frequency, so that if many replicate populations were surveyed the distribution of categories of effect would be negative exponential, whether mutations were counted when they appeared or when they became fixed. In the more interesting case that the effect of beneficial mutations is appreciable ($\Delta > 0$) those of larger effect are more likely to be fixed, and will be fixed more rapidly, once they appear. Consequently, the modal category will shift away from 0 to some intermediate value, that is, mutations of intermediate effect are most likely to be fixed first (Figure 3.16). Suppose that the mutation supply rate is very low, such that $t_{wait} \gg t_{pass}$. The category with minimal t_{sub} is then approximately $i^* = 1/k$, at which value $s^* = (i^* - 1)/\Delta$ and $u^* = u_0 \exp(-1)$. This category is simply the average of an exponential distribution with parameter k, so selection does not change the distribution of effect. To understand why this should be, suppose that mutation rates are given, so that the mutation supply rate depends only on N. Passage time becomes less important with smaller N, and reaches zero for $N = 1$; the distribution of fixed beneficial mutations among a large number of replicate populations of size $N = 1$ is clearly the same as the distribution of mutations counted when they appear. In larger populations the mutation supply rate is greater, and when passage time can no longer be neglected the category with minimal t_{sub} is $i^* = [1 + \surd(1 + 4kP)]/2k$, equivalent to $i^* = [1 + \surd(1 + 8k \ln N)]/2k$ for complete substitution. A rough general rule is then that $i^* = (2/k)\ln N$, showing that the first mutation fixed will have much greater effect than expected unless the population is very small.

In short, the initial step in adaptation may be larger than expected from mutation pressure alone. The average outcome depends on the mutation supply rate, and therefore on population size for any given mutation schedule. This average is very variable, however, so that if the decline of mutation rate with effect is neither very shallow nor very steep almost any intermediate outcome is quite likely to occur.

3.4.5 The extreme-value model

An elegant interpretation of the initial stages of adaptation can be developed from the assumption that the wild type is likely to be very well adapted, even to conditions that have recently changed. The distribution of effect among all mutations is unknown, but the great majority are deleterious

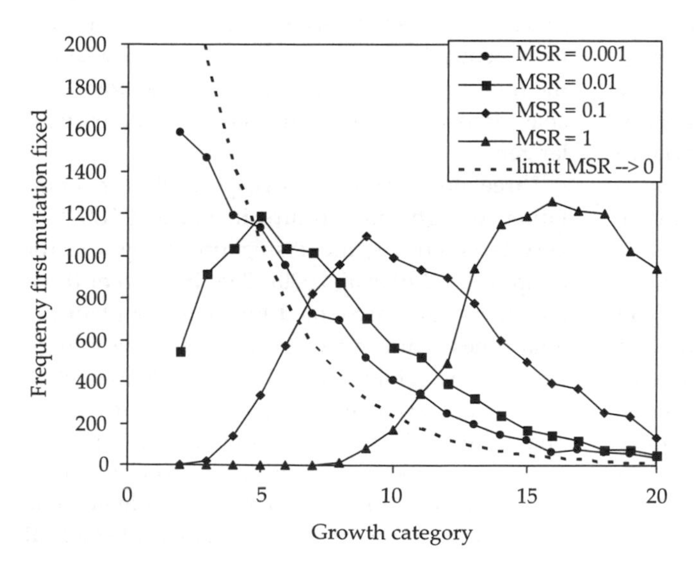

Figure 3.16 The effect on fitness of the first beneficial mutation fixed under a variable-mutation model. Simulation results using 20 equal increments of growth. Curves are distributions for 10 000 simulations at various mutation supply rates (MSR).

and will not contribute to adaptation. The distribution of effect among the very small minority of beneficial mutations that is alone relevant to adaptation is then predicted from general extreme-value theory, an approach pioneered by Gillespie (1984) and further pursued by Orr (2003). Suppose that we generate a very large number Λ of fitness values at random from some probability distribution, and then discard all but the largest λ values, where $\lambda \ll \Lambda$. These can be ranked from the largest value, representing the best possible type, to the λth, which represents the current wild type. Any two types i and j with adjacent ranks are then separated by an increment $\Delta_i = w_i - w_j$, so that the fitness of the type with rank i can be represented as $w_i = 1 + \Delta_{\lambda-1} + \ldots + \Delta_{i+1} + \Delta_i$. These fitness increments are random variables that grow steadily larger with distance from the wild type, as the probability of finding a yet larger value decreases. The expected values of successive increments follows the simple rule that $\Delta_i = \Delta_1/i$, which after the first few increments becomes nearly exponential. The effect of a mutation from an allele of lower rank i to one of higher rank j is $\Delta w = \Delta_{i-1} + \Delta_{i-2} + \ldots + \Delta_j$. Most mutations will have only a small effect on fitness, whereas the few best in a poorly adapted population may have a very large effect; and in general it can be shown that the probability of a given effect declines exponentially with its magnitude (Orr 2003). The effect of any given mutation will of course depend on its rank relative to λ, but the distribution of Δw for new beneficial mutations is exponential with average value Δ_1 regardless of λ (Orr 2003) (recall that the sum of increments for a mutation greatly superior to the wild type may exceed Δ_1). This invariance principle implies that the potential for adaptation of a population is independent of its current state of adaptedness.

The extreme-value model is clearly similar to a variable-mutation model, insofar as the supply of mutations falls as their effect increases, whether because the rate of mutation falls or because the fitness increments grow larger. It has the merit, however, of leading directly to a series of simple and striking conclusions, provided that we first assume as before that passage time can be ignored. Since $t_{wait} = 1/2Nus$ and u does not vary with effect then the probability that the ith type is the first to be established, and thereafter fixed, is simply s_i/S, where S is the sum of selection coefficients over all available beneficial mutations (Gillespie 1984). This can be expressed in terms of fitness ranks alone as the transition probability $P(\lambda,i) = [1/(\lambda - 1)]\Sigma(1/k)$ with the summation taken over all ranks from $\lambda - 1$ to i. It follows that the allele most likely to be substituted first is the fittest, the second most likely is the second fittest, and so forth: alleles of small effect are the least likely. The *average* rank of the first mutation fixed is simply $(\lambda+2)/4$ and the average effect on fitness is $\Delta w = [2(\lambda - 1)/\lambda]\,\Delta_1$ (Orr 2002). Unlike the potential for adaptation, the actual expected advance in fitness depends on λ, but unless the current wild type is very well adapted then $\Delta w = 2\Delta_1$ will be a good approximation, and in any case Δw will be bounded between Δ_1 and $2\Delta_1$. The full spectrum of selection coefficients is modal (Fig 3.17). The frequency of fixed mutations of given effect increases with the magnitude of the effect, up to a certain point, through the least-time principle. Beyond this point substitution time becomes insensitive to s, because the probability of establishment is so great and the passage time so short that neither can be substantially improved. The frequency now declines in a roughly exponential fashion because the mutation supply rate to any category of selection coefficients of fixed extent falls off in this way because of the roughly exponential decline in the size of fitness increments. The same result is obtained whether the fitness increments are treated as random variables or as fixed values characteristic of a particular gene.

Three point mutations responsible for adaptation to high temperature in phage ξX174 had very large effects, increasing fitness by a factor of up to 15 (Bull *et al.* 2000). The first two or three mutations fixed accounted for more than half the total fitness gain in two viruses adapting to high temperature (Holder and Bull 2001). Rokyta *et al.* (2005) transferred experimental lines of a DNA phage under low-M conditions and trapped 20 single beneficial mutations. These were found to represent nine distinct alleles by whole-genome sequencing, so the fitness rank of each beneficial

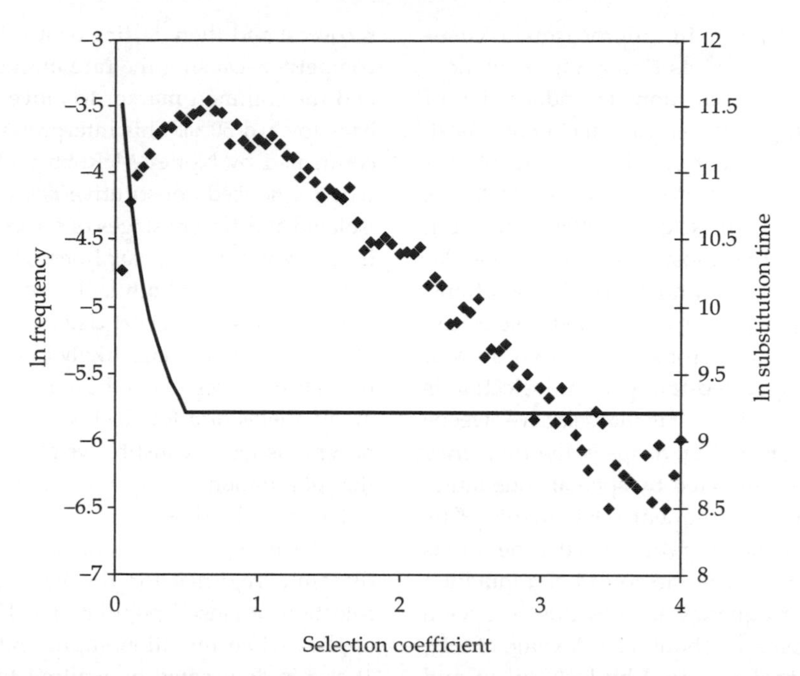

Figure 3.17 The effect on fitness of the first beneficial mutation fixed under an extreme-value model. Plotted points are frequencies of selection coefficients from 12 000 simulations. Line shows substitution time.

mutation could be estimated. The result failed to confirm Gillespie–Orr dynamics, mainly because the most fit allele was not fixed nearly as often as expected. To save the model, Rokyta *et al.* (2005) turned to a variable-mutation explanation. The fittest allele is generated by a G → T transversion, which rarely occurs. If the transition probability $P(\lambda,i)$ is adjusted to compensate for variable mutation rates, by replacing the term $1/(\lambda - 1)$ by $u_i/\bar{u}$ where $\bar{u}$ is the mean rate over all alleles, then the data can be made to fit the modified theory. It must be admitted, though, that the data clearly shows a modal distribution of effect that has been observed before at higher mutation supply rate.

3.4.6 Clonal interference

In practice, the effect of the first mutation fixed is even greater than these simple models predict, and the discrepancy increases with N. The reason is that competition among clones in large populations weakens the analogy of a race, because a slow runner nearing the finish line can be hauled back by a faster competitor who started out later. Gerrish and Lenski (1998) derive the expected value for the selection coefficient of a fixed beneficial mutation and show that when defined as a logarithm it takes a nearly constant value at low N and then increases linearly with ln N for $Nu > 0.01$. My own numerical simulations suggest that the average category of the first mutation fixed in variable-mutation models is a power function of population size, $\ln i = c + z \ln N$, where the intercept $c = \ln(1/k)$ and empirically $z \approx 0.1$–0.2. Consequently, in moderately large populations the effect of the first mutation fixed is much greater than expected; when the mutation supply rate is large the first mutation fixed may usually be one of the three or four best, whereas the least-time principle would suggest only a modest advance on the wild type. Likewise, the conclusions of extreme-value models are affected by clonal interference in large populations. The distribution of selection coefficients shifts to the right and the rank of the first mutation fixed increases.

Microbial evolution in microcosms is usually interpreted, at least as a first approximation, in terms of periodic selection. The adaptation of phage ξX174 to high temperature and a novel host, for example, follows the classical pattern of successive selective sweeps (Wichman *et al.* 1999). It is quite clear that periodic selection often occurs, but it is equally clear that most experimental populations of bacteria evolve under high-*M* conditions, where the simple process of periodic selection should break down (Korona 1996b). Two lines of evidence have suggested that periodic selection is far from universal. The first is that selective sweeps in chemostats, detected by drops in the frequency of a rare marker generated by spontaneous mutation, seem to be too frequent (Dykhuizen 1990). They often occur at intervals of 100 generations or less, and if they mark the spread of a mutation from mutation frequency to 99% this implies a selection coefficient of about 0.25. A value of 0.13, comparable to that reported by Dykhuizen and Dean (1980) as a maximum, would imply that the initial frequency was much greater (about 10^{-4}), that is, that the population is heterogeneous. The second line of evidence is that selective sweeps detected by the rapid spread of a neutral marker from about 50% to fixation, occurs too seldom. Dykhuizen (1990) commented that sweeps are common in batch culture but not in chemostats. We have described a case in which sweeps failed to occur in batch culture (MacLean *et al.* 2005): when cultured in rich complex medium, marker frequencies in *Pseudomonas* populations fluctuate widely through time, with a magnitude that cannot be explained by drift, but nearly always return towards 50% without becoming fixed. This pattern—too many sweeps according to a rare spontaneous marker but too few according to a common non-revertible marker—is explained by the continual flux of contending beneficial mutations in high-*M* conditions. These will undergo frequent excursions as a mutation rises to moderate or high frequency (causing the rare marker to drop or the common marker to move in one direction or the other) where it will remain for a while as its relative fitness becomes eroded by the establishment of a slightly superior mutant (while the rare marker recovers) and then decline as it is replaced by this competitor (causing the rare marker to drop again and the common marker to move, sooner or later, back toward 50%). This interpretation seems to be confirmed by Notley-McRobb and Ferenci (2000), who sequenced constitutive *mgl* and *mlc* mutants isolated at different stages of selective sweeps early in the evolution of *E. coli* lines. They found a wide range of alleles at both loci involving point mutations, frameshifts, short deletions, and insertions of IS elements. It seems likely that selection in chemostats often results in the simultaneous spread of many alleles of a few loci conferring similar phenotypes (such as constitutive glucose transport), so that populations are usually functionally uniform but genetically diverse.

Broadly speaking, one can recognize two dynamic regimes for the operation of natural selection in closed populations. The first (low-*M*) applies when mutation supply rate is low ($Nu \ll 1$) and is dominated by waiting time. The second (high-*M*) applies when mutation supply rate is high ($Nu > 1$) and is dominated by clonal interference. This distinction leads to several important predictions about the pattern of adaptation. First, a higher mutation rate will increase the rate of adaptation in small populations but not in large populations. Secondly, increasing population size for a given mutation rate will increase the rate of adaptation to a certain point, but will have little or no effect once the population has surpassed the threshold for high-*M* dynamics. Thirdly, this threshold will be higher for well-adapted populations because the rate of beneficial mutation will be lower. All of these predictions were confirmed in *E. coli* microcosms by de Visser *et al.* (1999). Miralles et al. (1999) confirmed that the fitness effect of fixed mutations tends to increase with population size in an RNA virus.

3.4.7 The distribution of fitness effects

To evaluate the predicted distribution of fitness effects we must distinguish between three categories of beneficial mutation: nascent mutations, that have only just appeared; contending mutations, that have appeared and are spreading; and fixed mutations, that have eliminated all others. Figure 3.18 shows how the distribution of effects on fitness is expected to shift from the appearance to the fixation of a mutation. The effect of nascent mutations should be exponentially distributed, but is difficult to measure because the mutations are in normal circumstances imperceptible before they have been amplified by selection. Kassen and Bataillon (2006) isolated a set of single mutations of *E. coli* resistant to nalidixic acid and then tested them against the ancestor in medium lacking nalidixic acid. They found that 28/665 mutations increased fitness, and their effects were consistent with an exponential distribution (Figure 3.19). It is noteworthy that so large a fraction of mutations were beneficial, which reinforces the impression that in some circumstances beneficial mutation is not very rare.

Contending mutations exist only under high-M conditions and should follow a gamma distribution through the stochastic loss of mutations of small effect (Rozen *et al.* 2002). Imhof and Schötterer (2001) used an unstable microsatellite marker to detect beneficial mutations spreading through populations of *E. coli* and found that their effects were exponentially distributed. It is difficult in practice to distinguish clearly between exponential and gamma distributions, however, unless a very large number of mutations has been tested.

The distribution of effect among fixed mutations depends on the dynamic regime. Under low-M conditions the distribution is roughly exponential except for mutations of very small effect, whereas under high-M conditions it is a conspicuously modal distribution (derived by Rozen *et al.* 2002) that shifts to the right as mutation supply rate increases. Distributions with this shape have been described for *E. coli* in minimal glucose medium (Rozen *et al.* 2002) and for *Pseudomonas* in serine medium (Barrett *et al.* 2005, Barrett and Bell 2006) by trapping mutations at or near the end of selective sweeps (Figure 3.20).

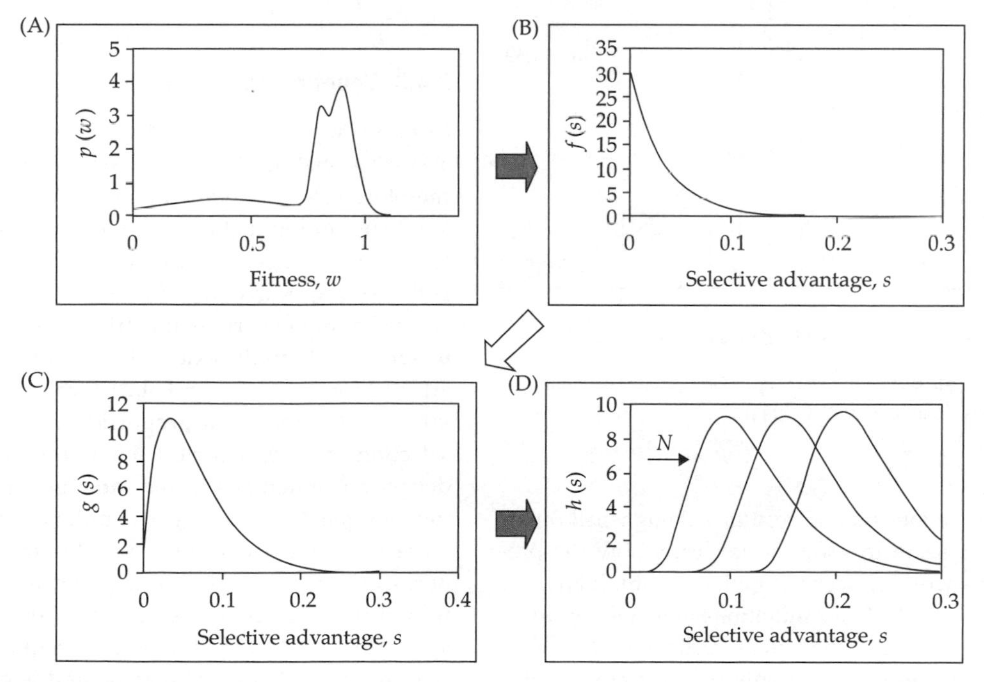

Figure 3.18 The expected distribution of fitness effects at different stages in adaptation. A. The full distribution of fitness among mutations might be arbitrarily complicated. B. The fitness effects of nascent beneficial mutations will follow an extreme-value distribution. C. Contending mutations will have a modal distribution of effect. D. The effect of fixed beneficial mutations will likewise be modal but shifted to higher values, the shift being more pronounced in larger populations. From Rozen *et al.* (2002).

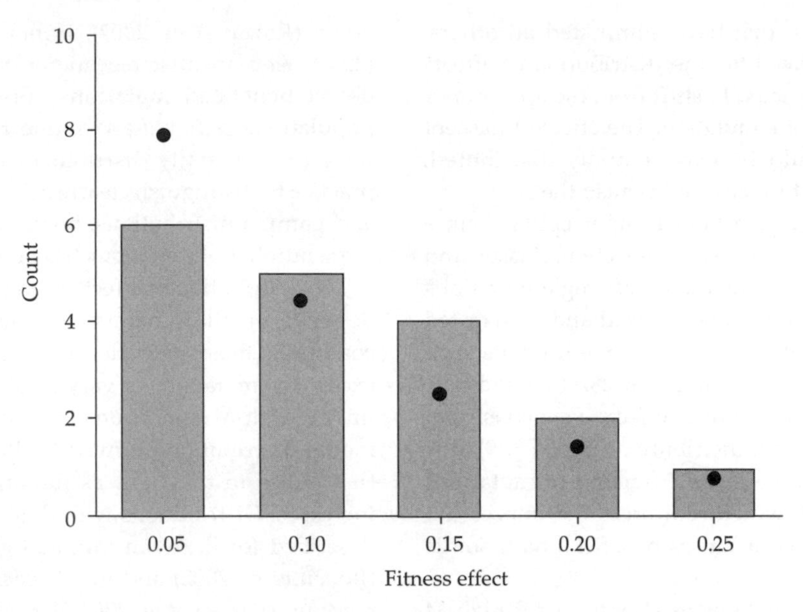

Figure 3.19 Distribution of fitness effects of nascent beneficial mutations. From Kassen and Bataillon (2006).

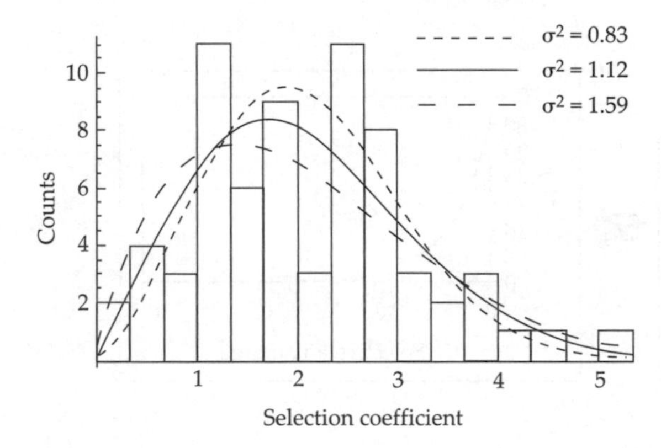

Figure 3.20 Distribution of fitness effects of fixed beneficial mutations. From Barrett *et al.* (2005).

These experimental results, although still rather meagre, seem to show quite clearly how the predominance of nascent mutations of small effect is translated into fixed mutations of generally much greater effect. It is the latter which imply that selection in novel environments will initially involve large increases in fitness. It should be emphasized that this conclusion depends on the high-M regime typical of bacterial microcosms, and may not hold for relatively small populations of animals and plants.

3.4.8 Genetic interference

In either low-M or high-M regimes, it has been assumed that mutations occur singly and are therefore selected independently. This is equivalent to assuming that the genomic mutation rate is low, $U \ll 1$. Otherwise, a new beneficial mutation is likely to appear in a genome that already bears one or more deleterious mutations, and will tend to spread only to the extent that it over-rides their effect. Moreover, the probability that it will survive will be smaller in proportion to the reduced selection coefficient attributable to the presence of deleterious mutations in the same genome. Using mutator genotypes or larger populations will both increase the mutation supply rate, but they will not necessarily produce the same patterns of adaptation. It will sometimes be useful to recognize the joint effects of M and U. Classic periodic selection requires a low-M, low-U regime, and is perturbed by clonal interference in high-M conditions and by genetic interference in high-U conditions. In a high-M, high-U regime selection operates in a highly diverse population where most individuals carry several mutations and produces complex dynamics (Figure 3.21).

3.4.9 The genetic basis of adaptation

The mutant alleles whose fitness effects are predicted by theory occur in particular genes and modify particular functions, and the next step in analysing adaptation is to ask what sorts of genes and functions are likely to be involved. In the past it has seldom been possible to identify the genetic basis of adaptation because the hunt for an unknown change in 1 of 10 000 genes is nearly hopeless, but technical advances and clever detective work have now tracked down the changes occurring in some experimental populations. I shall put them into three categories: predictable, repeatable, and idiosyncratic.

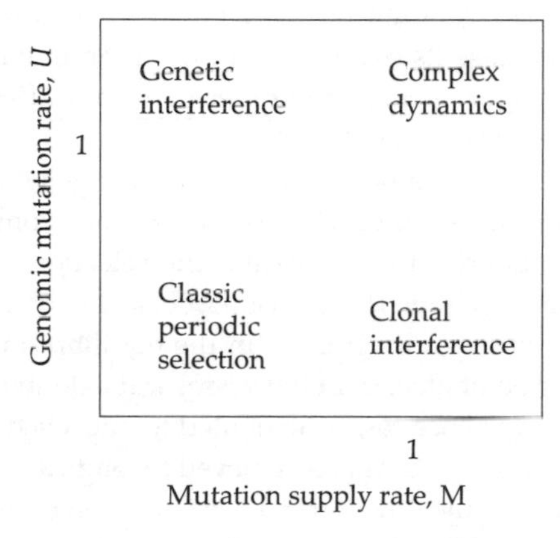

Figure 3.21 The process of adaptation under different dynamic regimes.

Laboratory conditions of growth differ in many ways from natural environments, but their crucial feature is simplification: at the extreme, a single limiting resource is provided. Provided that it can be metabolized, the most likely target of selection is then the rate at which the limiting substrate can be taken up from the environment. Glucose is the best example, because most substrates are converted into glucose before they are metabolized, so that short-term improvement in energy generation from glucose is unlikely whereas increased uptake can readily be achieved simply by loss-of-function mutations at regulatory loci. The growth of *E. coli* in glucose-based medium is achieved by a rather complex array of structural and regulatory proteins (Figure 3.22). The two outer membrane proteins OmpF and LamB bring glucose into the periplasmic space, from where two other protein systems in the cytoplasmic membrane, Mgl and PtsG, move it into

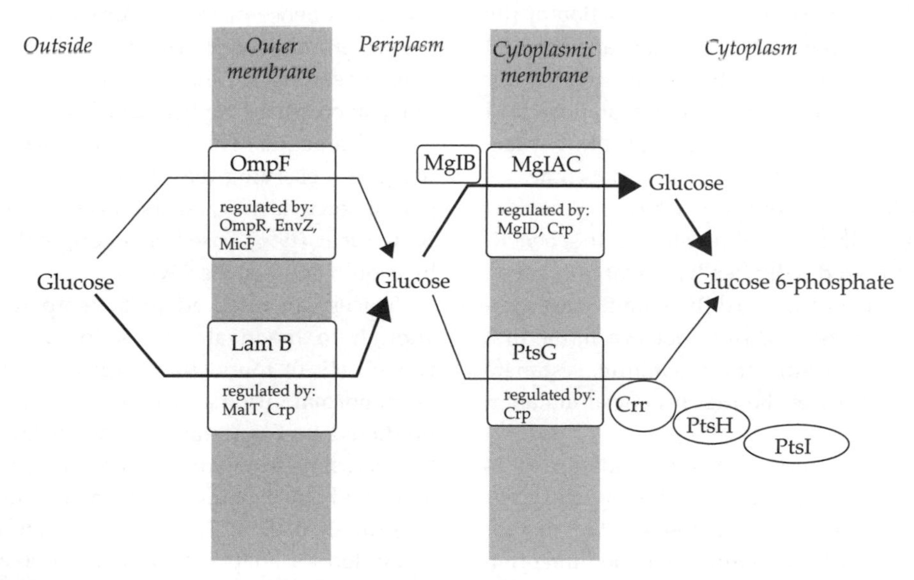

Figure 3.22 The glucose transport system of *E. coli*. (From Notley-McRobb and Ferenci 1999a.)

the cytoplasm. Chemostat selection increases glucose uptake (Dykhuizen and Hartl 1981) through an increase in the phosphotranferase system PtsG (Helling *et al.* 1987) which also regulates other transport systems, and the transcription of certain genes and may therefore have manifold consequences. A detailed description of the mutations responsible for enhanced uptake after about 300 generations of growth in replicated chemostat experiments has been given by Notley-McRobb and Ferenci (1999a, 1999b). In some cases phosphotransferase activity was elevated, apparently because of loss-of-function mutations at *mlc*, a gene that regulates sugar transport. The second inner membrane system Mgl is more important at low glucose concentration and was over-expressed in almost all lines, leading to very large increases in glucose uptake. The underlying genetic changes were substitutions, frameshifts and short insertions/deletions in both the *mgl* operator and the MglD repressor protein. At micromolar glucose concentrations uptake at the outer membrane is undertaken mainly by the LamB glycoporin, which is regulated by *mal*, which is in turn regulated by the global repressor *mlc*. LamB activity and *mal* expression were elevated in almost all lines, as the consequence of point mutations in the *mal* structural gene and by mutations in *mlc*. Thus, the basis of adaptation to glucose-limited chemostat conditions was constitutive production of the LamB protein on the outer membrane and the Mgl proteins on the inner membrane, causing greatly increased uptake of glucose. These mutations lead to an increase in the rate of glucose transport by factors of 8–15. In glucose-limited chemostats fitness is linearly related to glucose flux, which will depend primarily on uptake. Consequently, the first beneficial mutations fixed will often have a large effect on fitness. Dykhuizen and Hartl (1981) found that competitive fitness increased by about 13% in the first 40 generations of culture: this is a minimal estimate of the fitness effect of the first beneficial mutation to be fixed.

Constitutive expression has been identified as the initial step in adaptation in other systems. For example, cells taken from lactose-limited chemostats and spread on lactose-minimal medium often all die. This bizarre 'lactose killing' is attributable to mutants that have become constitutive for *lac* through mutations in its regulator *lacY* spreading through the population in the first 100 generations or so of chemostat culture (Dykhuizen and Hartl 1978). Adding a gratuitous inducer of *lac* abolishes the advantage of constitutive expression (Dykhuizen and Davies 1980). Other examples include serine in *E. coli* (Bloom and McFall 1975) and mandelate in *Pseudomonas* (Hegeman 1966). Acetate cross-feeding that evolves after a few hundred generations in *E. coli* cultured in glucose minimal medium always seems to involve mutations in the *acs* gene encoding acetyl CoA synthetase that increase its expression, rather than mutations in regulatory genes or in other pathways affecting acetate use (Treves *et al.* 1998).

Yeast that has been cultured for 450 generations in a glucose-limited chemostat becomes competitively superior to its ancestor, the selection coefficient being about 0.09. This is associated with a drop of one order of magnitude in the equilibrium concentration of glucose in the vessel and a doubling of yield, and hence can be attributed to improved utilization of glucose. This is achieved through a six-fold increase in the rate of uptake by the evolved strain. The physiological basis of this improvement was a doubling in the expression of two hexose transport *HXT* genes. There is good evidence that this was the consequence of several duplications involving the structural gene *HXT6* and the promoter of the related gene *HXT7*. This could have arisen through unequal crossing-over between *HXT6* and *HXT7* on sister chromatids. This remarkable chain of reasoning from observing the superiority of the evolved population to identifying its cause was described by Brown *et al.* (1998), based on an original experiment by Paquin and Adams (1983)

Viruses can often adapt to temperatures high enough to inactivate the wild type sequence. Dowell (1980) found that high temperature prevents normal virion assembly in phage ξX174, and predicted that mutations in capsid genes would be necessary to restore growth. Bull *et al.* (2000) identified three point mutations in lines selected for growth at 45 °C. The two most common, which accounted for adaptation in every line except one, were in a major capsid gene.

A very consistent feature of chemostat experiments is selection for wall growth or flocculation. This is usually regarded more as a nuisance than an adaptation, but from an ecological point of view it is at least as interesting as sugar transporters. Cells that adhere to a surface normally grow more slowly than cells moving freely in the medium, but their death rate is lower because they are not washed out. Cells that adhere to one another and settle in clumps to the bottom of the vessel enjoy a similar advantage. In either case reproduction will often produce a dispersive offspring that moves freely for a while, and may reproduce, before it or its descendants again attach to a surface. The evolutionary ecology of settlement and dispersal in chemostats does not yet seem to have received the attention that it deserves, although once it is reclassified as biofilm formation this situation may change. In batch culture wall growth is less pronounced because the inoculum is normally taken from the culture after mixing, so that attached cells have no advantage or are systematically less likely to be transferred. The interaction between cells in the surface film, in the medium, and in clumps at the bottom of the vial has provided an interesting system for studying social interactions in *Pseudomonas fluorescens*, and is described in Chapter 8.

These experiments are important for evolutionary theory because they show how evolution can be understood at all levels, from the initial genetic lesion to the final adaptive consequence, when the environment has been simplified to such a degree that the systems likely to be improvable can be predicted from our current knowledge of biochemistry. The deregulation and amplification of major transporters can be confidently expected to drive adaptation in simple microcosms. This gives us confidence that evolution is not mysterious: more complex systems and more subtle changes will become equally predictable when we understand evolutionary and biochemical processes better.

Other experiments have yielded repeatable results that are not clearly predictable. Lines of *E. coli* maintained in glucose-limited batch culture, for example, consistently lose the ability to utilize *D*-ribose as a substrate for growth (Cooper and Lenski 2000). This was associated with the presence of the transposable element *IS150* close to the *rbs* operon; when a copy of the element is inserted by transposition into *rbs* recombination between the old and new copies deletes the intervening sequence and is likely to create Rbs^- mutants (Cooper *et al.* 2001). These arise at a high rate (about 5×10^{-5} per replication) and have a slight advantage of 1–2% in glucose medium, so they are likely to be fixed in less than 1000 generations. In this case, it is not difficult to imagine some simple and general advantage of ribose auxotrophy, such as economy of material (but see §8.4). In other cases, repeatable changes are more difficult to understand. When *E. coli* is cultured at 41.5 °C for 2000 generations growth, yield, and fitness all increase (Bennett and Lenski 1993). The obvious candidates would be the eight heat-shock genes, which should be deactivated because their expression reduces growth, but these are still expressed at ancestral levels or higher. Instead, enhanced performance in several lines was associated with a large duplication involving about a dozen genes whose contribution to growth at high temperature is unclear; some are involved in stress resistance (such as *rpoS*, expressed in stationary phase), but there is no clear functional link between any of the genes involved and the observed adaptation (Riehle *et al.* 2001). Extensive genomic modification and rearrangement of uncertain function have been described in both *E. coli* (Bergthorsson and Ochman 1999) and yeast (Adams *et al.* 1992) and may be general features of microbial selection lines. Similar events were observed in experimental populations of yeast from glucose-limited chemostats, where several lines shared breakpoints for extensive deletions or duplications (Dunham *et al.* 2002). Two lines carried duplications on the right arm of chromosome 4, the location of the hexose transport gene *HXT6* that was locally amplified in a third line (see above), but the other rearrangements are not as yet so easily interpretable.

Repeatable genetic changes underline the potentially predictable course of evolutionary change; such cases will surely be moved into the 'predictable' category as knowledge advances. At present they have an important practical implication because they demonstrate how selection experiments can be used to identify genetic and biochemical mechanisms that would not otherwise be apparent. Long-term selection lines are valuable resources because they can be screened for repeated adaptive motifs that are candidates for poorly understood or unknown molecular processes.

Finally, there are idiosyncratic pathways of adaptation that may be peculiar to a particular system. Although *E. coli* cannot normally use propanediol as a food source, for example, Lin and Wu (1984) have shown that it is nevertheless possible to select variants that have evolved the ability to grow on propanediol as the only source of carbon and energy by modification of the pathway that is normally used to metabolize fucose (Figure 3.23). In normal *E. coli* cells growing anaerobically, the hexose L-fucose is eventually split to yield L-lactaldehyde and dihydroxyacetone phosphate. The lactaldehyde is then reduced to L-propanediol by an NAD-linked enzyme, and excreted from the cell via a 'facilitator' that promotes movement in either direction across the cell membrane. Discarding half the carbon skeleton of the original substrate in this way improves the efficiency with which the dihydroxyacetone phosphate can be used as a carbon source. When propanediol is supplied at high concentration as the only source of carbon and energy, some variants are able to utilize what was previously a waste product as a food. The crucial genetic change involved is the modification of the enzyme that normally reduces lactaldehyde to propanediol under anaerobic conditions into a new oxidoreductase that oxidizes

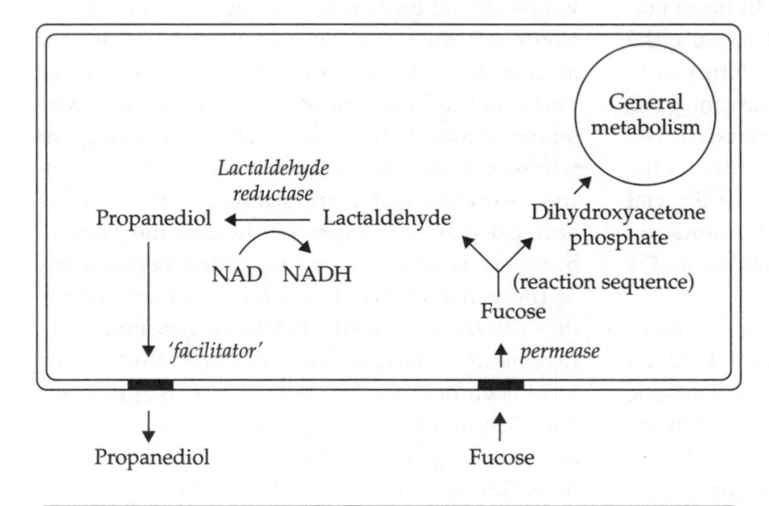

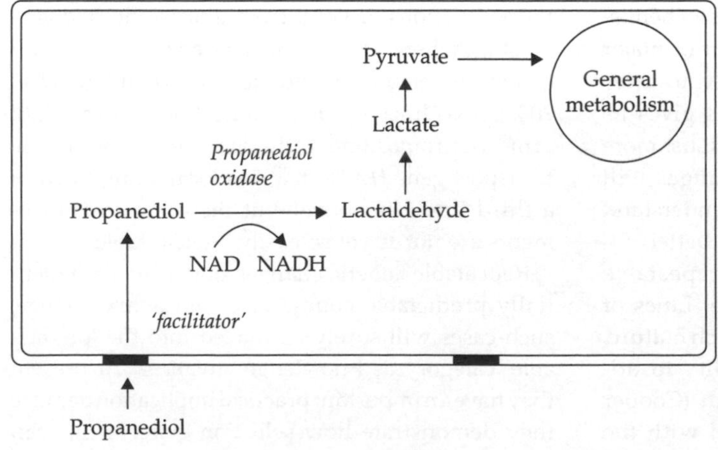

Figure 3.23 Experimental evolution of propanediol metabolism. This sketch shows the main features of the evolution of propanediol utilization in *E. coli*, from Lin and Wu (1984). In normal anaerobic metabolism, the fermentation of fucose produces lactaldehyde, which is reduced to propanediol and excreted from the cell. Selection on propanediol yields types in which the NAD-linked oxidoreductase that normally reduces lactaldehyde functions as a dehydrogenase that oxidizes propanediol; the lactaldehyde formed is further oxidized to lactate, a normal substrate for growth.

propanediol to lactaldehyde under aerobic conditions. Once this has been achieved, the lactaldehyde is readily oxidized to lactate, and enters general metabolism as a source of carbon and energy. Adaptation to a novel environment in which the original strain cannot live thus involves modifying the final enzyme in an anaerobic pathway so as to function as the first enzyme in an aerobic pathway. Perhaps cases such as this may eventually supply general principles that can be more broadly applied, but perhaps even laboratory microcosms have an irreducible natural history that must be mastered in order to understand a particular case.

As a very broad-brush conclusion, bacterial evolution in a simple microcosm often seems to involve a few themes and countless variations. The few themes are the major genes where beneficial mutations can occur. The course of adaptation can be predicted, in terms of the types of gene and protein likely to be responsible for improvements in growth and fitness, because the number of themes is limited. It cannot be completely predicted, however, because there is usually more than one theme, and this gives rise to genetic differences between lines. *HXT6/7* is duplicated in some lines of yeast cultured under glucose limitation, for example, but in other lines some different process must be responsible for adaptation. The variations are the alleles of the major genes, which may be exceedingly numerous and give rise to genetic diversity within lines. At this level, the course of adaptation is scarcely predictable at all. In the initial stage of adaptation replicate lines will discover a few broad themes, and will then build on these in subsequent evolution.

3.5 Successive substitution

When conditions change, a population may recover full adaptedness in a single step, but it is more likely that several changes will be necessary. The arguments used to analyse the first step can be extended to address the complete series of events leading to full recovery because they incorporate a property of self-similarity. The first move made by a population through Fisherian phenotype space will locate it closer to the optimum, but in all other respects the rules governing its future behaviour remain the same. The second and subsequent moves will therefore have the same properties as the first, save for the change of scale. In genetic models the first mutation fixed now constitutes the current wild type, and the series of beneficial mutations now available is constructed by advancing the wild-type rank or category λ and if necessary adjusting the mutation rate u_0. For the Gillespie–Orr model the invariance principle shows that successive episodes of substitution are self-similar, so that the results for the first mutation fixed can be applied to characterize the entire sequence of substitutions leading eventually to the fixation of the best allele.

3.5.1 Phenotypic evolution towards the optimum

Continued phenotypic evolution in n dimensions has been lucidly analysed by Orr (1998). As a Fisherian population recovers adaptation its distance to the optimum continually decreases by the constant proportion $(1 - 2x_1/n)^2(\frac{1}{2}\sqrt{n})$, where x_1 is the expected standard magnitude of the first move, expressing the self-similar nature of the process. As this happens, the magnitude of the largest possible beneficial effect necessarily decreases to the same extent. The second and subsequent moves are then expected to be shorter than the first move, by a factor equal to the proportion of the distance to the optimum expected to be still remaining. The length of moves thus tends to decrease in a geometric series as the optimum is approached. In any particular case, however, moves may be longer or shorter than expected, and an intermediate move may exceed the expected magnitude of the first. The distribution of effects for each successive move is modal, as it is for the first, with the mode shifting to the left in later moves. When these distributions are summed for all moves up to some point arbitrarily close to the optimum, the combined distribution has a peak at small values, but if very short moves are ignored (they will contribute little to adaptation) the distribution of the length of moves is roughly exponential. Thus, evolution towards an optimal phenotype is expected to involve many short steps, when the population is approaching the optimum, and a few long ones, usually near the beginning of the process.

3.5.2 Adaptive walks

The form of low-*M* models allows them to be extended in a simple way to second and subsequent substitutions by designating the first beneficial mutation fixed as the new wild type and waiting for the next. This can be chosen from the existing list of mutations, if any, superior to that just fixed, which is equivalent to assuming that any beneficial mutation could be reached in a single mutational step from the original wild type. This is a simple model that is often useful in practice. The progress of an evolving population through the DNA or protein space is more correctly simulated, however, by generating a new set of random values whenever a mutation is fixed. Those whose fitness exceeds that of the new wild type constitute the new series of beneficial mutations available to selection. This process is repeated until no superior type is found, at which point adaptation halts. This is the *mutational landscape* model of Gillespie (1984) that can be used to study the adaptive walk which, taken one step at a time, leads towards a local maximum of fitness. The most important feature of this walk is that it is short. The number of steps that must be taken, L, clearly depends on the fitness rank λ of the original wild type, but $L \propto \ln \lambda$ in the single-step model, and this seems to apply to a mutational landscape. As each step carries the population on average three-quarters of the way to the best currently available mutation, repeating this operation until $i = 1$ suggests $L \approx \ln(3\lambda)/\ln 4$; the argument is illegitimate but in practice the approximation works quite well for moderate values of λ between 10 and 100. More generally, the length of the walk is a power function of λ with a small exponent (Figure 3.24). Moreover, CV(L) is only about 0.4, so long walks are rare. Consequently, L increases only slowly with λ and adaptation normally requires surprisingly few steps: even for a poorly adapted wild type with $\lambda = 100$ the average adaptive walk is only about 4 or 5 steps.

These steps are correspondingly large in terms of fitness gains, with the first step alone

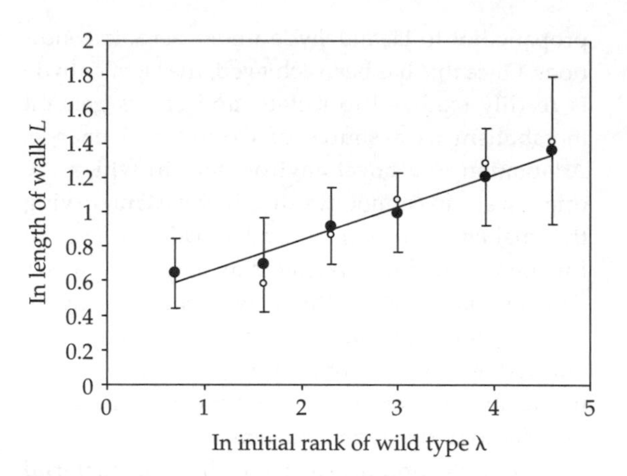

Figure 3.24 The length of adaptive walks. Output of a mutational landscape model with rate of appearance of each beneficial mutation set at u such that $N\lambda u = 0.01$. Solid points are means of simulations; the regression equation is $\ln L = 0.45 + 0.19 \ln \lambda$. Bars are standard errors. Open points are $\ln(3\lambda)/\ln 4$.

contributing on average more than 30% of the total gain (Orr 2002). The effect of subsequent steps falls off exponentially, albeit with a great deal of variation (Figure 3.25). In any given population, fitness increases irregularly in step-like fashion until a limit is reached and no further gain in fitness is possible. An ensemble of populations explores the landscape more thoroughly, and mean fitness continues to increase indefinitely through time. Simulations suggest that the increase of fitness over time t is described by a weak power law of the form $w = bt^z$, with z about 0.1 or less (Figure 3.26). This is highly non-linear and appears asymptotic when plotted on arithmetic axes, but $\ln w$ continues to increase at a diminishing rate for as long as the process continues.

3.5.3 Transitivity

When analysing selection experiments it is usually assumed that the fitnesses of successive wild types increase in an orderly and transitive fashion, that is if $w_C > w_B$ and $w_B > w_A$ then $w_C > w_A$. This is not often checked and may not always be true. The best-known exception was observed by Paquin and Adams (1983) in yeast cultures maintained in glucose-limited chemostats for about

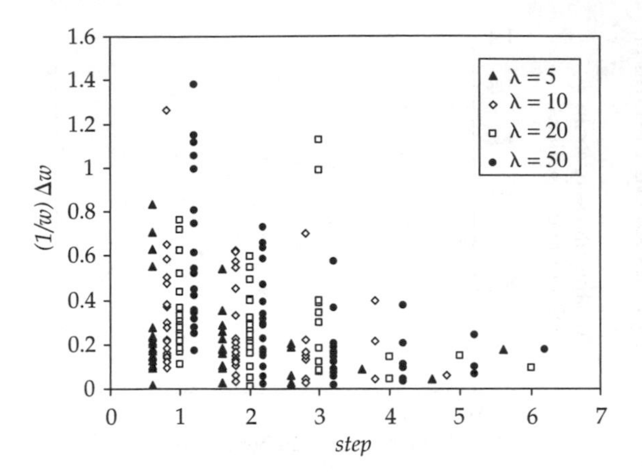

Figure 3.25 Progress during an adaptive walk. Relative increase in growth at successive steps for simulations over a range of λ. Points displaced along x-axis for clarity.

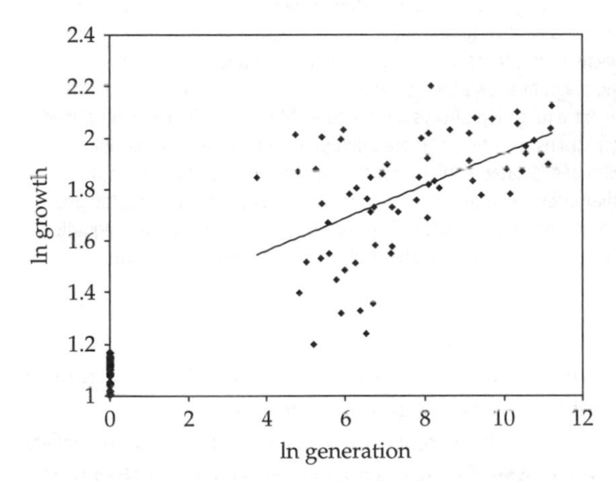

Figure 3.26 Increase of adaptedness over time. Data from 20 replicates of a mutational landscape model with $\lambda = 50$. The equation of the regression equation is $y = 1.31 + 0.063x$; growth of the initial type (points at extreme left) are not included.

250 generations. Isolates that differed by a single mutation always behaved consistently: the strain with one more mutation replaced the other. This was expected, since the strain with one more mutation had in fact replaced the other during the evolution of the population in the chemostat. Isolates that differed by several mutations behaved less consistently. They were always less fit, relative to the original strain, than would be expected from the separate increments in fitness caused by each successive new mutation. Fitness did not, then, simply cumulate through time. Although the difference in fitness between a given strain and the strain that it had just replaced remained about the same throughout the experiment, this difference represented a smaller and smaller advantage over earlier strains. Indeed, in some cases, strains that had accumulated five or six adaptive mutations were less fit than the original type; and in consequence, fitness in the diploid population at least actually decreased during the experiment (Figure 3.27). It is likely that these yeast populations were not merely adapting to a novel physical environment but were also responding to the social environment created by the currently prevalent strain, in such a way as to create an intransitive succession of new types (cf. §10.3). The observation has stimulated much discussion, but I cannot find another convincing example and think that it is probably a rare exception. The general transitivity of fitness among different species of *Drosophila* in competition experiments has been documented by Richmond *et al.* (1975) and Goodman (1979).

3.5.4 Clonal interference

Biologists working with large organisms with $N < 10^5$ or so find the low-*M* model natural, whereas those working with microbes where $N > 10^8$ or so are likely to think in terms of a high-*M* regime. The fundamental pattern of adaptation is not affected by clonal interference: substitutions still occur as discrete events separated by relatively long intervals of time, regardless of the mutation supply rate. The internal dynamics of the population, however, are quite different. Under low-*M* assumptions the population is genetically uniform almost all the time, barring recurrent mutations that fail to spread, whereas high-*M* populations always contain a high level of transient variation generated by the continual rise of superior but ultimately unsuccessful mutations. Competition among this crowd of contending mutations affects the progress of adaptation through two processes that affect the rate of substitution and the average selection coefficient respectively. The rate of substitution increases with mutation supply at low

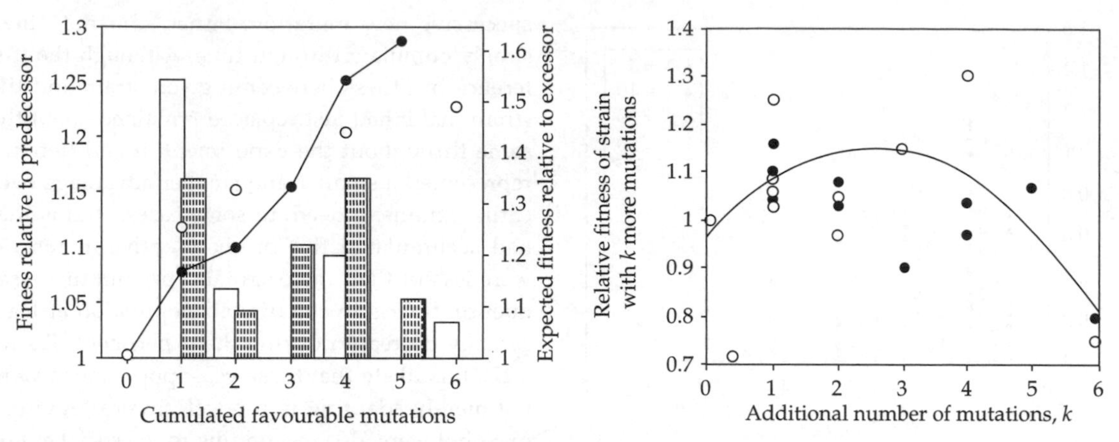

Figure 3.27 Non-transitive fitness. These diagrams show the outcome of competing strains evolved in the chemostat against their more or less remote ancestors, from the diploid series in the experiments on yeast by Paquin and Adams (1983). There are two sets of trials, differing only in the state of a 5-fluouracil resistance marker, distinguished by open bars and broken lines, or stippled bars and solid lines, in the left-hand diagram, and by open or solid symbols in the right-hand diagram. The histogram bars in the left-hand diagram show the fitness of successive mutant strains relative to their immediate ancestor; in all cases, a single additional mutation increases fitness, as expected. The lines show the expected fitness of strains with successively more mutations relative to the original strain, assuming that each mutation contributes independently to fitness; because each mutation causes an increase in fitness, relative to its immediate ancestor, expected fitness relative to the initial strain increases monotonically. The right-hand diagram shows the actual relative fitness of the evolved strains, relative to ancestors with up to six fewer mutations. Strains with a single extra mutation always have greater fitness; this is the result already shown in the upper diagram. However, there is no consistent tendency for strains with two or more additional mutations—for example, one isolated after the passage of five successive mutations, compared with one isolated earlier after the passage of three successive mutations—to display greater fitness. Indeed, there is some indication that after about 300 generations and the passage of six mutations, each superior to its predecessor, the evolved strains are actually inferior to the original inoculum. The solid line is a fitted quadratic equation ($y = -0.028x^2 + 0.144x + 0.97$, $r^2 = 0.44$), showing that relative fitness does not increase monotonically relative to distant ancestors.

N where low-M assumptions are satisfactory and greater mutation supply simply abbreviates the waiting time between successive substitutions. In larger populations the rate of substitution becomes insensitive to mutation supply, because the fixation of superior mutations is retarded more by competition with inferior types (passage time) rather than by mutation (waiting time). At the same time, the average selection coefficient, which is invariant at N sufficiently small in low-M conditions, increases at larger N because superior types must spread while competing against other mutants rather than against the wild type alone. Putting together the effects of high-M conditions on the rate of substitution and the average intensity of selection, the rate of fitness increase is greater in larger populations, as expected, but the effect becomes very small once Nu exceeds 0.1 or so (Figure 3.28). Gerrish and Lenski (1998) conclude that clonal interference imposes a 'speed limit' on the rate of fitness increase in large populations.

As well as restricting its speed, clonal interference may also impart a characteristic rhythm to the process of adaptation (Gerrish 2001). After a given period of time, a population will have accumulated a certain number M of mutations, where M is a random variable with variance Var(M). If Var(M) = 0 then adaptation is a perfectly clock-like process, with successive substitutions occurring at fixed intervals; as Var(M) increases this regularity breaks down and the schedule of adaptation becomes less predictable. The accumulation of neutral mutations is a Poisson process with Var(M)/M = 1, and low-M models share this unpredictability because the appearance of a successful mutation is a rare event independent of other such events. In high-M models, competition within the crowd of contending mutations creates to some extent a consistent

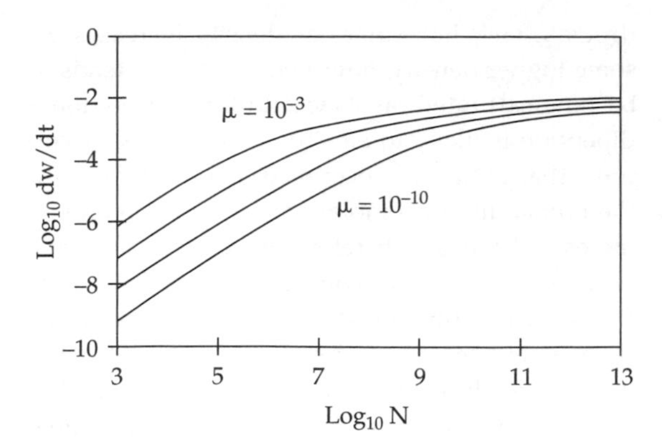

Figure 3.28 The 'speed limit' on adaptation in asexual populations. Rate of change in fitness d*w*/d*t* as a function of population size *N* and mutation rate *u*. From Gerrish and Lenski (1998).

selective environment from which successful mutations emerge in a more predictable pattern. Gerrish's rule is that $\mathrm{Var}(M)/M = 2\exp(-\gamma) - 1$, where γ is Euler's constant, with a numerical value of about 0.12 regardless of the biological details of the situation (so long as high-*M* rules apply). It is often thought that neutral mutations provide a reliable clock for dating phylogenetic divergence, in contrast to the unpredictable course of adaptation; Gerrish's rule suggests that the reverse is true, at least in large populations.

3.6 Cumulative adaptation

The elegant mathematical results from low-*M* and high-*M* models give a clear quantitative account of the dynamics of evolving populations. There are some important aspects of adaptation, however, that have not yet been made so accessible, and to explain these I have had to resort to toys and pegboard models, supplemented by guesses.

3.6.1 The protein matrix

The set of mutations and thus the set of proteins that successively replace one another in an evolving population constitutes an adaptive walk. When a single gene is involved, the route of the walk can be mapped out in an imaginary space defined by axes each representing the amino acid that occurs at a particular position in the protein. A typical protein consists of a sequence of, say, 100 amino acids of 20 different kinds. All 20^{100} possible sequences can then be represented as a 100-dimensional graph with axes of length 20. The axes are numbered consecutively axis 1, axis 2, . . ., axis *L* for a protein of length *L* amino acids, because the properties of a protein depend not only on its composition but crucially on its sequence. This is the *protein matrix* (Maynard Smith 1970), which is the second major concept that we can use to organize ideas about the process of adaptation. The corresponding DNA matrix has only four nucleotides in each row or column (Gillespie 1984). It is a more rigorous concept than the protein matrix, because it correctly represents evolution by single mutational steps. (The protein matrix is the clearer representation of functional change, however, and it can be rescued, if necessary, by including only those transitions from one amino acid to another that can be accomplished by a single mutation.)

The current wild type occupies a node within the matrix. When conditions change, adaptation will proceed through the stepwise substitution of one amino acid at a time, as it were by a series of rook's moves through the matrix. Most nodes of course represent proteins that will not function effectively in the new environment, or perhaps in any environment. A few nodes represent functional sequences that are at least as good as the wild type and that may therefore replace the present wild type through selection or drift. The simplest version of the model specifies the move rule without saying anything about the size of beneficial effects. By providing a way of visualizing the mechanisms at work in an evolving population, however, it leads directly to qualitative predictions that can be tested experimentally, and can also be used as the basis of more quantitative theory that aims to predict the rate of adaptive change, the effect of the mutations involved, and the length of adaptive walks. The extreme-value theory developed by Gillespie (1984 and Orr (2002) is based on the protein matrix (or DNA matrix) concept. Figure 3.30 illustrates the simplest version of the protein matrix, the 20×20

dipeptide matrix. The lines connecting the nodes of the matrix trace out possible routes for the adaptive walk. An adaptive walk ends with the fixation of the best allele, however, only if at each step the former sequence and the newly optimal sequence are connected by a single amino acid substitution. This requirement imposes two constraints on the degree of adaptation that can be achieved: connectance and reversibility.

3.6.2 Connectance

In the first place, functional sequences may not all be connected. If only single substitutions are allowed, the possible evolutionary transitions are rook's moves such as the CD→MD→MQ transitions shown in Figure 3.29A. Figure 3.29B shows one possible set of functional dipeptides. Because they can all be connected by rook's moves (Figure 3.29C), any one dipeptide can evolve into any other by successive single substitutions. Although the set at lower left (Figure 3.29D) looks superficially similar, it turns out on inspection to consist of two clusters of dipeptides (Figure 3.29E). Any dipeptide can evolve into any other within the same cluster, but cannot evolve into any dipeptide in the other cluster. The opportunity for and limits to selection are thus constrained by the inaccessibility of some functional states. The degree to which adaptation can be improved through selection thus depends on the connectance of the protein matrix—on how often changing a single amino acid at a given position results in a molecule that is functional in the new conditions of growth.

If any particular sequence is classified only as being either functional or not, the connectance of the matrix depends on the density of functional proteins, that is, the proportion of all possible sequences that are functional. If the amino acid at any given position has no consistent effect on function then functional proteins will be randomly located in the matrix. The probability that two dipeptides are connected is $(19 + 19)/399 = 0.0952$. If a third is added, the probability that all three are connected is $(19 + 19)/399 \times (19 + 19 + 18)/398 = 0.0134$, and continuing the product shows that the probability that all dipeptides are connected drops to very low values as density increases. At some higher density, however, the matrix tends to become saturated, as each additional functional dipeptide is likely to be situated on a row or column that is already occupied, and at this point the probability of complete connectance begins to increase. There are therefore two distinct dynamic regimes for protein evolution (Figure 3.30). The first is at low functional density, where evolution is severely constrained by the existence of many isolated evolutionary trackways, implying that the optimal sequence will often be inaccessible and the outcome of selection will often be only a modest level of adaptation. The second is at high density, where there are few trackways and the optimal sequence, or a nearly optimal sequence, can soon evolve. The dividing line between these two regimes seems to be at about 20 dipeptides, or 1/20 of those available. This is, of course, the same as the density represented by an unbroken row or column that guarantees connectance: the reason is that 20 is the maximum number of completely unconnected dipeptides that can be placed in the matrix. Below this density, the number of short isolated trackways increases with density and probability of complete connectance is very low; above this density, the number of isolated trackways decreases with density and the probability of complete connectance increases rather rapidly towards unity. In a matrix with more dimensions, representing a longer protein, the probability that random functional polypeptides are connected is necessarily lower, but the 1/20 limit seems also to apply in 3 dimensions and may be a general rule.

The DNA matrix is simple enough to solve completely If there are 4 or fewer functional dinucleotides the probability that the matrix is fully connected is 0.3–0.4, whereas if there are 5 or more the matrix is always fully connected. Thus, the DNA matrix is fully connected if the frequency of functional sequences exceeds ¼. Sanjuan *et al.* (2004) found that about 30% of random single-nucleotide substitutions in vesicular stomatitis virus (VSV) were neutral or beneficial when cultured in hamster kidney cells, presumably a novel environment for the ancestral virus. The RNA matrix may therefore be fully connected in this case.

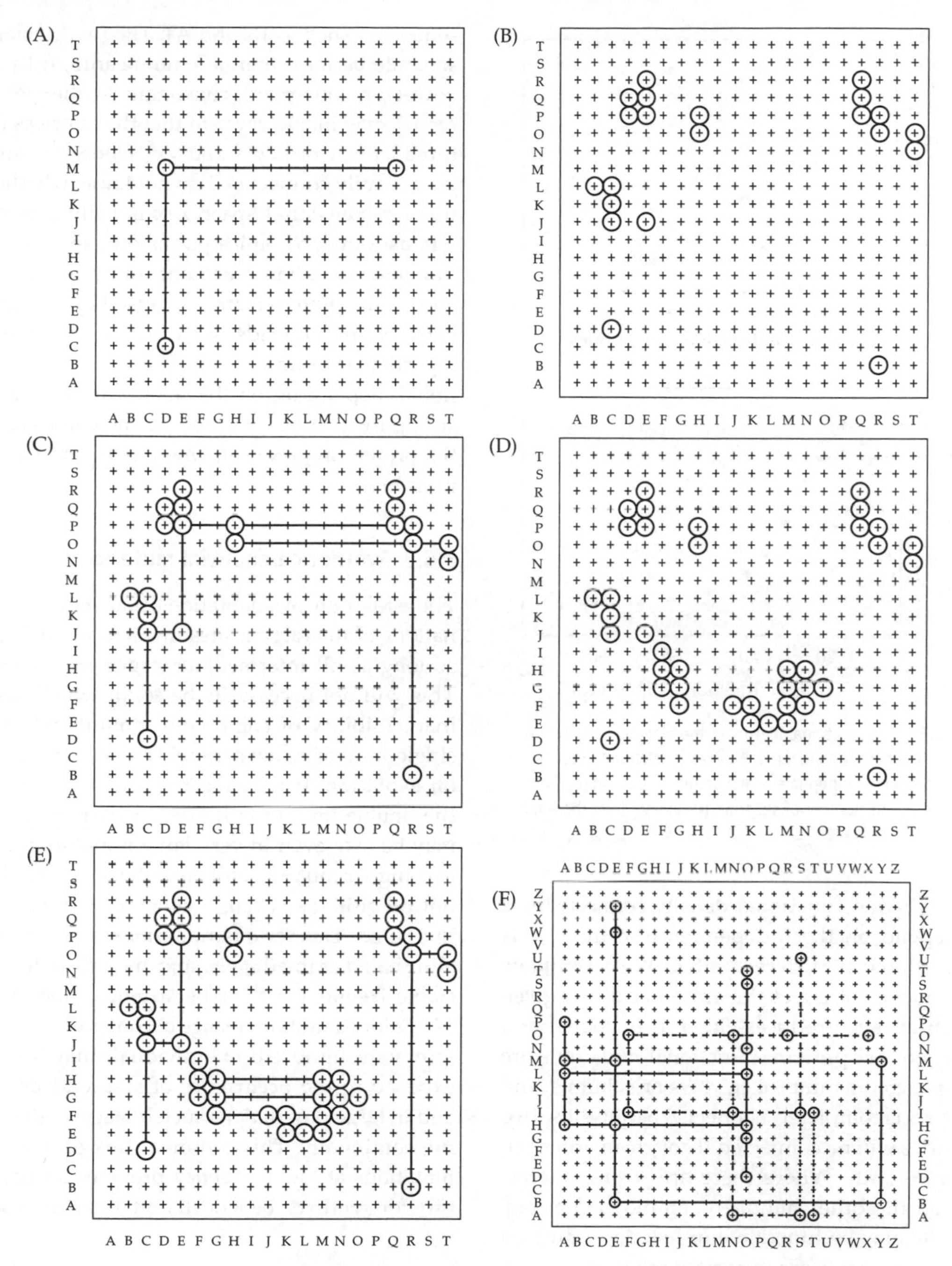

Figure 3.29 The dipeptide matrix. A. Accessible proteins are connected by rook's moves. B. A hypothetical map of functional dipeptides. C. This map is fully connected. D. Another map of functional dipeptides. E. This map is not fully connected. F. The map of two-letter English words; like E, these fall into two unconnected groups.

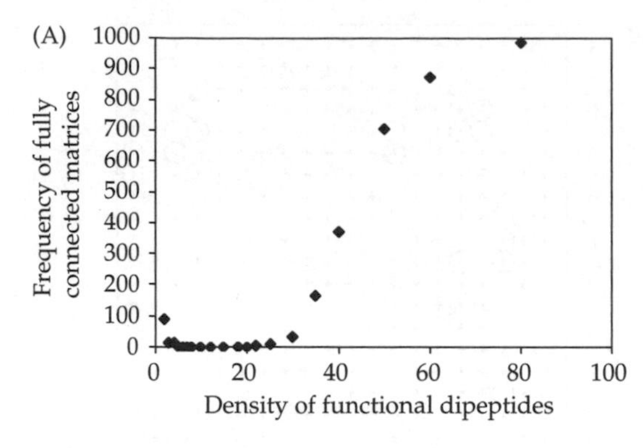

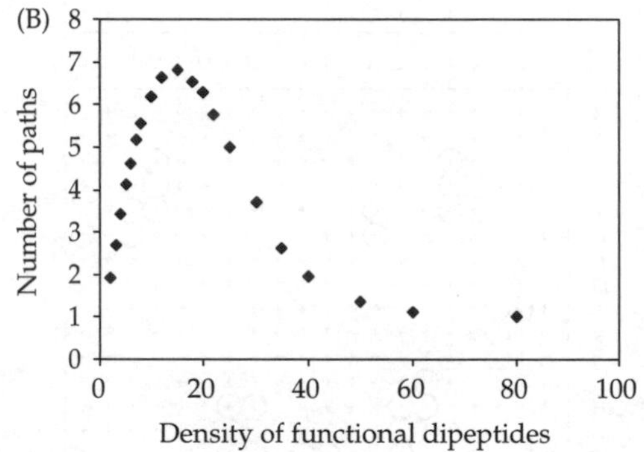

Figure 3.30 The functional density of the dipeptide matrix. A. The fraction of fully connected matrices in relation to the density of functional dipeptides. Each point is the mean of 1000 simulations. B. The number of separate trackways, from the same simulations.

At any given functional density connectedness will depend on the independence of the effects of mutations at different sites. With complete independence of effect the amino acid at a given position always produces either a functional or a non-functional protein. Functional mutations are then arranged as unbroken rows or columns, and the matrix is completely connected because any two dipeptides will be connected through an unbroken row or column. Where there are strong interactions, on the other hand, the amino acid at any given site may be functional or not, depending on which amino acids are present at other sites. As an example of strong interactions, Figure 3.29F shows the analogous 26 × 26 matrix of all meaningful two-letter words yx in English. There are two clusters, one comprising the consonant–vowel sequences such as DO and BY, the other the vowel–consonant sequences such as IN and AT. The two clusters are separate because there are no meaningful vowel–vowel or consonant–consonant sequences. The reader may like to confirm that the matrices for all three-letter words y*x and yx*, where * is any letter, are fully connected. I do not know whether the three-dimensional space zyx is fully connected, but, as Maynard Smith points out, spaces of four dimensions or more are not. Imperfect connectance of the protein matrix is caused by interactions between amino acids, such that at any given site a particular amino acid may or may not be functional, depending on the state of other sites, and strong interactions will reduce the connectance of the protein matrix even when its functional density is high.

3.6.3 Synthetic beneficial mutations

The evolution of adaptations that require combinations of interacting sites involves no difficulties, so long as all intermediate stages are connected. This will not necessarily be so, in which case the inaccessibility of compound mutants when each single mutant is deleterious is a severe constraint on evolution. If function requires two mutations the double mutant will arise at frequency u^2 and may be rare even in very large populations; triple and higher mutants will never be observed. It may not be quite as severe a constraint as is usually imagined. Deleterious mutations are not always eliminated immediately, and may drift to appreciable frequency: their persistence time is about $1/s$. Before a mutant lineage becomes extinct, therefore, some hundreds of individuals may have been exposed to the occurrence of a second mutation, and in large populations such lineages will be arising continually. This continual flux of deleterious mutations at low frequency provides an opportunity for synthetic beneficial mutations to arise.

3.6.4 Functional interaction in protein structure

The most stable shape for a RNA or protein molecule is one that minimizes its free energy, which in the case of proteins typically involves an arrangement of hydrophilic residues on the outside and hydrophobic residues on the inside of the molecule. Since these residues are arranged in a linear sequence along the peptide chain, the stable configuration is achieved through interactions between non-adjacent hydrophobic amino acids: the bond formed between two hydrophobic residues that are not neighbours will bend or fold the polypeptide chain because it reduces the free energy of the molecule as a whole. The classic protein-folding problem is to specify the shape of a molecule whose sequence is given; this is very difficult and has not yet been completely solved. The converse problem is to specify the sequence of a molecule whose shape is given, which is much easier. A randomly folded protein of given length is placed on a two-dimensional grid. The sequence is at first completely hydrophilic, but we allow mutation to substitute hydrophobic residues at random. Each pair of hydrophobic residues that are neighbours on the grid but not in the sequence lowers the free energy of the molecule by $e = -1$. Hydrophobic residues are assumed to be somewhat more costly than hydrophilic residues in order to prevent the sequence from becoming uniformly hydrophobic through mutation pressure alone. The fitness of an individual is determined by the difference between benefit and cost, so selection favours the substitution of locally interacting non-neighbouring hydrophobic residues. The difficulty is, of course, that any mutation will be deleterious unless it is combined with a second mutation at exactly the right site to create a local interaction.

Figure 3.31A shows a short (18 amino acid) peptide folded to make a central core and a projecting loop. The optimal configuration is for 6 hydrophobic residues in the core, with 12 hydrophilic residues on the outside and along the loop. This is readily found by generating mutations in an initially hydrophilic sequence and accepting them if they increase the stability of the molecule. If only single mutations are allowed then no progress can ever be made (because two hydrophobic sites are needed to interact) whereas if double mutants are allowed the sequence rapidly converges to the optimal state. An explicitly evolutionary model tracks the dynamics of a population of individuals whose

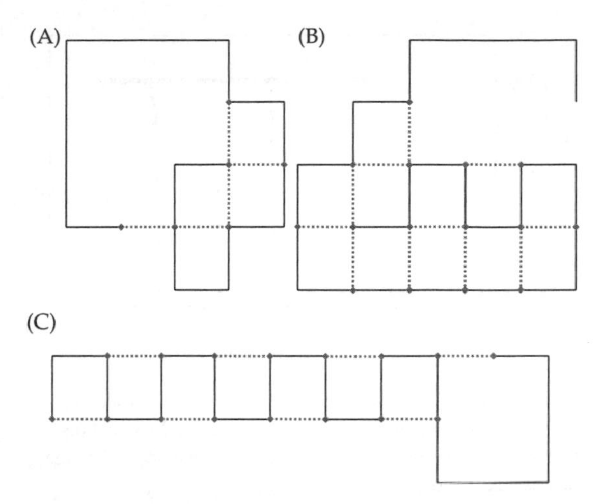

Figure 3.31 Randomly folded proteins. The amino acid skeleton is the continuous black line. Dots and broken lines indicate hydrophobic residues and the bonds between non-neighbouring residues.

offspring may carry a mutation. If the mutation rate is 10^{-5} the expected frequency of double mutants for an interacting pair of hydrophobic residues is 5 $\times$ 10^{-10}, suggesting that progress will be very slow when single mutants are neutral: a population of 1000 individuals will experience such an event at intervals of 2 000 000 generations. In fact, the optimal configuration evolves in about 2000 generations even when the mutation supply rate ($Nu = 0.01$) is low (Figure 3.32). The explanation is that the double mutants do not arise in a single step, but from the single mutants drifting in the population. A hydrophobic site will drift to fixation in about 1/ $(6 \times 10^{-5}) = 17\ 000$ generations, but a second mutation is likely to occur long before this while the first is still fluctuating at modest frequencies. The first mutation usually occurs at the terminal residue in this example, because it has more potential interactions than any other. Once the first pair has been fixed, a further mutation at any site that interacts with either of them will be favoured, setting off a cascade of fixation that rapidly establishes the optimal configuration at all the sites that interact directly or indirectly. In a highly integrated protein, such as the one illustrated here, all sites are epistatic and the whole molecule is quickly

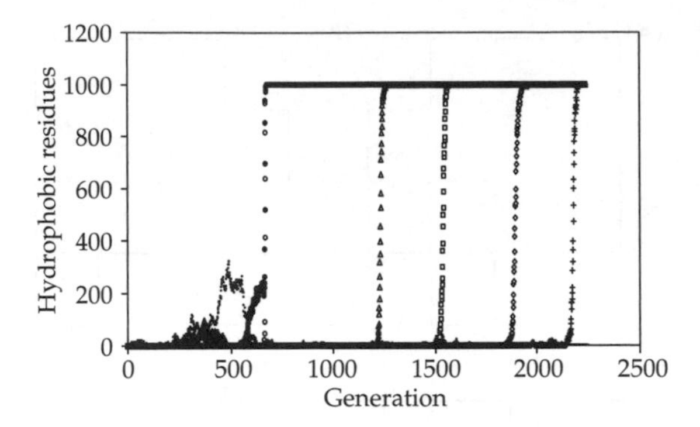

Figure 3.32 Successive substitution of hydrophobic residues in protein 3.31A in a population of 1000 individuals. Note that the first event is the capture of a second interacting residue when the first has drifted to appreciable frequency.

optimized by selection. More generally, a protein will consist of several clusters of sites, with direct or indirect interactions occurring within but not between clusters. In a moderately well-integrated protein (Figure 3.31B) sites in the major cluster are likely to be fixed first, and, within this cluster, sites with the greatest number of interactions. Poorly integrated proteins with many small independent clusters (Figure 3.31C) will evolve much more slowly.

3.6.5 Evolution of RNA sequences

Nucleic acids and proteins fold into shapes that minimize free energy given the primary sequence of nucleotides or amino acids. The secondary structure of RNA is determined by base pairing and therefore by interactions between nucleotides that are not adjacent in the sequence. It can be approximated using a fast algorithm that identifies a set of possible shapes and computes the probability of each. Cowperthwaite *et al.* (2005) defined the fitness of a random 76-nucleotide RNA molecule as its ability to bind a target molecule, estimated as the Hamming distance between the shape of the RNA and the shape of the target, averaged over all possible secondary structures. They confirmed that the fitness increments of the few fittest sequences are accurately predicted by extreme-value theory, and generated evolved sequences by simulated adaptive walks, repeatedly identifying the fittest single-nucleotide mutation and then generating all single-nucleotide mutant sequences. The effect on fitness of fixed mutations leading to these relatively well adapted molecules was roughly exponentially distributed for mutations of large effect, whereas mutations of small effect were much more abundant than expected. This was also true for very fit sequences generated by the inverse procedure of predicting the sequence likely to give rise to a specified shape. The most likely reason for this deviation from Gillespie–Orr dynamics is that related sequences are functionally correlated: the molecule produced by a single-mutation step will usually resemble its ancestor more closely in performance than a molecule resulting from several successive single-mutation steps (Cowperthwaite *et al.* 2005). Thus, the secondary structure of descendant molecules will become steadily more different from their ancestor until at some characteristic distance ancestor and descendant are little more similar than random sequences (Fontana *et al.* 1993). This characteristic distance increases with ln L from about 3 at $L = 20$ to about 7.5 at $L = 100$. This implies that the RNA matrix is more easily navigated than might be thought, as most molecules will have a 'shell' of neutral or nearly neutral states to any of which they can readily evolve.

3.6.6 Irreversibility

Even if the protein matrix is fully connected the population will simply drift from sequence to sequence if all functional nodes are equivalent. It is variation in fitness among functional nodes that provides the motive force for driving the population through the matrix. Selection will be effective only if functional density is high, otherwise the matrix is poorly connected and differences in fitness cannot be readily exploited. The operation of selection, however, will necessarily reduce the functional density and thereby the connectedness of the matrix. Any of the proteins connected to the current wild type by a rook's move may become fixed, because each is at least as fit. If there are functional interactions between sites then some of the connected proteins will be more fit than others and are more likely to be fixed. Once this has happened the matrix changes, because proteins that are less fit than the new wild type will no longer be functional. Thus, adaptation is a self-extinguishing process: the number of effectively functional nodes will shrink, and their fitness relative to the wild type will fall, as adaptation proceeds (Figure 3.35). This is the conceptual basis of continued evolution in the mutational landscape model. The immediate consequence of this process is that the adaptive walk cannot be retraced, so the requirement that adaptation must proceed in single steps, each leading to increased or at least equivalent fitness, means that evolution is likely to be irreversible if functional substitutions vary in their effect on fitness. Irreversibility in turn gives rise to two fundamental features of evolutionary change: cumulation and contingency.

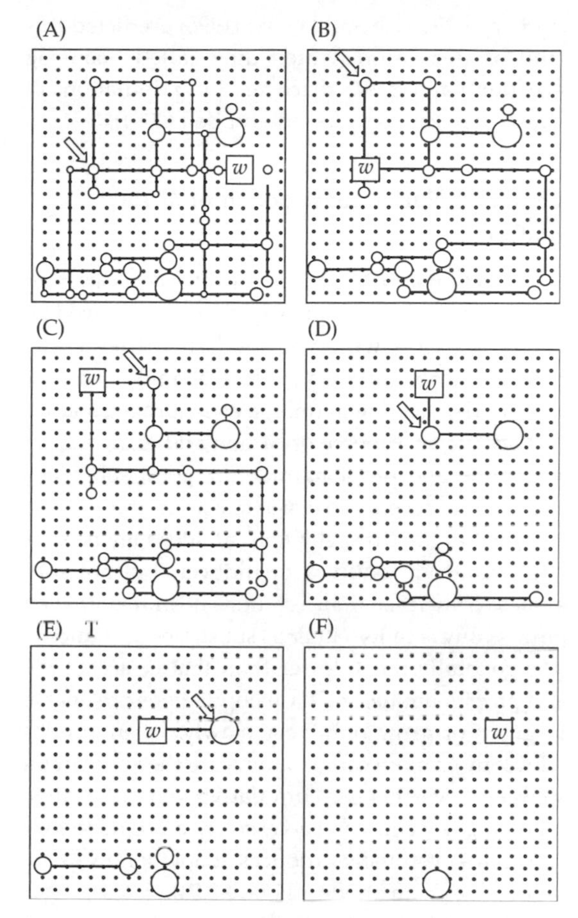

Figure 3.33 Pegboard diagrams of successive substitution in a mutational landscape. The diagrams trace the evolution of successive wild-type sequences (W) in the dipeptide space. Circles mark viable peptides; fitness is proportional to the size of the circle. In each diagram, an arrow indicates the position of the fittest accessible peptide. When this has become fixed, all less fit peptides are removed before the next substitution occurs. Eventually (F), no further evolution is possible under low-*M* conditions, although the wild type is still inferior to one possible peptide.

3.6.7 Cumulation

If the protein matrix were densely packed with sequences of equivalent fitness, except for a single superior sequence, evolution would almost always be reversible, and consequently the population would drift slowly through the matrix for a very long time before the optimal sequence were located. It is more likely that mutants will vary in fitness so that the next mutation substituted will have greater fitness than the current wild type. No protein of lesser fitness will ever again appear at appreciable frequencies in the population, so a large fraction of functional proteins is eliminated from future participation every time a mutation is fixed. Because the search space is continually narrowed in this way, the population threads its way rapidly through the matrix.

The pattern of cumulation depends on the dynamic regime of the population undergoing selection. In low-*M* conditions a single beneficial mutation will be present at appreciable frequency at any given time, so a lineage that is currently abundant will almost certainly be the ancestor of a lineage bearing a new beneficial mutation that becomes abundant in the distant future. In this case adaptation will be strictly cumulative. In high-*M* conditions, on the other hand, many beneficial mutations will become transitorily abundant before being overtaken by a superior genotype. So long as a lineage is abundant there is a good chance that a second beneficial mutation will arise in one of its members, and if it is not lost it will still be rare when its parent lineage has been replaced by the superior competitor. It will then continue to spread and eventually replace this competitor. A lineage that is successful in the future may then descend from a fossil lineage rather than from the current wild type. Gerrish and Lenski (1998) predicted this kind of succession (which they called 'leapfrog selection') and it was discovered in populations of digital replicators by Yedid and Bell (2001).

3.6.8 Cumulative construction of novel amidases

The selection of new amidases in *Pseudomonas* is an elegant example of cumulative change (reviewed by Clarke 1984). The wild-type amidase hydrolyses the two- and three-carbon amides acetamide and propionamide, yielding ammonia as a nitrogen source, and acetate as a source of carbon and energy. Both acetamide and propionamide are good substrates and good inducers. However, substrate and inducer specificities are distinct. Other amides may be substrates but not inducers, or inducers but not substrates, or neither inducers nor substrates. If they possess any activity, either as substrates or inducers, it is generally much lower than that of acetamide and propionamide, and base populations are unable to grow, or grow only very slowly, when amides with four or more carbon atoms are the sole carbon source. Nevertheless, cumulative genetic change following selection on successively more refractory substrates eventually produces adaptation to a new set of long-chain amides (Figure 3.34).

The simplest amide that is not normally used for growth is the four-carbon butyramide, which is hydrolysed only about 2% as fast as acetamide. It is easy to isolate mutants that overproduce the amidase and might therefore be able to grow on butyramide despite their inefficient utilization of the substrate. Some are regulatory mutants that express the amidase constitutively. Others appear to represent mutations in the promoter region of the amidase structural gene that cause increased rates of transcription. However, neither necessarily permits rapid growth, because butyramide, far from being an inducer, actually represses amidase synthesis. Effective adaptation requires one of two further changes. The first is another change in gene regulation, giving rise to the so-called CB strains, that causes a much higher rate of production of amidase. The second is the production of an altered amidase with higher activity towards butyramide.

These B mutants, because they are able to hydrolyse butyramide efficiently, remove it from the medium and thus prevent it from repressing amidase synthesis. One such mutant, B6, was studied in detail. It produced an amidase B that differed from the original amidase A in a single amino acid residue, the replacement of serine by phenylalanine in the seventh position from the N-terminus of the protein. This was presumably in turn caused by a single nucleotide change (UCU to UUU, or UCC to UUC) in the corresponding codon of the amidase gene.

The B6 strain was then selected on growth media containing more complex amides. A second mutation in the structural gene permitted growth on the five-carbon amide valeramide. These V mutants could in turn be used to select PhV mutants able to grow on phenylacetamide, which contains an aromatic ring, when this is supplied as a nitrogen source. These mutants have three alterations in the amidase. Phenylacetamide is neither a substrate nor an inducer for the original amidase system, so that by this point a genuinely new metabolic capacity had evolved. It had evolved through a cumulative process of successive substitution: the original B6 mutation is still present in the V strains, and both B and V mutations are present in the PhV strains, showing unequivocally how new capacities evolve through the stepwise modification of prior states.

This is not the only way in which the ability to metabolize phenylacetamide can evolve: another class of PhB mutants arose directly from the B6 strain, by a second change in the amidase structural gene. These are able to metabolize either phenylacetamide or butyramide. However, they have acquired one capacity at the expense of losing another: they are now unable to grow on the original substrate, acetamide. They are still constitutive; it is possible to select a strain that is induced by acetamide (although it cannot utilize it), however, by recombination with the basal stocks. A further regulatory mutation then produces a strain that is induced by butyramide. The result is a strain that both hydrolyses butyramide efficiently and is appropriately induced by it.

Finally, a quite different amidase was selected directly from the CB regulatory mutants. It was able to hydrolyse butyramide, valeramide, and

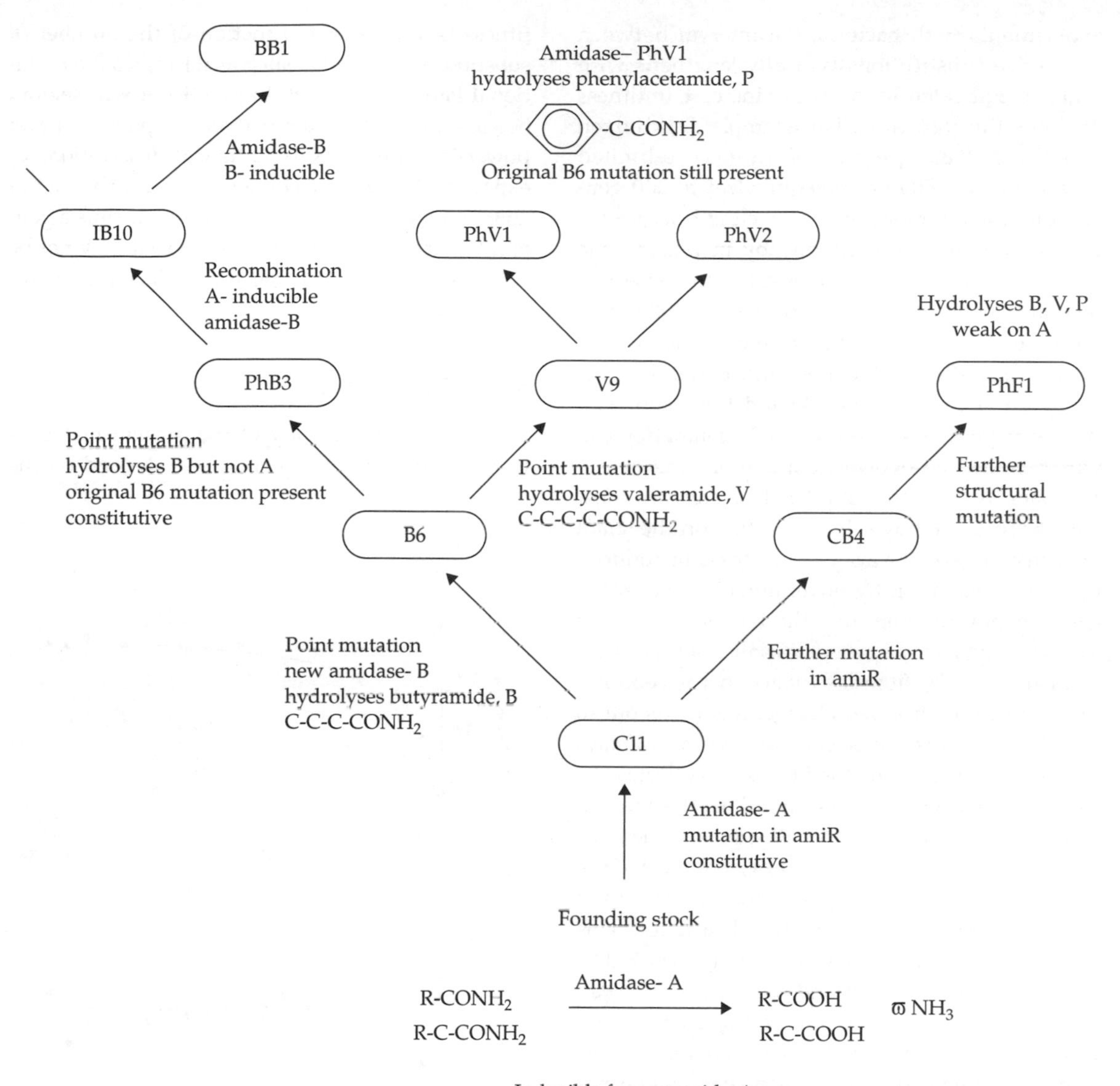

Figure 3.34 Experimental evolution of novel amidases.

phenylacetamide, but could not grow on acetamide. The capacity to utilize phenylacetamide, therefore, can arise in a variety of ways, involving different changes in regulatory and structural genes.

3.6.9 Diminishing returns

The simplest realization of long-continued change would be a long, regular succession of substitutions each involving genes of slight and more or less equal effect, driving a roughly linear increase of adaptedness through time, without any definite limit. Cumulation, on the other hand, produces a characteristic pattern of diminishing returns: progress is rapid at first but subsequently becomes slower. Thus, the first few moves through phenotype space are likely to be the longest and later moves are almost always much shorter. In chemostat experiments with bacteria, the interval between successive substitutions typically lengthens with time, or, equivalently, the rate of increase in fitness declines through time. For example, Dykhuizen and Hartl (1983) grew *E. coli* in glucose-limited chemostats for 500 hours (equivalent to 200 generations for the more rapidly cycling lines). Such populations evolve primarily by increasing the rate of glucose assimilation. Most of the observed response, however, occurred in the first 200 hours, with the rate of assimilation changing much more slowly thereafter. In the long-term *E. coli* lines set up by R.E. Lenski (see Lenski and Travisano 1994) fitness increased steadily for 2000 generations, at which point the evolved population was nearly 1.4 times as fit as the founder. This advance must have been caused by selection acting on the small quantity of genetic variance in fitness introduced in every generation by novel mutation. From the fundamental theorem that the rate of increase of fitness is equal to the genetic variance of fitness, we can calculate the fitness variance that is required to support the observed change; this turns out to be 2.7×10^{-4}. This can be compared with estimates obtained by measuring the fitness of replicate cells isolated from the same selection line; the median value of these estimates over all selection lines was 2.5×10^{-4}. It seems likely, therefore, that most, if not all, of the observed changes in fitness are caused by the selection of novel mutational variation. This experiment has now been extended to over 20,000 generations. Both fitness and cell volume followed the same sort of evolutionary trajectory: rapid change during the first 2000 generations of selection, with little further change in later generations (Figure 3.35).

This pattern of diminishing returns is predicted by both the geometrical and mutational-landscape models, and indeed is characteristic of any cumulative process with a fixed end point. Thus, suppose that we choose a uniform random number x in (0,1), then a second in (x,1), and so on. The distance remaining to the end point at 1 is halved at each step, and so fitness w approaches its upper limit as $wt = 1 - \exp(-kt)$, where t is the accumulated number of substitutions and $k = \ln(½)$. If the process has a random end point then (from numerical results) fitness is a power-law function of the number of substitutions, $w_t = w_0 t^z$, such as describes the mutational landscape model. In practice it will seldom be easy to discriminate between exponential and power-law functions as statistical descriptions of experimental results. The tendency for the rate of adaptation to diminish through time is thus a general feature of cumulative change that may not be very useful in identifying the underlying evolutionary mechanisms.

3.6.10 Contingency

If the functional density of the protein matrix is low, the outcome of evolution will depend on the

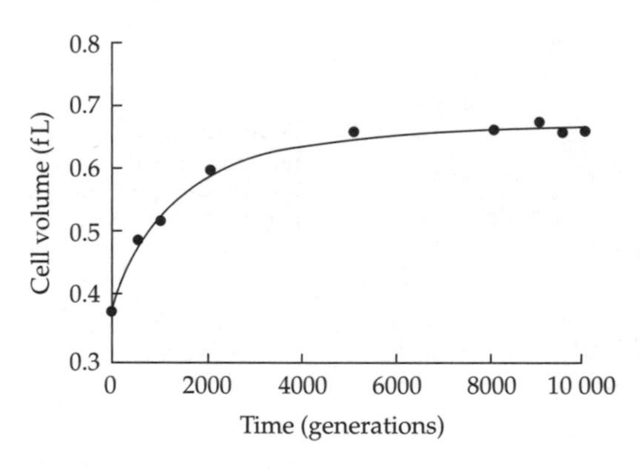

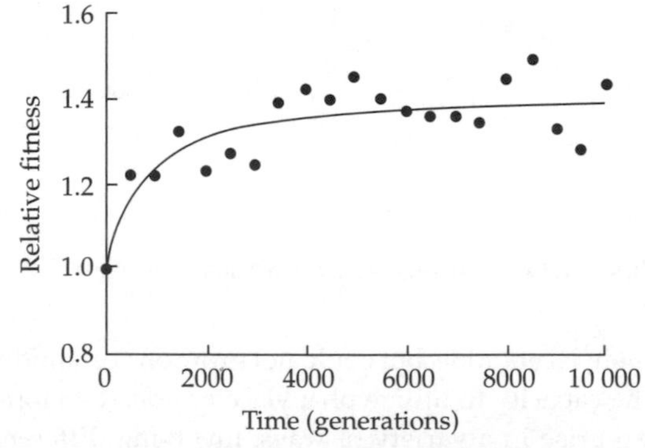

Figure 3.35 These figures show the response of cell size and fitness to natural selection in populations of *E. coli* maintained in glucose-limited chemostats for 10 000 generations (from Lenski and Travisano 1994 by permission of PNAS). The smooth curve fitted here presumably reflects the underlying discontinuous change caused by a parade of substitutions at progressively longer intervals, and indeed the authors interpret it in this way.

initial state of the population. This is one of the classical signatures of evolution: large grazing mammals evolved on the plains of all three southern continents, for example, but the three radiations produced different kinds of animal because they sprang from different ancestors. Even when functional density is high, however, adaptation will still bear the imprint of history because the stochastic fixation of one beneficial mutation rather than another may lead the population along a particular adaptive walk towards a distinctive end point. In this case there may be several or many outcomes of a uniform selection pressure applied to replicate isogenic populations. It is difficult to assess this possibility theoretically with conventional population genetic models because the constraints would have to be built into the model to start with. An alternative technique is to construct genetic algorithms able to replicate inside a computer. Such algorithms consist of a series of instructions for manipulating information, but no external rules directing how they change through time. Given a limiting resource (memory) and mutation (altering instructions by bit-flipping) they will evolve through natural selection and act as simulacra of microbial microcosms. They seldom yield pithy quantitative predictions, but can be used to explore general features of the long-term behaviour of self-replicating systems.

The Tierra creatures (mentioned in Chapter 1) are well-behaved insofar as they reproduce standard results such as the probability of fixation of beneficial mutations (Yedid and Bell 2001), so it is reasonable to use them as an extension of population genetics by other means. When the set of instructions defining a single creature is inoculated into a computer and allowed to proliferate, the population is not at first well adapted and undergoes extensive genetic modification. Fitness always increases, but replicate populations follow different trajectories and may evolve qualitatively different adaptations (Yedid and Bell 2002) (Figure 3.36). The reproduction of a sequence is controlled by a few key instructions that include *mal* (allocate a block of memory into which an offspring will be placed), *div* (divide by copying the parental sequence into the

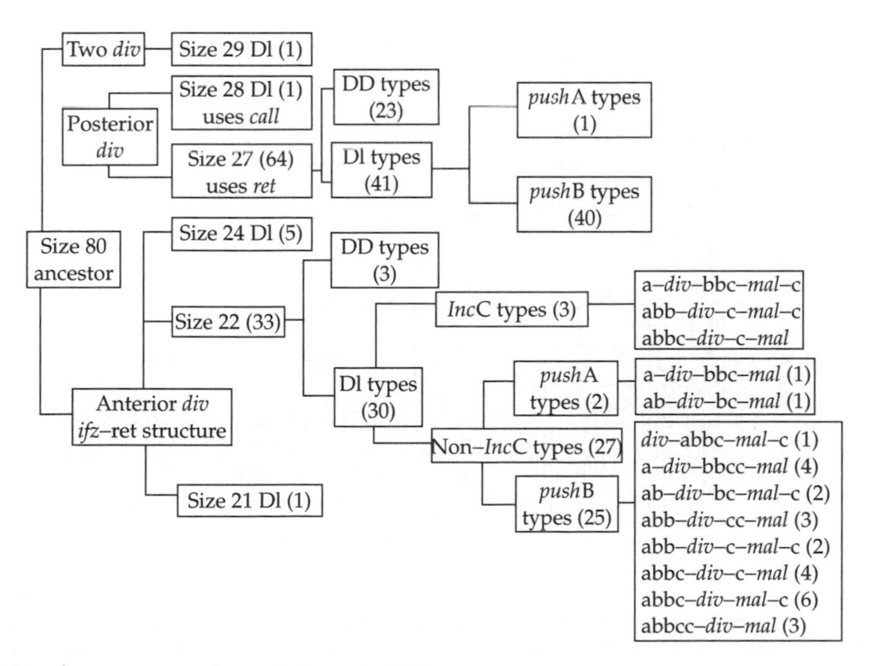

Figure 3.36 Variety of outcomes to a uniform selection regime in Tierra.

allocated space in memory), and *ret* (return from the copy operation and begin the life cycle again). Two main body plans evolved, one of 27 instructions with *div* close to the end and the other of 22 instructions with *div* near the beginning, linking *mal* and *div* in different ways. The crucial step that determines the fate of a lineage is the mutation that places *div* towards the front or towards the rear of the sequence, which in turn depends on the how *div* is affected by the pair of instructions *ifz*–*ret* in the copy procedure. (The *ifz* instruction may cause the instruction pointer to advance two places rather than one.) This strong genetic interaction creates two main pathways through the Tierran equivalent of the allele matrix, although there are many minor variants on either theme involving less important aspects of genetic architecture. A few outcomes could not be classified in this simple way: a 28-instruction sequence that used a surrogate for *ret*, a giant 39-instruction sequence unable to evolve further, and a bizarre creature with both anterior and posterior *div*. Thus, genetic interactions in this well-understood replicator channel evolution into two main themes, each with many variations, except on rare occasions when a hitherto unobserved kind of creature evolves unexpectedly. In a more complicated system, Lenski *et al.* (2003) imposed natural selection for algorithms capable of performing particular logical operations and recovered a diversity of outcomes, although they did not classify them.

A second approach is to construct genetic algorithms that are designed to simulate real organisms. These have biologically realistic properties, for example in the relationships between structure (such as leaf area) and function (such as photosynthesis rate). An algorithm is in this case an abstract representation of a plant or animal that is more or less successful in reproducing as a function of the joint state of many interacting characters. In this case the constraints are explicitly built into the model, but the model is so complex that the end point cannot be foreseen. In fact there appears to be no single end point, but rather a number of end points representing a range of creatures that have quite different designs but very similar rates of growth (Marks and Lechowicz 2006).

These models all predict that replicate selection lines should diverge through time because of functional interaction between the sites responsible for

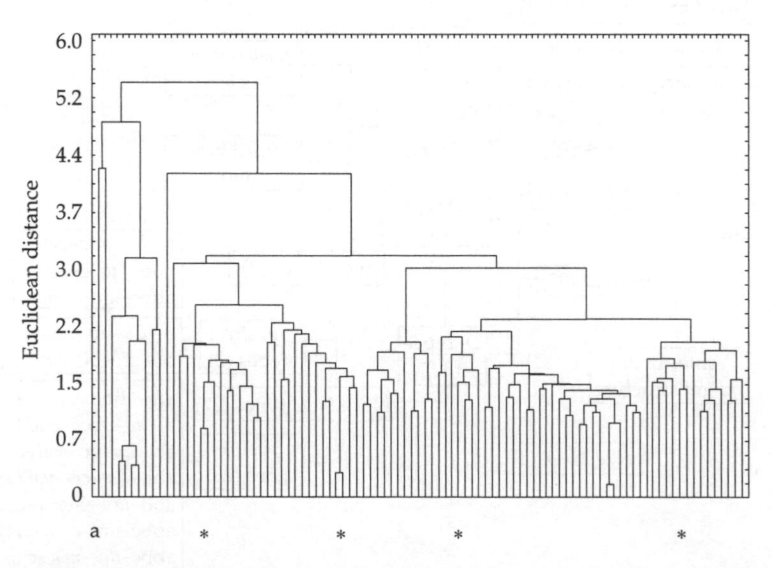

Figure 3.37 Phenogram of *Pseudomonas* lines selected on a range of substrates. The ancestor is marked A. Replicate lines that form sister taxa in this classification are marked with an asterisk; in all other cases sister taxa were selected on different substrates. From MacLean and Bell (2003).

adaptation. It is clear that the variance of fitness among lines increases through time, for example over several thousand generations in long-term *E. coli* lines (Lenski and Travisano 1994). This will be a feature of any cumulative process, however, including drift. Replicate lines of the soil bacterium *Comomonas* evolved similar levels of fitness and growth rate but had different yield and cell size after 1000 generations in laboratory conditions (Korona 1996a). MacLean and Bell (2003) propagated replicate *Pseudomonas* lines on a given carbon substrate for 1000 generations and then assayed growth on a range of substrates. They found that the pattern of utilization of substrates varied widely among replicate lines, to the extent that lines propagated on the same substrate were no more similar than lines propagated on different substrates (Figure 3.37). The pattern of substrate utilization reflects the expression of a large suite of genes, and the experiment conclusively demonstrates historical contingency. It does not necessarily show that contingency is an important factor in adaptation. If each line adapted through periodic selection, then each sweep would capture a beneficial mutation together with all the neutral or slightly deleterious mutations it was associated with. These hitchhiking mutations might then be revealed through their effect on growth on other substrates, without necessarily contributing to adaptation to the substrate of selection. This is an inescapable weakness of simple assays, whether genetic or phenotypic.

The protein matrix (or DNA matrix) serving as the basis for the successive substitution of alleles emphasizes the importance of functional interaction to the outcome of evolution. The primary constraint on evolution is the connectance of the protein matrix, which is governed by functional density and functional interaction. It is in turn modulated by reversibility, in such a way that connectance may decay as more substitutions are made, restricting each pathway to an increasingly isolated range of genotypes. This leads to a cumulative process of evolutionary change, marked by the approach to an asymptotic level of adaptation through the substitution of mutation of successively smaller effect. This process may also be historical, leading to a limited number of genetic combinations as the final outcomes of selection. The experimental evidence that we have so far obtained is consistent with this view, although it does not support it very strongly. To find stronger evidence I turn to studies that involve a broader range of genetic variation over longer periods of time.

3.7 Successive substitution at several loci

3.7.1 Genetic interactions

The mutational landscape, the protein matrix, and similar ideas are ways of describing directional selection at a single locus. Adaptation that involves mutations at several loci does not necessarily introduce any new principle, provided that the population remains asexual. The distinction between the substitution of alleles of the same locus and the substitution of mutations at different loci is thus a little artificial, and several of the examples described above probably involve more than one locus. Different mutations in a single gene typically affect a single protein and thereby often a single function, however, whereas mutations in different genes may affect very different functions. In practice, therefore, interactions between the proteins encoded by different loci are likely to be different in nature and in strength from the interactions between amino acids in the same protein. The interaction between alleles at different loci is called *epistasis*, because it originally referred to the between-locus equivalent of the dominance of alleles at the same locus; it is now used to describe any situation in which the combined effect of two or more alleles at different loci differs from its expected value based on their separate effects.

Dykhuizen and Hartl (1980) described a simple example of epistasis in chemostat populations of *E. coli*. The primary locus they studied was *gnd*, which encodes 6-phosphogluconate dehydrogenase (6PGD), an enzyme involved in the pathway by which glucose-6-phosphate is converted to a pentose phosphate. It is quite diverse, with 15 alleles being detected in about 100 natural isolates; Dykhuizen and Hartl (1980) studied four of these, with the unmemorable names of S4, S8, F2, and W^+.

The second locus involved in the experiments was *edd*, which encodes phosphogluconate dehydratase; this enzyme leads to an alternative pathway for the metabolism of 6-phosphogluconate, making it possible to bypass 6PGD. They also used genes conferring resistance ($T5^R$) or susceptibility ($T5^S$) to phage T5. To construct the initial strains, the appropriate combination of genes was inserted into the genome of a standard laboratory strain by viral transduction, so that they were as nearly as possible identical except for carefully-specified differences in *gnd*, *edd*, and T5. The strains bearing different *gnd* alleles could then be put into competition with one another in order to estimate the strength of selection. Each experiment involved four trials. Two were the experimental lines, in which changes in frequency at *gnd* could be followed by changes in the frequency of T5 alleles, measured by exposing the cultures to phage. The other two lines were controls, incorporating a point mutation that abolished 6PGD synthesis. There was strong selection against *gnd*$^-$ on gluconate medium, supplying a positive control ensuring that selection would be detected if it occurred. On the other hand, *gnd*$^+$ and *gnd*$^-$ alleles are nearly neutral in ribose–succinate medium that contains no gluconate. The growth of the experimental lines on ribose–succinate medium thus gives a negative control showing that any difference between the functional *gnd* alleles when growing on gluconate is indeed attributable to differences in 6PGD activity.

The general result of these experiments was that the four *gnd* alleles had indistinguishable effects on fitness when tested in an *edd*$^+$ background; the sensitivity of the assay is such that the result implies that any selection that occurs is very weak, with $s < 0.01$ per generation. There were, however, two complications. The first is that the S8 $T5^S$ strain was consistently selected over the S8 $T5^R$ strain in gluconate, although not in ribose–succinate. There is therefore some functional interaction between 6PGD and phage resistance, and the dynamics of selection of *gnd* alleles will depend to some extent on their genetic background. A more serious complication is that quite different results were obtained when the experiments were run with an *edd*$^-$ background. The details are complicated, because *edd*$^-$ strains do not grow well on gluconate, so that before the experiment is completed new *gnd* alleles have arisen by mutation. However, it was clearly established that S4 is consistently superior to F2. It seems that *edd*$^+$ genotypes can compensate for any small differences among *gnd* alleles in their ability to process gluconate, but these differences are expressed in *edd*$^-$ genotypes, where no compensation is possible.

Now, if we ask whether the *gnd* alleles are neutral, it will be difficult to give a straightforward answer. In most of the conditions where they were tested, they have very similar fitnesses, but differences appeared when they were tested on a particular genetic background (*edd*$^-$) in a particular environment (gluconate growth medium). I think it is very likely that many mutations will display this sort of conditional neutrality. Selection among alternative alleles is weak or non-existent across a broad range of conditions, but may be appreciable when the same alleles are expressed in a different genome or in a different environment.

3.7.2 The adaptive landscape

A well-known way of representing adaptation involving two loci is to arrange all possible alleles at each locus along the axis of a graph, so that any point on the plane is occupied by the two-locus genotype representing a combination of alleles at the two loci. The third dimension of the graph is the fitness of this genotype. The surface of the graph then resembles a landscape whose elevation differs from point to point, higher elevations corresponding to genotypes of greater fitness (Figure 3.38). It was devised by Sewall Wright (1931) in order to illustrate his thesis that fitness generally depended on complex interactions among genes. A second version of the concept is that the two basal axes represent allele frequencies while the third axis is population mean fitness. Polling my colleagues and students, I have found that several other versions are current, albeit unrecognized in the literature, including graphs with phenotypic values on the basal axes or with an environmental score on one axis (perhaps 'landscape' can be taken too literally). This is plainly a topic with the potential to

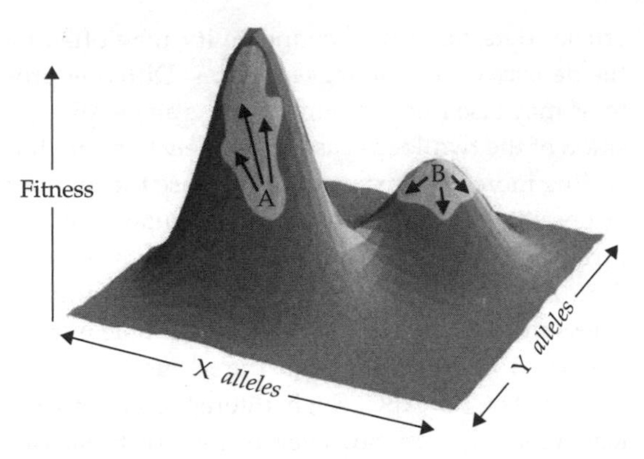

Figure 3.38 An adaptive landscape with two fitness peaks. The *x*- and *y*-axes are alleles ordered so that neighbouring sequences differ by a single substitution. Population A currently occupies the flanks of a fitness peak and is driven uphill by selection acting on beneficial mutations; population B occupies the summit of a (minor) fitness peak and is driven downhill by deleterious mutations. From Burch and Chao (2000).

confuse, along with the 'shifting balance' process of adaptation it is said to generate (§4.4.4). I shall use it here to refer to the fitness of individuals because this expresses most clearly the notion of gene interaction.

At any given time, a population that is dominated by a single genotype, or a small set of similar genotypes, occupies a certain place on this landscape. The selection of genotypes with greater fitness causes the population to move about the landscape in such a way that its mean fitness continually increases; that is, it always moves uphill. (It would be more appropriate to imagine the population as tending always to move downhill, but the uphill metaphor is too deeply entrenched to be changed.) Suppose that the alleles at either locus are arranged in order of increasing fitness. This will be possible only if they act independently, so that the fitness of an allele at one locus can be defined without reference to the allele present at the other locus. A combination of two low-fitness alleles thus gives a low-fitness genotype, and a combination of two high-fitness alleles a high-fitness genotype. The landscape then resembles a ramp, with its elevation increasing continually in one direction. Selection pushes the population up the ramp, until the fittest allele has been fixed at both loci. However, if there are epistatic interactions among loci, it will not be possible to arrange the alleles at either locus in a series of increasing fitness, because the fitness of an allele at one locus cannot be defined without reference to the state of the other locus. The topography of the adaptive landscape is now much more complicated than a simple ramp, and will instead resemble a hilly terrain of peaks and valleys. A population at any particular point will be pushed upward by selection until it reaches the top of a peak; the population will then cease to evolve, because movement in any direction is downhill, and is therefore prevented by selection against genotypes of lower fitness. However, this peak may not be the highest point in the landscape. There may well be higher peaks elsewhere, but they cannot be reached through the stepwise selection of unit genetic changes, because they are separated by valleys of low fitness. The simplest case would be if the two loci each had two alleles A,a and B,b. If AB and ab are fitter than Ab and aB, then a population fixed for AB will remain in this state even if the genotype ab has the greater fitness, because evolution from AB to ab through sequential substitution would involve passage through the valleys represented by the low-fitness genotypes Ab or aB. Suppose that the fitness of a normally distributed character with phenotypic variance σ^2 and additive genetic variance V_A is described by a normal fitness function with 'variance' τ^2. The population is currently located on the summit of a minor peak separated from the global optimum by a valley of depth Δ (the ratio of population mean fitness at the minor peak and in the floor of the valley). The expected time that will elapse before the population drifts across the valley and is driven by selection to the summit of the major peak is $(2\pi(\sigma^2 + \tau^2)/V_A)\ \Delta^2 N_e$ (Lande 1985). This is likely to be an exceedingly long time if N_e is not very small.

The simplest form of generalized epistasis is that combinations of alleles have random fitness, in which case the number of locally optimal genotypes increases exponentially with the number of loci (Kauffman and Levin 1987). This is unrealistic because similar genotypes will usually have similar fitness, so one could suppose instead that each of *N* loci interacts randomly with *K* others (the NK model of Kauffman and Levin). If $K = 0$ then alleles at different loci have additive effects on fitness, and as K increases the number of local optima increases. We have a very good estimate of $K = 30$ for strong interactions among the yeast double deletion set. This would certainly lead to a very broken landscape, but whether it applies to beneficial mutations is not known.

3.7.3 The allele matrix

A grave weakness of the adaptive landscape concept is that the shape of the landscape depends on the order in which alleles are placed along each axis, yet this order is arbitrary. An alternative concept that avoids this weakness is the *allele matrix*. Each axis represents all the possible alleles of a gene that are one mutation away from the current wild type (as in Figure 3.38). Thus, the first position on an axis might be occupied by the wild-type sequence; the second position by an allele in which the first nucleotide of the wild-type sequence was one of the three possible mutations; the third position by another of these three; and so forth to the final nucleotide of the $3L + 1$ encoding a protein of length L. Regardless of the order in which the alleles are placed, any can be reached from the wild-type sequence by a single mutation, so evolution proceeds by a series of rook's moves in a matrix with as many dimensions as there are genes in the genome. Once a mutant allele has become fixed at a locus it becomes the new wild type, and all $3L + 1$ sequences on this axis are adjusted accordingly. The allele matrix is then restricted to the neighbourhood of the current wild type, but gradually moves within the extremely large space represented by all possible sequences at all loci. This simple extension of the protein or DNA matrix allows us to ask whether all beneficial multilocus genotypes are accessible to selection. As in the single-locus case, the allele matrix is likely to be highly connected if genes have largely independent effects on fitness, or, should they be highly interactive, if the frequency of functional combinations exceeds $1/(3L + 1)$. Because the properties of a protein depend on its shape, mutations that affect amino acid sequence are likely to be highly interactive, and the crucial determinant of connectivity may often be the density of functional sequences. Different proteins may also interact, such that the ab → AB transition of the two-locus case is unlikely to occur, but adding more dimensions will increase the number of possible routes between double mutants. In the three-locus case, for example, AbC might be fitter than Abc even though Ab on other backgrounds is less fit than ab, in which case the double mutant AB could be fixed through the sequence abc → abC → AbC → ABC. Such interactions between loci are surely rare, however: in the great majority of cases, we expect that a beneficial mutation will tend to spread through a population regardless of the genetic background on which it occurs. It may sometimes happen that two mutant sequences are separately neutral or deleterious whereas they are beneficial when combined, although this is likely to be a rare exception rather than the rule. If this is the case, then the crucial determinant of connectivity for the allele matrix is the independence of mutational effects among loci, rather than the density of functional multilocus genotypes, and the allele matrix, unlike the protein or DNA matrices of a single gene, is likely to be highly connected.

Nevertheless, there are two very widespread kinds of situation in which different proteins are likely to interact strongly. The first includes compound structures whose components interact simultaneously, like the working parts of a machine. Transport systems, for example, involve complexes of several or many proteins that must interact precisely in order to provide normal function. The second includes processing chains whose components interact sequentially, like different machines in an assembly line. Metabolic pathways that degrade a substrate through a series of enzymatic reactions provide a familiar example. The allele matrix can be used to investigate these two situations through extensions of the single-locus Gillespie–Orr model.

3.7.4 Compensatory mutations

The compensation of genetic damage is an important special case of interaction. When an important gene is inactivated or lost, function can be restored either by reversion or by compensation. Reversion is simpler, but compensation seems to occur much more frequently. A compensatory mutation is beneficial only when a deleterious mutation is present at another locus (otherwise it would already have been selected as a straightforwardly beneficial mutation), and therefore by definition must involve genetic interaction. When fitness is degraded in *E. coli* lines by *Tn10* insertion it is usually recovered almost completely within 200 generations by selection for compensatory mutations at other loci (Moore *et al.* 2000). The effect is larger when the damage is greater because the relative fitness of the compensatory mutations is likely to be greater. Burch and Chao (1999) trapped a severely deleterious mutation in the RNA phage $\varphi 6$ by isolate culture and followed the sequential substitution of beneficial mutations that restored function. At high mutation supply rate fitness was wholly restored by single mutations of large effect in most cases. When effective population size was very small, however, fitness was only partly restored by a succession of mutations of small effect. Rokyta *et al.* (2002) deleted the ligase gene of DNA phage T7, which is severely deleterious in hosts that are themselves defective in ligase production. Function was partly restored by four point mutations, a short deletion, and a single-nucleotide insertion. Several of these mutations occurred in genes responsible for handling DNA and may have compensated for low levels of ligase by reducing the levels of functionally related proteins such as endonuclease, but the precise nature of the interactions is not clear.

The classical system for studying the evolution of compensation is the Lac operon of *E. coli*, where the *lac*Z structural gene encoding β-galactosidase can be completely deleted, while leaving the other elements of the system (the permease, for example) intact. The full story of the subsequent evolution of compensation by a well-defined alternative system has been reviewed by Hall (2003), who describes the chain of experiments leading to our current understanding. The initial observation involved streaking a culture of *lac*Z-deleted cells on to plates containing the normal nutrient broth supplemented with lactose. The broth is the primary source of nutrients, and allows colonies to develop; as it becomes depleted, any variant that can utilize lactose would have a large advantage. When the structural gene has been completely deleted, the only way of recovering the ability to ferment lactose is to use or modify the product of some other gene. Rather surprisingly, this turns out to be possible, and white papillae appearing on the surface of the colonies herald the spread of variants with a new β-galactosidase. It has been discovered subsequently that wild-type cells synthesize an enzyme that can hydrolyse β-galactosides, although too slowly to support growth in the absence of other substrates, and it is this enzyme, called Ebg for evolved beta-galactosidase, which is responsible for the growth on depleted plates. Both the structural gene and its repressor occur close together in the Ebg operon, which is situated far away from the Lac operon. The first mutation in the adaptive walk alters the structural gene *ebgA* and increases specificity for lactose about 10-fold. The second mutation occurs in the repressor gene *ebgR* of the Ebg operon and allows it to be induced by lactulose. A second mutation in *ebgA* permits the utilization of arabinose, and leads to the conversion of lactose to allolactose, the inducer of Lac permease. This strain thus represses the synthesis of Ebg and Lac permease in the absence of lactose but permits the induction of both Ebg and Lac operons when lactose is present. This allows growth on medium with lactose as sole source of carbon and energy, although the activity of Ebg is only about 1% that of the native Lac. The Ebg neatly answers two important questions about the evolution of novel metabolic systems. The first is the number of different ways in which a novel system compensating for the loss of an important gene can evolve. This is readily answered by the failure to find compensatory mutants able to rescue the double deletion of *lacZ* and *ebgA*: there is only a single evolutionary pathway in this case. The second question is the minimum number of mutations in the Ebg operon that lead to compensation. The answer is very few: two mutations in ebgA permit lactose hydrolysis, and a succession of three mutations is needed for a functional and appropriately regulated system. Evolution in this case seems to be operating within strictly confined boundaries.

3.7.5 Compound structures

An operation may be performed much more effectively by two or more units acting together than by any or all acting separately. A dimeric molecule consisting of two different proteins (such as an immunoglobulin) is an example of a compound structure: a reaction only weakly catalysed, if at all, by either component peptide is efficiently catalysed by the combination of both. A mutation at the locus encoding one peptide might then have little or no effect on function unless it were combined with some corresponding mutation at the locus encoding the second peptide; the resulting double mutant might be called a *synthetic vital*, by analogy with a synthetic lethal. We can imagine that any single mutation, at either locus, has some potential benefit, drawn as before from the extreme right-hand tail of a specified distribution, that is realized only when a mutation is also present at the other locus. This is then a two-locus version of the previous one-locus model. As the double-mutation rate is u^2, it might be thought at first that substitution would be extremely slow even in very large populations. It is not necessary, however, that both mutations occur at the same time. If the single mutants are neutral, however, one or the other locus will drift to fixation in about $1/2u$ generations, regardless of population size, and once this has happened, any mutation at the unfixed locus must occur on the single-mutant background. The dynamics are then the same as for the single-locus case. The second locus will usually be fixed much more rapidly than the first (given $Ns \gg 1$), and in practice $1/2u$ seems to be a good approximation provided that the mutation supply rate is very low. At higher mutation supply rates the fixation time of the synthetic vital is less than $1/2u$ and may be much less. This occurs when there is a substantial level of transient polymorphism, so that a second mutation often arises in a lineage which but for this would not have spread to fixation. Recall that under the conditions of most laboratory experiments with microbes, a lineage bearing a single mutation that increases to a frequency of 1% is likely experience *every* second-site mutation in every generation. Since the average frequency of a neutral allele at either locus will be $2Nu/(1 + 2Nu)$, the probability that a double mutant will appear and survive to become fixed is $4(Nu)^2s/(1 + 2Nu)$ per generation, and taking both loci into account the waiting time for the synthetic genotype is $(1 + 2Nu)/8(Nu)^2s$. This may amount to only a few hundred generations if $Nu > 0.1$ or so.

If the mutations are actually deleterious when they occur singly the assembly of a synthetic vital will take longer. In very small populations of 10–100 individuals a slight constant disadvantage of $s_1 =$ 1–2% will not greatly retard the process, since the persistence time of mildly deleterious mutations is roughly the reciprocal of the selection coefficient (Crow 1993), and does not prevent fixation within a time of order $1/u$ generations. In larger populations such alleles will seldom drift to fixation, but will persist at mutation-selection equilibrium with frequencies of about u/s_1. The waiting time for the appearance of a successful second mutation in this reservoir of slightly deleterious single mutants is therefore about $(s_1/s_2)/4Nu^2$, where s_2 is the advantage of the double mutant. If mutation supply rate is low such synthetic vitals will not be fixed within ecologically interesting periods of time, but if $Nu > 0.1$ or so then the waiting time will again be of the order of $1/u$ at most. An example is given in Figure 3.39.

The assembly of complex adaptations can be followed step by step in evolving populations of digital organisms. Lenski *et al.* (2003) gave greater rewards to individuals capable of performing more onerous logical operations on 32-bit strings using the Avida platform. The simplest operations are NOT and NAND (not and), with successively more onerous operations being AND, OR, NOR (not or), and EQU (equals). Any mutant capable of performing a new function was allocated supplementary resources, in the form of being allowed to execute more instructions. The base population comprised 3600 individuals each 50 instructions in length, which replicated with a mutation rate of 0.0025 per instruction copied, so $U = 0.225$ and $M = 9$. The most complex operation, EQU, requires many (about 20) changes to the ancestral sequence, which is capable of replicating but cannot perform any logical operation. If only EQU is rewarded it will not evolve because the requisite multiple mutant will never appear by chance. If

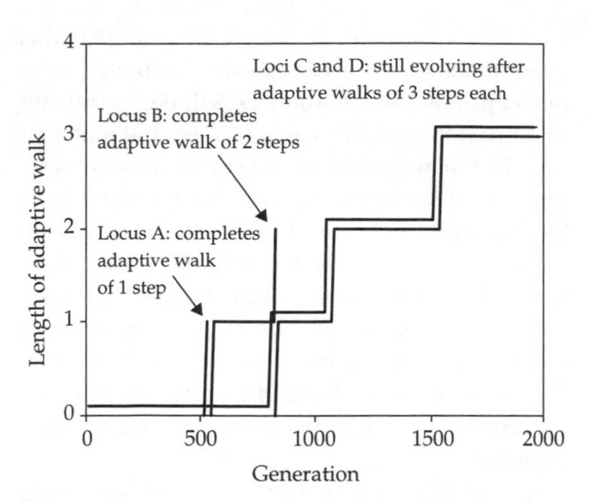

Figure 3.39 Assembly of synthetic vitals. This is output from an individual-based mutational landscape model with four interacting loci. Each locus is subject to mutation (NIu = 0.05), yielding alleles (initial I = 4) with extreme-value distribution of potential fitness. Single mutations are deleterious (s = 0.05) but the fitness of a double mutant is equal to the potential fitness of the fitter mutant allele. The process is refreshed when the double mutant is fixed. Two loci completed short adaptive walks; the other two advanced but still had a supply of beneficial mutations at the end.

most or all simpler operations are rewarded, however, EQU evolves in about one-half of all replicate lines. In one example that was analysed in detail the lineage that evolved EQU in about 5000 generations had substituted a total of 103 single mutations, 6 double, and 2 triple mutations. Most of these were beneficial (45) or neutral (48), but some were deleterious (18), and 2 of these were severely deleterious. The evolution of a complex novel ability in this system is thus based on the cumulation of changes that encode simpler functions. The appearance of highly improbable structures depends in part on the flux of neutral and deleterious mutations, typically at low frequency, that act epistatically with further mutations to produce superior genotypes. This seems to support the interpretation of complex adaptations in cellular systems.

3.7.6 Processing chains

In many cases some necessity is not produced in a single step but instead by a processing chain involving a number of steps which must occur in a certain order. A metabolic pathway is one example of a processing chain: a series of reactions taking place in a fixed order, each catalysed by a different protein, that results in energy production or biosynthesis. Each member i of a chain can be described in terms of three quantities: the input I_i (quantity of material passed on by the previous member of the chain), the processing capacity K_i (maximum quantity of material that can be transformed), and the output J_i (quantity of material actually transformed). The distinctive feature of a processing chain is that the output J_i of any member is the input I_i received by the next downstream member. The output itself may be limited either by input (if $I_i < K_i$) or by processing capacity (if $K_i < I_i$), and is therefore equal to the lesser of I_i and K_i. Fitness itself is affected solely through the final output J_n, which represents the total flux of material or energy through the chain. In this way, the output of any given member depends on the characteristics of all the upstream members, and the effect of this output on fitness depends on the characteristics of all the downstream members, so that each member of the chain interacts directly or indirectly with every other member. How will a processing chain evolve when conditions change such that a much greater quantity of some substrate becomes newly available?

This new source of supply means that the external input to the first member of the chain now exceeds its processing capacity (which had evolved previously to handle a lower input). Nevertheless, we assume as before that the ancestor is relatively well adapted and that the mutation supply rate is sufficiently low that low-M conditions hold. Each member of the chain is represented by a locus with a series of mutant alleles that are drawn as random values from the tail of a specified distribution and replaced in the same way when one of them becomes fixed. Adaptation proceeds by the successive substitution of beneficial mutations at all loci in the chain until no more beneficial mutations are available at any of them. To predict the sequence of substitutions in this system, it is helpful to draw the output profile, which shows the output from successive members of the chain, down to the final output J_n. The output from any given member is equal to or less than the output from the preceding member, so the profile descends to the right in a series of steps. If the input received by a given member is less than its processing capacity, the output must be the same as that from the previous member. If the processing capacity is less than the input, however, output will be less than output from the preceding member. Thus, each step is the result of a bottleneck attributable to the inadequate processing capacity of the member at this point in the chain. If there are several bottlenecks, a mutation that increases processing capacity in the bottleneck furthest upstream may increase flux through the first few members of the chain, but will not affect final output because inadequate processing capacity in later members of the chain prevents the increased flux from being transmitted to the end of the chain. The same argument applies to any intermediate step. The rule predicting the order in which substitutions occur is thus: the next successful mutation will involve the locus immediately upstream of the furthest downstream bottleneck. The consequence of this mutation is to remove the bottleneck and thus to flatten the output profile downstream. Selection now operates on the bottleneck next upstream, and this process continues until the output profile has become completely flat from beginning to end. At this point, the final output is limited by the processing capacity of the first step, since this is likely still to fall short of the newly increased rate of external input. The next event will then be a successful mutation at the first locus in the chain, which will give rise to a new irregular output profile, and the process of substitution from downstream bottlenecks upwards is resumed. In principle, adaptation will be complete when the output profile is flat, with the processing capacity of each member at least equal to the external input rate, but in practice it will almost always halt before this condition is reached because no new beneficial mutation is available at any locus.

Under low-M assumptions, therefore, selection effectively acts at a single locus at any given time. Since each locus will have Gillespie–Orr dynamics, the expected length of the adaptive walk for a processing chain of n members will be simply $L_{chain} = n \times L$. A minor complication is that this result holds only for functional mutations: mutations upstream of a bottleneck will be neutral and will therefore become fixed at the neutral mutation rate. The total number of substitutions will therefore exceed the number of functional substitutions by a quantity $(n - 1)ut$, where u is the mutation rate per locus and t the mean length of the adaptive walk in generations. These neutral mutations may become beneficial subsequently, however, as the output profile flattens towards the loci where they have occurred; for this reason, the observed number of functional mutations is always less than expected.

A successful mutation increases the processing capacity associated with a locus, but does not necessarily increase output to the same extent, because its elevated capacity may now exceed its input. Consequently, the effect of a successful mutation on fitness will be partly independent of (and on average less than) its effect on processing capacity. This will tend to reduce the advantage associated with early mutations, and simulations suggest that each successive substitution has on average about the same effect on final output. This does not affect the spacing of these substitutions, however, so that if fitness is directly proportional to final output it will increase linearly with ln time.

3.7.7 Effect of mutation in a simple processing chain

The evolution of a processing chain could then be predicted from the effects of mutation of each of its components. Dykhuizen and Dean (1990) extended this argument by showing how the effects of mutation could themselves be predicted from biochemical principles in the short processing chain of lactose assimilation (Figure 3.40). When lactose is the limiting resource, the fitness of a given type depends only on the rate at which it generates glucose and galactose from environmental lactose. Thus, relative fitness is an increasing linear function of the flux of lactose through the pore–permease–β-galactosidase system.

The rate V of an enzyme-catalysed reaction can be described in terms of the substrate

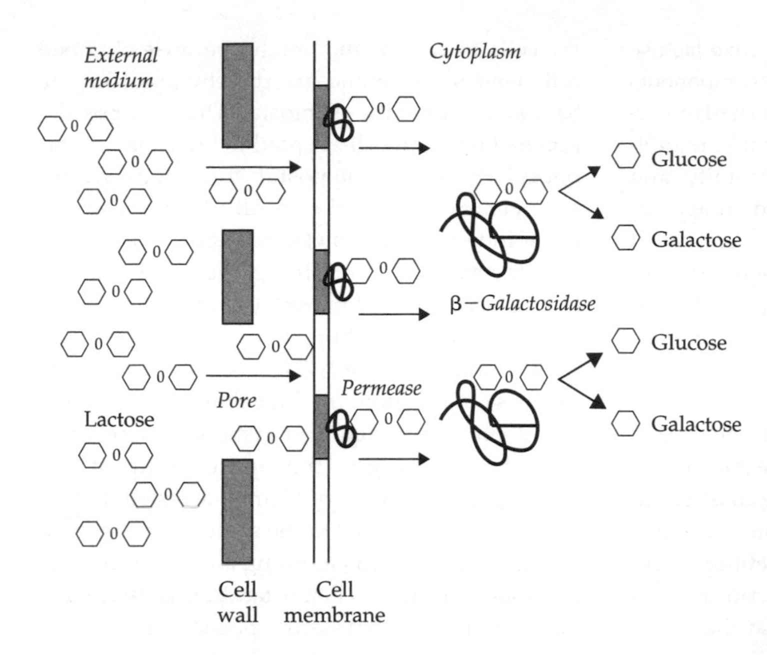

Figure 3.40 Lactose metabolism. Lactose present in the medium diffuses passively through pores in the cell wall into the periplastic space. It then passes across the cell membrane into the cytoplasm by an active process requiring the enzyme *lac* permease. In the cytoplasm, it is hydrolysed by a β-galactosidase, yielding glucose and galactose, which undergo further metabolic transformations.

concentration S by the Michaelis–Menton equation, $V = V_{max}S/(S + K_s)$, where V_{max} is the maximal rate at high substrate concentrations, and K_S is the half-saturation constant, i.e. the substrate concentration at which the rate of the reaction is half its maximal value V_{max}. In metabolic pathways, a series of reversible reactions are linked together by a series of enzymes with different Michaelis–Menton dynamics; the overall flux of material through the pathway depends on all these enzymes, and moreover the flux at each step depends on the flux at every other step in the pathway. The lactose system is relatively simple, because it is terminated by the irreversible hydrolysis of lactose by β-galactosidase. If the concentrations of substrates are low (less than their K_s) the system is unsaturated, and the flux J will be approximately

$$J = S/[(1/E_1) + (1/E_2K_{1,2}) + (1/E_3K_{1,3})].$$

The subscripts 1, 2, and 3 refer to the three components of the system—the cell-wall pores, lac permease, and β-galactosidase. E is a measure of enzyme activity, V_{max}/K_s; for the pores, this is equivalent to their diffusion constant. K is the thermodynamic equilibrium constant for the reaction (often symbolized K_{eq}, and not to be confused

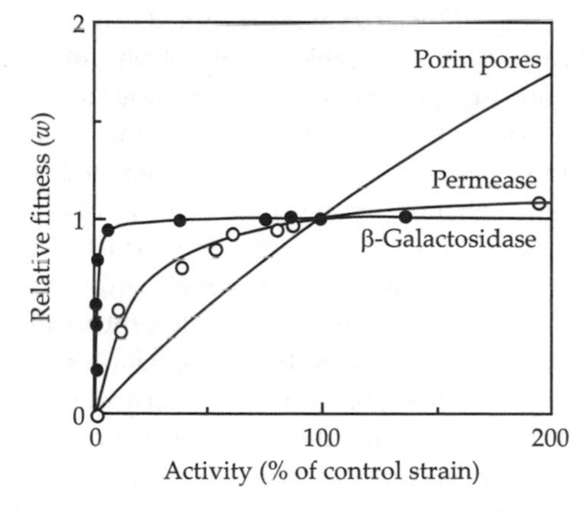

Figure 3.41 Predicted fitness of mutants in lactose pathway of *E. coli*. From Dykhuizen and Dean (1990).

with K_s). Writing out the equation in this way makes it clear how a change in any component of the system will affect flux through the pathway as a whole, and will thereby affect fitness. All of the parameters involved can be measured independently and then used to predict the outcome of selection in a lactose-limited culture. Figure 3.41 shows the predicted effects of alterations in the three components of the lactose system, relative to a control strain. Mutants unable to utilize lactose through a failure of any of the three components have zero fitness, there being no alternative carbon source; mutants that are equivalent to the reference strain have a relative fitness of unity; and in general relative fitness increases in an asymptotic fashion as the activity E of a given component increases. Note that fitness is much more sensitive to changes in the activity of some components than to comparable changes in others. A very modest level of β-galactosidase activity yields nearly wild-type performance, and further increases in activity give little increase in fitness. Minor changes in the permease are more likely to cause appreciable changes in fitness, and the system is most sensitive to pore diameter. The plotted points are actual relative fitnesses, measured in competition trials, of a series of mutant alleles of lac permease and β-galactosidase (no information is available for pore mutants). The outcome of selection seems to be adequately predictable from the biochemical properties of the enzymes concerned.

Yeast obtains a usable supply of phosphate by hydrolysing phosphate esters to release orthophosphate, which is then transported into the cell by a permease. The hydrolysis is mediated by an acid phosphatase located in the cell wall. Francis and Hansche (1972) set up chemostats so that phosphate utilization was limiting to growth. The only source of phosphate was β-glycerophosphate, which is hydrolysed by the phosphatase to inorganic phosphate. The optimal pH for this reaction is 4.2; by maintaining the culture medium at pH 6, phosphatase activity is made limiting to growth, and thereby exposed to intense selection. The experiment began with a single clone of brewer's yeast, *Saccharomyces cerevisiae*, with a population size of about 10^9 individuals. It grew slowly for about 180 generations, when there was a sudden increase in population density. This was attributable to more efficient assimilation of orthophosphate, presumably as the result of an improved permease. A second mutant appeared after about 400 generations. This was an altered phosphatase, with greater specific activity and a higher pH optimum than the original enzyme. A third mutant, appearing after about 800 generations, caused the cells to clump, and spread because clumped cells tend to settle and are thereby less likely to be washed out of the chemostat. Thus, the population had become well-adapted to utilizing organic phosphate under chemostat conditions within 1000 generations, as the result of a sequence of three independent genetic changes, affecting in turn the assimilation of inorganic phosphate, the hydrolysis of organic phosphate, and the retention of cells within the culture. In a second experiment (Francis and Hansche 1973), using the same strain as a base population, the first change affected the phosphatase. The mutant enzyme was more active than that evolving in the first experiment, but crossing the two evolved strains suggested that the mutation had occurred in the same gene. A second change again caused clumping and enhanced cell retention within the chemostat. In this case, independent lines evolved similar phenotypes through different mutations, but these mutations are likely to be variations on a common theme (Figure 3.42).

3.7.8 The pattern of adaptation

When bacteria are exposed to an environment so novel that no pre-existing system is adapted to deal with it, they can adapt only through the modification of enzymes that originally evolved in response to quite different selective regimes. However, this will rarely solve the problem. Some enzyme may possess minimal activity towards an exotic substance, but it would be a striking coincidence if it were regulated so as to be induced by the same substance. Even if such an enzyme exists, therefore, it will not allow growth on the novel substrate, because it will not be expressed. And even if it were expressed, its activity is likely to be too low to support growth, because enzyme specificity is an evolved character, and no enzyme is likely to be specifically active towards a substrate that the population has not encountered before. A population that is forced to depend on a novel substrate therefore encounters three formidable difficulties: no enzyme has the appropriate structure; no enzyme is appropriately regulated; and weakly-active enzymes, if expressed at all, are expressed too feebly to be effective.

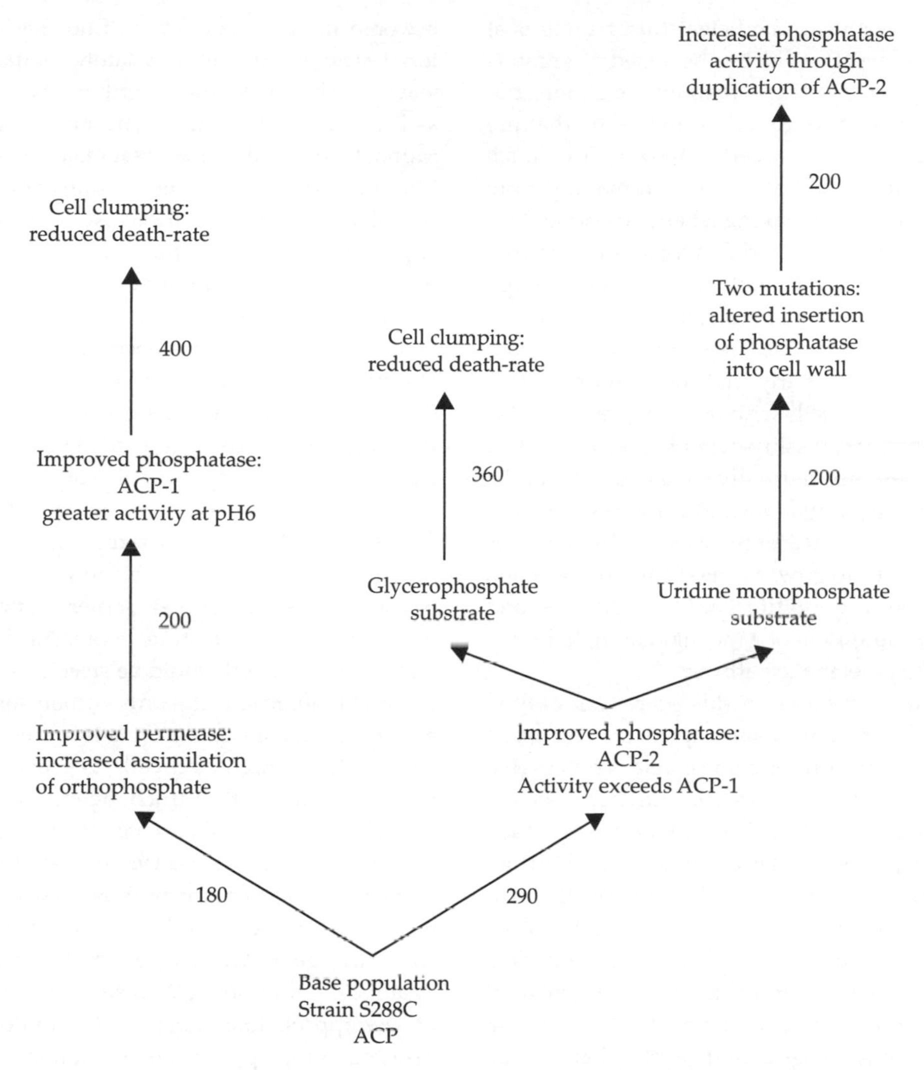

Figure 3.42 Experimental evolution of phosphate metabolism in yeast, from the studies by Francis and Hansche (1972, 1973) and Hansche (1975). The numbers by the lines are the number of generations between events.

Let us suppose that some enzyme exists that is able to metabolize the novel growth substrate slowly and inefficiently. Given the limited specificity of enzymes, this is not unreasonable; it will not be true for all substrates or for all bacterial strains, but it will often be true. The enzyme is not induced by the substrate, and before an appropriate regulatory system could evolve the population would be extinct. It is quite likely, however, that constitutive mutants in which transcription occurs whether or not the substrate of the gene product is present will be fairly frequent. For example, mutations that alter a regulatory locus so that it is no longer able to produce the protein that normally represses transcription of the structural loci are simply loss-of-function mutations, deleterious in normal circumstances, that are likely to occur quite often, and that will normally be present in large populations. We have seen that such mutations often provide the first large adaptive step in media containing easily metabolized substrates such as glucose or lactose. The gene is now expressed; but the enzyme is still so inefficient that growth is at best very slow; it could be increased if enzyme structure were modified through selection, but this requires gain-of-function mutations that are almost certain to be exceedingly rare. It is much more likely that selection will favour making more enzyme, rather than making a better enzyme. This could be achieved by modifications to the promotor, that controls the rate of transcription through RNA polymerase; much simpler, however, is simply to duplicate the gene. Gene duplication occurs spontaneously at a fairly high rate, and once two copies are present cells with many copies will arise through unequal crossing-over or some equivalent process. These are normally unstable, but can be retained in the population, and perhaps eventually stabilized, given sufficiently strong selection. This will enable cells to grow successfully, and selection can now begin to modify enzyme structure, and eventually the system of gene regulation, to form a new, exapted metabolic pathway.

There are four stages in this process of exaptation. The first is simply incomplete specificity, so that some pre-existing enzyme is able to process the new substrate, however inefficiently. The second is deregulation, so that the enzyme is expressed constitutively. The third is duplication, and further amplification, so that the total activity of the inefficient enzyme is adequate for growth. The final stage is the normal process of continued selection, causing cumulative improvement in the structural gene and its regulatory system. We can now ask whether the physiological and genetic changes that have been discovered in evolving microcosms correspond to this general framework.

3.7.9 Evolution of metabolic pathways

The importance of the initial breakdown of gene regulation to permit selection to modify inefficient enzymes has been very clearly established for the utilization of exotic five-carbon sugars by *Klebsiella* (*Aerobacter*) *aerogenes* (Lerner *et al.* 1964, Wu *et al.* 1968, Mortlock and Wood 1964, Mortlock *et al.* 1965, Wu 1976, Lin *et al.* 1976; see reviews in Mortlock 1984a, b). Some of these sugars, such as D-ribose or D-xylose, are relatively common in nature, and it would not be surprising to find that many bacterial strains have appropriately-regulated pathways for their catabolism. Others, however, such as D-arabinose and xylitol, are rare in nature and cannot be utilized by the great majority of bacteria. Mortlock's strategy was use a strain of *Klebsiella* that was able to metabolize these exotic sugars, after a lag indicating that selection was first necessary, to investigate how this ability evolves.

For D-arabinose to enter the main pathway of pentose metabolism, it is necessary to convert it to D-ribulose by an isomerase. The base populations do not possess any such enzyme, but L-fucose isomerase has a low activity towards D-arabinose. This enzyme is induced by its usual substrate, L-fucose, not by D-arabinose. Cultures are therefore unable to grow when D-arabinose is the sole carbon source present. Mutants that are able to grow turn out to be constitutive for L-fucose isomerase, producing it whether or not its substrate is present. This is normally wasteful, and would be selected against, but when D-arabinose is the only carbon source available, constitutive expression is favoured because it permits the deregulated cells to grow. The structural enzyme itself is unchanged. Further selection yields types with greater rates of growth on D-arabinose. These produce an altered isomerase. The course of events was thus, first a change in the regulatory gene, and then modification of the structural gene. Once viable cultures have been established, it is possible to select for inducibility by the appropriate substrate (D-ribulose in this case), restoring appropriate regulation.

Such evolved strains catabolize D-arabinose by constructing a new metabolic pathway from elements of two others. The first enzyme in the pathway is an isomerase borrowed from the L-fucose pathway that converts D-arabinose to D-ribulose. The second is a kinase from the pentose pathway that phosphorylates D-ribulose, so that it subsequently follows the normal course of pentose metabolism. It is in this way possible to acquire a new metabolic capacity by modifying pre-existing metabolic machinery.

Like D-arabinose, xylitol is not normally metabolized, and wild-type strains do not possess any enzyme specifically adapted to deal with it.

However, the ribitol dehydrogenase that converts ribitol to D-ribulose in the pentose pathway also has weak activity towards xylitol, converting it into D-xylulose. This can be phosphorylated by kinases of the pentose pathway, and proceeds through normal pentose metabolism. Mutants that are constitutive for ribitol dehydrogenase are therefore able to grow on xylitol. In chemostat culture, the rate of xylitol metabolism by constitutive strains often increases by a factor of 4 or so within 50 or 100 generations, and then after a time increases again, until specific activity may be 20 times as high as in the parental strain. The enzyme itself is structurally unchanged. The first advance seems to involve promotor mutants; the second is attributable to an increase in the number of copies of the gene. The result is massive overproduction of an unaltered ribitol dehydrogenase, which in evolved strains may have increased from less than 1% to nearly 20% of total cell protein. Under permissive conditions this is of course highly disadvantageous, and ribitol dehydrogenase activity falls rapidly, because supernumerary copies of a gene tend to be unstably inherited, and are quickly segregated out when they are no longer actively maintained by selection. The ability to grow on the novel substrate is, then, solely the consequence of changes in the regulation of gene activity: constitutive expression, promotor activity and the number of copies of the structural gene.

The exaptation of the fucose system for the utilization of propanediol (§3.4.9) can provide a basis for further modifications that extend the metabolic range of the strains. Most of these modifications involved changes in gene regulation. In some cases, little modification is required. If the membrane facilitator, the oxidoreductase and the lactaldehyde dehydrogenase are all expressed constitutively, ethylene glycol can be used as a substrate. It is converted first to glycoaldehyde and then to glycolate, which enters the gloxalate pathway. Xylitol can also be used, provided that it can enter the cell. Propanediol-using strains that are constitutive for the xylose pathway can use xylitol as the sole carbon and energy source. Xylitol is brought into the cell by the xylose permease, and there converted to D-xylose because the novel oxidoreductase can act as a xylitol dehydrogenase. The xylose is then processed in the normal way.

Variants that grow on D-arabitol arise after mutagenesis from propanediol-utilizing strains, but not from normal strains. In such variants, the propanediol membrane facilitator is constitutively expressed, and admits D-arabitol into the cell. Once inside, however, it can be used only if a third change occurs to convert it into a metabolic intermediate. This involves the constitutive expression of a dehydrogenase that has a high affinity for galactose, but is also capable of converting D-arabitol into D-xylulose, an intermediate in the xylose pathway. In this case, therefore, the use of a novel substrate depends on the prior acquisition of an unrelated mutation, permitting growth on propanediol, and then on regulatory changes in two other systems, that by switching them on permanently allow their specialized enzymes to process, rather inefficiently, the new substance.

The initial activity of ribitol dehydrogenase towards xylitol is rather low, and under the appropriate conditions can be limiting to growth. It is therefore possible to select for mutants with greater specific activity that have structurally modified dehydrogenases. At least one such mutant, when cultured at very low concentrations of xylitol, evolved a higher growth rate because it was able to acquire xylitol at a greater rate from the medium. This was because the ribitol transport system, normally induced by ribitol, had become constitutively expressed, and xylitol uptake increased as a consequence. In this case, then, we can trace the evolution of high growth-rates on an exotic carbon source through the constitutive expression of a catabolic enzyme, the capture of a pre-existing metabolic pathway, the structural modification of the enzyme, and the constitutive expression of a transport system (Figure 3.43).

These experiments were carried further by growing *Klebsiella* in xylitol chemostats (see Burleigh *et al.* 1974, Rigby *et al.* 1974, Hartley *et al.* 1976, Inderlied and Mortlock 1977 and Thompson and Krawiec 1983, reviewed by Hartley 1984). Cultures that were initiated either with strains that carried several copies of the ribitol dehydrogenase gene, or with strains bearing the modified

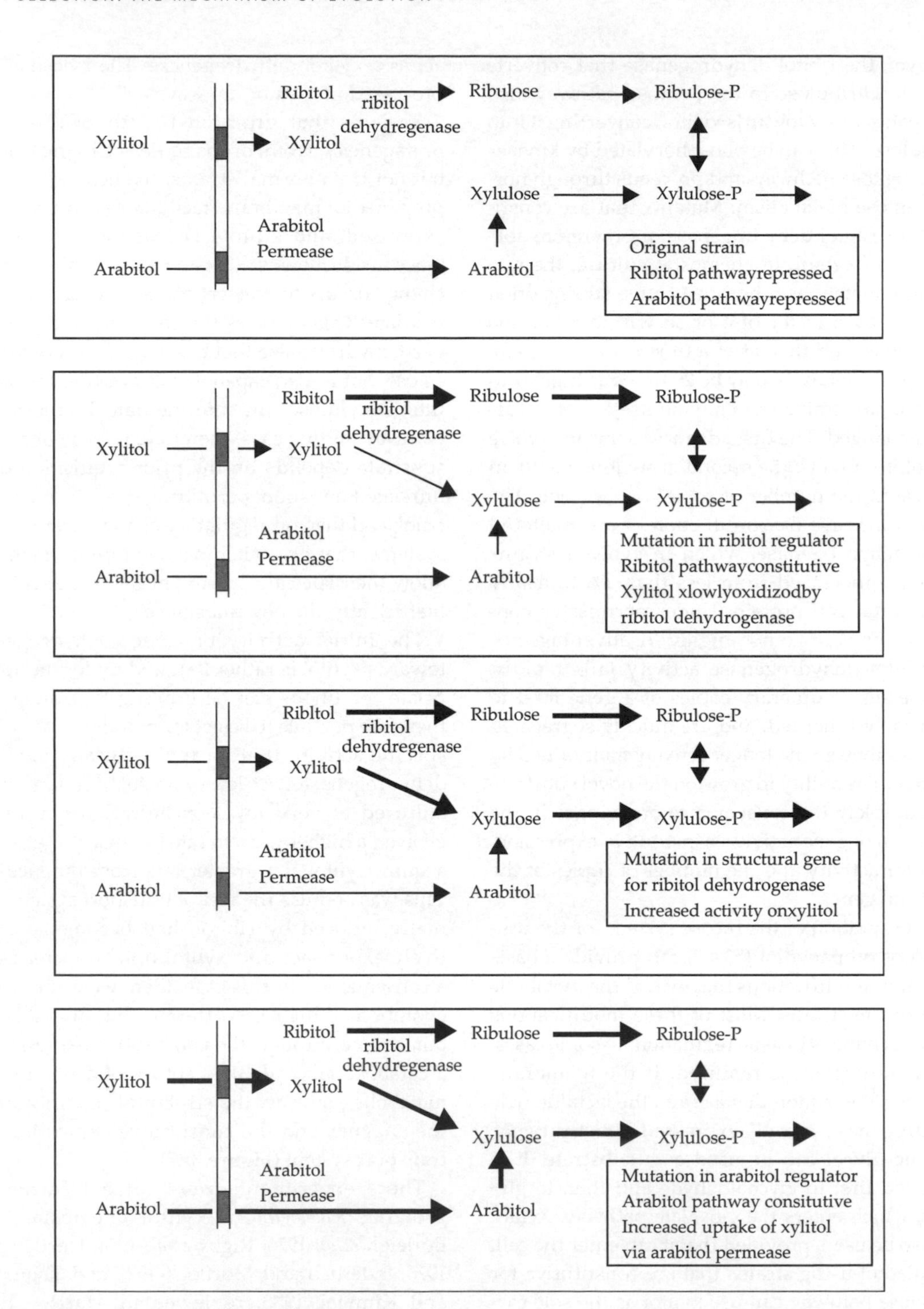

Figure 3.43 Experimental evolution of xylitol metabolism.

gene, were usually taken over by mutants with yet greater production of the same enzyme. However, when the cultures were exposed to rather severe mutagenesis, new strains with altered enzymes appeared. Most of these alterations appeared to involve single amino-acid substitutions that accumulated in a stepwise manner. Many different point mutations gave rise to strains with increased specific activity towards xylitol, leading to many different evolutionary outcomes. The genealogy of strains selected for growth on xylitol in several experiments is shown in Figure 3.44. The number of generations and the use of various genetic manipulations are shown alongside the branches: u signifies ultraviolet mutagenesis, n chemical mutagenesis with nitrosoguanidine, and p phage transduction of the *Klebsiella* ribitol and arabitol operons into *E. coli.* The ribulose dehydrogenase activity per unit total cell protein (RDH), the specific activity of the enzyme towards xylitol (SA), and the activity ratio (AR, activity on xylitol as a fraction of activity on ribulose) of the evolved strains were measured in different experiments. The uppermost tree shows the results of isolating strains from a xylitol-limited chemostat, after a drop in effluent xylitol concentration had signalled the passage of an improved mutant. These strains, evolving spontaneously or with mild ultraviolet mutagenesis,

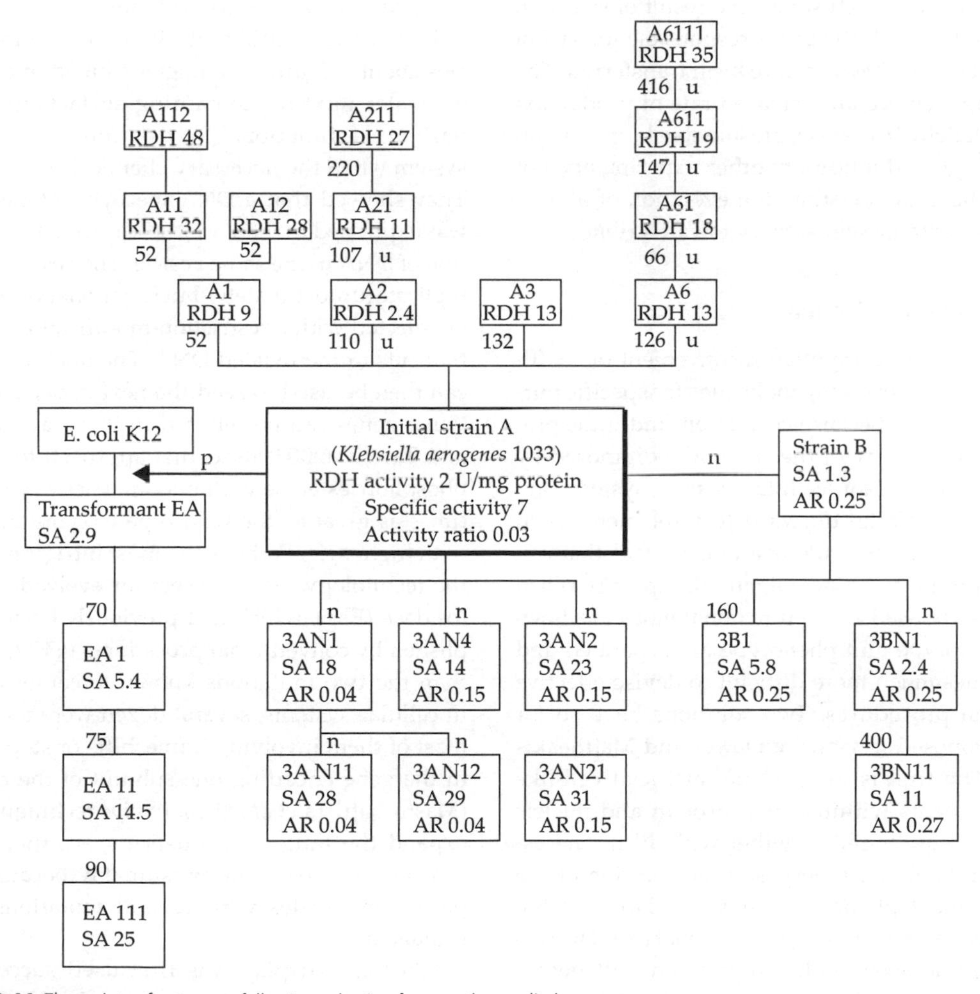

Figure 3.44 The variety of outcomes following selection for growth on xylitol

produce large quantities of ribitol dehydrogenase, which increases from 2% of total cell soluble protein to 30% or more. In some cases (such as strain A1), this increase in production seemed to be caused by more efficient transcription, but in most cases (such as strain A11) it was caused by gene duplication. The enzyme itself seemed to be altered little, if at all, having nearly the same K_m in all strains. Chemical mutagenesis resulted in strains with altered enzymes, as shown in the two genealogies on the lower right. These may have undergone several simultaneous changes in amino acid sequence; however, culturing such mutants often leads to the spontaneous appearance of strains such as 3B1 with even more active enzymes. The tree on the lower left shows the result of selection in *E. coli* to which the genes responsible for xylitol metabolism in *Klebsiella* have been transferred. The EA strains evolve an increased rate of production of ribitol dehydrogenase, presumably largely again through gene duplication; other experiments, not shown here, demonstrate the evolution of altered enzymes with greater activity towards ribitol.

3.7.10 *In vitro* selection

Cellular systems are often inconvenient or inefficient to use for evolving molecules for specific purposes, not least because cells often find unhelpful ways of dealing with the constraints imposed by experimenters. Non-cellular *in vitro* systems are in principle straightforward to implement when the functional molecule is a nucleic acid that can both replicate and bind a ligand (Chapter 1). When proteins are used as the functional molecule, however, genotype and phenotype are separated and it becomes much more difficult to devise effective selection procedures. Two solutions have so far been proposed (review by Dower and Mattheakis 2002). The first is to construct artificial cell-like compartments within which protein and nucleic acid are sequestered, together with all the necessary synthetic machinery, so that selection of the compartment phenotype allows isolation of the DNA sequences encoding this phenotype. The second is to do away with compartments altogether and instead to maintain the physical link between mRNA, ribosome, and nascent protein, usually by removing stop codons to make the ribosome stall. Mass selection of these complexes will then recover successful RNA sequences that can be reverse-transcribed into DNA. *In vitro* selection by compartmentalization has been used to select enzymes and *in vitro* selection by display to select binding proteins. It provides rather dramatic evidence that extensive modifications of protein structure involving several functional changes can be accomplished rapidly and repeatably in large populations. I describe here only a few of the major advances in the field; in the near future, they might occupy a large fraction of the text of books like this. Bull and Wichman (2001) give an elementary introduction to this and other aspects of applied evolution.

Tawfik and Griffiths (1998) discovered that vesicles about 2–3 μm in diameter formed in a stirred oil–water mixture containing surfactants would harbour a functional translation–transcription system when the necessary chemicals were added. They showed that a DNA methyltranferase gene was expressed in these vesicles, leading to methylation of DNA in the same vesicle. The vesicles do not replicate (unfortunately), but functional vesicles can be selected with a restriction-modification enzyme that cuts unmethylated DNA. The uncleaved DNA can then be used to seed the next cycle, and functional genes can be selected in this way. Griffiths and Tawfik (2003) used this approach to select a phosphotriesterase with maximal activity up to 60 times as great as the wild-type enzyme, although with higher K_m. Perhaps the most intriguing use of the technology was to select for evolved β-galactosidase (Ebg), which had previously been accomplished by conventional procedures (§3.7.4). Rather than the two mutations known to confer activity in cellular systems, several dozen were identified, most of them involving frameshifts or stop codons in the gene encoding one subunit of the enzyme (Mastrobattista *et al.* 2005). *In vivo* techniques thus expand the range of sequence space that can be screened by selection, presumably because they permit molecules with several mutations to be evaluated.

Ribosome display was first used successfully in a completely cell-free system by Hanes and

Plückthon (1997; see Mattheakis *et al.* 1994), who amplified a DNA library of antibody molecules by PCR and translated this into mRNA. The ribosome–mRNA–protein complexes were then washed over immobilized antigen to select the most tightly bound complexes, which were then dissociated to allow removal of the RNA and reverse transcription into DNA to close the cycle. After 5 cycles of selection the surviving sequences were identical with wild-type molecules or differed by 1–4 amino acids. Some showed greatly reduced activity (probably from mutation in the final PCR amplification) but most had more than 50% of wild-type activity and two molecules were equivalent to wild type or better. Hanes *et al.* (2000) used this technology to select molecules capable of binding insulin up to 15 times more strongly than the wild type. An even simpler technique is to bind the mRNA directly to its protein using puromycin as a linker (Roberts and Szostak 1997). Keefe and Szostak (2001) prepared a random library of 6×10^{12} 80-amino-acid proteins and selected for binding to ATP. After 8 cycles the frequency of ATP-binding proteins had risen from 0.001 to 0.06, with successful proteins falling into 4 unrelated classes—the main 'themes' of this phase of the experiment—none of which resembled any known native protein. After mutagenesis the frequency increased over the next 10 rounds to 0.34, and by this point all successful molecules belonged to a single family, one of the 4 identified earlier. Most of the variants on this surviving theme were characterized by the same 4 amino acid substitutions, with another dozen or so substitutions occurring at lower frequency; the crucial structure seemed to be a pair of cysteine residues able to facilitate the binding of zinc. Keefe and Szostak argues that only about 10% of the sequences initially present would fold correctly and thus be candidates for the first round of amplification, so that the 4 families of protein able to bind ATP that were recovered imply that the frequency of random 80-amino-acid proteins with substantial ATP-binding ability is about 10^{-11}, which is comparable to the frequency for ATP-binding RNAs (Wilson and Szostak 1999). So far as I am aware, this is the only empirical estimate of the functional density of the protein matrix. It is probably an underestimate, partly because neutral variants on the four themes were not catalogued, but mostly because the stochastic loss of sequences in early cycles of selection is probably underestimated.

3.7.11 Genetic changes during adaptation

According to the themes-and-variations notion there are few starting-points for an adaptive walk, and the examples that I have given above suggest that the same notion may apply to the walk as a whole, with replicate lines evolving in parallel through mutations in the same suite of genes. This is not necessarily the case when functional interaction between genes reduces the connectance of the allele matrix. In this case the first step will constrain the future course of the walk, because the second step must involve a gene with alleles that can interact favourably with the allele fixed at the first locus. This constraint applies equally to every subsequent step, so that adaptive walks progressively diverge towards end points involving quite different suites of genes. This would be an historical process of change in which each step is contingent on the previous step. Contingency implies that the products of evolution become steadily less predictable through time, so that the outcome of long-term evolution is essentially unpredictable and any particular realization of evolution represents only one of a large number of possible outcomes. This would clearly have a large effect on how we view the world in which we live.

Selection for 20 000 generations in glucose-limited medium was associated with shifts in the expression of 59 genes in two *E. coli* lines (Cooper *et al.* 2003). The same set of genes was affected in both lines and in every case expression had shifted in the same direction, a remarkable example of parallel evolution. Cooper *et al.* reasoned that expression was regulated primarily by guanosine tetraphosphate, whose concentration is controlled by the proteins encoded by *relA* and *spoT*, and by the cAMP receptor protein, controlled by the products of *cyaA* and *crp*. They sequenced these four candidate loci, found a point mutation in *spoT* in one of the lines and showed by allelic replacement that it was responsible for about 15% of the accrued gain in fitness. This mutation was not present in the other line, however. The two lines had thus evolved similar levels of fitness through similar changes in gene expression produced by different mutations in regulatory systems.

The adaptive walk of bacteria grown on exotic substrates seems to follow a predictable succession of events involving exaptation, deregulation, amplification, and modification (EDAM). Within these broad categories there is a great deal of functional variation from case to case, however, and it is seldom clear whether this reflects the evolution of a few characteristic genotypes involving strong genetic interactions or merely a large number of mutations with similar and independent effects on performance. To follow each step in the adaptive walk requires sequencing the whole genome, or at least the genes known to be solely responsible for adaptation, in a number of replicate populations. At the time of writing this is about to become technically feasible for cellular organisms through re-sequencing techniques, and in virus experiments the first results have already been published.

A single lineage of phage φX174 typically adapts to high temperature through mutations in 10–20 nucleotides of the 5400 in its genome. Replicate lines evolve similar levels of fitness through a large number of possible beneficial mutations, some of which are unique to a particular line whereas others recur in two or more lines. Pairs of replicate lines selected at high temperature on the same host species shared on average 20% of their beneficial mutations, and about 50% of all mutations were found in two or more of five replicate lines (Bull *et al.* 1997). When two replicate lines of φX174 were cultured on a novel host bacterium at high temperature 22 mutations were fixed of which 7 occurred in both (Wichman *et al.* 1999). These shared mutations were substituted in a completely different order in the two populations, however, suggesting that they act independently. By contrast, adaptation of phage G4 to high temperature consistently involved the same point mutations substituted in the same order, and go-back experiments in which stored intermediate sequences were re-run confirmed the repeatability of mutational order (Holder and Bull 2001). This does not necessarily imply strong genetic interactions, because beneficial mutations will tend to be fixed in more or less the same order if they vary in mutation rate or fitness. Mutations that typically appeared late in adaptation usually spread when they were introduced into the ancestral population, showing that their beneficial effect was not wholly conditional on the prior substitution of other mutations. Genetic interactions were much more clearly evident in the RNA virus VSV, where some pairs of point mutations were almost invariably found together in evolved lines (Cuevas *et al.* 2002). An even more extreme result was obtained by selecting the ribozyme L-20, a 400-nucleotide molecule derived from a self-excising group I intron, for the ability to cleave DNA (Hanczyc and Dorit 2000). Four replicate lines evolved a roughly 100-fold greater activity, whereas a fifth showed about half this improvement. A total of 25 beneficial mutations were fixed by the 5 lines, but 7 were fixed in all the 4 more successful lines but not in the fifth, whereas 9 were fixed in the less successful line but in none of the others. There was thus a clear dichotomy between two quantitatively different outcomes arising from two qualitatively different combinations of genes.

If all mutable sites are considered then the conclusion is plain: the mutations implicated in any episode of adaptation are always a small unrepresentative sample of possible mutations, even when many replicate lines are screened. This can be illustrated by re-sampling the data and calculating how the number of mutations detected rises as more lines are screened (Table 3.1). This soon shows diminishing returns because the same mutations are discovered repeatedly. For the ribozyme the fraction of point mutations that are beneficial is $26/(3 \times 414) = 0.021$. For viruses the estimates are much lower: 0.00093 for VSV and 0.0034 for φX174. In this sense, parallel evolution will always be found, but it remains possible that beneficial mutations have independent effects and are consequently uncorrelated, so that replicate lines do not evolve a restricted suite of combinations of mutations. To evaluate this possibility we confine our attention to those sites at which a mutation occurred in at least one line and construct the line × mutation matrix, in which each replicate line is a row and each mutable site is a column. The entry in each cell tells whether or not a mutation has occurred at that site in that line, $X_i = 0,1$. For any pair of mutations over all l lines, both mutations occur at n_{11} sites, the first but not the second at n_{10} sites, the second but not the first at n_{01} sites, and neither occur at n_{00} sites. The binary covariance is then $\mathrm{Cov}(X_i,X_j) = (n_{11}n_{00} - n_{10}n_{01})/l(l-1)$, which can be expressed in a standardized form as a correlation coefficient. If genetic interactions lead to distinctive combinations of mutations in evolved lines, any two mutations are likely to be more similar or more different than expected, so that there will be an unexpectedly large number of extreme correlations. This can be expressed by calculating the standard deviation of all pairwise correlations $s(r)$ and comparing this to randomized data (Bell 2005). A related procedure is to calculate the ratio of two variances: the variance of the number of mutations per line m and the sum of the variances $\mathrm{Var}(X_i)$ of mutations among lines. The two are related by: $\mathrm{Var}(m) = \Sigma\mathrm{Var}(X_i) + 2\,\Sigma\mathrm{Cov}(X_i,X_j)$, with expected value $V = 1$ (Piclou and Robson 1972). Estimates for the virus experiments are given in Table 3.2. In each case the distribution of correlations between mutations is broader than expected, especially for ribozyme L-20. This quantifies the visual impression that the evolved genotypes comprise relatively few combinations of mutations, presumably because of genetic interactions. These reduce the connectivity of the allele and protein matrices and create isolated paths leading to different evolutionarily stable outcomes. This is clearly true for the smallest molecule, ribozyme L-20, where adaptation is achieved largely through two non-overlapping sets of beneficial mutations. It applies to some pairwise combinations of mutations in VSV, but the effect is statistical rather than visual in φX174. It is conceivable that the matrices of larger genomes are more fully connected because their functional redundancy reduces the strength of genetic interaction, in which case cellular organisms will usually evolve towards the same theme.

3.7.12 Repeated adaptation

When a population completes its adaptive walk it has not become permanently adapted, for conditions will sooner or later change again. In broad terms, we can expect three possible kinds of

Table 3.1 The asymptotic number of beneficial mutations initially available in three evolving systems. Estimated by fitting $M = M_{max}[1 - \exp(-k_l)]$ to resampled data with varying numbers of lines l

Molecule	Length (nt)	Beneficial mutations M	Asymptotic M_{max}	Prob of beneficial mutation
DNA virus φX174	5386	39	55	0.0034
RNA virus VSV	11161	25	31	0.00093
Ribozyme L-20	414	25	26	0.021

Table 3.2 Analysis of the line × mutation matrix for beneficial mutations in three experiments. Observed values of $s(r)$ and V are compared with the distribution obtained from 1000 unconstrained randomizations of each data set. All three observed values of $s(r)$ exceed the mean of randomized data ($P < 0.001$). For details of analytical method see Bell (2005). Data sets from Holder and Bull (2001), Cuevas *et al.* (2002) and Hanczyc and Dorit (2000)

	$s(r)$	Random $s(r)$	SD	V	Random V	SD
DNA virus φX174	0.510	0.437	0.025	1.65	1.02	0.35
RNA virus VSV	0.296	0.213	0.014	0.13	1.27	0.82
Ribozyme L-20	0.705	0.472	0.033	0.18	1.29	0.83

behaviour. If the environment changes only rarely, with long intervening periods during which conditions remain the same, each adaptive walk will be completed, and the population will be well adapted most of the time. If the environment changes frequently there will seldom be enough time for the optimal allele to be fixed and consequently the population will be maladapted most of the time in the course of an infinite walk. Finally, the environment may deteriorate so rapidly that selection cannot prevent a chronic decline in mean fitness and the population sooner or later becomes extinct.

Recurrent environmental change can readily be simulated by shifting the distribution of growth to the left at prescribed intervals. The rank of each allele is preserved relative to that of other alleles, but the rank of all is increased relative to the optimal type, and their fitness reduced. In some cases it may be reasonable to think in terms of a simple model involving a series of alleles with fixed effects on fitness. If a single environmental factor is gradually changing, for example, perhaps the fitness increments in the series of beneficial mutations will not change with time. More generally, each shift in conditions is likely to involve many factors, so that while the fitness rank of an allele relative to that of others remains the same its relative fitness changes. This is contrived by drawing a new series of random values for each fitness rank, while shifting the rank of each allele upwards by a prescribed amount, and thereby creating a new series of mutants whose fitness exceeds that of any pre-existing type in the changed conditions. This is a mutational landscape model in which the fitness of alleles is adjusted when conditions change, while new beneficial alleles are revealed either by environmental change or by the fixation of a new wild type.

Suppose that growth rates remain the same for a certain period of time E, at the end of which they are abruptly reduced by a quantity proportional to Δ_1. With this formulation we can decompose the overall rate of deterioration into two components, the frequency of deterioration $1/E$ and the severity of deterioration Δ_1. Beginning with a well-adapted population, mean fitness at first drops and then reaches a rough equilibrium where episodes of deterioration punctuate periods of recovery under selection. The long-term average value of fitness, taken over many such episodes, is lower when deterioration is more rapid, as anyone would expect. Deterioration presses the fitness of alleles backwards, but also increases the mutation supply rate by exposing a longer series of potentially beneficial mutations; at some point the supply of mutations is sufficiently high to permit selection to act at a rate which counteracts deterioration, and at this point the population comes into equilibrium with its environment. Somewhat less expected is that mean fitness depends only on the frequency of deterioration, and not on its severity (Figure 3.45). In any particular episode a more severe change will produce a steeper decline in mean fitness, but this is then more effectively restored by selection before the next episode. This is reminiscent of the rule that the mean fitness of a population undergoing deleterious mutation depends only on the rate of mutation and not on the effect of mutation on fitness. Mutation and deterioration are not equivalent processes, however: mutation affects only a few individuals, or affects many individuals differently, whereas all individuals simultaneously experience the same deterioration. Consequently, an increase in the rate of deterioration does not affect mean fitness in the same way as an increase in the rate of mutation (Figure 3.46). Fitness is related to $\exp(-U)$, but to $\ln(1/E)$: this is because mean fitness depends on the rate of recovery through selection, and the rhythm of selection is ln time. At very low values of U or $1/E$ mutation and deterioration have similar effects, because the population has few mutants, or has time enough to recover fully after an episode of deterioration. There is also little difference at very high values of around 1, because at this point the whole population is affected in every generation by mutation in the one case or deterioration in the other. For all intermediate values a given rate of deterioration is more damaging than an equivalent rate of mutation. The reason is evidently that deleterious mutation creates variation in relative fitness that can be acted on immediately by selection, whereas relative fitness is unaffected by deterioration, and recovers only when beneficial mutations have appeared and spread.

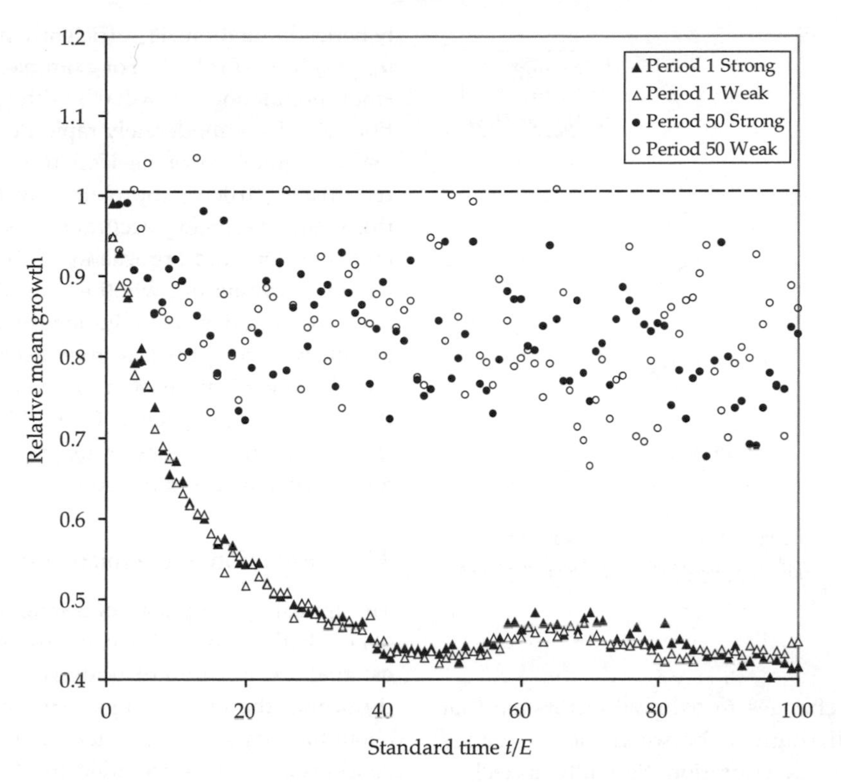

Figure 3.45 The effect of the frequency and severity of environmental change on adaptedness. Mutational landscape model with $\lambda = 100$; environmental change simulated by shifting back the wild type one rank after a specified period.

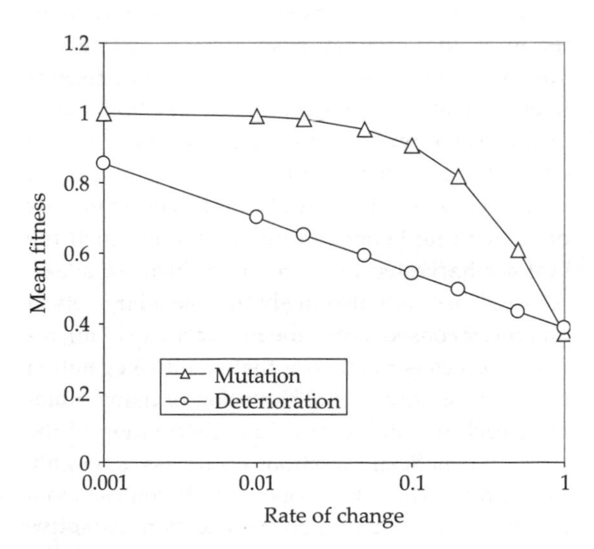

Figure 3.46 Effect of genetic and environmental deterioration on mean fitness.

Long-term mean fitness in a continually deteriorating environment can be expressed relative to a standard, the mean fitness of a comparable population whose environment does not change. Numerical simulation of a mutational landscape model suggests that $w_s = w_0 - b \ln(1/E)$, where $b \approx 0.1$ and w_0 depends on the mutation supply rate (Figure 3.47). This rule of thumb is not very sensitive to the details of the model: a simple (all alleles accessible) model and even a model with fitness increments of equal size yield similar results. It implies that populations are likely to be able to resist even rather rapid deterioration, especially if they are large enough to have a high mutation supply rate.

There are two main obstacles to relating these rules to real populations. The first is that the allelic series is seldom apparent, so that deterioration is

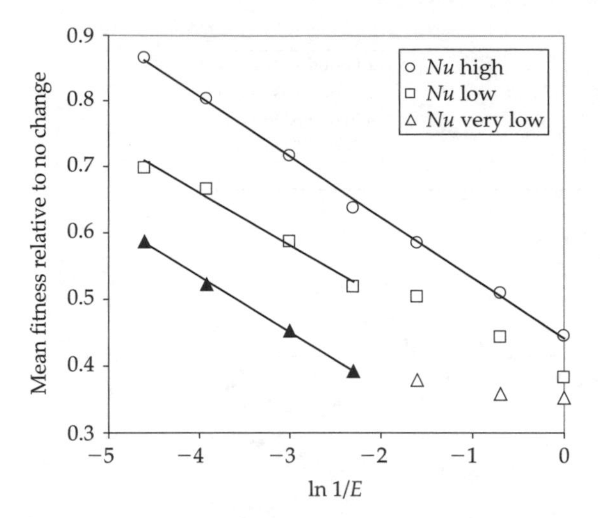

Figure 3.47 Effect of environmental change on mean fitness in relation to mutation supply rate. Based on a mutational landscape model with $\lambda = 100$.

perceived as changes in external factors, so that we cannot distinguish between the frequency and severity of deterioration. Secondly, a decline in mean fitness may reduce population size and thereby the mutation supply rate. Rapid deterioration may then force the growth rate of the population below the point at which it can fully replace itself. The lower mutation supply rate of a smaller population will further reduce the effectiveness of selection and thereby cause an accelerated decline in abundance, ending in extinction. Suppose that the environment deteriorates so that the resident type has rate of increase $r_0 < 0$ and consequently the population begins to decline from its carrying capacity of N_0 individuals. Before it becomes extinct, a total of $N_{tot} = -N_0/r_0$ individuals will have lived. The genomic rate of mutation is U, and a fraction φ of all mutations are beneficial, each with rate of increase $r_1 > 0$. When a beneficial mutation occurs, the probability that the lineage will spread is about $2s = 2(r_1 - r_0)$. The expected number of fixed beneficial mutations is then $B = 2N_0U\varphi(r_0 - r_1)/r_0$. With Poisson distribution, the probability that at least one beneficial mutation will spread is then $P = 1 - \exp(-B)$. The population will thus be rescued from extinction by adaptation if the frequency of beneficial mutations is sufficiently high, $\varphi > -[r_0/(r_0 - r_1)]\ln(1 - P)/2N_0U$. For example, for $P = ½$ in a small population ($N_0 = 1000$) with genomic mutation rate $U = 1$, moderately rapid decline ($r_0 = -0.1$) and moderately strong selection ($r_0 = -0.1$, $r_1 = 0.1$) requires $\varphi > 0.0002$, approximately. By simulation, this seems reasonably accurate for weakish decline and selection, and arguments of this sort may be useful in evaluating whether populations threatened by pollution, exploitation, or global change are likely to adapt before they become extinct. A more extended discussion of evolutionary rescue through sorting is given by Gomulkiewicz and Holt (1995), although in view of its potential utility the topic seems to have received rather little attention.

3.7.13 Evolution in the microcosm

The appealing simplicity of a small stoppered vial of cloudy fluid conceals a complex and subtle process that we still do not understand completely. I think that the main things that we have learned about the very simplest systems are as follows. The conceptual basis is provided by the protein matrix or its more formal descendant, the mutational landscape. Evolution occurs in one of two dynamic regimes according to the mutation supply rate M. In low-M conditions the population is genetically uniform or nearly so most of the time and adaptation is advanced by a succession of selective sweeps. The first beneficial mutation to be fixed is likely to have a large effect on fitness, and subsequent sweeps will involve mutations of decreasing effect. Most microbial populations live in high-M conditions where there is at first an abundant flux of contending beneficial mutations of which the best are harvested by selection to increase adaptedness. These are also likely to have a large effect on fitness; consequently, the first casualty of microcosm studies is the theory that adaptive evolution normally requires the substitution of many mutations each of small effect. The distribution of the effect of beneficial mutations on fitness is roughly exponential when they appear but becomes modal for those which are fixed by selection. Adaptive walks are short and often involve a predictable sequence of genetic changes. The first step is often a loss-of-function mutation causing constitutive expression of a gene, followed by increased expression and eventually structural change. Adaptation usually involves only a few major themes, corresponding to the major genes involved, albeit each with numerous variations, corresponding to allelic changes with nearly equivalent effects on function. The irreversibility of evolutionary change implies that the same ancestor exposed to the same conditions may give rise to a range of stable outcomes, although this has been convincingly demonstrated only for very small genomes. Adaptive change is cumulative but encounters diminishing returns as the supply of beneficial mutations shrinks. In many cases adaptation involves interactions between two or more genes but nevertheless evolves quite rapidly because quasi-neutral single mutants lurk at low frequency in the population.

CHAPTER 4

Prometheus Unbound: releasing the constraints on natural selection

Natural selection in the laboratory demonstrates the surprising ability of microbial populations to adapt to new conditions of growth, yet its operation is limited by the cramped confines of the microcosm. In an ideal world, any new adaptation would appear the moment the necessity for it arose and would spread without delay to all members of the population. In practice, the course of adaptation is impeded by the necessity for adaptation to await rare beneficial mutations that subsequently spread through the replacement of lineages. If most mutations are deleterious in well-adapted populations then the mutation rate should be reduced to a minimum, or, more precisely, to the point where the loss of fitness caused by the cost of repair just balances the gain associated with producing less heavily loaded offspring (see Liberman and Feldman 1986, Kondrashov 1995). The cost of repair is usually thought to be attributable to the metabolic burden of producing repair enzymes, but the extra time required for checking the fidelity of replication might also contribute. In any case, a minimal supply of beneficial mutations will be the primary constraint on adaptation, even in large populations. The simplest way of relaxing this constraint would to increase the mutation rate. The most straightforward but also the most expensive route would be to mount a constitutive genome-wide increase in mutation; it would be more economic to target particular loci for mutation, and most economic to increase mutation only at loci responsible for adaptation to a given stress, and only when that stress was experienced. It is not obvious, however, how elevated levels of mutation could be favoured by natural selection.

4.1 Increasing the mutation rate

4.1.1 Mutators

In fluctuating environments the generation of variation as a substrate for selection may over-ride other considerations (Leigh 1983). High rates of mutation can be caused very simply by loss-of-function alleles of the loci encoding polymerases or other components of the replication system (for a review of mutator genetics see Miller 1996). To the extent that the rate of evolution is limited by the rate of mutation, some degree of infidelity of replication might be favoured in the long term if it increases the rate of response to environmental change. Such selection is expected to be very weak (§2.2). Nevertheless, rather high rates of mutation may evolve in chemostat populations of bacteria. This is because mutator alleles, causing a high rate of errors during replication, can be autoselected in asexual populations through hitch-hiking (Taddei *et al.* 1997). Most of the mutations they induce will be deleterious; mutator frequency and mutation rate will therefore be low at equilibrium. In a novel environment, however, a mutator allele will occasionally cause a favourable mutation. As this spreads through the population, the mutator will spread with it, because in the absence of recombination the two are selected together. Once the new mutation has been fixed, the mutator frequency will decline, because more stable revertants that do not cause excess deleterious mutations will again be favoured. Mutator genes will continue to wax and wane in this fashion, until the population has reached evolutionary equilibrium, new favourable mutations become very unlikely, and mutator genes are almost always very rare. Thus, periodic selection (§3.4.2) implies the periodic spread of mutator genes, despite the fact that they reduce fitness in nearly all the individuals that bear them (for reviews see Giraud *et al.* 2001b, Denamur and Matic 2006).

The situation is quite different in sexual populations, where recombination separates mutator genes from the mutations they have caused. Selection is less effective in keeping mutation rates as low as possible, because it can act only through a fraction of the progeny of a mutant individual; however, it will be completely ineffective in causing any substantial increase in mutator frequency, because favourable mutations can be selected independently of the mutator genes that produced them. Mutation rates should therefore be somewhat higher on average in sexual populations, without displaying the extreme excursions that are liable to occur in asexual populations. I do not know whether this is in fact the case. Mutators are known from sexual organisms, however, and indeed were first described from *Drosophila*.

If the initial number of mutator cells N_m is rather small, so that $N_m u_m \ll 1$, only a small fraction of all possible mutations may occur in the population in every generation, and the mutator population is likely to dwindle to extinction before any beneficial mutation occurs. On the other hand, if $N_m u_m > 1$, almost all possible mutations will occur in every generation, including all possible beneficial mutations, and provided that the initial mutator inoculum is sufficiently large, it should tend to increase in frequency for as long as the population is not well-adapted to the chemostat environment (see discussion of dynamics by Tenaillon *et al.* 1999). Chao and Cox (1983) studied competition between isogenic wild-type *mut*$^+$ and mutator *mutT* strains in glucose-limited chemostats (see also Gibson *et al.* 1970, Nestmann and Hill 1973 and Cox and Gibson 1974). The *mutT* mutation increases the rate of AT → GC transversions by three or four orders of magnitude (see Cox 1976). Mixed populations in which mutator and wild-type strains were present in different proportions were made up from stationary batch cultures and used to inoculate the chemostat. There is an initial lag phase of about 60 generations, during which the frequency of *mutT* falls slightly but consistently. This represents the time taken for a beneficial mutation to arise and increase to appreciable frequency. The fall in the frequency of *mutT* during this period is caused by selection against the deleterious mutations that it induces and is necessarily linked to. If *mutT* is present at a sufficiently high frequency in the initial population, it is likely that the novel beneficial mutation will arise in a *mutT* cell, and that following the lag phase the *mutT* lineage will thereby spread. If *mutT* is very rare in the initial population, the increase in mutation rate that it causes is insufficient to counter the greater abundance of the *mut*$^+$ strain, the first beneficial mutation is more likely to arise in a *mut*$^+$ background, and consequently *mutT* is eliminated. This is what was observed: after the initial lag a mutator strain tends to spread when inoculated at low frequency into a chemostat occupied by an isogenic non-mutator strain. The main reason for believing that this is attributable to a higher mutation rate, rather than to some directly stimulating effect of the mutator gene product, is that replicate cultures often behave in quite different ways. The mutator gene may spread slowly or quickly, and sometimes fails to spread at all. The evolved populations may be changed in several ways: for example, they may adhere more readily to glass surfaces, reducing the rate at which they are washed out of the chemostat, or utilize novel carbon sources, or resist starvation more effectively; but different populations show different kinds of adaptation. Moreover, the superiority of the evolved population is usually retained even when the mutator gene is deleted. Thus, high mutation rates can evolve in chemostats because mutator alleles remain linked to the rare beneficial mutations that they cause.

Most mutators in natural isolates have defects in *mutS* and *mutL* of the mismatch repair system that increase both mutation and recombination rates. They have been found mostly in commensal and pathogenic bacteria, where they are selected during the recolonization of hosts (Giraud *et al.* 2001a). They also occur in long-term experimental lines, however: 3/12 of long-term *E. coli* lines had evolved high frequencies of mutators (Sniegowski *et al.* 1997). Roughly 1% of natural isolates of *E. coli* include mutators (1/408 from Gross and Siegel 1981; 4/110 from Jyssum 1960). They presumably spread when the population is challenged by a novel stress; thus, mutators are common among strains resistant to antibiotics, and indeed selecting for resistance to two antibiotics successively is an effective method for isolating mutator strains (Mao *et al.* 1997). Supposing that mutators have an indirect advantage (through the beneficial mutations they generate) of 10% when conditions change and a disadvantage of 1% once they reach a frequency of 99%, then they will be detectable at a frequency of 1% or more for about 1000 generations. This implies that natural conditions of growth change so as to evoke new adaptation about every 10^5 generations, which seems to imply a surprisingly low rate of environmental change.

4.1.2 Contingency loci

Mutators are more frequent in pathogens, where they can facilitate a faster co-evolutionary response to host defences. Loci whose products are displayed on the cell surface and which interact with host molecules, such as restriction-modification systems, pili, and iron-acquisition genes, are sometimes hypermutable (Bayliss *et al.* 2001). A high mutation rate is conferred by oligonucleotide repeats, similar to microsatellites, in the gene or its promoter (see Moxon *et al.* 2006). These *contingency loci* generate variation through slipped-strand mispairing, which causes a frameshift that switches the gene off (by introducing a stop codon within the open reading frame) or alters the transcript. The switching rate can be as high as 10^{-5}–10^{-2} per replication, exceeding that of a powerful mutator, without the concomitant cost of producing many deleterious mutations. Contingency loci have been interpreted as a specialized mutator system, possessed by about 10–100 loci in the genome, that reduces the apparency of a bacterium to the host immune system (Moxon *et al.* 1994, Field *et al.* 1999).

4.1.3 Mutation rate in stressful conditions

When bacteria are stressed they may de-repress error-prone DNA polymerases that have the effect of increasing mutation rates. The stressor may be a DNA-damaging agent such as UV light, but a general deterioration in conditions of growth in ageing cultures also induces the response (review by Foster 2005). Bjedov *et al.* (2003) found that the mutation rate to rifampicin resistance in *E. coli* increased from 6×10^{-9} in young colonies to 4×10^{-8} in ageing colonies experiencing stress through starvation and anoxia. Loewe *et al.* (2003) conducted a mutation-accumulation experiment using stationary phase cultures of E. coli and found a mutation rate about an order of magnitude greater than that reported for log phase growth by Kibota and Lynch (1996). A variety of chronic sublethal stresses were found to increase genetic variation in fitness in *Chlamydomonas* (Goho and Bell 2000). These responses are not Lamarckian, because there is no suggestion that the additional mutations are appropriately directed, but they would have the effect of increasing the mutation supply rate and thus the rate of adaptation in response to environmental deterioration. Mutation rates are usually estimated in benign laboratory conditions, and may be substantially greater in natural populations. Bjedov *et al.* (2003) make the remarkable observation that mutation rates measured in natural isolates of *E. coli* vary over an order of magnitude depending on the diet of the mammalian host: it seems to be stressful to live in the gut of a carnivore. An individual-based model of the evolution of protein structure in a fluctuating environment showed that more variable domains tended to be selected when the environment changed more often and more severely (Earl and Dee 2004), suggesting that 'evolvability' could be selected, at least in asexual populations (see Bell 2005).

4.1.4 Directed mutation

Combining contingency loci with stress-induced mutation raises the possibility that a particular stress might preferentially cause mutations at loci responsible for adaptation to that stress. Radman (1999) even proposed the existence of DNA mutases targeted to particular loci for highly error-prone replication, although a more reasonable interpretation is that the high rate of mutation is the cost of repairing lesions that the normal DNA polymerases cannot repair (Tenaillon *et al.* 2001). This would not necessarily be Lamarckian in the strictest sense, because deleterious as well as beneficial mutations might be generated, but it would be a marked departure from the conventional view of undirected mutation that would have the potential to accelerate adaptation. Cairns and Foster (1991; Foster and Cairns 1992; see Cairns *et al.* 1988) used a Lac⁻ strain unable to metabolize lactose, because of a chain-terminating mutation in the gene encoding β-galactosidase, the enzyme that hydrolyses lactose. Cultures of this strain were grown up in standard medium, which does not contain lactose, and then plated on to medium in which lactose was the sole source of carbon. The only colonies that could grow were those in which a mutation had restored β-galactosidase activity. These mutations might occur either during the preliminary growth of the cultures, or after they have been replated. There is good evidence that mutations occur at both times. However, mutations that restore Lac⁺ function are detected after replating only if the medium contains lactose, suggesting that the presence of lactose somehow causes the appearance of mutations enabling it to be used. These might be quite non-specific; it is conceivable that mutation rates generally increase when bacteria are stressed, and that the very small proportion of these that happen to restore the ability to hydrolyse lactose are then selected on lactose plates. It does not appear to explain Cairns's results, where the mutation to Lac⁺ seems to be specifically induced by lactose. The mutations responsible for these unusual results are small frameshifts generated by single nucleotide deletions in short mononucleotide repeats (Rosenberg *et al.* 1994). Mutations occur at many loci throughout the genome, however, rather than being specifically directed towards lac (Torkelson *et al.* 1997). Hall (1990) reported a similar experiment, in which loss-of-function mutations in the genes responsible for tryptophan synthesis reverted to their normal functional form in cultures starved for tryptophan (so that acquiring the ability to make it is strongly selected) much more often than in cultures where tryptophan was supplied in abundance (so that the ability to make tryptophan was only weakly selected, if at all). He suggested that nutritional stress induces some cells to enter a transient hypermutable state during which the beneficial mutations are generated. This is supported by the observation that unselected genes have high mutation rates among adaptive Lac⁺ revertants but not among Lac⁻ cells in the same lines (Rosenberg *et al.* 1994, Torkelson *et al.* 1997). Whether this is a general feature of metabolic evolution in bacteria is still unclear, as the *lac* allele used in the experiments is carried on an F plasmid and the process requires recombination, so it is possible that adaptive reversion occurs only or mainly on episomes, or on the particular episome used in the experiments (see Godoy *et al.* 2000). Regardless of the genetic details, however, it seems likely that bacterial populations adapt to novel conditions in ways that are not completely captured by simple models of fixed, uniform and undirected mutation. These are essentially Darwinian, so far as we know, but permit faster adaptation than would otherwise be possible. Similar mechanisms might occur in eukaryotes, for example in association with microsatellites or transposons, but their contribution to adaptive evolution has yet to be demonstrated (see Bridges 2001, Schmidt and Anderson 2006).

Although effectively elevated mutation rates will facilitate adaptation, there remain three constraints on the dynamics of evolutionary change that are imposed by the genetic and environmental context of selection:

- *Clonal interference.* Competition among contemporary beneficial mutations increases the passage time for single mutations.
- *Genetic interference.* Beneficial mutations may arise in combination with deleterious mutations, thereby increasing the passage time for compound mutants.
- *Incomplete connectance.* The stochastic fixation of beneficial mutations leads irreversibly to a state from which the optimal state cannot be reached through the selection of single mutations.

Natural conditions of growth may be more complex and stressful than glass and glucose, but they may also loosen these constraints. Three features of microbial life in the macrocosm are likely to quicken the pace of adaptation: horizontal transfer, sex, and dispersal. Each of these represents a different manifestation of the same general principle: infection.

4.2 Horizontal transmission

The single most important principle used to interpret the population genetics of asexual microbes in microcosms is that the entire genome is propagated as a single unit, so that adaptation is the outcome of the cumulation of genetic changes within a single lineage. One of the most striking discoveries of modern bacterial population biology, on the other hand, is that many bacterial genomes are mosaic, containing many genes that do not share a recent common ancestor. In the laboratory model *E. coli* 10–20% of the genome has not descended with modification from the ancestor of the remainder of the genome, but has instead been acquired from distantly related bacteria. Adaptation in natural populations often involves the acquisition rather than the modification of genes able to cope with new conditions of growth. There is a core genome governing processes such as ribosome assembly, transcription, and DNA metabolism that is transmitted from ancestor to descendant with little or no contamination, but such genes are rarely involved in specific adaptation. Genes responsible for ecologically sensitive functions such as transport, secondary metabolism and interaction with other species, on the other hand, are often located in regions that appear (from atypical G–C ratio, for example) to have been recently acquired from an unrelated lineage. This offers opportunities that are not available in microcosms and may leap the barriers set up by the paucity of mutations and the slowness of selection.

4.2.1 Gene cassettes

Natural environments may contain substantial quantities of DNA released by dying bacteria: about 1 μg/g in soil and 1–100 μg/L in seawater (cited by Thomas and Nielsen 2005). This can be taken up and integrated into the chromosome by prophage-like particles (*gene transfer agents*), where it may occasionally provide genetic variation that contributes to adaptation (Lang and Beatty 2000). It is difficult to assess the importance of this kind of macromutation and I do not know of any conclusive evidence that it contributes to adaptation. The *cassette–integron system*, on the other hand, is likely to provide an important source of new genes (reviewed by Rowe-Magnus and Mazel 2001, Michael *et al.* 2004). The cassette is a small circular molecule comprising an open reading frame (ORF) and a recognition sequence, the so-called 59-base element. The integron is a chromosomal region comprising the *intI* gene encoding an integrase mediating the insertion of the cassette at the recombination site *attI* downstream of a promoter. Once the ORF has become integrated into the bacterial chromosome it will be expressed and replicated and thus may immediately provide a new metabolic function to the cell. Gene cassettes were discovered through their participation in the evolution of resistance to antibiotics, but may bear genes encoding a broad range of catabolic functions (Figure 4.1). Michael *et al.* (2004) estimated that at least 2000 different cassettes could be found within a 50 m² plot of forest soil, so the repertoire

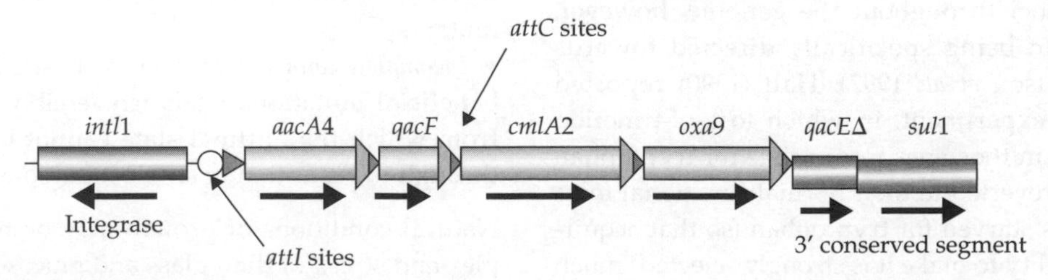

Figure 4.1 Structure of a mobile integron from *Vibrio cholerae*. The *aacA4*, *qacF*, *cmlA2*, and *oxa9* elements are resistance-gene cassettes. From Mazel (2006).

available from this source is equivalent to an entire bacterial genome. The recognition system, moreover, is quite permissive and allows many different ORFs to be harboured within the integron.

Gene cassettes accomplish the whole adaptive walk in a single step. Furthermore, they bear ORFs from a very wide range of bacteria and so the connectedness of the protein matrix becomes a moot point. The dynamics of the cassette–bacteria system have not yet been modelled, and will depend on the rate of uptake by cells, the rate of replenishment of the reservoir and the rate at which cassettes decay in the soil, none of which are known. The rapid spread of the 100 or so cassettes encoding antibiotic resistance to all parts of the world, however, is proof that they can be very effective in fuelling adaptive evolution. We have evaluated the effect of a small proportion of sterile soil water on the ability of *Pseudomonas* cultures to adapt to different carbon sources (Perron and Bell, unpublished). Soil water markedly stimulated growth, but in most cases this was a purely physiological effect. Soil from one site, however, caused a consistent increase in growth relative to unsupplemented controls over 20 serial transfers; this effect was diminished, although not completely effaced, by prior treatment with DNAse. This experiment raises the possibility that adaptation in natural populations of bacteria may involve processes that do not occur in most microcosms, although much more work will be needed to confirm this.

4.2.2 Conjugative plasmids

Gene cassettes can neither transfer themselves nor replicate. Other kinds of element encode their own transfer to another cell through conjugation (partial cell fusion) and then replicate in the recipient cell without necessarily becoming incorporated into the chromosome. This is a process of infection, by which genes can be transferred from one lineage to another and thereby spread through the population without depending on the replacement of lineages by competition. Whether or not infectious spread is responsible for the maintenance of conjugative plasmids in bacterial populations is discussed in §9.1; here I am concerned only with the consequences of plasmid carriage. The perfect plasmid would ensure conjugation and transmission whenever its bearer encountered an uninfected cell. If this happened after growth in every generation, the frequency of uninfected cells would fall from q to q^2 in each generation, and would therefore fall from q_0 to q_t in $t_1 = (1/\ln 2) \ln(\ln q_t/\ln q_0)$ generations. If the plasmid is introduced at low frequency, $q_0 = 10^{-5}$, then it will exceed $q_t = 0.999$ within 20 generations, so extreme transmission bias produces very rapid spread. In practice conjugation does not always occur and transfer is imperfect, so we can recognize that the overall probability v that an infected cell will meet an uninfected cell, conjugate, and transfer the plasmid will usually be much less than unity.

The specific transfer rate is usually estimated using the mass-action model of Levin *et al.* (1979), or one of its descendants, from the increase of transconjugants T in a population consisting of D donors (infected cells) and R recipients (uninfected cells), which is approximately $dT/dt = \gamma DR + rT$, where rT is the rate of growth of transconjugants and γ is a transfer coefficient. Estimates of γ range very broadly between 10^{-10} and 10^{-20} mL cell^{-1} h^{-1}. This is the appropriate parameter to describe the behaviour of bacterial populations with overlapping generations in continuous time, but it is not easy to grasp what these numbers mean in terms of evolutionary dynamics, which will be governed by the probability v that an infected cell will transfer the plasmid during its lifetime. If ΔT new transconjugants appear in a short interval of time corresponding to one bacterial generation, then $\gamma = \Delta T/DR$ and $v = \Delta T/D = \gamma R$ as a rough approximation when $T < D < R$. Licht *et al.* (1999) estimated $\gamma = 3 \times 10^{-11}$ for plasmid R1 in liquid cultures of *E. coli* containing about 10^{10} cells/mL, suggesting $v \approx 0.3$. In the more realistic environment of the mouse intestine, however, transfer was much slower with $v \approx 10^{-4}$. Sudarshana and Knudsen (1995) give 1.3×10^{-11} for R338 in liquid cultures of *Pseudomonas*, suggesting $v \approx 0.01$, whereas in soil microcosms estimates dropped to about 2×10^{-6}–5×10^{-3}. Bradley and Williams (1982) give direct estimates of 0.005–0.01 for the TOL plasmid of *Pseudomonas*. These figures are fairly representative and range so widely that no firm generalization is justified; the plasmid transfer rate is not a constant. Transfer is often faster early in growth, when the frequency of donors is low, and on solid surfaces such as agar and epithelial mucus. Later in growth and in viscous environments such as soil and gut contents, transfer seems to proceed very slowly. When observations are continued over several days, the plasmid is eventually almost fixed in the recipient population under favourable conditions such as chemostat growth.

In a discrete-generation model the recursion for the frequency of plasmid-bearing cells, ignoring any loss, is $p' = p(1 + vq)$, and the time required for fixation at given v, is approximately $t_v = (1/v)t_1$. It seems likely that $v \ll 1$ in most circumstances, in which case horizontal transmission will not greatly accelerate the fixation of a beneficial mutation of large effect. It may nevertheless affect the fate of a mutation in two respects. In the first place, if transfer rates are high in the initial stages of invasion plasmid carriage will boost the frequency of the mutation and thereby guard it against stochastic loss. This effect will be substantial only if $v > 0.1$ or so. Secondly, at each transfer the plasmid moves the beneficial mutation onto an arbitrary genetic background. Horizontal transmission thus screens the population for the best background, separating the beneficial mutation from any deleterious mutations with which it was initially associated, and reducing the level of genetic interference.

The most important consequence of plasmid transmission, however, is likely to be the rare acquisition of genetic material from very distantly related lineages. It might be very unlikely that a gene present in one bacterial lineage would have evolved through point mutation from an entirely dissimilar orthologue in a distantly related lineage, still less that it would evolve independently from a duplicated or redundant element. The very general nature of many of the mechanisms responsible for horizontal gene transfer, however, makes it possible for the gene to be transferred directly in a more or less fully functional state. This will happen only very rarely; but there is a world of difference between something that happens very rarely and something that never happens at all. In the case of chromosomal genes I used the concept of a *substitutional population* to define population size for bacteria: the set of individuals that will bear a beneficial allele once its spread is complete (§2.2.11). A conjugative plasmid is not necessarily limited to the substitutional population because it can be transferred to different kinds of bacteria living in different conditions. Hence for the plasmid, and any genes that it may convey, the appropriate population is the *conjugational population*, that is, the set of cells that can be infected by the plasmid. The conjugational population is not only potentially much larger than the substitutional population but is also phylogenetically diverse. Consequently, plasmid transfer makes it possible for optimal alleles to evolve despite a poorly connected protein matrix.

4.3 Sex

Plasmid transmission in bacteria is a rare event that results in a partial and asymmetrical transfer of genes between cells from distantly related lineages. Sex in eukaryotes is a frequent event that results in a complete and symmetrical transfer of genes between cells from closely related lineages. The nature of eukaryotic sex is discussed in §12.1. Here I shall take basic sexual biology for granted, defer any consideration of how sex itself evolves, and merely describe the distinctive features of purifying and directional selection in sexual populations. These rest on the two general consequences of sex, mixis, and the alternation of generations. *Mixis* is the mingling of genomes through gamete fusion and meiotic recombination. The *alternation of generations* is the shift from haploid to diploid state and back again as the consequence of sexual fusion followed by meiotic reduction in every sexual generation.

4.3.1 Dominance

Sex necessarily involves an alternation between haploid (gamete) and diploid (zygote) stages in the life cycle. I have assumed so far that growth and reproduction are confined to the haploid stage, which is the case in most algae and fungi. In many organisms, however, it is the diploid stage that grows and reproduces; such organisms are called *diplonts*. Since most plants and animals have life cycles of this sort, a great deal of evolutionary genetics has been developed for the special case of sexual diploids. Although this is understandable, it has had the unfortunate effect of introducing complications into the very simple arguments that apply to asexual haploid organisms, and thereby making it difficult for them to be presented clearly. The most serious complication concerns the interaction of allelic genes, recognized by Mendel and known as *dominance*. If the effects of genes are additive the fitness of a heterozygote will be the midpoint of the fitnesses of the homozygotes; otherwise the effect of one or the other allele will predominate.

There is a very simple reason for expecting most loss-of-function mutations that may be severely deleterious as homozygotes to have little effect on fitness in heterozygous combination with a normal functional allele. The rate of the reaction catalysed by an enzyme will rise asymptotically towards a maximum value, so it will not be increased much by additional synthesis of the enzyme beyond a certain level (Kacser and Burns 1973, Kacser 1988). Conversely, if the current rate of synthesis of the enzyme is sufficient to drive the reaction at somewhere near its maximal rate, a somewhat lower rate of synthesis will not decrease the rate of reaction very much. A heterozygote bearing a defective allele may synthesize the enzyme it encodes at little more than half the normal rate, but the reaction it catalyses may still proceed at a nearly normal rate, so loss-of-function mutations will tend to be recessive. This is consistent with the very slow loss of fitness in isolate cultures of asexual diploid organisms (§2.2). Direct estimates of the dominance of non-lethal deleterious mutations generally suggest h = 0.1–0.4 (Johnstone and Schoen 1995, Deng and Lynch 1996, Vassilieva *et al.* 2000).

One way of estimating the degree of dominance of severely defective genes that are lethal as homozygotes is to extract them from heterozygotes in samples from natural populations, and then test them in heterozygous combination with a normal gene. It has only recently become possible to extract single genes, and most of the information that we have comes from classical experiments with *Drosophila* in which individuals can be made homozygous for a particular chromosome through ingenious inbreeding procedures. Once a lethal chromosome has been identified, through the failure of homozygotes to appear in the progeny of a cross, it can be combined with a normal chromosome and its fitness measured. It is inevitably the fitness of the chromosome as a whole, rather than a single gene, which is scored in experiments of this kind, but in most cases the assumption that lethality is caused by a single gene of large effect is probably justified. The general result of experiments of this kind has been that heterozygotes for a lethal allele suffer only a minor handicap, amounting to a loss of 1–5% in viability. This is not unexpected. Natural populations will contain a biased sample of deleterious mutations: those that have the smallest effect in heterozygotes will accumulate at higher frequencies, and will be more likely to be sampled and tested. Arbitrary lethal mutations in yeast are usually almost completely recessive (Wloch *et al.* 2001), however.

4.3.2 Sorting in asexual diploid populations

In an asexual diplont a beneficial mutation will appear in a heterozygote whose fitness is $(1 + hs)$ relative to the fitness $(1 + s)$ of the homozygote. Completely dominant alleles, that direct the same phenotype regardless of their allelic partner, can be treated essentially as if they were haploid. The probability of fixation for a single new beneficial mutation is

$$\Phi = (1 - e^{-2hs})/(1 - e^{-4hsN}) \approx 2hs \ \{(N \text{ large})\},$$

and the passage time is the same as for a haplont with selection coefficient hs. It should be borne in mind that the mutation is 'fixed' in heterozygous state, and does not completely replace the previous wild-type allele until a second mutation occurs in the heterozygote lineage and spreads with a selection coefficient of $s(1 - h)/(1 + hs)$. More generally, the effect of diploidy on the rate of adaptation will depend on two opposed tendencies: the increased number of beneficial mutations (because of diploidy) and their reduced effect (because of dominance). The substitution time can be calculated by doubling the mutation supply rate and substituting hs for s: $t_{sub} = (1/4Nuhs) + (F/hs)$ where F is the ln frequency ratio used previously (§3.2). Consequently, diplonts will adapt more rapidly if $h > (Nu + 4F)/(2Nu + 4F)$, requiring $h > ½$ at least, which seems unlikely given the tendency for deleterious mutations to be recessive. Nevertheless, when Paquin and Adams (1983) counted beneficial mutations (detected by fluctuations in rare neutral markers) in haploid and diploid populations of yeast cultured in chemostats they found that the rate of fixation was 5.7×10^{-12} per cell per generation for the diploids and only 3.6×10^{-12} for the haploids. The diploid populations were thus evolving about 60% faster than the haploids, at least on a per-cell basis. This is consistent with the occurrence of partly dominant mutations at about twice the rate in diploids. However, it is offset by the greater abundance of the haploids: because they are smaller, they were about 40% more dense than the diploids. The overall rate of fixation per population was thus only slightly greater for the diploids.

4.3.3 Sorting in sexual diplonts

In sexual diplonts the situation is more complicated because each gamete carries only one allele, which can then be combined through sexual fusion with an allele from a different lineage. In an asexual diplont the appearance of a beneficial mutation would be followed simply by the vegetative propagation of the heterozygote, whereas in a sexual diplont each combination of alleles is broken up by meiotic segregation and restored by gamete fusion in every sexual generation. Consequently, the frequency of a diploid genotype AA among offspring f'_{AA} will not be directly related to its frequency among parents f_{AA} but rather to the frequency of the A allele among gametes f_A. If gamete fusion is unrestricted and the ratio of gamete frequencies at a locus with two alleles A, a is p:q then the ratio of the genotypes AA, Aa, aa in zygotes f'_{AA}: f'_{Aa}: f'_{aa} will be p^2:$2pq$:q^2, the well-known *Hardy–Weinberg ratio*. If fusion is restricted to gametes from the same individual then the ratio will be $(f_{AA} + ¼f_{Aa})$: f_{Aa}: $(f_{aa} + ¼f_{Aa})$. Thus, segregation and fusion generate a continuous supply of homozygotes, which is further increased by inbreeding. Consequently, recessive alleles concealed in heterozygotes may subsequently be exposed to selection by segregation into homozygote progeny. The initial stages of the spread of a beneficial recessive allele, or the final stages of the elimination of a deleterious recessive allele, will occur very slowly, depending as they do on the rare exposure of homozygotes to selection (Figure 4.2). Nevertheless, segregation will always increase the rate of response of rare recessive beneficial alleles to selection and thereby accelerate the initial stages of adaptation (see Kirkpatrick and Jenkins 1989).

Even aided by segregation, recessive alleles are fixed slowly; relative to the substitution time for a haploid t_{sub}(hap) a completely recessive mutation initially present at mutation-selection equilibrium will be fixed in $t_{sub}(\text{dip rec}) \approx t_{sub}(\text{hap}) + (su)^{-½}$ generations. Thus with $s = 0.1$ and $u = 10^{-6}$ the time to fixation will be about 3250 generations, almost 20 times as long as in a haplontic population. With increasing dominance this disparity grows less, and the passage time t_{sub}(dip dom) for a fully dominant allele is only a little more than for a haplont so long as it has not become too frequent: when it is very abundant selection is hindered because the few remaining recessives are sheltered in heterozygotes. For this reason, the time for complete substitution is minimal at $h = 0.5$, when $t_{sub}(\text{dip semidom}) = 2t_{sub}(\text{hap})$, and increases thereafter. Since we are primarily concerned with the establishment of a novel beneficial mutation, it is reasonable to calculate the passage time until it attains a frequency of 0.5, which is always a decreasing function of h. Alleles with very slight penetrance ($h < 0.001$) have nearly the same passage time as completely recessive alleles, whereas passage time is markedly reduced for any appreciable ($h > 0.01$) degree of dominance. There is no exact expression for arbitrary degrees of dominance, but numerically I find $\ln t_{sub}(\text{dip } h) \approx \ln t_{sub}(\text{dip dom}) - \ln h$ in the limit $p \to 0$. The convergence is very slow, and

$$\ln t_{sub}(\text{dip } h) \approx \ln t_{sub}(\text{dip dom}) - (1 + 2/\ln p_0) \ln h$$

is a fair approximation for reasonable values of initial frequency p_0. In short, incomplete dominance

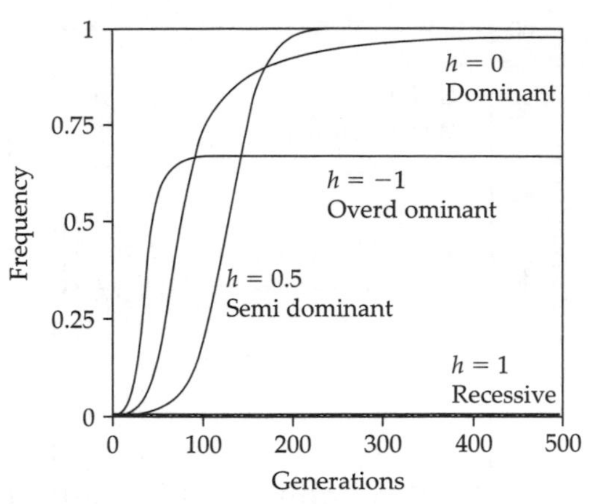

Figure 4.2 Selection in sexual diplonts. This diagram shows how dominance affects the spread of a rare beneficial mutation in a sexual diploid population. The fitnesses of the three genotypes at a locus with two alleles can be represented as: $w_{AA} = 1$; $w_{Aa} = 1 - hs$; $w_{aa} = 1 - s$. The degree of dominance is expressed by the value of h. When $h = 0$, Aa heterozygotes have the same fitness as AA homozygotes, i.e. the A allele is dominant. Conversely, when $h = 1$ the A allele is recessive. A value of $h = 0.5$ represents the situation when the A and a alleles combine additively, so that the heterozygote is intermediate between the two homozygotes. Values of h that fall outside the range 0,1 represent a genetic interaction that causes the heterozygotes to lie outside the range of the homozygotes; if $h < 0$ the heterozygote is fitter than either homozygote. The curves show the spread of the A allele from an initial frequency of 10^{-5} caused by selection with $s = 0.05$. Any new beneficial mutation occurs almost exclusively in heterozygotes so long as it is very rare. If it is recessive, selection is very weak, and its frequency increases only very slowly. If it is dominant, it is expressed in every individual it occurs in, and is selected as rapidly as it would be in a haploid population; when it becomes very frequent, however, selection does not eliminate the inferior allele as effectively (because at this stage it occurs mostly in heterozygotes, where it is not expressed) and the rate of spread of the superior allele falls off. A semidominant allele with $h=0.5$ is selected symmetrically, either allele being equally exposed to selection in homozygotes and heterozygotes, and a beneficial mutation is fixed quite rapidly. An overdominant allele with $h<0$ cannot spread beyond a certain frequency that represents a stable equilibrium. This is because the genotype with the greatest fitness, the heterozygote, cannot produce exclusively heterozygous progeny. This equilibrium frequency is in general $(w_{aa} - w_{Aa})/(w_{AA} - 2w_{Aa} + w_{aa})$, or for the model used here $(h - 1)/(2h - 1)$, which is equal to 2/3 for the case of $h = -1$ illustrated in the diagram. If $h > 1$ the heterozygote is inferior to either homozygote, and there is an unstable equilibrium: the A allele will become fixed or lost, depending on its initial frequency, and even if $w_{AA} > w_{aa}$ will not spread when rare.

increases the time required for a beneficial mutation to spread in an outcrossed population from t_{sub}(dip dom) to $(1/h)t_{sub}$(dip dom) or somewhat less. Recall that the heterozygote becomes fixed in a comparable length of time in an asexual population. Thus, sexual diplonts will adapt more quickly primarily because they fix beneficial mutations in a single episode, whereas asexual diplonts must fix first one and then the other copy of the allele. In terms of mean fitness, the difference will be greatest for alleles with recessive effects on fitness in low-M conditions.

An outcrossing individual heterozygous Aa for a deleterious mutation a with fitness $(1 - hs)$ is likely to mate with the wild-type homozygote and produce ½AA, ½Aa offspring. The offspring are more fit than one parent and less fit than the other: the mean fitness of the population is unchanged and the rate at which the mutation is removed is the same as for an asexual population insofar as the frequency of the mutant allele at equilibrium is u/hs. A self-fertilized individual produces ¼AA: ½Aa: ¼aa progeny whose mean fitness $¼ + ½(1 - hs) + ¼(1-s)$ exceeds their parent if $h > ½$. Thus inbreeding generates less fit progeny if deleterious mutations are recessive (*inbreeding depression*) and more fit progeny if they are dominant. The frequency of a deleterious mutation at equilibrium in an inbreeding population will be approximately $u/s[F + h(1-F)]$ (Agrawal and Chasnov 2001), where F is the probability that two alleles are identical by descent (Wright's *inbreeding coefficient*: $F = ½$ for selfing). Thus the load of deleterious mutations will always be reduced by inbreeding, provided that they are not completely dominant. The evidence for this is that the inbred progeny of parents from an outbred population almost always suffer reduced fitness. The probability of fixation of a completely recessive allele in an outcrossed population is $\Phi \approx \sqrt{(2s/\pi N)}$ (Kimura 1957), so that despite its lack of expression in heterozygotes it will still be substituted more frequently than a neutral allele if $s > 1/2N$.

4.3.4 Heterozygotes

Allelic interaction may be so pronounced that heterozygotes lie outside the range of homozygotes.

The example that everyone knows is provided by alleles specifying different sequences for haemoglobin A in human populations. One allele causes a fatal anaemia when homozygous; the other is associated with normal erythrocytes, which are unfortunately susceptible to infection by malarial parasites. The heterozygote is slightly anaemic, but is also resistant to malaria. Which allele is superior, given that malaria is endemic? There is no simple answer: it is the combination of the two alleles, in the heterozygote, that has the greatest fitness. In an asexual diploid the heterozygote will spread to fixation and resist invasion by either homozygote, so the population is genetically uniform at equilibrium. In sexual diplonts, however, selection for heterozygotes will actually maintain genetic variation because they will continually generate homozygous progeny. This will be the case whenever the heterozygote is fitter than either homozygote. For a diallelic locus in an outcrossed population, with the same fitnesses as in the previous paragraph, both alleles persist indefinitely at frequencies above mutation-selection balance if $h < 0$, the frequency of the A allele being $(h - 1)/(2h - 1)$. Allelic interaction, like mutation, may thus hamper adaptation and maintain a range of maladapted genotypes, although only in the special case of sexual diploid organisms.

There is abundant evidence that heterozygotes are generally fitter than homozygotes. The simplest experiment is simply to obtain inbred families from an outbred population; it is a very general result that these are less viable and less fecund than normal outbred families. However, this might well be caused by the homozygosity of a few recessive deleterious genes, without implying that genes with normal levels of activity generally confer higher fitness when they are matched with a somewhat different allele. A more sophisticated approach is to extract chromosomes from natural populations and compare the viability or fecundity of individuals that are homozygous or heterozygous for these chromosomes. Several such experiments, pioneered by Theodosius Dobzhansky in the 1960s, have given a consistent account of the difference between chromosomal homozygotes and heterozygotes (Figure 4.3). The distribution of viability or

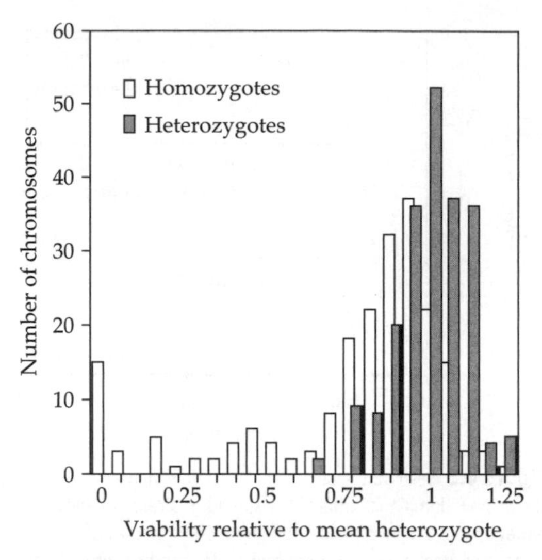

Figure 4.3 Viability of chromosomal homozygotes and heterozygotes in *Drosophila*. This diagram shows the distribution of viabilities among genotypes homozygous or heterozygous for wild-type chromosomes, using data on about 200 second chromosomes extracted from a South American population of *Drosophila pseudoobscura* by Dobzhansky *et al.* (1963). For the extraction procedure and the method of calculating viabilities, see Lewontin (1974), who gives an excellent account of the context in which these and similar results have been interpreted.

fecundity among homozygotes is bimodal: there is a lesser mode at zero, probably reflecting the segregation of a few recessive lethals, with the rest of the genotypes broadly distributed around a larger mode representing more or less normal performance. The heterozygote distribution is unimodal; it shows less variation than the homozygotes, and the mean of the heterozygotes exceeds the mean of the 'normal' homozygotes by about 10%. Heterozygotes are thus substantially more fit than homozygotes, at the level of whole chromosomes. However, it is not clear whether this difference is created by the slight superiority of heterozygotes at many loci, or by a few genes that are substantially inferior as homozygotes, and extending the result to single genes is not easy. Wallace (1958) created homozygous stocks of *Drosophila* and then induced mutations with low doses of X-rays. By crossing irradiated males with their non-irradiated sisters, he obtained flies that were heterozygous at only a few loci, rather than for a whole chromosome. These could then be compared with the progeny of crosses between non-irradiated partners, which should be homozygous at all loci, barring a very few spontaneous mutations. An alternative approach is to use spontaneous mutations that have been accumulated on inversions. Experiments of this sort have given rather ambiguous results. There is a general tendency for individuals that are heterozygous for an irradiated chromosome to be fitter, but the effect is so small that, despite the vast size of the experiments (Wallace scored over 2×10^6 flies), both its existence and its interpretation remain doubtful.

The search for heterozygote advantage was a very important theme in population genetics because it would provide a simple functional explanation for high levels of genetic variation as an alternative to neutral theory. It was, in fact, the only explanation that could readily be reconciled with the framework of population genetic theory, supplemented with laboratory experiments on *Drosophila* but lacking almost any reference to ecology. A few good examples of heterozygote advantage were discovered, but it is now clear that the bulk of variation that is not attributable to drift is maintained by ecological and behavioural interactions, which are discussed in Chapters 8 and 10.

4.3.5 Recombination

Segregation and fusion continually create new combinations of alleles at any given locus in sexual diplonts, without necessarily altering combinations of alleles at different loci. In practice, all eukaryotes have more than one chromosome, and the random segregation of whole chromosomes into the next generation of gametes creates new combinations of alleles at loci on different chromosomes. In most sexual organisms, moreover, crossing-over in the zygote or germline between DNA molecules derived from different gametes produces new combinations of alleles at loci on the same chromosome. The genetic recombination caused by this continual shuffling has profound effects on evolutionary dynamics. The genealogy of an asexual population is a strictly branching tree, whereas the genealogy of a sexual population is a net. A single individual in an asexual population will give rise to the entire population at some time in the future, whereas no such individual exists in a sexual population. The sexual genome is dissected by recombination in every generation so that it comes to consist of a number of segments within each of which recombination has not yet occurred. Each segment will have a most recent common ancestor, just like an asexual individual, but each segment is likely to have a different common ancestor. This is most obvious for segments in which recombination is systematically suppressed, such as the mitochondrial genome or the Y chromosome, which behave like asexual microbes embedded within a larger sexual genome. For ordinary autosomal regions the intact unrecombined segments will become shorter and more numerous as time goes on, until each is so short (a few hundred or a few thousand nucleotides) that it is very unlikely to experience a recombination event and will persist for many generations before eventually being broken up. Each segment, while it remains intact, will have a branching genealogy like an asexual genome. Each will have a different most recent common ancestor (mrca), however, so that the genealogy of one gene or segment is likely to be different from that of another. The interweaving of these independent branching phylogenies produces a net-like genealogy for the genome as a whole. We can treat a single gene as though it were an asexual replicator with a branching genealogy, at least as a first approximation. Combinations of genes or whole genomes, however, have distinctively different evolutionary dynamics in sexual populations.

In clonal populations the single clone with greatest fitness eventually replaces all others, which are maintained only by mutation. For example, different *lac* alleles flourish or dwindle in the bacterial populations of chemostats. However, this description is valid only because we are able to construct clones of *E. coli* that differ only at the *lac* operon. If we merely took *lac* variants at random from unrelated bacterial strains, they would probably differ at hundreds of loci. For example, two strains with different *lac* alleles might also differ at the *trp* operon, which is responsible for encoding the ability to synthesize the amino acid tryptophan. In minimal medium, a defective *trp* allele is lethal, and the clone that bears it will be eliminated, even if it has the superior *lac* allele. Hence, a superior allele may fail to spread in a population because the genome in which it occurs happens also to include a deleterious mutation at another locus. The converse may also occur. If a small amount of tryptophan is added to the medium, a mutant defective in *trp* will be able to grow, although it may grow more slowly than types able to synthesize their own tryptophan, and will tend to be eliminated through selection from mixtures. However, suppose that, as before, one clone bears the defective *trp* allele and a superior *lac* allele (it may be designated *trp*$^-$*lac*$^+$), whereas the other bears a functional *trp* allele but an inferior *lac* allele (*trp*$^+$*lac*$^-$). If tryptophan is present in the growth medium, it is perfectly possible that the *trp*$^-$*lac*$^+$ clone will become fixed, because its more efficient use of lactose gives it an overall competitive advantage. The *trp*$^-$ gene has thereby become fixed in the population, despite being an inferior variant. It has become fixed because it happens to be associated with a variant at another locus, whose superiority over-rides its own inferiority (*hitch-hiking*: Maynard Smith and Haigh 1974).

Selection has somewhat different effects in sexual populations, because sexual lineages are not independent. There are two essential components of the sexual process: the combination of genes from different lineages through gamete fusion, followed by the recombination of genes during meiosis. In asexual populations, lineages form a strictly branching tree; in sexual populations, lineages form a network. An important consequence of sex is that genomes are not perpetuated as units, because in every sexual generation the genes constituting the genome of one individual are recombined with those of its sexual partner. There are therefore no permanent associations between genes at different loci. Loci that are close together on the same chromosome are unlikely to be separated by crossing-over in any given generation, but otherwise genes in sexual populations are inherited nearly independently of one another. Thus, a beneficial *lac*$^+$ allele would not be permanently associated with a *trp*$^-$ allele, because mating between *lac*$^+$*trp*$^-$ and *lac*$^-$*trp*$^+$ individuals would produce some *lac*$^+$*trp*$^+$ progeny by recombination. The superior alleles at both loci can be fixed in sexual populations, because genes that are transmitted independently can be selected independently.

The linkage of genes in asexual lineages hampers selection, because the fate of an allele depends to some extent on the company it keeps. A mutation may on average tend to increase fitness, and yet fail to spread, because it happens by chance to arise in an inferior genome. More generally, the fate of a gene in an asexual lineage will depend, not only on its intrinsic properties, but also on the way it interacts with the rest of the genome. The intrinsic properties of a gene are those which it displays independently of genes at other loci with which it happens to be associated. The overall effects of a gene can be decomposed into two components: those which are expressed independently of other genes, and those which are modulated by alleles at other loci. Imagine a motile green alga in which genes at different loci control whether or not it is phototactic (swims towards light) and whether or not it is photosynthetic (uses light to fix carbon). Is the allele that directs phototaxis beneficial or not? The question cannot be answered by a simple yes or no. It may be advantageous if the cell can utilize the light it has found, but motility is only an expensive burden if the cell would be equally successful in the dark. The fitness of an allele affecting phototaxis, therefore, will depend on the state of loci that affect photosynthesis. This is a simple example of *epistasis*, the interaction of the effects of different genes. Epistasis would make no difference to the outcome of selection if the population were infinitely large, since in that case all possible combinations of genes would be present, and selection would sort out the best possible combination. In a finite population, only a sample, typically an extremely small sample, of all possible combinations of genes is present. There might, for example, be only non-phototactic photosynthetic strains and phototactic non-photosynthetic strains present. In an asexual population, one or the other would win, depending on the balance of advantage in given conditions of growth: in the light, the non-phototactic photosynthetic strains might win, and in the dark the phototactic non-photosynthetic strains might have the edge. Recombination produces strains that are superior either in the light (phototactic and photosynthetic) or in the dark (non-phototactic and non-photosynthetic) because it allows the independent effects of each locus to be expressed. Sex thereby permits selection to cause a greater degree of adaptedness, by permitting the independent effects of genes to be selected independently.

The converse is also true. In sexual populations, it is only the independent effects of genes that are sorted by selection. If adaptedness requires a particular combination of genes, then recombination may build up that combination, but it will destroy it just as efficiently, so long as other genotypes are present in the population. If the best combination already exists, then it will be selected more rapidly in an asexual population, because its increase in frequency under selection will not be opposed by its destruction through recombination. This principle implies that the leading features of adaptation may be different in sexual and asexual populations. One example is mutator alleles (above), which spread despite the damage they cause because they remain linked to the rare beneficial mutations they elicit. They are accordingly frequent in experimental and natural populations of bacteria. They should quickly become unlinked by recombination in sexual populations, where they are expected to be perennially deleterious.

Compensatory mutations are favoured because of their epistatic interaction with a loss-of-function mutation elsewhere in the genome. They have often been identified in experiments with bacteria, where they spread because they are tightly linked to the mutation they compensate. In sexual populations the situation is likely to be more complex. If the loss-of-function mutation is on balance beneficial (for example, because it confers antibiotic resistance despite reducing growth in permissive conditions) and spreads to fixation, then compensatory mutations will be favoured if they ameliorate the effect on growth because all gametes bear the loss-of-function mutation and recombination has no effect on the fitness of progeny. On the other hand, if the loss-of-function mutation is far from fixation (for example, because the antibiotic is no longer present) then it is readily unlinked from the compensatory mutation, which therefore should not spread in sexual populations.

4.3.6 Limits to adaptation

In an asexual culture of bacteria inoculated with *lac*$^+$*trp*$^-$ and *lac*$^-$*trp*$^+$ strains, the limit of sorting would be *lac*$^+$*trp*$^-$, whereas in a corresponding sexual culture the limit would be *lac*$^+$*trp*$^+$. A greater degree of adaptedness can evolve in the sexual culture, because recombination allows the two genes to be selected independently. The population will eventually be dominated by a genotype that is more highly adapted than either of the initial genotypes. This genotype was not originally present; but neither is it entirely new. Rather, it has been constructed by recombination from elements of the genotypes originally present. This principle is quite general. Fitness may be affected by genes at many loci, any of which may bear a mutation, either beneficial or deleterious. If we designate the wild type allele as 0, a beneficial mutation as + and a deleterious mutation as – ,then a sequence of wild-type, plus, and minus alleles defines a genotype, for example

$$\{0-0-0+0-00+0+00+0-0-\},$$

to exaggerate the likely number of mutations. The fitness of an individual is proportional to the count of + alleles in its genotype less the count of – alleles, 4 – 5 = –1 in this case. The population initially comprises a range of genotypes, such as

$$\{0-0-0+0-00+0+00+0-0-\}$$
$$\{-+000-00-00++0000--0\}$$
$$\{0000000000-000000000\}$$
$$\{00++0+0000-0000+0-0-\}$$
$$\{0000-00-0000+-0+0-00\}$$

and so forth. The fittest genotype in this set is the second from last, with a score of 4 – 3 = +1. This would be the limit of sorting in an asexual population; this genotype would become fixed despite bearing the deleterious—allele at several loci. If the population is sexual, it can transcend this limit. Imagine a mating between the second-from-last genotype and the last. This would produce recombinant progeny with a range of phenotypes. Some would by chance inherit mostly – alleles, and would have very low fitness: the worst is

$$\{0000000 - 00 - 00 - 0 + 0 - 0 - \},$$

with a score of –4. Others, however, would inherit mostly + alleles, and would have very high fitness: the best is

$$\{00 + + 0 + 000000 + 00 + 0 - 00\},$$

with a score of +4. Recombination has the effect of making all the allelic diversity present in the initial population available to selection, which then drives the population beyond the original limits of variation.

In short, recombination facilitates adaptation by resolving genetic interference. It frees beneficial mutations from the inferior background on which they might happen to arise, and causes the rapid appearance of beneficial combinations of epistatic mutations. It should therefore make both purifying and directional selection more effective. There are three biologically distinct situations in which this principle will apply: several deleterious mutations (*mutation clearance*), several beneficial mutations (*mutation assembly*), and a beneficial mutation with several deleterious mutations (*mutation liberation*).

4.3.7 Purifying selection in sexual populations: mutation clearance

The effect of sex on evolutionary dynamics hinges on a multilocus equivalent of the Hardy–Weinberg rule. For simplicity imagine a random-mating haplont bearing two loci each with two alleles A,a and B,b and thus four haploid genotypes AB, Ab, aB, ab whose frequencies are f_{AB}, etc, and whose fitnesses are w_{AB}, etc. (For diplonts the argument is slightly more complicated but the conclusions are essentially the same.) The crucial concept in understanding the effect of sex is the correlation between alleles at different loci, which is summarized by the coefficient of linkage disequilibrium $D = f_{AB} f_{ab} - f_{Ab} f_{aB}$. I shall use it first and interpret it later. The life cycle consists of selection during vegetative proliferation, followed by random fusion and meiosis. The recursions describing changes in genotype frequency under selection are $f'_{AB} = f_{AB}\, w_{AB}/\bar{w}$, etc. As a consequence of these changes, linkage disequilibrium is altered from D to D'. The recursions for genotype frequency after fusion and meiosis can be obtained by drawing up the table of all possible matings and writing in the frequency of each genotype among the progeny of each mating, given that the rate of recombination between the two loci is r. This yields:

$$f''_{AB} = f'_{AB} - rD'$$
$$f''_{Ab} = f'_{Ab} + rD'$$
$$f''_{aB} = f'_{aB} + rD'$$
$$f''_{ab} = f'_{ab} - rD'.$$

and

$$D'' = (1 - r)D'.$$

The last expression shows that random fusion and meiosis cause linkage disequilibrium to fall towards zero at a rate equal to the rate of recombination. The coefficient D is a binary covariance, expressing the correlation between alleles at the two loci. Suppose that the frequencies of A,a are p, $(1-p)$ and of B, b are q, $(1-q)$. When alleles are combined at random $f_{AB} = pq$, etc, and

$$D = pq(1 - p)(1 - q) - p(1 - q)q(1 - p) = 0.$$

Hence $D = 0$ means that the allelic state of one locus is uncorrelated with that at the other. What the expression for D'' means is that swapping partners through recombination in a zygote produced by the fusion of two random gametes destroys any correlation between the allelic states of loci. In a phrase, recombination randomizes the genome.

There are three processes that can prevent this randomization. The first is non-random gamete fusion, which will preserve any existing correlation among loci. The second is selection, which can restore the correlation eroded by recombination. Suppose that the mutations are deleterious and act independently, with selection coefficients s associated with allele a and t associated with allele b, so that they can be combined multiplicatively to form the genotypic fitnesses $w_{AB} = 1$, $w_{Ab} = (1 - t)$, $w_{aB} = (1 - s)$, $w_{ab} = (1 - s)(1 - t)$. Then $D' = (1/\bar{w}^2)\,(1 - s)(1 - t)D$, so if $D = 0$ it will not be altered by selection. Therefore selection can restore a correlation between loci only if fitnesses are non-multiplicative, that is, only if there are epistatic interactions between loci. Finally, the calculation is deterministic and assumes that the population is infinite. If the population were finite, some degree of correlation would be likely to arise in each generation through the stochastic outcome of selection or fusion. In brief, linkage disequilibrium will decline towards zero unless restored by epistasis for fitness, or by drift.

The recursions make it clear that when the population has reached linkage equilibrium ($D = 0$) recombination has no effect on genotype frequencies: once it has become randomized, further randomization has no effect. Recombination will therefore affect the response to selection only when the population is away from equilibrium. The effect that it has depends on the sign of disequilibrium. If $D > 0$ the coupling genotypes AB,ab are more frequent than expected from random combination. The excess of AB,ab gametes will produce Ab,aB recombinants. Hence, recombination in a population with positive linkage disequilibrium will reduce variation by generating genetically intermediate progeny. Conversely, when $D < 0$ there is an excess of Ab,aB gametes and recombination will increase variation by generating genetically extreme AB,ab progeny from intermediate parents.

Epistasis arises because individuals bearing both mutations are either more fit or less fit than they would be if the effects of the mutations were multiplicative. Suppose that they are more fit, $w_{ab} > (1 - s)(1 - t)$. The double mutant will be proportionately more abundant and linkage disequilibrium will be positive, $D > 0$. Conversely, if the double mutant is less fit than if effects were multiplicative then linkage disequilibrium will become negative, $D < 0$. The condition that sex should increase the variance of fitness is thus $w_{ab} < (1 - s)(1 - t)$, which has been called *synergistic epistasis* because the two mutations have a greater effect when combined.

4.3.8 Synthetic lethal mutations

The extreme case of synergistic epistasis is $s = t = 0$ whereas $w_{ab} = 0$. This is a *synthetic lethal interaction.* It has long been known from *Drosophila* genetics that such interactions occur, but until recently it has not been clear whether or not they are merely laboratory curiosities. A systematic screen of about 10^5 double deletions in yeast found that any given deletion had a synthetic lethal interaction with about 30 others, so about 3% of all pairwise combinations of viable deletions are synthetic lethal. Presumably the same would apply to all complete loss-of-function mutations. This is large enough to have a considerable effect on the population, which can be expressed in two ways (see § 2.2.6):

- The mean number of mutations per gamete W: the fraction of the haploid genome bearing loss-of-function mutations is then $m = W/G$. The corresponding number of loci H that are homozygous for loss-of-function mutations follows straightforwardly.
- The number of individuals killed by genetic impairment K, resulting in the loss of a fraction $k = K/N$ of the population.

The synthetic genotype is produced by the fusion of two double-mutant gametes, whose frequency at equilibrium is $\sqrt{u/s}$ (Christiansen and Frydenberg 1977), where s is the selection coefficient acting against the double homozygote. Under truncation selection, with $s = 1$, the frequency of the lethal zygote after fusion in each generation is equal to the mutation rate per locus, u. In a haplont, the frequency of each single-mutant 'carrier' gamete is $\sqrt{u/r}$, where r is the recombination rate (Phillips and Johnson 1998). The lethal genotype is thus created by recombination between opposite carrier gametes at a rate $2 \times [\frac{1}{2}r\,(\sqrt{u/r})^2] = u$, so haplontic and diplontic cases are equivalent.

The average number of loci homozygous for a loss-of-function mutation depends mainly on the frequency of zygotes homozygous for a single mutation, resulting from the fusion of carrier gametes each of which bears the mutation. For two loci, the frequency of single-mutant gametes is approximately $\sqrt{\ [u/r + \sqrt{u/s}]}$ in a random-making population (Phillips and Johnson 1998). With free recombination and truncation selection, the frequency of zygotes homozygous for a single mutation is then $2(y + \sqrt{uy})$, with $y = 2u + \sqrt{u}$. The mean frequency of loci homozygous is slightly greater than this, by the frequency of synthetic lethal zygotes. For a chromosome of arbitrary length, a reasonable empirical approximation for the mean number of homozygous loci is is $Z = y + \sqrt{y}U/2$, with $y = U + \sqrt{U}/2$. This approximation is difficult to simplify further, but a good numerical approximation for values of U between 0.01 and 1 is $Z = 2.44U^{0.707}$. The average number of loss-of-function mutations per gamete given random fusion is then $W = \sqrt{Z}G = G\sqrt{z}$, so the fraction of loci that are loss-of-function is $m = \sqrt{z}$. Note that this value depends on G directly, as well as on U: m increases with U, but for a given value of U it decreases with G. For a chromosome of arbitrary length, the frequency of lethal zygotes is $k = U/2$. Each death removes two loss-of-function mutations, so the total rate of removal is $2(U/2) = U$.

These results apply to the extreme case in which zygotes homozygous for a single loss-of-function mutation are completely unimpaired. If a single loss-of-function mutation causes reduced fitness when homozygous then the mean load of loss-of-function mutations per gamete will be reduced, because selection removes them more effectively. The fraction of zygotes that are killed by genetic impairment nevertheless remains close to $U/2$. When genome size (and thereby genomic mutation rate) is small, most genetic deaths are caused by the single-locus effect. As genome size increases, an increasing proportion of genetic deaths are attributable to synthetic effects, and in large genomes (with U approaching 1) synthetic effects are responsible for the great majority of deaths. Thus, in large genomes the single-locus effect on fitness has little effect on the load of loss-of-function mutations per gamete.

The load of potentially deleterious loss-of-function mutations might be reduced in two ways: redundancy and recombination. Suppose that all the individual tasks necessary for the operation of the cell are organized into a set of functional modules: one module might include transcription, for example, another cell–cell signalling and so forth. There are many strong interactions between genes in the same functional module, and relatively few between genes assigned to different modules. Each module incorporates sufficient functional redundancy that it will continue to operate normally if one of its genes is lost, although it cannot survive the loss of two. Thus, loss-of-function mutations are synthetic lethal only if they affect genes in the same module, and this is unlikely to happen because each module includes only a small fraction of the genome. Thus, dividing cellular labour into semi-isolated compartments might mitigate the severity of synthetic lethal effects.

In a structured genome, a zygote is inviable if it is homozygous for two or more deletions of genes in the same module, regardless of the total number of deletions in the genome. The mean homozygosity increases more or less linearly with the number of modules. The reason is that each module tends to acquire its characteristic load of deletions, depending on its size, after which the genome behaves in the same way as an unstructured genome that has become 'saturated' with deletions. A good approximation for the total load of mutations can be obtained informally by noting that each compartment will lose function at one locus without penalty. If the fraction of loci that are homozygous loss-of-function in an unstructured genome is z, the number in a diploid genome with C compartments will be Cz. Each gamete will bear $(C-1)z$ loss-of-function mutations, in addition to the $\sqrt{z}$ that it would bear with an unstructured genome. The total number of loss-of-function mutations per gamete is then $(C-1)z + \sqrt{z}$. Substituting the numerical approximation for Z described above yields $m = y + (C-1\ y^2$, with $y = (1.56U^{0.35})/\sqrt{G}$. The creation of new modules within the genome will always be associated with a temporary advantage, because it will offer some extra protection against synthetic lethals by reducing the effective genomic mutation rate. This applies only until the system comes to equilibrium, however, at which point the lethal fraction is restored and an additional load of deletions has been acquired. The fraction of lethal zygotes then depends mainly on the genomic mutation rate. It is reduced somewhat by modularity, but the effect is very small. Like diploidy, modularity purchases a short-term alleviation of mutational damage at the expense of a long-term elevation of mutational load.

The second possibility is that sex increases the efficacy of purifying selection. It can do this, rather paradoxically, by increasing the supply of double mutants. In an asexual population, every individual will bear a mutation at one of a pair of synthetic loci, as mutation is unchecked by selection. In a sexual population the fusion of gametes that bear mutations at different synthetic loci will produce recombinants that bear both or neither. Since those bearing both are killed, the survivors are a mixture of individuals bearing one mutation or none, so their mean load is lower. This will reduce the mortality caused by synthetic effects. With free recombination, we have seen that the lethal fraction depends on the overall genomic mutation rate, rather than on genome size itself. In non-recombining populations, however, the lethal fraction increases with genome size even when the overall genomic mutation rate is held constant. This is because in the absence of recombination there is an irreversible increase in the frequency of deletions, up to the point where all zygotes are homozygous for at least one deletion, the unloaded class being permanently lost from the zygote population. At this point the lethal fraction is about twice as great as in a freely recombining population. This situation persists over a broad range of genome size, but breaks down for large genomes, when the population enters a new phase. All gametes are now so heavily loaded that almost all fusions would give lethal zygotes if mutations were distributed among loci at random. The population therefore evolves a very high degree of negative linkage disequilibrium, and comes to consist almost entirely of two heavily loaded gamete genotypes. Fusion between gametes of different types are complementary, generate little homozygosity and usually give rise to viable zygotes; fusions between gametes of the same type yield highly homozygous lethal zygotes. Very high loads may be sustained in such balanced synthetic lethal systems, with 30% or more of zygotes being inviable. This is unlikely to happen very often (or ever) but underlines how recombination results in lower loads by creating variance among gametes and thereby restoring the unloaded class of zygotes.

4.3.9 Mean fitness under purifying selection

Similar conclusions hold for less extreme cases of synergistic epistasis (Figure 4.4). It seems reasonable to suppose (although it remains to be demonstrated) that fitness is a sigmoidal function of load, falling ever more steeply as load increases before levelling out towards an asymptotic value of zero for very large loads. At high average load a unit reduction in load will result in a more than proportional increase in fitness, so selection will push the population towards a reduced load. Within the region of synergistic epistasis sexual progeny will include an excess of very lightly loaded and very heavily loaded recombinants, and will therefore have lower mean fitness than asexual progeny because the reduced fitness of heavily loaded progeny more than outweighs the increased fitness of lightly loaded progeny. The death of most of the heavily loaded progeny, however, reverses this discrepancy, so that after selection the mean fitness of surviving progeny is greater for sexually produced than for asexually produced progeny. Thus, we expect that under synergistic epistasis the mean fitness of a sexual population will exceed that of a comparable asexual population, provided that they are censused after selection.

The most direct way of finding out whether epistasis is usually synergistic is to estimate the fitness of isogenic lines with different numbers of deleterious mutations. Some lines may be known to carry more mutations than others although the exact number of mutations is unknown. There is some evidence of an accelerated decline of viability in classical *Drosophila* mutation-accumulation experiments, although the existence and interpretation of this effect have been controversial (Fry 2004). Two episodes of mutagenesis does not cause a greater loss of fitness than a single episode in *Caenorhabditis* (Peters and Keightley 2000). There is some evidence for an accelerated decline of one fitness component (but not another) with dose of mutagen in a parasitic wasp (Rivero *et al.* 2003). In other experiments

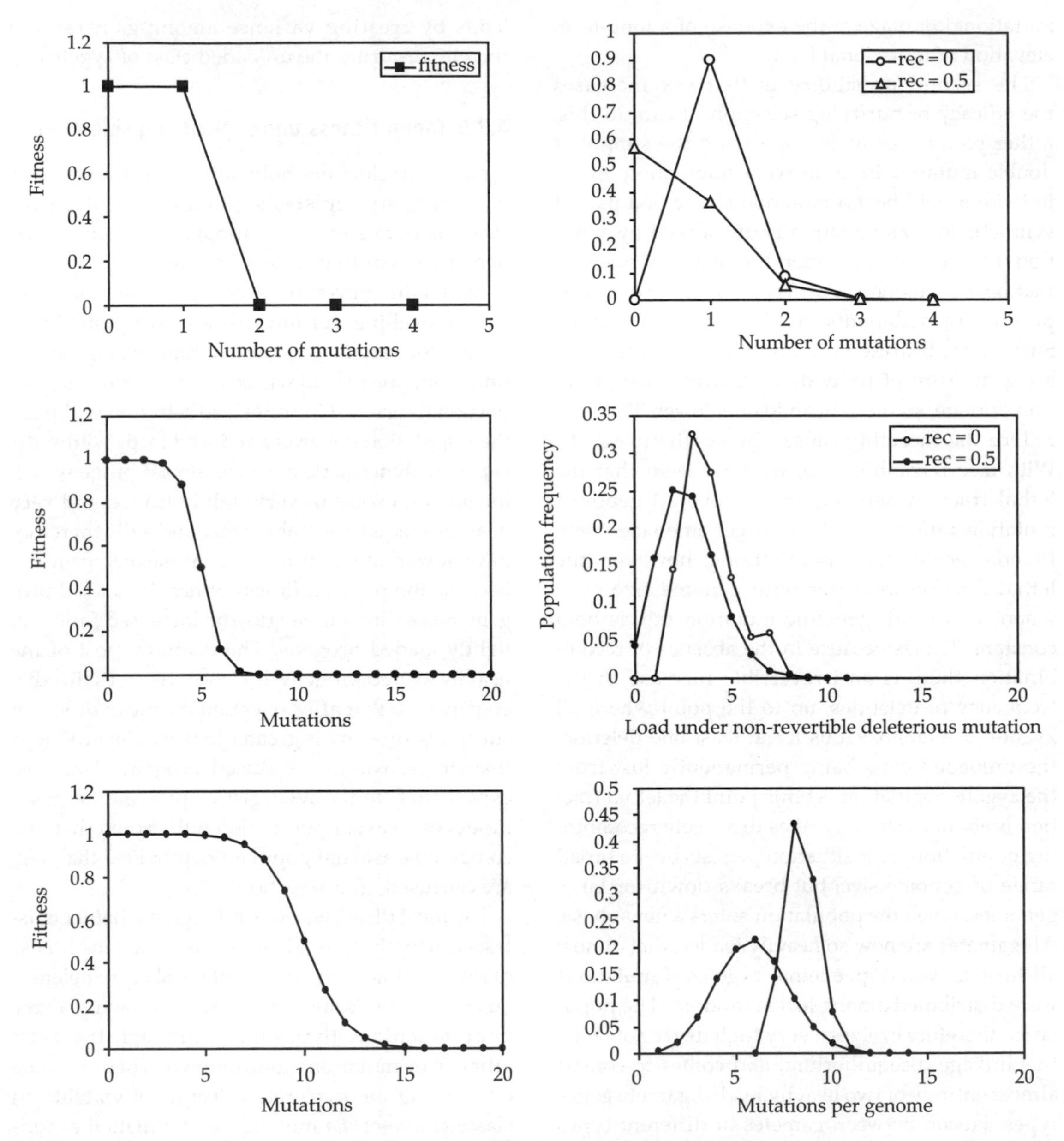

Figure 4.4. Effect of recombination on offspring fitness under different degrees of epistasis for deleterious mutations.

the number of mutations is known: none of these have shown a consistent excess of synergistic effects (de Visser and Hoekstra 1997, Elena and Lenski 1997, Elena 1999). Sanjuan *et al.* (2004) were able to compare pairs of known point mutations in an RNA virus and again found no excess of synergistic effects. On balance, studies such as these have shown that epistasis is widespread but also that antagonistic and synergistic epistasis are about equally common.

The second main approach is to measure the fitness of progeny immediately after meiosis in a haplont: with synergistic epistasis progeny mean fitness will be less than the midparent value. This has been observed (de Visser *et al.* 1996, de Visser and Hoekstra 1997) although the effect is slight and difficult to interpret (West *et al.* 1998). The approach has been criticized because it does not provide a quantitative estimate of the functional relationship between mutational load and fitness, but it has the advantage of showing the average effect over all loci and seems the most appropriate technique for evolutionary studies.

Once populations have reached equilibrium and remained under purifying selection for a considerable period of time, a sexual population should be better adapted, that is, should have greater mean fitness, than a comparable asexual population. This direct comparison is feasible only for organisms in which sex occurs intermittently, as it does in most eukaryotic microbes. I propagated sexual and asexual lines of *Chlamydomonas* for about 1000 generations in benign laboratory conditions, with some lines undergoing a sexual cycle every 10 generations or so, whereas others were perennially vegetative. There was no indication either from competition trials or from pure-culture growth that the sexual lines were superior. When spores were isolated from the sexual lines and crossed, the offspring were consistently less fit than their parents, indicating overall synergistic epistasis for vegetative growth (Renout et al 2007). The most reasonable interpretation of the long-term sexual lines is that after each sexual cycle they come to consist of several or many nearly equivalent genotypes whose high fitness is conferred by epistatic combinations of alleles. This diversity will be whittled away by selection, but before this happens the next sexual cycle disrupts these combinations and generates inferior offspring whose mean fitness is then again restored by selection.

Yeast grows on galactose at about half the rate on glucose. Selection lines maintained on galactose become better adapted whether sexual or asexual, probably through a single beneficial mutation. Lines maintained on glucose do not become better adapted, but after a few sexual cycles the sexual lines grow better on both glucose and galactose (Zeyl and Bell 1997). This is consistent with more effective purifying selection in sexual populations.

4.3.10 Directional selection in sexual populations: mutation assembly

In high-M conditions several beneficial mutations may arise independently soon after an environmental change, giving rise to clonal interference in an asexual population. This is resolved in a sexual population by the recombination of mutations carried by different gametes into the same genome. In an asexual population the optimal genotype AB must evolve sequentially through a second mutation in a lineage already carrying the first, whereas in a sexual population it will soon be generated by gamete fusion and meiosis (the *Fisher–Muller process*: Fisher 1930, Muller 1932). This process of mutation assembly is the mirror image of mutation clearance through the recombination of deleterious mutations, and equally requires epistasis for fitness in an infinite population. Thus, if A and B are beneficial mutations at two loci with positive selection coefficients s and t then synergistic epistasis is defined by $w_{AB} < (1 + s)(1 + t)$. (This may seem perverse, as the effect of a mutation in this case diminishes with the number already borne, but it is the standard usage. If the mutations are revertible, then the alleles produced by back-mutation are deleterious, and ln fitness falls off more steeply than linearly with the number of such back-mutations borne.) Recombination then increases the heritable variance of fitness and thereby enhances the increase of fitness under selection, an idea dating back to Weissmann (1889). Let us suppose that the population is currently dominated by – alleles, and that the environment changes so as to favour types with a larger number of + alleles, so that selection will move the population from types such as – – + – – – or – – – – + –towards types like + + – + + + or + + + + – +. In the simplest case, the loci act independently, in the sense that the fitness of any combination of alleles can be calculated by multiplying together their separate effects on fitness. Because the effects of the loci are independent, selection will not generate any correlation among them, and the population will remain in linkage equilibrium as it evolves. Recombination will have no effect on variation, and will therefore have no effect on the progress of directional selection. On the other hand, suppose that genes interact in such a way that a few + alleles cause a large increase in fitness, relative to having none, whereas the addition of more + alleles causes only a small further increase in fitness. Thus, the fitness of a genotype increases less steeply than multiplicatively as the number of + alleles it bears increases, rising asymptotically as it approaches the composition of the types favoured in the new environment. This kind of epistasis will lead to negative linkage disequilibrium, because types with intermediate numbers of + alleles will have much higher fitness than those with few + alleles, and will therefore increase swiftly in frequency, whereas extreme types with many + alleles are not much fitter than the intermediate types, and will spread more slowly. Recombination will therefore tend to increase variation, because genetically intermediate types will be disproportionately frequent and will produce extreme types when they mate. Conversely, nearly optimal genotypes will tend to be destroyed by recombination once they have been formed, so adaptation is likely to be accelerated only when epistasis is weak or sex infrequent.

4.3.11 Directional selection in sexual populations: mutation liberation

In a finite population linkage disequilibrium will arise by chance. It will at first be indifferently positive or negative, but positive disequilibrium will be consumed more rapidly by selection, leaving an excess of negative disequilibrium. This implies that beneficial alleles will often be found on low-fitness backgrounds, reducing the effectiveness of selection (Hill and Robertson 1966). Recombination will then allow these alleles to be selected more effectively by increasing the variance of fitness (Otto and Barton 2001). A particularly important situation is when a new beneficial mutation first appears in a population.

Given that most mutations are deleterious, a population under high-M conditions is likely to build up a load of deleterious mutations before the first beneficial mutation appears. In an asexual population any beneficial mutation is therefore likely to appear in an individual whose genome already bears several deleterious mutations. If the effect of the beneficial mutation is s_+ and that of each of d deleterious mutations is s_- then the overall fitness of the individual will be $(1 + s_+)(1 - s_-)^d$, assuming the deleterious mutations to act multiplicatively. There are then two possibilities. First, the beneficial effect might be too small to outweigh the combined deleterious effects, $s_+ < ds_-$, in which case the beneficial mutation will fail to spread. If the beneficial mutation is so feeble that its effect is less than that of a single deleterious mutation then it can spread only if it occurs in an unloaded individual. The distribution of deleterious mutations is Poisson with mean U/s_- (Haigh 1978), so that the frequency of the unloaded class is e^{-U/s_-} and the probability of fixation of a beneficial mutation will be $\Phi = \Phi_0 e^{-U/s_-}$ where Φ_0 is the probability of fixation of a solitary beneficial mutation calculated in the usual way (§3.3) (Peck 1994). The second possibility is that the beneficial effect is large relative to the combined effects of the deleterious mutations. In this case the beneficial mutation can be fixed even if it appears on a genome bearing several deleterious mutations, provided that its effect is sufficiently large to increase fitness above that of the unloaded type. Its probability of fixation is a complex function of U and s_- for which no useful approximation has yet been obtained (Johnson and Barton 2002), but will clearly be reduced by the occurrence of deleterious mutations.

The effect of sex is to redistribute the beneficial mutation over a range of backgrounds, some of which are likely to be lightly loaded. In this way it will become liberated from its initial background and concentrated into the least-loaded lines, where its net selection coefficient is largest, increasing the probability that it will become fixed and enhancing its eventual effect on population mean fitness. Recombination will then enhance the power of selection to cause the spread of beneficial mutations, provided that s_+ is neither too small nor too large. If $s_+ < 1/N$ then the fate of the mutation will be determined primarily by random sampling, so that further randomizing by recombination will have no effect. On the other hand, if s_+ exceeds some critical value then the lineage in which it arises is from the beginning the fittest in the population and will be fixed without recombination. This will always be the case if s_+ exceeds the range of heritable fitness effects, because even the most heavily loaded line would then become the fittest if it received the mutation. The ratio of the range to the standard deviation in a normally distributed population is about 3 for a population of 10 individuals; about 5 for 100 individuals; about 6.5 for 1000 individuals; about 7.75 for 10,000 individuals; and so forth. Thus, recombination will facilitate adaptation provided that $1/N < s_+ < K\sigma_w$ where σ_w is the genetic standard deviation of fitness and K is a coefficient depending on population size (Rice and Chippindale 2001). If only deleterious mutations have hitherto occurred then $\sigma_w \approx \sqrt{(U/s_-)}$ in a population that is about to undergo a sexual cycle after a long period of vegetative reproduction. This process of picking out the ruby from the rubbish (Peck 1994) does not require epistasis, and works even in relatively low-M condition where only a single beneficial mutation is likely to be segregating in the population at any particular time.

4.3.12 Effect of recombination in phage

Phage do not, of course, have any process resembling the sexuality of eukaryotes; however, when two different viral clones infect the same bacterial cell, many of the progeny phage will be recombinants, with a mixture of parental genes. Malmberg (1977) controlled the amount of recombination occurring in cultures of phage T4 simply by dilution. A fixed quantity of phage was added to a large or a small volume of medium containing a fixed density of bacteria. In the larger volume, the ratio of phage to bacteria is low, and almost all bacteria will be infected by a single phage; the phage population that grows up inside the cell will be a clone, and there will be no recombination. In the smaller volume, the ratio of phage to bacteria is higher, and any given bacterium is quite likely to be infected more or less simultaneously by two or more phage; this phage population is mixed, and will produce recombinant offspring. The phage were grown by serial transfer in batch culture for about 20 growth cycles, in medium containing proflavin, which is toxic to phage by interfering with DNA replication. The phage evolve resistance to proflavin during the course of the experiment, of course, and although the data are erratic high levels of resistance usually evolve more rapidly in the recombining lines.

The evolved resistance might be attributable to the independent effects of each new mutation, or might be dependent on particular combinations of genes that were much less effective when expressed separately. The genome of the evolved phage was broken into eight separate segments, any one of which might have borne a mutation conferring partial resistance to proflavin. Each segment was then inserted separately into an unevolved phage, and this construct tested for proflavin resistance. If the modified genes act independently, the resistance of the intact evolved phage should be equal to the sum of the partial resistances of the lines that were each transformed with a different segment of the evolved genome. The less dilute treatment, with a high ratio of phage to bacteria and thus a high rate of recombination, showed this additive relationship among the modified phage genes. In the more dilute treatment, however, the resistance of the intact evolved phage substantially exceeded the resistance that would be expected simply by adding up the independent effects of parts of the genome. Epistatic interactions among genes may thus evolve: they tend to be preserved when there is little recombination, so that combinations of genes are transmitted intact and selected as units.

4.3.13 Effect of sex in microbes

The most straightforward way of seeing whether these rather complex processes actually operate in simple populations is to document the progress of adaptation in comparable sexual and asexual populations. So far this has been done only in two unicellular eukaryotes, the chlorophyte *Chlamydomonas* and budding yeast *Saccharomyces*, in which mating and meiosis are readily manipulated although their life cycles are very different in many respects. A single sexual cycle followed by growth in a novel and stressful environment causes an immediate drop in mean fitness, relative to a parallel asexual line. This is accompanied by an increase in variance, after which mean fitness recovers as the variance is consumed by selection until the sexual line is fitter than the asexual line (Colegrave *et al.* 2002) (Figure 4.5). This process can be repeated, with successive sexual episodes consolidating the advantage of the sexual populations (Kaltz and Bell 2002) (Figure 4.6). The degree to which sex accelerates adaptation is greatest in large populations, where there is likely to be an abundant supply of beneficial mutations (Colegrave 2002) (Figure 4.7). The *Chlamydomonas* experiments provide a simple picture of adaptation in sexual populations (Figure 4.8). Under purifying selection in a constant

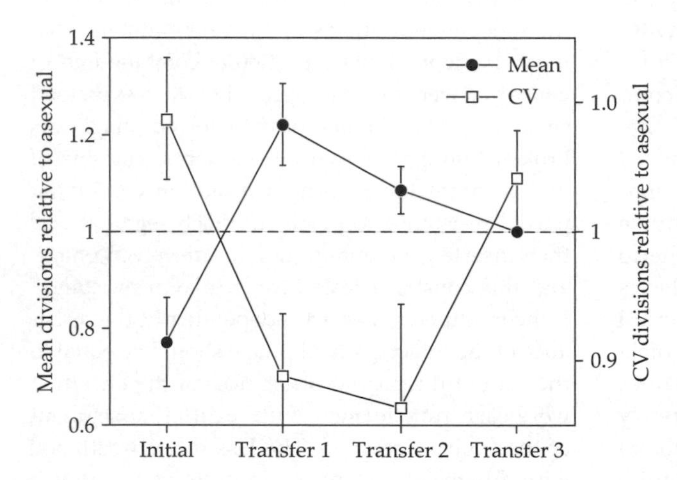

Figure 4.5 Effect of sex on the mean and variance of fitness in *Chlamydomonas*. 'Initial' is the state immediately following a sexual episode, relative to vegetative progeny (horizontal broken line at a value of 1). From Colegrave *et al.* (2002).

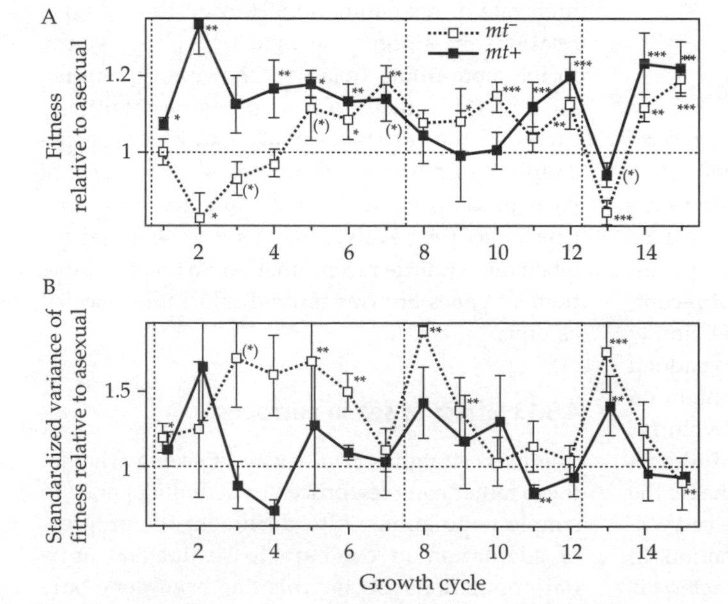

Figure 4.6 Effect of repeated sexual episodes on the mean and variance of fitness in *Chlamydomonas*. From Kaltz and Bell (2002).

benign environment a sexual cycle disrupts the well-adapted and nearly equivalent combinations of genes that have survived selection in preceding vegetative cycles and causes an immediate decline in fitness accompanied by an increase in variance, after which selection restores the previous level of fitness. Exposure to a novel environment has the same consequence, except that selection acting on the variance of fitness generated by sex increases mean fitness above that of an asexual line. Repeated sexual cycles thereby drive more rapid adaptation to novel conditions, although if the conditions persist for long enough the asexual population will eventually catch up.

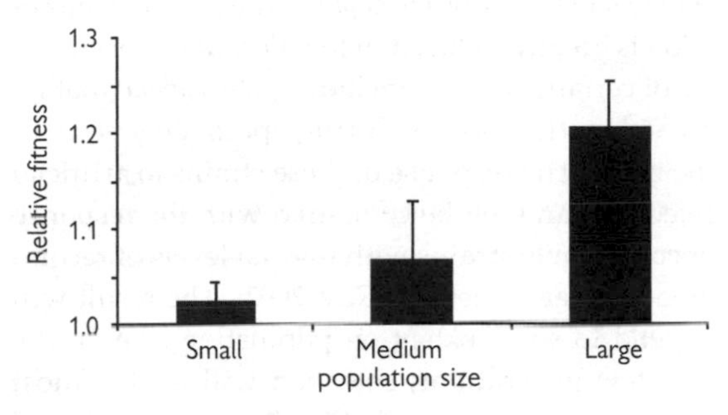

Figure 4.7 Sex accelerates adaptation in large populations. Histogram bars are fitness of sexual lines relative to asexual lines of *Chlamydomonas*. From Colegrave (2002).

One weakness of the *Chlamydomonas* experiments is that the sexual cycle is induced by nitrogen starvation, which may increase mutation rates and thereby confound the effect of recombination in generating selectable variation.. This can be avoided in yeast by constructing perennially asexual genotypes (Goddard *et al.* 2005). Yeast normally reproduces as a diploid, which on starvation undergoes meiosis to produce four haploid cells confined within a tough-walled ascus. These will normally fuse in pairs within the ascus when conditions improve, although outcrossing can be enforced by prematurely disrupting the ascus and mixing the haploid gametes. An asexual strain

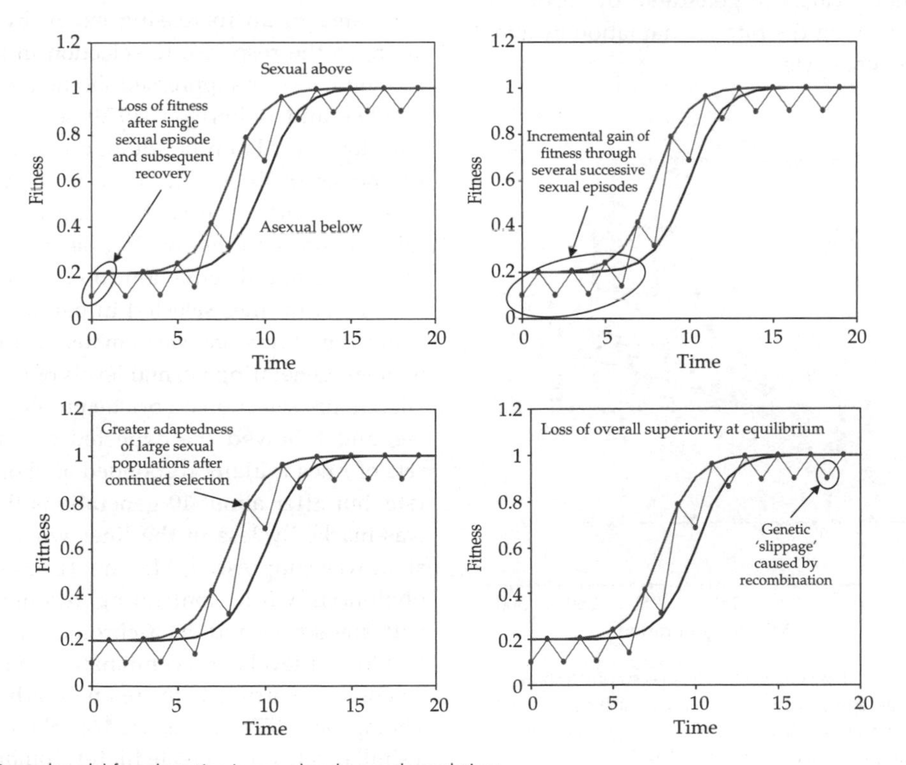

Figure 4.8 General model for adaptation in sexual and asexual populations.

can be constructed by deleting two genes: *SPO11*, which encodes an endonuclease that initiates crossing-over by causing double-strand breaks, and *SPO13*, which affects the pairing of sister chromatids and is necessary for initiation of the second meiotic division. As the second division does not reduce chromosome number, the double mutant produces non-recombinant diploid spores within the intact ascus. It can then be compared, under identical conditions, to an otherwise identical line in which meiotic reduction and recombination occur normally. When these lines are cultured in benign conditions there is little if any adaptation in either and no appreciable difference between them, so recombination has no effect on purifying selection. In harsh conditions (elevated temperature and salt) both types adapt, but the recombining lines adapt more rapidly (Figure 4.9). After 4 sexual cycles within 100 vegetative generations ln fitness relative to the ancestor had increased to 0.3 in the non-recombining and 0.4 in the recombining lines, showing that the additional genetic variance generated by recombination is increasing the rate of adaptation by about 2.5% per sexual cycle.

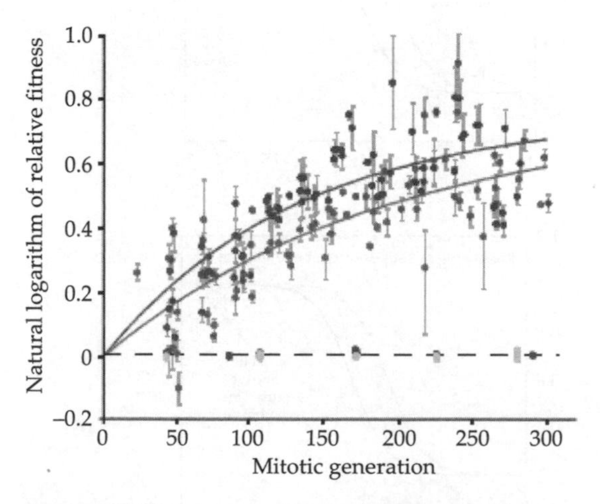

Figure 4.9 Effect of sex on adaptation in yeast. Graph shows sexual (upper line) and asexual (lower) lines in harsh environment; the lowermost broken line shows lack of response of either in benign environment. From Goddard *et al.* (2005).

4.3.14 Effect of recombination in *Drosophila*

The genetic consequences of sex depend on the rules for fusion and meiosis. Most theory and experimentation has focused on meiotic recombination, which can be manipulated even in obligately sexual organisms because recombination is suppressed in regions of the genome that are heterozygous for an inversion. This makes it possible to construct strains of *Drosophila* in which recombination is greatly reduced in females. (It is a peculiarity of certain Diptera, including *Drosophila*, that no recombination occurs during spermatogenesis in the male.) The response of these strains to artificial selection can then be compared with the response of comparable strains with normal levels of recombination (see review by Rice 2002). The result will depend to some extent on population size. In the first few generations, selection will for the most part sort the variation that is already present, and recombination will have little effect; the larger the population, and the more variation it contains, the longer this phase will be. Variation is subsequently generated to an increasing extent by recombination, so the response to selection in lines where recombination is suppressed should then be slower. McPhee and Robertson (1970) selected upwards and downwards on bristle number in *Drosophila*. In one set of lines, wild-type males were mated in every generation with females heterozygous for inversions on the two largest autosomes, causing a substantial reduction in crossing-over. In a second set of lines selected in parallel, males carrying the inversions were mated with wild-type females, generating normal levels of crossing-over. Directional selection on bristle number was effective, and followed the expected course: the two sets of lines initially responded at about the same rate, but after about 10 generations the response was markedly less in the lines where recombination was suppressed. Markow (1975) selected for phototaxis while controlling recombination on both autosomes and the X-chromosome and found that higher levels of recombination caused a faster response (according to the re-analysis by Rice 2002). Thompson (1977) also selected for phototaxis, using a similar scheme to create high-recombination and low-recombination lines, but could not confirm that recombination had any detectable effect on the response to selection after about 20 generations. It is possible that the number of flies maintained in the selection lines contributed to this disparity. McPhee and Robertson selected only 10 flies in each generation in each line, whereas Thompson selected about 50; it is possible that the initial phase of sorting pre-existing variation was prolonged in Thompson's experiment, so that the subsequent phase of selection on variation generated mainly by recombination was not witnessed.

The climax of chromosome manipulation in *Drosophila* has been the construction of *clone-generator* females that are able to transmit the autosomes of a random male unaltered to any number of offspring. This ability depends on a translocation of the two major autosomes to trap the male autosomes, and the combination of a doubled X chromosome and a Y chromosome to force all sons to inherit a Y from their mother. This can then be used to construct an almost completely non-recombining population to be compared with a population derived in the same way but lacking the translocation and thus transmitting random haplotypes and recombining through the segregation of intact chromosomes. This heady construct was used by Rice and Chippindale (2001) to follow the fate of alleles at the X-linked locus *w* (white). The w^+ (red eye) allele was introduced at a known rate by immigration to a base population otherwise fixed for the w^- (white eye) allele, to simulate a high rate of beneficial mutation, and a 10% surplus of w^+ males was added in each generation to simulate selection with $s = 0.1$. The base population itself was known to possess abundant genetic variance and so supplied a broad range of backgrounds for the simulated beneficial mutation w^+. For the first few generations w^+ spread at the same rate in recombining and non-recombining populations, presumably because may initially be associated either with heavily loaded or with lightly loaded autosomes. After about eight generations, however, its spread halted in the non-recombining lines, but continued in the recombining lines as it became concentrated in more lightly loaded lineages. This experiment provides us with a contrived but elegant demonstration of how recombination (through chromosome segregation) can facilitate adaptation by freeing a beneficial mutation from the inferior background in which it first arose.

4.3.15 Sex and the response to selection

So far as I know, there is currently little or no experimental evidence that sex facilitates the clearance of deleterious mutations, despite the enormous amount of theoretical effort that has been expended on this topic. The experiment by Zeyl and Bell (1997) stands out as a possible exception. It may be that any effect would be too small to be readily detectable. If we take at face value the estimate of $U = 0.003$ for microbes then the relative fitness of an asexual population will be $e^{-U} = 0.997$, which would certainly not be easy to measure. If so, this underlines the feebleness of the effect. On the other hand, attempts to demonstrate mutation assembly have usually been successful, and at this point we can be rather confident that sex will usually accelerate adaptation. The underlying mechanism is still not quite clear. The synergistic epistasis required by theories in which several beneficial mutations are combined into the same line of descent undoubtedly exists: we have a very good estimate that about 3% of interactions are synthetic lethal from the largest and most systematic experiment. It is not clearly true in a majority of cases for mutations of lesser effect, but most attempts to measure epistasis have used arbitrary pairs of loci or sites without regard to their functional interaction. The compensatory mutations in T4 and the results of crosses within long-term *Chlamydomonas* lines seem to show that epistasis will readily evolve, and it is the nature of epistasis in populations subject to some specified selection regime that is most relevant to its effect on adaptation. The most conclusive demonstration of an underlying evolutionary mechanism, however, is for mutation liberation in the Rice and Chippindale (2001) experiment. Granted that this was contrived to maximize the chance of observing the process, there are few if any experimental results that are inconsistent with a similar interpretation. The accelerated response to natural selection in microbes or to artificial selection in *Drosophila* could both be attributable to single beneficial mutations of fairly large effect. The immediate depression of mean fitness and elevation of the variance could reflect the temporary redistribution of a beneficial mutation to more heavily loaded lines. Moreover, the theory does not require the simultaneous presence of several beneficial mutations, and places no onerous conditions on their interaction.

4.4 Dispersal

4.4.1 Population structure

The open water of the sea or a large lake approximates a single large vial supporting a well-mixed plankton population, whereas organisms that live on surfaces or in sediments are likely to be aggregated into many small subpopulations or demes. Spatial structure may modulate the effect of selection, even if an allele has the same fitness in every deme. Selection in a large unstructured population acts exclusively through differences in fitness among individuals. In a subdivided population selection will act in the same way within each deme. The composition of demes may vary, however, even if they offer identical conditions of growth, because each will independently undergo genetic drift. This will create genetic variance among demes, as well as among individuals within demes, offering a new opportunity for selection. The independent drift of isolated subpopulations implies that a different lineage will proliferate in each deme, and in consequence the individuals in a deme will be more closely related to one another than they are to individuals in other demes. The overall probability that the alleles borne by a diploid individual (I) drawn at random from the total population (T) descend from different ancestors (A), A_{IT}, is the joint probability that they have different ancestry from other alleles in the same subpopulation (S), A_{IS}, and that two random alleles from the same subpopulation have different ancestry, A_{ST}: hence $A_{IT} = A_{IS}A_{ST}$ (Crow and Kimura 1970). This can be written in terms of the inbreeding coefficient $F = (1 - A)$, which is the probability that two alleles do share a common ancestor, as $F_{IT} = F_{ST} + F_{IS}(1 - F_{ST})$. The within-deme inbreeding coefficient $F_{ST} = (F_{IT} - F_{IS})/(1 - F_{IS})$ represents the correlation between two random gametes from the same deme relative to gametes drawn at random from the whole population (Wright 1951) and thus expresses the extent of genetic differences among demes If p is the population-wide frequency of an allele whose variance over demes is σ_p^2 then $F_{ST} = \sigma_p^2/p\ (1 - p)$ (Wright 1943), so the inbreeding coefficient is a standardized genetic variance among demes. The differentiation of demes by drift is opposed by dispersal, which inoculates each deme with unrelated individuals. If a fraction m of the individuals in each deme disperse to a random deme in each generation then $F_{ST} = 1/(1 + 4Nm)$ for neutral alleles, showing that genetic differentiation varies inversely with the scaled dispersal rate $D = Nm$. D is the migrant supply rate and plays a role in the dynamics of structured populations similar to that played by the mutation supply rate M in unstructured populations. High values of D homogenize demes through the frequent exchange of migrants, and for neutral alleles a spatially structured population will only develop substantial genetic differentiation if $D < 1$. The fate of neutral variation in structured populations is the subject of a very large literature (reviewed by Pannell and Charlesworth 2000) whereas the effect of spatial structure on a uniform process of adaptation seems to have been neglected.

How genetic variance among demes affects selection depends on how individuals compete with one another, say for some limiting resource. If they compete exclusively with other members of the same deme their density will be regulated at the level of the deme, so that each deme will produce in each generation a fixed proportion (in the simplest case, an equal fraction) of the total number of individuals in the population. If individuals in any given deme compete indifferently with all other individuals in the population, however, then their density will be regulated at the level of the population as a whole, and each deme will contribute a different number of individuals, in proportion to its mean fitness. Population-wide competition will enhance the response to selection, and is usually called *hard selection*. Within-deme competition is less effective because the local increase in frequency of a favoured allele is discounted by the limited production of its deme, so it is often called *soft selection*. As a broad generalization, any given area of land or water will support a fixed number of individuals of a given kind of organism (see Hubbell 2001). Moreover, individuals are often aggregated in rather small groups, such that most ecological interactions will be local (§2.3). Hence, local interactions, local density regulation, and soft selection are likely to be very common.

4.4.2 Subdivided asexual populations

Subdivided populations of asexual microbes are easily set up in the laboratory. One of the first decisions in any microcosm experiment is the size of the vial, which may vary over about four orders of magnitude from 50-μl cultures in serial transfer experiments to large chemostats holding 500 mL or more. This may seem mundane but is an important decision because large and small vials are likely to support large and small populations which may adapt at different rates and in different ways. If we wish to work with the same total number of individuals, what are the consequences of using a single large vial or many small ones? In practice the consequences might be quite complicated, because large and small vials offer different conditions of growth even they are the same shape, made of the same material and supplied with the same nutrient medium. I shall defer a discussion of selection in heterogeneous environments until §10.2 and for present purposes assume an ideal situation in which mean growth (absolute fitness) is the same for all vials, large or small.

The large vial supports N individuals , whereas each of V small vials supports $n = N/V$ individuals. At stated intervals v individuals are removed from each vial and replaced with m individuals from an equal mixture of all vials, so that a fraction $v/n = vV/N$ of the population of each vial at this point are immigrants. The probability that an individual will be born in one vial but grow in another is then a migration or dispersal rate $m = v(n-1)/n^2 \approx v/n$ $\{n \gg 1\}$. If $n = 1$ or $v = 0$ then vials are completely isolated, which is normally the case in microcosm experiments, where dispersal is called cross-contamination and prevented at all costs. If $n > 1$ while $v = 0$ then we are comparing a single large vial with many small isolated vials. In low-M conditions each small vial is less likely to fix a random beneficial mutation within a given period of time, but this is balanced by the greater number of small vials, so that once the first beneficial mutation has been fixed in the large vial it will also have been fixed in one of the small vials. In high-M conditions a beneficial mutation will be fixed in all the small vials at about the same time that it is fixed in the large vial. The most interesting case is when subdividing the vial switches a high-M regime in the large vial to a low-M regime within each small vial. The large vial will support a diverse mixture of competing beneficial mutations for some time before the best mutation prevails, whereas each mutation will arise singly and spread rapidly in the small vials. If clonal interference retards the substitution time of the best mutation to a greater extent than small vial size retards the waiting time for this mutation then it will be fixed first in a small vial. Naturally, it may be a very long time before it has been fixed in all the small vials. If there is some level of dispersal, however, emigrants from the vial in which the mutation first became established will inoculate other vials in the experiment, and this process of infection by the best mutation will become increasingly effective because its frequency among migrants is continually increasing.

The basic mutational-landscape model is readily extended to a subdivided population living in a uniform environment where conditions have recently changed, and some simulation results are shown in Figure 4.10. Under soft selection the length of time that elapses before the fixation of the optimal allele is very sensitive to the dispersal rate. In low-m conditions a beneficial mutation is for long trapped within the deme where it appears, and as dispersal increases it spreads more rapidly. At very high dispersal rates the fixation time actually increases again, because newly arisen beneficial mutations are apt to be swamped by poorly adapted immigrants, although this is a minor effect. Under hard selection the fixation time is much less sensitive to dispersal rates, because the advance of a beneficial mutation within a small

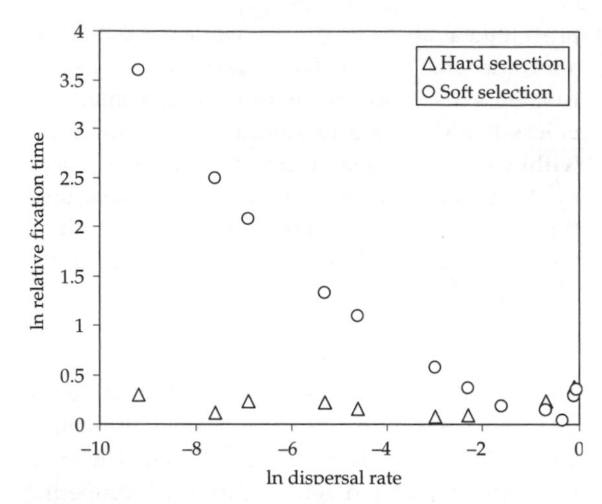

Figure 4.10 Fixation time in subdivided populations. Dependent variable is ln ratio of fixation time in subdivided populations to fixation time in undivided population. Mutational landscape model with $\lambda = 20$, $u = 0.0005$; metapopulation has 10 000 individuals in 100 demes. Points are means of 50 replicates.

deme actually increases the size of the deme. It is always less than the corresponding time under soft selection, and at low dispersal rates the optimal allele may be fixed 10–100 times more quickly than under soft selection. This is because under hard selection there can be wide variation in deme size, especially when dispersal is infrequent. The whole population may then come to consist of a few very large demes, removing the main impediment to the spread of beneficial mutations. Thus, subdivision generally impedes adaptation because it limits the possible advance of the best-adapted type during any given period of time. To put this in a slightly different way, individuals bearing a beneficial mutation in a subdivided population are likely to be competing against their relatives, whereas in a large homogeneous population they are mostly competing against other kinds of individuals. This limitation is most severe when density regulation is local, that is under soft selection, because the mutation is always spreading through a small rather than through a large population. It is mitigated by dispersal, which spreads it to other demes and so increases the scope for spread, or by global regulation, because this permits the demes in which it is abundant to grow. Nevertheless, the passage of the optimal mutation through a subdivided population is always longer, on average, than its passage through a homogeneous population of the same total size receiving the same total number of mutants in each generation.

4.4.3 Subdivided sexual diploid populations

In subdivided sexual populations the situation is different, because as well as competing with relatives individuals are also more likely to be mating with relatives. This will increase the frequency of homozygotes and thereby make selection on beneficial recessive mutations more effective. Indeed, hard selection is always more effective in subdivided populations for this reason. Whitlock (2002) shows that under hard selection the rate of change of gene frequency in a structured population is $\Delta q = spqH$, where q is the mean frequency over demes and $H = F_{ST} + (1 - F_{ST})[h(1\text{–}2q) + q]$. Here, F_{ST} expresses the genetic structure that is created by restricted dispersal. If this is not altered much by weak selection ($s \ll m$) then $H \approx (1 + 4hm)/(1 + 4m)$ for a rare mutation, and Δq is larger than it would be for an allele with the same dominance in an undivided population. There is then a close parallel between the effect of inbreeding in an undivided population and the effect of population subdivision under hard selection. Soft selection is more difficult to deal with, but Whitlock (2002) obtains $H_{soft} = (1 - r)H_{hard}$, where $r = 2F_{ST}/(1 + F_{ST})$. This will lead to more rapid adaptation (or reduced load) only for mutations that are sufficiently recessive.

4.4.4 The shifting balance

The *shifting balance* theory of Wright (1931, 1932) is a general theory of adaptive evolution that requires the division of the population into demes in order to maximize the response to uniform selection. The core of the theory is that adaptation will be facilitated when there are interactions among non-allelic mutations, as we have seen that it is when there is interaction between alleles. The process is considerably more complicated, however, than the production of homozygotes by inbreeding. Given that the adaptive landscape (§3.7.2) has a complex topography indicating strong interactions among genes a population may be unable to evolve from some state ab to a better adapted state AB because the intermediate states aB, Ab are inferior to ab. We now turn from an individual-based version of the adaptive landscape to a population-based version in which the mean fitness surface responds to allele frequencies. The distinction is important because given that an intermediate individual is poorly adapted it does not follow that an intermediate population is poorly adapted, as it might consist of a mixture of two well-adapted extreme types. With this in mind, this is now an evolutionary model in which selection drives a deme uphill to the nearest accessible local peak of fitness from where it cannot move to a still higher peak because of the intervening valley of low fitness. In a small deme there is a reasonable chance, however, that a deleterious combination will be fixed by drift. The best adapted genotype can then arise in this population by one further mutation and will then spread to other demes by dispersal, eventually moving each to the higher peak. In classical accounts of this process three distinct phases are recognized: in phase I maladaptive combinations are fixed within demes by drift, in phase II adaptive combinations are fixed by mutation and selection within demes, and in phase III the best adaptive combination is fixed in the population by selection among demes. The core concept is that alleles that interact with genes at other loci will have somewhat different effects on different genetic backgrounds. Where there are many small demes, drift will generate variation in composition so that these effects are expressed, so that the whole range of phenotypes that an allele can give rise to will be exposed to selection (Wade and Goodnight 1991). The differential success of those demes in which genes have their most favourable effects leads to the spread of particularly felicitous combinations by dispersal to other demes. In this way, complex novel adaptations become established much more readily than they could in a large homogeneous population.

This seductive proposal has ensnared many of the best minds in evolutionary biology and has led to a large and elaborate literature. A delicately sceptical review by Coyne and Orr (1997) argues that it provides a solution to a problem that does not exist: there are few if any examples of adaptation that are not explained by straightforward mass selection. They also point out difficulties in each phase of the process. In the first place, the process is intended to describe how complex adaptations involving many loci can evolve. The type of epistasis that is required is plausible for a simple two-locus model, but it is not clear that the interaction of many loci will give rise to many isolated fitness peaks; on the contrary, peaks are more likely to be connected by ridges of increasing fitness as the number of loci increases, as Fisher suggested. Secondly, the probability that a deme will become fixed for a deleterious genotype is about e^{-ns}, which is very small unless selection is very weak or each deme is very small. Finally, high rates of local dispersal are more likely to swamp an emerging adaptation than to spread it. Coyne and Orr (1997) conclude that the emperor is at best very lightly clad.

It is not difficult to see the shifting balance in operation, however, at least for a simple two-locus system. Let us populate a few hundred sites with a few thousand digital individuals and provide them with a deep ditch to cross (say $w_{ab} = 1$, $w_{aB} = w_{Ab} = 0.75$, $w_{AB} = 1.5$). The time elapsing before AB is fixed is shown for several selection schemes in Figure 4.11. The entire population is initially ab but thereafter a few single mutants arise in every generation. With such small demes it is not long before a few become fixed for aB or Ab, and provided that they are not quickly invaded by individuals dispersing from ab demes they will sooner or later experience a second mutation to AB. Local regulation (soft selection) favours the assembly of the double mutant, else the single-mutant demes will shrink. This deme will soon be fixed for AB and will proceed to inoculate nearby sites by dispersal—then the wave-front of AB propagates throughout the population until it is everywhere fixed. Global regulation (hard selection) favours the spread of the double mutant by enlarging the contribution of AB demes to the pool of dispersers. These events are readily visualized by writing an individual-based computer model and waiting for AB to become established in the whole population; in the model I used this time is log-normally distributed with a mean of a few thousand birth–death–dispersal

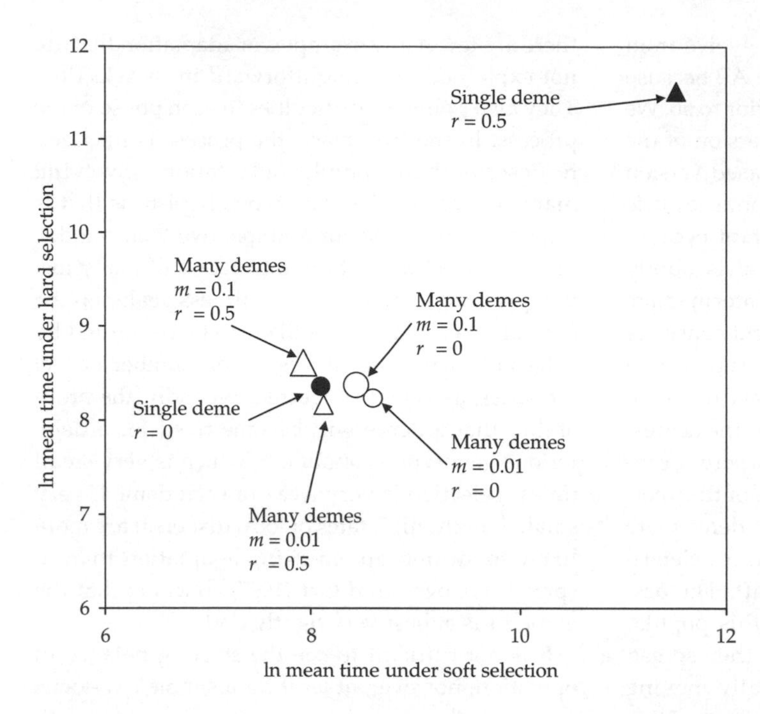

Figure 4.11 The shifting balance. Mutational landscape model showing the mean time (20 replicates) for the optimal genotype to attain a frequency of 0.5 in a metapopulation of 9000 individuals distributed among 900 demes, for high and low dispersal and recombination.

cycles. Contrary to some reports in the literature, shifting the balance does not seem to require any burdensome restrictions on deme size, dispersal, or epistasis, nor does it make much difference if selection is hard or soft. There is only one difficulty: the fixation of the double mutant will occur just as quickly in a single homogeneous population containing the same number of individuals. In this case, events proceed as I have described before: single mutants are always present at low frequency, after sufficient time has passed one of these experiences a second mutation, and sooner or later this will lead to the establishment of AB just as in the single-gene case. Provided the population is effectively asexual, subdividing it will not accelerate the progress of adaptation, and indeed will retard it when dispersal is low and density regulation local. Recombination changes this conclusion. Double mutants will usually appear in small demes where the single mutant is abundant, or fixed, and half their progeny will be double mutants whether or not recombination occurs. In a large homogeneous population any double mutant will almost certainly mate with the prevalent ab, and $r/2$ of its progeny will be low-fitness recombinants. In the presence of recombination, subdividing the population accelerates adaptation because it prevents the rapid break-up of beneficial combinations of genes. As in the single-locus diploid case, the effect of subdivision is equivalent to inbreeding. It might be better to say: recombination hinders adaptation when there is epistasis for fitness, but is not so harmful in subdivided populations.

There are two strong reasons for thinking that the shifting balance might provide a general description of how complex adaptations evolve. The first is that hybrids between species are often greatly inferior to either parent. On the face of it, this is strong evidence for multiple-peak epistasis, although it can also be caused by sexual or cytoplasmic incompatibility that has little if any relevance to ecological specialization. The second reason is that the populations of many species are subdivided with relatively low rates of dispersal between demes. The main evidence for this is that recently cleared sites are only slowly recolonized and that estimates of F_{ST} are often 0.1–0.3 for natural populations. There is at least one strong reason for scepticism: the inbreeding interpretation of the shifting balance that I have given here requires very small demes. This should lead to high levels of homozygosity, yet most outcrossing populations are close to Hardy–Weinberg equilibrium. Unfortunately there is no experimental system in which the operation of the shifting balance can be studied and evaluated. The closest approach is made by experimental studies of selection among demes (phase III), which are discussed in §10.4.

4.4.5 Partly unbound

Mass selection in large homogeneous populations of microbes is very effective, and horizontal gene transfer connects up the protein matrix by enabling bacteria to assemble complex adaptations from independent and widely separated lineages. In multicellular eukaryotes, however, the supply rate of beneficial mutations is a severe constraint on adaptation because their populations are relatively small and horizontal gene transfer is very rare. Sex and subdivision have traditionally been interpreted as ways in which this constraint can be overcome, but there is clearly a paradoxical element to the claim. If a beneficial genotype should arise, recombination will break it up, subdivision will restrict it, and dispersal will swamp it. In themselves, recombination and dispersal are randomizing processes that counteract selection, and they can facilitate adaptation only through a second-order effect on selectability. Sex resolves genetic interference and enables single beneficial mutations to be selected more effectively, but will usually make it more difficult to evolve adaptations that depend on sets of interacting genes. Subdivision will resolve clonal interference at a local level if demes are isolated, but dispersal is required for mutations to spread through the whole population. In short, it often requires special conditions or nicely adjusted rates for the second-order to overcome the ubiquitous first-order effect. This is why it has taken so long to develop theories of sexual and subdivided populations, why they remain so contentious, and why the experimental evidence lags so far behind the mathematics. To end on a hopeful note, future experiments will move the field forward by using clever molecular constructs to manipulate sexual biology and automated transfer systems to study selection in subdivided populations.

CHAPTER 5

Selection in multicellular organisms

5.1 Size matters

Larger organisms grow and reproduce more slowly. The maximal rate of population increase falls with body size as r (day^{-1}) = 1.5 $W^{-0.25}$ over 20 orders of magnitude of body size (W, in units of μm^3) from bacteria to large vertebrates (Blueweiss *et al.* 1978) (Figure 5.1). This is because metabolic rate mr does not increase proportionately with body size, $mr = aW^{0.75}$, which is in turn because surface area increases more slowly than volume. The allometric scaling of mr applies to eukaryotic microbes and continues across the microbe–multicell boundary at around 10^6 μm^3 (Bell 1985). Large size therefore exacts a high penalty and its evolution requires special explanation. Some possible advantages of deferring reproduction in favour of growth include continued occupancy of favourable sites in spatially varying environments, enhanced resistance to starvation in temporally varying environments, protection from predators for prey, and a greater range of prey for predators (Bell and Koufopanou 1991). When *Chlamydomonas* is exposed to rotifers it responds by aggregating into sheets or clumps of cells that are difficult for rotifers to ingest; the phenotype is irregular and quickly reverts to unicellularity when removed from the predator. When cultures of the unicellular chlorophyte *Chlorella* are grown in the presence of the predatory flagellate *Ochromonas*, however, they evolve into small, more or less regular colonies of eight cells through failure to separate completely after cell division (Boraas *et al.* 1998). This phenotype persists in the absence of predators and seems to be the only example of the experimental induction of permanent multicellularity.

Large size must nevertheless be mitigated by special adaptations to ensure an adequate level of metabolism. Large amoebas are polyploid or multinucleate, for example, and ciliates have evolved a specialized nuclear dimorphism. Most organisms larger than 10^6 μm^3, however, are multicellular. Extension in one dimension produces a filament, as in fungi and filamentous algae, which is specialized for absorption. Extension in two dimensions produces a sheet, as in macroalgae and tracheophytes, which is specialized for the interception of fluxes. Extension in three dimensions produces a massive body, as in ciliates and animals, which is specialized for ingestion. The smallest multicellular organisms of regular form are species of *Gonium* (Chlorophyta) and the gametophyte of *Syringoderma* (Phaeophyta), which are minimal two-cell organisms. Multicellular organisms with fewer than 100 cells are rare, and are usually undifferentiated, with all cells participating in reproduction. The smallest differentiated multicellular organisms are colonial chlorophytes (described below) and dicyemid mesozoans (internal parasites of cephalopods); the smallest free-living multicellular animal that can be routinely collected is probably the freshwater cnidarian *Microhydra*, which has 100–200 cells. The number of cell types C increases gradually with total cell number C_N as $C \approx 0.6C_N^{0.05}$, with complexity at given size increasing in the sequence Phaeophyta → Chlorophyta → plants → animals. These figures are from Bell and Mooers (1997).

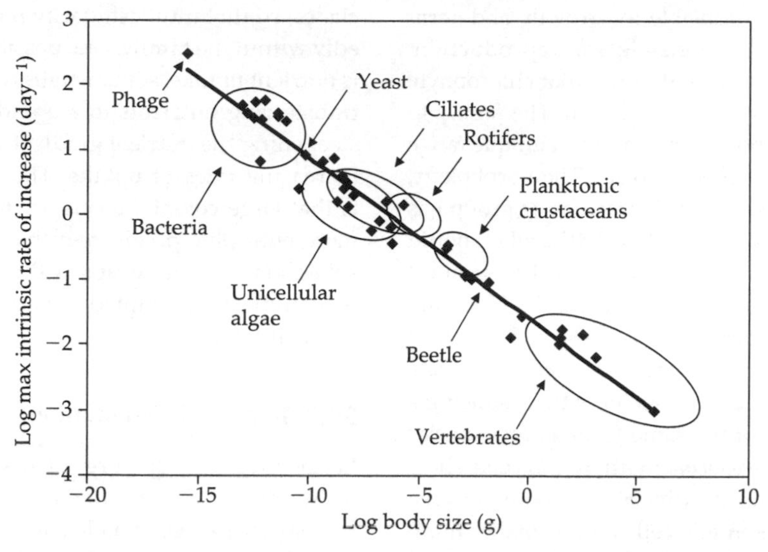

Figure 5.1 The allometry of *r*. This is the original diagram that showed the power law exponent to be about −0.25. From Blueweiss *et al.* (1978).

5.1.1 The universal spore

Many multicellular organisms reproduce vegetatively, forming new individuals from parental tissue by budding or fragmentation. To the best of my knowledge, however, no organism reproduces like this in perpetuity; all lineages are reduced sooner or later to a single cell or nucleus from which a new individual develops. One reason for this is that vegetative offspring will inevitably bear a load of somatic mutations. These are transmitted in a non-Mendelian fashion, with some progeny receiving more copies and some fewer, so that the lineage will become fixed for a succession of somatic mutations. Using a large number of cells to form offspring offers no solution, since although the assortment of previously acquired mutations will be slowed down, a greater number of new mutations will arise. Selection will oppose the accumulation of somatic mutations by eliminating heavily loaded individuals, but it will not be very effective because it can act only through variation among individuals in the mean number of mutations per cell. Exclusively vegetative lineages would in the course of time reproduce more slowly, show a greater incidence of developmental abnormalities, and at last become extinct, like isolate lines of ciliates (§12.1). When spores are produced, each cell is independently exposed to the action of selection. Genetic variance for the effects of deleterious mutations will be greater among spores than among vegetative offspring produced by budding or fission, and selection will be correspondingly more effective. At the same time, the elimination of within-body selection in the spore curtails the spread of selfish cell lines that would spread through vegetative reproduction at the expense of eroding individual adaptedness. Spore production thus sets up a 'selection arena' (Kozlowski and Stearns 1989) within which selection acts as an exogenous repair system for the lineage. Selection lines of multicellular organisms maintained solely from fragments or cuttings should therefore show a progressive decline in vigour and organization. I do not know of any deliberate experiments designed to test this possibility, although crops such as sugar cane and potato might provide some information.

5.1.2 The Volvocales

The fundamental differentiation within the bodies of multicellular organisms is between somatic cells, which are responsible for growth, and germ cells, which are responsible for reproduction. *Chlamydomonas* is a normal unicellular chlorophyte that reproduces by an equal fission. The family to which it belongs, however, contains a unique range of colonial and multicellular forms. The morphology, genetics, and ecology of this fascinating group has been reviewed by Kirk (1998, 2003), who should be consulted for detailed references. The smallest and simplest forms, like *Gonium* and *Pandorina*, are regular colonies consisting of 2–64 cells, each of which is similar in appearance to *Chlamydomonas*. Each cell in the colony reproduces by producing a miniature colony of the same form. Larger species of *Eudorina* and *Pleodorina*, with 32–128 cells, show the first indications of a division of labour between a sterile caste of somatic cells and a smaller number of germ cells, which alone reproduce. This tendency is taken further in *Volvox*, which contains 1000 or more cells. An asymmetric division early in colony growth creates the distinction between small somatic cells and much larger germ cells, or gonidia. The gonidia develop inside the parental spheroid into miniature *Volvox* (themselves with gonidia) before being released. At least three genes are involved in constructing a differentiated adult. The crucial asymmetrical division is regulated by *gls*, and if this is inactive all cells develop as somatic cells and the adult is sterile. Following this division, *regA* directs the expression of somatic genes in the small cells while *lag* directs the expression of gonidial genes in the large cells. *RegA* mutants at first develop normally but the somatic cells eventually resorb their flagella and redifferentiate as gonidia, so the colony consists entirely of small germ cells. *RegA* appears to regulate the expression of nuclear genes encoding chloroplast proteins and thereby keeps the somatic cells small by preventing them from synthesizing chloroplast material. *RegA* mutants with defects in colony matrix synthesis fall apart and can be cultured as unicells. The Volvocales are perhaps predisposed to the evolution of multicellularity because of the persistence of a cell wall within which division of the unicellular forms occurs. Phylogenetic analyses have shown that taxa like *Eudorina* and *Volvox* represent grades of construction rather than monophyletic clades, so that multicellularity has evolved repeatedly within the family. One possible selective agent is nutrient uptake: somatic cells may act as a source transferring nutrients to a gonidial sink, thereby steepening the nutrient gradient at the surface and increasing rates of uptake. The evidence for this is that large colonial forms are more abundant in more eutrophic ponds, and that gonidia in intact spheroids grow more rapidly than isolated gonidia, or gonidia in disrupted spheroids (Koufopanou and Bell 1993).

5.1.3 Somatic differentiation

Large multicellular organisms contain dozens or hundreds of specialized cell types organized into structures such as leaves and limbs responsible for processes such as photosynthesis and locomotion. Cells, structures, and processes are all clearly adapted, in the sense of being organized to perform a particular activity. Whether they are optimally adapted, in the sense of being able to perform it as well as possible, has not yet been established. Some are clearly suboptimal, such as the vertebrate optic disc or the human vas deferens (see Williams 1992b), but most field biologists, I believe, would view character state as being highly adapted in most cases. Optimal adaptation would be both accurate and precise. Accuracy is determined by directional selection that moves the population mean to the character state that maximizes fitness. The distance of the mean character value from its optimum is given by $|\beta/-\gamma|$, where β and γ are the linear and quadratic selection gradients (§7.3.8) (Estes and Arnold 2007). For the data set collated by Kingsolver *et al.* (2001) this distance is exponentially distributed (Figure 5.2) with mean 3.9. Thus, mean character state in about one-half of the populations surveyed was further than $1\sigma_p$ from the computed optimum, in about one-third more than $2\sigma_p$ and in about one-fifth more than $3\sigma_p$ from the optimum. In this quantitative sense, adaptation is often far from perfectly accurate. This is to be expected if selection fluctuates strongly over time so that the population always lags behind the changing environment.

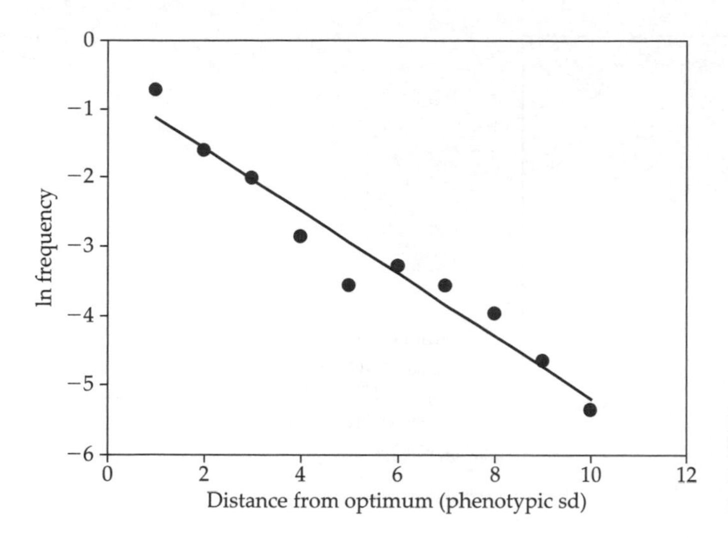

Figure 5.2 Distance of character values from their optimum in natural populations. The regression equation is $y = -0.68 - 0.45x$. Redrawn from Estes and Arnold (2007).

Precision is determined by purifying selection (or stabilizing selection, §3.3) that eliminates imperfectly adapted individuals. Linear skeletal elements of individuals collected from natural populations of mammals have coefficients of variation of about 0.04–0.10 (Simpson *et al.* 1960); for example craniofacial characters in deer *Odocoileus* had mean *CV* = 0.059. Extensive data on variation within inbred lines of mice (Mouse Phenome project at *http://phenome.jax.org/pub-cgi/phenome*) yields very similar values of *CV* = 0.04–0.10 (average 0.068). This is a little surprising, since given the rather high heritabilities found in natural populations one might expect *CV* to be about 50% greater than for inbred lines, but poor canalization might counter the lack of genetic variation. Other morphological characters seem to have similar average *CV* (e.g. ear length *CV* = 0.071), whereas organ weight in mice is more variable (brain 0.05, liver 0.08, kidney 0.55, atrium 0.23, ventricle 0.44). Overall body weight at various ages has *CV* = 0.2–0.3. The overall mean of about 30 000 estimates for about 700 characters is 0.35; the frequency of cases declines exponential with increasing *CV* (Figure 5.3). About 300 different physiological scores on human subjects had mean *CV* = 0.26 (*http://www.westgard.com/biodatabase1.htm*). Individuals may therefore be rather distant from the optimum with respect to many characters.

A very general justification for optimality is that the cost of investment exceeds the marginal return beyond a certain point. A bone that must support a given load, for example, has an optimal cross-sectional area; investing in a thicker bone would be wasteful, given that the resources involved could better be deployed elsewhere. The idea that structures evolve under natural selection until their capacity matches their load has been called *symmorphosis* (Taylor and Weibel 1981), and is supported by many observations of the loss of unused capacity, for example the eyes of cave-dwelling animals. In practice, however, many structures seem to have a much greater capacity than is normally necessary, as kidney donation proves. Bones usually break, in fact, at about twice the peak load that they would normally experience, so for both bones and kidneys the safety factor is an overcapacity of about twofold. Diamond and Hammond (1992) showed that the safety margin for uptake of amino acids and monosaccharides by the small intestine and for the glucose uptake capacity of the gut was in both cases a factor of about 3. This large reserve capacity might be necessary to meet loads that are rare in the laboratory but commonplace in nature, for example. The principle that regulates the evolution of safety margins is presumably the maximization of geometric mean fitness (§8.4), but I am not aware of any extensive theory or selection experiments.

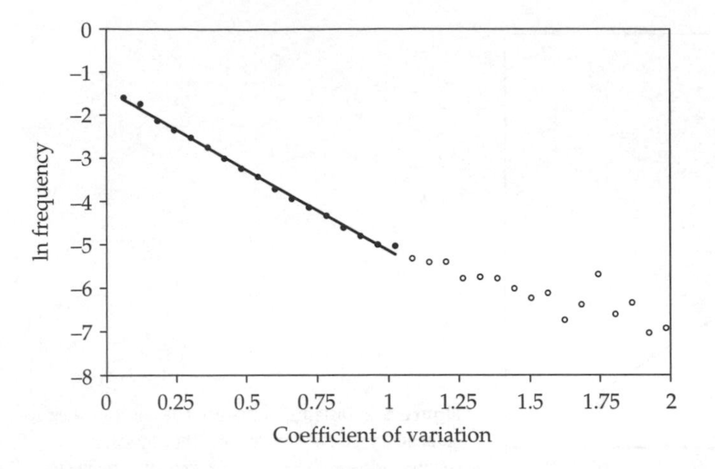

Figure 5.3 Phenotypic variation among individuals within inbred lines of mice. The regression equation is $y = -3.74x - 1.41$. Data from the Mouse Phenome Project at *http://phenome.jax.org/pub-cgi/phenome*.

A corollary of the idea that structures and processes are optimized is that all structures and processes are functional. There are two extreme points of view, laid out in the classical papers by Cain (1964) and Gould and Lewontin (1979). Curiously, both defend the primacy of the organism. The opening sentence of Cain's paper ('Listening to some physiologists, biochemists and karyologists, one might well get the impression that the phenotype is merely an inconveniently designed container for Krebs cycles and chromosomes.') chimes with the closing sentence of Gould and Lewontin ('A pluralistic view could well put organisms, with all their recalcitrant, yet intelligible, complexity, back into evolutionary theory'). Their views on organisms and phenotypes, however, are as different as their prose styles. Cain argues that features of body plans at all phyletic levels, from the pentadactyl limb of tetrapods to a minute bar of chitin on the limbs of certain specialized millipedes, are likely to contribute to normal levels of performance and have been selected, and continue to be selected, for that reason. Gould and Lewontin, on the other hand, argue that many features are unlikely to have any function at all, such as the colour of clams that live buried in the sediment, and are instead the outcome of history or constraint. This disagreement usually leads to philosophy or polemic, but should be settled experimentally by manipulating the phenotype: deleting or modifying the feature, or adding an arbitrary feature. Field experiments like this are done on a large scale by population ecologists who mark individuals to estimate survival or population size. The marks are, of course, intended to cause the least possible harm to the animals, but both deletions (e.g. toe-clipping; §2.2) and additions (e.g. bird-banding; see Burley 1986) have had unexpected effects on survival and reproduction.

The evolution of *Volvox* marks the appearance (within this group) of a clear distinction between soma and germ. It also marks the appearance of something else. When the daughter spheroids are mature they are released through rents in the parental spheroid, which then drifts on as a hulk of somatic cells that soon die. *Volvox* thus introduces two themes in the evolution of multicellular organisms. The first is the timing of reproduction and the allocation of resources between somatic and reproductive tissue. The second is the universal occurrence of senescence and death.

5.2 Reproductive allocation

A thrush sees a snail on the turf, captures it, and eats it; the fitness of the bird is somewhat increased while that of the snail is curtailed. The event will have been influenced by a multitude of characteristics of both predator and prey, all of which affect fitness in some way that could be described as a chain of physiological and ecological causes and effects. A particular mutation caused the synthesis of an altered protein, that modified the development of pigmentation so that the shell of the snail bore five bands rather than three, making an individual living in short turf more conspicuous to visual predators, and thereby reducing its chances of surviving to reproduce, and hence its fitness. It is often useful to think of the phenotype as being organized in this hierarchical way, as a series of levels intervening between the mutation itself and its eventual consequence in terms of fitness. These levels are successively genetic, physiological, developmental, and ecological. When a thrush kills the snail, this act is the manifestation of events at all of these levels, and determines, in this particular case, their effect on fitness. More precisely, the summation of many such acts determines the fitness of the allele in question, which will become more or less frequent depending on the rate of increase of the lineage that bears it. Individuals do not have a rate of increase; rather, each contributes through reproduction to the rate of increase of a lineage. The final phenotypic manifestation of a given genetic change is therefore its effect on the reproduction of an individual.

The schedule of reproduction during the life of an individual constitutes its life history. The life history of organisms such as *Volvox* or an annual plant is particularly simple: each surviving individual reproduces at a certain age and then dies. Its reproduction is therefore determined by whether or not it survives to reproduce; by the age at which it reproduces; and by the quantity and quality of offspring that it produces. These characters have directional effects on fitness: other things being equal, an allele that causes individuals to survive well, reach maturity early and produce numerous and well-provisioned offspring will spread, because the lineage that bears it will proliferate rapidly. Such characters represent the topmost level in the hierarchy of phenotypic effects, and are often termed *components of fitness*.

5.2.1 The correlated response to selection

Other things may not be equal, however, if components of fitness are not independent of one another. The evolution of a single phenotypic character is determined by its genetic covariance with fitness. Fitness can be regarded as a privileged character, in that it alone is directly selected. To analyse how a single character changes under natural selection thus involves considering two characters—the character itself, and fitness—and the relationship between them. The mean character state will change if character state is genetically correlated with fitness. This leads to the secondary theorem of natural selection, described further in §??. Adaptation generally involves coordinated changes in several or many characters, however, and to understand how complex adaptations such as the life history evolve we must ask how selection acts on several characters simultaneously. The basic principle is simply an extension of the argument for single characters: characters will evolve together if they are correlated genetically with one another and either or both are correlated genetically with fitness. This leads to the tertiary theorem of natural selection, described further in §6.2.7. The key concept is thus the genetic correlation between characters, which will govern their evolutionary dynamics. A genetic correlation may arise through linkage, epistasis, or pleiotropy.

In any finite population it is unlikely that genes will be associated entirely at random, if only because new mutations will arise by chance in some lineages and not in others. In asexual populations, such associations are permanent, and the characters that are expressed by the genes concerned will be genetically correlated because the genes themselves are transmitted together. Hitchhiking (§4.3.5) is an example of genetic correlation arising through linkage, but the phenomenon is quite general, and applies irrespective of the effects on fitness of the genes concerned. When different characters are encoded by different genes, they may become correlated through selection, either in sexual or in asexual populations, if their effects on fitness are not independent. This is because selection will cause some combinations of genes to increase in frequency and others to decrease. Suppose that there are two loci, A and B, each with two alleles. It happens that the genotypes A1B1 and A2B2 are more fit than the combinations A1B2 and A2B1. Selection will tend to increase the frequency of A1B1 and A2B2, creating a genetic correlation between the character states associated with the alleles at each locus. We might imagine, for example, that A controls the flower production of apple trees, with A1 producing many flowers and A2 few, whereas B controls fruit production, with B1 producing many fruit and B2 few. The genotype A1B1 is very fruitful in the short term, while A2B2 can invest the energy that it has not expended in reproduction on root growth, and thereby ensure that it will survive to reproduce in the following year; both genotypes may have high fitness. A1B2, on the other hand, wastefully produces many flowers that it cannot use to produce fruit, and A2B1 might mobilize resources for fruit production that cannot be achieved because of a paucity of flowers. The A and B loci will then evolve together if selection favours the combinations A1B1 and A2B2, creating a genetic correlation between flowering and fruiting. Conversely, a single gene may affect several characters, which in turn affect fitness in different ways. One allele of a gene affecting fruit production, for example, might direct the production of large apples, another of small. If the total mass of fruit produced is unaffected, then a tree producing large apples must produce few, whereas a tree producing small apples can produce many. The two characters, fruit size and fruit number, will therefore be correlated genetically because they are expressed differently by alleles of the same gene.

Characters that are genetically correlated through linkage, epistasis or pleiotropy will tend to evolve together. To describe the situation, it is convenient to recognize three types of character. One is fitness, which responds directly to selection. The second is the character that is the phenotypic criterion of fitness, having the effect of increasing adaptedness in a new environment; this responds indirectly to selection. The third is a character or characters that are not themselves responsible for differences in adaptedness, but that change because they are genetically correlated with the character that is. It is this third category of characters that represents the new element in the situation. The change through time of such characters can be termed a correlated (or indirect) response to selection. In practice, the distinction among the three categories is not as clear as I have made out. In the case of artificial selection, fitness is conflated with the character that is the target of selection, but there is a clear distinction between the direct and correlated responses. In the case of natural selection, the distinction between fitness and other characters is easily drawn, but often no one character can be designated as the target of selection, and there are instead a number of correlated characters, all contributing to fitness in some degree. Nevertheless, it may be useful to recognize three categories of character in order to emphasize the logical transition from self-replicators, through a character encoded by a gene and expressed by an organism, to a complex phenotype comprising many characters.

5.2.2 Antagonism of fitness components

The way in which characters such as components of fitness will change as a correlated response to selection is usually difficult to predict in detail for more than a few generations into the future. Nevertheless, it is possible to predict a general direction of change: the advance made by the primary character will cause other characters to regress. The notion that by improving one character we make others worse implies that 'worse' can be given an objective meaning: that any change in the value of a character can be expressed in terms of its partial effect on fitness. Mean fitness increases as the consequence of selection on the primary character, whereas the correlated response of other characters is such that, were these characters to change independently in the same way, fitness would be reduced. In this sense, the correlated response will generally be antagonistic to the direct response as the consequence of negative genetic correlation between components of fitness.

A general principle of antagonism is almost self-evident. If there were in general no deleterious consequences of selecting upward on a component of fitness, then it would be possible to select each component, simultaneously or sequentially, so as to produce an organism that could live for ever, and as soon as it was born would begin producing enormous numbers of large offspring as fast as resources became available. This is made impossible by the primary constraint responsible for selection, that the world is finite. Each individual can gather only a limited quantity of resources, and can therefore accomplish only a limited total quantity of reproduction. Insofar as different components of fitness impose separate demands on a common pool of resources, they must necessarily be antagonistic. This is the most general case of the principle of functional interference, which is created by the conflicting demands of the structures or behaviours contributing to any complex adaptation (§8.3.4).

The simplest case of competing demands made by components of fitness is the partition of resources among propagules. Other things being equal, producing more offspring will increase fitness. But producing larger offspring will also increase fitness, because larger offspring are more likely to survive. Given a fixed total quantity of resources, producing more offspring must mean that they will be smaller, whereas producing larger offspring involves producing fewer. This is a familiar problem in agronomy. In plants like cereals that are grown for grain, the total yield of an individual Y is determined by the number of infructescences per plant I, the number of fruits per infructescence F, the number of seeds per fruit S, and the mass per seed M:

$$Y = I \times F \times S \times M.$$

The characters I, F, S, and M are called components of yield representing successive partitions of total yield into different levels of structure. Because total reproduction is fixed for any given plant, this partition is necessarily antagonistic, with an increase in one component implying that other components must decrease. We therefore expect that components of yield will be negatively correlated. If a given genotype expresses an increase in one component, it will express a decrease in other components; there will be a negative genetic correlation among components of yield. Selection for advance in one component will thereby cause an antagonistic correlated response, reducing the value of other components, and will produce little if any increase in total yield per plant.

This commonsense argument is often justified. Characters such as seed number and seed mass in cereals or pulses are often negatively correlated. Moreover, the correlation can be altered by manipulating the total quantity of resources available to the plant: when the plant is stressed by low nutrient availability, the conflict between components of yield for resources is exacerbated, and the correlations among them become more highly negative. The same phenomenon occurs in other organisms: increasing the brood size of altricial birds by adding eggs to the nest, for example, causes a reduction in mean weight at fledging. Nevertheless, components of yield do not necessarily, or invariably, show negative correlation. The reason for this can be understood through the 'house–car paradox'. Household income is limited, and what is spent on one good is not available for purchasing another. We might therefore expect, following the same line of reasoning as before, that people living in grand mansions would drive small, cheap cars, whereas large and expensive models would be found outside modest tenements in less desirable areas of town. This is notoriously untrue. The reason is that although the income of any given household is limited, this income varies widely among households, and a rich family can afford to buy both a large house and an expensive car, while a poor family can afford neither. The negative correlation that would be created by variation in the allocation of resources is over-ridden by the positive correlation created by variation in the total quantity of resources.

The total quantity of resources available for reproduction may vary either because individuals have different opportunities, or because individuals have different capacities. A population dispersed in a heterogeneous environment will tend to display positive correlations between components of yield because some individuals will be growing in favourable sites, whereas others will have fallen on stony ground. This environmental variance can be eliminated experimentally by replication and randomization, but will always appear among natural populations of plants growing *in situ*. Even in a perfectly uniform environment, however, there may be genetic variance in the ability to gather resources. This will be particularly troublesome in two circumstances, involving poorly adapted populations in novel environments. The first is when individuals or families newly collected from the field are tested in the laboratory. Some genotypes will by chance be well-adapted to laboratory conditions of growth, and will express high values of all components of yield; others will be poorly adapted, and in all respects inferior. The second circumstance is when new lines are produced by mutation or inbreeding, so that some are likely to express unconditionally deleterious alleles. In either case, the result will be positive genetic covariance among components of yield.

The genetic covariance among components of yield, and hence the direction of the correlated response to selection, will thus depend on the ratio of two variances: the variance of total resource-gathering ability, or total yield, and the variance of allocation to different components of yield. When the variance of total yield is high, covariances will tend to be positive; this is likely to be the case in novel or heterogeneous environments. It is only when most of the variation of yield components among individuals is attributable to variance in patterns of allocation, rather than to variance in total yield, that components will be negatively correlated, directing an antagonistic correlated response to selection for any one of them.

The relationship between components of yield in natural populations of plants is illustrated by the comparative biology of *Plantago*, a very abundant and widely distributed herb, with over 200 species growing in different habitats throughout the world (Primack 1978, 1979; Primack and Antonovics 1981, 1982). This offers the opportunity to look first at variation among species. All the components of yield vary among species, although I and F vary much more than S and M, which is the usual case in plants. Thus, if all four components were independent, it is I and F that would be most important in regulating total yield. However, they are not independent: I and F are strongly correlated, as are S and M. Indeed, the regressions of F on I, and of M on S, have slopes that are close to -1, indicating a nearly perfect degree of compensation, with a unit increase in one character being accompanied by a unit decrease in the other. Thus, the total number of fruits per plant (fruit number, $I \times F$) and the total mass of seeds per fruit (fruit mass, $S \times M$) seem to represent more or less fixed quantities that do not vary much among species, and show very clear negative correlation between their components. The relationship between fruit number and fruit mass is more complicated (Figure 5.4). Both are good components of yield, in that both are correlated with yield, Y. However, fruit number is the better predictor of yield, because it is more variable than fruit mass. This is partly because annual and perennial species tend to produce different numbers of fruits: annuals have more fruits per plant than perennials. The correlation between fruit number and fruit mass is negative, as expected, although not very strongly so. It is the consequence of two different relationships. The first is that seed number varies according to life history: perennials bear more seeds per fruit than annuals, and have fewer fruits per plant. Species with similar life histories show no pronounced correlation between seed number and fruit number. The second is that seed mass varies independently of life history: annuals and perennials produce seeds of about the same size, but among species with similar life histories those with fewer fruits produce heavier seeds. The overall negative correlation between fruit mass and fruit number is thus a consequence of negative correlations between life histories, through seed number, and within life histories, through seed mass.

The pattern of negative correlations among yield components among species is consistent with a basic principle of antagonism. However, Primack found that correlations among individuals of the same species were usually positive. The reason for this is that there is less genetic variation within species than among species, so that a greater proportion of the variation within species is environmental. Natural populations growing in a heterogeneous environment will vary in total reproduction, and the positive covariance contributed by the environment will obscure any negative covariance among genotypes. This can be demonstrated by raising plants from a single species in the greenhouse. When fruit number and fruit mass of individuals growing in the field and in the greenhouse are

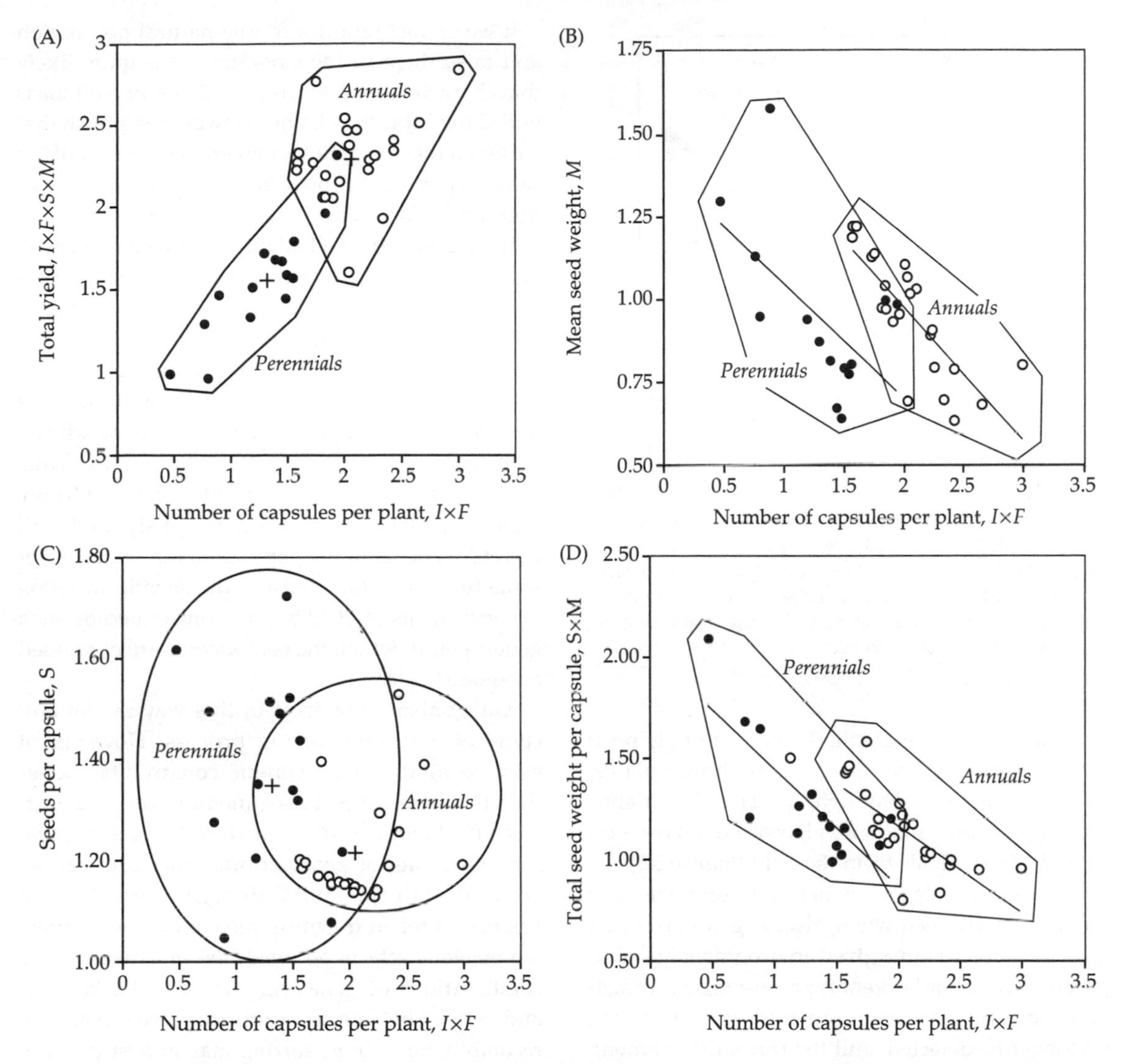

Figure 5.4 Comparative biology of reproductive allocation in *Plantago.* These figures are drawn from Primack's survey of reproduction in *Plantago*, used in the text to illustrate the antagonism of components of yield (Primack 1978, 1979; Primack and Antonovics 1981, 1982). They focus on the number of capsules per plant, $I \times F$; a similar analysis could be conducted for other components of yield. Each point is a different species of *Plantago.* A. Capsule number is a good component of yield, being strongly correlated with total yield. The correlation has two components: the difference between the mean values of the two groups of annual and perennial species, and the correlation among species within each group. B. Although annuals and perennials bear different numbers of capsules, their seeds are about the same size. Within either group, however, there is a pronounced negative correlation between seed weight and capsule number. C. There is no tendency for either annual or perennial species to form fewer seeds per capsule if they produce more capsules. However, annuals on average produce fewer seeds than perennials. Thus, there is a negative correlation of seed weight with seed number among species with a given lifestyle, but no variation between lifestyles, whereas seed number per capsule varies between annuals and perennials, but not among species within either group. D. The combination of these two effects causes a marked antagonism between the two components into which total seed yield can be partitioned, which is attributable in part to the difference between annuals and perennials, and in part to the differences among species in either group. The overall regression slope is about −0.5, so about half of any increase in capsule production is offset by a decrease in capsule weight.

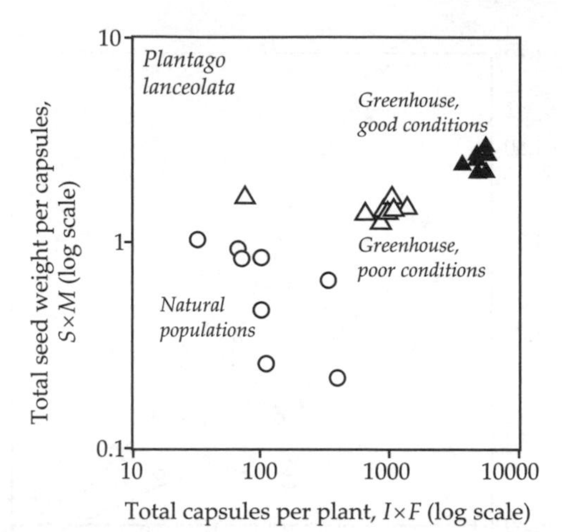

Figure 5.5 Environmental correlation of fitness components. The correlation between capsule weight and number is less marked among populations of a single species, *Plantago lanceolata*, and disappears entirely when plants from these populations are brought into the greenhouse. With more favourable conditions of growth, both capsule number and capsule weight increase greatly, creating a positive environmental correlation.

plotted on the same graph, they are strongly positively correlated, because the greenhouse plants, supplied with abundant light, water and nutrients, are simply larger and more vigorous in all respects than the individuals from the field (Figure 5.5).

In short, the antagonism between components of yield is clearly seen when diverse genotypes that have long evolved in a given environment are compared. When similar genotypes are grown in heterogeneous or novel environments, antagonism can no longer be detected, and the correlations among yield components are predominantly positive.

5.2.3 The evolution of genetic correlation

A straightforward genetic approach to the evolution of components of yield, by measuring correlations among genotypes in the laboratory or greenhouse, is therefore likely to fail. In order to display the underlying antagonism of such characters it is necessary to use an evolutionary approach, by showing that negative genetic correlations will evolve during the course of a selection experiment.

If we extract families from a natural population and raise them in the laboratory, it is quite likely that characters such as fruit number and fruit mass will show little correlation. However, suppose that we breed these families together, as members of the same experimental population, for many generations under laboratory conditions. Some genotypes will happen to be well adapted to this novel environment, and will produce large numbers of heavy fruits: these will increase in frequency through selection, and will quickly become fixed. Others will be very poorly adapted, producing a scanty crop of fruits with few seeds; these will decrease in frequency through selection, and will rapidly be eliminated. Genotypes that produce many fruits each with few seeds, or that produce few but heavy fruits, will be selected less rigorously, and will therefore persist in the population for longer. After some time, therefore, most of the genetic variation of components of yield will be contributed by such genotypes, in which the components are expressed antagonistically.

Antagonism may arise in this way as the outcome of a simple sorting process. However, it may be hindered by genetic constraints. Genes that by themselves have unequivocally favourable effects may be associated in the base population with inferior genes at other loci, and cannot be fixed until this initial linkage disequilibrium has been broken down by recombination. If there are epistatic effects on yield, the most productive combinations of genes may initially be lacking, and cannot be selected until they are created by recombination. Thus, sorting may at first produce positive correlations among yield components, by enhancing genetic integration. Nevertheless, as favourable combinations of genes spread and inferior combinations decline, continued selection will at last result in the emergence of antagonism. The force of this argument is that yield components will be consistently antagonistic only in well-adapted populations that are close to equilibrium under selection. It enables us to postulate that the evolution of yield is constrained by the antagonism of its components, despite the positive correlations that are often observed in novel or heterogeneous environments.

Despite the plenitude of studies of the phenotypic and genetic correlations among components of yield in crop plants, few selection experiments have been reported. McNeal *et al.* (1978) selected yield components of wheat for eight generations. Selection for mean kernel weight was only marginally effective, but caused a decrease in the number of kernels per spike, as expected. Selection for the number of kernels per spike, on the other hand, was completely ineffective, and actually caused an increase in mean kernel weight. In neither case was there any change in total grain yield per plant, so that selecting a given component of yield would not be likely to be an effective way of increasing yield itself. Perhaps the most general result, indeed, is that selecting yield components in order to increase total yield, on the grounds that yield itself has low heritability, is usually unsuccessful. For example, Lopez-Reynoso and Hallauer (1998) selected longer ears in maize for 27 generations. This increased ear length by nearly 40%, but yield was unaffected, perhaps because ear diameter and the number of kernel rows were reduced, while development time was increased.

5.2.4 Optimal allocation

Because selection will not favour the indefinite exaggeration of any given component of yield, the physiological and developmental machinery that underlies its expression will not continue to be enhanced beyond a certain point. The antagonism of fitness components prevents organisms from being perfectible. Instead of being perfectly adapted in all respects, organisms will instead evolve as compromises, expressing intermediate values of characters such as fruit size or fruit number. Nevertheless, selection will always favour an increase in total yield (or, more generally, fitness). The combination of values of yield components that emerges through continued selection is therefore an optimal compromise, in the sense that it maximizes total yield. Optimality models have been very useful for interpreting and predicting reproductive allocation. There are extensive reviews of theory and comparative data by Roff (1992) and Stearns (1992).

The pattern of reproduction should therefore vary in a predictable fashion among organisms with qualitatively different ways of life, even if they are closely related species sharing a recent common ancestor. Annual and perennial species of *Plantago* allocate resources to seeds in different ways: annuals produce a smaller number of larger seeds, while perennials produce a larger number of smaller seeds. Why should they differ in this way? Perennials occupy their local site by growth and proliferation over several years, and a seed, even a large and well-provisioned seed, has little chance of succeeding in competition with a well-established vegetative plant. The seeds of perennials will thus be adapted primarily to seeking out new sites. Because sites that are favourable to growth but currently unoccupied are likely to be rare, only individuals that produce a large number of seeds are likely to be successful; but because their total reproductive capital is limited, the seeds they produce must be small. Annuals have no vegetative propagation, and must recolonize the local site each year by reseeding. Larger seeds with more stored reserves and thus faster initial growth are likely to succeed in competition with smaller, less well-endowed seeds; but plants that produce larger seeds must produce fewer. The contrasted patterns of allocation that are found in annual and perennial species of *Plantago* show how the common constraint imposed by a limit on total reproductive output is modulated by the ecological context in which it is expressed.

It is in principle possible to predict the quantitative allocation of resources between components of yield if the relevant constraints are known. Parents which lay very small clutches obviously have low fitness, because they can produce only a few fledglings. Less obviously, parents which produce very large clutches may also have low fitness, because they are unable to provide enough food for so large a brood, and many of their offspring are small, fledge in poor condition, and die in their first winter. The number of surviving offspring is therefore maximized by an intermediate clutch size that represents an optimal compromise between antagonistic components of yield. This is one of the most famous hypotheses of evolutionary ecology: the mean clutch size in a population of birds should be its optimal value under natural selection, that is, the clutch size that maximizes the production of independent young (Lack 1966). The great tit, a passerine bird of deciduous woodland, has been intensively studied by ornithologists in England and the Netherlands. It can lay a clutch of up to about 20 eggs, but most clutches are much smaller: the mean is about 8.5, and the mode between 8 and 10. Why do most birds lay many fewer eggs than they are capable of doing? The data on clutch size and the survival of young birds in the Wytham population of great tits is taken from Lack's book (Lack 1966, Table 7); a great deal more work has been done at Wytham since then, but there is a certain appeal in using Lack's original data to test his own hypothesis. The data set includes the number of clutches of given size, and the number of young birds recovered 3 months or more after leaving the nest from clutches of given size, for 17 consecutive years. The mean clutch size in the whole sample of 1098 nests was about 8.5 eggs, but this cannot represent a physiological limit because some birds produced clutches of as many as 15 or 16 eggs. Both sampling effort and the overall survival of nestlings and fledglings varied from year to year, so I have calculated a standard value of survival by multiplying the observed survival of clutches of given size in a given year by a quantity (P_{mean}/P_{year}), where P_{year} is the mean survival rate over all clutches in a given year, and P_{mean} is the mean survival rate over all years. Other ways of dealing with year-to-year variation are possible, of course. This standard survival rate decreases with clutch size B; the regression is survival rate = $C-kB$, with $C = 0.177$ and $k = 0.0098$. (Very small clutches of 1–3 eggs and very large clutches of 16 eggs do not fit this relationship, but were found only rarely.) Thus, both small and large clutches are relatively unproductive, and an intermediate clutch size may be optimal. More precisely, the total productivity of a brood, that is the total number of surviving chicks, is the product of clutch size and survival rate, or $B(C - kB)$. The optimal clutch size is that which

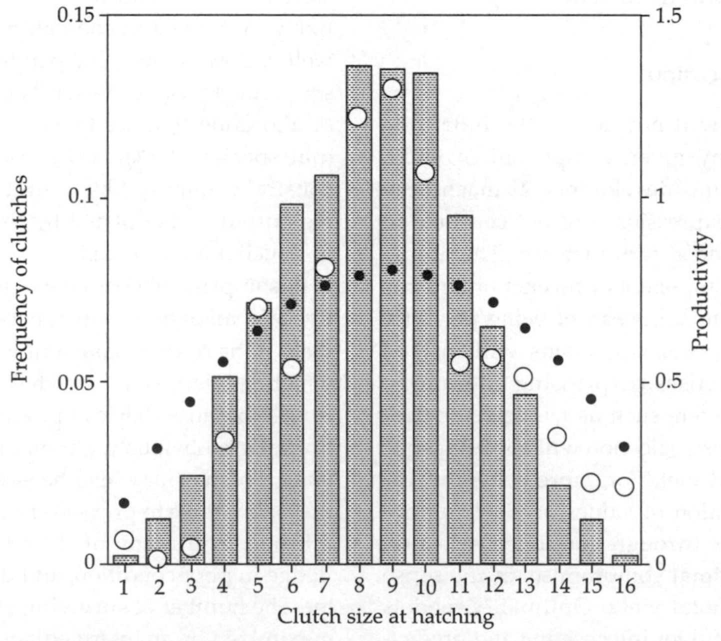

Figure 5.6 Optimal clutch size in the great tit. The histogram bars are observed values, a nearly normal frequency distribution with mean 8.48 and standard deviation 1.75. The large open circles show the actual productivity of clutches, in terms of survival to 3 months of age. This curve is also unimodal and approximately normal, with a mode at 9 eggs. The small filled circles give the productivity as calculated from the survival regression, as $B(C - sB)$, yielding an optimal value of $C/2s = 9.03$ eggs.

maximizes this quantity: by simple calculus, this is $B = C/2k$. Observed clutch sizes follow a nearly normal frequency distribution with mean 8.48 and standard deviation 1.75. The actual productivity of clutches, in terms of survival to 3 months of age, is also unimodal and approximately normal, with a mode at 9 eggs. Optimal productivity as calculated from the survival regression is $C/2k = 9.03$ eggs (Figure 5.6). A straightforward optimization approach seems to give a satisfactory quantitative account of adaptation for characters such as clutch size.

5.3 Life histories

The compromise between seed size and seed number is adjusted differently in annual and perennial species of *Plantago*, but this does not explain why some species should be annual and others perennial. The difference between the two lifestyles is that annuals die immediately after reproducing for the first time, whereas individuals of perennial species often survive to reproduce in the following season. Reproduction requires resources that are exported as seeds or chicks; but survival equally requires resources that must be retained in order to grow, to accumulate storage tissue, to maintain defences against predators and pathogens, and so forth. In discussing components of yield, I have emphasized the antagonism of current components of reproduction, arising from a limited pool of available resources. These resources, however, may more generally be spent on present reproduction, or hoarded in order to support survival and reproduction in the future. To the extent that individuals can garner a finite supply of resources at any one time, these resources must be allocated between present and future reproduction, and what is spent in the present will not be available in the future. There may then be antagonism between prospective components of fitness: if present reproduction is increased, the prospect of future reproduction may be threatened. This concept has been called the *cost of reproduction*. An increase in the current level of reproduction may represent an additional metabolic drain that reduces the quantity of resources available for reproduction later in life: this is the *fecundity cost*. It may also decrease the likelihood of surviving to reproduce at all, either because individuals whose stored reserves have been depleted are more vulnerable to stress, or because activities associated with reproduction, such as courtship, may be inherently risky: this is the *survival cost*.

Studying the cost of reproduction by measuring phenotypic or genetic correlations between prospective components of fitness in experimental populations encounters the same difficulties as a similar approach to components of yield. Indeed, they are more severe. The longevity of different species of *Daphnia* and related genera is negatively correlated with the rate at which they produce offspring, as we might expect. However, if we raise clones in the laboratory using newly captured individuals from natural populations, the correlation between longevity and reproduction is usually positive. Part of this effect is straightforwardly environmental; individuals from the same clone also show a positive correlation between longevity and reproduction, no doubt because some culture vessels receive more food than others, or are subtly different in some other way. Another part is because some clones flourish better in laboratory culture than others. This pattern of positive correlation has been observed by myself (Bell 1984a, b) and by several other biologists; it applies to many organisms other than water fleas; and it also applies to other prospective components of fitness, such as the number of offspring produced in different instars during adult life. Various statistical devices for manipulating the data so as to reveal an underlying pattern of negative genetic correlation have been suggested, but none are very convincing, and the only satisfactory way of demonstrating the fundamental antagonism among components of fitness is, as before, the selection experiment.

5.3.1 Selection in age-structured populations

Pacific salmon (*Oncorhynchus*) migrate as smolts from the streams in which they were born to the depths of the North Pacific Ocean, 2000 km or more away. Having fattened in the pastures of the sea, they then swim all the way back to their natal stream to spawn. This was a very hazardous journey even before the first gill-net was set on the Fraser river; few would survive it, and the probability of surviving a second round trip to the Pacific and back would be very low indeed. An adult salmon that has reached the spawning stream has therefore nothing to gain by restraint, and should commit all its reserves to reproduction, rather than holding some back in a futile attempt to survive to breed for a second time. The suicidal reproduction of these fish is the consequence of a survival cost that makes any attempt to reproduce at all nearly certain to be fatal. Atlantic salmon (*Salmo*) undertake a similar but somewhat less arduous migration to the north-west Atlantic; they spawn with great vigour, and many die, but a few survive to repeat the voyage and spawn for a second time. Sea trout move into coastal waters and back, without venturing into the open sea, and routinely survive for two or more spawning seasons. Exclusively freshwater salmonids such as lake trout, and many related fish such as chars and graylings, make at most very limited journeys along the lake shore to find their spawning grounds. Post-reproductive survival in this group of fish thus appears to be governed by a prospective cost, the risk entailed by the spawning migration.

The complete collapse of somatic maintenance after reproducing is an extreme case. More generally, current reproduction may be somewhat risky, or may reduce the reserves available for future growth and reproduction, without leading inevitably to death. The timing and intensity of reproduction will evolve through selection towards an optimal schedule that will differ among species or populations, depending on how it is constrained by the current and prospective costs of reproduction in different ecological contexts. The expected reproductive output at age x is the product of fecundity at that age $b(x)$ and the probability of surviving to that age $l(x)$. The expected lifetime output is the sum of this quantity over all ages from birth to death, $\Sigma l(x)b(x)$ in discrete time. Reproduction at any age must be discounted by the population growth rate r, because population growth will progressively dilute the future descendants of offspring born to older parents. The fitness associated with a particular schedule of reproduction is thus $w = \Sigma e^{-rx} l(x)b(x)$ (Charlesworth 1980).

In some cases it is possible either to identify a particularly important cost of reproduction in a natural population, or to impose a specified cost of reproduction on an experimental population. The result is a shift in the schedule of reproduction that is predictable from the age-specific incidence of selection. Edley and Law (1988) passed experimental populations of *Daphnia* through a series of sieves. By removing small individuals from some lines and large individuals from others they acted themselves as size-specific predators. The base population was a heterogeneous mixture of clones from different localities, and lines of several hundred individuals were selected through about 40 4-day cycles of growth and culling, representing about a dozen generations of asexual reproduction. The outcome was no doubt attributable primarily to sorting the variation originally present. Lines from which small individuals were culled evolved more rapid growth, because genotypes that escape more rapidly from the vulnerable size-classes will increase in frequency. Individuals from lines in which large individuals were culled grew more slowly, but reproduced at smaller sizes, because selection would favour genotypes that could reproduce before being sieved out from the population.

5.3.2 Depreciation of later reproduction

Selection will generally favour more intense reproduction in early life, even when no deliberate age-specific selection is applied. The reason for this can readily be appreciated through an analogy with insurance and investment.

Suppose that you wish to purchase an annuity that becomes payable when you reach a certain age. This future income has a present value that depends on the probability that you will live to enjoy it: obviously, it might be very valuable if you are confident of living well beyond the redemption date, but almost worthless if you are likely to die beforehand. This value is reflected by the premium that you pay to the insurance company, which represents a bet that you will survive for long enough to profit from the transaction. At any given age, the amount of the premium that you must pay in order to purchase a given annuity depends on the insurance company's estimate of your probability of survival. The premium will be lower when your chances of survival are lower, because the present value of the annuity is less. The present value of a future return on investment thus declines with the probability that you will survive to collect it. Thus, the later in life an annuity commences, the less valuable it is at present.

Alternatively, suppose that you wish to invest a sum of money at compound interest in order to realize the greatest possible capital gain at some fixed date in the future. It is assumed that you will still be alive on this date. Clearly, it is better to invest the money sooner than later, because the sooner the investment is made the sooner the interest that it earns will itself begin to earn interest. This effect will be larger when the rate of interest is greater. Thus, the later in life an investment is made, the less valuable it will be.

Putting these two arguments together, we can see that postponing an investment will reduce its value in proportion to the rate of mortality and the rate of interest. Because populations grow at compound interest, there is a direct analogy between investment and reproduction. Postponing reproduction will reduce the number of your descendants alive at any given date in the future, because it postpones the time at which your offspring will themselves begin reproducing, and makes it less likely that you will survive to reproduce at all. This principle was first expressed precisely by Hamilton (1964a, b). Fitness has been defined as a rate of increase, which is wholly determined by the rates of survival and fecundity throughout life. A genotype whose rate of increase exceeds that of the population as a whole will tend to spread. Any age-specific change in survival or fecundity will therefore be selected according to its effect on the rate of increase of the population, given that the schedule of reproduction is otherwise held constant. More formally, the manner in which selection modifies the schedule of reproduction will depend on the partial derivative of the rate of increase with respect to a given age-specific change. These partial derivatives for the rate of survival $p(x)$ and the rate of fecundity $b(x)$ at age x are as follows:

$$\partial r/\partial \ln p(x) = (1/T)\, \Sigma_{x+1}^{\text{death}}\, e^{-ry}\, l(y)\, b(y)$$
$$\partial r/\partial\, b(x) = (1/T)\, e^{-rx}\, l(x),$$

where T is mean generation time. For any age within the reproductive span (any age x for which $b(x) > 0$), the right-hand sides of these equations decrease as age x increases. What these equations say, therefore, is that a given small change in survival or fecundity will have a greater effect on fitness, and will therefore be selected more strongly, if it is expressed earlier in life: that is, the rate of selection on any component of fitness decreases with age. The effect of a gene on survival or fecundity is discounted with increasing age by a factor $e^{-rx}\, l(x)$, or in other words by the rate of mortality (which decreases $l(x)$), and by the rate of population increase (which decreases e^{-rx}). This is the formal justification of the intuitive argument that I began with.

The implication of this conclusion is that selection will generally act so as to make the age at first reproduction as early as possible. However, this is true only for microbes; in multicellular organisms, a period of growth supervenes before reproduction. The weakening of the force of selection with age means that a special explanation is required for multicellularity, and for somatic growth and delayed reproduction generally. A larger individual can produce more offspring, and if it grows fast enough the increase in fecundity through growth may exceed the discounting factor $e^{-rx}l(x)$. However, growth alone cannot account for delayed reproduction, because, on purely demographic grounds, there is no reason that an individual should not grow and reproduce at the same time. Reproduction should be delayed, therefore, only if future growth is retarded by reproduction more than future reproduction is discounted. This will only be the case if the virgins grow so fast that their discounted fecundity $e^{-rx}l(x)b(x)$ increases with age early in life. This seems likely to be true in many cases: in very small, young individuals, even a very few propagules will represent a large fraction of somatic mass, and reproduction must severely inhibit growth because it uses up most of the tissues on which future growth would have been based. It is thus the prospective cost of reproduction in terms of a loss of potential fecundity later in life that reduces the value of early reproduction, and in some organisms may cause selection for a prolonged period of pre-reproductive growth.

5.3.3 Artificial selection for early reproduction

A culling programme that preferentially removes larger and older individuals selects for increased reproduction among smaller and younger individuals. Conversely, selection for early reproduction may have the indirect effect of reducing survival and fecundity later in life. For example, it is slow and inconvenient to select for total lifetime productivity in domestic fowl, which may live for several years; a much more effective method would be to select for egg production early in life, provided that early vigour is a reliable guide to lifetime egg output. However, it is known that the correlation between first-year output and output in subsequent years falls steadily throughout life, so that egg production by birds 5 or 6 years of age is almost independent of their output in the first year of life; moreover, longevity is negatively correlated with early production. This suggests that selection for early vigour will expose and exacerbate the antagonism of fitness components expressed at different ages. Morris (1963) selected egg production during the first 6 months of life for 12 years in a closed flock of White Leghorns (Figure 5.7). There was a steep and continued response that amounted to a gain of about 35 eggs (from about 45 to about 80) by the end of the experiment. This was achieved largely through a decrease of nearly a month in the age at which the first egg was laid, and to a lesser extent through an increase in the rate of lay thereafter. Residual egg production, in this case the number of eggs laid between 6 and 18 months of age, at first increased along with early production. This was consistent with the high positive genetic correlation

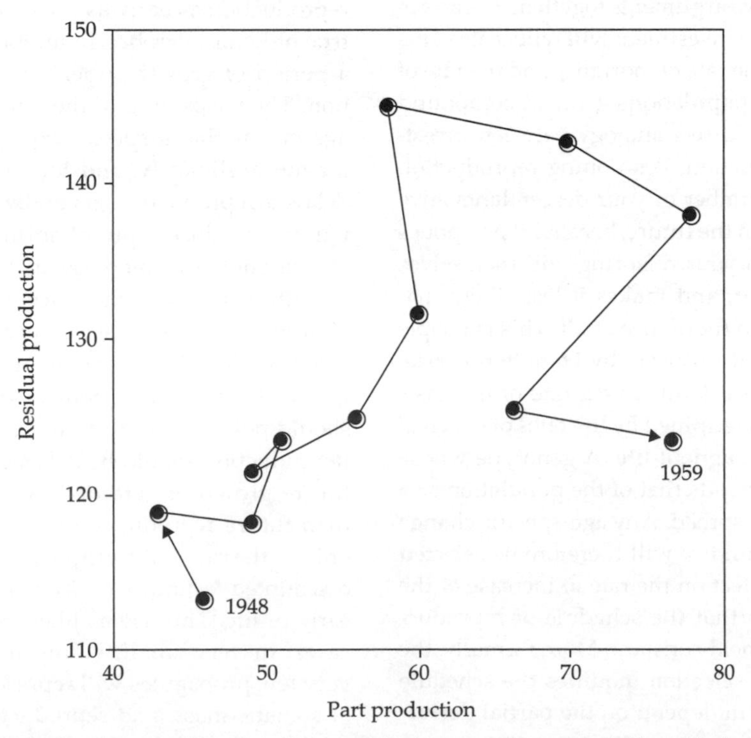

Figure 5.7 Effect of selection for egg production early in life (part production) on production later in life (residual production). After Morris (1963).

between early and residual production in the base population. However, during the last 5 years of the experiment residual production fell steeply, while early production continued to increase; during this period there was a gain of 20 eggs in early production, but a correlated loss of 19 eggs in residual production. At this point, part-record selection had ceased to be effective, with early gains being annulled by later losses. It had first caused an overall increase in productivity, offset by a decrease in egg weight as a correlated response to selection for egg number. It later failed to generate any overall gain in egg number, because of a decrease in egg number later in life as a correlated response to selection for early egg output.

Wallinga and Bakker (1978) selected a laboratory population of mice for the number of offspring in the first litter for 25 generations. They obtained a direct response of about six offspring. This effect extended to subsequent litters, although by the fourth or fifth litter the difference between selected and unselected lines was very small, and older animals in the unselected line may have produced larger litters. Selecting for the size of the first litter involves selecting for mice that survive to reproduce for the first time, and early survival was greater in the selected line than in the unselected control line. However, animals in the selected line survived much more poorly after their fourth litter. In this case, the antagonistic response to selection for early vigour seems to involve reduced survival among older individuals. Similar results have been obtained in experimental populations of beetle and flies. Mertz (1975) selected populations of the flour beetle *Tribolium* for increased reproduction early in life by simply discarding individuals after a few days of egg-laying. He did not observe any effect of selection on longevity, but fecundity later in life was less in lines selected for early reproduction. The correlated response to selection for early reproduction may thus involve either a decrease in subsequent survival or a decrease in subsequent fecundity.

The general implication of experiments such as these is that selection for age-specific components of fitness may be no more effective than selection for components of yield. In the first few generations of selection overall fitness may advance, but if selection is continued an antagonistic correlated response is induced, and although the character under selection may continue to advance, this advance is more or less completely countered by the regress of other fitness components.

5.3.4 Senescence

Most individuals in most species die young, but a few linger on into extreme old age. The limit to lifespan is poorly defined because any estimate depends on sample size and the conditions of growth, but it varies so greatly among species that a few generalizations are possible. Maximum lifespan L_{max} of metazoans was found to scale with body size as $L_{max} = 1570W^{0.15}$ (L_{max} days, W g) (Blueweiss *et al.* 1978), although regressions for mammals, birds, and fish separately have larger exponents of about 0.2–0.25 (Speakman 2005). This discrepancy arises because the longevity of smaller animals is greatly overstated by the Blueweiss regression and instead follows a rule of roughly $L_{max} = 250W^{0.25}$. It is not very far out to say that maximum lifespan increases as $W^{0.25}$ with an elevation of 1 year at 1 g, so it increases from about 10^1 days in the smallest animals to about $10^{4.5}$ days in the largest and longest-lived (Figure 5.8). Age at maturity L_{mat} can be estimated more reliably than lifespan and scales in a similar way, $53W^{0.27}$ (L in days, W in g) (Blueweiss *et al.* 1978), so a ¼-power allometry may apply to all epochs of life.

Lifespan would be limited even if old individuals were fully as vigorous as their younger comrades, as all would sooner or later fall victim to accident or disease. In many organisms, however, the probability of dying eventually increases with age. The increase in the probability of dying, from whatever cause, defines the process of senescence. It usually begins at or shortly after the age at first reproduction (Promislow 1991). If older individuals die more readily than younger individuals by some factor, but from similar causes, there will be an exponential increase in mortality d at rate γ: $d(x) = d(0)\exp(\gamma x)$, the *Gompertz equation*. If this process operates on only a part of the juvenile mortality rate then $d(x) = d(0) + a\exp(-\gamma x)$, where a is the senescent component of overall juvenile mortality $d(0) + a$. This is the *Gompertz–Makeham equation*, which Promislow (1991)

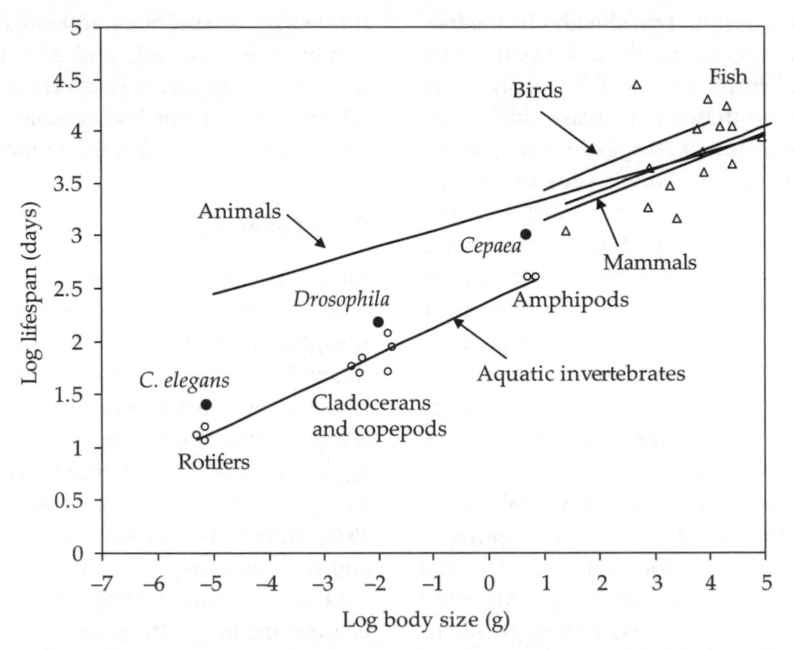

Figure 5.8 Animal lifespan. Data and regressions from Blueweiss *et al.* (1978) for animals, Peters (1983) for fish, Gillooly *et al.* (2002) for aquatic invertebrates, Speakman (2005) for mammals and birds.

fitted to demographic data for wild populations of mammals and found that adult mortality generally increases with age, showing that senescence occurs in natural conditions of life. Different causes of mortality in juveniles and adults can be represented by $d(x) = d(0) + ax^b$, or equivalently the excess adult mortality $d(x) - d(0) = ax^b$, the *Weibull equation*. Estimates of the Weibull parameters for birds and mammals are given by Ricklefs (1998) and show that mortality increases sharply with age, estimates of b falling between 1 and 5. The fraction of deaths attributable to senescence is greater when the fixed rates $d(0)$ and a are small and the exponent b is large. Thus, if $d(0) = a = 0.1$, the fraction of deaths attributable to senescence rises from about 5% for $b = 1$ to 14% for $b = 3$, and to 23% for $b = 5$.

5.3.5 Soma and germ

Reproduction in ideal unicells that reproduce by a perfectly equal fission will be discounted by the rate of mortality and the rate of population increase, as in multicellular organisms, and they will generally evolve so as to reproduce as soon as possible. They will not senesce, however, because there is no difference in age between the products of fission. If fission is imperfect, however, this conclusion does not hold. This is clear in budding yeast, where parent and offspring can be distinguished and the accumulation of bud scars eventually terminates the life of a cell. Even in *E. coli*, however, one daughter cell will inherit a greater share of used structures and this creates an ageing lineage whose last representative dies of old age (Stewart *et al.* 2005). Somatic lineages of multicellular organisms have a finite lifespan in tissue culture, amounting to about 50 divisions for human fibroblasts (Hayflick and Moorhead 1961; review by Shay and Wright 2000), caused by the progressive shortening of telomeres (Bodnar *et al.* 1998).

The conventional view of animal development, dating from August Weissman, is that hereditary transmission occurs solely through a special caste of germ cells, while the sterile vegetative tissues of the body, the soma, represent a temporary vessel for the germ cells that is rebuilt in every generation.

The germ cells thus proliferate within bodies just as clones of unicellular algae or yeasts proliferate outside bodies: and in the same way, they form lineages of dividing cells that do not senesce. The germline is, so to speak, the generating lineage of the body, with the important qualification that somatic lineages cannot usually re-enter the germline through mutation. The activity of the soma in supporting the proliferation of germ cells is discounted by mortality and population increase, but the persistence of the soma as a distinct physical entity through the lifetime of an individual implies that genes can be expressed at different ages during that lifetime, and therefore that deleterious mutations, or the deleterious effects of mutations, can be expressed later in life. It is thus the soma that undergoes senescence.

It may be useful to distinguish between two kinds of development (with many intermediate cases). The first occurs in organisms such as plants or cnidarians, where a population of stem cells continues to divide freely throughout life, giving rise both to germ cells and to somatic tissues. In the extreme case, a multicellular organism may reproduce vegetatively by dividing into two identical fragments, each of which regenerates the adult form. This would be equivalent to the binary fission of a unicell, and such organisms should not senesce. There is, indeed, some evidence that worms that reproduce by a nearly equal binary fission do not senesce, or senesce very slowly, and this supports the evolutionary view that senescence in organisms such as mice or flies is directed by the weakening of selection with age (Bell 1984a). In practice, fission is rarely perfectly equal. In vegetatively reproducing worms, for example, cells in the hind region of the body proliferate to form a new large offspring, while the tissues of the head region persist; successive head fragments might then senesce whereas tail fragments would not. In *Hydra*, a band of cells near the base of the trunk region divides continually to produce tissues that move upwards, differentiating as they do so, before eventually being shed from the body at the tips of the tentacles. It is not clear that a living fountain of proliferating stem cells should senesce at all.

The second pathway of development involves the early segregation of the germline and the re-differentiation of somatic cells to form different tissues. In the extreme case the number of cell divisions is fixed and the number of germ cells is likewise fixed and small. In the gastrotrich *Lepidodermella*, for example, exactly four internally self-fertilized eggs are produced (Sacks 1964). Once these have been shed there is no longer any effective selection for prolonging life, and the animal dies some random number of days afterwards (Bell 1984b). Nematodes have this type of development, although the number of eggs is larger. In flies there are no further mitotic divisions in the adult, but the eggs are proliferated by a germinal band and it is not clear whether their number is effectively limited. In most vertebrates reproduction continues throughout adult life, either because eggs are continuously produced or because the number initially segregated in the germline is much greater than could ever be matured. Gastrotrichs, nematodes, flies, and mice all senesce, but it is not obvious that the underlying mechanisms will be the same.

5.3.6 The nature of senescence

There are two general physiological theories of senescence. One treats it as a discrete process with a single predominant cause, whereas the second regards it as the consequence of many independent underlying pathologies. The dichotomy has profound consequences for the evolution of senescence and the potential for extending the lifespan. If there is a single cause then this can be identified and mitigated, and the small number of genes that govern the physiological mechanism will be those primarily responsible for normal senescence. If there are many causes, on the other hand, mitigating any one of these will merely produce a trifling extension of lifespan that will then be limited by a disease with a slightly later age of onset.

The simplest theory of senescence is that it is caused by the wearing-out of tissues, and that the amount of wear depends on the amount of use. Since basal metabolic rate scales with $W^{0.75}$ (Peters 1983) the mass-specific rate scales with $W^{-0.25}$. Lifespan, or components of lifespan, scales with $W^{0.25}$, and the metabolic expenditure per unit weight per lifespan is therefore approximately a constant: one gram of tissue performs the same amount of work (about 5 kJ) in any organism. Hence, it is inferred, life ends when this fixed amount of work has been accomplished. This is the rate-of-living hypothesis (reviewed by Speakman 2005). Tissues are usually thought to wear out through the irreversible accumulation of damage. The source of the damage caused by use has been vigorously debated, with the prevalent view being that reactive oxygen species that attack organic molecules, including DNA, are primarily responsible. Whatever the underlying cause, the theory suggests that there are three ways of extending lifespan. The first is dormancy: a stage in the life history that is resistant to stress requires severely reduced metabolism and could therefore be very long-lived. The second is reduced input: a low diet will necessarily reduce the rate of living. The third is reduced output: slowing down the rate of physiological processes should reduce the rate of wear. All of these ways are effective. Small animals like rotifers, for example, reconfigure their body as a resistant structure in response to stresses such as desiccation, and may survive in this form for many times longer than their normal span. In active animals caloric restriction is often effective in extending life beyond its usual limits (see Sohal and Weindruch 1996). This suggests that genes governing these processes act to restrict lifespan by inducing senescent changes in normal individuals. This has been strongly supported by the isolation of very long-lived mutants of the model nematode *Caenorhabditis* (reviewed by Hekimi *et al.* 2001). Mutations in *clk* genes extend lifespan, often in an epistatic fashion: thus, the normal median lifespan of about 15 days is increased to 18–19 days in *clk-1* and *clk-2* mutants and to 28 days in the *clk-1 clk-2* double mutant. These animals have slow physiological rates. Mutations in *eat* genes affect the development of the pharynx and thus the rate at which food is pumped into the gut; they extend lifespan by about the same amount as the *clk* single mutants, perhaps through the caloric restriction they impose. The *daf* genes are involved in metabolic regulation through the insulin-signalling pathway and regulate development as a dauer larva, the dormant resistant phase of the life cycle, and *daf* single mutants may live twice as long as the wild type. Combining mutations in different pathways can dramatically extend lifespan: thus, *daf-2 clk-1* individuals may live for almost 80 days, whereas few if any wild-type individuals survive as long as 20 days. All these mutations involve loss of function; hence, the normal activity of genes like *clk*, *eat*, and *daf* brings about senescence and death. This seems to be the consequence of normal levels of activity and metabolism; loss-of-function mutations extend lifespan as an indirect effect of reducing the rate of living.

Whether or not this conclusion can be extended from *Caenorhabditis*, where the adult soma consists of a fixed number of non-dividing cells, to organisms with different modes of development, such as mice, is still unclear. There are no known mutations in mice that extend lifespan as much as 50%, with the possible exception of some dwarf strains. Lifespan is doubled, however, by caloric restriction, suggesting a common physiological basis for senescence. Moreover, the mouse orthologue of *clk-1*, *mclk1*, may modestly extend lifespan in heterozygotes (Liu *et al.* 2005).

5.3.7 The evolution of senescence

Instead of having a fixed effect at a single age, mutations may be expressed differently in younger and older individuals. Mutations that are deleterious at all ages will be rapidly eliminated, of course, and those with beneficial effects at all ages will rapidly spread. The remainder will have opposed effects on fitness at different ages, increasing fitness at some ages but decreasing it at others. Those that increase survival or fecundity in younger individuals, at the expense of reducing survival or fecundity in older individuals, will be selected more strongly than those with the converse effect of benefiting older individuals at the expense of younger ones, and will be more likely to spread, or will spread more rapidly. Indeed, the greater effect of early vigour on overall fitness implies that a rather small increase in survival or fecundity early in life will be favourably selected, even if it has severely deleterious consequences later in life. Mutations that increase fecundity early in reproductive life will spread more quickly than mutations with similar effects later in life; thus, selection will generally tend to increase early vigour. The converse is also true: selection will indirectly reduce vigour later in life, causing a senescent decline in rates of survival and fecundity. Senescence then evolves because the expression of prospective costs of reproduction is biased by the general weakening of selection with age. This idea, originally due to Williams (1966), is the *antagonistic pleiotropy* theory of ageing. It seems to be completely compatible with the rate-of-living physiological theory of senescence, because a prospective cost of reproduction is necessarily entailed by the finite capacity of tissue to perform work. Increased metabolism early in life will incur a greater amount of damage and will thereby accelerate senescence and curtail lifespan.

An alternative evolutionary theory of senescence is that the expression of deleterious mutations will increase with age. Suppose that deleterious loss-of-function mutations with variable ages of expression occur at many loci. The accumulation of deleterious mutations will be checked by selection at all ages, of course, but countervailing selection is more effective in younger than in older individuals. The equilibrium level of mutational load will therefore increase with age: deleterious mutations will accumulate to a greater extent in older age classes. In the simplest case, different deleterious mutations will have different fixed ages of expression, and those that are expressed later in life will be maintained at higher frequencies in the population. It is also possible that the age of expression of a deleterious mutation is itself a character that can be selected, in which case selection will favour shifting expression to later ages, although in this case selection on modifiers that govern the age of expression will be very weak (Partridge and Barton 1993). This is the mutation-accumulation theory of ageing, first suggested by Medawar (1952). It corresponds to a polygenic theory in which senescence is the outcome of many independent sources of mortality, and it therefore predicts that the variance of fitness components will increase with age. A large experiment using lines from an outbred population of *Drosophila* adapted to laboratory conditions reported that the genetic variance of age-specific reproductive success increased sharply with age (Hughes *et al.* 2002, who review earlier experiments). Moreover, the severity of inbreeding depression also increased with age, suggesting that late-onset recessive deleterious mutations were segregating in the population. This is the strongest evidence that such mutations contribute to senescence within the 250 or so generations since the ancestral population was founded.

5.3.8 Selection for delayed senescence in *Drosophila*

If senescence reflects the weakening of selection with age, then not only can it be accelerated by selecting for early reproduction, as the experiments described above show, but it can also be delayed by selecting for reproduction late in life. A procedure that makes old individuals more likely to live is, of course, a more impressive testimonial to the theory than one that merely makes younger individuals more likely to die. The simplest procedure is to breed solely from older parents. This artificially opposes the discounting of reproduction with increasing age, and should lead to an increase in fecundity late in life, with a correlated increase in longevity. (It necessarily creates direct selection for increased longevity, too, because only surviving individuals can breed.) Rose (1984; see also Rose and Charlesworth 1980, 1981) used a base population that had been maintained in the laboratory for over 100 generations on a 14-day cycle. This was selected in two ways: one set of lines was transferred on the same schedule, while another set was transferred at longer and longer intervals—at first 28 days, then 35, 42, 56, and finally 70 days. Eggs were collected only from the oldest females, so that only females that survive into old age are able to reproduce at all in the latter treatment. The effect of selecting for reproduction late in life is to increase late fecundity, at the expense of reducing fecundity early in life. Reverse selection increases early fecundity at the expense of accelerating senescence (Service 1987). These results were broadly confirmed by an independent series of experiments run at about the same time by Luckinbill (Luckinbill *et al.* 1984). Both research groups have continued to maintain and analyse these selection lines. They have reported morphological and physiological correlates of delayed senescence (Rose *et al.* 1984, Service *et al.* 1985, Service 1987, Graves *et al.* 1988, Service 1993) that include increased resistance to starvation and desiccation. They have also investigated the genetic basis of the response to selection (Clare and Luckinbill 1985, Luckinbill *et al.* 1987, 1988, Hutchinson *et al.* 1991). These and some earlier experiments are critically reviewed by Rose (1991). There are some discrepant results. Mueller (1987), Engström *et al.* (1992) and Roper *et al.* (1993) all obtained a direct response to selection for increased longevity, but failed to observe any substantial reduction in early fecundity. This may be attributable in part to differences in culture conditions, but may also reflect the different contributions of mutation accumulation and antagonistic pleiotropy in different experiments. Most experiments using outbred stocks maintained in the laboratory for many generations before conducting the experiment seem to give the expected result: lines maintained by reproduction late in life evolve greater longevity, and fecundity early in life often falls as a correlated response. The physiological mechanism responsible seems to involve costs of reproduction. In the first place, the response occurs too rapidly to be explained by the accumulation of late-onset deleterious mutations. Secondly, the loss of early fecundity is not restored in crosses between selection lines (Hutchinson and Rose 1991). Curiously, Rose *et al.* (2005) have recently repudiated their own research programme, chiefly on the grounds that the results are sensitive to culture environment and genetic background. Since genes can act only by modulating physiological processes whose effects vary with the conditions of growth, this is not unexpected, and even the major genes affecting lifespan in *Caenorhabditis* show strong interactions with environment and genotype. Despite the protests of their authors, these *Drosophila* selection experiments provide the strongest direct evidence for the evolutionary theory of senescence.

5.3.9 Endogenous evolution

According to the account I have given, senescence is the consequence of internal deterioration. There is a second and less familiar possibility: that the external environment may systematically deteriorate during the lifetime of every individual, so that individuals will senesce even if they are able to maintain a constant level of repair. It is obviously out of the question that the physical environment should continually deteriorate during each individual's lifetime, since if this were true the Earth would no longer be capable of supporting life. It is not inconceivable, however, that the biotic environment should continually deteriorate as the consequence of the evolution of increased virulence within populations of pathogens. When a pathogen infects a host individual, it may proliferate so successfully that the host is quickly killed by the disease it causes. More often, the growth of the pathogen population is suppressed by host defences, and disease symptoms cease, or never appear. The absence of symptoms is normally taken to imply elimination of the pathogen. Suppose, however, that a small population of the pathogen, made up of somewhat more resistant forms, persists in host tissues. This population continues to reproduce, but is prevented from increasing in numbers, and thereby causing overt symptoms, by host defences. As time goes on, however, variants which are able to replicate somewhat more successfully will be selected in the pathogen population resident in the host. These variants are those pathogen genotypes which happen to be best able to exploit the genotype of the particular host in which they are evolving, and may differ from individual to individual. However, although the pathogen population within the host can evolve adaptation to the host genotype, the host itself cannot evolve, being an individual and not a population. As the pathogen population continues to evolve more effective ways of circumventing host defences, the rate of replication and the abundance of the pathogen will tend to increase, causing a progressive increase in the level of damage sustained by the host, which is macroscopically apparent as a senescent decline in survival or fecundity. Eventually the host dies, either because the pathogen finally eludes host defences entirely and breaks out to cause acute disease, or because the indirect effects of increased levels of damage caused by evolving pathogens result in the host starving to death or being eaten by a predator. A long list of conditions must be met for this theory to be plausible, the most important being that all individuals do indeed harbour persistent populations of pathogens, that selection for increased virulence occurs within the host's body, that selection is host-specific, and that persistent infections cause progressive deterioration. There are well-documented cases, however, in which all of these conditions are met (Bell 1994). The best-known example is human immunodeficiency virus, but several other viruses, such as hepatitis A and B, are also candidates, and the phenomenon may be widespread in metazoans.

Most plants are colonies of actively proliferating stem cells, the meristems, connected by somatic tissue. To most botanists 'senescence' refers to plant parts, such as leaves, rather than to the individual, and there has been remarkably little interest in the ageing of the whole plant (see Thomas 2002). Long-lived perennial plants such as trees and ferns do not appear to suffer from the degenerative diseases typical of old animals, and although cancer is common it is seldom fatal because it cannot metastasize. Old plants do suffer visibly from pathogens, however, and I speculate that old age and death in plants are normally caused by infection. Whether they bear persistent internal pathogen populations has not been investigated so far as I know; I have cored healthy-looking trees using sterile techniques and failed to find any.

One internal pathogen is certainly an important agent of mortality: the cancer cell derived from the soma itself. In microbes a continued high rate of cell division is favoured by selection, whereas in multicellular organisms stable development is possible only if cell division is strictly regulated. A cell lineage that escapes from this regulation, however, tends to proliferate because it has a replicative advantage over other somatic cells within the body. Consequently, we expect to find two processes during the development of multicellular organisms: the evolution of features that either prevent the uncontrolled division of somatic cells or mitigate its consequences, and selection to evade these controls during the growth of a tumour. Cancer cells arise through somatic mutations in proto-oncogenes (encoding proteins that initiate normal cell division and differentiation) or tumour suppressor genes (which repress cell division). An embryo developing from an unmutated zygote and consisting of C diploid cells when fully formed has undergone $C-1$ divisions and with somatic mutation rate to malignancy of u the frequency of cells that have received at least one mutation will be $1 - \exp[-4(C-1)u]$. Roughly speaking, most individuals with 1000 cells or more will have at least one malignant cell (and will therefore die of cancer while young) if $u > 5 \times 10^{-7}$; consequently, the induction of malignancy must require mutations in two or more genes for individuality to be viable (Nunney 1999). If the double mutant appears, its spread can be blocked by the arrangement of tissues into many small isolated compartments: thus, a malignant lineage arising in the rapidly proliferating epithelium of a crypt in the small intestine will expand as it passes up the villus (the projecting ridge of epithelium) but it will then be shed into the lumen of the gut without being able to invade the neighbouring crypt (Cairns 1975). The aspects of tissue architecture and competition among cell lineages that inhibit the spread of tumours are discussed by Frank and Nowak (2004). Complete eradication of malignancy is impracticable, unfortunately, especially in tissues that must proliferate rapidly in order to function normally, such as the epithelia of skin, intestine, lung, or cervix. The growing tumour is a clone that will evolve like the bacterial population of a microcosm, through the spread of beneficial mutations that will, in this case, confer the ability to divide faster, evade the defences of the rest of the soma, and invade new sites in the body (Nowell 1975). The generation of variation and the effects of selection in tumour populations are described by Shackney and Shankey (1998) and Snijders *et al.* (2003).

The onset and rapidity of senescent decline depend on the hazards of life. Individuals in lineages that are unlikely to survive long will be selected to reproduce early and therefore age soon and rapidly (Ricklefs 1998). Large animals resist stress more effectively and live longer, but they too will eventually senesce. Kirkwood (1977) expresses this vividly as the 'disposable soma'. Like the dying hulk of *Volvox* somatic tissue, all multicellular bodies are eventually discarded while the germline makes a new shelter for itself.

CHAPTER 6

Artificial selection

The founding document of evolutionary biology begins with an example of experimental evolution, the diversification of domestic varieties of pigeon through artificial selection, in order to establish an unarguable mechanism for modification through descent. This failed to create an experimental tradition in evolutionary biology: the next 50 years are almost bereft of experimental work, and although a book on *Experimental Evolution* was published by de Varigny in 1892, it does not describe a single selection experiment. The experimental tradition was instead founded as a consequence of the rediscovery of Mendelism and the consequent development of the pure-line theory of Johannsen (1903; English account by Johannsen 1911). This led to many attempts to produce permanent modifications within asexual clones by artificial selection on morphological characters, which on Johannsen's view will always fail. Most experiments did indeed end in failure: Hanel (1908) and Lashley (1916) attempted to select for increased tentacle number in *Hydra;* Jennings (1908) for size in *Paramecium;* East (1910) for various characters of potatoes; Agar (1913) for the rate of reproduction in the cladoceran *Simocephalus;* Ewing (1914) for rate of reproduction in *Aphis* (see also Agar 1914); Surface and Pearl (1915) for yield in oats; Mendiola (1919) for the shape and rate of growth of *Lemna* (see Jennings 1910 for a review of selection experiments in relation to the theory). All these attempts were in vain, although Jennings (1916) succeeded in producing a diversity of lineages with more or fewer spines from a single clonal stock of the rhizopod *Difflugia,* and Middleton (1915) obtained a response to selection for higher fission rate in clones of the ciliate *Stylonichia.*

This line of investigation petered out before 1920 as the mechanism of inheritance became better understood. It was replaced by a much more important tradition in artificial selection, the development of systematic programmes of selection to improve crop plants and livestock at dedicated agricultural research stations. The comparable tradition in university laboratories was founded by selection for coloration in rats by Castle and Phillips (1914) and for bristle number and eye facet number in *Drosophila* by MacDowell (1919) and Zeleny (1922; see also Zeleny and Mattoon 1915, May 1917, Zeleny 1921). There is a good contemporary review by Pearl (1917), who cites a number of other early references, and emphasizes the lack of any evidence for the cumulative effect of selection on slight novel variations. Indeed, the ease of selecting in sexual populations, and the lack of progress in asexual populations, was generally interpreted as a *falsification* of Darwinism. The Mendelian elements were fixed and finite in number; they could be reassorted and recombined, but not altered nor destroyed nor created. Consequently, selection would sort any variation initially present in the population, but could never produce new kinds of organism. With hindsight, it is not difficult to understand the situation. The attempts to select within clones involved populations that were small, and characters (such as the number of tentacles in *Hydra*) that were sensitive to environmental or developmental variation. The failure of these experiments was a decisive influence on the history of evolutionary biology for the next 50 years. It led to the neglect of experimental work in favour of an increasingly elaborate theory of sorting in sexual populations. This did not begin to be remedied until the need to investigate the effect of artificial selection on quantitative characters over short periods of time in the laboratory led to the choice of *Drosophila* as a rapidly reproducing model organism because it already served as a model in transmission genetics. The *Drosophila* programme, heralded by L'Héritier and Tessier in the 1930s, began to gather pace in the 1950s, and in the following decades hundreds of experiments were reported and collectively made a large contribution to understanding how selection works. Fifty years on, the basic principles of population genetics have been illuminated by a considerable body of experimental work; but the lasting scar left by the failure to develop experimental programmes in the early years of Darwinism has been, as we shall see, the continued obscurity of the effects of selection in the long term.

6.1 Selection acting on quantitative variation

6.1.1 Inheritance of quantitative characters

Once we move outside the laboratory microcosm we are likely to be interested in the response of some trait Z that is not equivalent to fitness—the grain yield of corn, the beak shape of finches or the coloration of moths, for example. In most cases these characters will vary over a continuous range of values rather than falling into a small number of discrete categories corresponding to genotypes. When the genetic basis of variation is cryptic we need a theory of quantitative genetics to predict the response to selection. I can give only the barest details of this large and technically difficult field: the reader should consult Falconer (1981) and Roff (1997) for extensive accounts of the classical theory and Mackay (2001) for a review of the genetic basis of quantitative variation.

The variation of individuals around the mean of the population may be caused by genetic or environmental factors. The two can be distinguished by raising families in isolation and measuring the degree of resemblance among relatives. Roughly speaking, the variation among families is genetic, and the variation within families is environmental. The genetic variation is itself compound: some part is attributable to variation of the independent effects of genes on the character, another part to variation of the effects of genes in combination. The independent effects are usually defined as *additive*, in the sense that the independent effects of genes sum to the character value; thus, if one gene causes the character value to change by an amount x_1 and another gene causes it to change by an amount x_2, then the total change in individuals bearing both genes is $x_1 + x_2$. The interaction between genes is measured as the deviation from this strictly additive relationship, and it can in turn be decomposed into a component arising from *dominance*, the interaction between allelic genes, and a component arising from *epistasis*, the interaction among genes at different loci. These different genetic effects can be estimated from the resemblance among different kinds of relatives. However, the response to selection in sexual populations depends primarily on the independent effects of genes, and therefore on the additive genetic variance. This can be estimated directly from the resemblance between parents and offspring, since this necessarily represents the differences among individuals that are transmitted from generation to generation, and which can therefore be selected so as to cause a permanent genetic change in the population. When several pairs of adults are mated and their offspring reared, the average value of a brood of offspring can be plotted against the average value of their two parents; the slope of this graph is the fraction of the total variance among offspring attributable to additive genetic variance. This quantity is called the *heritability*, denoted by h^2. We can modify the fundamental theorem to state that the rate of change in fitness in a sexual population will be equal to its additive genetic variance. In artificial selection experiments, we can use this result directly to predict character evolution, because the character that we are selecting deliberately is a surrogate for fitness.

Quantitative characters are often inherited as though their state depended on the action of many genes, each having only a very small effect relative to environmental variation. Taking this to an extreme, we can imagine them to be affected by an infinite number of genes, each with an infinitesimally small effect. This is the *infinitesimal hypothesis*, originally developed by Fisher, which forms the basis for the mathematical theory of quantitative genetics. It cannot be literally true, of course, and was proposed as a device to facilitate calculation rather than as a realistic description of inheritance. Nevertheless, it would be interesting to know whether it is roughly true: whether quantitative variation is normally attributable to many genes each with small and nearly equal effects (Mather 1941), or to a few genes of large effect which are de-Mendelized by small-effect genes or environmental variation (Robertson and Reeve 1952). A number of hypothetical 'polygenes' were identified through linkage with major genes, but it was only after 1980 that it was possible to use cheap, highly polymorphic marker systems to locate the genes responsible for quantitative variation routinely and with reasonable precision. They were re-dubbed *quantitative trait loci* (QTLs), and thereupon became a fashionable research theme. Identifying QTLs can in principle estimate the number of genes responsible for quantitative variation and the size of their effects, although in practice these estimates may be biased in several ways (see Erickson *et al.* 2004). Kearsey and Farquhar (1998) reviewed the literature and found that the average number of QTLs reported was 4, with almost all studies reporting 8 or fewer. They explained an average of about 50% of the genetic variance of the character, independently of the number of QTLs. A typical QTL is thus associated with about 10% of the genetic variance, showing that the variation of quantitative characters is often attributable to a few genes with rather large effects.

In large experiments the distribution of QTL effects is highly skewed, with a long tail containing a few QTL of large effect (Edwards *et al.* 1987 for maize; Mackay 1996 for *Drosophila*; Hayes and Goddard 2001 for livestock; Xu 2003 for barley). There may be several uninteresting reasons for this (Bost *et al.* 2001). QTL effects are often reported in terms of the fraction of phenotypic variance explained, which will be proportional to the square of the genetic effect, and the size of effect is biased upward because small studies will only discover large effects (and because studies that fail to detect QTLs might not be published at all). QTLs are not really loci, but rather long and variable chromosome segments between flanking markers, so QTLs of large effect might be long segments containing many genes of small effect. Nevertheless, orthologous QTLs are often detected in related species, showing that their effects have been accurately estimated, and the effect of a QTL is unrelated to its length. Effect size can be fitted to negative exponential (Otto and Jones 2000) or gamma (Xu 2003) distributions. There is clearly an interesting parallel between the number and effect of QTLs contributing to a quantitative trait chosen for deliberate selection and the number and effect of beneficial mutations available in an altered environment. In both cases extreme-value theory provides a reasonable starting point, though I am not aware of any attempts to apply it to morphological or physiological traits. The drawback is of course that the ancestors of modern cultivars did not necessarily express nearly maximal values of the characters we attempt to exaggerate, such as the grain yield of corn or the back-fat thickness of swine. The distribution of QTL effects in published studies seems to be roughly consistent with an extreme-value model, however, and this would provide an unexpected link between natural and artificial selection.

In natural populations the function of a character, that is, the relationship between character state and fitness, is often obscure and finding out what it is may be the whole point of an extensive research programme. Applying artificial selection, on the other hand, apes the part of nature by making Z equivalent to fitness. For this purpose, any character can be chosen on the grounds of utility or convenience, and equated with fitness (as far as possible, at least) by making it the criterion for choosing the parents of the next generation. That settles the question of function, at least in the short term: the experimenter defines the function.

The rapid modifications that can be caused through the deliberate generation and sorting of variation are the basis of modern varieties of crop plants and livestock, and the wide differences between modern strains of corn, cattle, or dogs and their wild progenitors are familiar to everybody. From the innumerable examples available, I have chosen a sorting experiment on maize by Genter (1976). The object of the experiment was to broaden the genetic basis of standard Corn Belt cultivars by developing entirely unrelated strains with which they could be crossed. The initial material for the experiment was 25 Mexican races, including indigenous, pre-Columbian, and prehistoric stocks, that varied widely but were generally unsuited to cultivation in Virginia because of low yield, late maturity, high ears, and susceptibility to corn smut. The base population comprised material from over 200 crosses among these races. This was then sorted for 10 generations; in every generation, 10 000 seeds were planted at Blacksburg, and about 600 ears selected at the end of the growing season from erect, healthy, productive plants. Because the base population was poorly adapted to the climate and soil of Virginia, 10 generations of selection caused rapid and substantial change. The selected population was much more productive, with yield increasing from 1.2 to 3.2 tons per hectare. It comprised shorter plants (232 cm vs 289 cm) that bore ears lower down (at a height of 115 cm vs 179 cm), developed more rapidly (from 90 to 78 days at silk), and were healthier (the proportion of diseased plants dropped from 44% to 19%).

The genetic basis of sorting in asexual populations is straightforward, and it is the physiology and biochemistry of the characters under selection that receives most attention. In sexual diploid populations, on the other hand, the genetics of sorting are more complicated, and the aim of most experiments is to understand the genetics while ignoring the function of the characters under selection. We can now ask whether the evolution of the population through artificial selection can be predicted from knowing how the character is inherited; whether the genetic dynamics can be predicted from the genetic statics.

6.1.2 Directional and stabilizing selection

In discussing natural selection I made a distinction between purifying and directional selection. Both processes are directional, however, insofar as they favour types with high fitness. Selection acting on fitness is always directional, in this sense, because variants with greater rates of replication always tend to increase in frequency. This is not true for the structure and biochemistry that underlies fitness. If any character is exaggerated too far in any direction, it will have the effect of reducing fitness. This is because any character affects fitness in many ways; some effects increase fitness when the character is altered in one direction, others increase fitness when it is altered in another direction. If it is altered too far in any direction, the effects that reduce fitness come to outweigh those which increase fitness. A resistant spore produced when conditions are unfavourable for growth must persist for a considerable period of time, and then germinate when conditions improve. It must have a thick wall in order to resist desiccation, abrasion, and the attacks of fungi and grazing animals. The thicker its wall, the more likely it is to survive. It must also have a thin wall, in order to be able to germinate easily. The thinner its wall, the more likely it is to germinate successfully. These effects are incompatible. A spore with a very thin wall will have nearly zero fitness, because it is almost certain to dry out or rot. A spore with a very thick wall also has nearly zero fitness, because it is almost certain to be unable to germinate. The optimal value for the thickness of the wall is therefore intermediate between these extremes. Because all characters will have intermediate optima, the fitness function will in general be a humped curve. We can then draw the general case of the various fitness functions shown in Figure 6.1. The effect of selection on a character depends on the relationship between the population frequency distribution and the fitness function. If the mean character value does not coincide with the maximum of fitness (for example, if the environment has recently changed), directional selection will drive the population so as to increase mean fitness. This process slows down and finally halts as the population mean approaches the maximum of fitness, at which point selection is stabilizing.

The theory of stabilizing selection acting on quantitative characters has been reviewed by Falconer (1981) and Roff (1997). The leading prediction is

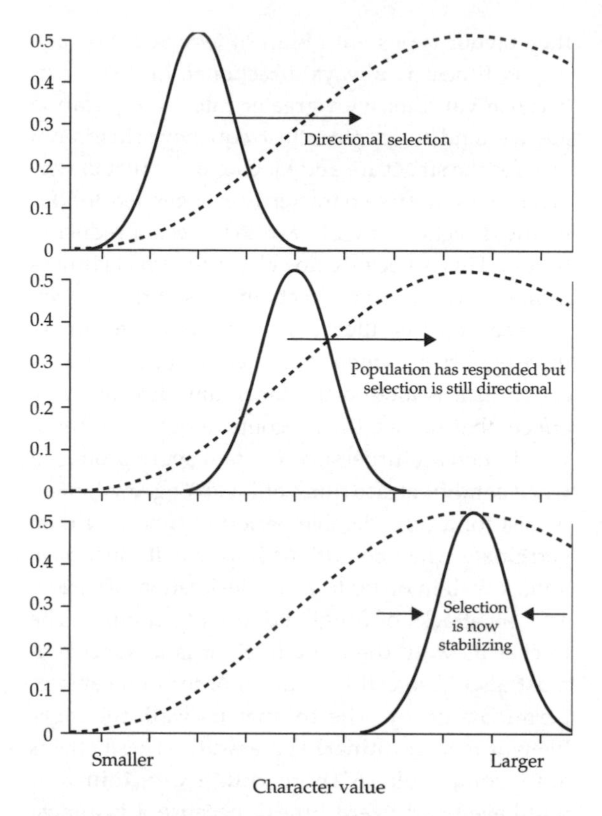

Figure 6.1 Response of the population (solid line) to a Gaussian fitness function (broken line) by directional and stabilizing selection.

that it will reduce both phenotypic and genetic variance, which has been demonstrated experimentally for bristle number in *Drosophila* (Gibson and Bradley 1974) and for pupal weight in *Tribolium* (Kaufman *et al.* 1977). The conventional interpretation of these results is that stabilizing selection acting on a character influenced by a large number of loci of small effect will favour individuals with roughly equal numbers of alleles increasing and decreasing character value. An alternative interpretation, however, is that normal development is perturbed by mutations at one or a few loci which produce extreme phenotypes in one direction or the other and are therefore selected against. In this case, stabilizing selection reduces to purifying selection and the concept is unnecessary. So far as I am aware, this dilemma has not yet been satisfactorily resolved.

6.1.3 Directional selection of quantitative characters

Selection is usually applied to experimental populations, whether of flies or of crop plants, in one of two ways. The first is to choose as parents individuals in which the character being selected exceeds a certain value, and rejecting all the rest. This is called *threshold selection* or *truncation selection*. The second way is to choose a certain fraction of the population, decided in advance, made up of those individuals with the most extreme values of the character. This is called *mass selection*. The rate of selection can be expressed in several ways. The relative fitness of distinct types, usually in the context of natural selection, is expressed by the selection coefficient *s*, whereas in experimental situations we are more likely to be interested in continuously varying characters such as seed yield or fleece quality. In this case the experimenter creates two types, the selected and the rejected individuals, according to some phenotypic criterion that imposes stronger or weaker selection. This criterion can be expressed as the selection differential, which I shall symbolize as D: the selection differential is the difference between the mean of the population before selection and the mean of the selected individuals. The selection differential is clearly related to the fraction of the population that is accepted in a mass selection scheme: the greater the selection differential, the smaller the fraction that are chosen. In truncation selection, setting the threshold for acceptance further from the population mean increases the selection differential. To compare the response of different characters, it is often convenient to use a common scale for the selection differential. This is usually supplied by the phenotypic standard deviation of the base population, σ_p: the standardized intensity of selection is thus $i = D/\sigma_p$, or in other words a selection differential in units of phenotypic standard deviations. If an allele has an effect x on a character (that is, the two homozygotes differ by $2x$) then the intensity of selection i and the selection coefficient s acting on this allele are related by: $s \approx (2x/\sigma_p)i$ (Falconer 1981).

If all the differences among individuals in the population were heritable, then the response to selection R would be equal to the selection differential D. Because only a fraction h^2 of the phenotypic variation will be heritable, the response to selection will be the corresponding fraction of the selection differential: $R = h^2D$, or equivalently $R/\sigma_p = h^2i$. This modification of the fundamental theorem, the 'breeder's equation' of Lush (1947a, b), is the basic predictive equation for the change in population mean value under artificial selection. The selection differential is under the direct control of the experimenter, who may choose to select the top 1% or the top 90% of the population, applying powerful selection in the former case and weak selection in the latter. The heritability is a property of the base population, and is not under the direct control of the experimenter, although the quantity of genetic variation initially available for selection can be specified to some extent through an appropriate choice of the number and relatedness of individuals constituting the base population. The theory of sorting in a sexual population can therefore be tested by using estimates of the heritability, derived from breeding trials, to predict the change in population mean, estimated from selection experiments.

6.2 Generations 1–10: the short-term response

6.2.1 A bristle experiment

Clayton *et al.* (1957a, b) and Clayton and Robertson (1957) carried out a classical experiment to test the ability of this elementary theory to predict the response to selection in terms of the heritable variation of a character in the base population. The character they used was the number of bristles on the abdomen of *Drosophila*. The bristles are sensory chaetae borne on the thorax and abdomen of the fly; their main virtue for experimental work is that are easy to count, and they have been the subject of dozens, if not hundreds, of selection experiments. I shall call this the 'Edinburgh bristle experiment'. The base population was a large outbred population maintained in a cage in the laboratory for many generations. The mean bristle number was about 30 or 40 (it differs between males and females), with a standard deviation of 3 or 4, and showed no tendency to change through time. They first estimated the heritability of bristle number in this base population, using a variety of techniques that consistently gave values of about 0.5. They then established upward and downward selection lines, in each of which a different proportion of individuals was selected. They selected 40 flies—20 males and 20 females—in each line, but the 20 of each sex were chosen as the most extreme individuals among 25, 50, 75, or 100 scored, in order of increasing intensity of selection, in different lines. The difference among lines in the rate of selection gave a range of predictions of the expected response. The response to selection after five generations was accurately

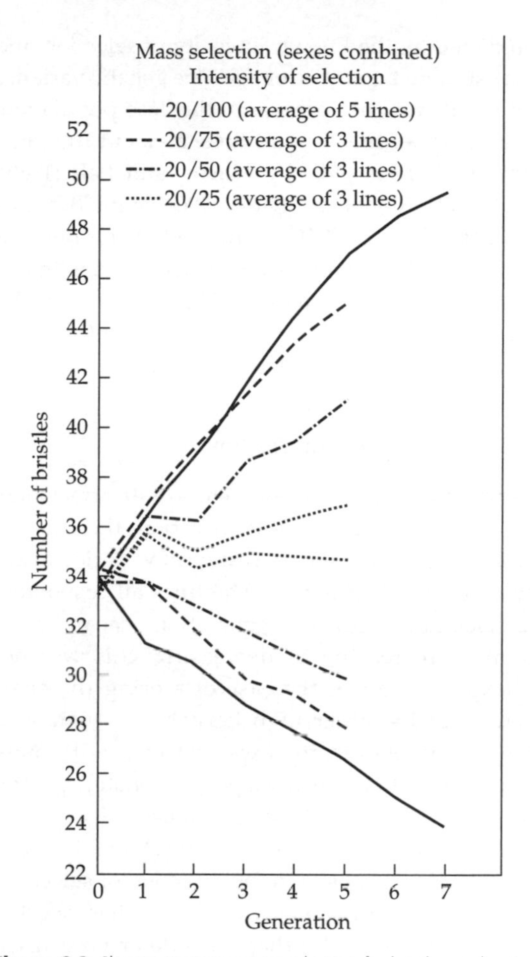

Figure 6.2 Short-term response to selection for bristle number in *Drosophila*. From Clayton and Robertson (1957).

predicted by the known intensity of selection and the estimated quantity of additive genetic variance for bristle number present in the base population. Similar experiments were done at about the same time with *Drosophila* by Martin and Bell (1960), Sheldon (1963a, b) and Latter (1964); with *Tribolium* by Enfield *et al.* (1966); with fowl by Lerner and Hazel (1947); with mice by Falconer (1953); with rats by Chung and Chapman (1958); and with swine by Hetzer (1954). A wide range of *Drosophila* experiments is reviewed by Mather (1983).

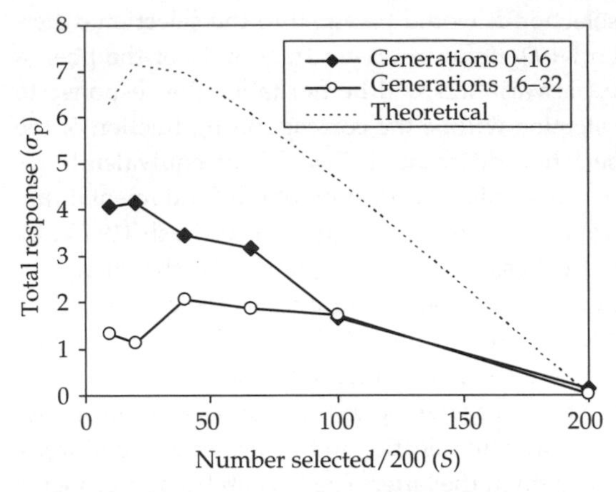

Figure 6.3 Response to selection for egg production in *Tribolium*. These are overall responses for early and late periods of selection; the broken line is a theoretical expectation. From Ruano *et al.* (1975).

6.2.2 The short-term response

In the short term, three factors will affect how rapidly a line responds to selection: the fraction of individuals chosen, the range of variation available and the heritability. The lines all responded to selection within five generations, mean bristle number increasing or decreasing; this was not unexpected, given the ease of scoring the character and its rather high heritability. Moreover, they responded in the expected way, with more intense selection causing a proportionately greater response (Figure 6.2). With intense selection (20 flies selected from a sample of 100), there is a steep and regular response in the direction of selection; when selection is much weaker (20 flies selected from a sample of 25) there is little or no consistent response in the short term. Another example from a different system, fecundity in the beetle *Tribolium*, introduces some complications: the total response falls with selection intensity, but is always less than expected, and is less still in later generations of selection (Figure 6.3). The reasons for this are considered in the next section. Altering the intensity of selection, in the short term at least, causes a regular and predictable change in the response of the population.

The range of variation is manipulated most easily by maintaining selection lines with different numbers of individuals. Frankham *et al.* (1968a, b) set up a similar experiment in which they began with a large outbred cage population and selected at different rates on abdominal bristle number. For each selection intensity they set lines of different size, made up in each generation of 10, 20 or 40 pairs of flies. As expected, the response during the first 10 generations of selection increased with the rate of selection. However, it was also affected by population size. For any given rate of selection, the larger populations tended to respond more rapidly, presumably because a larger sample is likely to include a greater diversity of genotypes. This effect can be large enough to overcome a substantial difference in selection intensity. For example, after 10 generations, the response of lines with 40 pairs, selected by choosing the top 40% of the population, was as great as that of lines with only 10 pairs, selected more intensely by choosing the top 20% of the population. Moreover, population size influenced the repeatability of the response. Replicate lines, at a given population size and selection intensity, diverged through time, as in the Edinburgh bristle experiment. However, the larger lines diverged less than the smaller lines, because sampling error will be less in larger populations. Thus, larger populations tend to give both a greater and a more reliable response to selection in the short term.

The responses in the two fly experiments were consistent with the heritabilities estimated in the base population, which was much greater in the Edinburgh bristle experiment (0.51) than in Frankham's experiment (0.15). A range of characters that vary in different degrees and are selected at different intensities can be compared by scaling the response to selection on the selection differential, to form the realized heritability, R/D. The simple theory I have outlined above suggests that this should be equal to the heritability h^2 as estimated by a breeding trial in the base population. In Frankham's experiment, for example, the average realized heritability was 0.16, compared with 0.15 estimated in the base population. This is a matter of some importance to animal and plant breeders, who can manipulate the relative magnitude of response by adjusting selection intensities or (within limits) population sizes, but who would like to be able to predict the absolute value of the advance they are likely to make. In practice, such predictions are not very successful when collated for a range of unrelated studies (Figure 6.4). There is a tendency for the scaled response to increase with heritability, but in many cases the response has been much greater or much less than was anticipated. There are two reasons for this. The first is that the intensity and duration of selection varies among studies, although in practice most studies are rather similar. The second is that the intensity of selection and the size of the selected population can be specified precisely, whereas the heritability is a property of the population whose estimate is subject to a considerable sampling error.

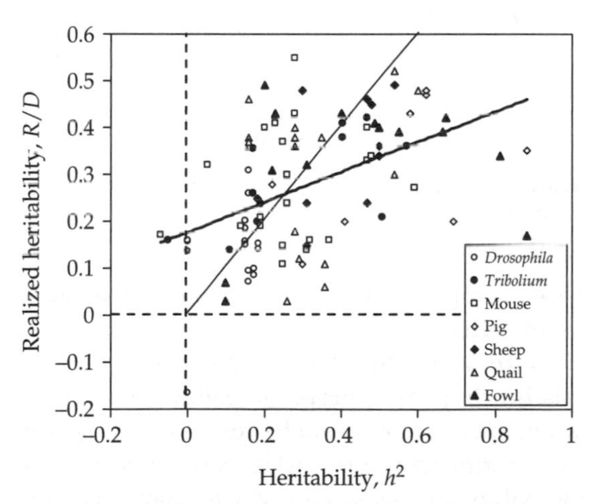

Figure 6.4 The short-term response to selection in relation to heritability. Data from Sheridan (1988). In each generation, the selection differential is the difference between the mean value of the character before selection and the mean after selection. Cumulating this difference over all preceding generations gives the cumulative selection differential in a given generation. The regression of response on cumulative selection differential is the *realized heritability*, which in simple mass selection experiments should be equal to the proportion of additive genetic variance in the base population. The expected result is the steeper line in the diagram. The regression of the data is $y = 0.18 + 0.32x$, $r^2 = 0.22$.

In these *Drosophila* experiments, the observed response to selection usually corresponds quite closely to the value predicted from basic theory. However, such predictions are rather limited. They are made on the basis of an estimate of heritability in the base population, and are therefore strictly speaking valid for only the first generation of selection; when this has been completed, selection will not only have changed the mean, but will also have reduced the additive genetic variance, and the heritability must be estimated anew. In practice, the heritability is unlikely to change substantially in the first five or six generations of selection, so that the original value can be used as the basis for predicting short-term sorting on timescales of 10 generations or less. On longer timescales the theory often breaks down, as we shall see.

6.2.3 Asymmetry

There were two unexpected trends. The first was the asymmetry of the response: the upward selection lines behaved in almost precisely the way anticipated by theory, except at the lowest intensity of selection, but the downward lines showed consistently less response than expected. In most of the downward lines, indeed, the response was only about half the expected response; at the lowest selection intensity, there was no response at all. Asymmetrical responses, which are often met in experiments involving divergent selection lines, can be produced either by the genetic composition of the base population, or by the nature of gene action. A general interpretation of asymmetric responses to selection is that they are caused by an underlying genetic asymmetry that can be represented as a non-linear regression of offspring on parent. Prior natural selection will tend to have elevated the frequency of genes associated with higher fitness. The offspring–parent regression should therefore be shallower for the range of character values associated with higher fitness under natural selection, and steeper for character values associated with lower fitness. The response will then be greater for artificial selection applied in the direction of greater heritability, that is, for character values associated with lower fitness under natural selection. In this case, more rapid development will lead to earlier reproduction and thus greater fitness (§5.3.2); consequently, lines should respond more quickly to selection for slower development than to selection for faster development, which is what is observed. This interpretation of asymmetric responses has been reviewed critically by Frankham (1990), who lists about 30 studies of bidirectional selection in various organisms on characters which contribute directly to fitness. These include female fecundity (Falconer 1955, 1971, Narain *et al.* 1960, Richardson *et al.* 1968, Land and Falconer 1968, Joakimsen and Baker 1977, Lambio 1981, and de la Fuente and San Primitivo 1985); male mating (Manning 1961, and 1963, Siegel 1965, Tindell and Arze 1965, Kessler 1969, Sherwin 1975, and Spuhler *et al.* 1978); and rate of development (Sang and Clayton 1957, Marien 1958, Hunter 1959, Clarke *et al.* 1961, Moriwaki and Fuyama 1963, Dawson 1965, Bakker 1969, Englert and Bell 1970, Smith and Bohren 1974, Drickamer 1981, Soliman 1982, and Hudak and Gromko 1989). Frankham concludes that on balance the experimental evidence favours the interpretation of asymmetric responses to selection given here.

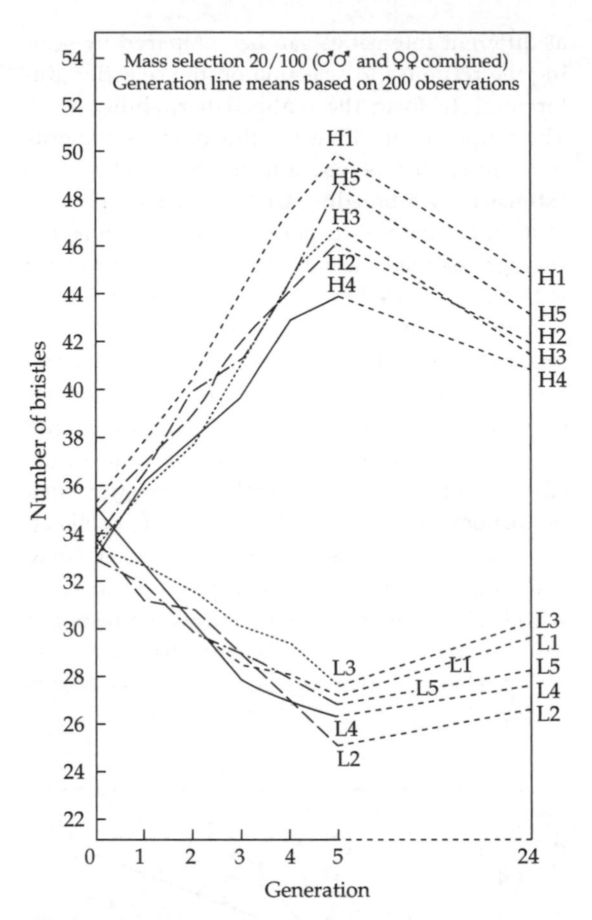

Figure 6.5 Divergence of lines similarly selected. From Clayton and Robertson (1957).

6.2.4 Divergence

The second trend was the variability of the response: lines that were selected identically responded at different rates. The lines all responded in the same direction, and the differences in response are not very great. Nevertheless, they are consistent, with any given line tending to remain above or below average in successive generations (Figure 6.5). This is caused by sampling error. The lines are set up as rather small random samples from the base population, and differ somewhat in their initial composition. This difference will subsequently tend to increase, as the lines are propagated, as biased samples, from generation to generation. If the experimenter could scrutinize the genotypes of the flies directly, choosing only those individuals bearing the appropriate genes, these errors would not exist, or would be very slight. However, the experimenter must instead select phenotypes, and will often choose flies which express extreme phenotypes because they have obtained somewhat more or less food than average, which have developed from somewhat larger or smaller eggs, or which vary for any of a multitude of reasons attributable to the unique circumstances of their individual development. Consequently, any two selected samples will be somewhat different genetically, and because only genetic differences are transmitted independent selection lines will tend to diverge. A single selection line thus represents a unique historical process that cannot be precisely repeated. Unreplicated selection experiments are therefore difficult to interpret: in this case, the predictions of the theory were borne out on average, but observations, however detailed, of a single, unreplicated selection line would have been much less informative. As selection proceeds, the supply of genes causing variation in the direction of selection is exhausted more quickly in the lines that respond more quickly, and replicate lines should then begin to converge on the phenotype that represents the limit to selection; whether this is actually the case is discussed below.

6.2.5 Selection of heritable merit

The schemes used by animal and plant breeders are typically much more complex than simple mass selection. They have two objectives. The first is to reduce the uncertainty associated with phenotypic selection by estimating the genetic value of individuals. The second is to make the best use of genetic interactions between alleles and between loci.

The genetic value of an individual can be evaluated either by screening its genome directly or by measuring its offspring. If pedigrees are available it may be possible to find traits that are simply inherited and strongly associated with the phenotype of interest: allozymes and restriction fragments, for example, have been used for this purpose. Indirectly selecting these linked loci (*marker-assisted selection*) may then be easier and more reliable than phenotypic selection on the trait itself (Lande and Thompson 1990). If QTL have been identified, they can be used as a surrogate for the trait itself (Dekkers and Hospital 2002). The main drawback of these methods is that once the marker or QTL has been fixed then another must be found before any further progress can be made. A more traditional approach (which has parallels in nature) is to apply selection to families rather than to individuals. The reason for doing so is that the phenotype expressed by an individual depends on both genetic and environmental sources of variation, and if the most extreme individuals are predominantly those that happen to have experienced unusually favourable conditions of growth then mass selection will make little progress. It would, of course, be preferable to select solely on the basis of heritable differences, if these could be discerned. Roughly speaking, when offspring are reared separately, or when the offspring of all families are reared together as a single group, the variance of family means represents genetic variance, and the variance of individuals around the mean of their family represents environmental variance. The breeder may choose, therefore, to select individuals from the most extreme families, rather than selecting the most extreme individuals in the population. The concept of family selection was first extensively developed by Lush (1947a, 1947b). This technique is particularly useful for characters whose variation is largely environmental, because the mean value of a family, being based on a number of similar individuals experiencing a range of different environments, is a more reliable measure of heritable superiority than the value of a single individual. Conversely, if all the members of a family are kept together, like the litter of a pig, the variance among families may be largely environmental; but the variance among individuals from the same family, growing in very similar conditions, will be largely genetic. In these circumstances, it may be most effective to select the best individuals in each family, disregarding the average value of the family as a whole. Mass selection can thus be thought of as having two components, one represented by selection among families, the other by selection among individuals within families. Each is of practical use in different circumstances. The efficacy of family selection relative to individual selection has been investigated experimentally by Kinney *et al.* (1970), Wilson (1975), Campo and Tagarro (1977) and Garwood *et al.* (1980).

Family selection works because the mean value of a character in a family is heritable: families, as well as individuals, can transmit qualities to their descendants. This is because sibs inherit copies of the same ancestral genes from their parents, and therefore form part of the lineage of those genes. This is obvious enough in an asexual population, in which the families are clones, but it remains true in an outcrossed sexual population. Family selection is less effective in sexual organisms, because a given gene is transmitted to only half the full-sib progeny: whereas the family as a whole is selected, only half are part of the lineage of the gene. The response to family selection can be predicted from the same equation as the response to individual selection, except that all of the terms refer to families rather than individuals: $R_f = i_f h_f^2 \sigma_f$. In general the intensity of selection i_f will be somewhat less, because fewer families than individuals are available for selection. The heritability h_f^2 of family means depends on the relatedness of family members r and their phenotypic correlation t: $h_f^2 \approx (r/t)h^2$ (Falconer 1981). If family size is large then a very rough approximation is $R_f \approx R\sqrt{(r/h^2)}$. Full-sib and half-sib family selection was included in the Edinburgh bristle experiment, and followed the predicted outcome quite closely: selecting full-sib families is not very different from mass selection, whereas selecting half-sib families is markedly less effective. The same principle applies, not only to sibs, but to any set of related individuals. Selecting sets of individuals that are related to one another in a given way is equivalent to selecting lineages of different degree, that is, lineages that extend backwards in time for the corresponding number of generations. In principle, selection can be applied to lineages of any degree, but the response to selection among lineages, relative to the response to the mass selection of individuals, will be less when their members are less closely related; very roughly speaking, the response will decline with the square root of relatedness. In addition, as more and more generations are added to the lineage, selection can be applied only at longer and longer intervals. In practice, therefore, family selection is hardly ever used except for full-sib and half-sib families.

6.2.6 The indirect response to selection

The target of selection may be a single character, such as bristle number or oil content, yet many other characters may respond. The countervailing natural selection that may limit the desired response to selection is one example of this. More generally, characters will evolve together if they are correlated genetically with one another and either or both are correlated genetically with fitness. A genetic correlation between characters can arise in three ways. First, they may be linked: in any finite population it is unlikely that genes will be associated entirely at random, if only because new mutations will arise by chance in some lineages and not in others. In asexual populations, such associations are permanent, and the characters that are expressed by the genes concerned will be genetically correlated because the genes themselves are transmitted together. Secondly, when different characters are encoded by different genes, they may become correlated through selection, either in sexual or in asexual populations, if their effects on fitness are not independent. This is because selection will cause some epistatic combinations of genes to increase in frequency and others to decrease. Finally, a single gene may affect several characters, which in turn affect fitness in different ways. One allele of a gene affecting fruit production, for example, might direct the production of large apples, another of small. If the total mass of fruit produced is unaffected, then a tree producing large apples must produce few, whereas a tree producing small apples can produce many. The two characters, fruit size and fruit number, will therefore be correlated genetically because they are expressed differently by alleles of the same gene. Characters that are genetically correlated through linkage, epistasis, or pleiotropy will tend to evolve together. To describe the situation, it is convenient to recognize three types of character (see §5.2.1). One is fitness, which responds directly to selection. The second is the character that is the phenotypic criterion of fitness, having the effect of increasing adaptedness in a new environment; this responds indirectly to selection. The third is a character or characters that are not themselves responsible for differences in adaptedness, but that change because they are genetically correlated with the character that is. It is this third category of characters that represents the new element in the situation. The change through time of such characters can be termed an indirect or correlated response to selection. In practice, the distinction among the three categories is not as clear as I have made out. In the case of artificial selection, fitness is conflated with the character that is the target of selection, but there is a clear distinction between the direct and correlated responses. In the case of natural selection, the distinction between fitness and other characters is easily drawn, but often no one character can be designated as the target of selection, and there are instead a number of correlated characters, all contributing to fitness in some degree. Nevertheless, it may be useful to recognize three categories of character in order to emphasize the logical transition from self-replicators, through a character encoded by a gene and expressed by an organism, to a complex phenotype comprising many characters.

Individuals that differ in overall body size inevitably differ in the size of some, or many, of their body parts. A very simple example of a correlated response is thus the increase in the size of a particular body part as the consequence of selection for overall size. Atchley *et al.* (1982) conducted an unusually thorough experiment of this sort by measuring a series of skull characters in lines of rats that had been selected for about 20 generations for increased or decreased body weight. There was a pronounced, although asymmetrical, direct response, the upward lines increasing in weight by about 3.5 standard deviations, relative to randomly selected controls, while the downward lines decreased by about 1.3 standard deviations. This caused a correlated response of most of the skull characters. Not surprisingly, the most conspicuous feature of the correlated response was a change in the value of each skull measurement corresponding to the change in overall size: all of the characters increased in the upward lines and decreased in the downward lines. These correlated responses were also asymmetrical, averaging (for males) about +2 standard deviations in the upward lines and about –0.5 standard deviations in the downward lines. As is usually the case, the correlated response is smaller than the direct response, because of the imperfect genetic correlation between the skull measurements and body weight. However, although the correlated response was dominated by a general increase or decrease in character value, it also involved some changes in shape. In the upward selection lines, for example, the zygomatic (cheek-bone) length increased, but did not increase in proportion to the increase in total skull length. These rats thus evolved skulls whose middle region (corresponding roughly to the tooth-row) was short relative to either the snout or the brain-case. Straightforward selection for size may thus have subtle and unexpected effects on the conformation of body parts.

The occurrence of a correlated response can often be understood qualitatively because the primary and secondary characters are both particular cases of a more general category of character. A simple example is the evolution of resistance to a particular stress, such as starvation, high temperature, or the presence of a toxin. A part of the response to selection may be specific to the stress applied, but part may also be attributable to a more general response, such as lowered activity or metabolic rate. This general response is likely to give some protection against a broad range of stresses, and thus resistance to other sources of stress will evolve as a correlated response. Hoffman and Parsons (1989) selected *Drosophila* for resistance to desiccation by placing vials of flies in a desiccating chamber, waiting until most have died, then breeding from the survivors. There is a strong response to selection: within three generations, the proportion of females surviving 24 hours in the desiccator has risen from 10–25% in the control lines to about 85% in the selection lines. The physiological basis of the response is a decrease in water loss through the spiracles, which can be viewed as an appropriate and rather specific adaptation to desiccation. However, decreased spiracle activity seems to be in turn the consequence of lower metabolic rate: less demand for oxygen leads to decreased spiracle activity, which reduces water loss. It seems likely, therefore, that the response to selection was achieved simply by reducing resting metabolic rate. Now, this is a rather generalized kind of change, which might be effective against many different sources of stress. When the selection lines are tested, they are found to be resistant to temperature shock, ethanol and acetic acid fumes, and gamma radiation. Resistance to these factors is a correlated response rooted in the rather non-specific physiological mechanism evolving as an adaptation to desiccation.

6.2.7 The tertiary theorem of natural selection

It may not surprising to learn that selecting flies with longer wings yields lines that not only have longer wings but also have a longer thorax, and vice versa; selection for longer wings is likely to favour larger flies, that also have a longer thorax. It is more interesting that the correlated response of thorax length to selection for wing length (or vice versa) can be predicted in the base population, before selection is applied. I shall describe this below (§7.3.7) as the secondary theorem of natural selection: the rate of change of character value is equal to the standardized genetic covariance of the character with fitness. It follows from this that the response to selection of a character Z can be expressed as $(1/\bar{z})\Delta\bar{z} = \beta_{wz}' SV_A$, where β_{wz}' is the selection gradient (phenotypic regression of fitness on character value) standardized by the mean, and SV_A is the standardized additive genetic variance of the character Z (see §3.2). A second character Y will be affected by selection if it is genetically correlated with Z. This indirect response will be $(1/\bar{y})\Delta\bar{y} = \beta_{A(yz)} \beta_{wz}' SV_A$, where $\beta_{A(yz)}$ is the genetic regression (regression of breeding values) of Y on Z (see Falconer 1981 p. 286). Perhaps a more convenient way to express this argument is to say that the direct response to selection R will be depreciated by the genetic regression of y on z, such that the correlated response is equal to $\mathcal{R} = r_{yz}(\sigma_y/\sigma_z)R$. The genetic correlation r_{yz} and the genetic standard deviations σ_y and σ_z can be estimated in the base population, and the equation then used to predict the change in $\bar{y}$.

This procedure is subject to the same constraints as predicting the outcome of selection on a single character, where the quantity of genetic variance that determines the rate of response to selection itself changes as the result of selection. Where more than one character is concerned the difficulty is worse, because both genetic variances and genetic covariances may change because of selection. The predictions are therefore unlikely to be useful for more than a very few generations. Moreover, even in the short term the correlated response will be predicted less precisely than the direct response, because of the additional error involved in estimating genetic covariances. Sheridan and Barker (1974a, b) selected for bristle number on two segments in *Drosophila,* measuring the correlated response of coxal bristle number to selection for sternopleural bristles, and vice versa. The predictions were reasonably successful, except that they consistently overstated the response that was actually achieved. Thus, the predicted response of coxals to selection for sternopleurals, over the first 10 generations, was 2.6 bristles, whereas the actual response was only 2.3 bristles; the disparity was greater in the converse experiment, where the correlated response of sternopleurals to selection for coxals was only 0.9 bristles, against a prediction of 1.8 bristles. Another way of expressing this result is to say that measuring the correlated response to selection gives a lower estimate of genetic correlation than measuring the resemblance among relatives in the base population: in this experiment, the realized genetic correlation was only about 0.24, against an estimate of 0.48 in the base population.

Although it may seem paradoxical, the correlated response of the secondary character may exceed the response of the primary character. This may happen if there is much more genetic variance for the secondary than for the primary character, and the two are highly correlated genetically. This has occasionally been put to use in agronomy. Selecting directly on the primary character might be ineffective if most of its variance were environmental. In this case it is much easier to select individuals that are genetically superior with respect to the secondary character, using them to get leverage, so to speak, on the primary character. This technique, which has been called *indirect selection,* is especially useful when the real object of selection is very difficult or very expensive to measure precisely, so that genetic variance would be obscured by measurement error, whereas an easily measured surrogate character is available. Its relevance to natural populations is that selection may often cause unexpectedly large responses by characters that seem to have little to do with adaptation.

6.2.8 Stabilizing selection

There are far fewer experimental studies of stabilizing selection than of directional selection, because it is less relevant to the improvement of crops and livestock. McGill and Mather (1972) crossed two wild-type strains of *Drosophila*, obtaining an F2 in which bristle number was approximately normally distributed with a mean of about 19 and a mode at 18 bristles. A pair of these flies with a given number of bristles and a pair of flies from the 6CL strain, which bear many more bristles, were permitted to mate freely, and their progeny were then reared together. The proportion of wild-type flies emerging from this F3 was then used as a measure of the fitness of the test flies. When fitness was plotted as a function of bristle number, it was maximal at 18 bristles and decreased on either side of this value. This suggests that the distribution of bristle number in the population is maintained through stabilizing selection. Other experiments on stabilizing selection in *Drosophila* include Falconer (1957; abdominal bristles), Thoday (1959; sternopleural bristles), Mather and Cooke (1962; sternopleural bristles), Barnes (1968; sternopleural bristles), Gibson and Bradley (1974; sternopleural bristles), Prout (1962; development time), Scharloo *et al.* (1967b; length of wing vein), Bos and Scharloo (1973a, b, 1974; thorax length), Tantawy and Tayel (1970; thorax length) and Grant and Mettler (1969; escape behaviour). The underlying process may be stabilizing or purifying selection in all these experiments (§6.1.2).

6.2.9 Relaxed selection

The broken lines in Figure 6.5, after generation 5, show the outcome of 19 generations in which selection was relaxed, the flies being chosen at random in each generation. These relaxed lines tend to return towards the phenotype of the base population, although rather slowly. This shows that the lines still contain genetic variation for bristle number (not surprising, after only five generations of selection), and that natural selection in the culture vials is antagonistic to artificial selection. You may expel nature with a fork but she always returns: the experimenter does not completely determine fitness by choosing how to select, because flies die for many reasons before they pass under the lens. This must eventually limit how much populations can be altered.

6.3 Generations 10–100: the limits to selection

6.3.1 Surpassing the ancestor

In outbred populations where mating is more or less at random, the selection of genes that increase (or decrease) a character such as bristle number will not be appreciably hindered by their linkage with genes at other loci that have the opposite effect, and the population will immediately respond smoothly to selection in either direction. However, if variation is constrained by linkage, there may be little or no response to selection for a few generations, until these constraints are loosened by recombination. For example, suppose that the base population is set up by crossing two unrelated, highly inbred lines. The F1 progeny are of course genetically uniform. The F2 are highly diverse, but retain parental combinations of genes. The early response to selection thus depends on the segregation of large blocks of linked genes, or whole chromosomes, and is likely to be slow. It will accelerate only when recombination breaks up these blocks, allowing genes to be selected independently. The same principle applies during selection. If lines are deliberately inbred, by mating relatives, the number of different combinations of genes is reduced, and the response to selection will be retarded.

Alleles that enhance the character under selection may be associated in the ancestral population with alleles at other loci having the opposite effect. When they are released by recombination the response to selection may take the population far beyond the limit of variation that was originally expressed. This principle is clearly demonstrated by the classic experiment involving artificial selection on bristle number in *Drosophila*, the initial phase of which I described above. In the base population, the mean bristle number in female flies was about 39, with a standard deviation of about 3.5; virtually all the flies in this population would bear between 25 and 55 bristles. After 35 generations of upward selection,

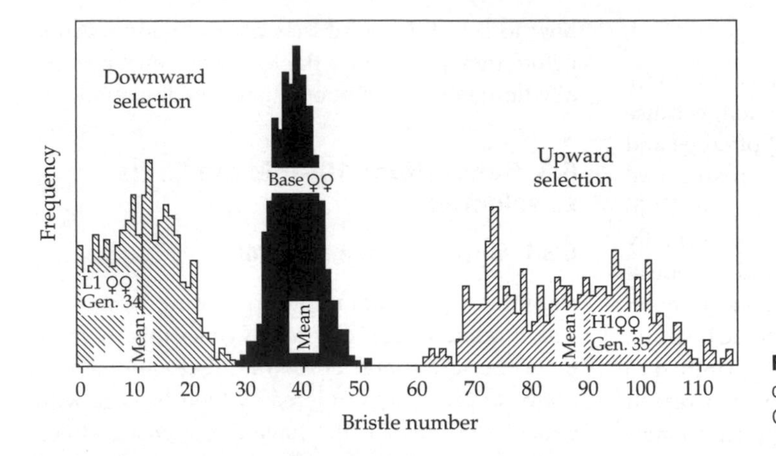

Figure 6.6 The outcome of 35 generations of divergent selection on bristle number. From Clayton and Robertson (1957).

mean bristle number had increased to about 87; in the downward selection lines, mean bristle number had decreased to about 11. After 30 or 40 generations of selection the population mean can shift by more than 10 phenotypic standard deviations—equivalent to about 20 genotypic standard deviations—from its original value; no population, however large, would initially contain individuals so extreme (Figure 6.6). D.S. Falconer (1981 p. 121) remarks that this fact 'still surprises many undergraduate students'. So it should. The ability of selection to conjure up apparently non-existent kinds of individual would scarcely be believed, were it not for the concrete demonstration that experiments provide.

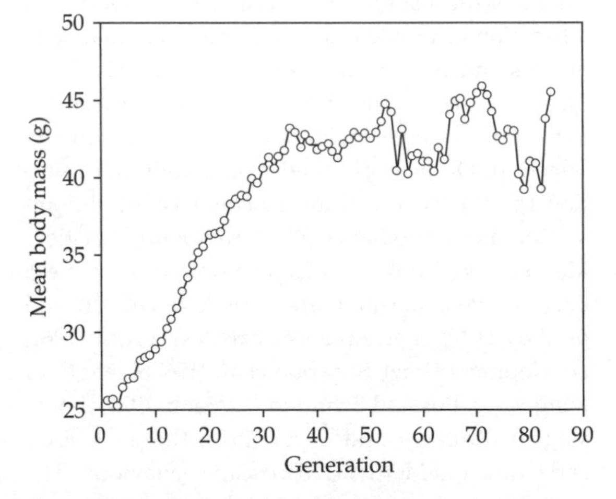

Figure 6.7 Response to selection for large size in mice. From Wilson *et al.* (1971).

6.3.2 The selection limit

After 35 generations, however, the population had reached a plateau and there was little if any response to further selection. This has been observed in many other experiments. For example, H.D. Goodale tried to breed mice as large as rats (Figure 6.7) (Wilson *et al.* 1971). Selection began in 1930 and was continued for over 30 years, spanning over 80 generations of mice. The base population comprised 8 males and 28 females, obtained from a commercial breeder. The weight of the mice initially averaged about 25 g, with a phenotypic standard deviation of about 2.5 g. After 35 generations of selection with an average intensity of about $i = 1$, mean weight had risen to about 43 g, for a total advance of about 7 phenotypic standard deviations.

More systematic experiments by MacArthur (1949) and Falconer (1953, 1960) quickly reached a limit at about 30 g for an advance of 4–6 σ_p (see reviews by Roberts 1966a, b, 1967). Garland (2003) selected mice for voluntary wheel-running and obtained a large response of about 20 σ_p in 24 generations, largely through increases in running speed, but at this point the population seemed to have reached a limit. Most mouse experiments have encountered a well-defined selection limit within 10–25 generations. This is not unexpected. The sorting of recombinants produces surprisingly rapid and extensive modification of quantitative characters, and selection lines soon move well beyond the limits of variation of their ancestors. Under the infinitesimal hypothesis this response should be sustained for a considerable length of time. If quantitative variation is largely attributable to a few genes of large effect, on the other hand, these genes will quickly attain high frequency because the individuals carrying them will readily be identified by the experimenter, most of the selectable variance will thereby be lost, and the response will peter out.

6.3.3 Heroic experiments

Natural selection can be applied to asexual populations of microbes for hundreds or even thousands of generations in the laboratory. It is seldom, if ever, possible to study sexual populations at such length, because the sexual cycle is much more time-consuming than the asexual cycle; the complete sequence from gamete production or induction, through sporulation, to the eventual release of new vegetative spores takes several days in organisms such as yeast or *Chlamydomonas*, which go through an asexual cycle of reproduction in a few hours. Using artificial selection rather than natural selection also slows down the experiment simply because of the time taken by the experimenter to carry out the selection. A number of ingenious devices have been invented to speed up selection experiments, usually by automating the process of selection. Weber (1990) lowered a glue-covered screen on to a platform where anaesthetized flies had been strewn; the flies with the longest wings stuck to the screen first and could thus be selected readily in large numbers. Applying selection for tolerance to ethanol fumes, he also invented the 'inebriometer' (Weber and Diggins 1990). Nevertheless, the time and labour involved in applying artificial selection to organisms with lifespans of several days or weeks means that few such experiments extend beyond 20 or 30 generations. In a few cases, a deliberate attempt has been made to continue selection for as long as possible. Some well-described long-term experiments are listed in Table 6.1: volume 24 of *Plant Breeding Reviews* (2004) is a mine of information about such experiments and provides much more information than there is space for here. Figure 6.8 displays the outcome of these experiments by plotting the total response to selection against the 'extent' of the experiment (individuals selected × generations of selection). Roughly speaking, the data show that selecting upwards will eventually double the population mean, while selecting downwards will halve it. More precisely, the figure shows that extensive experiments are capable of shifting the population mean by 20 phenotypic standard deviations. This conclusion is not restricted to some particular system: *Drosophila* bristles and mouse weight follow much the same rule. So does poultry body size, implying that laboratory models will be useful in anticipating the response to selection shown by crops and livestock. There are some exceptions. Increasing ear length in corn or decreasing body size in *Drosophila* has been ineffective, presumably because there is too little selectable variation. Selecting for increased oil content in maize or pupal weight in *Tribolium*, on the other hand, has shifted experimental populations by about 40 standard deviations. These heroic experiments, some of which extend over the working lifetime of the experimenter, or even of several generations of experimenters, provide a striking demonstration of the effectiveness of simple mass selection, even when applied to relatively small populations over geologically trivial periods of time.

Goodale's attempt to select mice the size of rats continued for over 30 years, until 84 generations had been completed. The response ceased after about 35 generations at a size of about 42 g, and during the last 50 generations there was little or no consistent increase in size, although there were marked fluctuations. Bunger and Hill (1999) obtained mice of about 50 g in weight after 50 generations, again with an erratic response after generation 35 or so. After 100 generations the upward line of the Dummerstorf experiment had advanced at a constant rate to about 85 g without any pronounced fluctuations (Renne *et al.* 2003). Eisen (1980), Bunger *et al.* (2001) and Hill and Bunger (2004) have reviewed these and other long-term mouse experiments. There seem to be two kinds of outcome. The first is a rapid advance for 30–50 generations, after which strange things happen and there is little if any further progress. The second is

Table 6.1 Some medium-term and long-term selection experiments. The final response to selection R is given as an a positive value in terms of the phenotypic standard deviation σ_p and the mean value X_0 of the base population. G is the number of generations, S the number selected from P candidates. For complex experiments, representative lines or averages are given. Where data for males and females are available, the data for females is given; the data for males is always very similar. In a few cases missing data has been supplied by a plausible guess, for example that the coefficient of variation of body weight for turkeys is the same as that for chickens

Organism	Trait	↑↓	G	S	N	R/σ_p	R/X_0	Reference
Drosophila	Bristles	↑	30	40	200	12.8	1.15	Clayton and Robertson (1957)
		↓	30	40	200	8.7	0.78	
Drosophila	Bristles	↑	50	80	800	15.7	1.47	Frankham *et al.* (1968b)
Drosophila	Bristles	↑	82	100	500	16.1	3.15	Yoo (1980a–c)
Drosophila	Bristles	↑	33	40	200	5.6	0.39	Sheldon (1963a)
		↓	33	40	200	7.6	0.52	
Drosophila	Body wt	↑	39	40	200	4.0	0.45	Sheldon (1963b)
		↓	39	40	200	0.3	0.03	
Tribolium	Pupa wt	↑	131	50	200	17.0	1.30	Enfield (1977)
Tribolium	Pupa wt	↑	340	50	200	40.1	3.07	Muir *et al.* (2004)
Mouse	Body wt	↑	21	74	336	4.4	0.49	MacArthur (1949)
Mouse	Body wt	↓	21	74	336	4.8	0.53	
Mouse	Body wt	↑	52	15		3.4	0.29	Falconer (1953)
Mouse	Body wt	↓	42	15		5.6	0.49	
Mouse	Body wt	↑	84	108	650	7.7	0.79	Wilson *et al.* (1971)
Mouse	Growth	↑	26	19		5.1	0.82	Eisen (1975)
Mouse	Body wt	↑	43	43		8.2	1.02	Bakker (1974)
Mouse	Body wt	↑	52	50		7.6	0.57	Sharp *et al.* (1984)
Mouse	Body wt	↑	100	160	307	18.3	1.20	Renne *et al.* (2003)
Mouse	Body wt	↑	122	50	100	4.6	0.93	Holt *et al.* (2005)
Mouse	Protein	↑	70	160	306	9.0	0.79	Bunger *et al.* (1998)
Mouse	Activity	↑	24	20	240	19.9	1.80	Garland (2003)
Mouse	Nesting	↑	15	16	96	4.3	1.79	Lynch (1980)
Quail	Body wt	↑	97	60	200	21.7	1.88	Marks (1996)
Quail	Body wt	↑	30	60	200	12.2	1.06	Aggrey *et al.* (2003)
Quail	Body wt	↓	30	60	200	4.9	0.43	
Chicken	Body wt	↑	38	60		4.9	0.63	Dunnington and Siegel (1996)
Chicken	Body wt	↓	38	60		6.1	0.80	
Chicken	Eggs	↑	30	109		2.1	0.64	Gowe and Fairfull (1985)
Turkey	Body wt	↑	28	72	1094	6.9	0.89	Nestor *et al.* (1996)
Turkey	Eggs	↑	34	144	1094		0.88	Nestor *et al.* (1996)
Maize	Oil	↑	100	48	240	37.2	3.34	Dudley and Lambert (2004)
Maize	Oil	↓	80	48	240	10.8	0.97	
Maize	Protein	↑	100	48	240	17.3	1.66	
Maize	Protein	↓	95	48	240	6.4	0.61	
Maize	Ear l	↑	27	300	4000	0.5	0.38	Hallauer *et al.* (2004)
Maize	Ear l	↓	27	300	4000	1.4	0.52	

a smooth and uninterrupted rise well beyond the limits experienced by other experiments. The same dichotomy can be seen in the other main model system, the abdominal bristle count of *Drosophila*. Mather and Harrison (1949) and Clayton and Robertson (1957) both encountered a plateau at about generation 25, whereas the lines selected by Frankham *et al.* (1968a, b) advanced smoothly

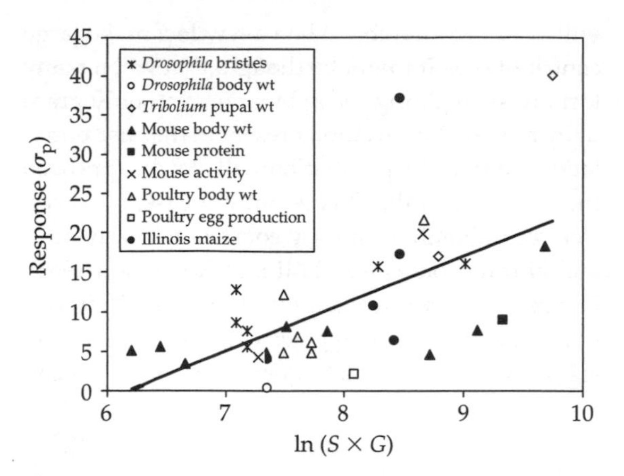

Figure 6.8 Long-term response to selection. See Table 6.1.

for 50 generations. Yoo (1980a–c) selected several lines for nearly 90 generations and most were still responding when the experiment was ended, although at a somewhat lower rate. The pattern of response, however, was quite variable. Some lines showed a rather smooth increase in bristle number, but in others bristle number seemed to reach a limit that was maintained for 10 or 20 generations before abruptly resuming their response to selection. Other systems have shown one pattern or the other. Enfield *et al.* (1966) selected increased pupal weight in flour beetles for 120 generations. Two replicate lines continued to respond throughout the experiment, although the response was slow from about generation 75 onwards. Again, the pattern of response was not the same for the two lines; in particular, one appeared to reach a plateau for several generations at about generation 50, but subsequently resumed its advance. Selection was applied intermittently to these lines for over 300 generations, with pupal weight falling when selection was relaxed and then increasing again when it was re-applied, with no real indication of a permanent limit.

Flour beetles and fruit flies have generation times of weeks, so that selection experiments lasting for 100 generations represent an enormous commitment of time and labour. The generation time of mice is months, so that Goodale's experiment required the whole of a research lifetime to complete; annual plants represent the limit of 100-generation experiments, which must span the lifetimes of a succession of experimenters, few of whom will see the outcome of the experiment they are conducting. The Corn-Oil Experiment, begun in 1896 and still being continued at the University of Illinois, is probably unique. Corn seeds have been selected for higher and lower oil content, beginning from a base population with an oil content of about 5%. After 87 generations, seeds from the low line had substantially less than 1% oil, and selection was discontinued because of the technical difficulty of measuring such low oil contents. Oil content in the high line has increased more or less linearly to about 20%, and when last reported, after over 100 generations of selection, the response showed no sign of slackening (Dudley and Lambert 2004, Moose *et al.* 2004). After about 40 generations, however, the fluctuations in oil content from generation to generation have become more marked.

Not all experiments fit neatly into two categories, an early plateau or a long-continued advance. Litter size in one mouse experiment increased for 20 generations, then actually decreased for the next 20 generations, then increased rapidly following line crossing, and finally seemed to be almost stationary after 122 generations (Holt *et al.* 2005). These two outcomes, however, lead directly to two simple questions: what sets the selection limit, and how can some lines continue to advance well beyond it?

6.3.4 Limits to selection: loss of useful variation

To follow the effects of artificial selection in an idealized population, imagine that the character being selected (which we may as well call bristle number) is increased by alleles at L loci on a single chromosome. Each locus has a wild-type allele and a 'beneficial' allele that increases bristle number and whose effect (the difference between the two homozygotes) is an exponentially distributed random variable with mean x. Gametes fuse at random to form the zygotes, whose bristle number is found by summing the effects at all the loci, assuming that the dominance of the beneficial allele is h, and adding a normally distributed environmental deviation with mean zero and variance V_E. From the total population of P zygotes the S zygotes with most bristles are selected as parents. For each zygote, a single random gamete is chosen from each of P/S meioses in which a single crossover occurs at a random position with probability r, and these gametes then repeat the cycle. This is not a very realistic account of fly biology, but it picks out the basic parameters that might affect the response to selection: L, x, V_E, P, S, and r. We wish to increase bristle number, and begin with a population of randomly assembled gametes: this is not very realistic either, because a previous period of stabilizing selection would create linkage disequilibrium, but I shall disregard this complication. An omniscient experimenter would select individuals on the basis of their additive genotype (i.e. $V_E = 0$), and the initial rate of response will depend on the strength of selection S/P. If $S = P$ then there is no selection, and the population mean fluctuates without direction through drift alone. If $S < P$ then selection will drive the population mean upwards, and smaller values of S will accelerate the initial response. In principle this could continue until beneficial alleles are fixed at all loci for a total advance of Lx bristles. In practice this does not happen because S is much smaller than the potential number of combinations of alleles, provided that L is fairly large, so the beneficial alleles at some loci are lost by chance early in the process. The actual selection limit is thus likely to be less than the ultimate limit Lx, because the probability of fixation of a beneficial allele depends on its frequency and its effect on the character under selection. The response of the population will therefore be governed partly by the underlying genetic basis of the character and partly by the selection protocol itself.

The ultimate limit Lx depends to some extent on L and x separately. When there are fewer alleles of larger effect the response is faster and the limit reached sooner. The limit itself is not much different in terms of bristles, but it is much less in terms of phenotypic standard deviations (Figure 6.9). This is because of the law of large numbers: individuals will be very similar when many loci each contribute a small quantity to the overall character value, but will vary considerably when a few loci make large contributions. It might be thought that when many loci are segregating recombination will only gradually release the variation present in the base population, so that the population will slowly approach the selection limit. This is not the case, however: free recombination quickly constructs nearly optimal chromosomes (see Hill and Robertson 1966). Figure 6.10 shows an example of selecting 25 from a population of $P = 100$ zygotes where beneficial alleles at $L = 100$ loci have a frequency of 0.5 and a mean

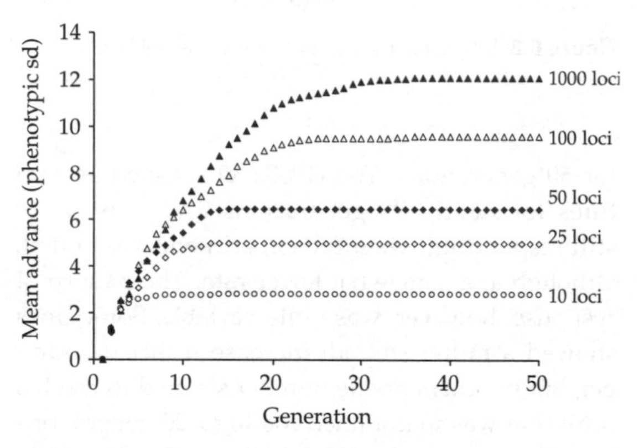

Figure 6.9 Effect of number of loci on response to selection. The total effect Lx is fixed, so the average effect x falls with the number of loci L. This figure and those immediately following give typical output from individual-based models with default parameters $L = 100$ equally frequent alleles ($p_0 = 0.5$), $x = 0.1$, $h = 0.5$, $V_E = 0$, $P = 100$, $S = 25$, $r = 0.5$.

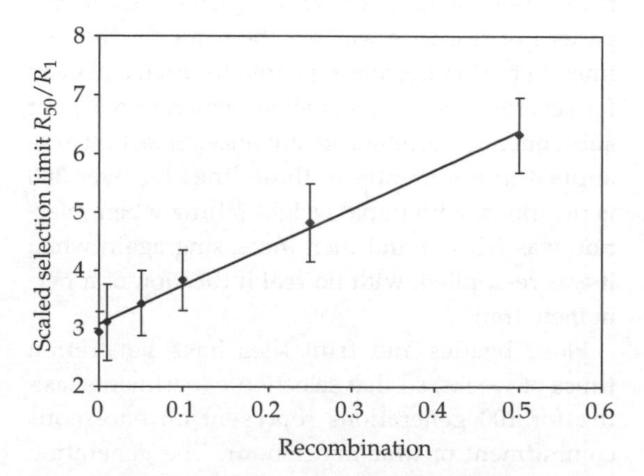

Figure 6.10 The effect of recombination on the response to artificial selection. The regression equation is $y = 6.64x + 3.1$.

effect of $x = 0.1$ bristles. With no recombination, the selection limit is set by the most extreme gamete in the base population. The maximum score increases almost linearly with ln P for $P > 10$, and numerically its average value for $P = 200$ gametes is 6.4; after 50 generations of selection the mean advance was 6.8 bristles. With free recombination, $r = 0.5$, the theoretical limit is $Lx = 10$ and the theoretical linear effect of recombination is $(10 - 6.4)/0.5 = 7.2$. Simulation shows a nearly linear response with a slope of 6.6, so this simple argument is not far out. A more complex scenario involves several or many loci bearing deleterious alleles that reduce bristle number, so that recombination must separate beneficial from deleterious alleles before the potential advance can be realized. This does not make much difference either. If the beneficial alleles are dominant the bristle number at the selection limit will be almost unchanged, but the advance in terms of σ_p is much reduced because the base population has a greater mean. If recessive, the initial mean is lower but more alleles are lost to drift and the outcome is much the same as co-dominance (Figure 6.11). If bristle number were increased by rare recessive alleles at many loci then the advance is prolonged and the limit (in terms of the initial phenotypic standard deviation) extended if S is large enough to prevent most of them being lost to drift. In this case, however, there would be an increasing rate of response in the first 20–30 generations of selection, which is never seen in outbred stocks. Thus, the genetic architecture of a quantitative character does not make much difference to its overall response to selection, except in ways that are obvious from single-locus theory.

Smaller values of S will at first mitigate the loss of beneficial alleles, because selection is then much stronger than drift, and will consequently extend the selection limit. Beyond some limit, however, the increasing loss of alleles caused by sampling a small number of individuals overbalances the effect of stronger selection, and the limit is correspondingly reduced. In a finite population, therefore, the selection limit is at first extended and then curtailed by reducing the fraction selected, and the total advance is maximized by an intermediate value (Figure 6.12). The optimal strength of selection is $S/P = 0.5$ (Robertson 1960). This conclusion is based on the loss of alleles by drift and is therefore governed by S rather than by P: that is, for fixed P the limit is sensitive to S, whereas for fixed S the limit is not sensitive to P. An experimenter who is less than omniscient will not screen the whole population, however, but will take a random sample of N individuals and choose the best S from these. In this case the selection limit depends on the fraction selected S/N and is sensitive both to S and to N.

A formal theory of the selection limit based on the infinitesimal model was developed by Robertson

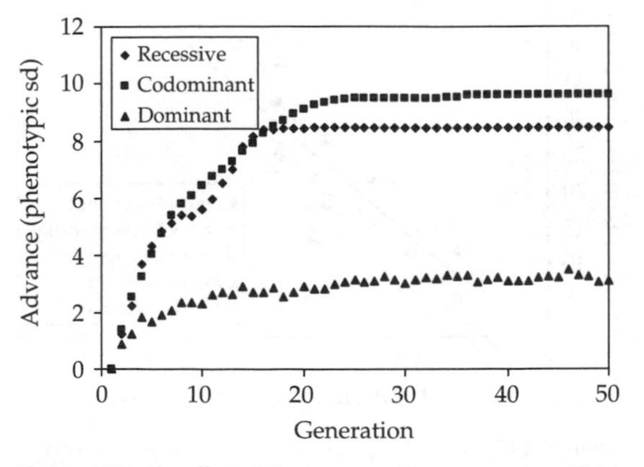

Figure 6.11 The effect of dominance on the response to artificial selection.

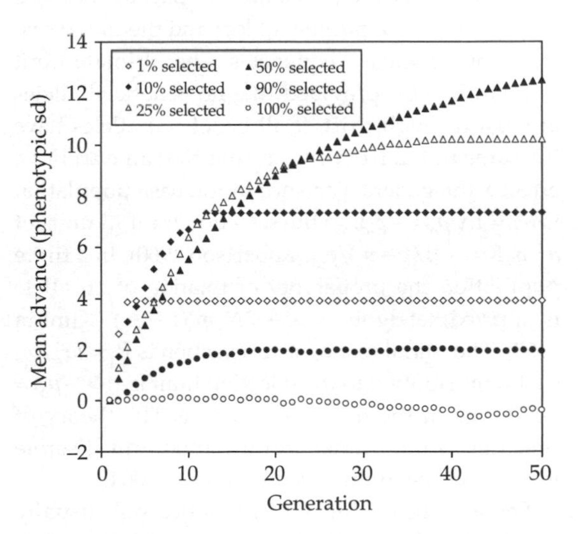

Figure 6.12 The effect of selection intensity on the response to artificial selection.

(1960). It is framed in terms of the effective population size N_e, which is approximately equal to S in most circumstances, although it may be much less when lines become highly inbred. The fraction of genetic variance lost through drift in each generation is $(1 - 1/2N_e)$, and the response to selection is reduced in proportion. The response in generation t is thus only $(1 - 1/2N_e)t$ times as great as the initial response, and summing this quantity for all the intervening generations gives the cumulative response as $R_t = 2N_e[1 - \exp(-t/2N_e)]R_1$, with the limit $R = 2N_eR_1$. If we define the intensity of selection i as the difference between the general population and the selected parents, then the total response is $R = 2ih^2\sigma_pN_e$ (Robertson 1960), or $R = 2iN_e$ in units of phenotypic standard deviations if we continue to ignore environmental variation. The intensity of selection is inversely related to the fraction selected, and $i = 2(1 - S/P)$ can be taken as a rough approximation provided that S is not too small. Taking $N_e = S$ the selection limit attained by $S/P = 0.5$ is $R = 2S$: selection will shift the mean of the line a maximum of $2S$ phenotypic standard deviations from the mean of the base population. The half-life of this process (the time required to achieve an advance equal to half the selection limit) is got from $1 - \exp(-t/2N_e) = ½$, that is, $t_{0.5} = N_e \ln 2$. The selection limit for a character governed by a few loci of large effect is more complicated because it depends on the number of loci and the initial distribution of allele frequencies. The ultimate limit is again $R = Lx$, provided that no beneficial alleles are lost through drift. If all beneficial alleles have the same initial frequency p_0 and the same additive effect x the genetic variance in the base population is $\sigma^2_A = Lx^2p_0(1 - p_0)$, so the selection limit in units of σ^2_A is $R = \sqrt{[L(1 - p_0)/p_0]}$ (Robertson 1960). In a finite population the probability of fixation of an allele is approximately $\Phi_p = p_0 + 2N_esp_0(1 - p_0)$ (Kimura 1957). The initial response to selection is $R_1 = i\sigma^2_A/\sigma_p$, so its contribution to the selection limit is $x(\Phi_p - p_0) = 2iN_eR_1$, as in the infinitesimal case. The theory of selection limits under infinitesimal and discrete models has been reviewed by Walsh (2004).

The advance achieved in practice will usually fall short of the ideal Robertsonian limit. The selection limit is a linear function of N_e only when there are infinitely many loci; otherwise it is asymptotic because adding more individuals does not add a corresponding number of alleles. Figure 6.13 shows data summarized and analysed by Weber and Diggins (1990) from 10 studies of 4 organisms. The limit is taken to be the total response at generation 50, R_{50}, although some lines were still responding at this time, and the figures are thus underestimates. It is scaled by the response in the first generation of selection, R_1; this is a measure of the quantity of selectable variation in the base population, and is equivalent to scaling by the initial quantity of additive genetic variance. The effective population size was calculated by the original authors, or recalculated by Weber and Diggins (1990), using standard approximations for *Drosophila* populations. The limit is clearly a decelerating function of N_e: there is little further extension of the limit above an effective size of about 100, but the relationship is clearly curvilinear for smaller values. Weber and Diggins (1990) use an empirical argument to fit the expected limit, but it seems simpler to conclude that the data falsify the theory. Empirically, the limit is a linear function of $\ln N_e$: $R_{50}/R_1 = 8.0 \ln N_e - 3.7$. An unusually well-replicated experiment selecting for postweaning weight gain in mice over 16 generations also shows that response increases with $\ln N_e$ (Kownacki 1979), with $R_{16}/R_1 = 8.3 \ln N_e - 16.4$, assuming from other mouse experiments

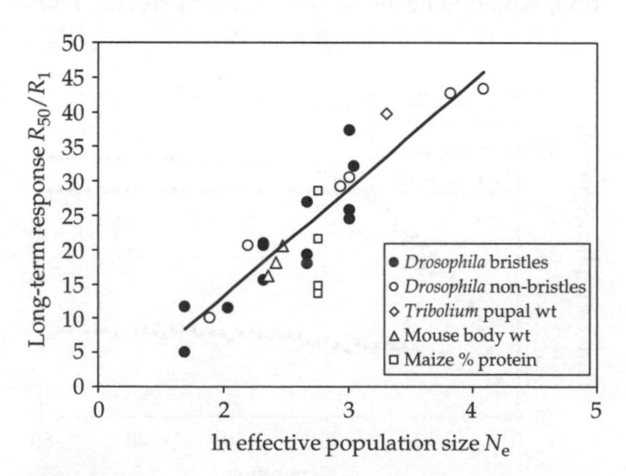

Figure 6.13 The effect of population size on the selection limit. Data from Weber and Diggins (1990) Table 2. The regression equation is $y = 8.0x - 3.7$.

that $R_1 = 0.25$ g. A simple sorting model in which beneficial alleles have an initial frequency of 0.5 leads to a regression of R_{50}/R_1 on $\ln S$ with a slope of about 1 and is clearly discordant with the data. If beneficial alleles are less common the slope increases, because adding more individuals adds relatively more alleles and thereby increases the ultimate response. If the initial frequency of beneficial alleles is only 1% then the slope is about x and the model lies close to the data (Figure 6.14). Moreover, the response in terms of σ_p also compares well with observed values. In other respects the model is less satisfactory: the initial coefficient of variation is very small, because many beneficial alleles are lacking in the first generation of parents, and consequently the absolute advance, relative to the mean value of the base population, is modest. This example proves nothing, but it is certainly not easy to reproduce all features of experimental outcomes with a simple sorting model.

In brief, reasonable models of a pure sorting process of artificial selection applied to an outbred population show that lines can advance far beyond the extreme limit of variation in the base population and that this process will end when genetic variance has been exhausted because the alleles responsible have been fixed, which usually occurs in 20–30 or so generations. This seems to be a good description of some experiments. Falconer (1960) increased the weight of an upward-selected line of mice from 25 g to 35 g in 50 generations. Breeding trials showed little if any remaining genetic variance, and reversing or relaxing selection were without effect. Roberts (1966a) re-analysed this and earlier experiments in which limits of about 2–6σ_p were attained in 10–30 generations and concluded that exhaustion of selectable variance was the simplest explanation. Brown and Bell (1961) and Scowcroft (1968) reported similar results in *Drosophila*. However, two outcomes have been observed repeatedly that are inconsistent with this simple interpretation: some lines continue to respond at about the same rate for more than a hundred generations, and others retain abundant genetic variance despite having reached a selection limit. We need to invoke some additional process in order to explain these results: the most likely candidate is recurrent mutation.

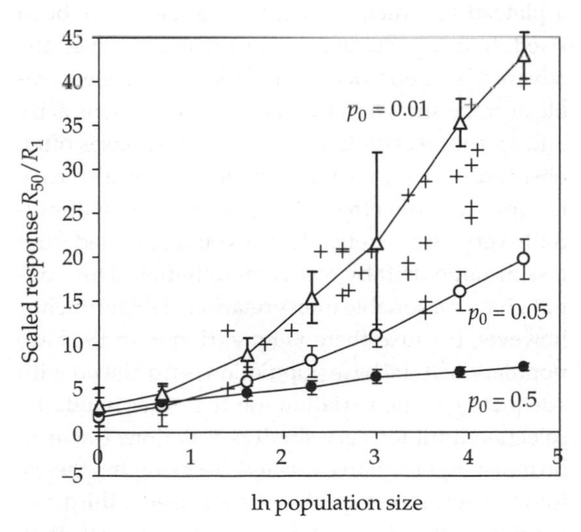

Figure 6.14 How the selection limit R_{50}/R_1 depends on the initial frequency of beneficial alleles p_0. Plotted points are means of 20 replicates + 2 sd. The crosses are data from Weber and Diggins (1990) (see Figure 6.13).

6.3.5 Long-continued response: recurrent mutation

In some long-term experiments the response to selection does not reach a limit but continues unabated for 100 generations or more. There are two current explanations: that the character is affected by very many loci in the base population and the favoured alleles have not yet been fixed, or that variation is continually replenished by mutation (Barton and Keightley 2002). The infinitesimal model is a Ptolemaic theory capable of producing, in expert hands, a passable simulacrum of reality, but it is based on assumptions known to be incorrect. Perhaps it should be emphasized that an ensemble of 100 loci each contributing on average 1% of the variance of a character does not behave like an infinitesimal model—most are quickly fixed by drift in the small populations characteristic of artificial selection experiments. And yet 100 loci is clearly an upper bound for the number of loci with appreciable effects. The infinitesimal model has been falsified as a description of the genetic basis of variation by the success of the QTL programme. It is likely that estimates of QTL effects are biased upwards, but the leading fact is that loci of large effect often contribute to quantitative variation and adaptation. It is also true that more QTL are uncovered as experiments become more extensive and more precise. Both oil and protein content in maize were found (by non-QTL techniques) to be affected by many loci—between 10 and 70 for oil and more than 100 for protein (Dudley and Lambert 2004)—and bristle number seems to be affected by about 100 QTLs (Mackay and Lyman 2005). A few loci of large effect are nonetheless likely to be responsible for a large fraction of the response to selection. The alternative is that variation is replenished by recurrent mutation.

Mutation supply rate may constrain adaptation even in microbial populations of hundreds of millions of individuals, so at first sight it seems unlikely that mutation can support fairly strong artificial selection in populations of a few hundred individuals. This is why mutation was ignored in the early development of the theory of selection limits. The population exposed to mutation (P), however, is much larger than the population exposed to drift (S). Moreover, a mutation that increases character value by a few per cent does not have to spread slowly through the population: if it occurs on a favourable background and is detected by the experimenter (still supposed to be omniscient) then it will be recruited immediately to the elite corps of parents. Although selection is most effective when applied to a diverse base population, highly inbred populations may respond slowly to selection (Mather and Wigan 1942, López and López-Fanjul 1993), and this is probably largely attributable to the slow generation of selectable variation by mutation. The rate of response is typically about one-tenth of that seen in outbred populations (Keightley 2004). Clayton and Robertson (1955) estimated mutational heritabilities for bristle number in *Drosophila* of $V_M = 2–7 \times 10^{-3}$ per generation, Weber and Diggins (1990) obtained $V_M = 9 \times 10^{-4}$ per generation for alcohol resistance, and Martorell *et al.* (1998) estimated $V_M = 3 \times 10^{-3}$ for fecundity and viability, values comparable to that found in microbes (§2.2). We would then expect the response to decline from the initial value of $ih^2\sigma_P = iV_A/\sigma_P$ fuelled by the initial quantity of additive genetic variance V_A towards a persistent level of $2iN_eV_M/\sigma_P$ fuelled by recurrent mutation (Hill and Mbaga 1998). Experimental lines should thus continue to respond indefinitely, although with small effective population size the advance would be very slow. Large inbred populations respond more rapidly than small populations, and the response is sustained for at least 50 generations (López and López-Fanjul 1993). In large-scale experiments, however, the response is often very erratic (especially Mackay *et al.* 1994 and Keightley 1998). An exceptionally rapid advance can readily be attributed to the passage of a large-effect mutation, but occasional precipitate falls in value are more difficult to explain.

A simple model with a few loci, each capable of mutating to a single beneficial allele, is shown in Figure 6.15. With mutational input $V_M = 10^{-3}$ there is a steady rise over 100 generations, even though this simulates an experiment of modest size with $P = 100$ and $S = 25$. In contrast with sorting in the infinitesimal model, it is very easy to find reasonable parameter values that produce a sustained increase through 100 generations or so before approaching a plateau at which all available alleles have been fixed. In any particular experiment, moreover, the advance is a good deal more jerky than the ensemble average shown in the figure, as mutations arise and spread, resembling the erratic responses often observed in long-term experiments. It is also easy to reproduce the effect of population size with initially very rare beneficial alleles being rescued from loss through drift by recurrent mutation. This does not give a reasonable interpretation of experiments, however, because there is no variation in the base population. If the base population is furnished with adequate genetic variation, on the other hand, the selection limit for very small populations seems to be much higher than expected. To bring the theory into line with the experiments we need a third factor that will reduce the effectiveness of selection: the most likely candidate is imperfect heritability, caused by environmental variance.

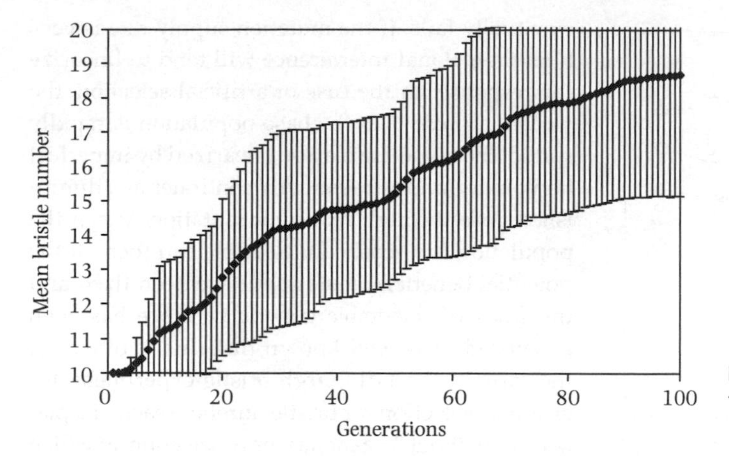

Figure 6.15 Sustained response to selection through recurrent mutation. Mutation rate per locus = 10^{-4}. Plot shows mean + 1 sd of 20 replicate runs.

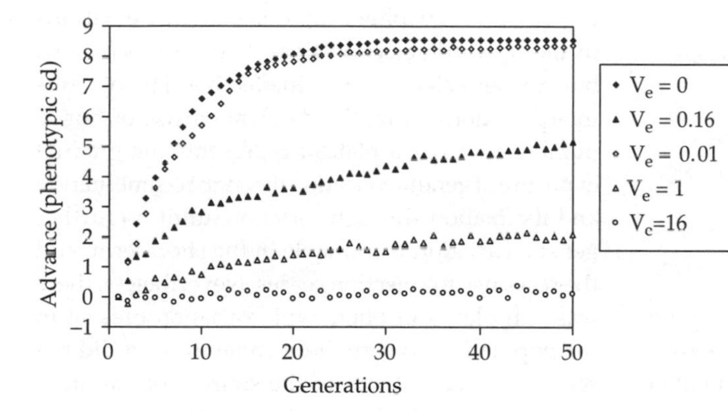

Figure 6.16 How the response to selection is flattened by environmental variance.

6.3.6 Long-continued response: environmental variance

All of these arguments apply to an ideal situation in which the experimenter is omniscient. This is never the case in practice, so the theory of artificial selection has always been closely concerned with the extent to which the phenotype is a reliable guide to the genotype. From the point of view of its relationship to natural selection, perhaps the most interesting constraint on artificial selection is that the experimenter perceives only the phenotype, by counting bristles, without knowing the underlying allelic states, and must select on the basis of this imperfect information. Environmental variance reduces the heritability and thereby the initial rate of response. Moreover, it also curtails the selection limit, because individuals are selected in part on the basis of effects that are uncorrelated with their genotype. This reduces the effect of selection and allows more loci to drift to fixation. A very small amount of environmental variance has little effect on the outcome, which is still characterized by a rapid increase to a clear plateau, and a very large amount simply prevents the population from responding appreciably to selection. An intermediate amount, however, converts the normal decelerated response into a more or less linear advance which may continue for a considerable time (Figure 6.16). The population will eventually reach a plateau, and this will be lower than if environmental variance were less, but the population may give the impression for some time that it is advancing at a constant rate with no plateau in sight.

Imperfect heritability reinforces the effect of drift because the experimenter may unwittingly discard

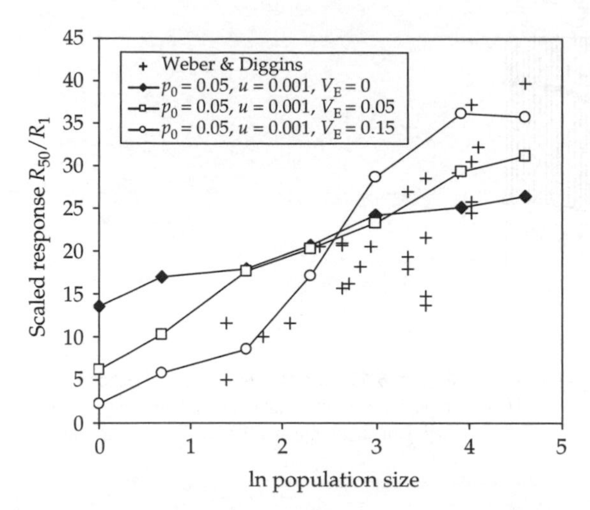

Figure 6.17 Imperfect heritability caused by environmental variance, combined with recurrent mutation, provides a reasonable explanation for observed selection limits. Heritabilities are 1 ($V_E = 0$), 0.5 ($V_E = 0.05$) and 0.25 ($V_E = 0.15$). The regression for a heritability of 0.25 (in combination with all the other parameter values used) has a slope of 8.4. Crosses are data from Weber and Diggins (1990) (see Figure 6.13).

superior genotypes. This bears more strongly on small populations and thereby steepens the regression of R_{50}/R_1 on ln N_e (Figure 6.17). Simulated populations in which heritability is initially intermediate ($h^2 = 0.25 - 0.5$), mutation provides a mutational heritability $V_m = 10^{-3}$ per generation and the frequency of beneficial alleles generates a realistic level of phenotypic variation seem to evolve in much the same way as experimental populations, in terms of the initial response, the scaled selection limit (in terms of R_1 or σ_p) and the absolute number of bristles added. This proves nothing, but all the values used are all based on reliable estimates which are likely to apply quite broadly.

6.3.7 Limits to selection: countervailing natural selection

Both natural selection and artificial selection lead to a non-linear, decelerating response to selection (see §3.7). In the case of natural selection, this is because the supply of beneficial mutations to an initially isogenic population falls through time and eventually fails; if the mutation supply rate is very high then clonal interference will tend to linearize the response. In the case of artificial selection, the variation present in the base population is rapidly sorted, the response may be linearized by imperfect heritability, and subsequently continues at a diminished pace through recurrent mutation. When the population has reached a clear plateau then all the potential beneficial mutations have been fixed and the stock of selectable genetic variance has been exhausted. It is well known that this is often not the case. In the Edinburgh bristle experiment, for example, selection for bristle number reached a plateau after about 35 generations of selection, at which time an enormous shift in character value, amounting to some 20 phenotypic standard deviations in the upward selection lines, had been achieved, but further selection was ineffective. The obvious interpretation is that the decelerated rise of bristle number towards a plateau represents the gradual exposure of genetic variation through recombination and its fixation through selection, until no further genetic variation is available in the short term, and the response to selection ceases. Nevertheless, there was still plenty of phenotypic variation present in the population. Where did it come from? It did not seem to represent any of the sources of variation that were identified in the base population: it was not developmental variation, because asymmetry had not increased, and it was not environmental variation, because conditions of culture were more or less the same throughout the experiment. It did not seem to represent selectable genetic variation, because no further response to upward selection could be obtained; but there was an immediate and rather rapid response to back-selection, when flies with fewer bristles were selected in lines that had previously been under upward selection. The retention of large amounts of genetic variation at the selection limit is a common feature of *Drosophila* experiments: the main lines of evidence are that phenotypic variance often does not decrease (and may increase), that breeding trials reveal substantial heritable variation, that relaxed selection causes a regression towards the ancestral state, and that reverse selection is often effective (e.g. Frankham *et al.* 1968a, b, Yoo 1980a–c).

The limit is more likely to be set in these cases by natural selection acting in the opposite direction to artificial selection. The character chosen by the experimenter is a surrogate for fitness, but only up to a certain point: the normal causes of mortality and morbidity continue to operate. If the individuals chosen by the experimenter are systematically less vigorous or fecund than average, then it will eventually be impossible to advance further because of countervailing natural selection. It is a very common problem in laboratory experiments that selection lines become difficult to propagate because most individuals are sickly or infertile; most of the experiments I have referred to suffered from this phenomenon. The simplest explanation of it is that artificial selection relies on genotypes that are rare in the base population (otherwise the desired phenotype would already be available) and therefore have less than average fecundity or viability (otherwise they would be common). In other words, the base population is roughly at equilibrium under stabilizing natural selection, but is driven away from its optimum by directional artificial selection. At some point the forces of artificial selection, provided by the experimenter, and natural selection, provided by the distance between the optimum and the mean of the selection line, will become equal: individuals with the most extreme phenotypes now have low fitness under natural selection and those with the least extreme have low fitness under artificial selection. At this point the population is at equilibrium, with the total response to selection of about $(i/y)\sigma_p$ where $y = \sigma^2_p/\omega^2$ is a quantity akin to a selection coefficient that expresses the strength of stabilizing selection in terms of the width ω (calculated like a variance) of the fitness function (Zeng and Hill 1986). The plateau is thus proportional to the relative intensity of artificial and natural selection. This is reasonable but artificial, and the response of a real population will depend on the properties of genes, and in particular on the selection of alleles that increase the value of the selected character but reduce fitness. ('Fitness' here means the effect of the allele in the absence of artificial selection.) If the genetic effect of such an allele on the character is x and its deleterious effect on fitness is s then its net selection coefficient is $i(x/\sigma_p) - s$ and the usual consequences will follow. The response of the selected character Z is $iV_A/\sigma_p + \mathrm{Cov}(Z,s)$, where V_A is the additive genetic variance of the character in the population and $\mathrm{Cov}(Z,s)$ is its genetic covariance with s. Thus, the response is reduced by an amount equal to the pleiotropic effect of the allele on fitness. The response of fitness itself is $(i/\sigma_p)\mathrm{Cov}(Z,s) + V_s$, where V_s is the genetic variance of fitness. Since this is negative, fitness will drop as the consequence of artificial selection. These results and their attendant complications were derived by Hill and Mgaba (1998). Pleiotropic gene effects obstruct the response to selection but do not in themselves bring it to a halt—there might be an endless supply of mutations whose effect on character value more than compensates for their effect on fitness. In practice, fitness must eventually decline to a level where the lines can only just be propagated, which imposes an effective limit on their improvement.

The classic experimental paper on the correlated response of vigour to artificial selection is by Latter and Robertson (1962). (I am using 'vigour' in the sense of fitness under natural selection, as distinct from the fitness entailed by artificial selection.) They used lines propagated by only 10 pairs per generation, so that some loss of vigour was caused by inbreeding depression; after 25 generations, the vigour of randomly selected control lines, measured as competitive ability, had fallen to about 60% that of the base population. However, the vigour of lines selected for bristle number was only 20–40% that of the base population, showing a substantial additional effect of artificial selection on vigour. It has been suggested, on the basis of experience in the poultry industry, that this decline in vigour could be halted by a special type of two-trait selection, selecting simultaneously for the desired phenotype and for high vigour. In practice, this would amount to culling weak, sickly, or infertile individuals from the selection lines. This idea was tested by Frankham *et al.* (1988a, b), who selected *Drosophila* for resistance to ethanol vapour. There was a straightforward direct response, with 25 generations of selection doubling the inebriation time of the flies, but vigour, again measured as competitive ability, was halved as a consequence. In comparable lines that were selected simultaneously for vigour, by discarding the 20% of females who produced the fewest offspring, there was no detectable loss of vigour. From the explanation that I have given in the text, this result would follow from the antagonistic effects of pleiotropic genes, the culling having the effect of removing genotypes with the most strongly antagonistic effects. This can be expected to maintain vigour, although at the expense of retarding the response of the selected character. This was not the case—the culled lines responded as fast as the others. Frankham *et al.* explain their result as the outcome of a non-linear heritability of vigour, the heritability being greater for lower vigour, because most deleterious alleles are recessive.

Countervailing natural selection not only contributes to setting the selection limit in many experiments, perhaps in all, but can also explain how genetic variance persists once the limit has been reached. Beneficial alleles that increase bristle number are likely to be systematically deleterious in terms of natural selection, because otherwise they would already have been fixed in the base population. In the simplest case the effect of such an allele on fitness is a random variable with mean s, while its dominance is a random variable with mean h_s. Any given allele may cause a large increase in bristle number at a small cost in terms of reduced fitness, and is likely to be fixed; or it may cause a relatively small increase with a large cost, and is likely to be lost. Such alleles will soon cease to contribute to phenotypic variation, which will become increasingly dominated by alleles with evenly balanced effects. In some cases the opposed effects on bristle number and fitness will make the heterozygote superior to both homozygotes, one of which has many bristles but low fitness whereas the other has few bristles but high fitness. The condition for this is approximately

$$\tfrac{1}{4}(h_s/h_x)(s/d) < x < \tfrac{1}{4}[(1 - h_s)/(1 - h_x)](s/d),$$

where d is the fraction discarded by the experimenter. This is most readily satisfied when the effect on bristles is dominant and the effect on fitness is recessive. Such loci will continue to segregate indefinitely, unless fixed by drift, and will in time contribute most of the genetic variance. Thus, the population will evolve so that effects on bristles tend to become more dominant while effects on fitness tend to become more recessive, until further progress is blocked because most of the successful parents are heterozygotes. The population has then reached a plateau, even though it will readily respond to relaxed selection or reverse selection, which favour the wild-type homozygote. In this way, countervailing natural selection automatically sets a selection limit that cannot be transcended despite the presence of ample genetic variation.

This elegant idea does not work very well in small populations with many loci involved. The selection limit will indeed be curtailed, but the phenotypic variance at the limit will not greatly exceed the environmental variance. The reason is that the heterozygous effect of each beneficial allele will be small, on average, so most recessive lethals will be lost through drift. Even when there are few loci it is implausible because the combination of a large positive effect on bristles, a large negative effect on viability and recessive action is not very likely when these parameters are assigned at random. If it be granted that there is a direct link between gain of bristles and loss of viability, however, then flies with many more bristles are necessarily much less viable. This will curtail the selection limit, without much affecting the amount of variance at the limit. Only if there are few loci of large effect governing bristle number, several of which are recessive lethals, can we expect to see a strong restriction of the selection limit combined with abundant genetic variance still expressed.

The agent responsible for the plateau reached in the Edinburgh bristle experiment was, in fact, an allele that increased bristle number in heterozygotes but was lethal when homozygous. Artificial selection favoured the mutant form, because the flies chosen by the experimenters were likely to be heterozygotes; but when these mated, a quarter of their progeny died, so that natural selection favoured the normal gene. There was no further response to artificial selection, because it was opposed by countervailing natural selection: any advance created by the choice of flies with more bristles by the experimenters was annulled because, among this selected sample, flies with fewer bristles (normal homozygotes) produced more progeny than flies with more bristles (mutant heterozygotes). This allele probably arose during the course of selection, as an unforeseen consequence of the selection regime itself, in which case no survey could have detected this nascent limit to selection in the base population. Similar recessive lethal mutations have appeared in several other cases (see Frankham *et al.* 1968a, Yoo 1980b, Garcia-Dorado and López-Fanjul 1983), and seem to be a routine feature of long-term selection experiments. A reasonable interpretation of the selection limit in these experiments is that bristle number is mainly governed by a small number of loci, whose action to increase bristle number is offset by a corresponding reduction in viability, so that alleles which produce the most bristles are likely to be lethal.

The evolution of reduced fitness in lines selected for up to about 100 generations is by now well established. In the very long term, however, compensatory mutations might restore fitness without compromising the character being selected. The most characteristic manifestation of the effect is the tendency to evolve back towards the character state of the base population when selection is relaxed. In one experiment, however, the opposite effect was observed. This happened in lines of *Drosophila* selected for positive geotaxis which were then selected in the reverse direction, for negative geotaxis, for several generations before selection was relaxed. The response to relaxed selection was increased positive geotaxis, suggesting that the extreme behaviour evolved through artificial selection was now favoured by natural selection. The observation is unique, perhaps because the experiment is unique: Ricker and Hirsch (1988) selected these lines for 25 years, or about 500 generations. It is possible, then, that the harmful side effects of mutations enhancing a selected character, which cause such mutations to accumulate in the population, may themselves be modified and eventually ameliorated by selection extending over hundreds of generations.

6.3.8 The limits to stabilizing selection

Artificial stabilizing selection, practised by removing extreme individuals, should tend to reduce phenotypic and genetic variance, and will reach a limit when the genetic variance has been exhausted. The phenotypic variance should then be equal to the environmental variance of the base population. Enfield (1977) investigated the limits to stabilizing selection by selecting pupae closest to the median weight within half-sib families of *Tribolium*. The experiment continued for 95 generations, and is the only long-term experiment on stabilizing selection that I have been able to find. Genetic variance decreased, but not dramatically, and the heritability of the character fell only slightly, from about 0.25 to about 0.2. As with directional selection, this may have been caused by the accumulation of genes with antagonistic effects on pupal weight and fitness, although there is no direct evidence for this. It should be noted that stabilizing selection does not cause the general reduction in fitness associated with directional selection: viability and fecundity were both greater in the selection lines than in control (randomly selected) lines. However, another complication in such experiments is that the environmental variance may itself decline, because selection will favour genotypes that are insensitive to fluctuations in conditions of culture. The evolution of the environmental variance is considered in §8.1.

6.3.9 Transcending the limit

If the limit to mass selection were set by the loss of beneficial alleles through drift, replicate lines would lose different alleles and the limit would readily be transcended by crossing them. The most effective selection procedure would then be to set up a subdivided population whose demes were merged at intervals. The procedure of selecting and crossing the most responsive lines has been tried on a number of occasions with *Drosophila* and *Tribolium*, but such experiments have generally failed to demonstrate any clear advantage in subdividing the population, and in most cases, indeed, a single large population showed the greater response. Katz and

Enfield (1977) selected upwards for pupa weight in *Tribolium* over 40 generations in a single 'large' population propagated by 24 males and 48 females in each generation, and a set of 6 small populations, each comprising 4 males and 8 females. At certain intervals the best 2 lines were selected and crossed, the set of 6 lines being reconstituted from their progeny. In this case, population subdivision seems to retard progress under selection. A similar experiment, with a similar outcome, was reported by Goodwill (1974). Madalena and Robertson (1975) tried a similar but more complex scheme when selecting for bristle number in *Drosophila*, without finding any clear-cut benefits in subdividing the population.

A second possibility is that replicate lines may evolve different epistatic combinations of alleles. Sewall Wright's 'shifting balance' theory stemmed from his experiences as an animal breeder, and his ideas may have important practical consequences. Breeds of livestock are maintained as a series of herds, with a certain degree of migration, physical or genetic, among the herds. How should the herds be managed so as to maximize response to selection for some desirable character? If epistasis produces a rather hilly adaptive landscape (§4.3), through which any given selection line takes a unique and essentially unpredictable path that may lead to a peak of only modest elevation, a set of selection lines, comprising the same total number of individuals, will set off in different directions and climb different peaks. Perhaps these peaks will be on average lower than that climbed by the single line, because of the smaller number of individuals, and thus the smaller quantity of variation, in each line. However, the *tallest* peak climbed by one of these lines is likely to surpass the peak climbed by the single large line. The limit of response for a structured population, consisting of several or many herds, might thereby exceed the limit for a single large herd under mass selection. This argument suggests two ways in which population structure might be manipulated so as to take advantage of multiple-peak epistasis. The first is to select at intervals among lines that have been propagated independently for a number of generations, retaining those lines that have shown the greatest response and discarding the others. The set of independent lines can then be reconstituted either by expanding and subdividing each of the selected lines, or by crossing them and re-extracting lines from the progeny. As I have mentioned, crossing seems to be ineffective: it is possible that crossing the selected populations breaks up any epistatic combinations of genes, thereby frustrating the process that the experiments are designed to exploit. Experiments in which the more responsive lines are expanded, rather than crossed, have been performed on *Tribolium* by Wade's group in Chicago; these demonstrate group selection and are described in §10.4. The second possibility is to allow a certain low level of migration in each generation, so that lines of greater mean fitness are able to infect, so to speak, lines of lower fitness, causing all lines to converge on the highest available peak. Katz and Young (1975) selected for body weight in *Drosophila* while allowing some restricted movement of individuals among lines; in these experiments, subdivided populations showed an appreciably greater response to selection. The subject has been reviewed by Barker (1988).

6.4 Generations 100 up: new kinds of creatures

6.4.1 Selection for yield in crop plants

Artificial selection is of great practical interest through its use in the improvement of crop plants and livestock, and the purpose of many of the laboratory experiments was to predict the response to selection in agricultural situations by using shorter-lived and more convenient models. It must be borne in mind, however, that the objective of practical plant breeding is not to increase the average value of the population but rather to isolate the best line. An outbreeding population such as the open-pollinated maize varieties of the early 1900s may not realize its full potential because it might include many superior genotypes that would be disrupted by recombination. Ideally these would be isolated and propagated asexually, but it is not yet possible to incorporate apomixis into most commercial crops, although some (such as sugar cane) can be propagated vegetatively by cuttings. The alternative is inbreeding to obtain homozygous lines that will produce nearly identical offspring which retain epistatic combinations of parental genes. This has two severe drawbacks: recessive deleterious alleles will create inbreeding depression, and superior heterozygotes cannot be utilized. Furthermore, the genetic homogeneity that is the goal of inbreeding is incompatible with the genetic diversity that is necessary for continued selection. The primary objective of most of the ingenious combinations of crossing and inbreeding devised by plant breeders is to mitigate these difficulties. A common procedure is to estimate the *combining ability* of an inbred line by crossing it with another line (*specific combining ability*) or a diverse population (*general combining ability*), selecting lines through the merit of their outbred progeny. Most widely planted maize cultivars, for example, have been derived by crossing inbred lines of high combining ability, selfing the progeny of the most productive crosses to obtain a range of new inbreds, and repeating the process. The outcome of selection is a collection of inbred and highly homozygous lines; the seed that is planted is a vigorous and highly heterozygous hybrid from the cross between two or more of these lines. A more complex alternative is to maintain two populations, using each to supply testers for the other. Individuals from one population are each at the same time selfed and crossed with a number of individuals from the other population. The selfed progeny of the individuals from this population that gave the best crossed progeny are chosen, and then crossed among themselves to reconstitute the population. This scheme, called *reciprocal recurrent selection*, is designed to maximize both general combining ability (through evaluating the crosses with a separate population) and specific combining ability (through crosses among the selected inbred lines from the same population). There are laboratory studies of the efficacy of reciprocal recurrent selection in *Drosophila* by Bell *et al.* (1955), Rasmuson (1956) and Kojima and Kelleher (1963); in *Tribolium* by Bell and Moore (1972); and in fowl by Saadeh *et al.* (1968) and Calhoon and Bohren (1974). There are many variations on these themes in the agronomic literature, but I shall not describe them further because there are few if any parallels in nature.

The first agricultural research stations were founded in the nineteenth century and were responsible for pioneering large-scale systematic attempts to increase crop yield through selective breeding. After 1920 there was increasing participation by commercial institutions, resulting in the appearance of modern cultivars based on controlled crossing in the 1930s and 1940s. About this time crop yield began to increase dramatically, and has sustained this advance to the present day. Wheat and maize yields, for example, changed very little between 1870 and 1930, but thereafter have advanced at a rate of about 100 kg ha^{-1} y^{-1}, so that within a single lifetime these crops have become four or five times as productive (Figure 6.18). It is interesting to compare the agricultural data with the results of long-term laboratory selection experiments, which suggest that the advance to be expected after 50 generations of selection in terms of the state of the base population is about $(X_{50} - X_1)/X_1 = 0.8$ (with a great deal of variation), equivalent to nearly doubling the ancestral value. All of the major crops fell well below this benchmark in the early period of scientific agriculture, whereas all have surpassed it

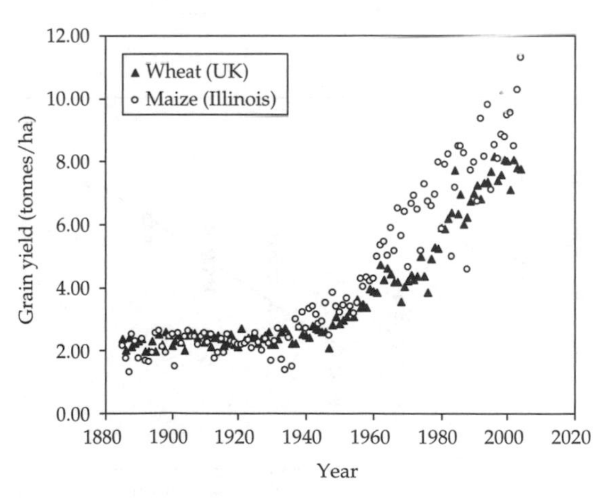

Figure 6.18 Increase in grain yield 1885–2004. Sources: DEFRA for UK wheat (*http://statistics.defra.gov.uk/esg/datasets/cyield_c.xls*), USDA for Illinois maize (*http://www.nass.usda.gov:8080/QuickStats/PullData_US*). Regression slopes for 1950–1999 are 0.104 (wheat) and 0.110 (maize) tonnes ha^{-1} y^{-1}.

since 1930 (Figure 6.19). Agricultural practices have changed greatly over this period, however, and it is not immediately clear whether improvements in yield should be credited to the selection of superior varieties, or to greater inputs of fertilizer and pesticide. Fortunately, seed of commercial varieties has been kept in storage, so that obsolescent cultivars of corn or cotton can now be woken from their long sleep, and planted out in company with the modern varieties that have replaced them. These trials still show an advance of about 80 kg ha^{-1} y^{-1} when modern and obsolete cultivars are grown together in the same conditions (Castleberry *et al.* 1984) (Figure 6.20). It remains possible that modern cultivars flourish only under modern conditions, yet they retain much of their superiority in stressful conditions. The total advance, from obsolete cultivars in relatively poor conditions to modern cultivars in better conditions, has an environmental component attributable to improved practices and a genetic component attributable to selective breeding (Figure 6.21). These data from the early 1980s show improved practices added 3.2 tonnes ha^{-1} and improved cultivars added 4.1 tonnes ha^{-1} for a total response of 7.3 tonnes ha^{-1} in about 50 years. These figures are not quite conclusive, for two reasons. First, modern cultivars are to some degree adapted to modern conditions, which is why the two regressions in Figure 6.21 are not parallel. In particular, modern cultivars are able to produce grain at very high densities while as individual spaced plants they are often little superior, if at all, to old open-pollinated varieties (see §10.1). Secondly, the superiority of modern cultivars could be attributable in part to a strictly temporary advance, especially resistance to contemporary pathogens. Nevertheless, it is reasonable to claim that selection has been responsible for at least half the great improvement in grain yield since 1950. The same is true for other crops, such as cotton (Bridge and Meredith 1983), soybean (Specht *et al.* 1999), and barley (Wych and Rasmusson 1983).

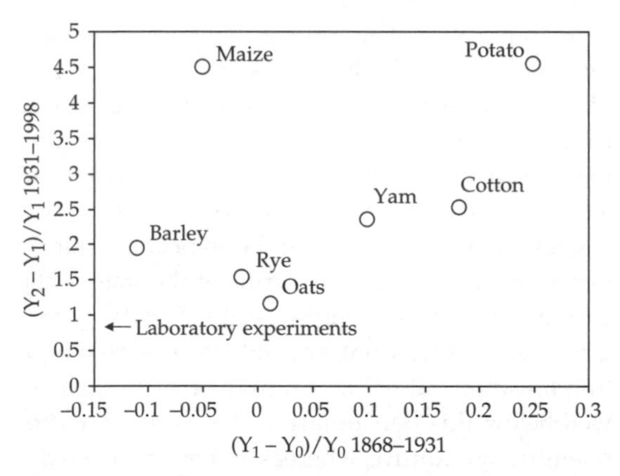

Figure 6.19 Advance of crop yield in two epochs, 1870–1931 and 1931–1998. Data from Tracy *et al.* (2004), Table 3.1.

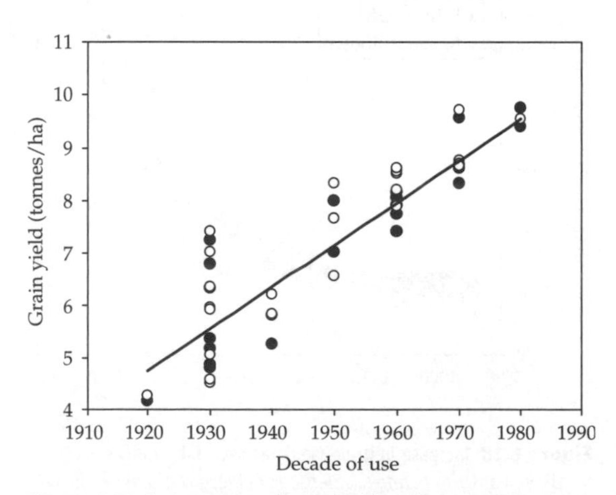

Figure 6.20 Improvement of maize cultivars in common-garden trials. Symbols are two trial years. The slope of the regression is 0.08 tonnes ha^{-1} y^{-1}. Data from Castleberry *et al.* (1984).

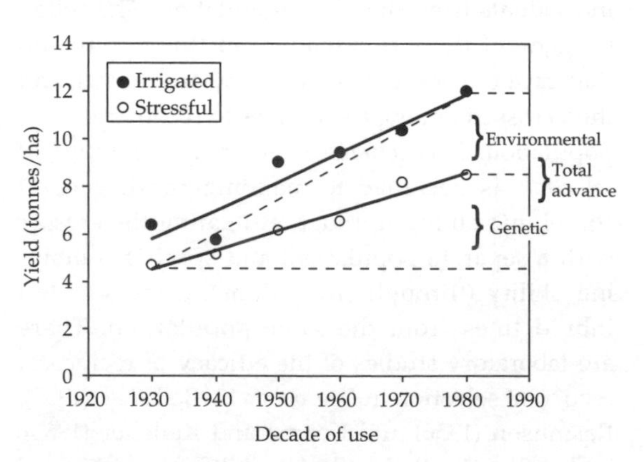

Figure 6.21 Performance of obsolete and current cultivars under obsolete and current management regimes. Data from Castleberry *et al.* (1984).

6.4.2 Historical improvement

The advance of grain yield in field crops since 1950 has been matched by equivalent advances in livestock. For example: broiler weight has increased from 1500 g to over 4000 g (Emmerson 1997), pig growth has increased from about 600 g/day to nearly 1000 g/day (Rauw *et al.* 1998), and milk production has increased from 2400 kg to about 8250 kg per cow (Blayney 2002). The degree of improvement per generation relative to breeds of 50 years ago is similar to that reported from laboratory selection experiments with much shorter-lived model organisms (Figure 6.22). There is no reason to doubt that a large fraction of the improvement is attributable to selection, and in several cases there is good evidence of genetic change (see Hill and Bunger 2004). There have also been a few failures. The daily egg mass of laying chickens increased modestly from 34.2 g in 1950 to 49.3 g in 1993 (Jones *et al.* 2001), perhaps because there is a physiological upper limit of one egg per day that is difficult to transcend. The best-known example of complete failure to advance is the speed of racehorses, where the trend has been precisely the opposite of that in crops and livestock: a steady fall in winning times until about 1950, since when no advance has been made. This is puzzling because racing ability seems to be heritable and there is strong artificial selection, especially on males (Gaffney and Cunningham 1988). Thoroughbred horses are the most valuable animals in the world and their breeding is taken very seriously, but winning times in unhandicapped races such as the Derby (Epsom or Kentucky versions) have nevertheless changed little in 50 years. Racing greyhounds show the same lack of progress (Hill and Bunger 2004). Taken at face value, this implies that training methods have steadily deteriorated, which is surely unlikely. It is possible that heritability is overestimated because offspring of the best sires are likely to receive the best treatment (Hill and Mackay 1988), and the effective population size may be very small because these sires are responsible for a large fraction of recruitment. I think that the most plausible explanation is that there is very little if any selection for speed in itself: what is selected is primarily the ability to win handicaps. The purpose of handicapping is to give all contestants an equal chance, so that if it were done perfectly the winner would be a random entrant and there would be no systematic selection for speed. I doubt that this explanation will be widely accepted, however, and leave the slowness of horses and greyhounds as a curious anomaly in the annals of artificial selection.

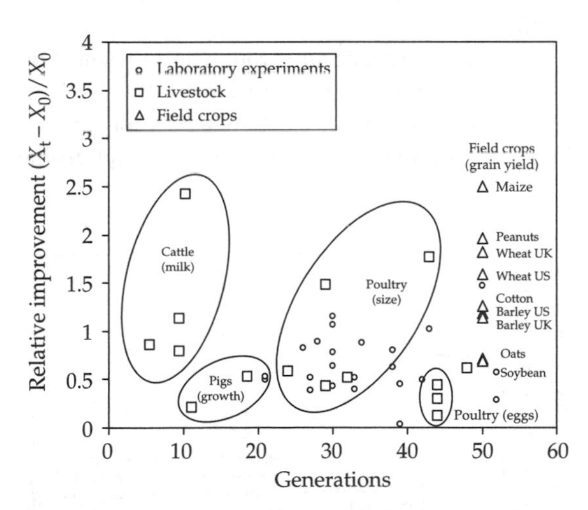

Figure 6.22 Improvement of crops and livestock 1950–2000. Data from Rauw *et al.* (1998), Etches (1998), Emmerson (1997), Jones *et al.* (2001) and the USDA and DEFRA websites. Generation time taken to be 1 y for crops and chickens, 1.5 y for turkeys, 2 y for pigs, and 5 y for cattle. Plotted points represent different characters within a category (e.g. daily egg mass and mean egg size), different regions or different data series. The points from laboratory selection experiments are from Table 6.1.

6.4.3 Domestication

The landraces grown as field crops in the nineteenth century were themselves highly modified versions of wild ancestors. How modern crops and livestock were derived from wild ancestors is a large subject that I can only glance at: there are extensive accounts by Hancock (1992) and Harlan (1992). The processes that are involved in domestication, and in particular the role of deliberate selection, have not been finally resolved, but I think that it is reasonable to recognize three types of character. The first comprises those that would evolve in cultivation whether or not any deliberate selection were applied. This category would include the retention of seed in persistent infructescences and the loss of seed dormancy, because plants which lacked these characteristics would not for long persist in cultivation. The second category includes traits that are readily identified, generally found to be desirable, and would be likely to be deliberately selected by early cultivators. Seed size and ease of harvesting would be examples. The third category would be population traits that are intrinsically desirable but difficult to select. Grain yield per unit area is a character of this sort: it is often the principal object of modern selection programmes, but cultivators down to the late middle ages would fid it difficult to disentangle the merits of different varieties, even if they were distinguishable, against the background of environmental noise created by widely separated holdings in a variable landscape. The early stages of domestication, once the habit of cultivating plants had become established, would then proceed automatically: genotypes with non-shattering fruits and lack of seed dormancy would accumulate. The intermediate stages would be more problematic because advances in single-plant traits such as increased seed size do not necessarily translate into increased yields. It would only be in the final stage of scientific agriculture that the ability to perceive and efficiently select for increased yield would appear. The history of wheat yield can be viewed in this perspective. Wild wheats were gathered in the eastern Mediterranean basin 12 000–20 000 BP and would yield 500–1000 kg/ha when sown, according to Evans (1993). They bear ears which at maturity disarticulate through the formation of an abscission layer between the spikelets and thereby scatter the seed. The first domesticated forms were hulled emmer and einkorn wheats that were first cultivated 9000–10 000 years ago. These have a tougher rachis (central axis of the spike) and the abscission layer is repressed so the seeds are retained on the ear until they are gathered and threshed. This is clearly a desirable characteristic that enables the grain to be harvested easily and might be deliberately selected; but it is not necessary to assume this, as plants with a fragile rachis would lose their seeds before harvest and consequently would not be re-sown. Similarly, the seeds of most wild grasses lie dormant for a variable period of time. Loss of dormancy leads to rapid and uniform germination that is desirable to the cultivator, but need not be deliberately selected because dormant seed will be unable to compete with established plants if it germinates at all. Consequently, there is no need to invoke any process other than the annual cycle of sowing and harvest to account for the early evolution of domesticated strains. To the extent that these could be improved by deliberate selection, it would be applied to easily distinguishable traits of whole plants, such as seed size. Traits of populations are much more difficult to discern, and until the beginning of scientific agriculture it seems likely that no consistent increase in grain yield occurred for 8000 years following the first domestication. Indeed, yields in medieval times may not have exceeded the yields achieved by the first cultivators of the soil, and yield increased only slowly in the 500 years 1350–1850, probably because the climate grew warmer and methods of cultivation improved (Figure 6.23). It is only in the last 0.5% of the Holocene that the systematic application of artificial selection has driven large increases in yield.

The changes that occur through natural selection in the environment created by primitive agriculture constitute a 'domestication syndrome' shared by most crop plants. In crop plants the syndrome

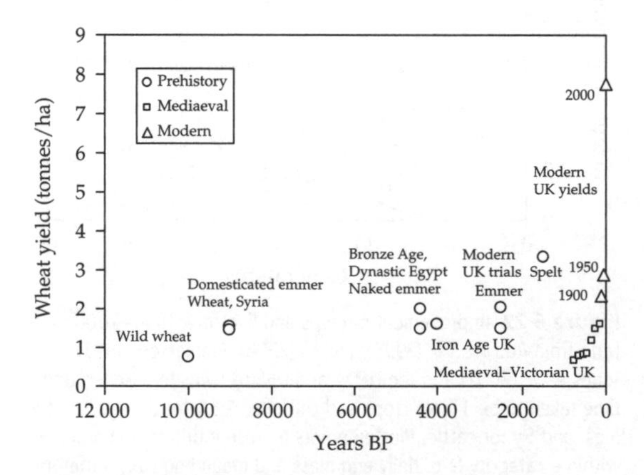

Figure 6.23 Historical wheat yields. Estimates of ancient yields from Araus *et al.* (2003); medieval yields from Clark (1991).

includes seed retention, loss of seed dormancy, ease of processing, compact growth habit, increased levels of selfing, and reduction of toxic secondary compounds. These characters are largely determined by major genes. In hulled wheats (einkorn, emmer, and spelt) the rachis is less fragile than in wild wheats but is still quite brittle, whereas the grain is enclosed by tough glumes, so that when the ear is threshed it breaks up into spikelets that require further processing to release the grain. Durum and bread wheats are free-threshing: the rachis is very tough but the glumes are fragile, so threshing releases the grains directly. The free-threshing phenotype is produced by a single point mutation in the *Q* gene, now known to be a transcription factor homologous to a gene that regulates flower development in *Arabidopsis* (Simons *et al.* 2006). The best-known set of major genes involved in domestication is that involved in the evolution of maize from wild teosinte in Mexico (reviewed by Doebley 2004). Some affect the infructescence (cob) which has become modified like the ear of wheat. Thus, in teosinte each grain is protected by a tough covering consisting of a rachis segment and a glume, whereas in maize these are reduced in size and form the cob on which the naked grains are borne. The gene responsible is *teosinte glume architecture1* (*tga1*), which is thought to be a regulatory gene. Furthermore, the ears of teosinte disarticulate at maturity whereas those of maize remain intact and are easily gathered, and the size of the ear has been greatly increased. Moreover, there has also been a radical change in plant architecture: teosinte has long lateral branches bearing tassels (male flowers) whereas maize has short branches bearing ears. The gene primarily responsible for this change is *teosinte branched1* (*tb1*) which like *Q* is a transcription factor.

If the interpretation I have given is broadly correct, there are three main epochs in crop cultivation. The first is rapid natural selection of major genes for fruit and seed characters in the early stages of agriculture. This is followed by a long period during which plant architecture and the size of individual fruits and seeds are modified through desultory artificial selection, whereas the yield of the crop remains much the same. The third is the epoch of systematic selection and breeding, which has quadrupled yields in a single human generation. This is a simplistic account, but is intended to distinguish between the three processes of natural selection in the humanized environment, selection by people without consistent procedure or direction, and systematic artificial selection.

At about the same time that the first crops were being domesticated from wild grasses, dogs were beginning to be domesticated from the grey wolf, a large and rather variable canid that struck up an acquaintance with human bands towards the end of the last glacial advance, some 10 000 years ago. They were for long selected, no doubt largely unconsciously, for several useful characteristics, particularly for hunting and guarding, and have more recently been deliberately selected and inbred to create the vast range of modern breeds. Plant breeders generally select for a particular phenotype, typically high yield. Naturally there is some differentiation among varieties in terms of local adaptation and quality, but in general different dietary needs are met by different species of crop plant. Dog breeders, by contrast, have selected for specialization. In body size alone, the variation among domestic dogs not only exceeds that among wolves, but actually exceeds that among all members of the family Canidae—wolves, jackals, wild dogs and foxes (Figure 6.24). There is subfossil evidence of domesticated dogs from China, Syria, and arctic North America about 8000–10 000 BP. The subsequent evolution of dog breeds has been worked out by Parker *et al.* (2004). The deepest-branching lineages are spitz dogs, with thick fur, pointed ears and muzzle, and a tail curled on to the back. No radical morphological changes separate these dogs from wolves, and the domestication syndrome is instead mainly behavioural, involving affective behaviour towards humans and tolerance of restraint (very imperfectly developed in some breeds). In the next few thousand years more specialized types seem to have been bred, although gene flow has erased most of the phylogenetic signature. These include gaze hounds (saluki), guarding dogs (mastiffs), and herding dogs (shepherd dogs, collies). Finally, a wide variety of hunting dogs began to be bred in Europe a few hundred

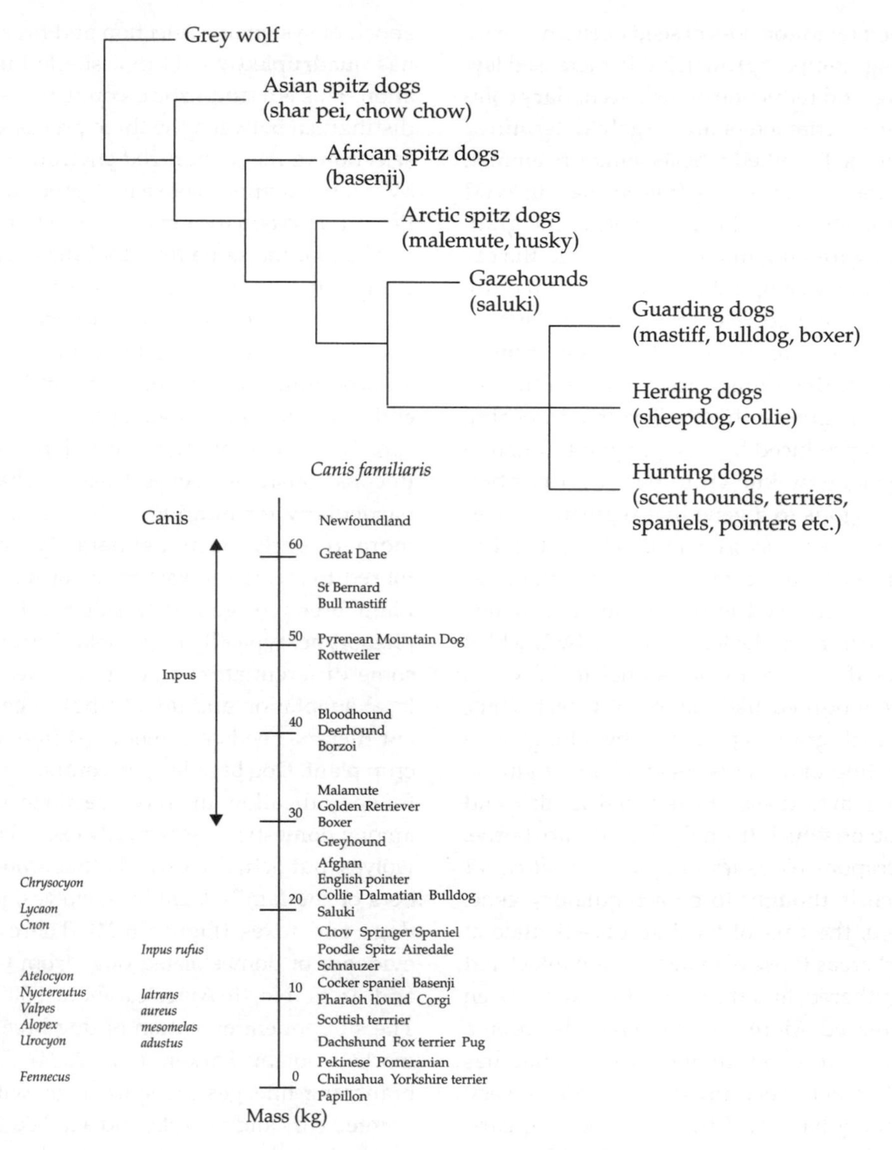

Figure 6.24 Dogs. Schematic phylogeny of dog breeds is adapted from Parker *et al.* (2004).

years ago. The capture of prey by wild canids may involve a whole sequence of behaviours: flushing, tracking by scent, detection by sight, pursuit, either individually or in a pack, herding, crippling, killing, and finally carrying the prey back to be eaten. Domestic dogs have been selected to excel at one of these tasks, while often suppressing all others. Spaniels will flush game from low undergrowth; bloodhounds follow a scent trail; gaze hounds such as salukis have exceptional visual acuity, while pointers will actually detect prey and then freeze without proceeding to pursue it; foxhounds and beagles hunt in packs; sheepdogs will herd flocks of sheep (and almost anything else) without attacking them; bulldogs were bred to grip their prey and hang on; mastiffs are fighting and killing dogs; retrievers will fetch dead or wounded prey without (in theory) damaging it; every component of the strategy of wild canids for capturing prey is represented by a specialist breed. Moreover, different breeds tackle different prey: deerhounds and otterhounds for large errant animals pursued in the open, the whole tribe of terriers for small game hiding in burrows and crevices. In the nineteenth century these strains were adopted by breed clubs, each with the intention of preserving a breed through a closed pedigree, inbreeding and artificial selection for defined criteria. The result has been extremely successful: after all, no palaeontologist who unearthed the fossil bones of a wolf, a bulldog, and a Yorkshire terrier would classify them in the same species. The exuberant diversity of dogs is a striking testimonial to the power of selection to direct adaptive change far beyond the limits of the original population within a few hundred generations.

CHAPTER 7

Natural selection in open populations

W.F.R. Weldon

The extreme slowness with which evolution was presumed to proceed inhibited any attempt to observe the operation of selection in natural populations, still less to measure its strength, for the first 40 years of Darwinism. In the later decades of the nineteenth century, however, the biometrical school founded by Francis Galton—a cousin of Charles Darwin—supplemented the traditional belief in the importance of minute variations with a growing ability to analyse and interpret quantitative data. In 1893 the Royal Society of London established a 'Committee for Conducting Statistical Inquiries into the Measurable Characteristics of Plants and Animals' chaired by Galton. Its secretary was W.F.R. Weldon, a rising young biometrician who later went on to found *Biometrika* with Karl Pearson. The report of the committee was presented on 28 February 1895 and consisted of a statistical analysis of the measurements of the carapace of about 8000 shore crabs, *Carcinus maenas*, prepared by Weldon. He was most concerned to show how the variability of frontal breadth changed with age, and for this purpose used the quartile deviation Q, half the difference between the third and first quartile. He was able to distinguish between adult and juvenile crabs but otherwise could not estimate their age directly, so used carapace length as a surrogate. In the very smallest crabs, with a carapace 7.5 mm long, $Q = 9.42$ mm and this value increased steadily to a maximum of $Q = 10.79$ mm at a carapace length of 12.5 mm. Weldon took this to be the consequence of allometric growth. In adults, however, Q had fallen to 9.96 mm, and Weldon attributed this decline in variation to the greater mortality of extreme individuals. With brilliant insight, he realized that the selective mortality required to produce this effect is just $(10.79 - 9.96)/10.79 = 0.077$, or 77 individuals per thousand. At a stroke, Weldon had established the study of evolution in natural populations and given it the strongly quantitative and mathematical flavour it has retained to the present day.

Weldon spent the summers of 1900 and 1901 at Gremsmühlen in northern Germany, where at that time an extensive beech forest grew on the shores of the Diek See. Here he collected specimens of the land snail *Clausilia laminata*. The shell of a snail is a cylinder wrapped helically around a central axis, the columella, that increases in diameter as the snail grows. It is formed incrementally, each episode of growth adding to the existing shell, so that every snail carries its own history on its back. Weldon realized how this could be used to demonstrate the operation of selection: the shells of young snails which had completed only a few turns of the helix could be compared with the apical portion of the shells of older snails, the persistent record of their state when young. By grinding the shell along its longitudinal axis through the plane of the columella one can measure the geometrical parameters determining its shape, such as the peripheral radius of each whorl in relation to its position along the columella. The mean value of the radius did not differ between juvenile and adult individuals, so no net directional selection had acted. The standard deviation, however, was always greater in the juveniles: consequently, extreme individuals must have been removed by selection (Weldon 1901) (Figure 7.1).

The reduction of variation in the youngest whorls suggests a selective removal of 5–6% of individuals. Weldon next repeated these observations on snails collected in the spring of 1902 at Brescia—clearly another holiday—but was surprised to find no consistent differences between juvenile and adult states. He published the results anyway (Weldon 1904). Meanwhile, he had inspired A.P. di Cesnola, a young student at Queen's College Oxford, to make the same observations on a very different snail, *Arianta arbustorum* (then *Helix arbustorum*). He collected these from the river path between Godstow and the Perch Inn, no doubt a pleasant labour, and followed Weldon's procedures of specimen preparation and data analysis to the letter. The results were in every respect similar to those for *Clausilia laminata*, the adults being consistently less variable than the juveniles, both for columellar and for peripheral radii (di Cesnola 1907). At this point, then, it had been clearly established that natural selection is acting in contemporary populations, and that its force could be precisely estimated. The stage seemed set for a rapid expansion of field studies of natural selection in action. It never happened: Weldon died unexpectedly of pneumonia in April 1906 at the tragically early age of 46, and left no intellectual heirs. During the next decade the biometrical and Mendelian schools wrangled about genetics, with evolution pushed into the background. When their debate had been resolved, it led to the development of theoretical population genetics, not to empirical studies of selection. There were a few disconnected observations of selection at the time and thereafter, including the well-known case of storm-blown sparrows described by Bumpus (1899), but these do not compare with Weldon's careful and systematic studies. The defining and much-discussed document of the time, R.A. Fisher's *The Genetical Theory of Natural Selection* (1930), does not cite Weldon and despite including much matter of peripheral interest it nowhere describes a field study of selection. It was to be nearly 50 years before systematic observations of selection in natural populations were mounted again, and by then Weldon had been almost completely forgotten.

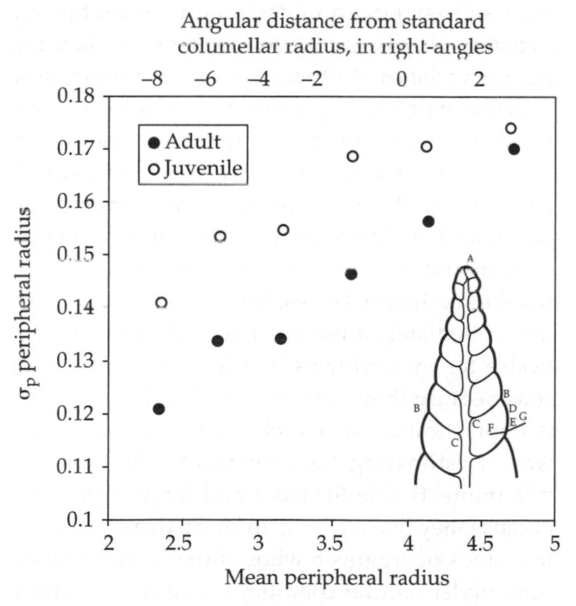

Figure 7.1 Variation and selection in *Clausilia laminata*. Age is represented by angular distance from a standard columellar axis, increasing from left to right. In the cross-section of a shell, taken from Weldon's paper, this is the number of whorls measured from the standard whorl at 5 mm from the apex. The peripheral radius vector is AB. The mean radius increases as the snail grows, but its variation (as phenotypic standard deviation) is always greater for the shells of juvenile snails than for the remnant juvenile shells at the apex of the shells of adult snails. Data from Weldon (1901).

7.1 Fitness in natural populations

It is difficult to draw a clear line between laboratory microcosms, artificial selection programmes, and what goes on in the lake or the forest. It is common to refer to 'natural' populations and I shall often do so, but it might be objected that we are part of nature, or, alternatively, that no site on Earth has been left untouched by human activity. Sometimes one speaks of evolution 'in the wild', but when this turns out to involve snails in an English meadow the term seems a little absurd. But outside the microcosm or the cloistered selection line, populations are at any rate open: open, that is, to unconstrained gene flow and environmental perturbation. No new principle need be introduced, but studies of open populations bring us a step closer to understanding the evolutionary dynamics underlying adaptation.

The routine operation of selection in natural populations can be analysed through the fundamental theorem, which shows how the intensity of selection can be estimated by measuring either the change in fitness over a generation or the standing genetic variance of fitness. To understand how this can be done I revisit the fundamental theorem, following the accounts in Price (1972) and Ewens (1989). The overall change in mean fitness from one generation to the next can be expressed as

$$\Delta\overline{w} = (\overline{w} \text{ in } \epsilon') - (\overline{w} \text{ in } \epsilon),$$

where ϵ stands for the state of the environment and the changed state is designated by a prime. This introduces a difficulty: from one generation to the next, mean fitness will be affected by two sources of change, the change in the environment and the change in the population. In principle, these components can be separated:

$$\Delta\overline{w} = [(\overline{w} \text{ in } \epsilon') - (\overline{w} \text{ in } \epsilon)] + [(\overline{w} \text{ in } \epsilon) - (\overline{w} \text{ in } \epsilon)].$$

The first component represents the change in the environment (from the frame of reference of the next generation) and the second the change in the population (from the frame of reference of the previous environment). This is equivalent to distinguishing between changes in the frequencies of alleles and changes in their effects. Any given allele has some average effect on fitness which is conventionally symbolized as α, so that the fitness of an individual bearing alleles i and j is $w_{ij} = \overline{w} + \alpha_i + \alpha_j$. If the frequency of these individuals is p_{ij} then mean fitness is $\Sigma\Sigma p_{ij}w_{ij} = \Sigma\Sigma p_{ij}(\overline{w} + \alpha_i + \alpha_j)$. Consequently, mean fitness may change either because allele frequencies (p_{ij}) change or because average effects (the αs) change. Allele frequencies will be altered by mutation and immigration, which will create genetic variance that causes selection, the overall change in mean fitness being given by $(1/\overline{w})\Delta\overline{w} = V_A/\overline{w}^2$. This is the form in which the fundamental theorem is usually expressed and used. Because the right-hand side must be positive it might be inferred that mean fitness is continually increasing, which on a geological timescale is clearly absurd. In fact, mean fitness will usually change little if at all. This is partly because the expression refers to a single generation, in which variance is generated primarily by deleterious alleles introduced into the population by mutation or immigration. Another reason, however, is that the environment may not remain constant. Average effects will be altered by changes in the environment, so that an allele having a positive effect on fitness when population density is low and resources abundant may be an inferior competitor in crowded sites, for example. The response of mean fitness to environmental degradation is not addressed by the fundamental theorem and no general rule for predicting its magnitude has yet been established. This section discusses the effects of changes in gene frequency, the day-to-day operation of selection in a constant environment. Renewed adaptation to a changing environment is discussed in later sections (§7.5).

7.1.1 The variance of fitness

At first glance, the fundamental theorem seems to offer an easy key to understanding evolutionary dynamics: it will suffice to measure the standing genetic variance of fitness in order to predict how selection will act. In practice this is not very easy to do. To measure the phenotypic variance of fitness it is necessary to follow a cohort of individuals through the whole of their lives from birth to death, as their fitness may be strongly affected by early mortality. This is impracticable for insects or planktonic invertebrates, because they will scatter immediately they are released; it is impracticable for invertebrates that live in sediments or soils because they are too cryptic to be studied; it is impracticable for microbes because there is no way of estimating the growth of a lineage; and it is impracticable for trees and large vertebrates because they live too long. In short, there are only a few kinds of organism whose fitness can be measured under natural conditions. The two that have been used most often are annual plants, because they stand still and will yield estimates after a few years of study; and vertebrates that can be individually marked and scored, such as hole-nesting birds.

However difficult and laborious to measure, the phenotypic variance does little more than set an upper limit to the flux of mean fitness. Estimating the genetic variance (let alone the additive component of the genetic variance) is even more challenging. Without manipulating the population, it is necessary to know both the fitness and the relatedness of individuals. This requires long-term studies extending over at least two generations and has been achieved only for a few animals. One drawback is that the mating system is uncontrolled, hence mating may not occur at random, it may not correspond to observed family groups, and only simple comparisons (such as presumptive parents and offspring) can be used. Alternatively, individuals can be crossed in the laboratory or experimental garden and their offspring put back into the field. This requires large-scale studies because most of the offspring will die almost immediately, but it has been achieved for a few short-lived plants. One drawback is that planting seeds does not correspond to natural seed dispersal and may not produce typical rates of establishment. It is much easier to germinate plants in the greenhouse and either rear them to maturity or plant them out once they are well established. This provides much fuller and better-balanced data sets, but the environment is far from natural and the estimate of genetic variance is inaccurate to an unknown degree.

As an example of surveys of natural variation I shall use a plant I know well, the annual herb *Impatiens*. It grows as dense stands along watercourses or by seeps in the forest. The plants bear both open (chasmogamous) flowers pollinated by bumble bees and closed, self-fertilizing (cleistogamous) flowers; the chasmogams make a colourful display in sunny sites, but most reproduction under the canopy is through cleistogams. The seeds are dispersed ballistically about 0.5m from the parent plant and germinate in the following year, so the species is strictly annual, with no seed bank. Dozens of investigators have used this attractive organism to study pollination biology, seed dispersal, and inbreeding. It can also be used to measure fitness and study selection, which I have done in collaboration with colleagues at McGill (Bell *et al.* 1991, Schoen *et al.* 1994). We marked out a quarter-hectare plot within old-growth woodland occupied by a more or less continuous population of *Impatiens pallida*. In the spring we removed 100 random seedlings and reared them in a screened exclosure so that we could construct outcrossed families when they flowered in the summer. These seed families were then carefully sown in random locations in the same quarter-hectare site, so that we could follow the fate of every individual. Each family was sown into each site, so that we could estimate both genetic and environmental components of variation. We estimated fitness from the sum of chasmogam and cleistogam fruits (through seed) and chasmogam flowers (through pollen). This is probably as close as one can come to studying the genetics of fitness of resident plants in an undisturbed natural setting. The first result was that most individuals have zero fitness, usually because they die in the first few weeks of life: only about 10% succeeded in producing any flowers at all (Figure 7.2). Mean fitness is therefore low (0.58) and variance high (5.36), so the standardized phenotypic variance is $SV_P = 15.7$. Almost all reproduction was through cleistogams, with an average of 0.86 seeds, variance 6.55, giving $SV_P = 7.6$. Either figure implies is that most of the seed produced by the population comes from a very few individuals: 1% of the initial seeds generated 30% of the total reproduction. This creates a very large opportunity for selection (and a rather small effective population size). It is often supposed that nearly

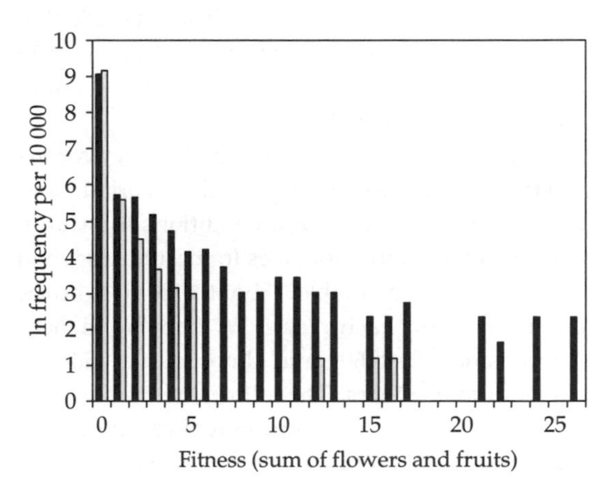

Figure 7.2 Distribution of fitness of *Impatiens* individuals in natural conditions. Solid and open columns are two sites. Data from Bell *et al.* (1991).

all of this early mortality is random and therefore non-selective, mainly because seedling survival rarely shows consistent local adaptation in transplant experiments (e.g. Andel 1998), but one might argue that this is the stage at which many deleterious mutations might be eliminated. If we neglect plants with zero fitness the mean and variance become 4.6 and 23.8, so that SV_P falls to 1.13. Much of the variance is clearly environmental. This is in part attributable to a NW–SE gradient across the plot generated by unidentified environmental factors, but also in part to heterogeneity at much smaller scales comparable with the dispersal range of the plants. Consequently, the correlation between the performance of plants from the same family falls off steadily with the distance between sites (Figure 7.3). The variation among sites and the variation among families give estimates of environmental and genetic variances in seed production, $V_E = 0.366$ and $V_A = 0.022$. Thus, the bulk of variation in fitness is caused by unidentified hazards killing plants early in life, a smaller but substantial fraction is contributed by systematic variation from site to site, and the very small remainder is genetic. The standardized genetic variance is $SV_A = 0.022/0.862 = 0.03$. Bennington and McGraw (1995a) carried out a similar study in the same species growing in optimal (floodplain) and stressful (hillside) sites, and their data lead to estimates of $SV_A = 0.008$ on the floodplain and 0.06 on the hillside. It seems reasonable to conclude that in natural populations of *Impatiens pallida* $SV_P \approx 10$ in offspring, falls as the cohort ages to $SV_P \approx 1$, and that SV_A lies between 0.01 and 0.1.

The great tit *Parus major* is a small insectivorous bird that has been intensively studied in Britain and the Netherlands. In natural conditions it nests in holes which it readily forsakes for nest-boxes when these are made available. About 1000 nest-boxes have been set out each year in Wytham Wood, near Oxford, since the 1960s, and almost all the 100–500 resident tits use them. This allows the nestlings to be banded, so that each individual can be recognized and subsequently recorded for the rest of its life. Each adult bird will produce offspring, some of which will survive to be recruited into the breeding population a generation hence: the fitness of an individual is the number of recruits it produces during its lifetime. Forty or fifty years of continuous effort then provides pedigrees from which fitness can be followed generation after generation (McCleery *et al.* 2004). The mean number of recruits is 1.11 with variance 2.36 for females (males are similar), so $SV_P = 1.9$. Analysing the pedigrees yields $V_A = 0.004$ for females and 0.031 for males, leading to $SV_A = 0.003$ and 0.025 respectively. Another long-term study involved a different species of

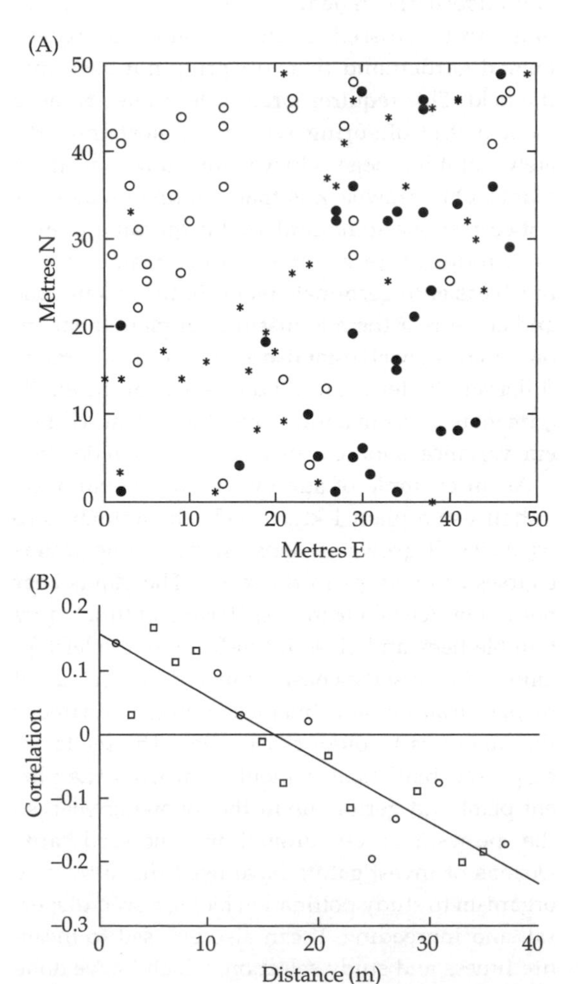

Figure 7.3 Variation of fitness in natural conditions. Families with above-average fitness (filled circles) are not randomly distributed among sites (upper diagram), revealing environmental heterogeneity that causes the correlation between families to fall with their distance apart (lower diagram).

insectivorous bird, the collared flycatcher *Ficedula albicollis*, on the island of Gotland, off the coast of Sweden (see Merilä and Sheldon 2000). Mean lifetime reproductive success was 2.21 with variance 1.93 for females, so $SV_P = 0.4$. The pedigree data gave $SV_A = 0.085$ for females, considerably greater than the Wytham tits, albeit the estimate for males ($SV_A = 0.029$) was almost the same. In small insectivorous birds it seems reasonable to conclude that $SV_P \approx 1$ and SV_A lies between 0.01 and 0.1.

Estimates of SV_P in other intensively studied vertebrates are shown in Figure 7.4. These tend to fall from values of 1–2 in birds to about 0.5 in large mammals (deer, lions) and about 0.1 in humans. This is largely attributable to variation in juvenile survival. Suppose that each female produces B offspring of whom 2 survive and themselves each produce B offspring. Then $SV_P = ½B - 1$ is simply a linear function of B. This accurately reflects the potential for selection but clearly gives no information about the genetic consequences of selection.

Impatiens and *Parus* both live in temperate deciduous forest but in other respects their biology is different: the contrast between a rooted photosynthesizer whose seeds disperse 1m and a motile predator whose offspring disperse 1km could hardly be stronger. Moreover, a successful adult *Impatiens* may bear 1000 seeds, whereas

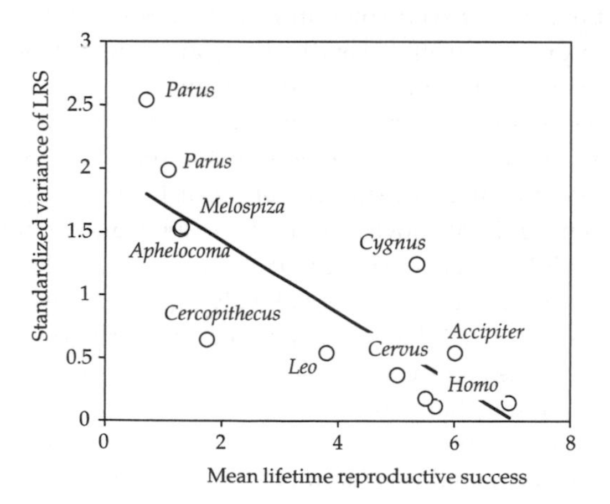

Figure 7.4 Phenotypic variance of lifetime reproductive success in vertebrates. The regression equation is about $y = 2 - 0.3x$. Data from chapters in Clutton Brock (1988).

a successful *Parus* incubates only about 10 eggs. Consequently, juvenile survival is much lower in *Impatiens* (1% in the Yellow Valley plot) than in *Parus* (25% at Wytham), and this creates a large difference in SV_P. Nevertheless, estimates of SV_A fall in the same range of 0.01–0.1 for both organisms. One serious reservation should be made: despite the labour involved in these studies, estimates of V_A are seldom significantly different from zero. Nevertheless, they are the best available, and suggest that SV_A may fall within a surprisingly narrow range in natural populations.

7.1.2 Immigration pressure

The individuals who survive and succeed in reproducing in a given site will be adapted in some degree to the specific conditions of growth at that site. The next generation to occupy the site will be their descendants together with immigrants from nearby sites. The immigrants are likely to be somewhat less well adapted and will reduce the mean fitness of the population, which is then restored by selection acting to change gene frequencies back to their values in the previous generation. The intensity of selection involved depends on the rate of immigration and the difference in relative fitness between sites. Consider an individual growing in a given site and producing pollen and seed. If it had been moved as a newborn individual to a nearby site it would have fitness $1 - s$ there relative to the resident adults. Consequently, its offspring through pollen will have fitness $1 - s/2$ for additive gene effects. If a fraction φ_p of all newborn individuals develop from resident seed fertilized by immigrant pollen then mean fitness will be reduced, relative to the resident adults, by approximately $\Delta \ln \bar{w} = \varphi_p \ln(1 - s/2)$. If immigrants arrive as seed and constitute a fraction φ_s of the total seed then they will cause a larger reduction $\Delta \ln \bar{w} = \varphi_s \ln(1 - s)$, assuming that they are not carrying pollen genomes back to their home site. The overall loss of ln fitness is thus $\varphi_s \ln(1 - s) + (1 - \varphi_s)\, \varphi_p \ln(1 - s/2)$. Alternatively, the fitness of offspring will be reduced by emigration. Suppose that individuals invest equally in male and female gametes, so that the average parent has one surviving outcrossed offspring through male and one through female function. A fraction φ_p of ovules at the home site are fertilized by pollen from the same site, so the fraction of successful pollen produced by a plant that fertilizes home ovules is also φ_p. The resulting seed may either grow at the home site or emigrate. Ovules may be fertilized by home or by foreign pollen and may subsequently remain or disperse. Adding up these contributions, the change in fitness of the offspring is

$$(1/\overline{w})\Delta\overline{w} = (1 - \varphi_p)(1 - \varphi_s) + \varphi_s(1 - \varphi_p/2)(1 - s) + \varphi_p(1 - \varphi_s/2)(1 - s/2) - 1.$$

If only pollen moves ($\varphi_s = 0$), for example,

$$(1/\overline{w})\Delta\overline{w} = -\varphi_p s/2 \approx \varphi_p \ln(1 - s/2),$$

as before. Thus, the movement of gametes or zygotes creates selection acting to restore adaptedness in every generation.

7.1.3 Local selection coefficients

The selection coefficient acting against immigrants has been estimated by implanting seeds or ramets from the surrounding region into the home site and measuring their subsequent success relative to re-planted residents (Table 7.1). There have been four short-range studies of *Impatiens* where seed has been moved 100m or less within more or less continuous populations (Schemske 1984, Schmitt and Gamble 1990, Bennington and McGraw 1995a, Donohue *et al.* 2001), yielding 11 estimates of relative fitness involving somewhat different fitness measures. The mean value is 0.34, but the experiments involved implants gathered at different distances from the home site and greater separation is equivalent to more generations of dispersal. Estimates of s from these studies increase with distance: $s = 0.1 \times \ln$ distance (m) + 0.0152. *Impatiens* has ballistic dispersal with a mean dispersal distance of 0.5m, so the home site can be reasonably taken to be the area within 1 m of the parent plant. One estimate of s for a single generation of dispersal is thus 0.015. Schmitt and Gamble (1990) used plots at 3m and 12m from the home site which showed an exponential rate of fitness decline of 0.03m^{-1}. Another estimate can be got by arguing that the mean squared dispersal distance (0.45m^2 for *Impatiens*) is additive over generations for random-walk dispersal (Burt 1995), so that immigrants derived 3m from the home site are separated from it by $3^2/0.45 = 20$ generations. Their production of outcrossed seed gave $s = 0.1$, so the per-generation equivalent is $s = 0.005$. The calculation is dubious: vacant sites 100m away will not remain vacant for 20 000 years before being colonized, and in practice a linear cumulative dispersal distance may be more realistic. This gives $s = 0.017$ for a single generation of dispersal. Seed from sites 100m away give estimates of $s = 0.5$, or 0.0035 per generation (Bennington and McGraw 1995a, Donohue *et al.* 2001) These *Impatiens* experiments suggest that $s = 0.01$–0.02 is a reasonable estimate of the consequences of a single generation of dispersal. Similar studies of other species have given comparable results: for example, Waser and Price (1991) obtained mean estimates of $s = 0.33$ at 50m (0.008 per generation) for *Delphinium*, and Antonovics (1976) reported $s = 0.6$–0.7 for *Anthoxanthum* sites 14m apart. Nevertheless, such estimates must be treated with some reserve. In the first place, some refer to sites chosen for their ecological distinctiveness and may not be representative of average values: Antonovics's samples were taken from forest and grassland sites, for example. It is also possible that the home site is likely to be chosen because it offers superior conditions of growth and therefore supports an unusually vigorous population, so that seed sown in nearby sites is unlikely to grow as well. Secondly, there are many inconsistent and contradictory observations. We moved *Impatiens* seed about 500m in old-growth forest and found $s = 0.75$ for one combination of sites but $s < 0$ for another (Schoen *et al.* 1986). Galloway and Fenster (2000) moved *Chamaecrista* 100m and found $s = 0.3$–0.4 for two experiments and $s < 0$ for three others. Some experiments involving much greater distances of 1–100km have found that immigrants are consistently inferior, but others, at least as numerous, have found that they are at no disadvantage (§7.4). My feeling is that s may be strongly skewed: small displacements, corresponding to a single generation of dispersal, will usually have no discernible effect, but occasionally they will cross some sharp ecological boundary, which may or

Table 7.1 Estimates of selection coefficients from short-range transplant experiments. Selection coefficient is [*w*(resident) − *w*(immigrant)]/*w*(resident)

w(resident)	*w*(immigrant)	*s*	Distance	Organism	Character	Reference
33.7	23.5	0.303	100	*Impatiens*	Seed production	Bennington and McGraw (1995b)
7.5	3.1	0.587	100			
16.7	10.4	0.377	32	*Impatiens*	Survival to maturity	Schemske (1984)
20.2	13.8	0.317	32			
4	1.4	0.650	100	*Impatiens*	Seeds/plant, high density	Donohue *et al.* (2001)
413	328	0.206	100		Seeds/plant, low density	
7.7	3	0.610	100			
9.65	8.69	0.099	3	*Impatiens*	Cleistogamous flowers, outbred	Schmitt and Gamble (1990)
9.65	8.3	0.140	12			
8.53	6.96	0.184	3	*Impatiens*	Cleistogamous flowers, inbred	
8.53	5.81	0.319	12			
0.34	0.1	0.706	500	*Impatiens*	Chasmogamous fruit per survivor	Schoen *et al.* (1994)
0.9	0.15	0.833	500			
4.6	5.9	−0.283	500		Cleistogamous fruit per survivor	
3.29	3.87	−0.176	500			
0.39	0.16	0.590	1500	*Polemonium*	survival	Galen (1996)
0.46	0	1.000	1500			
1.81	1.07	0.409	100	*Chamaecrista*	seeds/plant	Galloway and Fenster (2000)
1.94	2.02	−0.041	100			
1.81	1.94	−0.072	100			
1.61	1.73	−0.075	100			
0.62	0.41	0.339	100			
0.79	0.67	0.152	50	*Delphinium*	finite rate of increase	Waser and Price (1991)
0.91	0	1.000	50			
0.75	0.66	0.120	50			
1.08	1.03	0.046	50			
155.63	39.22	0.748	14	*Anthoxanthum*	seeds/survivor	Antonovics (1976)
53.32	21.24	0.602	14			

may not be apparent to the experimenter, but which produces radical shifts in relative fitness. There are no sufficiently detailed observations available to resolve this issue.

7.1.4 The field gradient

Adaptation is degraded in every generation by the difference in gene frequencies between the home site and the surrounding region because a random individual moved from one to the other would suffer a loss in relative fitness. I shall call the reduction in relative fitness of immigrants the *field gradient* ('selection gradient' would be preferable, but has already been extensively used for the linear partial regression of fitness on character value, discussed below). The steepness of this field gradient, which governs the fitness degradation $(1/\overline{w})\Delta\overline{w}$ at the home site, depends on the selection coefficient s and the rate of immigration m: greater differentiation and dispersal will set up steeper gradients. As both s and m can vary across the whole range of possible values it might seem impossible to make any generalization about selection gradients and the restoration of adaptation. The degree of differentiation that can be maintained, however, depends on the rate of immigration. If s is small then beneficial alleles will be eliminated by modest immigration pressure so the fitness degradation must be small. It will also be small when s is large but m small despite the steep field gradient because the few immigrants do not greatly reduce mean fitness. Strong selection can sustain beneficial alleles in the face of high immigration rates, but then the field gradient will be shallow, because the average individual on the home site will not be greatly different from the average individual in the region, and the fitness degradation will again be small. The inverse relationship between the immigration rate and the selection gradient leads to a smaller range of fitness degradation than would otherwise be the case.

Suppose that immigrants arrive as zygotes; selection then acts on survival or fecundity and adults outcross within the home site. The basic theory is given by Maynard Smith (1966), with extension to quantitative variation by Bulmer (1985, p. 180). Fitness at the home site is affected by co-dominant alleles at a single locus; the favoured allele has frequency p_h at the home site and p in the surrounding region. The regional frequency is fixed and does not respond to selection within the home site, as would be approximately true if the home site were small compared with the region. After immigration mean fitness is $\bar{w}' = 1 - (1 - m)sq_h - msq$, so $(1/\bar{w})\Delta\bar{w} = ms(p - p_h)/\bar{w}$. The fitness degradation thus depends on the product of the immigration rate, the selection coefficient, and the difference in gene frequency. The equilibrium frequency at the home site can be found in the usual way; the interesting case is $p = 0$, so that we are considering an allele favoured only at the home site, in which case $p_h = 1 - 2[m/(1 - m)]/[s/(1 - s)]$. The allele persists if $m < s/(2 - s)$; this defines a critical immigration rate, which if exceeded prevents the maintenance of adaptation. Substituting, at evolutionary equilibrium $(1/\bar{w})\Delta\bar{w} = -m(s - 2m + ms)/(1 - 3m + 2ms)$. The fitness degradation is small when m is small (because immigrants are rare), grows (becomes more negative) as m increases, but then again becomes small at the largest values of m compatible with the persistence of the allele (because residents are rare) (Figure 7.5). There is therefore an immigration rate m^* that creates the maximum fitness degradation for given s. For weak selection $m^* = s/4$, increasing to $m^* \approx s/3$ for values up to $s = 0.25$, and described roughly by $m^* \approx 0.05 + s/5 + s^2$ for values of s between 0.2 and 0.8. Substituting m^* gives the maximum fitness degradation as a function of the selection coefficient. For small s, therefore, the maximum value of $(1/\bar{w})\Delta\bar{w}$ is about $s^2/8$. As s increases the selection gradient generated by immigration at rate m^* becomes steeper and the fitness degradation increases (Figure 7.6). The response of the fitness degradation is sluggish, however, because the frequency of the favoured allele is kept low by immigration when selection is weak. Even with selection coefficients as large as 0.5 the maximum value of $(1/\bar{w})\Delta\bar{w}$ is less than 0.1. Hence, despite the very large potential range of m and s the fitness degradation created by immigration pressure is unlikely to exceed 10%, and can usually be expected to be much less over all selected loci.

The most precise estimate of a field gradient is for the soil bacterium *Bacillus mycoides* (Belotte *et al.*

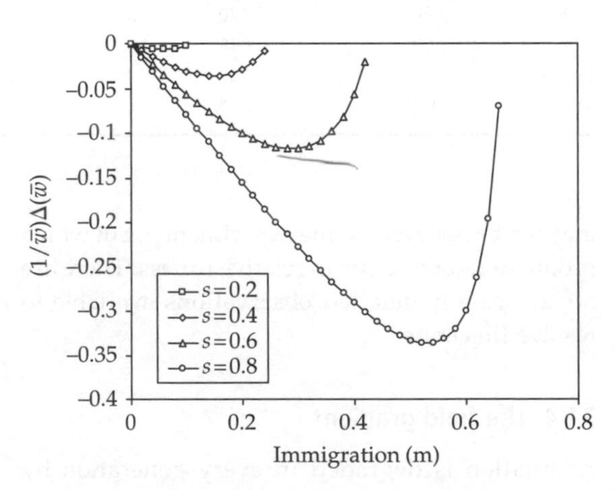

Figure 7.5 Effect of immigration pressure on rate of selection.

2002). Strains were isolated from soil samples from 1 ha of forest floor, and the same soil samples used to prepare growth medium. Growth in the medium prepared from a given sample then reflects some attributes of the corresponding site, although much potentially important variation may be lost. Average growth declined with distance from the home site such that comparing home site fitness w_h with overall mean fitness gave $(1/\bar{w})\Delta\bar{w} = 0.086$. Relative fitness dropped about 40% in the first 10 m away, compared with 40–50% in the perennial herb *Ipomopsis* (Waser and Price 1989). Thereafter fitness w_d declined exponentially with distance d towards some asymptotic value w_∞ such that $w_d = w_\infty + \alpha \exp(-\beta d)$ with $\beta = 0.06\ \mathrm{m}^{-1}$. The corresponding estimate from the study of *Impatiens* by Schmitt and Gamble (1990) is $0.03\ \mathrm{m}^{-1}$. The similarity between the bacterial and plant estimates is unexpected and striking. Dispersal rates are unknown and of course bacteria and plants have radically different life styles, but if bacterial growth responds to soluble nutrients that also influence the growth of plants then this may explain the resemblance.

Although estimates of genetic variance and local selection in natural populations are uncertain and difficult to obtain, they seem to be consistent with values of $\Delta \ln \bar{w}$ in the range 0.01–0.1: selection acts to increase fitness by 1–10% in each generation. This also seems consistent with some rough theoretical arguments. The main reservation that should be made is that many transplant experiments fail to detect any superiority of home isolates, so selection may often be weaker than 1%.

7.2 Phenotypic selection

So long as conditions of growth do not change, the opposition of mutation and immigration by selection will direct the predictable ebb and flow of fitness. They do perpetually change, however, and this will usually cause further degradation of fitness because change is usually for the worse (§11.3). The amount of degradation cannot be predicted in any axiomatic way. Moreover, it will be countered by specific and appropriate adaptations: evolving resistance if a new disease appears, or a thicker pelt if winters become more severe, for example. Most field studies are intended to evaluate the response of particular characters, or the effects of particular agents of change. Their focus thus moves from abstract considerations of fitness based on axiomatic theory to more concrete studies of phenotypes directed towards finding empirical rules.

7.2.1 Environmental variance of fitness

Displacing a single genotype to different sites or years will show how fitness is degraded by changes in the average effects of alleles. Fitness changes

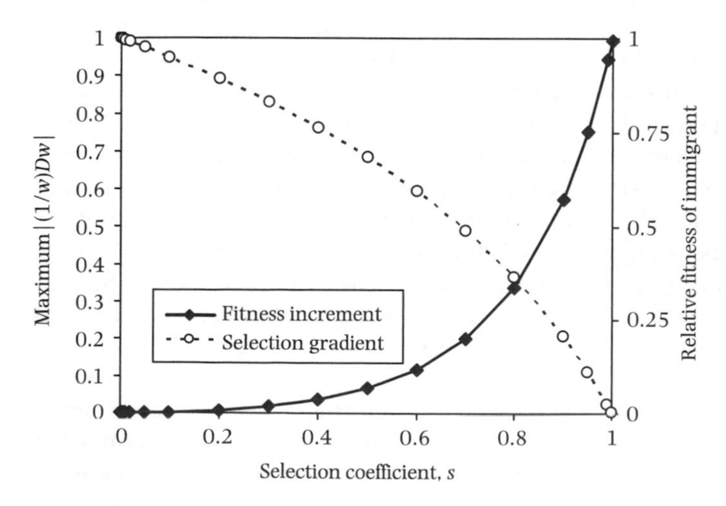

Figure 7.6 The maximum fitness increment generated by a selection gradient.

from w_h at the home site to w_d at the site to which offspring are displaced, so $\Delta w = w_h - w_d$. From bioassays we know that the variance of absolute fitness obeys a weak power law $\mathrm{Var}(w) = Cd^z$ with $z \approx 0.1$, where d is the displacement of the offspring (§2.3). If we rescale d in units of mean dispersal distance per generation φ then $\delta = d/\varphi$. Since $\mathrm{Var}(w) = \frac{1}{4}(w_h - w_d)^2$, $\ln \Delta w = \frac{1}{2}(\ln 4C + z \ln \varphi + z \ln \delta)$, or $(\ln 4C + z \ln \varphi)$ over a single generation. Sedges *Carex* have feeble powers of dispersal, the seeds usually falling close to the parental culm unless they are dispersed by ants or fall into flowing water: 1 m would be a generous estimate of the average dispersal distance for most species. From the *Carex arctata* explant bioassay (§2.3.9), $\ln C = -7.98$ and $z = 0.0835$, so $\Delta w = 0.037$. The *Carex prasina* data give a similar estimate of 0.044. Because the exponents are small, these values are not very sensitive to φ, so Δw for *C. arctata* is only slightly reduced to 0.035 at 0.2 m and only slightly increased to 0.041 at 10 m. These values suggest that natural dispersal in these plants involves a loss of fitness of a few per cent per generation.

7.2.2 Cost of selection

When conditions change, fitness may be abruptly and severely reduced. This can be observed easily enough by stressing natural populations, and happens spontaneously from time to time after droughts, floods, or fires. Catastrophes will often create high rates of selection that will cause rapid allelic substitution. Although such events are very instructive, because they allow us to study evolution on a human timescale, they are surely exceptional. No population can sustain powerful selection acting simultaneously on many characters. If a bacterial population is exposed to a novel antibiotic, 99.999% of cells may be killed, but the tiny fraction of resistant cells will build the population up again. If the same population were exposed simultaneously to two novel antibiotics, only a tiny fraction of a tiny fraction would bear mutations giving them resistance to both, and very few cells indeed, if any, would survive. It would probably be impossible to adapt to three new antibiotics presented simultaneously: there would be no triply resistant mutants already present in the population, and no opportunity to build up the triply resistant genotype in a stepwise fashion over a number of generations. Instead of adapting, the population would become extinct. There is therefore a limit to the amount of evolution that can occur in a given period of time because there is a limit to the rate of selection that a population can sustain. Haldane (1957) expressed this principle by referring to a 'cost of natural selection', by which he meant the excess mortality or sterility needed to drive evolutionary change. The term may be misleading. Selection is not in itself costly: it does not damage the population any more to have 99% of its members killed selectively than it does to have them killed at random. It would be more precise to say that the opportunity for selection is limited by the capacity of the population to regenerate itself. However, though imprecise, the term is probably too familiar to discard, and the principle that it embodies is an important constraint on the rate and pattern of evolutionary change.

While a newly favoured allele is spreading after a change in the environment, some proportion of the population must be eliminated in every generation in order for selection to continue. At first, when the newly favoured allele is still rare, a fraction s of individuals bearing the common gene must be eliminated, because their fitness is $1 - s$. When the newly favoured allele has become very common, selective mortality will be negligible, because it now affects only a very few individuals. One might guess that on average the proportion of the population that has to be eliminated in every generation during the substitution of the allele is about $s/2$. The process is complete in t generations, so the total selective elimination required by the gene substitution is $st/2$. From the equation describing the length of time required for the substitution, this is equivalent to $-\ln (F_t/F_0)$ as defined in §3.2.2. The cost of selection is therefore independent of the rate of selection. Very weak selection acting over a very long period of time, or stronger selection over a shorter period of time, will involve similar total costs per allele substitution. (This is approximately correct only when selection is not too strong; if $s = 0.1$ or more, the cost is greater than these expressions suggest.) The cost is directly proportional to the initial frequency of the allele, because most of the cost is incurred early in the process of substitution, while the newly favoured allele is rare. Alleles that are only mildly deleterious, and that are therefore maintained at fairly high frequencies at mutation–selection equilibrium, are more likely to be substituted when the environment changes than are severely deleterious and therefore extremely rare mutations. Even so, the cost is substantial. As a rough rule of thumb, the number of genetic deaths required by the substitution of a single mildly beneficial allele is about 10 times the number of reproducing individuals in the population at any one time. These genetic deaths represent individuals that die or fail to reproduce because of their genotype at this locus; most death or sterility is likely to be accidental, or caused by other genetic effects, and the actual death rate is certain to be much greater then this figure suggests. A population of 1000 individuals, exposed to a novel environment, can proceed to substitute alleles at a rate equivalent to one new substitution every generation, at the cost of 10 000 genetic deaths per generation. This is not impossible, because the selective elimination could take place among young offspring. It does, however, severely restrict the rate of evolution in slow-growing organisms that produce few offspring where the opportunity for selection SV_P is modest. Even in fecund and fast-growing organisms, excess production will never be sufficient to permit selection to act strongly on several independent characters at the same time.

7.2.3 Lack of response: genostasis

This implies that the rate of input of genetic variance may be inadequate to sustain adaptedness. Any population consists of a finite number of types that represents only a very small fraction of possible combinations of genes. When the environment changes, some possible types will be fitter than others, and these will tend to increase in frequency provided that they occur in the population. Sorting will increase adaptedness, until the sorting limit is reached. However, if the environment changes radically, those types that would be well adapted to the changed conditions of life may simply not exist in the population; the new adaptation required to cope with environmental change may exceed the sorting limit. Selection is then powerless to maintain adaptedness. It might be that even poorly adapted individuals can continue to survive and reproduce, so that the population is able to replace itself and persist. If the rate of replacement is inadequate, with adults giving rise on average to fewer than one descendant (two descendants, in an outcrossed sexual population) in the next generation, the population will dwindle and eventually disappear. Or the change may be so drastic that the population is wiped out immediately; when a lake dries up, the fish will not evolve into amphibians, and most have no means of surviving even short periods in dry mud. This is the extreme case of the cost of selection, or of limited opportunity for selection; when the relevant variation is not available during the lifespan of the population, the result of environmental change is not adaptation, but extinction.

The limits to adaptation are seen most clearly in extreme environments such as hot springs, salt marshes, and mountain summits. Even where there is unoccupied ground, unexploited resources, and ample time, very few species succeed in adapting to these harsh conditions, and the great majority fail to do so even though they occupy nearby sites in great abundance. Species do not continually expand their ranges. This may be attributable to the swamping of adaptation at the edge of the range by poorly adapted immigrants from central populations, but the simplest explanation is that the relevant genetic variation does not exist. Some of the classic demonstrations of selection in the field involve adaptation to novel anthropogenic stresses such as pesticides and pollutants, but even in these situations many or most species fail to adapt. Bradshaw (1991) has convincingly documented the limits of adaptation to soil pollution by heavy metals in the vicinity of smelters, refineries, and mines (§7.5.2). Some plant species possess heritable variation for the ability to grow at high concentrations of copper, lead, or zinc, even in pristine sites, and they rapidly evolve populations capable of living in heavily polluted areas. Others do not, and fail to evolve resistance. Bradshaw argues that *genostasis*, the exhaustion of variation capable of sustaining adaptation, is the primary reason that everything is not found everywhere.

7.2.4 Field studies of selection in *Cepaea*

The cost of selection, or limited opportunity for selection, leads to a very simple conclusion. A specific stress, eliciting a specific response, may from time to time generate intense selection leading to rapid genetic change; but most evolution must involve weak selection. Because the rate of evolution depends on the rate of selection, most evolution must be slow. This conventional view of extreme gradualism discourages attempts to study selection in open populations, because it is likely to be imperceptible or rare. There were indeed very few systematic field studies for 40 years after Weldon's death. The modern history of field studies can be divided into two phases. The first began in 1950 with the publication of the first *Cepaea* paper (Cain and Sheppard 1950), which inspired a generation of researchers whose results were summarized by Endler (1986). The main goal of these studies was to measure the selection coefficient operating on polymorphic characters such as colour and pattern that take discrete values. At about the time that Endler's book appeared, statistical procedures for describing selection on quantitative characters were refined (Lande 1979, Lande and Arnold 1983, Arnold and Wade 1984), and this launched a cohort of graduate theses whose main goal was to measure selection differentials operating on continuous characters. The results have now been summarized by Hoekstra *et al.* (2001) and Kingsolver *et al.* (2001). In combination, these two research programmes, extending over 50 years, provide the basis for evaluating the strength of directional selection acting on phenotypes in natural populations.

When di Cesnola (1907) was collecting snails along the Isis towpath he noticed that they varied considerably in colour, and that some were marked with a prominent dark band. Moreover, they had enemies: 'Fragments of broken shell were quite frequent and on one occasion I observed a blackbird... trying to break up the shell on the gravel of the towpath in order to devour the inmate.' This piece of natural history lay dormant for half a century before forming the basis for a classic investigation of colour polymorphism in another snail, *Cepaea nemoralis*, which is often found in the same places, is similarly marked, and often suffers the same fate. *Cepaea nemoralis* is about 1 cm in height, lives 2–3 years, disperses 10–50 m from its natal site, and typically forms populations with $N \approx 100$–$10\,000$ and $Nm \approx 1$. It has a very variable shell; the background colour ranges from pale yellow through pink to dark brown, and it may bear a number of dark bands running around the whorls. Background colour, the presence or absence of bands, and certain other characters are determined by a series of tightly linked loci, while an unlinked locus controls the number of bands. There are an enormous number of possible variants, which might be (and were) dismissed as being of no functional importance. They might equally be regarded as characters that conceal the snail more or less effectively against visual predators in different kinds of habitat.

Thrushes eat snails. They cannot swallow whole any but the smallest; larger ones they carry to a nearby stone, hammering them until the shell is broken and the snail can be extracted. The same stone is used over and over again, so that the preferred anvil is surrounded by the broken remains of the thrushes' prey. Biologists are more effective predators than thrushes, and can probably find almost all the adult snails in a small area, all those that have been missed by the thrushes. This gave Cain and Sheppard (1950, 1952) the opportunity to compare the snails that had been detected and eaten by the thrushes in a small area of woodland with the remnant population that had escaped predation. In early April, when the floor of the wood was covered by sodden dark leaves, snails with yellow shells were disproportionately frequent on the anvils; as the season advanced and the understorey greened, the preference of the thrushes changed, and yellow shells became less frequent on the anvils than in the population at large. The obvious conclusion, that the thrushes, visual predators, were choosing the more conspicuous snails, was checked experimentally by releasing marked snails in the wood. Those with yellow shells were more at risk earlier in the season, and survived better later in the season, confirming that selection favoured the more cryptic individuals. Visual predation, moreover, explained the geographical distribution of colour and banding morphs in sites around Oxford. There were often large differences between nearby sites that were associated with background: mostly banded shells in the coarse grass of hedgerows, yellow unbanded shells in short turf, and so forth. The *Cepaea* studies thus made use of a striking, easily scored character controlled by a few genes to demonstrate the operation of natural selection, to measure its intensity, to identify the agent responsible, and to link these observations to the seasonal and spatial pattern of variation.

The discrimination exercised by the birds can be considerable. The frequency of unbanded snails living in a small fen was about 0.53, whereas their frequency among the victims of thrushes was only 0.44, a difference of nearly 20% (Cain and Sheppard 1954). The value of the selection coefficient, however, depends not only on the selectivity of the birds but also on the total mortality they inflict on the snail population. Suppose that the favoured type survives thrush predation with probability s_A and has frequency p in the population and p' among the snails eaten by thrushes. The selection coefficient is then

$$s = [(1 - s_A)/s_A]\,[(p/p')(q'/q) - 1].$$

Obviously, if a snail is very unlikely to be eaten by a thrush then the discrimination shown by the birds is inconsequential. A very careful study of a downland population showed that dark brown shells were favoured over paler yellow and pink shells in tussock grass: the frequency of browns was 0.53 in the population and 0.46 among those killed by thrushes. The overall mortality caused by the thrushes, estimated by releasing marked snails, was only 0.067. However, thrushes only eat snails when their preferred prey (earthworms) are difficult to find. Consequently selection was rather weak, $s = 0.023$, despite the strong discrimination shown by the birds. In adjacent nettle beds, where the birds dislike foraging, selection was even weaker: $s = 0.011$.

As more researchers were attracted to the *Cepaea* programme it became clear that visual predation is by no means the only selective agent acting on shell colour and pattern. The polymorphism is expressed where there are no thrushes; at the downland site, for example, the birds were killed by a hard winter and never returned in numbers. In upland areas an extensive area might be dominated by a single morph, which then abruptly gave way to another with no sign of ecological discontinuity. There are geographical and altitudinal clines that reflect the selective effect of climate. Where selective predation did occur, it might depend on the composition of the population as well as on the vegetation in which the snails lived. The multiplicity of factors moulding the polymorphism was reviewed by Jones *et al.* (1977), who concluded that the multifarious consequences of variation in natural settings require each population to be interpreted separately, as the historically unique outcome of a complex process of selection involving many agents.

The *Cepaea* project is probably the largest exercise in scientific natural history yet attempted, led by some of the best evolutionary biologists of their generation and encompassing the collection of millions of individuals from thousands of sites. It initially succeeded in demonstrating that simple and powerful processes of selection can mould adaptation in natural populations. As time went on, it also succeeded in demonstrating that several or many different processes may act on a single well-defined character, depending on circumstances and scale. This is the distinctive feature of selection in open populations. Laboratory microcosms are ecologically simplified and the selective agents known, and in artificial selection the agents and their effects are defined by the experimenter; in both cases the object of an experiment is usually to understand the genetic basis of the response to selection. In open populations many agents, most of them unknown or poorly understood, may be acting simultaneously in combinations that vary idiosyncratically from one side of a stream to the other. Concepts such as 'deleterious mutation', 'selection coefficient', and even 'genetic variance', which serve so well in the laboratory, must be treated with reserve when taken out of doors.

7.2.5 Selection coefficients

After 1950 the number of publications describing direct demonstrations of natural selection in the field grew exponentially at a rate of more than 10% per year according to Endler's review (Figure 7.7). This early growth was in large part due to the Oxford school of ecological genetics led by E.B. Ford, which the *Cepaea* project was initially affiliated with (Ford 1964). Other projects, such as polymorphism in *Panaxia dominula* and wing-spotting in *Maniola jurtina*, were less successful, perhaps because the functional basis of the variation was never clearly identified, and with one exception (industrial melanism) they have faded from the textbooks. After the mid-1960s the enthusiasm for working in the field became much more widely diffused and several hundred studies were published in the following 20 years. These have been collated and epitomized by Endler (1986). Studies of polymorphism, where a selection coefficient is estimated for each discrete type, yielded mean values of $s = 0.33$ for undisturbed situations and $s = 0.30$ for situations that involved recent human disturbance. In either case, s was roughly exponentially distributed so that for about 1000 cases the frequency of estimates of s followed $f(s) \approx \exp[-(2.5s + 2)]$ (Figure 7.8). The simplest way to summarize this result is that selection is seldom weak; about three-quarters of cases involve $s > 0.1$. For several reasons, this conclusion cannot be taken quite at face value. In the first place, many studies measured only one component of fitness, such as mortality, and if equally strong but countervailing selection acted on other components then net selection might be weak. Secondly, an episode of strong selection might contribute only a small fraction of overall variation in survival or fecundity and would then have little genetic effect. Finally, there may be a bias towards selecting, analysing, and publishing cases that are interesting because they are likely to involve strong selection. All these reservations being admitted, however, there is no doubt that field studies of genetic polymorphism have convincingly demonstrated that natural selection is easily detected in natural populations and is often found to be strong.

Changes in the frequency of species are not different in principle from changes in the frequency of morphs or alleles and can be used in the same way to calculate selection coefficients, although this may strike ecologists as an unconventional procedure. I have calculated values for British birds, an exceptionally well-surveyed community for which unusually extensive and accessible information is available through the Breeding Birds Survey (*www.bto.org/bbs*). The data are records of the abundance of all 200+ species of breeding bird in 1500–3000 1-km² quadrats throughout Britain over the 12 years 1994–2005. Some caution needs to be used in interpreting these figures, as noted on the website, but they nevertheless present an exceptionally detailed

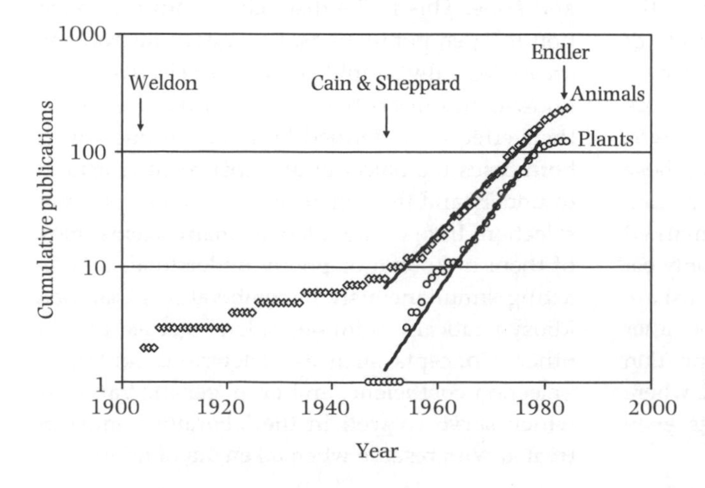

Figure 7.7 Cumulative number of direct demonstrations of natural selection. Data from Endler (1986) Table 5.1. Exponential rates of increase 1950–1979 are 0.11 y^{-1} for animals and 0.16 y^{-1} for plants.

and accurate record of the composition of a diverse community in a large region. I have included only the 106 species that were recorded in every year, so that fluctuations in frequency are likely to be caused by differential reproductive success rather than by immigration. The average over all species and years is $|s| = 0.18$ with standard deviation 0.25. Estimates are exponentially distributed as $f(s) = \exp[-(7s + 1)]$, so weak selection of $|s| < 0.1$ is more frequent than for morphs and consequently the frequency of successively more intense episodes of selection falls off more rapidly (Figure 7.9). The two distributions cannot be compared directly, however, since the estimates in Endler's collation were obtained in a variety of different ways. Moreover, the estimates shown refer to annual changes in frequency, whereas small passerines generally live 2–3 years and other common birds, such as ducks and crows, somewhat longer. Reducing them to units of generations would increase the average and thus reduce the slope of the exponential distribution, making it more similar to that for morphs. Regardless of the exact correspondence of the two distributions, however, their qualitative similarity is quite striking.

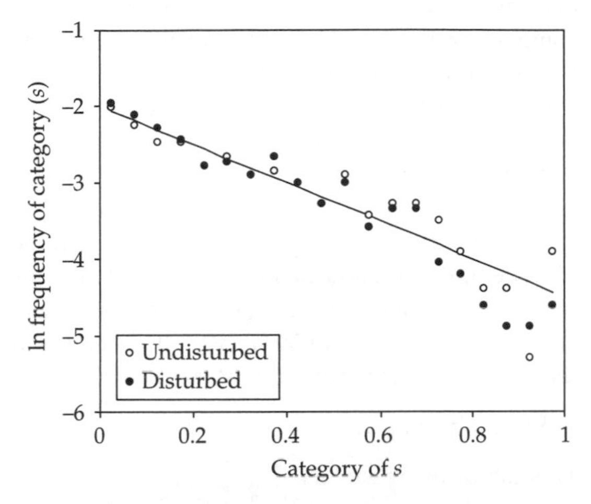

Figure 7.8 Distribution of selection coefficients acting on discrete (polymorphic) variation in open populations. Open circles, undisturbed situations (566 cases); filled circles, disturbed situations (394 cases); line is $y = -2.5x - 2$. Data taken from Endler (1986) Figure 7.1.

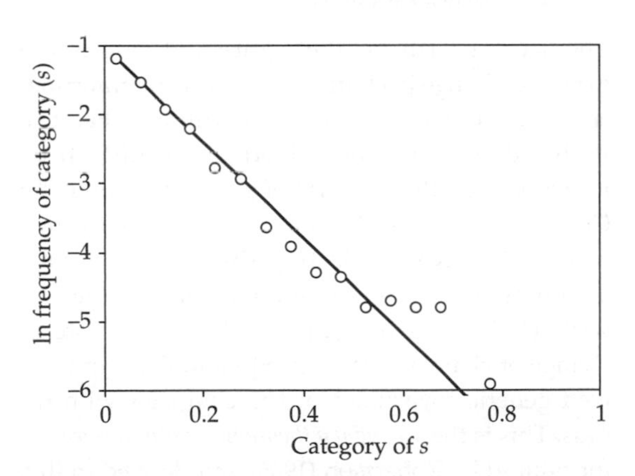

Figure 7.9 Distribution of selection coefficients calculated from annual changes in frequency of species of British birds. Data are mean abundance in 1 km^2 plots throughout UK in 1994–2005. Only the 106 species that were recorded in every year are plotted. The records were taken from the Breeding Bird Survey of the British Trust for Ornithology (*http://blx1.bto.org/bbs-results/results/reg_lists/bbsdenslist-20.html*). The website describes several potential shortcomings of the dataset, but it remains among the best available. Line is $y = -7x - 1$.

7.2.6 Heritability

Polymorphism has a special place in evolutionary biology because it supplies such clear illustrations of natural selection in open populations. Selection acting on continuous variation is less apparent but perhaps even more pervasive. As with fitness itself, the two main approaches to quantitative variation in natural populations are to estimate variation and to measure selection directly. Artificial selection changes the mean of a quantitative character by an amount equal to the product of the heritability and the selection differential in every generation (§6.2), so the heritability has been widely used to express the potential for selection in natural populations. To relate this usage to the standardized genetic variance appropriate to express the dynamics of fitness it can be recalled that $SV_A = h^2 SV_P$, so that provided the phenotypic variance is recorded the two are easily interconvertible. Catalogues of the heritabilities of traits in laboratory and open populations have been collated by Mousseau and Roff (1987) and Weigensberg and Roff (1996). The leading feature of these data is simply that heritabilities estimated in the field are on average rather large, with an average of about 0.5 and standard deviation about 0.25 (Figure 7.10). (The heritability of the columellar radius in the *Arianta arbustorum* that di Cesnola 1907 studied, for example, is 0.54; Cook 1965.) It is not really clear why this should be the case. In principle it need not imply that selection would be very effective, since low values of both genetic and phenotypic variance could lead to high heritabilities. In practice, however, phenotypic variance is usually substantial and the observed heritabilities imply that most characters have substantial genetic variance and will readily respond to directional selection. Another possibility is that selection on a character is usually weak relative to mutation, although if this be so it is strange that selection should be so easy to observe. Alternatively, selection might be strong but highly specific, whereas dispersal rates are high, so that most populations have high frequencies of poorly adapted immigrants, although what we know about fitness in nature populations does not make this very plausible as a general rule. If the conditions of growth are poorly correlated from one generation to the next then selection might be strong but fluctuating, so its effects do not cumulate and genetic variance is less rapidly depleted. Finally, selection might act strongly against extreme individuals of all types, preserving the mean and permitting only very gradual directional change. Stabilizing selection of this sort will tend to deplete genetic variance, but this could plausibly be restored by mutation.

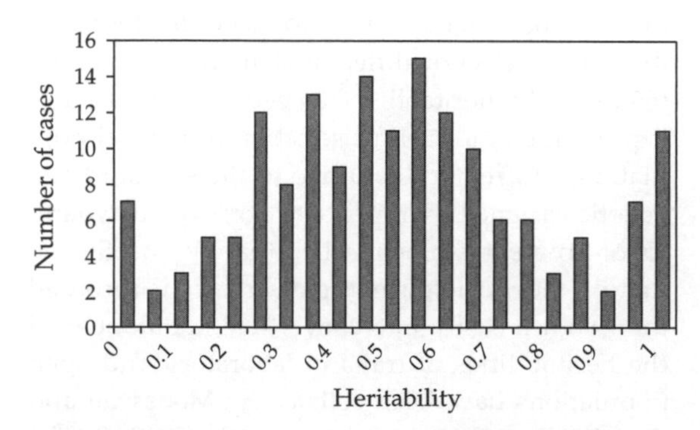

Figure 7.10 Distribution of estimates of heritability of quantitative characters obtained from open populations in the field. Data from Weigensberg and Roff (1996).

7.2.7 Secondary theorem of natural selection

Selection cannot act directly on quantitative characters; it acts only indirectly, through the effects that they have on fitness. The response to natural selection, therefore, is governed by the genetic covariance of a character with fitness. Let us compare two populations, P_1 and P_2: P_1 contains the ancestors of all individuals in P_2, and P_2 comprises all the descendants of individuals in P_1. Each population contains the same range of types that differ with respect to some character z. In P_1 the ith type has frequency p_i and character value z_i. The fraction of individuals in P_2 that descends from the ith type in P_1 is p_i', and this will be equal to $p_i w_i/\bar{w}$, where $\bar{w}$ is the mean fitness of P_1. The mean character value of these individuals is z_i'. The change in the mean value of z from P_1 to P_2 is thus

$$\Delta\bar{z} = \Sigma p_i' z_i' - \Sigma p_i z_i.$$

This can be expressed as

$$\Delta\bar{z} = \Sigma p_i[(w_i - \bar{w})/\bar{w}]z_i + \Sigma p_i(w_i/\bar{w})\Delta z_i.$$

From the definition of a covariance, this is equivalent to

$$\bar{w}\Delta z = \mathrm{Cov}(w,z) + \mathrm{E}(w\Delta z).$$

The second term on the right-hand side is the expected change in character state during transmission, weighted by fitness. This will arise because the breeding value of an individual may differ from its genotypic value because of gene interaction (see Crow and Nagylaki 1976) or other processes such as meiotic drive. Neglecting this term, and thus assuming strictly additive allelic effects, we can write $(1/\bar{z})\Delta\bar{z} = [\mathrm{Cov}(w,z)]/\bar{w}\bar{z}$. That is, the rate of change of character value is equal to the standardized genetic covariance of the character with fitness. This is the *secondary theorem of natural selection* introduced by Robertson (1966) and derived in this form by Price (1972). If the character concerned is fitness itself, Fisher's fundamental theorem appears as a special case of the secondary theorem. Price's argument and its implications have been explained at length by Frank (1998). Price himself emphasized that the expression for Δz is perfectly general and holds for any situation: change through time in business practices, religious beliefs, or military technology will all follow the same general rule. He was, indeed, working on a general theory of selection shortly before his untimely death in 1975. This project has yet to be achieved.

7.2.8 Selection gradients

In studies of wild populations genetic covariances are likely to be unknown, but the phenotypic regression of fitness on character value is often measured. Suppose that we plot the fitness of individuals expressing a given character as a function of character value. If the plot is flat, with zero slope, individuals have the same fitness, regardless of the value of the character. In this case, selection will have no effect on the mean value of the character. On the other hand, if individuals expressing different values of the character have different fitnesses, selection will tend to cause a change in the mean value of the character. In some of the chemostat experiments that I have described, for example, it was possible to define the relationship between the value of a phenotypic character—such as the rate of lactose utilization—and fitness, and therefore to predict how selection would drive phenotypic change. More generally, we can define β_{wz} (or more simply β) as the slope of the graph of fitness w on character value z, or the 'selection gradient'. This approach was introduced by Lande (1979; Lande and Arnold 1983). The complications involved in defining the relationship between character value and fitness have been discussed at length by de Jong (1994). As a regression coefficient, β is the ratio of the phenotypic covariance Cov(w,z) to the phenotypic variance V_P; that is, β is a selection differential D standardized by V_P. Consequently, the response to selection will be $\Delta z = h^2 D = V_A\beta$. If the character value is expressed in units of σ_p then β is a standardized selection gradient and when mean fitness is set to unity (as it usually is) the relative fitness of an individual with character value one standard deviation in excess of the mean is $+\beta$. The steeper the gradient the greater the effect that variation in character value has on fitness, and the faster the mean of the character will change under selection: the fractional change in the mean value of a character during a single episode of selection, from generation to generation, is $(1/\bar{z})\Delta\bar{z} = \beta h^2 V_p / \overline{w}\bar{z}$. This is a purely phenotypic description: the change caused by selection is established permanently in the population only to the extent that the character is inherited. Houle *et al.* (2002) have suggested that it is more useful to standardize Cov(w,z) by the mean rather than the phenotypic variance: $\beta' = \bar{z}\beta$. This leads to the simple relation $(1/\bar{z})\Delta\bar{z} = \beta' S V_A$, so that $\beta' = 1$ when the character is fitness itself. This provides a useful benchmark against which estimates of β' for different kinds of character can be evaluated.

Any natural situation is more complicated. Individuals express many characters that are correlated with one another to a greater or lesser extent. Selection acting directly on a primary character may then induce changes in other characters with which the primary character is correlated. To separate the direct effect of selection from its indirect effects we must know the phenotypic correlations between characters and then calculate the selection gradient β as the partial regression of fitness on the value of a given character when the value of other characters is held constant. There may also be several episodes of selection during a single generation so that the strength and the direction of selection (measured by β) vary within the lifetime of an individual. The estimation of β as a partial regression coefficient and its partition among episodes of selection was described by Arnold and Wade (1984). This approach was used to describe selection on a range of characters in *Impatiens pallida* at Mont St-Hilaire by Stewart and Schoen (1987), who studied the fate of groups of seedlings emerging in a small area (about 0.1 ha) of forest floor. They first scored 7 morphological and phenological characters and calculated their partial regression on viability, the first episode of selection. Since there were 24 groups of seedlings there were $7 \times 24 = 168$ opportunities for selection to occur, in 36 of which selection was detected in the form of a significant selection gradient. The surviving plants then flowered, and fecundity as a second component of fitness was regressed on 9 characters in the 22 plots that had enough survivors, yielding significant selection gradients in 47 cases. Thus, a random character in a random plot experienced selection through differential viability in 21% of cases, and selection through differential fecundity in an additional 24% of cases. In general, selection favoured large plants with many leaves, and was often rather strong, with standardized $|\beta| > 0.2$. It was also heterogeneous, however, with selection differing in strength and even in direction between sites separated by a few metres. This appeared to be caused by consistent differences in the conditions at nearby microsites: for example, seedlings that produced leaves later survived better in sunny sites, perhaps because their more precocious neighbours succumbed to water stress. This very detailed study showed that selection was frequent and often strong, but also varied substantially between sites within dispersal range.

The many estimates of selection gradients published in the ensuing 15 years have been collated and epitomized by Hoekstra *et al.* (2001) and Kingsolver *et al.* (2001), who have made these data available at *www.bio.unc.edu/faculty/kingsolver*. For about 1000 cases involving many kinds of characters and organisms about a quarter of all estimates of β were formally significant, with at mean value over all cases of 0.22, median 0.15, in line with Stewart and Schoen's early study. Estimates from small-scale studies may be inflated by sampling error, and the largest studies (1000 individuals or more) suggest somewhat more modest values of about 0.1. Kingsolver *et al.* (2001) regard this as rather weak selection, but it is much stronger than was anticipated when field studies began in earnest in the 1950s; perhaps the success of the ecological genetic school has raised expectations. The selection gradient is roughly log-normally distributed, with a mode of ln s between -1 and -2 (i.e. s between 0.15 and 0.35) and a pronounced skew toward low values (Figure 7.11). On an arithmetic scale the selection gradient is exponentially distributed (Figure 7.12), like the selection coefficient. If a trait is governed by additive alleles at a single locus then the fitness of the favoured homozygote would be $1 + s$ or $1 + 2\beta$, so if $f(s) = c\exp(-ks)$ then $f(\beta) = c\exp(-2k\beta)$. This is roughly true: the distribution of β is reasonably well fitted by $f(\beta) = \exp[-(5\beta + 1.5)]$ (Figure 7.12). The distribution of selection intensity i is necessarily similar to that for β.

Hereford *et al.* (2004) calculated β' for about 300 cases and found the same kind of distribution with an even larger mean of about 0.5: that is, the average strength of selection acting on a character is about half as great as selection acting on fitness itself. This was reduced on appeal to about 0.3 by removing statistical biases, but even so we have clearly moved from thinking that selection is too weak to be readily detectable to finding that it is too strong to be credible. There are several reasons for treating the Endler, Kingsolver, and Hereford collations with circumspection, as the authors themselves recognize (Hendry 2005). Perhaps the most troubling is

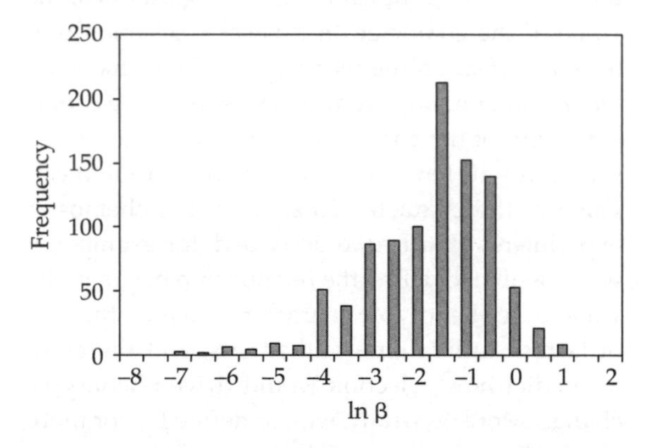

Figure 7.11 Distribution of estimates of the selection gradient β. Data from Kingsolver *et al.* (2001) and *www.bio.unc.edu/faculty/kingsolver.*

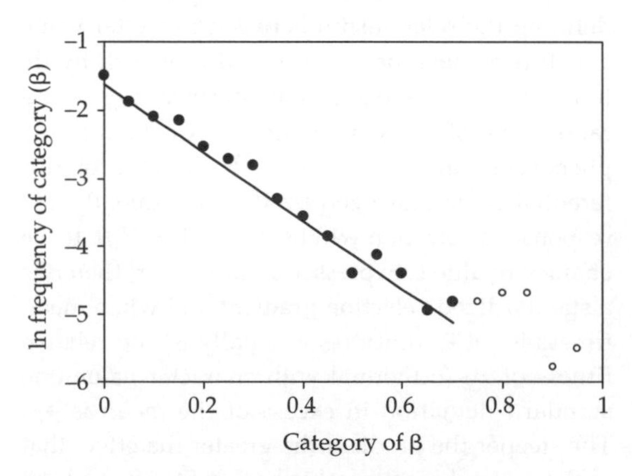

Figure 7.12 Distribution of selection gradients acting on continuous variation in open populations. The line is $y = -5x - 1.5$. Data from *www.bio.unc.edu/faculty/kingsolver.*

the environmental correlation between trait values and fitness (Rausher 1992). Seed production is often highly correlated with plant size, which in turn depends on nutrient supply and insolation. In a heterogeneous environment plants will be large and fecund in good sites but small and barren in poor sites, giving rise to a steep selection gradient even in the absence of genetic variation. Furthermore, most studies estimate the covariance of character state with the state of some other character thought to be correlated with fitness rather than with fitness itself. If the correlation is in fact rather weak then the real selection gradient (relative to fitness) will be much shallower than the apparent gradient. Allowing for this effect reduces the average of estimates of β' based on survival to 0.4 and those based on fecundity to 0.06 (Hereford *et al.* 2004). Finally, the investigators themselves might be biased, being more likely to study characters expected to experience strong selection or more likely to publish observations of strong selection. Most studies in any case have insufficient power to detect even moderately strong selection (Hersch and Phillips 2004). For all these reasons, the averages of published selection coefficients and selection gradients may overestimate the strength of selection typically acting on natural variation. Nevertheless, I find it impossible to reconcile the dozens of studies reporting strong selection with the view that selection in natural populations is weak or infrequent. The very different data sets for selection coefficients acting on discrete characters and for selection gradients or selection intensities acting on continuous characters seem to agree quite well, and it seems to me inescapable that selection in open populations is commonplace and often rather strong. As a rough rule of thumb, the force of selection on an arbitrary character in an open population is exponentially distributed with an average value corresponding to a difference in fitness of at least 0.2 between individuals of recognizably different phenotype. Whether or not this will cause any permanent change in the population is another matter, because in most cases we do not know whether these individual differences are heritable. We can resolve this point by reversing the argument: instead of estimating selection in order to predict the magnitude of changes in frequencies, we can measure changes in frequencies in order to infer the strength of selection responsible for causing them.

7.2.9 Stabilizing selection

Extreme individuals might tend to be removed selectively for two reasons. The first is that extreme phenotypes are functionally inferior and reduce fitness relative to the average, as expected in a population that has become well adapted through long continuance in a constant environment. Stabilizing selection will then favour individuals with nearly average phenotypes. Secondly, genotypes which are systematically inferior, for example because they are heavily loaded with deleterious mutations, may express extreme phenotypes as the result of their general debility. Purifying selection will then remove these inferior types. These two processes can be distinguished by crossing individuals from opposite extremes: under stabilizing selection these will have high fitness, whereas under purifying selection they will have low fitness. Selection against extreme individuals is normally referred to simply as *stabilizing selection*, although whether the mean phenotype is generally preserved through stabilizing or purifying selection, or both, does not seem to have been clearly established yet.

Weldon's study of *Clausilia* remains one of the best examples of stabilizing selection. A decline in the variability of size or shape during the lifespan of a cohort caused by a single episode of stabilizing selection has been reported from sparrows (Bumpus 1899; see O'Donald 1973, Lande and Arnold 1983), ground finches (B.R. Grant 1985, P.R. Grant *et al.* 1976), ducks (Rendel 1943), snakes (Inger 1942), lizards (Fox 1975), sticklebacks (Gross 1978, Hagen and Gilbertson 1973), whelks (Berry and Crothers 1970), bivalves (Palenzona *et al.* 1971), beetles (Mason 1964, Scheiring 1977), and corals (Potts 1984). As a graduate student, I studied the morphology of a cohort of newt larvae in a pond near Oxford. By measuring a variety of skeletal characters from the time when the larvae hatched in early spring to the time when most of them left the water in midsummer, I was able to show that there was a general and continuous decline in morphological variation during this period (Figure 7.13). This did not happen when the larvae were carefully reared in the laboratory, supplied with adequate food and protected from predators, so it was probably caused in the field by stabilizing selection. Selection gradient studies evaluate the strength of stabilizing selection through the partial regression coefficient of the square of character value on fitness. This can be either negative, indicating stabilizing selection, or positive, indicating disruptive selection that favours extreme individuals. Kingsolver *et al.* (2001) have collated the available estimates of this coefficient and report a double-exponential distribution with a mean near zero and a standard deviation of about 0.1. This leads to the surprising conclusion that disruptive selection is just as common as stabilizing selection—that is, in about half of all cases nearly average phenotypes have the lowest fitness. This seems difficult to believe, and perhaps the difficulties of estimating second-order coefficients will justify declining to interpret it further.

7.3.10 Fluctuating selection

The two leading conclusions from studies of open populations are that selection is often strong and that heritability is often high. This combination is unexpected because it is very difficult to reconcile with simple population genetics theory. The most likely explanation is that selection often changes direction, because offspring grow up in a different place and at a different time from their parents. They will experience different conditions of growth because their environment varies in space and time. A dispersing propagule will experience both sources of change simultaneously, through the particular characteristics of the site where it settles and through the average state of sites at that time. Spatial and temporal variation will have quite different effects on adaptation, however, essentially because populations may readily become adapted to a place but not to a time. The evolution of specialization and diversity in relation to spatial variation is discussed in the next chapter. In this section I am concerned mainly with how the strength and direction of selection fluctuate through time, although as I have pointed out this will often be combined with spatial variation.

If a lineage increases by a factor of $\lambda_1, \lambda_2, \ldots, \lambda_t$ in t successive generations, then the number of individuals descending from a single founder is $\Pi\lambda_i$ and the mean rate of increase of the lineage is $(\Pi\lambda_i)^{1/t}$. This is a *geometric mean* (GM), which is related to the

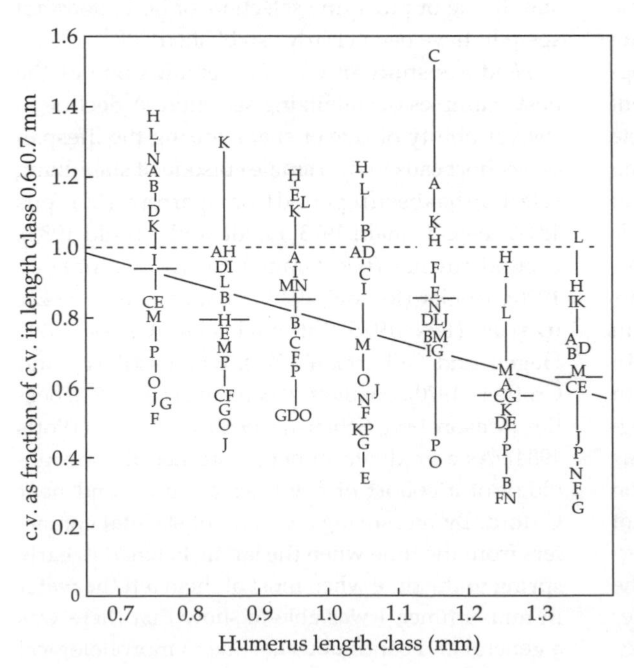

Figure 7.13 The fate of morphological variation in a cohort of newt larvae. Humerus length is a surrogate for age. Letters denote different characters, chiefly measurements of skull and teeth. From Bell (1978).

arithmetic mean (AM) as: GM ≈ AM − ½σ^2. Thus, the long-term proliferation of a lineage is reduced by environmental variance; indeed, if the variance is so large that λ is occasionally close to zero the variance has a greater effect than the arithmetic average on the fate of the lineage. The same principle applies to relative fitness: selection in a varying environment will favour the type with greatest geometric mean fitness (Gillespie 1977). In general, therefore, fluctuating selection in simple haploid models with discrete generations will not permanently protect allelic diversity, nor will it protect species diversity in ecological communities. Some exceptions to this rule are discussed in §8.3. Fluctuating selection may retard the loss of diversity under directional selection, however, and some simple simulations to illustrate this are shown in Figure 7.14. The fitness of the *i*th type is defined as $1 + v_i + \epsilon$, where $s = v_i$ is the basic component and ϵ is a random variable with mean zero and standard deviation σ_s, the environmental standard deviation of fitness over generations. With 20 alleles at intervals of $v = 0.01$ the range of fitnesses from worst to best is $V = 0.2$. Under pure sorting diversity is always eventually lost, but we can census the population after a few hundred generations when diversity has almost been eliminated in a constant environment in order to evaluate the effect of fluctuating selection. Weak fluctuations of $\sigma_s = v$ have little effect, but as σ_s increases diversity is conserved for longer. This is mostly because competition between the best few alleles takes longer to resolve. Strong fluctuations of $\sigma_s > V$ cause a rapid elimination of diversity, however, because alleles are liable to become extinct after an episode of very low fitness; one allele is soon fixed, although this is not necessarily the best allele. When a high rate of mutation is allowed this stochastic extinction is prevented and diversity continues to increase up to unrealistically high values of σ_s. If field situations are intermediate between these extremes, it is reasonable to expect that environmental fluctuations of σ_s = 0.05–0.2 will preserve an appreciable amount of genetic diversity. Sexual diploid models are somewhat more complicated because diversity is protected if the heterozygote has the greatest geometric mean fitness (Haldane and Jayakar 1963). If the heterozygote is intermediate in that it has basic fitness $1+s/2$ this is equivalent to requiring that the difference between the geometric mean fitnesses of the homozygotes is less than ¼ σ_s^2 (Gillespie and Langley 1974).

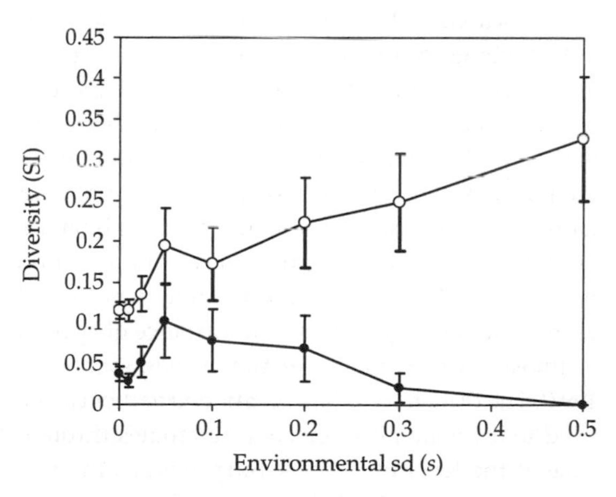

Figure 7.14 Effect of fluctuating selection on allelic diversity. Based on a model with 20 alleles with basic fitnesses 1, 1.01, . . ., 1.2 and a population of 10 000 individuals. In each generation a random normal deviate with mean zero and variance σ^2 is added to the basic fitness of each allele: σ is the environmental standard deviation of fitness in the plot. Error bars are + 2 se of 50 simulations per point. Solid symbols: pure sorting with no mutation. Open symbols: mutation rate to each allele 10^{-4} per individual per generation. SI is Simpson's index.

Temporal variation is more difficult to study than spatial variation, so we know less about it. The most direct evidence would be provided by long-term studies of the relative fitness of types in a population, but these are very rare. Cain *et al.* (1990) continued their census of the downland population of *Cepaea* for 20 years after the thrushes had disappeared and thus in the absence of substantial visual predation. In each year they conducted a mark–recapture census to estimate the survival rates of the morphs, providing direct evidence of variation in selection coefficients over time. For banded vs unbanded brown snails the mean selection coefficient was 0.053 with no general trend but a standard deviation of $\sigma_s = 0.53$. Comparisons of pink with yellow, or brown with non-brown, gave similar results. This is a very large annual variation, but it includes drift (whose contribution was probably minor) and experimental error (likely to be large) and so represents only an upper bound on the true variance of s, probably an extreme upper bound. A more amenable semi-natural system was provided by barley Composite Cross V, which was founded by intercrossing 30 barley cultivars from different regions of the world and subsequently propagated for many years without conscious selection under standard agricultural conditions in California. Clegg *et al.* (1978) grew stored seed from populations 10 generations apart in uniform conditions and estimated the fitnesses of single-locus isozyme genotypes. The standard deviation of s between populations for homozygotes was $\sigma_s = 0.087$, again an upper bound because the populations being compared are 10 generations apart.

Time series of genotype frequencies provide a much larger source of data. If the frequency of the favoured type is p then $s = \Delta/q(1 + \Delta)$, where $\Delta = (1/p)\Delta p$. Applying this to the *Cepaea* series gives a mean of -0.015 with $\sigma_s = 0.66$, probably an unrealistically large value. Lynch (1987) developed sampling theory for this kind of estimate and conducted a survey of isozyme genotypes in *Daphnia* populations that gave an average of $\sigma_s = 0.195$ for homozygotes. He also analysed similar data from the literature which gave an average of $\sigma_s = 0.11$. These are the best estimates currently available. O'Hara (2005) estimated selection coefficients for a spotting polymorphism in the Cothill population of the moth *Panaxia dominula*, first studied by Fisher and Ford, and found on average selection against the *medionigra* type of $s = -0.103$ with $\sigma_s = 0.075$. The correlation between successive years is low (0.15) so the strength of selection in any year is nearly independent of selection in the previous year. Selection favoured *medionigra* in 2 of the 50 years for which estimates are available, so despite considerable fluctuation in the strength of selection its direction was seldom reversed. O'Hara calculated that only about 30% of the variation in frequency was caused to selection, however, the rest being attributable to drift. These data, with some other values, are plotted in Figure 7.15. All are to some extent upper bounds, especially because they are inflated to different degrees by sampling error. Nevertheless, it seems likely that the standard deviation of s is generally about 0.1. This implies

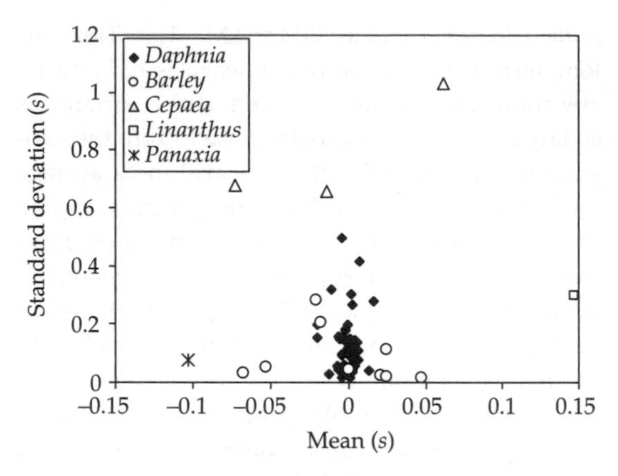

Figure 7.15 Fluctuating selection. The standard deviation of selection coefficients over time for several organisms; references in text.

that fitness fluctuates markedly through time, so that a genotype that is favoured in one generation will often be deleterious shortly afterwards.

Long-term surveys of community composition can also be used to calculate s, which in this case refers to the relative fitness of species. The abundance of a species N often fluctuates widely through time, with ln variance increasing over ln time with a slope of about 0.4 (§2.2), the coefficient of variation of ln N being about 1 in long-term surveys regardless of taxon (Inchausti and Halley 2002). This need not necessarily imply a corresponding degree of variability in relative fitness, since if abundances are positively correlated frequencies might remain more or less constant. In practice, the total biomass or abundance of individuals in a community does not usually fluctuate nearly as widely as that of each component species, so the variance of species composition increases over time (Bengtsson *et al.* 1997). The British bird data mentioned earlier can be used to estimate how selection fluctuates through time at the level of species. Surprisingly, the values for species of bird throughout Britain are not very different from those for PGM genotypes of *Daphnia pulex* in Busey Pond: the average standard deviations of the annual coefficients cluster around 0.2 (Figure 7.16). Using longer intervals of time to calculate s does not substantially alter this conclusion: all intervals from 1–8 years yield an average standard deviation over time of 0.2–0.3.

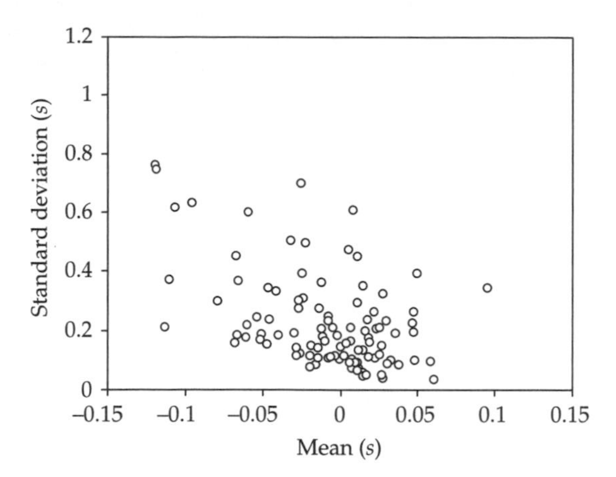

Figure 7.16 Fluctuating selection. Selection coefficients calculated from change in species composition of British birds. Data are mean abundance in 1 km² plots throughout UK in 1994–2005. Only the 106 species that were recorded in every year are plotted. The records were taken from the Breeding Bird Survey of the British Trust for Ornithology (*http://blx1.bto.org/bbs-results/results/reg_lists/bbsdenslist-20.html*).

7.2.11 Historical change

A standardized difference in mean character state can be used to express rates of historical change in the medium term of 10^2–10^5 generations. The simplest measure is the darwin (Haldane 1949): the rate of change of the mean value of a character x in darwins (Dar) is

$$(\ln x_t - \ln x_0)/t \approx (1/t)(1/\bar{x})\Delta\bar{x}.$$

This expresses the proportionate change per unit time and is conventionally expressed per million years. In contemporary samples it may be more appropriate to use the kilodarwin (kDar) as the proportionate change per 1000 years. An alternative is the haldane: the rate of change in haldanes (Hal) is $(1/t)(x_t - x_0)/\sigma_x$ per generation. This expresses the inferred intensity of selection. There is no compelling reason to use a mean-standardized rate in terms of calendar time and a variance-standardized rate in terms of generations, and no objection to varying this usage when appropriate, but I shall retain the traditional definitions here.

The shore crab *Carcinas maenas* that Weldon measured was introduced accidentally to the eastern coast of North America in the nineteenth century and subsequently spread northwards from New England to Nova Scotia. It feeds on marine snails such as *Littorina* by crushing the shell and extracting the animal, a marine analogue of thrushes and *Cepaea*. The way of a crab with a snail depends on the shape of the snail, because the adult shell preserves its juvenile whorls. This can be expressed by the exponent z of the allometric relation (spire height) = (shell width)^z, a high value of z expressing a high-spired shell whose whorls have a large angle of rotation about the columellar axis. In low-spired shells the thin-walled juvenile whorls are protected by the thicker adult whorl, whereas the juvenile whorls remain exposed and vulnerable in high-spired shells. Seeley (1986) compared museum specimens collected before invasion by *Carcinas* with contemporary material from the same localities and found a trend from high-spired shells ($z = 0.34$ at one locality) in the 1870s to low-spired shells ($z = -0.35$) in the 1980s, consistent with elevated rates of predation by crabs. As z is already a logarithmic variate it does not need further transformation and its rate of evolution is $1000 \times [0.34 - (-0.35)]/84 = 8.2$ kDar. The equivalent variance-scaled measure was 0.12 Hal. Museum collections dating from 1915, however, suggest that most of this change may have taken place in the first 40 years after the appearance of *Carcinas*, in which case the time-averaged value is much less than the rate experienced by the population during the period over which most selection occurred. This also shows that the estimate depends on the arbitrary length of time elapsing between the selective event and the subsequent census, a serious impediment to interpreting historical differences in terms of the strength of selection responsible for creating them.

The best-known example of an historical process of selection driven by a known selective agent is the change of beak shape in the large ground finch (Darwin's finch) *Geospiza fortis* on the island of Daphne Major in the Galapagos (Figure 7.17). A prolonged drought in 1976–1977 caused a change in the composition of the vegetation, by favouring plants with large, tough-shelled seeds. These could be consumed only by finches with unusually large and powerful beaks, and between 1976 and 1978

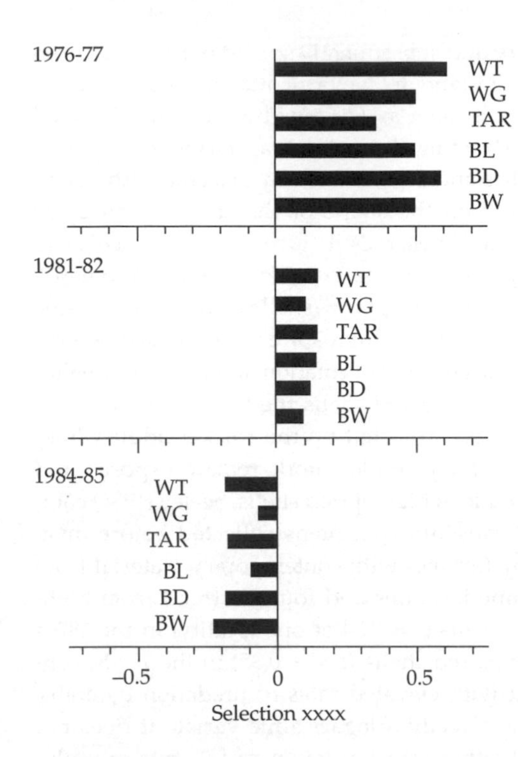

Figure 7.17 Reversal of selection on body and beak morphology in *Geospiza fortis* over time on Daphne Major. 1976–77 and 1981–82 were dry periods; 1984–85 followed an El Niño event causing very heavy and prolonged rainfall. Characters measured are: WT, body weight; WG, wing length; TAR, tarsus length; BL, beak length; BD, beak depth; BW, beak width. From Gibbs and Grant (1987b).

beak depth increased at a rate of 26.1 kDar (0.66 Hal) (Boag and Grant 1981). Heavy rain in 1983 reversed the trend in the vegetation by favouring plants with smaller, softer seeds that germinated more readily and thereby favoured birds with smaller beaks that were more adept at processing them (Gibbs and Grant 1987a, b). Within a few years the response to reversed selection at a rate of 8.8 kDar (0.37 Hal) had more or less restored the *status quo* (Grant and Grant 1995). This study has become a classic example of selection in a nearly pristine environment, the thoroughness of the fieldwork being buttressed by detailed knowledge of the ecology of the populations and the genetics of beak shape (the romantic location may also help). It is particularly noteworthy that selection is episodic and fluctuating, so that an opportune study would reveal strong natural selection, as indeed it did, whereas less fortunately scheduled surveys in (say) 1972 and 1992, however carefully executed, would have shown little if any change. Beak shape is modulated by *Bmp4*, whose product is a bone morphogen, which is strongly expressed early in the development of *Geospiza* species with deep beaks but not in those with long thin beaks (Abzhanov *et al.* 2004, Grant *et al.* 2006). Thus, selection on this quantitative character may act primarily through alleles of a single gene to produce adaptation.

M.A. Bell *et al.* (1985) measured morphological characters in a lake population of sticklebacks (*Gasterosteus*) that lived in the Miocene. The fish were preserved in varved sediments that allowed samples to be taken at roughly 5000-year intervals over a total span of about 110 000 years; the generation time was probably about 1 year. The fluctuations in average morphology are shown in Figure 7.18. Two characters shifted radically and abruptly during the census period, representing a shift from weakly armoured forms with a single dorsal spine to strongly armoured forms with up to three spines. Pelvic armour and dorsal spines are defences against predators and suggest a brief incursion of predatory fish or birds that produced temporary but powerful selection for enhanced body armour. Variation in plates and spines is largely attributable to a few major genes (see §13.3.1). Other characters, such as fin-ray number, fluctuated with lower amplitude and often showed slight trends

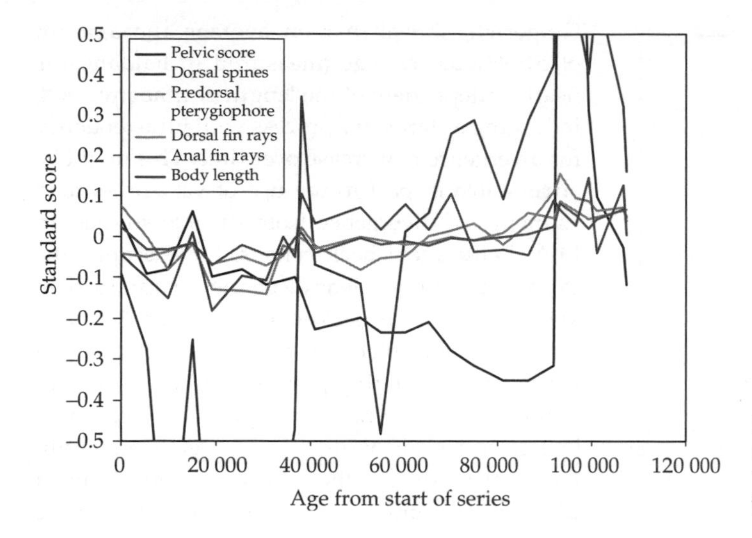

Figure 7.18 Morphological changes in a stickleback (*Gasterosteus*) population over 100 000 years. Standard score is $(z_t - \bar{z})/\bar{z}$. Data from M.A. Bell *et al.* (1985).

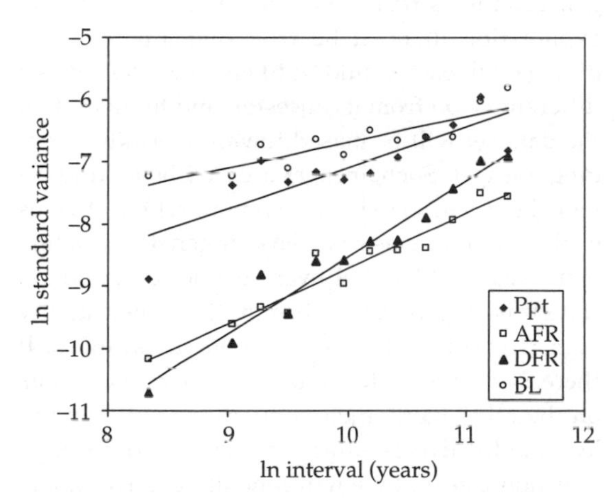

Figure 7.19 Variance plot for stickleback morphology spanning about 10^4–10^5 years. Standard variance is the variance of standard score = ln(score/mean). Slopes (exponents) of plots calculated from raw data (not pooled into distance classes, as here) are: PPt predorsal pterygiophores 0.60; AFR anal fin rays 0.68; DFR dorsal fin rays 1.19; BL body length 0.22. Data from M.A. Bell *et al.* (1985).

over the census period, or at shorter periods. The increase of variation with time followed a power law quite well (Figure 7.19). Body length and pterygiophores have exponents of about 0.5, and thus show $1/f$ noise: samples separated by longer periods of time are more different, but with little indication of a secular trend on the scale of the census. Anal and dorsal fin rays have larger exponents of about 1, in this case because of a consistent tendency to increase over the census period. Similar plots of ln (absolute difference) on ln time have exponents one-half as great. Thus, morphology is shifting on relatively short timescales of less than 10^4 years (body armour), intermediate timescales (body size), and long timescales of more than 10^5 years (fin rays).

In many cases populations are known to have diverged from a common ancestor whose state is not accurately known. Accidental or deliberate introductions of organisms into new areas provide a rich source of data, although much of this concerns arbitrary morphological characters responding to unidentified agents of selection. Well-known examples include the introduction of rabbits into Australia (Williams and Moore 1989), sparrows into North America (Johnston and Selander 1964) and mosquitofish into Hawaii (Stearns 1983), all of which were evaluated after a lapse of about 100 years. According to Hendry and Kinnison (1999), body size in all these cases changed at a rate of roughly 0.1–1 kDar. Other characters might show much greater rates of change, of course. Following the introduction of myxoma virus the remnant population of rabbits evolves resistance, with ln survival increasing at a rate of 0.26 per year, or 260 kDar (Ross 1982).

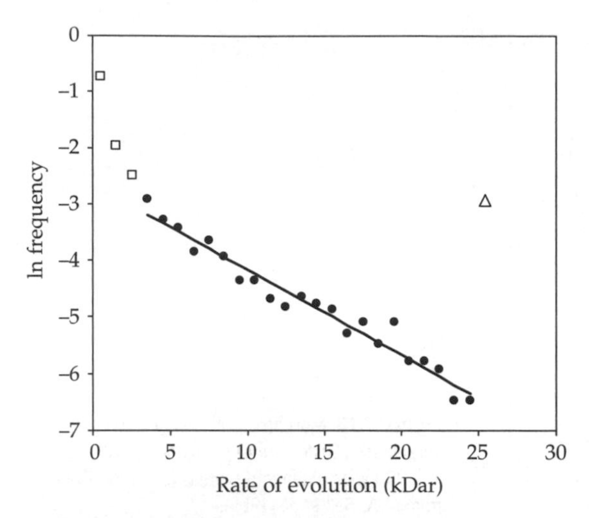

Figure 7.20 Rates of evolution. The regression of the intermediate rates (solid circles) is $y = -0.15x - 2.67$. The triangle represents all cases with rates exceeding 25 kDar. Database from Kinnison and Hendry (2001) was kindly provided by M.T. Kinnison.

Kinnison and Hendry (2001) collated about 2000 estimates of rates of evolution for periods of up to about 100 years. The distribution of rates is roughly exponential, ln frequency = − 0.15 rate (kDar) − 2.7 (Figure 7.20). On a log scale these rates are roughly normally distributed, although with a long left tail of small values, as expected for an exponential random variable. There are indeed too many small values for the exponential to fit the lowest rates very well, and there are also far too many exceptionally high rates. These discrepancies probably arise because of the pattern of research (for example, many of the long-term estimates come from a very few studies) but there are too few studies available to be sure. The average rate over all studies is 5.3 kDar, or 0.02 Hal (these values differ from those given by Kinnison and Hendry). Unfortunately, both rates are functions of time. Since estimates of the average rate of evolution will be sensitive to the length of census the values of 5 kDar or 0.02 Hal should be taken as no more than a rough guide to rates of change over 10–100 generations.

Regressions of ln rate on ln time have slopes very close to −1 (Gingerich 2001) (Figure 7.21). The negative slope is due to autocorrelation (since x/t is regressed on t) but its value, according to Gingerich, shows that on average the amount of directional change (measured in haldanes) is nearly independent of the length of time involved, indicating a stationary process with little tendency for divergence to increase over time. This might be attributable in part to the use of variance-scaled estimates of divergence from diverse studies. In M.A. Bell's stickleback census, for example, the exponent for body length change (expressed in kDar) is −0.88: this is close to −1 but the exponent for the variance plot is nevertheless 0.22, and that for the corresponding plot of absolute difference is thus 0.11, showing that populations separated by longer periods of time are on average more different. Nevertheless, the very slow average rate of divergence, representing an increase of no more than one phenotypic standard deviation over 10^6 generations, is truly remarkable and calls for an explanation. It must be true that any strongly divergent lineage would be likely to be classed as a different taxon from its ancestor, and to this extent the data set will be biased towards weakly diverging lineages. Such lineages are clearly not uncommon, however, so the puzzle remains. One obvious explanation for stasis is a lack of genetic variation (genostasis, §7.3.3), but over very long periods of time this is implausible if mutational heritability is in the region of $10^{-3}\sigma_p$ per generation (§2.2). If there is an adequate supply of genetic variation, on the other hand, then populations will more or less rapidly diverge either through drift or through selection caused by environmental change. Long-term stasis therefore requires stabilizing selection around a constrained optimum (Charlesworth *et al.* 1982). Estes and Arnold (2007) investigated the long-term outcomes of a variety of quantitative models of stabilizing and directional selection and compared them with Gingerich's data. If the optimal character state follows a random walk from one generation to the next then the expected divergence from the ancestral state is a linear function of time, and if this divergence is appreciable over short periods of time, as it is known to be, then it is much too large over long periods of time. If the optimum merely changes randomly around a fixed mean, on the other hand, the divergence is much smaller than observed over long periods of time.

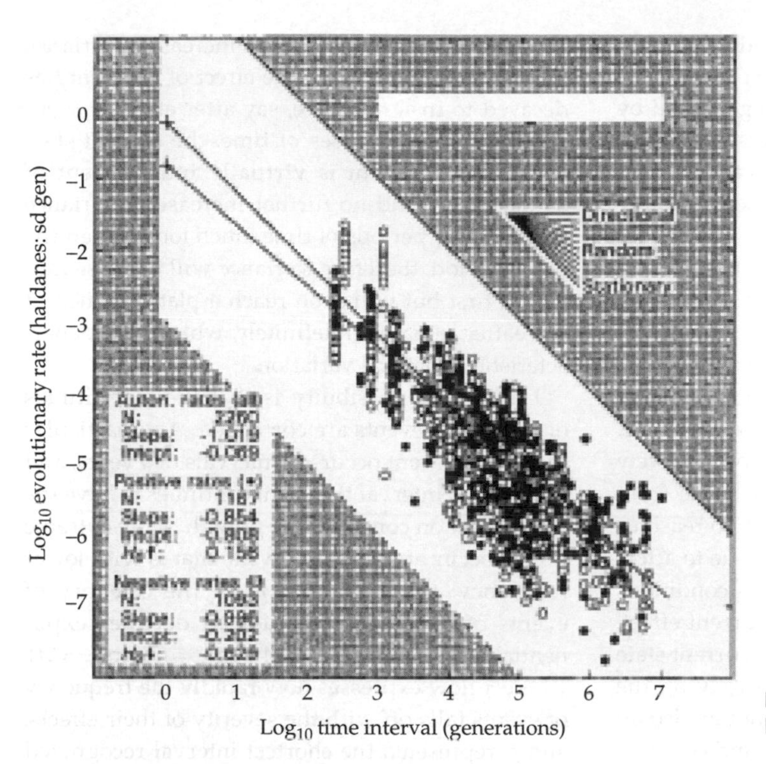

Figure 7.21 Rates of evolution on short and long time scales. From Gingerich (1983).

If the mean itself shifts consistently through time, the long-term divergence is again much too large for any reasonable value of short-term change. The only model that provided a reasonable fit to the data required that the optimum is fixed throughout any time period, however long, except that on a single occasion somewhere within that period it changes abruptly and thereafter maintains its new position. This generates a certain amount of divergence that is independent of the elapsed time. However, none of these models are very realistic. Environments do not follow a random walk, as replicators do; environmental variance is not constant through time; and imposing a single shift within a time period of any duration is clearly artificial. The physical environment, at least, varies from year to year and is autocorrelated so that successive years tend to be alike and consequently the variance of environmental factors increases over time (§2.3 and below). If environmental change causes selection the population mean will change from generation to generation in proportion to the change in the environment. Over longer periods of time more extreme excursions will occur, and if the population is observed at this time a greater divergence will be recorded. This will inflate the low level of divergence associated with white-noise environmental variation, giving a better fit of theory to data. Estes and Arnold (2007) speculate that a series of small abrupt changes might produce nearly the same outcome as a single large displacement, and as changes of greater magnitude are more likely to occur over longer periods of time this would bring the data on divergence into line with what we know about environmental variation through time. The process that is responsible, however, is directional selection fluctuating in direction and magnitude over time, rather than stabilizing selection as usually understood.

7.2.12 Multiscale temporal variation

Field studies of selection support two broad generalizations: the first is that selection coefficients are exponentially distributed, and the second is that the variance (or absolute difference) of population means increases with elapsed time as a power law. What is the link between them? The conditions of growth at a site will be governed by a series of events occurring in successive periods of time, which for simplicity I shall call years. If the magnitude of these events were exponentially distributed this would provide a physical basis for the exponential distribution of selection coefficients. Longer periods of time are then more likely to include more extreme events, but this does not in itself generate a power-law increase of variance: if the magnitudes of successive events are independent then variance will not increase over time. The variance of successive years will be relatively small if conditions remain much the same from year to year, which might happen for two reasons. In the first place, an event might continue to affect conditions in the future. If its effect on conditions in the following year, relative to its current effect, has decayed by a factor $\gamma < 1$ then the current state of the environment will depend not only on the magnitude M_t of the most recent event but also on the discounted effects of past events, and could be represented in some such form as $1 + M_t + \Sigma\gamma^{t-T}M_{t-T}$. This will generate a power-law increase of variance over time, but only until the effect of an event has decayed to insignificance, say after about $(1-\gamma)^{-2}$ years. For longer lapses of time, the current state of the environment is virtually independent of its past state, and no further increase of variance occurs. Over periods of time much longer than the decay period, therefore, variance will increase rapidly at first but will soon reach a plateau where it thereafter remains indefinitely, which is not characteristic of natural variation.

The second possibility is that the magnitudes of successive events are correlated. Any particular category of event occurs at intervals of y years, and during this interval the event continues to have the same effect on conditions of growth. More extreme events occur at longer intervals, that is with lower frequency: the frequency f_M of the category of events of average magnitude M declines exponentially with increasing M as $f_M = f_0 \exp(-\gamma M)$, where γ now expresses how rapidly the frequency of events falls off with the severity of their effects, and f_0 represents the shortest interval recognized in a survey. Events of small magnitude are more

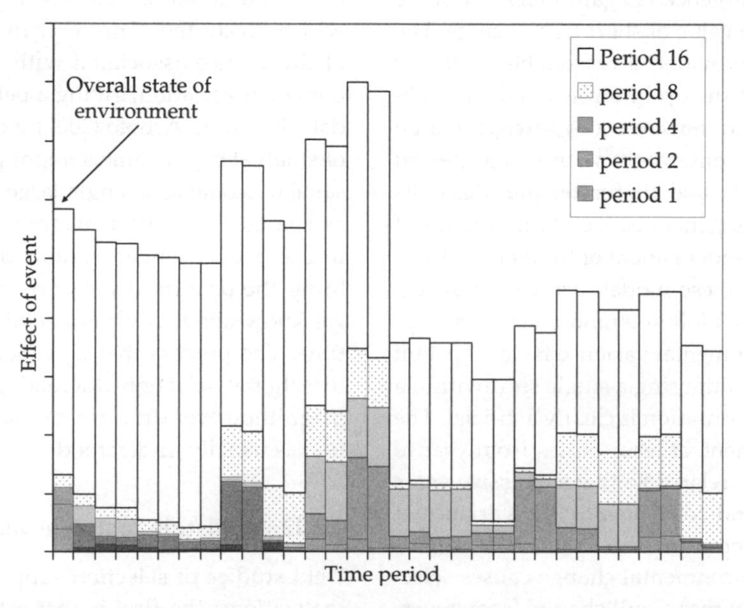

Figure 7.22 Multiscale model of environment. Events occur on all timescales and their effect is represented by the stacked bars for each time period. The effect of an event is an exponentially distributed random variable, whose mean is an exponentially decreasing function of its frequency. The overall state of the environment is the sum of effects at all scales.

likely to occur while a survey is being conducted, so that the selection coefficient, representing the response to these events, will be exponentially distributed. An event in the category of events that occur every year will be an exponential random variable with some small average value; an event in the category that occur every 2 years will likewise be exponentially distributed but with somewhat greater average, and so forth, the average magnitude of events that occur with frequency f_M being $M = -(1/\gamma)\ln(f_M/f_0)$. The current state of the environment is then the sum of the effects of events in all categories. Its absolute value is largely determined by rare events of large magnitude, whereas its variation from year to year is usually attributable to smaller and more frequent kinds of event (Figure 7.22). Conditions will thereby change on all timescales, and will fluctuate around an overall trend, no matter what period of time is chosen for a survey. This generates a power-law increase of variance over time that continues indefinitely. An hypothetical example might be a population of annual understorey herbs such as *Impatiens* which is affected every year by the severity of the winter, every 10 years by a tree-fall that opens a gap in the canopy, every 25 years by a major tempest or ice storm that breaks the trees, and every 100 years by forest fire. In a more realistic model, the effects of each of these events will decay over time, but the recurrence of progressively more severe events at longer intervals will still generate a power-law increase in variance, provided that effects of each category persist for an appreciable fraction of the corresponding interval. I think that this multiscale model of the physical environment is a reasonable way of linking the exponential distribution of selection coefficients with the power-law increase of variance over time.

In brief, the conditions of growth vary on all timescales, with rarer events of greater magnitude making a progressively smaller contribution to the overall variance (§2.3). Because the environment changes on all timescales, allele frequencies will also show trends on all scales of calendar time. At the shortest scales there is strong fluctuating selection that can cause appreciable shifts in allele frequency within a generation, or a few generations. Longer-term changes in conditions will create fluctuating selection over longer periods, which if the period is long enough will be perceived as a directional trend. Thus, morph frequencies in long-term fossil series frequently show pronounced fluctuations over short intervals of time combined with gradual change over longer periods of time, so that estimates of the rate of selection decrease with the length of the census period. Long-term evolution is usually depicted in two alternative ways: either as gradual change driven by chronic weak selection, or as abrupt change following a long interval of stasis. I am presenting here a third interpretation that seems to me more consistent with observations of natural selection in open populations: selection is generally rather strong and fluctuates on all timescales such that abrupt changes can occur over short periods of time and gradual directional change occurs over long periods of time. This process is brought to a halt when the species becomes extinct, at which point it expresses directional change corresponding to the long-term environmental change on the timescale of its longevity.

7.2.13 Genetic revolutions

Environmental change will cause adaptation provided that there is appropriate genetic variation for selection to act on. Having been exhausted by selection, new variation will be generated by mutation, but single or even multiple point mutations may not be sufficient to cause any appreciable advance. Renewed adaptation then requires a 'genetic revolution' of some kind in which a radical reorganization of the genome creates a new basis for variation and selection. The best-documented product of a genetic revolution may be *Spartina anglica*, a grass living at the lower levels of salt marshes (Gray *et al.* 1991). Angiosperms have seldom been able to adapt to saline conditions, and both the European *Spartina maritima* and the North American *S. alterniflora* are restricted to the less-saline upper reaches of the marsh, despite having had the opportunity to extend their ranges for millennia. When *S. alterniflora* was introduced into Europe a century ago in ballast it hybridized with *S. maritima* to produce the diploid *S.* × *townsendii*, which subsequently gave rise to the sexually sterile polyploid *S. anglica*. In this case the new combinations of genes created through hybridization made it possible to adapt very rapidly to conditions that had previously been completely inaccessible.

7.3 Selection experiments in the field

The course of adaptation to a new environment by a natural population can be studied in two ways: either by modifying the site that a population occupies or by introducing it into a different site. Selection experiments extending over several generations are very seldom attempted under field conditions, however: they are too long for student theses and too uncertain for funded research programmes. Almost all experimental studies of natural selection simply take advantage of an existing situation by measuring adaptation in populations that have recently occupied different sites. In most cases these sites merely reflect unmanipulated environmental variation, and the purpose of the assay is to determine whether adaptation can be maintained in the face of dispersal and immigration; such experiments are discussed below. If the sites have been manipulated deliberately, however, in the course of an experiment with some other objective, they provide an opportunity to document adaptation to a known environmental factor in an open population.

7.3.1 Habitat modification: the Rothamsted Park Grass Experiment

In 1856 an old hay meadow of uniform appearance in the grounds of the Rothamsted Experimental Station at Harpenden, England, was divided into 20 plots that received different fertilizer treatments in order to determine their effects on the production of hay. The experiment is still continuing, and despite some changes in treatment and incomplete data recording it provides a 150-year record of biological changes attributable to specific environmental manipulations. Selection has produced changes at two levels: by the sorting of species from the original vegetation of the meadow to alter the species composition of each plot, and by the adaptation of each surviving species within a plot. Many plots have received high inputs of specific nutrients, such as calcium and phosphate, that can influence growth and yield. The vegetative yield of the outcrossed annual grass *Anthoxanthum odoratum* is much greater in plots fertilized with calcium and phosphate, whereas other nutrients, such as magnesium, have no effect. This is an environmental effect, caused by the differences in soil nutrient concentrations between the plots, but there are also genetic changes in the populations. This can be demonstrated by extracting clones from the plots—easy to do, because most grasses proliferate vegetatively, and can be broken up into individual ramets that can be planted separately—and growing them under standard conditions in the greenhouse. Ramets from limed plots are more responsive to calcium than ramets of the same species from unlimed plots: that is, when high levels of calcium are applied to plants in the greenhouse, they can be utilized for growth much more effectively by the plants from the limed plots (Davies and Snaydon 1973a) (Figure 7.23). Similarly, the plants extracted from plots receiving phosphate addition are much more responsive to phosphate addition than are those from untreated plots (Davies and Snaydon 1973b), although responses to potassium

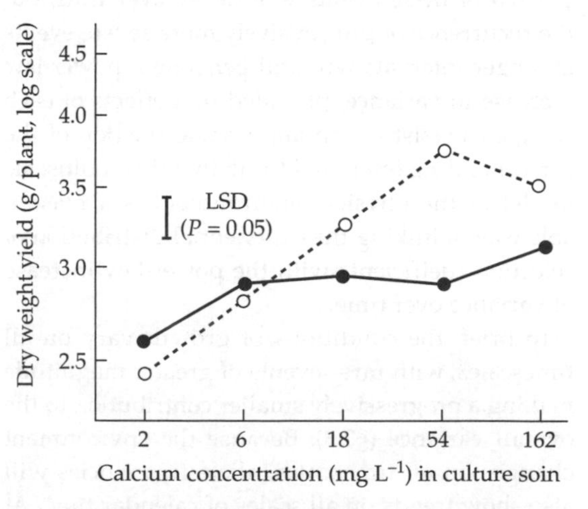

Figure 7.23 Adaptation of *Anthoxanthum* to liming in the Rothamsted Park Grass Experiment. Broken line: ramets from limed plots; solid line, ramets from unlimed plots. From Davies and Snaydon (1973a).

were less consistent and no adaptation to magnesium addition could be found (Davies 1975). During the few hundred ramet generations since the beginning of the experiment, *Anthoxanthum* has adapted to the nutrient treatments that most effect its growth. The time course of adaptation is unknown, but populations on newly limed plots evolve heritable differences in yield and morphological characters within 6 years (Snaydon and Davies 1982). These results show that the intense natural selection that can be imposed on laboratory populations also acts to drive adaptation in open populations.

7.3.2 Introduction: guppies in Trinidad

Streams that drain adjacent valleys in a steeply cliffed coastline are often more or less independent replicates of natural experiments for organisms such as fishes that cannot disperse by sea or land. On the north shore of Trinidad small rivers a few kilometres apart are inhabited by a fish community that includes guppies (*Poecilia*). The topography is rugged and the course of the rivers is often broken by waterfalls that impede the upstream dispersal of fish and filter out different components of the community: the whole community, including relatively large predators, lives in the broader downstream reaches, whereas further upstream only smaller fishes such as guppies are able to pass the barrier waterfalls, and the highest reaches beyond the steepest falls contain only *Rivulus*, a small fish able to squirm a short distance over the wet forest floor. Where there are large predators the guppies tend to be small, fecund, and dull because large and conspicuous individuals are most likely to be consumed; above the barrier waterfalls males are more brightly coloured and both males and females are older and larger when they mature (Endler 1978, Reznick and Endler 1982). The effect of visual predation can be checked by moving guppies upstream, from sites with large predators to sites that lack either predators or guppies. Males of an introduced population were more colourful after only two years, representing 3 or 4 generations (Endler 1980). Age and size at maturity increased for both males and females in two implant experiments, the rate of evolution being about 5–10 kDar measured over 10–20 generations (Reznick *et al.* 1997) (Figure 7.24).

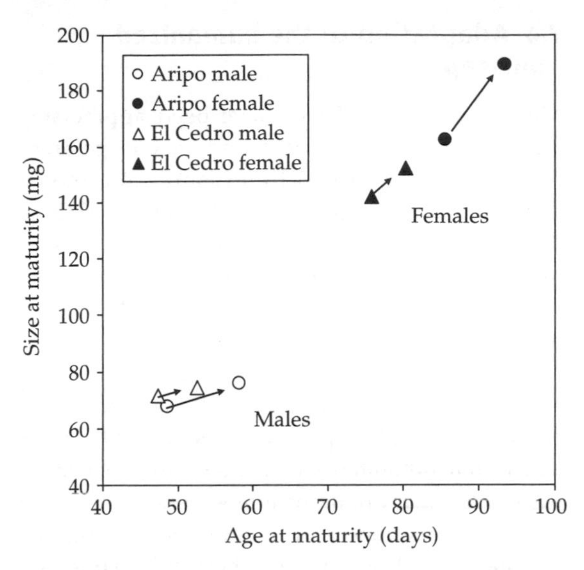

Figure 7.24 Effect of transplanting guppies to low-predation sites after 11 (Aripo) or 7.5 (El Cedro) years. Data from Reznick *et al.* (1997).

7.3.3 Perturbation: cactophilic *Drosophila*

If selection is primarily responsible for maintaining the genetic composition of natural populations, gene frequencies should tend to return toward their original values after they have been experimentally perturbed. Barker and East (1980) used isofemale lines of *Drosophila buzzattii* homozygous for alleles of esterase, pyranosidase, and alcohol dehydrogenase loci to increase the frequency of a given allele in an isolated natural population of these cactus-feeding flies. This caused sustained and substantial shifts in gene frequency over a period of about 250 days. When the introductions were discontinued, gene frequencies returned to their original values within about 300 days. This was not caused by immigration, as the nearest cactus patch was 3 km away; in any case, the three loci responded at different rates. It was therefore attributable to natural selection acting on these loci or on loci linked to them. Comparable experiments, with less positive results, were reported by Halkka *et al.* (1975) and Jones and Parkin (1977).

7.4 Adaptation to the humanized landscape

Most areas of the Earth have been appreciably affected by human activity and very large areas have been transformed by urbanization, industrialization, or agriculture. In the course of these changes some animals and plants have disappeared completely while others have prospered, but all survivors live in a humanized landscape to which they will adapt. In some cases a completely new environment has been created, such as a ploughland or a sewage plant, and only a few fortunately exapted species will be sorted from the aboriginal inhabitants of the site. In other cases the environment remains almost unaltered except for the consequences of a single perturbing factor, such as a trawl net or a herbicide. These are particularly interesting because they provide a quasi-experimental situation in which evolutionary change can be attributed with near-certainty to a particular selective agent. Indeed, they have provided some of the classical examples of natural selection that will already be familiar to most readers of this book.

7.4.1 Unexpected consequences of harvesting

When Pacific salmon, *Oncorhynchus*, return from their feeding grounds in the ocean to the streams where they spawn off the coast of British Columbia they must run the gauntlet of the coastal fishery, where they are caught mainly with seines, gill-nets, and trawls. A seine is a close-meshed net supported by a cork-line that is drawn around a group of fish in shallow water; even the smallest fish cannot pass through the meshes, so seines probably catch fish with the same efficiency, irrespective of their size. Seine-caught fish are thus taken to be a random sample of the population, and the selectivity of other kinds of gear is estimated by comparing the fish they take with the seine-caught sample. (There is an interesting problem here: if the selective effect of a particular procedure is to be established by comparing the selected sample with a random sample from the population, how can we be sure that the methods used to capture the 'random' sample are not themselves selective? The short answer is that we cannot be sure, and a selection differential that is estimated in this way always represents the difference between two sampling techniques, both of which may be selective.) Gill-nets are passively suspended in the water, and capture fish which are too large to pass through entirely, but which are prevented by their gill-covers from withdrawing their head. The selectivity of gill-nets depends on their mesh size: a large mesh allows small fish to pass through, while retaining larger individuals, and therefore selects for small size. A very small-mesh net would select for large size, because large individuals would be unable to insert their heads; but in practice gill-nets with such small mesh are never used. The most profitable mesh-size depends in part on how the fishery is conducted. In the salmon fishery, payment was originally by the piece; this encouraged fishermen to use small-mesh nets, which are rather unselective with respect to size, in order to capture the greatest possible number of individuals. In 1945, the method of payment was changed to payment by weight. This made it more profitable to use large-mesh nets that maximize the total weight of fish caught, by increasing the take of large individuals. For the same reason, the trawl fishery specialized in capturing, or retaining, the larger fish. The selection against large size caused by this change in economic policy had two sorts of effect on the salmon populations. Most of the species of salmon—chum, sockeye, chinook, and coho—return from the sea at different ages, and therefore at different sizes, and selection for smaller fish will cause selection for earlier maturity, as well as for smaller size at any given age. The situation is simpler for pink salmon, which always return after 2 years at sea. The mean size of pinks decreased between 1945 and 1975 in almost every fishing area along the coast. The rate of decrease has averaged about 20–30 g per year, so that over 30 years the average size of the fish caught decreased from about 2.3 kg to about 1.7 kg, a very considerable change, from either a biological or a commercial point of view (Ricker 1981). This change was almost certainly caused by the selectivity of the fishing gear, rather than by any trend in environmental factors such as sea temperature or salinity. For example, the rate of decrease in size was greater in fishing areas where a larger proportion of the population is captured. This decline continued between 1975 and the early 1990s, although Bigler *et al.* (1996) attribute it mainly to increased abundance.

While working for the government of Alberta, I found that gill-nets were selective, not only for size, but also for body form: they capture individuals that are relatively shorter and thicker than those caught by trawls, which because of their fine-meshed cod-end are much less selective (Handford *et al.* 1977). In lakes where gill-nets were set, whitefish (*Coregonus*) tended to become longer and thinner at given age, and reproduced earlier in life and at smaller sizes. Changes in growth rate caused by the selective effects of harvesting have also been documented in cod *Gadus* by Borisov (1978) and in *Oncorhynchus* (Hamon *et al.* 2000). Such evolutionary trends driven by size-selective predation are economically important, because they upset the steady-state assumptions on which fishery management is based. Heritable shifts in growth and size at reproduction are irreversible in the short term, and mean that populations are more likely to become extinct as fishing pressure shifts towards smaller and less valuable individuals, and less likely to recover if fishing is suspended. Despite the very widespread occurrence of such trends, evolutionary considerations are given little or no weight in the management of fisheries.

7.4.2 Unintended consequences of pollution

Mining

The metal ores found in the igneous rocks of North Wales were being mined when the Romans came to Britain, but the pace of exploitation increased sharply when the industrial revolution created new markets for copper, lead, and zinc. The deposits were relatively small and had been worked out by the late nineteenth century, leaving a fenced patch of rubble and spoil surrounding each old mineshaft. This patch was sterile; nothing grew on it, because high concentrations of heavy metals were leached from the low-grade ore that had been discarded there into the soil, creating a bare circle in the surrounding sheep pasture. After a while, a few plants from the pasture succeeded in growing on the old mine area, which eventually became tufted with grasses such as *Agrostis* and *Anthoxanthum*. By taking the progeny (either ramets or seeds) of plants growing on pasture or mine, McNeilly and Bradshaw (1968) showed that the mine plants had evolved a heritable resistance to high levels of heavy metals, and were able to grow fairly well at concentrations that were lethal to most pasture plants. The resistance is often highly specific: resistant plants grow within the mine boundary, but are replaced by susceptible plants in the surrounding pasture (Bradshaw 1952; Hickey and McNeilly 1975) (Figure 7.25). It can also evolve quite quickly: it has been little more than a century since the mines were abandoned, but resistant populations can appear on spoil heaps within 50 years or so. It has even been reported that linear populations of plants resistant to zinc, a metre or so in width, appear beneath galvanized-iron fences within a decade of their construction, or in the small area beneath electricity transmission pylons within about 20 years (Al-Hiyaly *et al.* 1988). In most cases resistance is conferred by a few major genes of large effect, the same genes often being involved in different species (Macair 1991, Schat *et al.* 1996). The evolution of metal tolerance on mine sites has been reviewed by Macnair (1987).

Smoke

About the middle of the nineteenth century, black varieties of several moths, formerly rarities, began

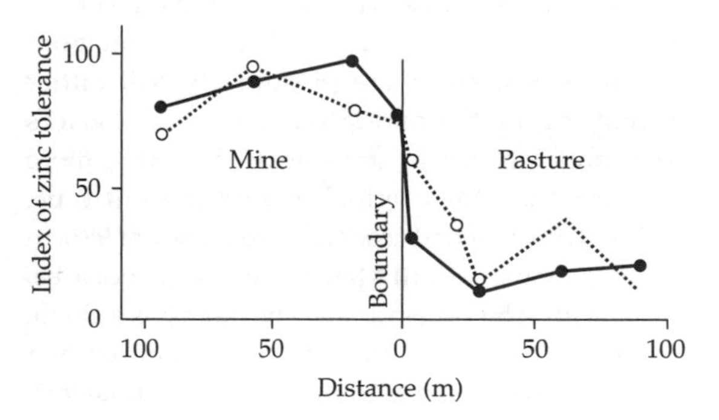

Figure 7.25 Evolution of resistance to heavy metals. Abrupt replacement of susceptible by resistant types at the mine boundary in two species of grass. From Macnair (1987).

to spread to a remarkable extent in the industrial regions of midland and northern England. The canonical example is the peppered moth, *Biston betularia* , in which a marked increase in melanic pigmentation is caused by a single mutation. Their spread followed the appearance of clouds of coal smoke above the newly industrialized towns, smoke that was washed down by the rain to blacken walls and tree trunks with soot. It is easy with hindsight to appreciate that the pepper-and-salt markings of the type originally common blended with the lichen-covered tree trunks of the unpolluted countryside, whereas the black wings and abdomen of the melanic variant were inconspicuous when the moth was resting on the soot-encrusted bark of urban trees. Visual predators such as birds would detect the pepper-and-salt variety more easily against a blackened background, creating selection that favoured the melanics and caused their spread. Kettlewell (1955) showed that birds preyed on moths resting on tree trunks, taking the more conspicuous kind first. Their effect on the frequency of melanics during a single episode of selection was demonstrated by following the fate of a relatively small number of moths, melanic and pepper-and-salt, released into natural populations. In the countryside, where lichens still grew thickly on tree-trunks, the pepper-and-salt variant increased in frequency over the course of a few days; in polluted areas, it was the melanics that increased. A single episode of selection, therefore, showed that melanic variants tended to spread in the novel environment furnished by sooty towns. Melanic variants of the peppered moth were first recorded in 1849. In 1875 they were still listed in catalogues as curiosities, and were presumably still rather infrequent. By the mid-1880s, however, collectors were noticing that in some areas they were more common than the original pepper-and-salt type, and by 1898 they had reached a frequency of 98% in the (highly industrial) Manchester–Liverpool area. Thus, within 50 generations—the moth has a single generation per year—the melanic phenotype had increased in frequency from about 1% to about 95% (Kettlewell 1973). Since in this case melanism is caused by a single dominant allele, the frequency of the melanic allele must have increased more than a hundredfold during this period, from about 0.005 to about 0.775. This implies intense selection, with $s = 0.2$ or so (see Haldane 1924). Selection favouring the melanic form in the city reverses in rural areas, and this is reflected in a cline of decreasing melanic frequency from the city into the countryside (Bishop 1972) (Figure 7.26). Since the abatement of smoke in the 1960s urban trees have became cleaner, their lichen flora has returned, and the frequency of melanics has decreased (Cook *et al.* 1999) (Figure 7.27). In recent years some scepticism has been voiced (e.g. Sargent *et al.* 1998), including a popular book claiming to debunk the whole story (Hooper 2002). The evolution of industrial melanism in the peppered moth and other species is certainly more complicated than simple selective removal by birds; there is a monographic treatment by Majerus (1998). The technical quibbles are overstated, however (Cook 2000), and Hooper's book is thoroughly unreliable (Clarke 2003). The crypsis of melanic moths in smoky cities remains a classical example of natural selection.

Carbon dioxide

The effects of industrialization have grown from point-source pollution to perturbation of the planetary economy as a whole. The most general and best-characterized effect is an increase in carbon dioxide (CO_2) concentration in the atmosphere, accompanied by a rise in temperature. Pre-industrial levels of about 200 ppm have now risen to about 380 ppm, and are expected to rise further to 700–1000 ppm about a century hence. There has been much research into the ecological consequences of elevated CO_2, especially its effect on plant productivity, and some speculation about the changes in community composition that are likely to occur, but very little attention has been paid to evolutionary changes within populations. Global change will inevitably lead to adaptation, however, and is a scientifically and socially valuable opportunity for evolutionary biologists to predict the qualitative changes that will occur in the next two or three human generations.

The first and rate-limiting step in photosynthesis is mediated by the enzyme Rubisco (ribulose bisphosphate carboxylase/oxygenase), which

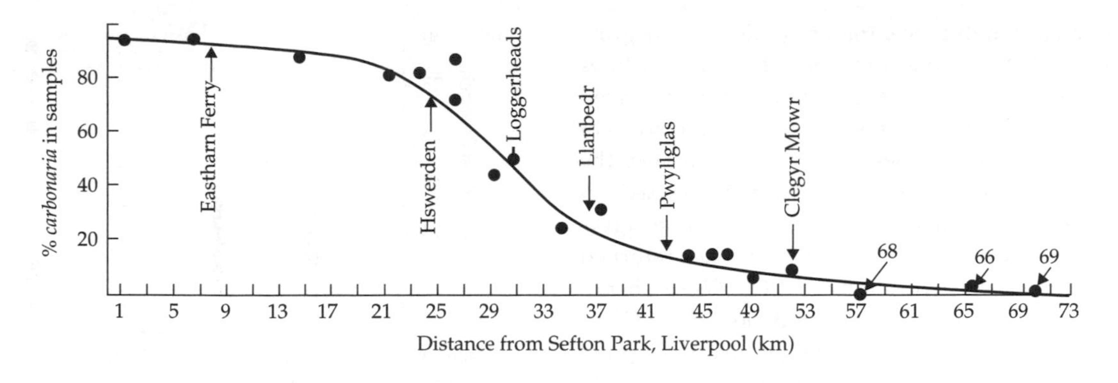

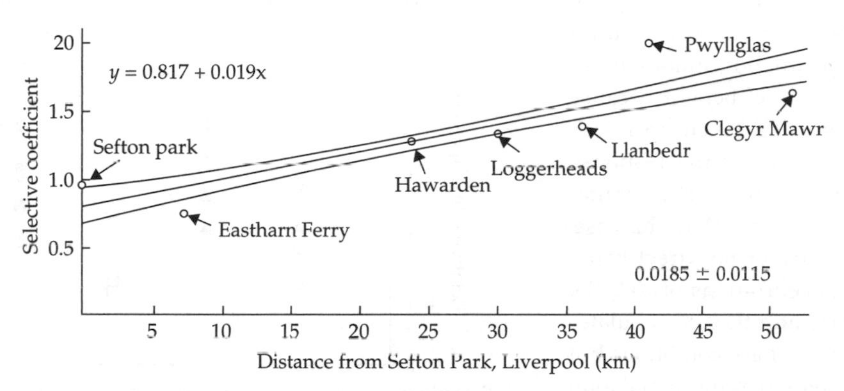

Figure 7.26 Industrial melanism in the peppered moth, *Biston betularia.* The upper figure shows the cline in the frequency of the melanic form *carbonaria* from the centre of Liverpool (Industrial) to North Wales (rural). The lower figure shows the selection coefficient along the same transect, estimated by capture-recapture experiments. From Bishop (1972).

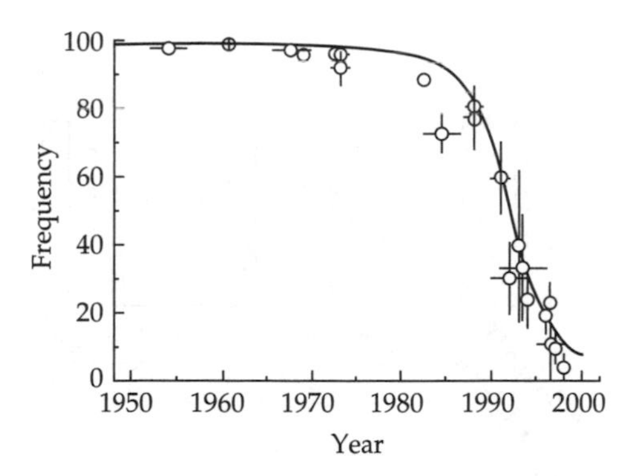

Figure 7.27 Industrial melanism in the peppered moth, *Biston betularia.* The steep decline in the frequency of carbonaria following smoke abatement. From Cook *et al.* (1999).

attaches a molecule of CO_2 to ribulose bisphosphate and then cleaves the lengthened sugar into two identical three-carbon phosphoglycerates that can be fed into central metabolism. Rubisco is an unusually inefficient enzyme that operates two or three orders of magnitude more slowly than most others, and to achieve high rates of photosynthesis green plants must either increase the concentration of Rubisco in the vicinity of CO_2 (by packing the chloroplast with large amounts of the enzyme) or increase the concentration of CO_2 in the vicinity of Rubisco. Green algae concentrate CO_2 around Rubisco with an array of external and internal carbonic anhydrases that are induced by low concentrations of CO_2. This inducible CCM (carbon-concentrating mechanism) is able to maintain high rates of photosynthesis even at low concentrations (<200 ppm) of CO_2. Collins and Bell (2004) cultured lines of *Chlamydomonas* at increasing concentrations of CO_2, rising from 400 ppm to 1000 ppm to simulate the atmospheric change expected to occur in the next century. They found to their surprise that there was no direct response to selection: lines selected at high concentrations of CO_2 did not grow faster at these concentrations than control lines maintained in air. Some selection lines had evolved elevated levels of photosynthesis, but this was not translated into higher growth rates because the additional carbon fixed was respired or leaked out of the cell (Collins *et al.* 2006a). Some lines grew very slowly or were unable to grow when returned to ambient concentrations of CO_2. These phenotypes are caused by the degradation of the CCM: the selection lines are no longer able to respond to changes in CO_2 concentration (Figure 7.28). In experiments where bacteria are cultured with a limiting substrate, loss-of-function mutations in regulatory systems are of direct benefit because they enable the substrate to be taken up faster. In the *Chlamydomonas* experiment the response to selection also involves loss-of-function mutations in a regulatory system, the CCM. In this case, however, these mutations are of no direct benefit: at perennially high concentrations of CO_2 the CCM is simply unnecessary and thus accumulates mutations that are neutral in these conditions but become severely deleterious when the population is returned to normal air.

Similar phenotypes have been described from mutants where components of the CCM have been inactivated. Curiosity led us to isolate soil algae from natural CO_2 springs, bizarre sites where subterranean sources of CO_2 seep to the surface and create pools of air with extremely high concentrations of CO_2 that have been in existence for hundreds or thousands of years (Collins and Bell 2006). These often showed the same syndrome as our laboratory selection lines: no specific adaptation to high CO_2 coupled with poor growth at normal levels of CO_2 and inability to induce a functional CCM. Several lines were lethal at normal concentrations of CO_2. Induction of the CCM seems to be particularly vulnerable to mutational degradation, leading to a large cost of adaptation when populations that have long lived at high concentrations of CO_2 are returned to normal atmospheres. The broader implication of this result is that algal growth is likely to become less efficient, so that the future increase in the productivity of the oceans, and their ability to act as a carbon sink,

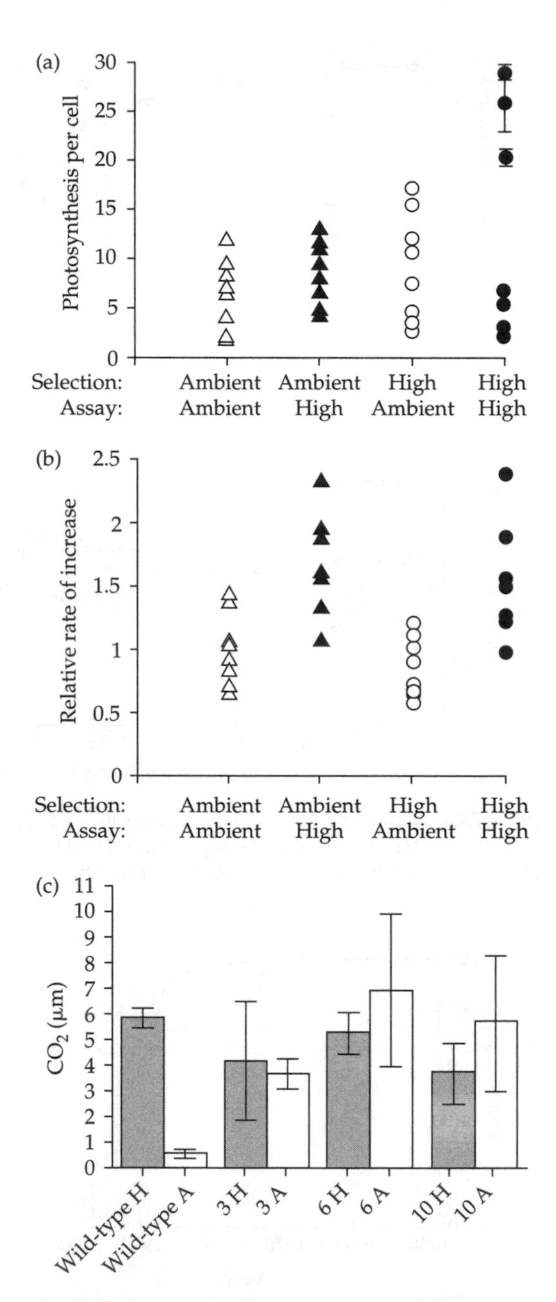

Figure 7.28 Adaptation of *Chlamydomonas* to elevated CO_2. A. Some selection lines, but not all, evolve higher rates of photosynthesis. B. This is not translated into growth, which does not differ from controls. C. At high CO_2 (filled bars) $K_{0.5}$ for net CO_2 uptake is the same in wild type (left-hand pair of bars) and three selection lines; in air the wild type is able to induce a carbon-concentrating mechanism (reducing $k_{0.5}$) whereas the selection lines are unable to do so. From Collins and Bell (2004) and Collins *et al.* (2006b).

will be less than expected on purely physiological grounds.

7.4.3 Unwelcome effects of eradication: herbicides, pesticides, and antibiotics

The outcome of selection caused by human activity is sometimes not merely unintended, but—from our point of view—highly undesirable, because it tends to frustrate our efforts (see Palumbi 2002 for a broad review). A classical example, with enormous social and economic importance, is the evolution of pesticide resistance in insect populations. It is generally caused by simple changes involving one gene, or a few genes. The first case was recorded in 1908; by 1948 there were some 14 species resistant to one chemical or another; after the widespread use of pesticides in the next two decades over 200 species had evolved resistance, and the number continues to increase. Almost equally dramatic processes occasionally occur in natural populations exposed to similar stresses. Black grass (*Alopecurus myosuroides*) is a common weed of cereal crops, which is controlled by herbicides that inactivate a crucial enzyme in fatty acid biosynthesis. Délye *et al.* (2004) surveyed populations in northern France and found high frequencies of variant sequences in a highly conserved region of the gene targeted by the herbicide, although it had been used for the first time only 15 years previously. Assuming that these mutations were initially very rare, this suggests that black grass populations in France are experiencing selection coefficients of at least $s = 0.85$ and evolving very rapidly in consequence.

The possibility that bacteria would evolve resistance was foreseen but discounted at the time when modern antibiotics were being developed. β-Lactams such as penicillins and cephalosporins kill bacteria by inhibiting cell-wall synthesis. They were introduced into general practice in the early 1940s, and were at first dramatically successful in suppressing infections by *Staphylococcus*, *Streptococcus*, and other bacteria. However, the first cases of resistance began to be reported only a few years after their introduction, and most populations throughout the world are now resistant to doses hundreds or thousands times larger than those that were once effective. There are various sources of resistance. Penicillin normally acts by binding to the proteins that link peptidoglycans in the cell wall; the wall is then weakened, and ruptures under the osmotic pressure of the cytoplasm. Changes in the structure of these cell-wall proteins reduces their tendency to bind penicillin, and enables the cell to survive. Some bacteria are resistant because they produce a β-lactamase that cleaves the antibiotic and renders it harmless. More simply, some strains are just less permeable to antibiotics, as Gram-negative bacteria are generally less sensitive than Gram-positive bacteria. Unfortunately, the response to the appearance of strains resistant to currently prescribed doses is very often to increase the dose until the infection is just brought under control; this is an efficient procedure for selecting increased levels of resistance. The various sources of resistance are sometimes encoded by chromosomal genes; in other cases, however, they are borne by plasmids or transposons, and a gene selected in one lineage of bacteria can quickly be transferred to a large proportion of the bacterial community. Despite the ingenuity of pharmacologists in devising new kinds of antibiotic, every hospital in the world now harbours bacterial strains resistant to a range of antibiotics, and whenever a new antibiotic is released the evolution and spread of resistance follows, sometimes with bewildering rapidity (Figure 7.29).

The evolution of antibiotic resistance is, sadly, a familiar story. Other than epidemiological issues, the topic adds little in terms of fundamental principles to natural selection in microcosms and plasmid carriage. It has been studied from the point of view of an evolutionary geneticist by B. Levin and others (see Levin *et al.* 1997). The point that I would like to emphasize here is that understanding, and ultimately controlling, the evolution of resistance has been greatly hindered because physiologists (interpreted broadly to include pharmacologists, biochemists, and molecular biologists) and evolutionary biologists have fundamentally different and sometimes incompatible views of natural mechanisms. This was brought home to me by the case of ribosomally assembled antimicrobial

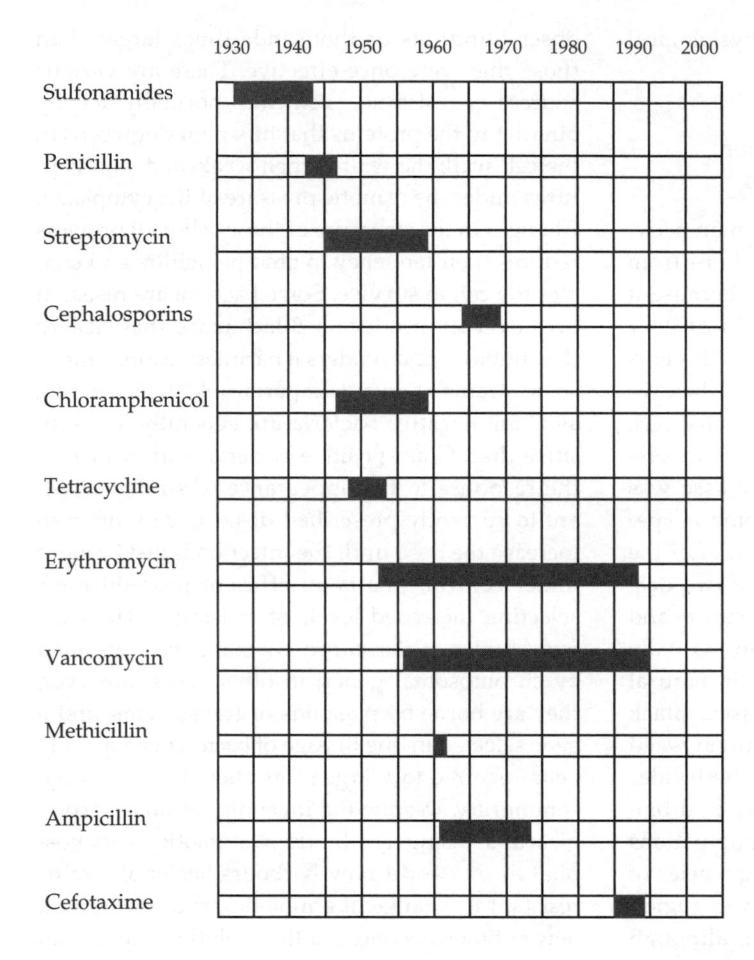

Figure 7.29 Evolution of antibiotic resistance. Bars indicate the period between the introduction of an antibiotic and the first appearance of resistance. From Palumbi (2002).

peptides (RAMPs). The RAMPs are gene-encoded polypeptides composed of fewer than 100 amino acids (Huang 2000) and have a net positive charge conferred by lysine and arginine residues (Hancock 2001). The selective toxicity of RAMPs toward bacterial cells is primarily due to an initial electrostatic interaction between the peptide and the anionic phospholipid head groups in the outer leaflet of the bacterial cytoplasmic membrane (Zasloff 2002, Yeaman and Yount 2003). The point of view of physiologists was that resistance is unlikely to evolve because it would require fundamental changes in membrane structure (Zasloff 2002). Host defence peptides have remained effective against bacterial infections for at least 10^8 years, despite the continual presence of these peptides in bacterial environments, and we may infer that resistance is very unlikely to evolve in the short term (Zasloff 2002). The great diversity of RAMPs, the presence of several different kinds at the infection site, and their different modes of action might have impeded the evolution of resistance in natural bacterial populations. Evolutionary biologists argued, however, that the therapeutic use of particular RAMPs would alter natural environments by creating a source of specific and continued selection that might readily lead to the evolution of resistance (Bell and Gouyon 2003). We resolved this disagreement by exposing lines of *E. coli* and *Pseudomonas* to gradually increasing dosages of pexiganan, an artificial analogue of a natural RAMP of frog skin (Perron *et al.* 2006). The outcome was decisive: high levels of resistance evolved in almost all selection lines within 700 generations. There is no doubt that widespread therapeutic use of RAMPs would lead to the evolution of resistance, in much the same way and within the same time frame as conventional antibiotics. The wider implications of this conclusion are somewhat more complex. It should not be used, in my view, to discourage or retard the development of potentially useful antimicrobial agents. Rather, as we develop RAMPs for use as human and veterinary anti-infectives we should also seriously consider the consequences of the emergence of resistant organisms. Will organisms that emerge resistant to a synthetic RAMP such as pexiganan also exhibit resistance to endogenous, natural antimicrobial peptides? Would these resistant organisms be more pathogenic than non-resistant strains? Could resistance against a synthetic peptide evolve *in vivo*, for example, in the setting of the long-term exposure of an animal? How should these concerns be translated into practice by the governmental regulatory bodies that ultimately approve new anti-infectives? Thoughtful analysis, informed by evolutionary principles, of the potential emergence of resistance against such novel anti-infective molecules before they are in widespread use, will, in my opinion, help maximize their ultimate benefit to society.

7.4.4 Human evolution in the humanized environment

Humans certainly evolved in the past, and indeed our present state is in large measure due to a sustained increase of brain size in our lineage from about 400 cm^3 in the late Pliocene to about 1400 cm^3 at a rate of about 400 kDar. In the last 10 000 years, however, we have created our own conditions of life so successfully that it has been questioned whether natural selection continues to operate at all in modern human societies. There would indeed be little opportunity for selection if everyone survived to reproduce and then limited their family to two children. As this is not the case selection may occur, and if we are like other animals it will often be rather strong and capable of producing substantial and rather rapid adaptation. The ability to resist starvation or avoid large predators may no longer be necessary in most modern societies, but the agents of selection have changed rather than disappeared. There is no clear reason to suppose that there should be less variation in the attributes and skills required for agriculture, commerce, and industry than in those required for hunting and gathering, nor that they should be systematically less heritable. Indeed, the novelty of the humanized landscape that elicits strong selection in other animals should have a similar effect in human populations.

The phenotypic variance of lifetime reproductive success in modern but pre-industrial human populations is $SV_P \approx 0.1$ (see Figure 7.4). Historical records from eighteenth–nineteenth century Lapp communities in Finland which lived by fishing, hunting, and reindeer herding gave $SV_P = 0.18$ (Kaar *et al.* 1996). A contemporary sample of Australian women with a modern Western lifestyle gave a somewhat higher value of $SV_P = 0.4$, with a heritability of 0.4 (Kirk *et al.* 2001). The standardized genetic variance is then $SV_A = 0.16$, around the upper limit of estimates from natural populations of annual plants and small birds. For both Finnish and Australian women the most important determinant of fitness was the age at first reproduction; the selection gradient for the Finnish sample was $\beta \approx -0.5$.

7.5 The ghost of selection past

The main purpose of studying how selection acts in the laboratory or the field is to be able to predict its outcome. Perhaps this logical sequence can be inverted, in order to infer how selection acts from observations of its outcome. Indeed, this has been the preferred method for making large-scale generalizations about selection since extensive standardized estimates of protein variation first became available in the 1960s, and has been given additional impetus in the past decade by the advent of genomics. The outcome of any genetic survey is information about the state of genes in a series of individuals, ideally the DNA sequence of a large fraction of the genome, or of the whole genome. The differences between individuals constitute the information that has been printed on to the population by the operation of selection during the last $4N$ generations. If this information can be decoded, then we can reconstruct the history of the population and discover how it has been moulded by selection. In particular, we can estimate the frequency and intensity of different kinds of selection acting on different kinds of genes. This is potentially a very powerful approach, since it is capable of synthesizing evolutionary processes over the whole genome, and thereby enabling us to make generalizations that would otherwise require hundreds of experiments. The methods that have been used to do this fall into two broad categories. The first uses population-genetic models that compare observed patterns of variation within species against the prediction made by neutral theory, whereas the second uses the properties of different kinds of mutation.

7.5.1 The analysis of allele frequencies

The simplest survey will report the frequencies of alleles at a single locus. If the alleles are neutral the heterozygosity is expected to be $1/(1 + \theta)$, with $\theta = 4N_e u$, and the number of different alleles in a sample of N diploid individuals is $\Sigma\ [\theta/(\theta + i)]$, with the summation from $i = 0$ to $i = 2N - 1$ (Ewens 1972). Given the observed number of alleles, therefore, the corresponding heterozygosity can be calculated and compared with the observed value; this is the widely used Ewens–Watterson test (Watterson 1978). Heterozygosity may be elevated or reduced by selection. Balancing selection will lead to greater heterozygosity if allele frequencies are more even than expected under neutrality. Heterosis will produce this result, unless the homozygotes are very different. Frequency-dependent selection will also have this effect, provided that the ecological or sexual opportunities sustaining the polymorphism are correspondingly even. Sexual opportunities, such as the availability of gametes of different gender, have an inherent tendency towards evenness; ecological opportunities do not, but if they are not fairly even it is unlikely that a polymorphism will be sustained. On the other hand, directional selection will reduce heterozygosity. If a beneficial mutation has recently become fixed, all other alleles will be rare, the distribution of allele frequency will be markedly uneven, and heterozygosity will be unexpectedly low.

The distribution of allele frequency will determine the relationship between the number of segregating sites and the average number of differences between individuals. Thus, if there are 2 equally frequent alleles differing at 10 nucleotide sites the number of segregating sites is much greater than the difference between 2 random individuals, whereas if there are 10 equally frequent alleles each differing at a single site the number of segregating sites is the same but the average difference between individuals is much greater. For neutral substitutions the expected number of differences between individuals is θ, while the expected number of segregating sites in a sample of n alleles is $i\theta$, where $i = 1 + 1/2 + 1/3 + \ldots + 1/(n-1)$. A survey of any set of loci, or any DNA sequence, will thus provide two estimates of θ, which will be approximately equal for neutral alleles. Directional selection will cause them to differ because it reduces the coalescence time and thereby reduces variation among individuals relative to the number of segregating sites; this is Tajima's D test (Tajima 1989). Because the test can easily be applied to genomic data it has been widely employed to identify sequences where variation has been strongly reduced by recent directional selection.

When two lineages become sexually isolated they will diverge at neutral sites at a rate equal to the neutral mutation rate. But heterozygosity in either lineage is also proportional to the neutral mutation rate, given that alleles are selectively equivalent, and therefore the ratio of the number of segregating sites P in either lineage to the number of fixed differences between the lineages D should be the same for all loci. If directional selection is acting on some loci but not on others, however, the variance of P/D among loci will be unexpectedly large; this is the Hudson–Kreitman–Aguarde (HKA) test (Hudson *et al.* 1987).

The collation and analysis of genomic data has been pursued enthusiastically for the last 20 years and has given rise to a large literature (see Kreitman 2000, Nei 2005). As a very broad-brush summary, statistics such as Tajima's D suggest pervasive purifying selection affecting a substantial proportion of loci, with occasional examples of strong directional or balancing selection. For example, about 40% of genes surveyed in *Arabidopsis thaliana* gave a signal of recent selection, and four of seven plants that have been intensively studied have D statistics suggesting a prevalence of purifying or directional selection (Wright and Gaut 2004). In some cases there is convincing circumstantial evidence to support the statistics, for example low diversity in the region of the *tb1* (*teosinte-branched*) gene of maize, which is implicated in domestication (Clark *et al.* 2004) (§6.4.3). Moreover, most of the genes in non-domesticated plants such as *Arabidopsis* that give the clearest signal of recent selection are concerned with resistance to pathogens or herbivores (e.g. Tian *et al.* 2002) (§11.3). In general, however, it is difficult to place much trust in most analyses. The reason is simply that selection and demographic events can have similar effects on allele frequencies; thus, a selective sweep and a population bottleneck will leave similar signatures of reduced diversity and more recent coalescence. It should be possible to distinguish the two processes, as demographic changes affect the whole genome whereas selection affects genes differently, but in practice it is far from easy (Nielsen 2001). Most studies do not take demography into account, and very few explicitly investigate its effects. Haddrill *et al.* (2007) surveyed 10 X-linked loci in *Drosophila melanogaster* populations and found that standard statistics rejected the neutral model in many cases, but after detailed analysis were unable to exclude demographic fluctuations as the cause. There are probably very few patterns that cannot be explained either by drift or by demography, so that allele frequency distributions offer a less than convincing evaluation of selection in natural populations. One response is to call for more extensive data and more powerful analytical procedures, but I am pessimistic about the chances of success. My experience with models of species abundance and co-distribution has persuaded me that the bulk properties of populations or communities, such as allele frequency distributions, will usually be dominated by stochastic processes even when selection is quite strong and frequent. Lewontin (1974) described how attempts to distinguish adaptive from neutral interpretations of enzyme polymorphism by inspecting patterns of variation were doomed to failure, and I suspect that history is repeating itself in the analysis of genomic surveys.

7.5.2 The analysis of divergence

An alternative method is to use information about the phenotypic consequences of mutations, in particular the difference between replacement mutations, which alter the encoded amino acid, and synonymous mutations, which do not. The fundamental assumption is that synonymous mutations are neutral. For homologous sequences from any two species the fixed differences can be counted and characterized. The number of replacement substitutions per replacement site (k_A) and synonymous substitutions per synonymous site (k_S) should be equal if replacements have no effect on fitness, whereas purifying selection will yield $k_A < k_S$, and conversely directional selection will yield $k_A > k_S$ (reviewed by Fay and Wu 2003). The ratio k_A/k_S is 0.08 for *E. coli* vs other bacteria (Jordan *et al.* 2002), 0.1–0.2 for *Drosophila melanogaster* vs *D. simulans* (Eyre-Walker *et al.* 2002), and 0.12 for human vs mouse (Waterston *et al.* 2002). For single-nucleotide polymorphisms in the human genome $k_A/k_S = 0.2$ (Fay *et al.* 2001). This is unambiguous evidence for purifying selection and shows that on average 80–90% of replacement mutations are deleterious. Even in non-coding regions, however, a large fraction of nucleotides may be under purifying selection (Halligan and Keightley 2006).

There are a few examples of genes involved in disease resistance (§11.3.4) or mating (§12.4) for which $k_A/k_S > 1$. Endo *et al.* (1996) found 17 candidate genes among nearly 4000 sequences of viruses, bacteria, and eukaryotes, of which 10 were genes governing infection or virulence in pathogens; 2 others were neomycin resistance and snake venom, which contains peptides related to those of the innate immune system. This is a very severe condition for detecting directional selection, however, as the occasional substitution of a beneficial mutation will be veiled by the mundane elimination of deleterious mutations. The McDonald–Kreitman criterion detects directional selection, even in the presence of purifying selection, by comparing the ratio of replacement polymorphisms (P_A) to replacement fixed differences (D_A) to the ratio of synonymous polymorphisms (P_S) to synonymous fixed differences (D_S) for two fairly closely related species. The neutral expectation is that $P_A/D_A = P_S/D_S$, and any departure from equality indicates selection, regardless of demographic history (McDonald and Kreitman 1991). The fraction of substitutions attributable to directional selection is then $1 - [(D_S/D_A)(P_A/P_S)]$. This argument shows that a large fraction of gene substitutions are adaptive: about 35% in the human–platyrrhine divergence (Fay *et al.* 2001) and about 45% between *Drosophila* species (Smith and Eyre-Walker 2002). These figures are equivalent to one adaptive substitution per 200 years in the human lineage and one every 45 years in the *D. simulans* lineage.

At the point where two lineages begin to diverge they will share all their polymorphisms, but the standing neutral variation will soon dissipate and be replaced by new polymorphisms arising independently in the two lineages. Transgressive (or trans-species) polymorphisms that are shared by well-separated species are thus unambiguous signals of balancing selection (see Charlesworth 2006). As with strong directional selection, cases have been reported from genes governing disease resistance and mating; MHC alleles, for example, may be shared by mouse and rat (Figueroa *et al.* 1988). In general, however, transgressive polymorphism is rare. A detailed analysis of human and chimp genomes failed to uncover any shared single-nucleotide polymorphisms in excess of the number expected to occur by coincidental mutation (Asthana *et al.* 2005). This shows that almost all single-nucleotide polymorphisms are neutral, or that sister species such as human and chimp quickly become divergently specialized.

7.5.3 Commonplace, strong, fluctuating, oligogenic

The prevailing account of adaptation is the Fisherian process of gradual change through the occasional substitution of alleles of small effect under weak selection at many loci over long periods of time. This has served well as a theoretical framework for thinking about selection, and in some cases it may provide a realistic description of evolutionary change. To this point, however, a very different picture seems to be emerging. Adaptation in the laboratory usually occurs swiftly, driven by strong selection acting on a series of beneficial mutations of rather large effect. Artificial selection soon produces marked shifts in phenotype attributable in great part to genes of large effect. Populations in the field often experience strong selection that fluctuates in direction over time. Not all loci can be strongly selected all the time, of course. Adaptation may often occur, however, within a non-Fisherian context in which selection is commonplace, often acts strongly on genes of major effect, and fluctuates over time. This is the view I shall carry forward in describing selection in more realistic circumstances involving ecological complexity, genetic parasitism, competitive neighbours, unrelenting enemies, and ambivalent mates.

CHAPTER 8

Adaptive radiation: diversity and specialization

8.1 Adaptive and non-adaptive radiation

8.1.1 Wrinkly spreaders

A culture of *Pseudomonas fluorescens* that is vigorously shaken becomes turbid and milky. When a sample is spread on agar, the colonies that grow are circular and have a characteristic mottled appearance with a smooth surface. If the culture is not shaken, however, it develops a tough skin, rather like neglected custard. This apparently mundane observation set off one of the classical case studies in experimental evolution (Rainey and Travisano 1998). Cells isolated from the surface biofilm form dryish, wrinkled colonies on agar quite unlike the smooth colonies formed by cells from the broth. These *wrinkly spreaders* secrete large quantities of a cellulose-like polymer as the result of a mutation in the 10-gene operon *wss*: loss-of-function mutations in different *wss* genes give wrinkly spreaders with somewhat different phenotypes. When the culture vial is continuously stirred these mutants are at a disadvantage, but if it is left undisturbed they colonize the surface of the medium and thereby monopolize the oxygen supply. The broth itself then becomes depleted of oxygen, but the smooth cells remain abundant and a third type, the *fuzzy spreader*, may appear in the abyssal depths as a low-oxygen specialist. Thus, the existence of sites offering potentially different conditions of growth induces a rapid miniature adaptive radiation in unstirred vials, with the occupation of these sites by specialists such as the wrinkly spreaders exacerbating the ecological heterogeneity of the virgin vial. In this situation a cost of adaptation is demonstrated by invasion: any of the three types will invade a microcosm currently occupied by only one of the others, because it is a superior competitor in one niche but cannot displace the resident from the other. The unstirred vial has become a model system for studying ecological and social interactions among microbes, and several other studies that have used it are described below.

8.1.2 Measurement and interpretation of diversity

There is thus a natural tendency for lineages to diversify because replication is usually but not invariably accurate and the errors that are occasionally made will be propagated provided that they are not too damaging. This tendency is opposed by recombination, because entities that recombine cannot remain distinct. Consequently, biological diversity is studied primarily in situations where recombination is rare enough to be neglected: among asexual taxa (especially bacteria and archaea), among sexual species, and among alleles. These situations represent disciplines—microbiology, community ecology, and population genetics—with little overlap in membership, separate journals and conferences, and more or less independent histories. To some extent, their isolation reflects technical differences in measuring diversity. Most species of animals and plants can readily be identified by specialists using little more than a lens; systematic attempts to catalogue species date back to the mid-eighteenth century, and generalizations about species diversity to the mid-nineteenth. Genetic diversity is more difficult to measure, because the variation that is readily apparent is only a partial and unrepresentative sample of the whole range of alleles, and it was not until the introduction of protein electrophoresis in the 1960s that the true extent of allelic diversity could begin to be explored. The physiological diversity of bacteria has been known for more than a century, but their small size and lack of convenient morphology hindered attempts to measure their phylogenetic diversity until the introduction of DNA techniques in the 1980s.

The foundation of all knowledge about diversity is the biological survey: samples are taken from a number of sites and the individuals found there assigned to the appropriate categories, such as alleles or species. The output of a survey is thus the abundance or occurrence of each type at each site, and the distribution of abundance is the fundamental statistic of diversity. For most ecologists the most important aspect of diversity is the total number of species, or *species richness*, because every species is felt to be important. This simple approach has two serious drawbacks. The first is that we have no theory that predicts species richness. The second is that when the underlying distribution of abundance is given the estimate of species richness depends on sample size: when more individuals are collected, more species will be found, so the estimate is meaningless unless sample size is standardized by rarefaction. Population geneticists sometimes use an equally blunt approach by recognizing a locus as polymorphic if it has at least two alleles with frequencies of more than 5%, an arbitrary practice that has historical roots in the study of visual polymorphisms, where each morph has almost the same cachet as a species. It is more usual, however, to calculate the variance-like quantity $1 - \Sigma p_i^2$ from the frequencies p_i, which is the expected heterozygosity in genetics and the inverse of Simpson's index in ecology. This is independent of sample size and depends solely on the distribution of abundance; moreover, it is directly related to theory. Its main drawback is that it does not tell you how many different types there are. The parameters that can be estimated from survey data and the relationships between them have been described by Bell (2003).

More important than differences in the measurement of diversity are differences in its interpretation. The simplest interpretation is provided by the neutral theory: all types are equivalent, or nearly equivalent. Neutral communities have many predictable properties: the amount of variation is proportional to the number of individuals of novel type appearing in each generation, the rate of fixation is equal to the neutral mutation rate, and isolated communities diverge linearly over time. The theory was developed by population geneticists in response to the high levels of allelic diversity uncovered by the first allozyme surveys. It sparked a vigorous controversy that was the central theme of the field for 20 years and still continues in various guises. The arguments of the population geneticists had little effect, however, on their colleagues in community ecology. Neutral models of species diversity were developed by Caswell (1976) and Hubbell (1995), but had little influence, and it was not until much later that neutral theory was re-introduced to ecology by Hubbell (2001) and Bell (2000, 2002). It has been very successful in providing a mechanistic explanation for many ecological patterns, such as the distribution of abundance, the increase in species richness with area, and the decay of community similarity with distance. The alternative to neutral theory is that each type is specialized, in some sense, so that variation is actively maintained by selection. This has always been the prevalent theory in ecology, where the consistent and distinctive differences between species have led ecologists to suppose that each species occupies a 'niche' to which it is specifically adapted, and since the beginnings of scientific ecology in the early twentieth century one of its principal goals has been to relate the distribution and abundance of organisms to the features of the landscape they inhabit.

8.2 G × E

A plant growing on the forest floor occupies a definite volume of space, in the soil and in the air above it. At any instant in time, conditions vary throughout this volume. The most obvious and important distinction is between soil and aerial conditions, which is reflected in the gross architectural design of the plant. But neither soil nor aerial conditions are uniform: some leaves will be receiving more light than others; some roots will be adequately supplied with water and others not. Moreover, this pattern of variation changes through time. As sunflecks moves across the plant, leaves will pass from light to dark and back again; roots will be wetted after rain and then dry out. An individual plant thus has a certain extension in space and time, and during its life experiences a corresponding variation in conditions of growth. Its ability to cope with this variation may be called *versatility*. The more versatile the individual, the broader the range of conditions over which it can maintain growth. In order to do so, its physiological systems must be apt to accommodate themselves to changing circumstances, shifting in space and time as conditions dictate. The outcome of this underlying physiological variability, however, is to maintain components of fitness as nearly constant as possible: versatility implies the *stability* of survival and reproduction, despite environmental variation and because of physiological variability. All organisms must be versatile to some degree, although the amount of environmental variation that they experience, and the different contributions of spatial and temporal change, depend on their organization. A bacterial cell growing in favourable conditions has a spatial dimension of a few micrometres and a temporal dimension of a few hours; it is likely to be exposed to brief pulses of different nutrients, and has inducible metabolic systems that switch the activity of the whole cell in order to exploit them. A tree has a spatial dimension of tens of metres and a temporal dimension of tens of years; although it must respond to temporal variation, it will also experience substantial spatial variation, and must be able to adjust light-harvesting or water-uptake systems to different levels in different parts of its body. Any particular organism, then, has a certain individual scale that represents its extension in space and time, and must display the versatility appropriate to this scale. Large, motile, and long-lived organisms operate at larger scales, and require greater versatility.

The seeds produced by an individual plant will be dispersed to a greater or lesser distance within the area occupied by the population, and will germinate after a longer or shorter period of time. Each will grow in a different place from its parent, and at a different time. Over and above the variation experienced by an individual during its lifetime, parents and offspring will experience different conditions of growth. Thus, the lineage has an extension in space, determined by the distance that propagules are dispersed, and an extension in time, determined by the time that elapses before the propagules germinate and grow. These constitute the characteristic scale of the lineage. In addition to their ability to accommodate changes in conditions during the lifetime of an individual, organisms must also adjust to the average change in conditions between one generation and the next. Insofar as this is possible, the same genotype will be able to express different phenotypes in different environments. Its ability to do so may be called *plasticity*. A more plastic genotype has a more variable developmental programme that enables it to exhibit greater stability with respect to components of fitness over a broader range of environments.

The distinction that I have made between versatility and plasticity is essentially the difference between reversible (versatile) and irreversible (plastic) developmental shifts. It is not a conventional usage: the ability of individuals to change their behaviour from day to day, or even from minute to minute, is called plasticity by physiologists and ignored by geneticists, whereas the ability of the same genotype to develop into different kinds of individual is called plasticity by geneticists and ignored by physiologists. In this part of the book I am concerned with the evolutionary response of populations to selection in heterogeneous environments, and therefore primarily concerned with the average differences among individuals. I have therefore adopted a geneticist's definition of plasticity, while being careful to distinguish it from the equally important phenomenon studied by physiologists. Plasticity is thus the property of a genotype, not of an individual, and is measured by the environmental variance, that is, by the variance among individuals with the same genotype raised in different environments. Microbes in a laboratory microcosm live in a very simple world whose environment is uniform and unchanging, unless it changes once to remain the same thereafter, and in which genes have straightforwardly beneficial or deleterious effects on characters that affect fitness. The natural world, on the other hand, varies on all scales of space and time (§2.3), and consequently organisms must adapt, not merely to a given set of conditions, but to diverse and changeable conditions. This might conceivably lead to the evolution of a perfectly plastic super-microbe that flourished in all conditions, but in practice almost all sites support a diverse community of microbial and multicellular organisms. Diversity is the second-order theme of evolutionary biology, adaptation being the first-order theme.

The fact that fitness varies in time or space is an ecological principle that does not necessarily have any evolutionary consequences. If all genotypes vary in the same way, and to the same extent, over a given range of environments, then selection will not cause any change in their frequencies. It is only if relative fitness varies among environments that selection will act differently according to circumstances. This constitutes a third possible source of variation among individuals. If several genotypes are each tested in several environments, the variance of the average scores of genotypes represents genetic variance, and the variance of the average scores in each environment represents environmental variance. However, if genotypes do not respond consistently to environmental variation, the score of a particular genotype in a particular environment may deviate from the score that would be expected on the basis of the average properties of that genotype and that environment. The variance represented by these deviations is attributable neither to purely genetic sources, nor to purely environmental sources, but rather to the interaction between genotype and environment: the genotype–environment interaction, or G × E. This is the variance of relative fitness over environments.

There are two equally valid ways of interpreting G × E. On the one hand, it expresses the extent to which genetic variation is expressed differently in different environments. The quantity of genetic variation is not a fixed property of a population, but may vary according to the environment in which the population is living. There may therefore be greater opportunity for selection in some environments than in others. Moreover, part of this variation among environments in the quantity of genetic variation that is expressed in them arises because genotypes that have very high relative fitness in one environment are mediocre or even inferior in others. Thus, genotypes may be selected as specialists in particular environments. On the other hand, G × E expresses the extent to which environmental variation is expressed differently by different genotypes. Some genotypes may express very different phenotypes in different environments, and therefore possess a large quantity of environmental variance; others will be less responsive, and express more or less the same phenotype regardless of the environment in which they are raised. More plastic genotypes are more stable; that is, they display *less* variation for components of fitness as the result of displaying *more* variation for the physiological mechanisms that are responsible for maintaining survival and fecundity. Thus, the quantity of environmental variance, or plasticity, is itself a character, and may vary among genotypes, which may be selected as generalists over a range of environments. The main issue raised by G × E is the balance between generalization and specialization that should evolve in populations that live in a heterogeneous environment.

8.2.1 The ecogenetic landscape

The interaction of genotype with environment might be visualized as a landscape, akin to the adaptive landscape used to represent interaction among genotypes (§3.7.2). I shall call it the ecogenetic landscape (Figure 8.1). Latitude represents a series of strains, as in the adaptive landscape; but longitude now represents a series of sites. The strains, which may be alleles at a single locus, or clones that differ to any extent, or species, are arranged from the least to the greatest mean fitness; the longitudinal relief is genetic variation. Likewise, the sites are arranged from the least productive—that in which mean fitness is lowest—to the most productive,

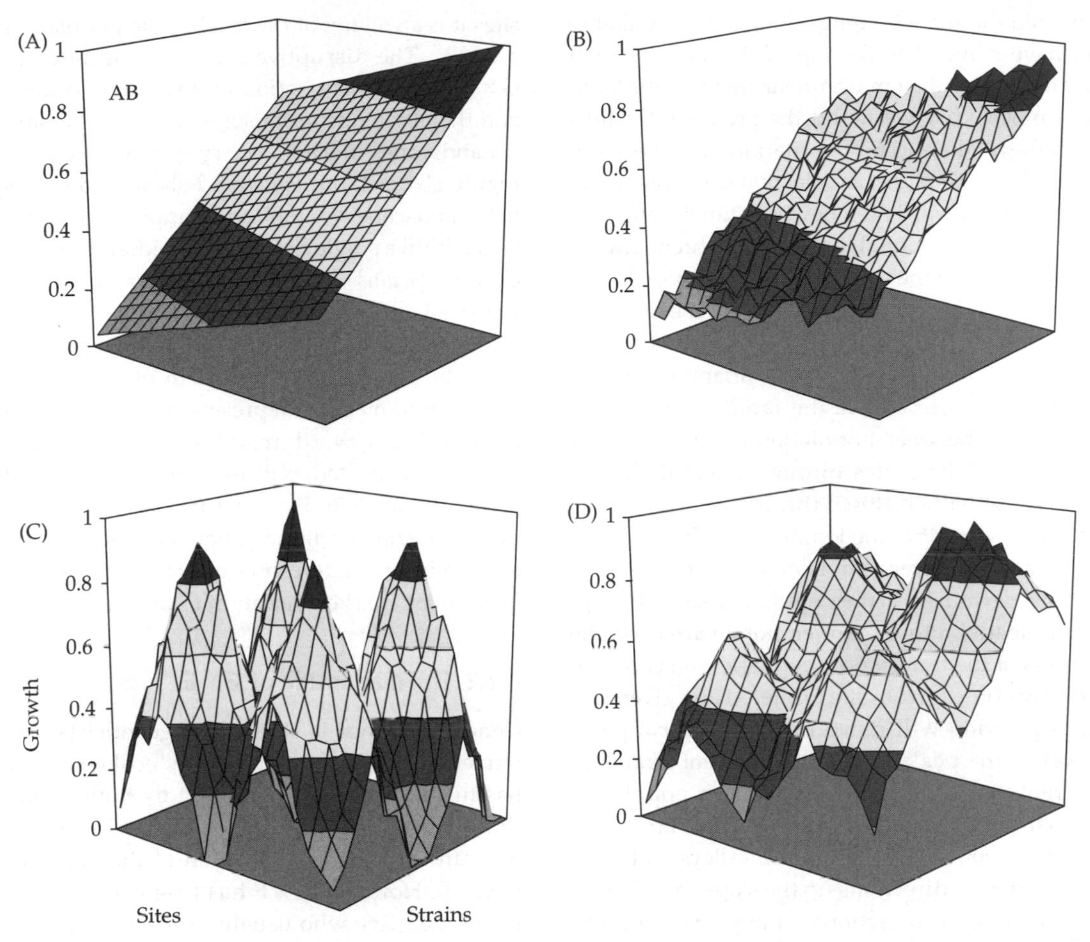

Figure 8.1 Ecogenetic landscapes. A. A purely additive landscape. B. Additive + random. C. Additive + G×E. D. Additive + G×E + random.

and the latitudinal relief is environmental variation. If we can arrange both strains and sites in this way, then the landscape will be a ramp, sloping smoothly upwards to the north-east. The same strain will be selected, regardless of the state of the environment, because the same strain is the most fit in all sites. However, it may not be possible to arrange either genotypes or environments in this way. If strains are ranked differently in different sites, then they cannot be unequivocally arranged from least to most fit. Consequently, rather than being a smooth ramp, the selective landscape will have a more or less complex topography of hills and valleys. This roughness of this topography is produced by the interaction between genotype and environment, with the height of the hills and the depth of the valleys representing the magnitude of G × E.

A landscape of this sort can be used to visualize how the population responds to environmental change in time or in space. Imagine that the environments are arranged in a temporal sequence, so that the landscape moves in a east–west plane past some line representing the present, like a conveyor belt. At any given time, the population occupies a small region of the map, represented by the prevailing strain and its variants. The fitness of the whole range of possible strains at this time is a south–north section through the landscape, with the hills and valleys seen in profile. If the environment has remained the same for long enough, the population will have climbed to the top of the hill, or one of the hills. When the environment changes, the landscape moves east–west, and the profile of the hills and valleys alters. The population now moves in the north–south plane, as selection shifts the population uphill towards the newly optimal state. This represents directional selection, which always tends to increase the mean fitness of the population. When adaptedness has been restored, mean fitness is maximized, and the population stands at the top of the hill, with a few mutational variants straggling down the slopes. This is the familiar process by which the fit between population and environment continually deteriorates through external change, and is continually restored through selection.

Now imagine that the longitudinal dimension is the range of different kinds of site that are simultaneously available to the population, so that it represents spatial rather than temporal variation. The hilliness of the terrain now expresses the variation of relative fitness from site to site. At any given site, the population will move uphill to a local peak. However, the peak represents different strains in different sites. Viewed from above, a population that is initially dispersed over the whole landscape would be seen as evacuating the valleys and moving in different directions up the slopes of different hills. Selection is directional at any given site, but because it acts in different directions at different sites it is *disruptive* at the level of the population as a whole. The disruptive selection engendered by G × E retards the fixation of any single genotype, and thereby tends to conserve genetic diversity.

Landscapes in evolutionary genetics are usually highly abstract, but Figure 8.2 shows a real ecogenetic landscape. We took soil samples at 10 m intervals within a single hectare of woodland, isolated a clone of *Bacillus mycoides* from each sample and also prepared a soil-water medium from it (Belotte *et al.* 2002). We could then estimate the ecogenetic landscape by measuring the growth of each clone in the batch of medium representing each site. There is a clearly a upward trend towards the north-east corner, as expected, but the surface is deeply dissected (Figure 8.2). Thus, natural isolates grown in quasi-natural conditions show consistent genetic and environmental responses that are nevertheless overlain by marked variation representing G × E.

8.2.2 The magnitude of G × E

Genetic variance is studied by geneticists; environmental variance is studied by ecologists; G × E has historically been neglected by both. There is thus no synthetic literature on the subject to which one can turn for an evaluation of the importance of G × E. However, G × E has long been of interest to agronomists, who usually wish to develop cultivars that will perform well over as broad a range

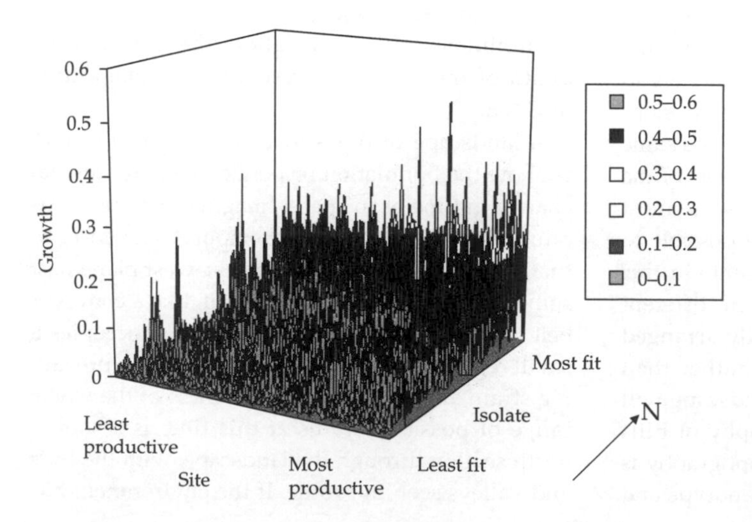

Figure 8.2 Adaptive landscape for natural isolates of *Bacillus mycoides*.

of environments as possible, and the agronomic literature is replete with accounts of trials in which a series of cultivars are grown at several localities for two or three years. Such trials are by far the largest available source of information about how genotypes respond to environments, and they provide very extensive documentation of the magnitude and generality of G × E. For a representative series of these trials, I surveyed the journal *Crop Science* from 1978 to 1986, and abstracted all papers from which satisfactory estimates of sources of variation can be obtained. The data can be summarized in one of two ways. The first is to calculate the three components of phenotypic variation: the genetic variance σ^2_G, the environmental variance σ^2_E, and the genotype–environment variance σ^2_{GE}. Methods for doing this can be found in any elementary book on parametric statistics. The proportion of the genotypic variance contributed by straightforward genetic variance is then $t_G = \sigma^2_G/(\sigma^2_G + \sigma^2_{GE})$, an intraclass correlation coefficient. If this is close to unity, genotypes perform consistently over environments, the best or worst in one place being the best or worst everywhere else; that is, adaptation is very general. If t_G has a value close to zero, then genotypes are very inconsistent, with fitness in one environment providing little if any information about its fitness in other circumstances; that is, adaptation is very specific. The survey yielded 269 estimates of t_G from about 80 studies, with a mean value of 0.602 (standard error 0.016). Further examination of the data tends to reduce this value. Thus, the 102 estimates from studies conducted over more than one year at more than one locality gave a mean value of 0.584. All these estimates refer to a very broad range of agronomic characters, from seed weight to milling quality; restricting the analysis to 30 or so studies dealing exclusively with seed yield, the character closest to Darwinian fitness gives a mean value of 0.35. Moreover, these estimates are not very sensitive to the nature of the source material: random selfed lines and outcrossed populations give similar results. In short, some 40% or more—say a half—of genotypic variance, especially for characters close to Darwinian fitness, is attributable to the variation of relative performance among environments.

Another way of analysing crop trials is to compute interclass genetic correlations, As an example, the largest single survey that I have found was a yield trial conducted by Campbell and Lafever (1980) involving 19–30 cultivars of winter wheat scored at 12 localities over 7 years. How can we express the degree to which the relative fitness of these genotypes, measured as their seed yield, varied among the environmental conditions prevailing at different times and places? Perhaps the most straightforward way is to consider the variation of a series of genotypes in just two environments. The growth of a genotype in one environment could then be plotted against its growth in the other. This graph would then represent a genetic correlation; if the relative fitness of genotypes is similar in the two environments the correlation is high, whereas if relative fitness in one environment is unrelated to relative fitness in the other the correlation will be zero, and if high fitness in one environment is consistently associated with low fitness in the other the correlation will be negative. The advantage of this approach is that the genetic correlation coefficient is related directly to the correlated response of one character to selection applied to another. It amounts to considering the different rates of growth expressed in the two environments as being two different characters; and more generally, any attribute measured in several different environments can be treated as though it were as many different characters. The problem of how populations evolve in an heterogeneous environment can then be viewed in the wider context of how selection acts on several different characters simultaneously. In Campbell and Lafever's trial, the average of the 464 genetic correlations that can be calculated from their data was 0.29; half the estimates fell between 0.09 and 0.51, and 15% of them were negative (Figure 8.3). The relative fitness of the genotypes they studied thus varied extensively from place to place, and from year to year, giving the impression of a selective landscape with a very rough and broken topography. A second survey, from the same literature source, turned up 12 other papers that supplied 255 estimates of genetic correlations between environments, either localities or years. These did not overlap with the first data set.

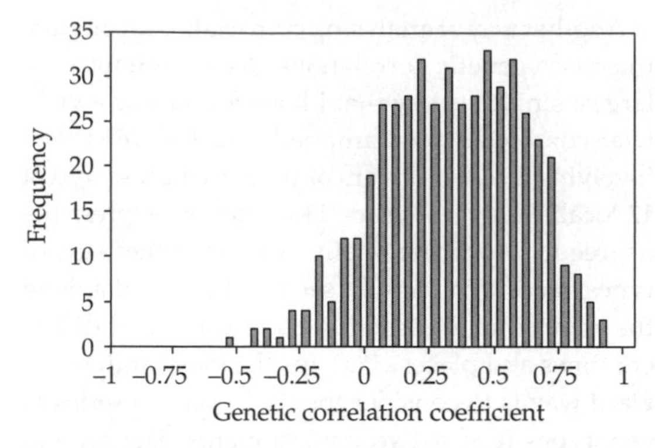

Figure 8.3 Distribution of genetic correlation of years and localities for grain yield in wheat. *Source*: Campbell and Lafever (1980).

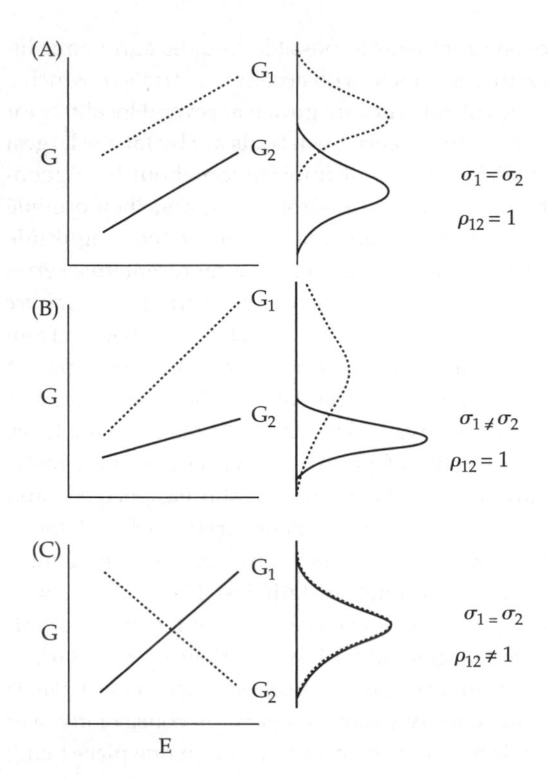

Figure 8.4 Components of G×E. A, no G×E. B, G×E caused by variation in responsiveness. C, G×E caused by inconsistency.

The mean value of the genetic correlation coefficient was 0.486 (standard error 0.023).

There are few comparable studies in non-agronomic situations. However, I have made extensive measurements of the growth of *Chlamydomonas* in the laboratory that were intended to supplement the results from crop trials (Bell 1991a, b, 1992a). I used either clonal isolates of different species, or spores (clones) isolated from crosses within the single species *Chlamydomonas reinhardtii*. These were grown with different combinations of nitrate, phosphate, and bicarbonate concentrations, so that genetic correlations could be calculated between media differing in one, two, or all three of these macronutrients. For both the limiting rate of increase r_{max} and the asymptotic density K (see §10.1.5), genetic correlations lay between 0 and +0.5; when two or three components of the growth medium were adjusted at the same time, the genetic correlation among environments was only about +0.1 or less.

8.2.3 Inconsistency and responsiveness

The variance of a genotype over environments confounds two sources of G × E that it is interesting to separate (Falconer 1952, Robertson 1959a). To illustrate the approach, Figure 8.4 shows how two genotypes might respond to a range of environments. If there is no G × E these reaction norms are parallel, implying both that the same amount of genetic variance is expressed in each environment ($\sigma^2_{1G} = \sigma^2_{2G}$) and that the cross-environmental genetic correlation is perfect ($r_{1G2G} = +1$). Either condition may fail. In the first place, the genotypes may differ more in some environments than in others, so that the rate of response to selection will vary among environments. This factor can be called *responsiveness*. Secondly, the cross-environmental genetic correlation may be imperfect, which can be called *inconsistency*. If the reaction norms cross, genetic correlation is negative and relative fitness changes with conditions of growth, so that inconsistency leads to differences in the direction of selection among environments. The overall quantity of G × E can be decomposed into two components, one representing responsiveness and the other inconsistency (Robertson 1959b):

$$\sigma^2_{GE} = ½(\sigma_{1G} - \sigma_{2G})^2 + \sigma_{1G}\sigma_{2G}(1 - r_{1G2G}).$$

Inconsistency is the more important component, because it governs the evolution of specialization.

If fitness is positively correlated across environments then some genotypes will have high fitness in all environments, and these generalists will be selected. If the cross-environmental correlation is negative, on the other hand, any genotype with superior performance in some environments is likely to be inferior in others, so that the best available genotypes are specialists. Thus, specialization is not promoted merely by the quantity of G × E in relation to genotypic variance, but also by the quantity of inconsistency relative to G × E. For random *Chlamydomonas* strains, inconsistency constituted 80–90% of G × E (Bell 1990a). This reflected genetic correlations that were modest but nevertheless almost all positive.

8.2.4 The genetic correlation in relation to environmental variance

The main axis of the argument so far has been that environments are heterogeneous, and that this heterogeneity causes variation in relative fitness. However, it will be very difficult to use these facts to interpret populations unless there is a regular relationship between the extent of heterogeneity, measured as the variation in average fitness of a given set of genotypes, and the extent of variation in relative fitness. If the most trifling environmental shift may often cause a complete revolution in the ranking of genotypes, then it will be impossible either to predict or to explain evolutionary change in terms of environmental change. Rather, we might imagine that a very slight change in the environment will cause only very slight alterations in relative fitnesses; and that greater environmental change will cause progressively more change in relative fitnesses. This amounts to saying that G × E should become larger and more inconsistent as environmental variance increases. In crop plants and in *Chlamydomonas*, I have found that the genetic correlation among environments decreases linearly as environmental variance increases (Figure 8.5). In very similar environments it is nearly unity; in the most dissimilar environments represented in the trials, it approaches zero. A similar pattern was displayed by clonal ramets of species of *Carex* implanted into an old-growth forest (Bell *et al.* 2000). Conversely, for any given set of environments we would expect closely related types to have similar ecological reactions whereas distantly related types might respond quite differently and display more G × E. Taking both genetics and ecology into consideration, closely related types growing in similar conditions should express large positive genetic correlations, which should decay towards zero as the relatedness of types or the similarity of conditions lessen, and eventually become negative beyond some combination of genetic and ecological distance. We have used a phylogenetically ordered set of *Chlamydomonas* species to demonstrate this pattern experimentally (Kassen and Bell 2000) (Figure 8.6).

These large-scale assays show that the growth and fitness of arbitrary strains or species varies widely among conditions of culture: G × E is usually substantial, with a large contribution from inconsistency reflecting low or even negative cross-environmental genetic correlations. In general, any given genotype does not possess some fixed fitness or selection coefficient, but rather may be superior, inferior, mediocre, or neutral depending on circumstances. This inconsistency is an essential part of the basic theory of selection—when the environment changes, the relative fitnesses of genotypes changes, causing some that were previously rare to increase in frequency through selection. However, change in this model is rare, and at any particular time a genotype expresses a fixed fitness in a uniform environment. This representation of the world, though convenient for many purposes, is evidently wrong. Instead of a world in which populations that have reached equilibrium under selection are occasionally disturbed by some abrupt perturbation, the surveys that I have described suggest that populations of all kinds of organisms, regardless of the scale on which they live, are kept in flux by continual change in the force and direction of selection. This is a very important change in perspective.

8.2.5 The outcome of selection in different environments

To the extent that the phenotype expressed by a given genotype varies with circumstances,

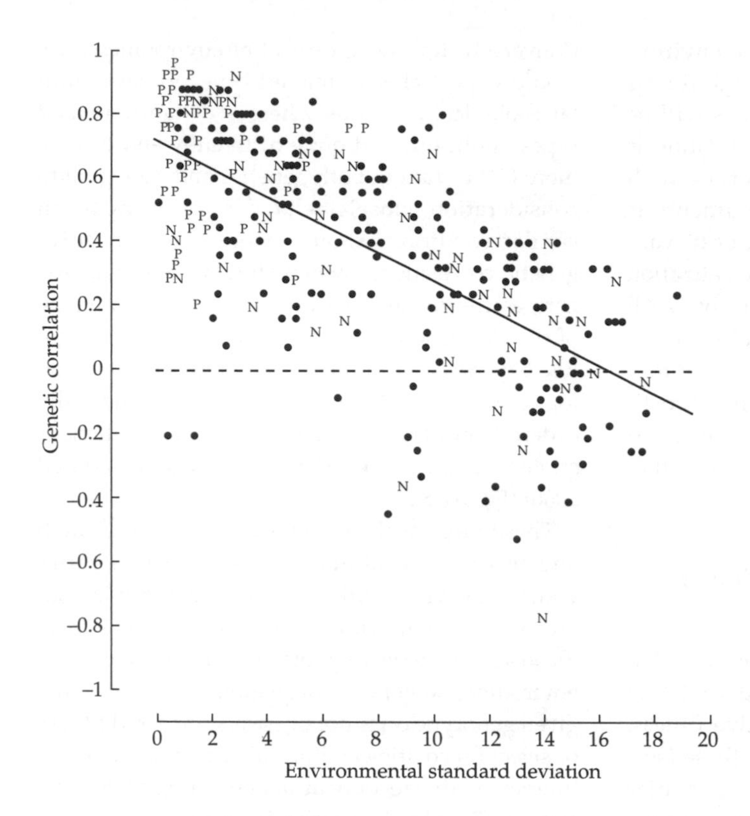

Figure 8.5 The relationship between genetic correlation and environmental variance. The genotypes were 12 species of *Chlamydomonas*. The environments used varied in the concentration of nitrate (N), phosphate (P) or both (filled circles). From Bell (1992a).

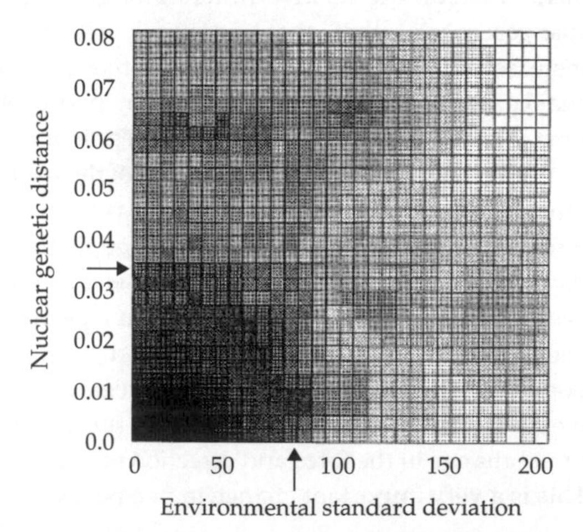

Figure 8.6 The genetic correlation in relation to the similarity among genotypes and environments. Dark regions represent positive correlations; light regions are negative correlations. The arrows mark the points at which the isocline of zero genetic correlation crosses the environmental and genetic axes. From Kassen and Bell (2000).

selection for a given character in different environments will favour different genotypes. In the case of artificial selection, this implies that the response of a character to selection will depend on the conditions of culture; when the character is fitness, natural selection will produce different results in different environments.

This argument leads to an important practical question: in selecting improved strains of crop plants or livestock, how should one choose the environment in which selection takes place? Falconer (1960) investigated this problem by selecting for body size in mice, as a model or surrogate for attempts to increase carcass weight in larger mammals. The mice were reared either on the normal laboratory diet, or on the same food diluted with a large quantity of indigestible fibre. There were thus two characters being selected in different lines: 'body size on a high plane of nutrition', and 'body size on a low plane of nutrition'. Both characters were selected upwards and downwards for about a dozen generations, and, as expected, there was in every case a direct response to selection in the appropriate direction. It is the correlated responses to selection that were the main point of the experiment. We can ask three questions about these responses.

First, is the outcome of selection in either environment completely reproducible in the other? If this is the case, the environments are not really different, with respect to evolutionary change, and it does not matter which of them we choose to select in. This is equivalent to saying that the genetic correlation between the two environments is unity. This was not invariably the case. The correlated responses to selection for increased growth on high plane (i.e. growth on high plane of the line selected for increased growth on low plane) and for reduced growth on low plane (i.e. growth on low plane of the line selected for reduced growth on high plane) were indeed comparable with the direct responses. However, although selection for increased growth on the low plane produced only a modest direct response, there was no consistent correlated response (i.e. the line selected for increased growth on a high plane showed no response when tested on the low plane). Moreover, there was a pronounced direct response to selection for decreased growth on a high plane, but little if any correlated response (i.e. the line selected for decreased growth on a low plane showed no response when tested on a high plane). The response to selection in one environment is thus not necessarily reproducible in another environment.

The second question is whether the correlated responses to selection can be predicted, by treating rates of growth in different environments as different characters and estimating their genetic variances and covariances (§ 7.3.7). This is similar to asking whether the direct response can be predicted from estimates of genetic variances, and the answer is much the same. For the first few generations, the genetic correlation is a good predictor of the correlated response, but it is of little value in later generations, presumably because the genetic variances and covariances involved have themselves changed under selection.

Finally, we can ask which of the two planes of nutrition is preferable as an environment for selection. The answer depends on the direction of selection. For upward selection the low plane is better, because the direct response achieved in this environment was reproducible in the other environment. Conversely, if the object is to reduce body size would be better to select on the high plane.

Harlan and Martini (1938) mixed or crossed 11 barley cultivars to create diverse populations that were sown in several different widely separated localities spanning the continental USA. For about a decade in the 1920s and 1930s each population was harvested without conscious selection to collect seed for re-sowing, while at the same time monitoring its genotypic composition . Some localities were used for only 4 or 5 years, but longer-term records covering all 12 years 1925–1936 are available for stations in Montana, Idaho, and California. All the cultivars could grow reasonably well as pure stands at all the localities, but when they were sown together as a mixture their frequencies changed rather rapidly through selection. Some cultivars seemed to be unequivocally inferior competitors, and were quickly reduced to low frequencies at all localities. Others were more successful, but no cultivar increased in frequency at all localities; rather, the genotypic composition of the population changed in different directions in different populations. Composite Cross I was created by making 32 crosses among the 11 parental cultivars, and was sown at the same localities as the mixture. Although its varietal composition can no longer be discerned directly, the frequency of distinct characters can be scored instead. Again, selection produced different results at different localities. *Black seed coat*, for example, was evidently nearly neutral at most localities, remaining in the population at about the same frequency for up to 12 years; at the California station, however, it declined from about 10% to about 1% within 5 years. *Two-rowed* plants increased in frequency somewhat at the Idaho and Montana stations, from about 20% to about 30%, but decreased to 5% in California; *smooth-awned* plants doubled in frequency in California and Montana, but showed no change in Idaho. All of these characters are attributable to single Mendelian genes, although of course selection acting at linked loci may have been responsible for some of the changes. These experimental populations of barley provide the best evidence that the G × E shown by cultivars when grown in isolation translates into differences in the direction of change in gene frequency within populations grown in different environments.

Schizophyllum is a basidiomycete in which growth can be measured as the linear extension of hyphae in solid medium (Jinks and Connolly 1973, 1975). High and low growth rate were selected by choosing the 2 most extreme of 50 sexual progeny of the previous cycle. Both upward and downward lines were established at 20 °C and 30 °C; the growth of unselected lines is considerably greater at 30 °C, which is thus the more favourable environment for growth, corresponding to the high plane of nutrition in Falconer's experiment. Despite the brevity of the experiment, the lines evolved specific adaptation to 20 °C and 30 °C conditions (Figure 8.7). Thus, at 20 °C the lower of the two low lines and the higher of the two high lines were those selected at 20 °C, while the lines selected at 30 °C were the more extreme when tested at 30 °C. The effect was nearly symmetrical, in that the difference between the direct and correlated responses was about the same at 20 °C and at 30 °C, in proportion to the growth of unselected lines.

By testing the selection lines over a range of temperatures it was possible to make a point that is often neglected, that while adaptation may be specific to a particular environment it is not confined to that environment. Selection at 20 °C specifically causes adaptation to 20 °C; but this does not imply that at 20.1 °C the line would be equivalent to an unselected line, or to a line selected at 30 °C. The effects of adaptation will be reproducible to a greater or lesser extent in similar environments. To investigate the breadth of adaptation associated with selection in a specific environment, the lines were tested at 2.5 °C intervals between 15 °C and 35 °C after 8 cycles of selection (Figure 8.8). The high line selected at 20 °C had a higher growth rate than the unselected control at all except the highest temperature; it exceeded the 30 °C line between 15 °C and 25 °C, but the 30 °C line grew faster at temperatures of 27.5 °C or more. If this were not the case, then the effects of selection could not cumulate, because any

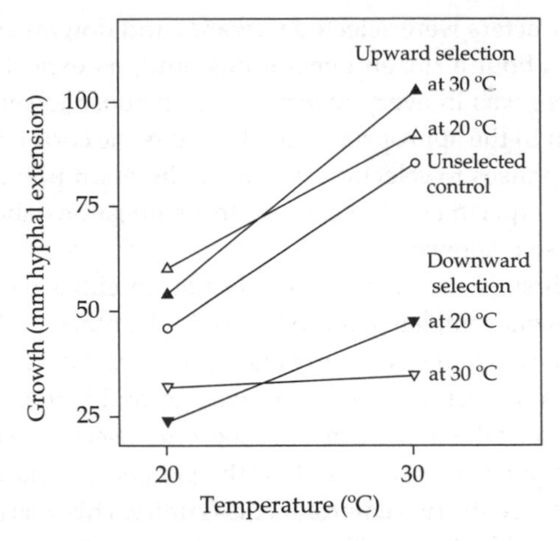

Figure 8.7 Direct and correlated responses to selection. This diagram shows the growth of *Schizophyllum* selected upwards and downwards at two temperatures. The lines drawn here to represent the character value of a genotype, or a selection line, over a range of environments is sometimes called a *norm of reaction*, or *reaction norm*. The direct response is the difference between a selection line, assayed in the environment of selection, and the unselected control. For example, the direct response to downward selection at 20 °C is 24.25 – 46.25 = –22 mm. The correlated response is the difference in growth at a given temperature between the response of the line selected at the other temperature and the unselected control. Thus, the correlated response at 30 °C to downward selection at 20 °C is (32 – 46.5) = –14.25 mm. The full pattern of direct and correlated responses was as follows:

response to selection		direct	correlated
upward selection	at 20 °C	+14.5	+7.25
	at 30 °C	+17.75	+8.25
downward selection	at 20 °C	–22	–37.75
	at 30 °C	–50.25	–14.25

Thus, the correlated response may exceed the direct response. This is because the reaction norm of the unselected control is not parallel to the reaction norms of the selection lines: selection is more effective in increasing growth at 20 °C and more effective in reducing growth at 30 °C. However, it is useful to distinguish between the correlated response of lines (in a given test environment) and the correlated response in environments (of lines selected in another environment). In all cases, the line selected at a given temperature deviated more from the control than did the line selected at the other temperature. That is, in either environment the direct response (of the line selected in that environment) always exceeded the correlated response (of the line selected in the other environment). For example, at 30 °C the direct response was –50.25 mm, and the correlated response –37.75 mm. This is why the reaction norms cross; this represents genotype-environment interaction caused by selection. From Jinks and Connolly (1973; see also Jinks and Connolly 1975).

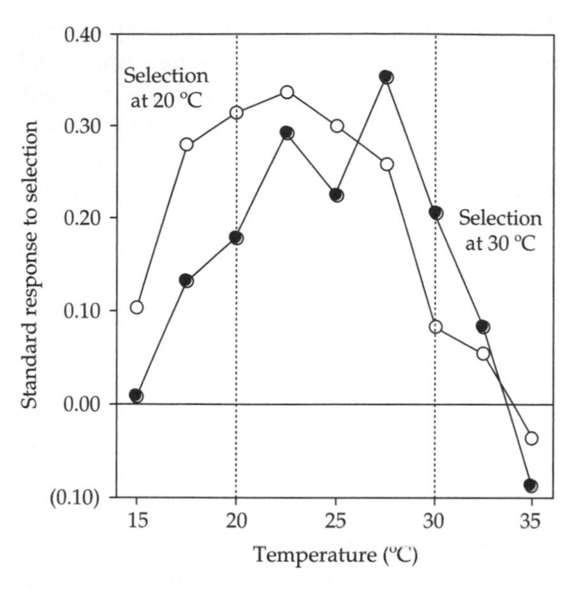

Figure 8.8 Response of lines differently selected over the whole range of environments. From Jinks and Connolly (1973).

trifling change in the environment would cause selection for a different set of genotypes.

The evolution of growth at different temperatures has been studied with larger populations over longer periods of time by using natural selection in bacterial systems. The *E. coli* lines discussed in §3.3 had been maintained in the laboratory for 2000 generations, and had by this time become reasonably well adapted, both to general laboratory conditions and to a temperature of 37 °C. Two sets of lines extracted from this population were then grown for a further 2000 generations, one at 32 °C and one at 42 °C. The effect of selection at a particular temperature, on adaptation over a range of temperatures, could then be assessed by competing the selection lines against their common ancestor. The results were similar to those obtained with *Schizophyllum*: the 32 °C lines had greater fitness at temperatures around 30 °C, and the 42 °C lines at temperatures above 40 °C (Bennett *et al.* 1992).

8.2.6 Stability

A population may become locally adapted to any feature of the environment, even a feature as exotic as severe lead pollution. However, one universal source of environmental heterogeneity is variation in productivity from site to site. I have already emphasized that the growth of a test organism is itself the most appropriate measure of the state of the environment. The average growth of a particular set of genotypes thereby defines the value of the environment at a given site, supplying a common scale that enables us to compare sites within any sort of environment. The behaviour of a genotype can then be represented by plotting its growth at a series of sites as a function of their environmental value (Figure 8.9). This graph is usually linear, and can be used to partition the environmental variance expressed by a genotype into two components. The first is the variance in growth of a particular genotype that is accounted for by variance in average growth, or in other words that part of its environmental variance that is attributable to the regression of genotypic value on environmental value. The second part is the variance in growth expressed by a particular genotype at a fixed environmental value, because of factors that affect the relative fitness of genotypes without affecting mean fitness; this part of its environmental variance is represented by deviations from the regression line. This regression analysis of G × E is useful because it isolates the response to one source of environmental variance—the availability of resources for growth—that can be studied in any organism in any kind of environment.

The slope of the regression represents the responsiveness of a genotype to improved conditions of growth. A shallow regression slope characterizes an unresponsive, or plastic, genotype, whose growth is not consistently affected by a general amelioration of the environment. This might be merely an unconditionally inferior genotype that is unable to grow well anywhere; or it might grow well in some conditions and poorly in others, but in a way that cannot be explained by variation in the general conditions of growth, in which case the regression analysis is uninformative. More interestingly, it might be a genotype that grows relatively well in unproductive conditions, but relatively poorly at more productive sites. Because its fitness does not vary much with general productivity, it might be regarded as a generalist that grows equally well anywhere. However, this is not necessarily the case, because its growth may vary widely with features of the environment that are not consistently related to overall productivity. It is better interpreted as a type that is specialized for growth in unproductive sites. Conversely, a steep regression slope characterizes a highly responsive type that grows poorly in generally unproductive sites, but is able to exploit productive sites more effectively. Where we can recognize these two types of specialist, their regression lines will cross at some intermediate environmental value at which they have equal fitness. A generalist will exhibit nearly average growth at all sites, being inferior to one specialist and superior to the other. The regression of its growth on environmental value, which is defined as the mean growth of all genotypes, will therefore have a slope of unity. Consequently, provided that we exclude genotypes that are everywhere inferior, or everywhere superior, we can define a type specialized for growth in poor conditions as one whose regression on environmental value has a slope of less than unity, and a specialist for rich conditions as a type whose regression slope is greater than unity.

Hillesheim and Stearns (1991) selected for body size in *Drosophila* in rich and poor growth environments. Within a single family, the ratio of the body sizes of individuals tested in rich and poor environments is an estimate of plasticity; a high ratio is equivalent to a steep regression slope, indicating a responsive genotype, whereas a low ratio indicates stability. By selecting the families with the largest and the smallest ratios in each generation, they produced lines that were more or less responsive to environmental variation. Phenotypic plasticity is therefore a character that can evolve through selection in the short term. Falconer's experiments with mice, and Jinks' experiments with *Schizophyllum*, had the same basic design, measuring the direct and correlated responses of growth rate in two environments, one of which was generally more favourable to growth than the other. The Hillesheim–Stearns experiment, besides involving family selection on plasticity, also involved individual mass selection on body size in favourable and unfavourable environments. All three experiments yielded similar results: lines that were selected for increased growth in the more favourable environment grew faster in that environment than did lines selected in the less favourable environment, but they grew less well in the less favourable environment than did lines selected in that environment. Thus, the environmental variance of lines selected for increased growth in a favourable environment was greater than that of lines selected for increased growth in an unfavourable environment. In terms of the regression analysis of stability, the regression slope is steeper for the lines selected in the favourable environment, indicating that they were more responsive to environmental variation. The converse was true for lines selected for decreased growth: lines selected in the unfavourable environment had greater environmental variance, or in other words were more responsive, than lines selected in the favourable environment. We may summarize these results by saying that selection will favour individuals that are highly responsive to environmental variation when selection is applied in the same direction as the general effect of the environment.

This will be the outcome of selection whenever the correlated response is not as great as the direct response in either environment, as will usually be the case. Jinks interpreted this pattern in genetic terms as follows. Selection for high growth rate at 30 °C will fix some alleles that increase growth regardless of temperature. However, it will also favour alleles that have antagonistic effects on growth at high and low temperature, increasing growth at 30 °C while decreasing growth at 20 °C, and will thereby increase the responsiveness of growth rate to temperature. Selection for high growth rate at 20 °C will likewise fix alleles that have the same effect on growth at all temperatures; and will also favour alleles that increase growth at 20 °C while reducing it at 30 °C, thereby reducing the responsiveness of growth to temperature. This interpretation posits two classes of genes. One includes genes that affect growth to the same extent in all environments, will contribute equally to the direct and correlated responses to selection, and do not affect plasticity. The other includes genes whose effect varies according to conditions, producing an antagonistic correlated response in conditions that are sufficiently different from the conditions in which selection is carried out, because of some unidentified source of functional interference (§ 8.3.4). The combined effect of these two kinds of genetic effect is that the correlated response will often be in the same direction as the direct response (because of genes in the first category), but will be less pronounced (because of genes in the second category). It follows that selection in more favourable environments will increase responsiveness and decrease plasticity, whereas selection in more stressful environments will decrease responsiveness and increase plasticity. Selection for reduced growth rate will produce the same results, bearing in mind that 30 °C is the more stressful and 20 °C the more favourable environment for the expression of low rates of growth.

Growth was measured in each generation of selection in the *Schizophyllum* experiment, so that the joint evolution of mean growth rate and responsiveness can be followed. Regardless of the environment in which selection was carried out, responsiveness declined in lines selected for low growth rate, and increased in lines selected for high growth rate. However, responsiveness declined more steeply in lines selected for low growth at 30 °C, and increased more steeply in lines selected for high growth at 30 °C. This is the clearest experimental demonstration that the extent of plasticity or responsiveness depends on the environment in which selection is practised, and that plasticity can be increased or reduced in a predictable manner as a correlated response to selection in different kinds of environment.

8.2.7 Evolution of stability and responsiveness

The *Schizophyllum* experiments also included lines that were selected on the basis of their average performance at 20 °C and 30 °C. Not surprisingly, these lines evolved intermediate growth rates when tested at either temperature. Their plasticity and responsiveness were therefore also intermediate: when selected upwards, they were less responsive than the 30 °C line but more responsive than the 20 °C line, and when selected downwards this relationship was reversed. These lines were thus more stable than either the 20 °C or the 30 °C lines, in the sense of showing less G × E, and therefore a regression slope of close to unity.

In view of the great interest in broad adaptability among agronomists, it is surprising that there does not appear to have been any attempt to select for stability in crop plants, beyond choosing the more stable varieties from a trial. There are, however, several descriptions of how stability changes during the course of a conventional programme of selection for increased yield (Figure 8.9).

The selection of new crop varieties is generally carried out at an experimental station, and the evolved lines then tested on farms under more or less normal agricultural conditions. Experimental lines are often grown as spaced plants, carefully nourished and protected, so that conditions of growth are usually more favourable at the experimental station than on the farm. Upward selection for yield is then expected to increase responsiveness. This has been demonstrated in several experiments with maize. For example, Mareck and Gardner (1979) increased grain yield by about 14% over 15 cycles of mass selection at the Nebraska experimental station. The base population was a cultivar called Hay's Golden, an open-pollinated variety developed during the drought years of the 1930s that is said to be resistant to high temperature and low water availability. When tested under modern conditions, it is low-yielding and plastic, as expected: it yields reasonably well in the poorest environments, but shows little response to improved conditions. The selection lines are much more responsive: they are about equally productive, or perhaps a little less productive, in the poorest environments, but yield 20–25% more in the most favourable environments. Selection with the grain of the environment, so to speak, thus produces less plastic, more responsive genotypes, as we expect from elementary theory and the experiments with mice and *Schizophyllum*.

It follows from this argument that yield and responsiveness will usually be positively correlated, provided that selection for yield is practised in favourable environments. To put this in another way, plasticity will decrease as a correlated

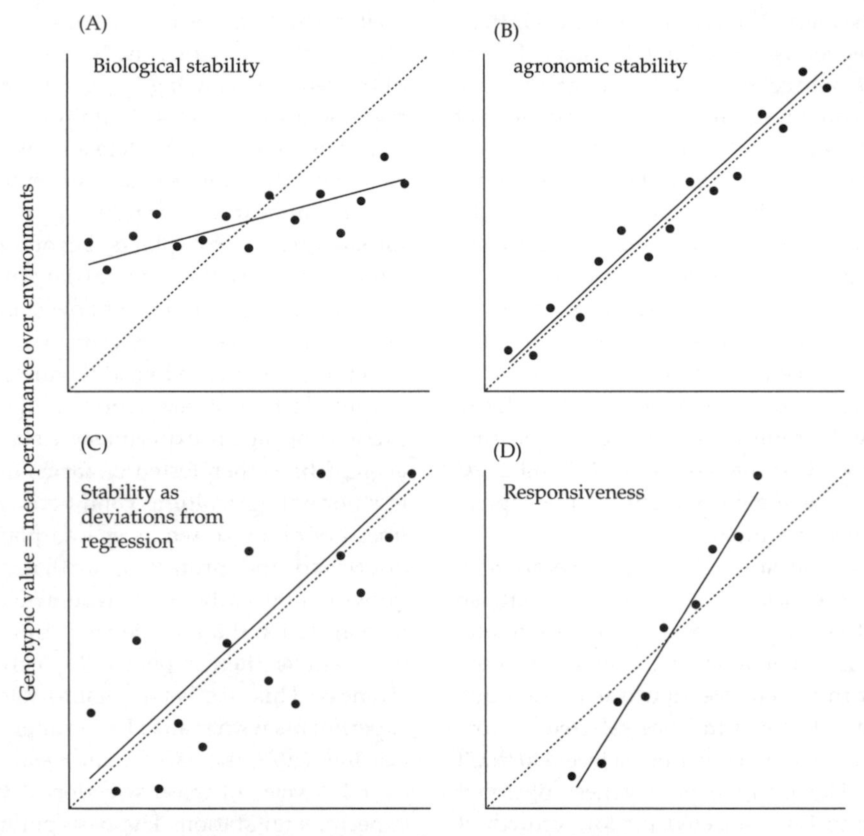

Figure 8.9 Phenotypic stability. These figures illustrate the types of response that might be shown by different cultivars in a crop trial conducted at several localities or in several years. Environments may be evaluated by the mean performance of all genotypes; by the mean performance of all genotypes except the one whose response is being scored; or by the mean performance of a separate series of check cultivars. Four types of genotypic response may be found. A, low environmental variance, giving a shallow regression slope; B, low G×E variance, giving a regression slope close to unity; C, a pronounced scatter of genotypic values around the regression line, showing that G×E is not consistently related to the mean environmental value; D, high environmental variance, giving a steep regression slope.

response to selection for increased yield. This is the general finding from agronomic trials, and is attributable to the superior management usually practised at experimental stations. Falconer (1990) has pointed out that increased responsiveness will be a very general consequence of directional selection, even when lines are tested in the same environment that they were selected in. The conditions of growth experienced by individuals in a population will always vary somewhat, even when they are raised in the laboratory, or at a single location, and those that are selected will tend to be those that have experienced more favourable conditions (with respect to the expression of the character under selection) and are able to respond to them. Under stabilizing selection, on the other hand, individuals that express phenotypes close to the optimal value are likely to be those that are insensitive to environmental variation. Responsiveness should therefore increase under directional selection and decrease under stabilizing selection.

Agronomists have long argued whether selection should be practised under favourable or stressful conditions. On the one hand, favourable conditions might facilitate selection by allowing a greater range of heritable variation to be expressed; on the other hand, most farm environments are stressful in at least some respects. It will be evident from the preceding sections that no general answer can be given. Selection in favourable conditions will increase responsiveness, maximizing yield in good years, at good localities or on well-managed farms; selection in stressful conditions will increase plasticity, minimizing losses in poor years and broadening the range of localities in which a cultivar can be profitably grown. Which procedure is preferable is determined by the subjective priorities of the breeding programme. One might add that any attempt to select in stressful environments must take into account the possibility that plasticity with respect to one source of stress is not necessarily expressed when a different stress is experienced. The only sound general rule is that selection is best practised in the environment where the plants will eventually be cultivated, because the direct response to selection will almost always exceed the correlated response. (This advice is not very practicable if it requires selecting independently at a great number of localities; and it cannot be applied to yearly variation at all.) Where the crop is grown in a range of environments, it is best to select for average performance over this range, if the object is to increase mean yield. The *Schizophyllum* experiments suggest that this will also enhance stability. If it is impracticable to select in several environments, the selection environment should approximate the average conditions that will be experienced by the commercial crop.

8.3 Specialization and generalization

8.3.1 Niche separation

This chapter is mostly about the ecological theory of adaptive radiation, and neutral variation will not be discussed at length. Schluter (2000) has written a synthetic account of the ecology of adaptive radiations that should be consulted for a more detailed treatment of the field. The theory makes three general propositions. The first is that most environmental factors have an optimum: increasing values are beneficial up to a certain point and deleterious thereafter. Resources and the conditions required to use resources (such as temperature and water availability) perhaps always follow this rule: beyond certain limits, above and below the optimum, the net reproductive rate is negative and populations are unable to persist. Secondly, these optima vary among species (or strains): in any given conditions, the birth and death of individuals depends on which species they belong to. This is not as clearly or universally true; the alternative is the neutral theory. Each species has an optimal value and a characteristic range of tolerance for any given factor, and taking all factors together defines a multidimensional space that constitutes the fundamental niche. Thirdly, the sites actually occupied by natural populations (the realized niche) correspond approximately to the fundamental niche because populations are more likely to persist where their net rate of reproduction is greater. In this way, a continuous process of sorting moulds the community to fit the structure of the landscape. This is most likely to be true when the differences between species or strains are so pronounced that the field gradient (§7.2.4) is steep. Temporal change or dispersal, however, may loosen the fit so much as to make it difficult to detect.

Niche separation may arise through two processes. The first is divergent selection, with populations descending from a common ancestor becoming adapted to different conditions of growth and thus evolving different fundamental niches. The location of the fundamental niches in multidimensional environmental space that is expressed by a community then reflects the pre-existing opportunities in the landscape it occupies. The second process is competition, by which the realized niche of a species is moulded through its interaction with other species. This distinction is not very easy to maintain in practice, but is adopted so that this section on divergent selection follows naturally from previous discussion of directional selection and adaptation. The consequences of competition are described in the chapter on social selection.

8.3.2 The cost of adaptation

When it encounters changed conditions a population is likely to adapt; but what if they should change back again? If ancestral population was well adapted to the ancestral environment then its descendants are unlikely to have been further improved by selection in a different environment. Clearly, we expect the descendants to have the same or a lower fitness when cultured under ancestral conditions. Labelling the ancestral and descendant populations as Anc and Des, and the corresponding environments by *Anc* and *Des*, the direct response to selection is $R = w(\mathrm{Des}, Des) - w(\mathrm{Anc}, Des)$, which we expect to be positive. The indirect response is $\mathcal{R} = w(\mathrm{Des}, Anc) - w(\mathrm{Anc}, Anc)$. If $\mathcal{R} < 0$ there is a cost of adaptation in terms of impaired growth in the ancestral environment, as illustrated in Figure 8.10. This can be expressed in terms of the genetic correlation of ancestor with descendant in their two environments, which will be negative if the indirect response is negative. Without a cost, the successful lineage would evolve superiority over an expanding range of conditions until it was well adapted to any turn of events and evolution would stop. The existence of a cost of adaptation puts a limit on the evolution of plasticity and thereby creates the necessity for continued adaptation in changing environments.

The cost of adaptation was first demonstrated in Dallinger's primordial experiment (§3.1.1). After years of selection, he had obtained microbial strains capable of growing at 70 °C: when these were exposed to the original temperature of 15 °C they all died. This is the extreme case, in which $w(\mathrm{Anc}, Des)$ and $w(\mathrm{Des}, Anc)$ are both zero.

Chlamydomonas normally grows autotrophically by photosynthesis in illuminated cultures, but it will also grow heterotrophically in the dark on medium supplemented with acetate. Lines maintained in the light for several hundred generations show no improvement, being already well adapted for photosynthetic growth. They rapidly become better adapted to growth in the dark, but at the cost of reduced growth in the light (Bell and Reboud 1997) (Figure 8.11). In this case both ancestor and descendant populations were viable in both environments, but the indirect response (regress in the light) was about as great as the direct response (advance in the dark), so the cost of adaptation was considerable.

8.3.3 Divergent specialization

Adaptive radiation involves divergent specialization: lines selected in different environments not only become different from their common ancestor

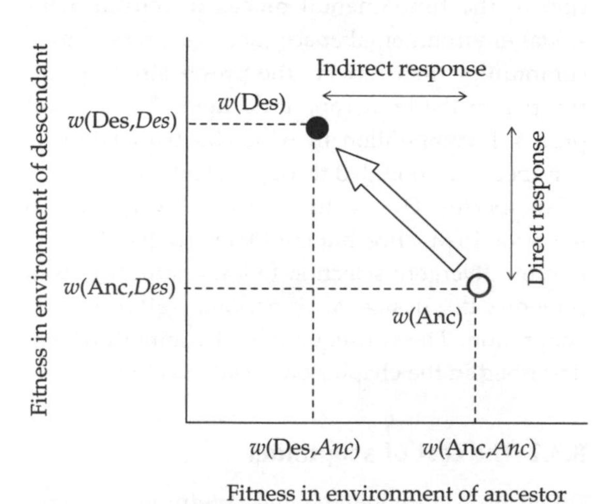

Figure 8.10 The indirect response to selection and the cost of adaptation.

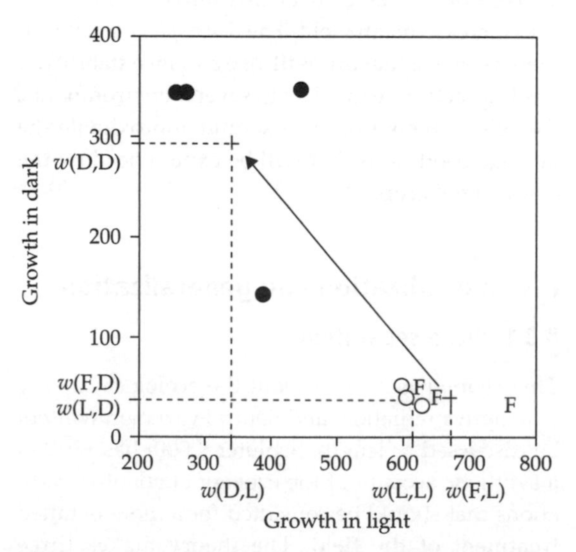

Figure 8.11 Response to selection for growth in light (L) and dark (D) in *Chlamydomonas.* Open circles, spores from line selected in the ancestral Light environment; filled circles, spores from line selected in Dark environment; F is the stored founder (ancestor). From Bell and Reboud (1997).

but also become different from one another. Suppose that the ancestral population is introduced to two 'sites' with different conditions of growth that are completely isolated from one another. The sites may be thought of as localities, or simply as conditions of growth resulting from different substrates, physical factors, or toxins. The outcome of selection might be a single superior type capable of flourishing in both sites. This would be obstructed if the indirect effect of selection in one environment is to reduce fitness in the other, that is, if there is a cost of adaptation, assessed in this case relative to an alternatively adapted specialist rather than the ancestor. We can use this concept to describe how organisms adapt to complex environments by an extension of the approach followed in previous sections. Characters such as (say) fecundity and wing length are clearly distinct, because they are measured in different ways, and these measurements are imperfectly correlated. The nature and magnitude of the correlation between them directs their joint evolution. However, fecundity need not be regarded as a single monolithic character; in the preceding sections, values of fecundity at different ages have been treated as different characters. The correlation between such components of fitness then directs the evolution of the life history. In a similar way, we can treat the values of fecundity (or of any other component of fitness) expressed in different conditions as different characters (Falconer 1990). The response of a population to a complex environment can then be described in terms of the nature and magnitude of the correlations between the site-specific values of components of fitness. If fitness in two sites be taken as the two characters Y and Z, then the indirect response of Y to selection on Z will be $\mathcal{R}_y = r_{yz} (\sigma_y/\sigma_z)R_z$, where R_z is the direct response of Z and r_{yz} is the genetic correlation between sites (tertiary theorem of natural selection, §6.2.7). If there is a cost of adaptation then $r_{yx} < 0$, and the indirect response to selection in one site will be regress in the other. This prevents the evolution of a universally superior type, since in either site the resident specialist will be better adapted than an alternative specialist from the other site.

It is possible for a population to be selected in both sites, for example if conditions alternate between generations. The cost of adaptation implies that it will not become as well adapted to either as will a population exposed to only one of them. It will be better adapted to either, however, than a population that never experiences that site, provided that the direct response exceeds the indirect response. This condition is likely to be met unless the genetic variance is grossly unequal in the two sites. In either site a generalist will then be intermediate in fitness between the resident specialist and one from elsewhere. Generalists are thus less likely to evolve when there is a cost of adaptation, although the outcome of selection in any particular situation will also depend on ecological factors such as site frequency, immigration, and population regulation that are discussed below.

When the ancestral population first invades two new sites the initial response to divergent selection will depend on the pre-existing genetic correlation. The conditions of growth at each site can be thought of as having three components. The first is common to the ancestral site and both new sites; for example, all might have the same temperature. The second differs from the ancestral site but is shared by both new sites; if the ancestor is a wild isolate, for example, selection lines will both experience common features of the lab environment such as high nutrient levels or constant illumination. The third is unique features of each new site; for example, different limiting carbon substrates. The only mutations that matter are those which are beneficial in at least one of the two sites. If the ancestor is poorly adapted to the shared distinct features of the new sites there will be many mutations that improve fitness in both sites, and the genetic correlation will be positive. This leads to synclinal selection where the response is in the same direction in both environments and hence each line will evolve higher fitness than the ancestor in both environments. Provided that the direct response exceeds the indirect response, this will nevertheless lead to adaptive divergence and the appearance of negative cross-environmental genetic correlation between the selection lines (Figure 8.12). Adaptation to different temperatures in glucose-minimal medium in *E. coli* is an example of this: adaptation to 42 °C was not accompanied by regress at lower temperatures,

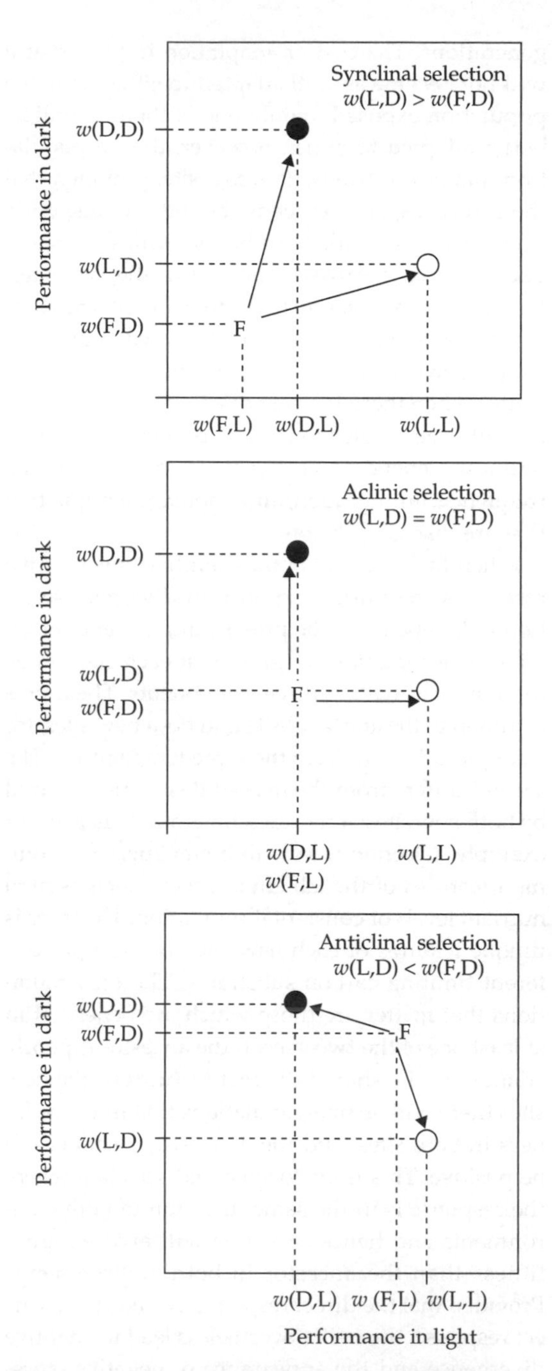

Figure 8.12 The concepts of synclinal, aclinic, and anticlinal selection. The open and solid circles represent Light and Dark selection lines, respectively; F indicates the founder.

but the direct response exceeded the indirect response (Bennett and Lenski 1993). This may be attributable to parallel changes in glucose transport (Bennett and Lenski 1996). Conversely, the ancestor may be well adapted to the shared features of the new sites so that most mutations that improve adaptation to the unique features of one site will degrade performance in the other. The genetic correlation is then negative to begin with, and anticlinal selection will convert this into a negative cross-environmental correlation. In intermediate situations where both shared and unique distinctive features of the new sites provide an opportunity for adaptation the initial genetic correlation may be close to zero. Selection is then aclinic, with a direct response but little or no indirect response at both sites, which necessarily leads to a negative cross-environmental genetic correlation. The general rule is that the cross-environmental genetic correlation will become negative under divergent selection regardless of its initial value, provided that the direct response in each environment exceeds the indirect response.

Most experiments can therefore be expected to demonstrate a cost of adaptation, in the sense of a negative correlation between selection lines, but this provides little information about whether or how mutations in the ancestor cause antagonistic effects on fitness. Moreover, the genetic correlation within either selection line is likely to be close to zero and may be positive. Price and Schluter (1991) illustrate this for a continuous character z that affects fitness differently in two sites. The ancestor is adapted to one site, where it is under stabilizing selection, and a derived population becomes adapted to the other through an increase in z. As the mean character value z in the derived population increases it traverses the range within which increased values are associated with reduced fitness in the ancestral site but increased fitness in the new site. Since these effects are in opposite directions the genetic correlation within this population will become negative. Once the mean of the derived population has attained the optimal value for the new site, however, a mutation that reduces z will reduce fitness in the new site but increase fitness in the ancestral site, whereas a mutation that increases z will reduce fitness in both sites. The positive and negative effects will cancel to leave a net correlation of zero. This argument ignores the effect of selection in complex environments and of immigration in structured environments (§8.4), both of which will tend to create negative within-population genetic correlations. It is nevertheless important to distinguish between the within-line or within-population genetic correlation and the among-line or cross-environmental genetic correlation in evaluating the consequences of divergent selection.

Changes in the fitness rank of genotypes is reflected by crossing reaction norms, requiring $w(A,A) > w(B,A)$ and $w(B,B) > w(A,B)$, using $w(X,Y)$ to signify the fitness of a strain adapted to site X when tested in site Y. Precisely the same conditions hold for the cross-environmental genetic correlation to be negative. Thus, a negative cross-environmental genetic correlation and crossing reaction norms are equivalent ways of expressing the same result, as implied by the definition of the component of $G \times E$ attributable to inconsistency.

8.3.4 Sources of antagonism

Functional interference

Selection will be anticlinal when mutations tend to be beneficial in one environment and deleterious in another. This will follow when the character state causing enhanced adaptation to one environment necessarily reduces adaptedness in the other. This very broadly applicable principle of functional interference is familiar in everyday life from the design of tools. A single tool may serve diverse purposes, as a combination pocket-knife may include large and small blades, screwdriver, tweezers, and so forth. Nevertheless, surgeons or electricians do no usually rely on a Swiss Army knife; they have instead a battery of specialized implements, each of which is well suited to a particular task but of little use for others. A scalpel and a spoon are each apt for a particular task, but the features that fit them for one task disqualify them for the other, and an implement that combined these features would be effective for neither. This applies equally to different types of individual. Snails and birds provide simple examples. A beak able to crack hard nuts is unlikely to be well suited to picking small insects out of cracks. Dark brown, unbanded shells are difficult to see against the sodden leaf-litter of the woodland floor; yellow, banded shells are more likely to escape detection in rough grassland. What is cryptic at one site will necessarily be conspicuous at the other; better adaptation to one type of site can be achieved only at the expense of poorer adaptation to the other. Functional interference is often referred to as *antagonistic pleiotropy* and the negative correlation it creates as a *trade-off.*

When *E coli* lines are cultured in glucose-limited minimal medium they often lose the ability to utilize ribose within 500–2000 generations (Cooper *et al.* 2001). The main reason is that spontaneous Rbs^- mutants have a 1–2% advantage over wild type. They arise at high frequency because transposition of an upstream insertion element, IS150, element into the *rbs* operon causes deletions of varying length, through recombinational excision, that lead to complete loss of function. Restoration of the intact *rbs* partially restores function and reduces fitness. The uniform loss of ribose catabolism and the consistent advantage of Rbs^- mutants in minimal medium strongly suggest that silencing *rbs* is directly responsible for elevating fitness when glucose but not ribose is available as a carbon and energy source. This is a clear example of functional interference, although its physiological basis is not yet understood.

Outside the laboratory, most examples of adaptive radiation involve morphological characters that affect feeding, moving, or mating. These often provide clear examples of functional interference. Benthic sticklebacks, for example, are relatively powerful predators with large gape that can readily capture insect larvae but would be incapable of the agile chase needed to hunt copepods. A long thin beak that is ideal for eating insects would be unable to crush thick-shelled fruits. Other examples will readily come to mind.

Mutational degradation

When selection is synclinal, selection in one environment will actually increase fitness in the other.

This correlation will tend to be degraded and eventually reversed by the effects of disuse. These are also familiar from everyday life, where tools that are not kept in use and maintained will rust or rot. In Labrador a vehicle needs a heater, but its air-conditioner may be superfluous; in Australia the priorities would be reversed. We expect that air-conditioners in Labrador and heaters in Australia will tend to accumulate unrepaired faults, so that a vehicle transferred from one location to the other might perform poorly. In a similar way, mutations that reduce fitness in a given environment will be held at low frequency by selection, whereas mutations that reduce fitness in other environments will be nearly neutral in effect, and will tend to accumulate. Long-continued selection in one site will thereby lead to some loss of adaptedness to other kinds of site by mutational degradation. It is not necessary that such conditionally deleterious mutations should be somehow entailed by increased adaptedness to local conditions. They will accumulate merely because adaptedness tends to break down unless it is actively and continually maintained by selection. I have already described the degradation of the carbon-concentrating mechanism in *Chlamydomonas* after selection at elevated CO_2, with evolved lines becoming unable to up-regulate photosynthesis at low CO_2 (§7.5.2). Consequently, some evolved lines are unable to grow at contemporary concentrations, and there was on average a fall of about 40% in net CO_2 uptake (Collins *et al.* 2006a). Back-selection restored the ability to grow at low CO_2 levels, but did not restore the carbon-concentrating mechanism (Collins *et al.* 2006b).

8.3.5 Consequences of interference and degradation

The two sources of the cost of adaptation have somewhat different consequences. If enhanced growth in one site necessarily leads to regress in others, the amount of the regress should be proportional to the amount of the advance. Moreover, the cost is paid immediately, as once a beneficial mutation is fixed at one site it will express lower growth at other sites. Conditionally deleterious mutations, on the other hand, should accumulate continuously through time, and the cost they incur is not necessarily related to the amount of adaptation that has been achieved. Mutational degradation should be accelerated by elevating the mutation rate, whereas it will be retarded by exposure to both (or all) environments. These consequences have been used to evaluate the contributions of functional interference and mutational degradation to the cost of adaptation.

Long-term *E. coli* lines cultured in glucose minimal medium permanently lost the ability to utilize a range of other carbon substrates. Cooper and Lenski (2000) concluded that this was primarily attributable to functional interference, because most losses occurred early in the history of the lines, and because a restricted set of substrates was involved in most cases. Moreover, some lines had evolved higher mutation rates (through the spread of mutators; Chapter 4) and yet suffered no more damage. Funchain *et al.* (2000), however, found that mutation-accumulation lines of *E. coli* often evolved catabolic defects. In some cases the substrates affected seemed to be predictable; small indels in a long GC repeat, for example, caused frameshifts that knocked out xylose catabolism.

Independently evolved wrinkly spreaders appearing early in the history of *Pseudomonas* cultures also express partial or complete loss of ability to utilize substrates available to the ancestral smooth type (MacLean and Bell 2004). These defects reduce the fitness of the strains that bear them, relative to their smooth ancestor, and thus constitute a severe cost of adaptation comparable to that observed in *E. coli.* If a strain is re-inoculated into fresh medium smooth types will spontaneously arise and spread; these lack the catabolic defects of the ancestor, showing that the mutation causing the wrinkly phenotype is also responsible for the pattern of defects associated with it. If the wrinkly continues to be propagated, however, these defects are greatly attenuated or lost although biofilm formation and wrinkly colony morphology are fully retained. This might happen in two ways. If many wrinklies arise independently in the early history of the culture, competition will favour *wss* mutations associated with fewer defects and will therefore ameliorate catabolic damage in the long term. Alternatively, if a single mutation becomes established early on, compensatory mutations at other loci that restore catabolic function will be favoured. It is not known which of these two mechanisms is acting, but in either case the cost of adaptation itself evolves, being severe at first but fading over time until it has almost completely disappeared.

The difference between the *E.coli* and *Pseudomonas* experiments may be that some loss-of-function mutations contribute directly to success in novel conditions whereas others do not. The loss of catabolic function in wrinkly spreaders is incidental to biofilm formation and will be attenuated by selection, whereas *rbs*$^-$ mutations are directly responsible for elevated fitness in glucose minimal medium and are thereby maintained by selection (MacLean *et al.* 2005).

We analysed the consequences of dark growth in *Chlamydomonas* by using four treatments to extend the experiment mentioned above, using the lines previously selected in Light and Dark sites for about 1000 generations as base populations (Reboud and Bell 1997) (Figure 8.13). One set of lines continued to be maintained in the same conditions for a further 100 generations: these changed little, showing that the base populations had approached evolutionary equilibrium. A second set of lines experienced reverse selection: Light lines were transferred to the Dark and maintained there during the course of the experiment, while Dark lines were transferred to the Light. Each line will then become adapted to the novel reverse site to which it has been transferred. If the cost of adaptation displayed by the base populations were attributable to functional interference there should be a continual increase in adaptation in the reverse site, accompanied by a continual decrease in adaptation to the base site. On the other hand, if the cost of adaptation is caused by mutational degradation, adaptation to the reverse site will at first be achieved without any regress when tested in the base site, because it will involve only the replacement of the conditionally deleterious mutations which accumulated during selection in the base site. The reverse lines from the Dark base population rapidly improved, to reach levels of growth in the Light comparable with those of the continued Light lines, with no loss of ability to grow in the Dark. The reverse lines from the Light base population were heterogeneous; some spores had evolved the ability to grow well in the Dark, whereas others showed little or no improvement. Those which grew well in the Dark grew slightly less well in the Light, presumably because of functional interference. A third set of lines experienced temporal variability: Light lines were transferred to the Dark for one cycle of growth, then back to Light, then to Dark again, and so forth. Adaptation to either site will cause regress in the other through functional interference, so these lines will be inferior to the specialist continued line in both Light and Dark. On the other hand, mutations which are deleterious in either site will be selected against very effectively, so conditionally deleterious mutations will not accumulate, and those which have accumulated in the base populations will tend to be eliminated. The theory of mutational degradation therefore predicts that generalists will evolve which are as fit in both Light and Dark as the specialist continued lines. The most surprising result of the experiment was the evolution of these superior generalists in a temporally varying environment. It is most likely attributable to the clearance of conditionally deleterious mutations that had accumulated in the base populations. In short, divergent selection in Light and Dark conditions involves a cost of adaptation that is mostly attributable to mutational degradation.

Both functional interference and mutational degradation, then, may contribute to the cost of adaptation in different circumstances. The relative importance of mutational degradation may depend on how well-adapted the ancestor is and on how many genes are involved. The photosynthetic apparatus of *Chlamydomonas* is operated by a large number of genes (including much of a large chloroplast genome) and seems to be nearly optimally configured for normal conditions of growth, so it presents a large target for mutation when it becomes redundant. It is only the transport and catabolism of other sugars that is made redundant by growth in glucose minimal medium, and these may be fewer and less highly integrated. No firm generalization can yet be made, however.

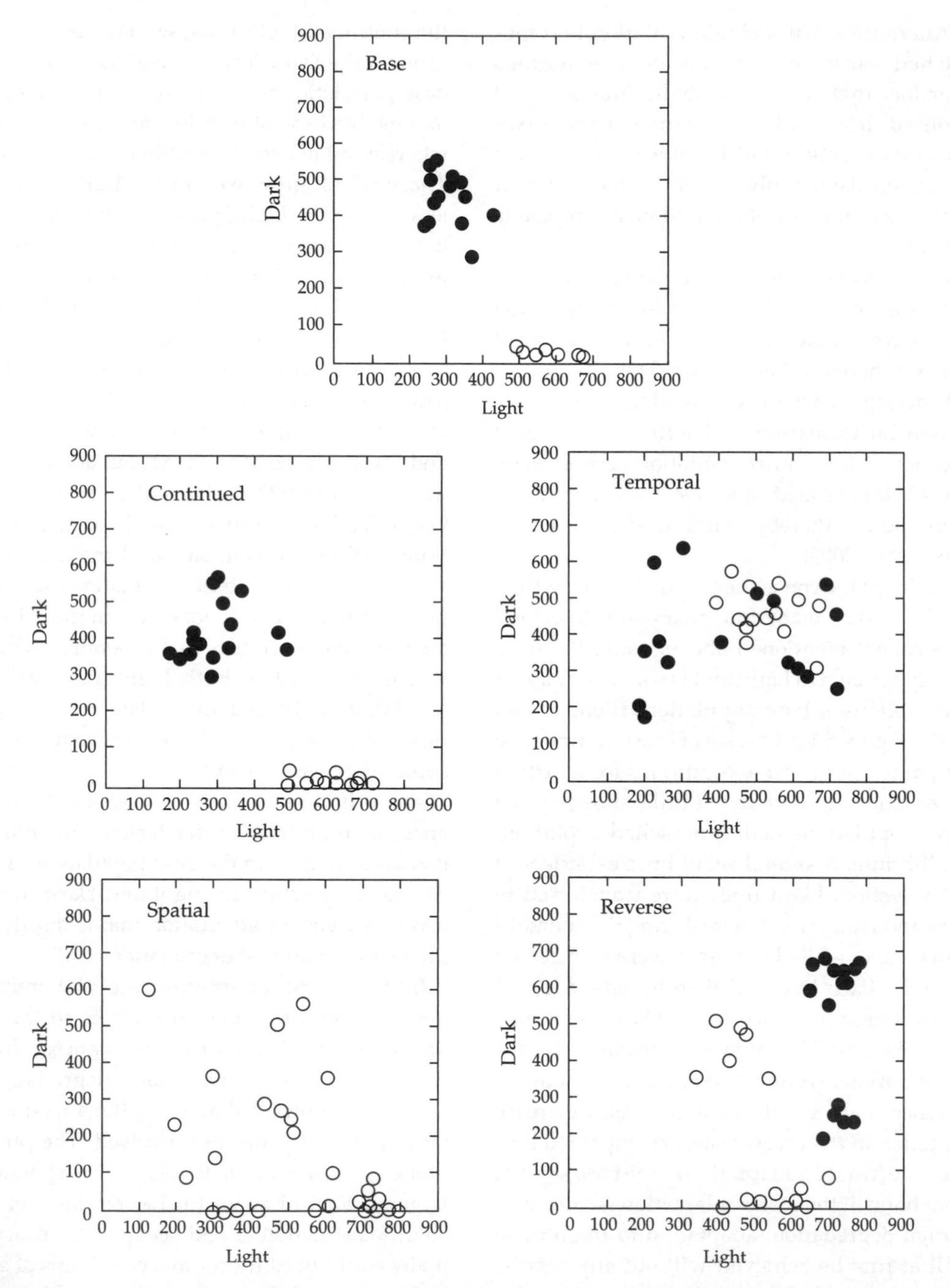

Figure 8.13 Selection for Light and Dark growth in *Chlamydomonas*. Open circles are lines from Light base populations, filled circles from Dark base populations. From Reboud and Bell (1997).

8.3.6 An experimental adaptive radiation: *Pseudomonas*

We instigated an extensive adaptive radiation by inoculating *Pseudomonas fluorescens* into Biolog plates—microtitre plates in which each well contains a single unique carbon substrate (MacLean and Bell 2002). It is easy to transfer a whole plate at once using a 96-pin replicator, so that a large number of separate selection lines can be propagated at the same time. We initially supplemented the wells with complete medium, gradually reducing this supplement to permit the populations to become adapted to each substrate. This initiated an adaptive radiation in which hundreds of specialized lines evolved. The ancestral clone that we used could grow on 46 of the 95 substrates available, which constitute its ancestral niche. Growth on any one of these substrates usually improved when it was the only source of energy available, although the improvement was usually modest. The more interesting result was that about 70% of the lines became capable of growing on substrates outside the ancestral niche. The improvement in adaptation, indeed, was negatively related to the performance of the ancestor: the advance was greatest for substrates where the ancestor grew poorly if at all, and declined to zero for substrates such as glucose where the ancestor grew well (Figure 8.14). For 13 substrates, however, at least one of four replicate lines failed to respond: adaptation does not always occur. Indeed, for one substrate (α-cyclodextrin) no lines responded, so in some kinds of site adaptation may occur seldom if ever.

Some of these novel specialists showed no qualitative cost of adaptation, since they were still capable of growth on all the substrates of the ancestral niche, but about 80% had lost the ability to grow on at least one of them, and on average they dropped two or three substrates. The main cost of adaptation was thus a contraction of about 5% in the extent of the ancestral niche. Some substrates were lost much more often than others, as Cooper and Lenski (2000) and Funchain *et al.* (2000) had previously found for *E. coli*. For each substrate, the indirect response could be regressed on the direct response over the four replicates. A negative

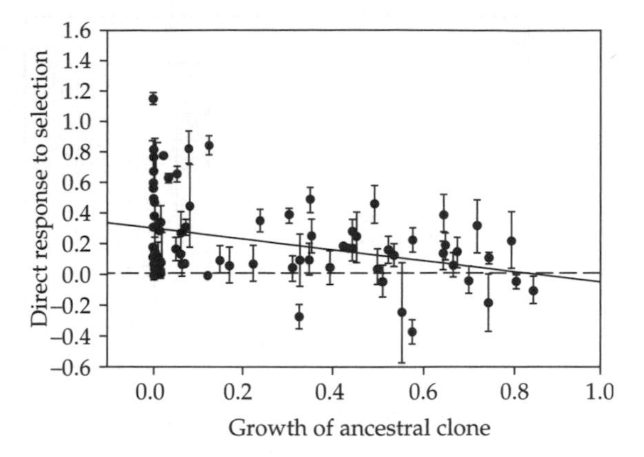

Figure 8.14 The direct response to selection as a function of ancestral performance in an experimental adaptive radiation. Plotted points are the mean direct response to selection of a selection line and the mean growth of the ancestral clone on the substrate on which the line was selected. The units of growth are standardized measures of the increase in turbidity. The dashed line is the zero response corresponding to lack of adaptation. From MacLean and Bell (2002).

slope indicates functional interference, a greater advance entailing a greater regress, but the slope was negative less often than expected by chance. Conversely, a negative intercept indicates mutational degradation, with lower fitness for other substrates despite a failure to adapt, which was seen more often than expected. On balance, then, the side effects of the radiation were dominated by mutational degradation.

The most surprising feature of the radiation, however, was that over all substrates the indirect response was positive: selection lines actually acquired a broader niche, evolving the ability to utilize about 30 more novel substrates than the ancestor and consequently being able to grow on 70–85 of the 95 available (Figure 8.15). Indeed, the lines evolved the ability to utilize every substrate as an indirect response, and in several instances lines could grow strongly on a particular substrate as an indirect response when the direct response was weak or lacking. We called this 'roundabout' selection because such lines take a roundabout route to adaptation. We do not know the reason for this, but the ancestor is a wild strain and a widespread breakdown of regulation resulting in the

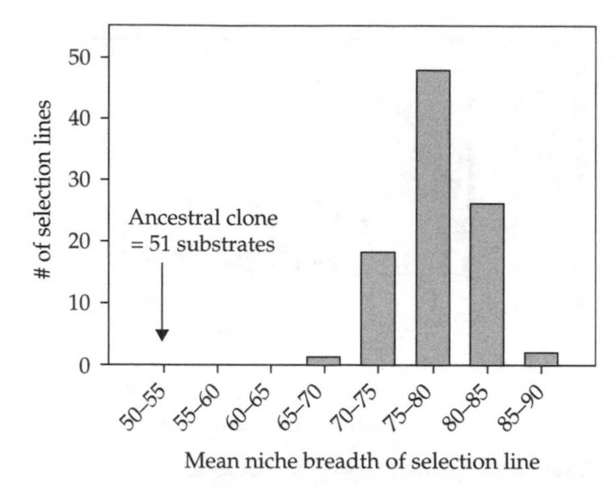

Figure 8.15 Frequency distribution of niche breadth after selection in an experimental adaptive radiation. From MacLean and Bell (2002).

constitutive expression of catabolic genes is a possible explanation. Whatever the reason, it implies that the invasion of a complex environment with many novel sites might proceed much more rapidly than expected on the assumption that adaptation had to evolve separately at each site.

8.3.7 An historical adaptive radiation: *Anolis*

Like isolated streams, an archipelago is a series of more or less independent replicates for organisms that cannot readily disperse across water. Lizards of the genus *Anolis* are found on all the larger Caribbean islands, each of which supports a characteristic suite of *ecomorphs* living in different kinds of site, from the smallest species found on twigs to larger species with highly developed toe-pads that frequent trunks and large branches, and the largest species that live high in the canopy (Williams 1972). Different ecomorphs on the same island are more closely related to one another than each is to the corresponding ecomorph on other islands, strongly suggesting that the whole suite has evolved in a similar fashion on many occasions (Losos *et al.* 1998). The Greater Antilles date back to the Eocene, providing a considerable period of time for repeated dispersal and radiation to occur. Nevertheless, Losos *et al.* (1997, 2001) introduced *Anolis sagrei* from a single source population to several islands in the Bahamas that lacked lizards and reported adaptive changes after only about 10 years.

8.4 Opportunities in space: obligations in time

Essential resources such as carbon and nitrogen must be available in order for organisms to grow, but other resources may or may not be present: a carbon source is required, for example, but it might be glucose, galactose, or any of a thousand other possibilities. A simple environment receives a constant supply of one such substrate, as in a basic chemostat experiment. Natural conditions of growth, however, vary widely in time and space (§2.3, §8.2). Complex environments provide a range of substrates simultaneously available to all individuals. In structured environments this range differs among sites, whereas in variable environments it changes over time. Selection may then favour either plasticity or specialization. If the cost of adaptation prevents any single genotype from being superior in all conditions, a community of specialists may evolve, each superior in different times and places; alternatively, a mediocre generalist may have the greatest overall fitness. This section describes the outcome of divergent selection in complex, structured and variable environments as the consequence of G × E and the cost of adaptation. For the time being I shall ignore how resource supply is affected by the density and composition of the population (as it almost always will be), which is the subject of Chapter 10.

8.4.1 Simple environments

An ideal simple environment contains a single limiting substrate that is freely available to all individuals. As this substrate is depleted, types able to grow on lower concentrations will be favoured, until at last only the single most frugal type remains (§10.2).The classical sorting experiments of Gause (1934) led to the formulation of the competitive exclusion principle (Hardin 1968): in a closed environment with a single limiting substrate only one strain can persist indefinitely. This is a restatement of the fundamental theorem in the context of ecological competition among species. It has been very extensively confirmed by classical microcosm experiments in which there is little if any genetic variance of fitness most of the time in most cases, at least for 1000 generations or so. There are nonetheless some circumstances in which it has to be qualified or abandoned.

Neutral or slightly deleterious alleles will drift in the population, with a slow accretion of mutational heritability (§2.2). Many mutations may be conditionally deleterious and can be detected by growing the population in a different medium, which shows that there is initially a linear increase in genetic variance even in chemostat populations exposed to the full force of natural selection (Goho and Bell 2000).

There may also be cryptic variance attributable to nearly neutral epistatic combinations of alleles in sexual populations. We propagated sexual and asexual lines of *Chlamydomonas* for about 100 sexual episodes and 1000 vegetative generations and then measured the fitness of individual spores (Renaut *et al.* 2006). The asexual lines were devoid of genetic variance for fitness, as expected, but the heritability of fitness was 20–25% in the sexual lines. This was largely contributed by a conspicuous excess of inferior spores. Mating within each selection line generated a large amount of genetic variance among the sexual progeny, but these progeny were consistently less fit than their parents: their limiting density was about 8% lower, corresponding to $\Delta \ln \bar{w} \approx 0.005$. Hence, recombinational load can preserve genetic variance in sexual populations.

In the long term the competitive exclusion principle breaks down even in completely asexual populations. There is now abundant evidence to show that asexual lines which are maintained for hundreds or thousands of generations in simple environments become polymorphic, that spores isolated from the lines differ substantially in fitness, and that fitness is frequency dependent (Helling *et al.* 1987, Rosenzweig *et al.* 1994, Turner *et al.* 1996, Elena and Lenski 1997, Treves *et al.* 1998). The principle breaks down because the simple environment provided by the experimenter is modulated by the activities of the organisms themselves, who collectively create a complex environment within which different kinds of specialists can evolve. This may involve utilizing the incompletely respired metabolites of other strains or succumbing to toxins and allelopathic substances (§10.3.2).

When a range of potentially limiting substrates is provided by the experimenter (or by nature) there is greater potential for maintaining diversity. Competitive exclusion continues to serve as a central organizing principle, generally in the form that the number of distinct types that can be maintained is equal to the number of potentially limiting resources. The details depend on how the resources are presented to the population: complex environments in which all resources are simultaneously available at the same site, structured environments in which each resource is available at a different site, and variable environments in which each resource is intermittently available.

8.4.2 Complex environments

It is convenient to conduct experiments in simple environments, but it may also be misleading. Natural environments are seldom simple, and often comprise a large variety of resources none of which are sufficiently abundant to support growth alone. Individuals must then simultaneously consume several or many different kinds of resource, and most microbes are able to use a variety of sugars, organic acids, and other compounds to provide energy or carbon skeletons, for example. Such substrates are said to be *substitutable*, because growth can be supported by any one of them or any mixture of them and is not prevented by the absence of any one of them. The simplest case is a mixture of two substrates A and B consumed by two strains 1 and 2. The fitness of type 1 (frequency p) relative to type 2 is $1 + s$, where $s = (r_1 - r_2)/r$ is defined in terms of growth rate on either substrate separately (s_A and s_B) or on a mixture of the two (s_{AB}). It can then be shown that

$$s_{AB} = [R_A s_A(1 + ps_B) + R_B s_B(1 + ps_A)]/[R_A(1 + ps_B) + R_B(1 + ps_A)],$$

where R is the resource supply rate (Lunzer *et al.* 2002). Hence, the outcome of selection in a complex environment can be predicted from the outcome in each component simple environment, together with the resource supply rates, which are under the control of the experimenter. When type 1 is rare, $s_{AB} \rightarrow s_A F_A + s_B F_B$, where F_A is the resource ratio $R_A/(R_A + R_B)$: it spreads if its arithmetic mean fitness (fitness on each substrate, weighted by the contribution of that substrate to the complex environment) exceeds unity. Conversely, when type 1 is almost fixed,

$$1 + s_{AB} \rightarrow 1/\{[F_A/(1 + S_A)] + [F_B/(1 + s_B)]\},$$

and type 1 becomes fixed if its harmonic mean fitness exceeds unity. Consequently, both strains can be maintained if each has a higher growth rate on one of the substrates, such that its arithmetic mean fitness is greater than unity and its harmonic mean is less than unity. This is because the harmonic mean fitness of the common strain is by definition the reciprocal of the arithmetic mean fitness of the rare strain. Lunzer *et al.* (2002) used two strains of *E. coli* differing in growth rate on methylgalactoside and lactulose as a consequence of a mutation in the *lac* operon to show that fitness at any given resource ratio is frequency dependent, as expected. Hence, there exists a range of resource ratios within which the arithmetic mean fitness of either type exceeds unity whereas its harmonic mean falls short, so that both types will persist. The width of this 'window of coexistence' depends on the strength of selection on the single substrates. In Lunzer *et al.*'s experiment selection is intense ($s_A = +0.3$, $s_B = -0.1$) but the window is nevertheless quite narrow (23–30.5% methylgalactoside), and they conclude that resource competition is unlikely to be responsible for maintaining the bulk of allelic variation at allozyme loci in natural populations.

Organisms may evolve as generalists that are able to consume a large proportion of the substitutable substrates in a given category or as specialists that are able to grow on only one substrate or a very few. Communities might thus comprise a diversity of specialists at one extreme or a single universal generalist at the other (see Harder and Dijkhuizen 1982, Gottschal 1986). The theory of coexistence in mixtures of substitutable substrates has been exhaustively analysed in the ecological literature, especially from the point of view of how many species can be maintained on a given number of substrates (MacArthur and Levins 1964; MacArthur 1969; Stewart and Levin 1973). The most general result is an extension of Gause's exclusion principle that the number of species stably coexisting cannot exceed the number of distinct substrates. This holds for competition in chemostats (Taylor and Williams 1975), for non-substitutable resources (Tilman 1977), and for spatially structured environments (Strobeck 1975). Thus, when a single genotype is cultured in a complex medium, we might expect to observe either the evolution of a single broadly adapted generalist or an adaptive radiation involving a diversity of specialized types.

One afternoon in the summer of 2003 we set up some *Pseudomonas* cultures and then monitored marker frequencies with the intention of demonstrating selective sweeps. This turned out to be more difficult than we had expected: a marker might become very abundant at some point, but instead of becoming fixed by the passage of the first beneficial mutation it would then retreat to lower frequency. Rather than a waiting period during which the markers remained about equally frequent followed by rapid fixation and subsequent uniformity, we saw wide and persistent fluctuations in frequency (MacLean and Bell 2004) (Figure 8.16). This was because the cultures were growing in rich complete medium that supplied hundreds of substrates at low concentration. In minimal medium with a single limiting carbon source, a selective sweep announcing the arrival of the first successful specialist will usually occur fairly soon. If there are two substrates at high concentration one will often be utilized exclusively to begin with and the other only once the first has been exhausted (*diauxic growth*). In complex media where there is not enough of any single substrate to support growth alone, cultures will grow readily if the substrates are substitutable, but each cell must consume several or many substrates simultaneously. In these circumstances the concept of the limiting factor breaks down (see Harder and Dijkhuizen 1982), and with it the paradigm of periodic selection. Rather than being genetically uniform most of the time, cultures instead evolve a high level of

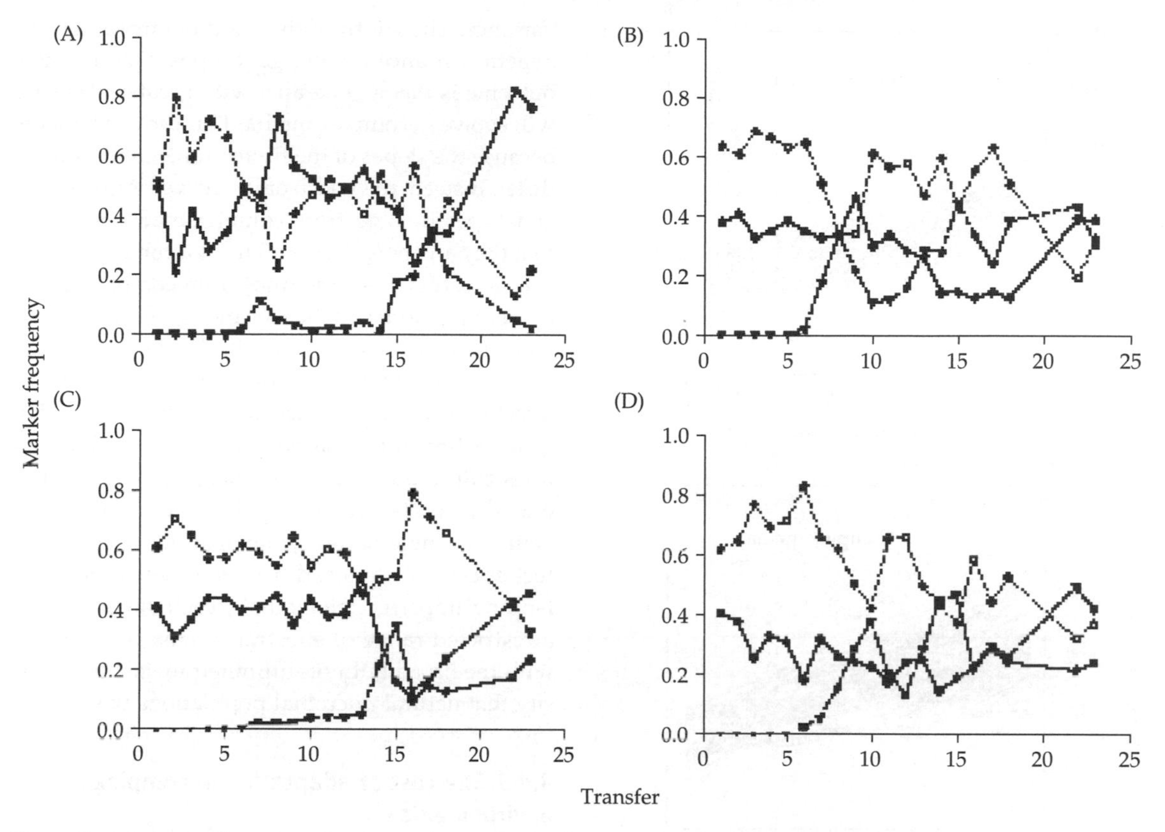

Figure 8.16 Dynamics of genetic markers in *Pseudomonas* batch cultures maintained in rich complete medium. Each graph (a–d) shows the dynamics of the three genetic markers in an independent selection line. Initially, all lines consisted of smooth genotypes carrying the pan+ (open circle) or pan– (filled circle) marker. During the course of the experiment, colony morph variants evolved, providing a third genetic marker (triangle). From MacLean *et al.* (2005).

genetic diversity that can be visualized by isolating genotypes and recording their metabolic profiles on Biolog plates. We found that most genotypes fell into two clusters: the first resembled the ancestor and had a broad metabolic range, whereas the second had a narrower and qualitatively different range. The stationary density attained by a genotype increased with the range of substrates it was able to consume, but genotypes with a relatively narrow range nonetheless persisted in culture. This was because genotypes from either cluster were able to invade cultures dominated by the other. The most likely explanation for the maintenance of diversity in rich medium is that costs of adaptation prevent the evolution of a single superior generalist and instead facilitate the evolution of a series of specialists each efficiently consuming a restricted range of the substrates available.

The rich medium used in these experiments is an undefined mixture of many organic compounds. We can document adaptive radiation in complex environments more precisely by using chemically defined mixtures of relatively few substrates (Barrett *et al.* 2005). To illustrate the outcome of selection in simple and complex environments, we constructed a fitness rank curve for each genotype isolated from the lines after 900 generations of culture by ranking the substrates in order of growth from most to least. Combining these curves for several genotypes gives a useful visual summary of the differences between lines or treatments (Figure 8.17). Note that the identity of the

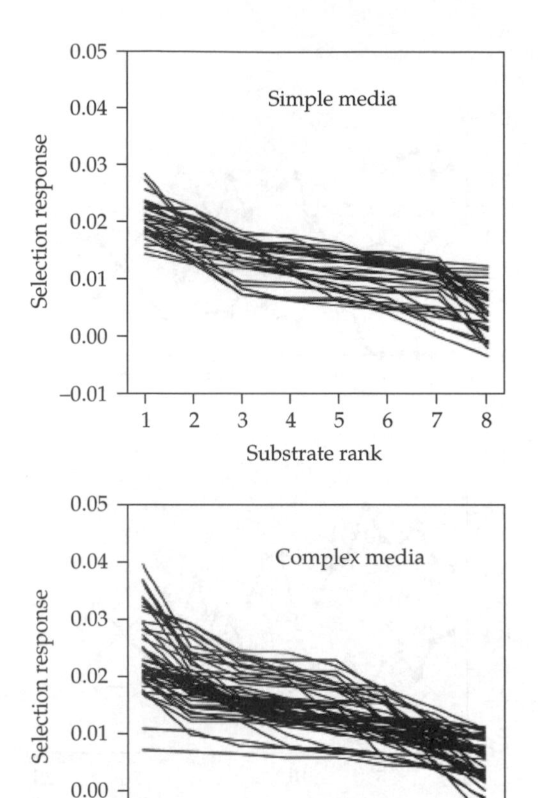

Figure 8.17 Fitness rank curves in simple and complex media. Each line represents the fitness of a genotype across different substrates. Substrates are ranked by decreasing fitness of each individual genotype, so the substrate at each rank may be different for different genotypes. Increased separation between responses indicates greater genetic variance among genotypes. Increased slope indicates greater environmental variance in response. From Barrett *et al.* (2005).

substrate corresponding to a particular rank may, and usually will, differ from genotype to genotype. The curve for each genotype, and the band of curves from a group of genotypes, necessarily slopes downwards from the highest-ranking to the lowest-ranking substrate. The slope of the curve or band of curves represents environmental variance: a steeper slope reflects a greater difference between the highest-ranked and the lowest-ranked substrates and thus greater environmental variance. The width of the band of curves reflects genetic variance among genotypes. One possible outcome is that a generalist with greater plasticity will evolve in complex media. This did not happen, because the slopes of the fitness rank curves do not differ. Instead, the width of the curves is greater for genotypes isolated from complex media, showing that they are more diverse. This is a consequence of specialization, with the lines from complex media expressing more genetic variance, more $G \times E$, and more inconsistency. Surprisingly, the curves are higher for complex than for simple media; that is, selection in a complex medium results in types that grow better in its component simple media than lines cultured on each component separately. The outcome of selection in complex media is therefore neither a single perfect generalist nor a set of perfect specialists; instead, the genotypes are overlapping imperfect generalists, each superior over a restricted range of substrates. This is consistent with the broad but not unlimited metabolic diversity that natural microbial populations possess.

8.4.3 The cost of adaptation in complex environments

Negative genetic correlations in complex media will arise through disruptive sorting alone. Suppose that a genetically diverse population is growing on a mixture of two substitutable resources. The genetic correlation between growth on either resource separately is zero, so the plot of growth on the two resources is a circular scatter of genotypes. Any genotype that grows poorly on both resources will be quickly eliminated by selection, however, thus removing those genotypes in the lower left corner of the plot. As time goes on, selection will shave away variants that have little ability to utilize one or both resources, generating a negative genetic correlation in the population. This process of disruptive sorting is expected to occur even when there is no cost of adaptation. The eventual outcome of selection should then be a single superior generalist, but in practice negatively correlated genotypes of more or less equivalent overall fitness may persist for a long time. Even at equilibrium, mutation that produces uncorrelated variants will generate some negative correlation through the subsequent operation of selection.

The costs that that may be generated by functional interference can be evaluated by reverse-genetic experiments. Auxotrophic mutants often spread in chemostat populations of bacteria, when the substance they are unable to synthesize is supplied in the growth medium. This would be easy to understand if a metabolic pathway that is no longer necessary represents a metabolic burden, so that strains possessing the pathway reproduce less rapidly and are lost through selection. Such a theory of energy conservation would provide a general rationale for functional interference, and would explain why anabolic pathways are usually repressed by their end-products. For example, it is known that tryptophan auxotrophs often increase rapidly in frequency in glucose-limited chemostats, when tryptophan is supplied in excess. The *trp* operon of *E. coli* comprises five linked genes that encode the biosynthetic enzymes, controlled by an upstream promoter and an unlinked repressor. The pathway they form can be represented crudely like this : chorismate → anthranilate → → → indole → tryptophan. Dykhuizen (1978) used *trp* mutants to find out whether auxotrophs tend to spread in permissive conditions of growth because of the energy cost of maintaining redundant metabolic machinery. A mutation accumulation hypothesis was not even considered as an alternative. Neutral mutations certainly increase in frequency in chemostats—this provided much of the impetus for the early chemostat work—but the *trp*⁻ mutants that Dykhuizen studied increased in frequency when competing against isogenic wild-type *trp*⁺ strain thousands of times more rapidly than mutation could explain. However, an explanation in terms of energy conservation seemed almost as unappealing. Only about 1% of the total energy budget of the cell is used in making tryptophan; moreover, when inhibited by the presence of excess tryptophan in the medium, the level of tryptophan synthesis is only about 1% of normal, and so constitutes only about 0.01% of the total energy budget. The selection coefficients involved in the spread of *trp*⁻ strains, on the other hand, are roughly 0.05–0.10. There is thus a substantial disproportion—amounting to about three orders of magnitude—between the amount of energy saved and the competitive advantage that this is supposed to create. It might be argued that very slight metabolic economies are somehow magnified into large differences in growth rate, but when auxotrophic and prototrophic strains are cultured separately there is little if any difference in their rates of growth. It is also possible that the major expense of operating the pathway lies in making the enzymes involved, rather than tryptophan itself. However, mutants in which these enzymes are not synthesized have no greater advantage than otherwise isogenic mutants in which they are. Finally, selection should be weaker when using mutants in which only the first enzyme of the pathway is dysfunctional if indole is supplied in the place of tryptophan (because both mutant and wild-type strains would have to perform the final reaction in the pathway), and still less if anthranilate is supplied. However, this prediction could not be confirmed either. Dykhuizen and Davies (1980) found much the same result for *lac*⁻ strains in maltose chemostats, where energy conservation again seemed an inadequate source of selection. They attributed the success of *lac*⁻ to a type of functional interference they called *resource interference*: strains are less efficient in using a resource when they are simultaneously using other resources. The source of this interference may have been competition between lactose and maltose permeases for the limited number of uptake sites in the cell membrane.

Dykhuizen and Davies (1980) competed specialist and generalist genotypes of *E. coli* in medium containing maltose and lactose; the specialist was simply deleted for the *lac* operon. With only maltose in the medium, the Δ*lac*⁻ specialist grew faster and out-competed the generalist, with a selection coefficient of about 0.05. In medium containing both substrates the specialist and generalist coexisted at frequencies governed by the proportion of lactose: the more lactose, the greater the frequency of the generalist. Above a certain proportion of lactose the specialist could not persist, showing that genotypes specializing on rare resources are unlikely to evolve when a much more abundant resource is available.

8.4.4 Structured environments

Not even culture vials are perfectly homogeneous, however, and natural environments are heterogeneous at all spatial scales (§2.3). There is an extensive body of theory in population genetics dealing with *multiple-niche polymorphism* in metapopulations inhabiting discrete sites providing different conditions of growth (Levene 1953, Dempster 1955, Maynard Smith 1962, Clarke 1979, Maynard Smith and Hoekstra 1980, Via 1984). At a single site, local adaptation is driven by selection and obstructed by immigration from an external source of constant composition (§7.2). The simplest case of a spatially structured environment is two alleles A1, A2 in a sexual diploid population inhabiting two sites that exchange migrants, so that each is affected by the other. The sites can be imagined to present different substrates, or mixtures of substrates, using the term in the broadest sense. (It is not necessary that there be literally a single site for each substrate; there may be many randomly distributed sites for either.) Immigrants make up a fraction I_1 and I_2 of all newborn individuals in the two sites respectively (adults die after reproduction). If one homozygote has fitness $w_{11,1} = 1 - s_1$, $w_{11,2} = 1 - s_2$ relative to the heterozygote in the two sites then it can invade if s_1 and s_2 have opposite sign and $(I_1/s_1) + (I_2/s_2) < 1$ (Bulmer 1972). If the migration rate (probability that a newborn individual will successfully disperse to another site) is m and the two sites contribute fixed fractions K_1 and K_2 to the overall production of the environment, then the condition is equivalent to $m[(K_2/s_1) + (K_1/s_2)] < 1$. The condition that the other homozygote, with relative fitness $w_{22,1} = 1 - t_1$, $w_{22,2} = 1 - t_2$ in the two sites, can also invade is likewise $m[(K_2/t_1) + (K_1/t_2)] < 1$, provided that t_1 and t_2 are differently signed in the opposite sense to the first strain. If there is no dispersal then divergent selection, each homozygote being fitter in one of the two sites, is sufficient to maintain diversity. If all newborn individuals disperse (the original Levene model) then the fitness of a rare dominant A1 allele will be simply the arithmetic mean of its fitness on each substrate weighted by the productivity of the substrate. The relative fitness of a rare A2A2 homozygote must be the reciprocal of this quantity, that is, the weighted harmonic mean. The condition for maintaining diversity is thus that the arithmetic mean fitness of individuals bearing A1 is greater than unity and the harmonic mean is less, that is, $\Sigma K_i w_{11,i} > 1$ and $\Sigma K_i/w_{11,i} > 1$ where i signifies the ith substrate (Maynard Smith 1962). This is precisely the same condition that holds in a complex environment (above): hence, selection in complex environments and in structured environments with complete dispersal has precisely the same outcome. For arbitrary dominance the condition is that the harmonic mean fitness of both homozygotes weighted by substrate productivity exceeds the fitness of the heterozygote, that is, $\Sigma K_i/w_{11,i} > 1$ and $\Sigma K_i/w_{22,i} > 1$ (Levene 1953). These conditions are rather severe, and diversity is maintained only if selection is intense (lethality of each type in one site is of course equivalent to lack of dispersal) or the two sites are about equally productive (Maynard Smith and Hoekstra 1980), as in complex environments. With intermediate rates of dispersal the situation is less easy to summarize, but with symmetrical selection (each strain having an advantage s in one site equal to its disadvantage in the other) the condition for the maintenance of variation is just $m/s < 1$, that is, selection must overturn the effect of immigration in either site. A rough indication of how selection, dispersal and productivity combine to protect diversity is given in Figure 8.18. There are five ecological factors that we have to take into account: the diversity of substrates, the frequency of substrates, the rate of dispersal, the pattern of dispersal, and the regulation of productivity.

The simple argument given above is readily extended to any number of substrates, and it follows from the principle of competitive exclusion that the number of distinct alleles or strains maintained by divergent selection cannot exceed the number of distinct substrates, or kinds of site (Strobeck 1975). Now, we have seen that natural environments are heterogeneous at all scales, but there is no rule that tells us how many different kinds of site exist within a given region, and the theory, despite its clarity, has no predictive power. At some risk of circularity, we might recognize two substrates as being different if the genetic correlation between

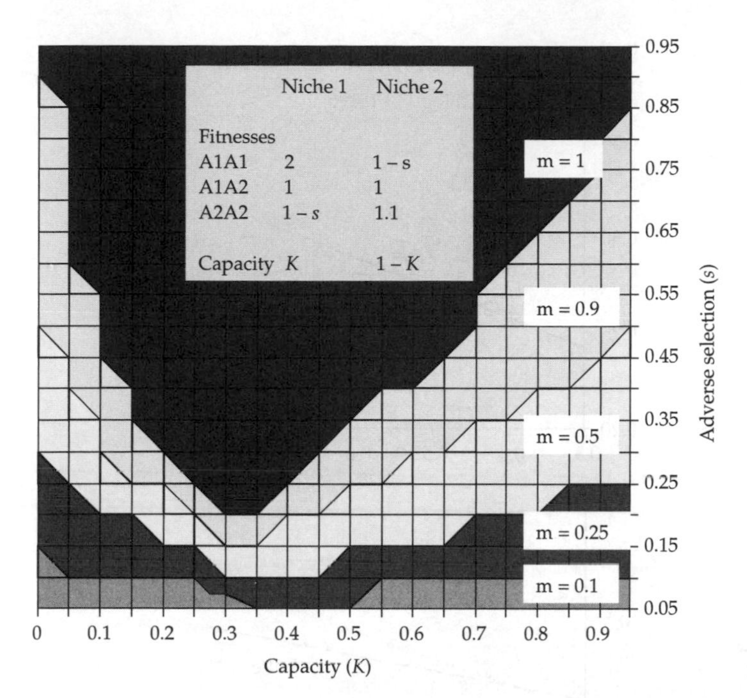

Figure 8.18 Conditions for protected polymorphism in a structured environment.

them $r_{ij} = -1$, so that divergent selection will conserve diversity (Via and Lande 1985). This leads to two difficulties. The first is merely that partitioning a region into different substrates would require a very large number of experiments. The second is that the definition applies only to the set of strains or species included in the calculation of r_{ij}, and will not necessarily hold for any other set. Consequently, no uniform classification of sites into substrates is possible, either in principle or in practice, and information about ecological factors, however detailed and extensive, is not sufficient to predict the level of diversity that will be maintained. The best that can be done is to predict that diversity will increase with heterogeneity, since r_{ij} tends to decrease as the environmental variance increases (§8.2). The environmental variance itself is not straightforward to interpret: if only a single factor is measured, it may not have a large effect on fitness, whereas if several factors are measured it is not clear how to combine them into a single estimate of the variance. There have nevertheless been several attempts to demonstrate this effect, although most rely on very crude intuitive assessments of environmental heterogeneity (reviewed in Rosenzweig 1995).

We assessed the state of the environment for species of *Carex* (sedges) living in old-growth forest in southern Quebec by measuring the survival of clonally propagated ramets of 15 species implanted at 10-m intervals along three 1-km transect lines, relying on the plants themselves to integrate all the relevant environmental factors (Bell *et al.* 2000). For any pair of sites we could calculate the environmental variance (from the difference in survival of the experimental implants) and measure the species diversity (from a survey of the resident vegetation). Species diversity increased with heterogeneity, both within and among survey lines (Figure 8.19). This is consistent with niche separation among these closely related and visually similar species: the effect is weak, but this could be attributed to the small size of the sites. This experiment can also be used to estimate the grain of the environment: as the distance between two sites increases the genetic correlation of fitness will tend to fall, and will eventually become negative. If there are M distinct kinds of site then diversity will be maintained

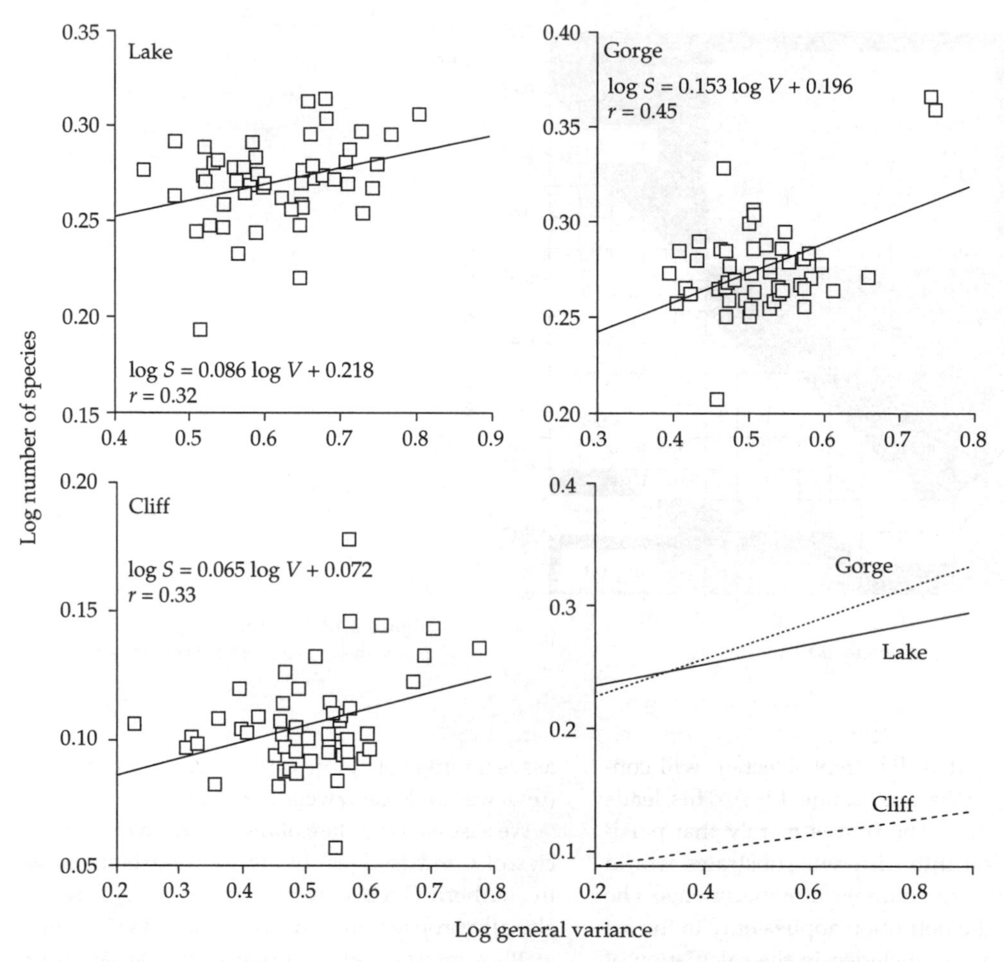

Figure 8.19 Environmental heterogeneity and species diversity of *Carex* in old-growth forest. Heterogeneity (general variance) is estimated as the variance of survival of clonal implants of 15 species of *Carex* between pairs of sites at 10 m intervals along three 1-km transect lines (Gorge, Lake, and Cliff). The number of resident species was recorded at each site. From Bell *et al.* (2000).

if on average $r_{ij} < -1/(M-1)$ (Dickerson 1955), so that if we assume M is fairly large, $r_{ij} < 0$ is a reasonable criterion for identifying the transition from one kind of site to another. We found that the correlation among closely related species of *Carex* fell as environmental variance increased, but became negative only at distances of about 1 km, at the limit of our experiments. This suggests that the rather dry upland forest that we were sampling constitutes a single kind of site from the point of view of the resident *Carex* species, at least so far as survival is concerned. This is consistent with the ecophylogenetic analysis (§13.3.3). Wetland sites are occupied by a completely different suite of species, and had our transect lines spanned both dry and wet sites we would have detected at least two kinds of site. We could not confirm, however, that the fine-scale heterogeneity detected by measuring environmental factors translated into the fine-scale response of the plants themselves that would result in the maintenance of diversity through divergent selection.

If all substrates are equally productive then a given set of alleles will be conserved if the harmonic mean of selection coefficients exceeds the dispersal rate m for each strain. If they differ in productivity then the less productive sites will receive a proportionately larger number of immigrants, and specialized strains can persist only if selection is more intense. In the simple two-allele two-substrate model a specialist adapted to substrate 2 will invade only if the combined productivity of these sites, as a fraction of total productivity, exceeds a certain level: $K_2 > (ms_1 - s_1s_2)/(ms_1 - ms_2)$ (Figure 8.20). For example, suppose that a mutation arises that halves fitness in one substrate but increases it by a factor of three in the other, with 70% of offspring dispersing; it will fail to spread unless the second substrate contributes more than about 8% of overall production. Thus, genotypes that specialize on rare substrates will seldom evolve.

Conversely, high rates of dispersal will inhibit adaptation to either substrate. If specialized populations are completely isolated from one another then the within-population genetic correlation will fall to around zero, as shown by Price and Schluter (1991). If there is dispersal among populations, however, most immigrants will be relatively poorly adapted to local conditions, while being relatively well-adapted to at least some conditions prevailing elsewhere. The within-population genetic correlation should then be negative when the appropriate comparison is made. The effect of dispersal may be mitigated when sites are aggregated and local dispersal is limited. If the sites providing a given substrate are randomly distributed, they will donate and receive migrants from other sites at rates proportional to their overall frequency. If these sites are aggregated, however, they will exchange migrants mostly among themselves provided that the typical dispersal distance is not much greater than the distance between neighbouring sites. The level of aggregation is expressed by the variance plot (§2.3), which generally indicates that viscous media such as soil tend to be coarse-grained. Organisms with short-range dispersal (such as *Carex*) should be more likely to radiate into a large number of specialized types, whereas long-range dispersers should evolve as generalists.

The simple two-substrate model used above assumes that each substrate contributes a fixed proportion of the overall population. This implies

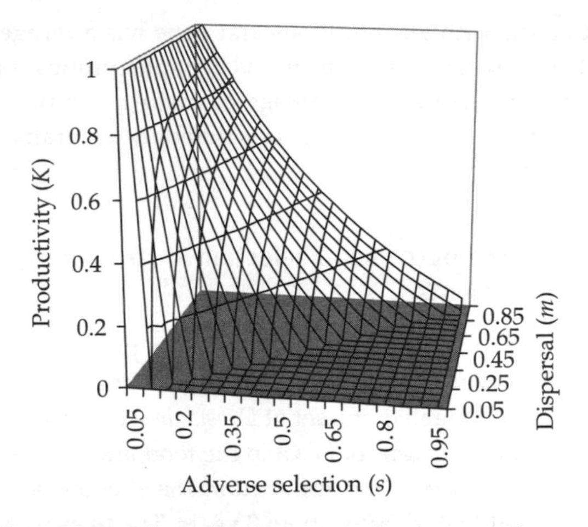

Figure 8.20 Polymorphism in a structured environment. The minimum productivity of sites for a new specialist to invade when the resident generalist has fitness 1 – *s* in these sites, fitness 2 in other sites, and the dispersal rate is *m*. The new specialist can invade if the productivity of the sites where it is superior lies above the plane of the plot.

that production is regulated at each site rather than at the level of the population as a whole. This is an essential condition for variation to be conserved by selection. If the production of each site depends on the mean fitness of its inhabitants, then one allele or strain will have the greatest overall fitness and will become fixed regardless of spatial variation in relative fitness (Dempster 1955). The effect of local regulation is best understood through the concept of a *refuge*: diversity is likely to be maintained if each specialized type has a refuge that protects it from extinction. A refuge is a resource of any kind that the specialist can exploit more effectively than any of its competitors, so that once established there it cannot be dislodged: it may be a physical location such as suitable patches of soil, but it can just as well be a substrate, a way of avoiding predators, or anything else that contributes to growth. A refuge guarantees that the specialist can persist, or that it can invade if it has become temporarily extinct. When many kinds of site or substrate are simultaneously available each is a potential refuge, and there are consequently many opportunities for specialists to evolve. When resources change over time, however, all lineages are obliged to cope with all conditions and none has a refuge. Thus, spatial variation provides opportunities for specialists to take advantage of whereas temporal variation imposes obligations that only generalists can meet.

8.4.5 The outcome of selection in structured environments

Dispersal is expected to reduce the precision of local adaptation. Wallace (1982) and Ehrman *et al.* (1991) propagated one set of *Drosphila* with increasing concentrations of NaCl in the food and another set with $CuSO_4$ and found that the flies became resistant to both after about 3 years. The lines were then mixed in cages where both types of toxic food were available and thereafter the level of resistance fell. Forbes Robertson (1966) set up cages with or without EDTA in the food. In isolated cages resistance soon evolved; in cages that were connected by narrow tubes, permitting a low level of dispersal, selection was less effective. In general, *Drosophila* experiments have shown that marked local adaptation occurs only when the frequency of immigrants is less than the fraction selected (Mather 1983). Endler (1977) set up a linear series of vials representing 4% increments of artificial selection in favour of Bar, a deleterious but easily scored chromosome duplication. When the vials were completely isolated the frequency of Bar increased sharply with the intensity of selection, in vials above the point in the series where artificial selection counterbalanced natural selection. The effect of dispersing 20% of flies in every vial to each of the two adjacent vials in the series was to make the experimental cline shallower, as expected.

Interest in structured environments usually concerns the maintenance of diversity rather than the precision of adaptation. The diversity evolving in *Pseudomonas fluorescens* populations maintained in unshaken vials (§8.1) is a simple demonstration of diversity maintained by balancing selection in a structured environment. Otherwise, surprisingly few experiments designed to test this extensive body of theory have been reported. Korona *et al.* (1994) described a pure effect of structure, without divergent selection. They isolated a soil bacterium capable of growing on the pesticide 2,4-D as sole carbon source and cultured it in liquid and solid medium. In liquid medium, each cell competes against every other and replicate lines converged on the same fitness. On the agar surface, by contrast, only neighbours can compete, and the outcome of selection was more variable. It is possible that many different combinations of mutations would be adaptive in the novel conditions of laboratory culture, many of which would persist in the structured environment because selection is local and hence less effective.

One approach to the maintenance of allelic diversity is to follow the frequency of specified alleles in different kinds of environment. Haley and Birley (1983) set up *Drosophila* population cages to represent environments with different degrees of heterogeneity. The simplest environment included vials of a single standard food medium made up of oatmeal and molasses. In more complex environments, two types of medium were available: the standard recipe and either fig or potato, in separate vials. The most complex environments contained all three media. Larvae would thus develop in a vial containing one of the three media, later emerging and mating with flies that in the more complex environments might have developed, and been selected, in a different food medium. They began their experiments with large populations that were segregating for alleles at several enzyme loci, and monitored the frequency of these alleles at intervals for about 30 generations. Allele frequencies shifted substantially at most of these loci, almost certainly through selection, and mean heterozygosity also changed over time. However, there was no simple relationship between genetic heterozygosity and ecological heterogeneity. The highest levels of heterozygosity were found in the most heterogeneous environments—those containing all three food media—but there was as much or more heterozygosity in the simplest environments as in those with two types of medium.

I set up a structured environment for *Chlamydomonas* by making culture media with different combinations of non-substitutable resources: nitrate, phosphate, and bicarbonate (Bell 1997a). The algae were grown in each medium separately for several generations before being mixed and redistributed to fresh vials to simulate free dispersal ($m = 1$). The corresponding uniform environment consisted of the same number of vials each filled with an equal mixture of all the resource combinations. Production was regulated globally by taking 1 mL from each tube at the end of growth, and locally by taking a fixed number of cells. The base population consisted of progeny from a set of crosses and contained a substantial amount of genetic variance for fitness. After 7 cycles (about 50 generations) genetic variance had been eliminated from the uniform environments but conserved at about the same level in the structured environments. There was no difference between environments with local and global regulation. This experiment seems to correspond most closely to a Levene-type model, and shows straightforwardly that selection is obstructed by environmental heterogeneity, effectively preventing the loss of genetic variance in fitness. It was probably too brief, however, to identify any effect of hard vs soft selection.

The experiment involving light and dark growth in *Chlamydomonas* described above included a fourth set of lines that experienced spatial variability with dispersal between sites: the Dark and Light base lines developed from the same founder were mixed, and this mixture used to inoculate Dark and Light sites. This process of mixture and redistribution was then repeated after each growth cycle. Because the appropriate conditions are always available for the growth of specialized types, they will tend to persist, and the composition of the population should be roughly the same as that of the mixture of the two base populations. This was confirmed: the tight clustering of specialized types in the base populations and continued lines becomes less pronounced, but genetic variance is conserved and the negative cross-environmental correlation is maintained (see Figure 8.13).

8.4.6 Fluctuating fitness

The loss of diversity through selection is retarded if the environment changes from generation to generation, so that first one type and then another has the greater fitness. If the fitness of a type over a sequence of t generations is $w_1, w_2, \ldots, w_t$ then its overall fitness is the product of these values, and hence the appropriate mean fitness is the geometric mean $\hat{w}=(\Pi w_i)^{1/t}$. If fitness does not vary through time then the geometric and arithmetic mean fitnesses are identical, which is the justification for using the arithmetic mean fitness when the environment is assumed to remain constant. When fitnesses vary through time, the geometric mean is less than the arithmetic mean by a quantity roughly equal to half the variance: $\hat{w} \approx \overline{w} - ½\, \sigma^2_w/\overline{\mathrm{w}}$. In this case, the population becomes fixed for the type with the greatest geometric mean fitness (Gillespie 1991). If all types have equal arithmetic mean fitness then this is the most plastic type, that is, the type expressing the least environmental variance. More generally, temporal variability will favour plastic generalists, because any specialist will occasionally encounter hostile conditions that greatly depress its numbers. Fluctuation of fitness over time cannot permanently maintain diversity in asexual populations. There are two major exceptions to this rule. The first is that allelic diversity will be preserved in a sexual diploid if the heterozygote has the largest $\hat{w}$ (Dempster 1955). The second is that at least two types can coexist in density-regulated populations living in a seasonal environment (§10.1).

In an infinitesimal model with two types that differ in plasticity, neither is completely eliminated; the frequency distribution becomes steadily more widely bimodal, but never becoming absorbed at fixation. This is an artefact of the model: the frequencies are bounded away from fixation because they will increase from time to time in favourable states of the environment, even though the types have become so rare that they would almost certainly have become extinct in a finite population.

8.4.7 Outcome of selection in variable environments

The results of the few experiments that have been reported support the general principle that generalists will evolve in variable environments. I have already described how generalists evolve when

Chlamydomonas lines are cultured in alternating light and dark conditions, and similar results have been reported in other systems. Bennett *et al.* (1992) propagated *E. coli* at constant or variable temperature and found that the variable-temperature lines were more plastic.

We exposed *Chlamydomonas* populations to different schedules of alternation between light and dark growth (Kassen and Bell 1998). With a photoperiod of 1 hour, each individual will experience about five changes in its lifetime: this should select for versatile generalists capable of mounting a rapid physiological response to environmental variation. With a photoperiod of 24 hours, on the other hand, conditions will change only every five generations, most individuals will experience no change, and selection should favour plastic generalists. Lines that were maintained in a constant environment became specialized, as in previous experiments, with a negative cross-environmental genetic correlation. After 250 generations of selection in a variable environment we obtained generalists that grew better than the light specialists in the dark and better than the dark specialists in the light. These generalists were insensitive to the time period of environmental change, however, so we failed to observe versatility and plasticity as distinct adaptations. Leroi *et al.* (1994) alternated *E. coli* lines between 32 °C and 42 °C at intervals of six or seven generations and found that they became superior to the ancestor at both extremes but actually became worse at making the transition between them. Thus, the lines evolved plasticity at the expense of versatility, perhaps because any given individual was unlikely to experience any change.

8.5 Local adaptation

8.5.1 The precision of local adaptation

When different treatments are applied to adjacent plots of land, whether intentionally or not, divergent selection may produce sharp differentiation between populations over very small distances. The Rothamsted Park Grass Experiment (§7.4.1) has provided some classical examples. The strength of selection involved in maintaining adaptation despite unrestricted gene flow between the plots was estimated by reciprocal transplants of *Anthoxanthum* tillers between comparable plots differing in macronutrient concentration or pH (Davies and Snaydon 1976). For survival, the performance of incomers, relative to residents, was about 55% for transplants from unlimed plots (low pH) to limed plots (high pH), and about 85% for transplants in the reverse direction; for vegetative performance, the corresponding relative growth rates were about 65% and 75%. Overall, the relative fitness of incomers relative to residents is about 60–70%, and this rate of selection maintains the sharp distinctions in physiological characteristics between the populations on adjacent plots.

Resistance to high concentrations of heavy metals may evolve rapidly among plants growing on old mine sites (§7.5.2), but even more remarkable than the rapid appearance of adaptation is its extreme specificity. At most sites, metal concentration drops off steeply at the edge of the old working, so that there is an abrupt change, often marked by a wall or a ditch, from the scanty vegetation of the mine to the much lusher pasture. Mine and pasture populations are genetically differentiated, with tolerance increasing steeply as one passes from pasture to mine, the frequency of tolerant individuals paralleling the concentration of metal in the soil. This differentiation is maintained despite the unrestricted movement of seed and pollen across the boundary. The mine site is often quite small and completely surrounded by pasture, so that a large proportion of new seedlings are incomers, bringing pasture genes on to the mine site. The increase in frequency of non-tolerant genotypes caused by immigration must be opposed by intense selection against these genotypes, in order to account for the maintenance of tolerance on the mine. This is most clearly demonstrated by reciprocal transplants between mine and pasture soils, which show that pasture plants are often almost incapable of growth on the mine, and seldom have fitnesses, relative to mine plants, of more than about 0.3 (Jain and Bradshaw 1966).

The erosion of adaptation by immigration was very clearly demonstrated by McNeilly (1968), who studied a mine situated in a narrow glaciated valley down which the prevailing wind is funnelled. At the upwind end of the mine, where pollen and seeds are blown from the pasture population on to the mine, an abrupt increase in tolerance occurred over less than 20 m at the boundary between pasture and mine. Adult plants from the mine (tested as tillers) were much more tolerant than seedlings, grown from the seed that the same adult plants produced in the field; the difference between tillers and seedlings represents the extent to which adaptation is broken down in every sexual generation by pollen from the pasture, and is therefore a minimal estimate of the selection that must be acting in every sexual generation to restore adaptation. In this case, that selection is sufficient to keep mine and pasture populations distinct. At the downwind end of the site, the wind blows pollen and seeds from mine to pasture. The evolution of tolerance on the mine involves a cost of adaptation: the growth rate of tolerant plants in uncontaminated soil is less than that of non-tolerant plants, and reciprocal transfer experiments show that the fitness of mine plants on pasture, relative to the resident individuals, is usually about 0.7–0.9. There is thus rather strong selection against mine plants on the pasture (if this were not the case, of course, metal-tolerant plants would often be common even in uncontaminated pasture), but it is not nearly as strong as selection against pasture plants on the mine. Consequently, there is only a gradual decline in the frequency of tolerant adults, over a distance of about 200 m, as one passes from mine to pasture at the downwind end of the site. As before, seedlings are less well adapted than adults, showing that local adaptation tends to be restored in every generation, but less effectively because selection is weaker and gene flow stronger.

Abrupt ecological transitions in natural systems can also lead to local adaptation. Soils derived from serpentine rocks have high magnesium/calcium ratios, are poor in nutrients, and have high levels of nickel and chrome. They are toxic to most plants and are consequently unproductive and tend to be occupied by a restricted set of species forming a distinctive plant community. Some species that grow on both serpentine and non-serpentine soils show pronounced local adaptation, seeds from the non-serpentine populations performing poorly when sown into serpentine soil (Kruckeberg 1954). Clones of *Agrostis* collected from serpentine populations were much more tolerant of high magnesium and nickel concentrations in liquid culture than clones collected from normal calcareous soil (Proctor 1971). Other sharp transitions to bog, fen, heath, and so forth are likewise associated with characteristic shifts in community composition, and the species involved are likely to show a high degree of local adaptation.

Where edaphic heterogeneity is less clearly apparent, within a region of forest or grassland for instance, populations may nevertheless be adapted to local conditions of growth. Where adequate surveys are available, the distribution of any given species, with respect to any given environmental factor, can be compared with the range of conditions available throughout the region: more highly specialized species will be found in a narrower range. Thus, if the variance of a given factor among the sites occupied by a species is σ^2_{range} whereas the variance of all sites in the region is σ^2_{region} then the precision of local adaptation can be expressed as $-\log(\sigma^2_{range}/\sigma^2_{region})$. The distribution of precision for a survey at 1-ha grain of native plants growing in 1000 ha of old-growth forest at Mont St-Hilaire, Québec, resembled a left-skewed normal distribution on the log scale, with a mode at small positive values, for several physical factors (Bell *et al.* 2001) (Figure 8.21). The overall mean for these factors was +0.085, corresponding to 82% of the overall environmental variance, so that most species occupy a large fraction of the available range of growth conditions. Moreover, this figure does not unambiguously represent local adaptation alone. A neutral community model readily gives rise to spatial pattern through limited local dispersal, and when the output of neutral models is projected on to a real landscape it generates a surprisingly similar estimate of precision. The hallmark of local adaptation is not restricted distribution in itself, but rather a repeatable association of species with the same narrow range of conditions. This can be evaluated by aggregating contiguous sites into larger blocks, so that if a species occupies similar

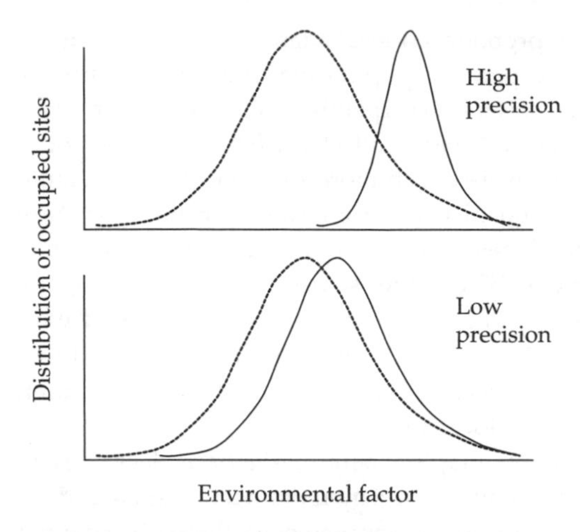

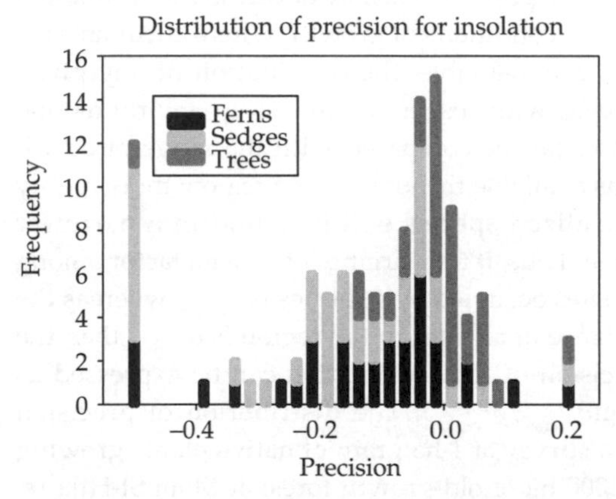

Figure 8.21 Precision of local adaptation in forest plants. The upper figures illustrate the concept of high and low precision. The lower diagram shows the distribution of estimates of the precision of adaptation relative to insolation; other factors give similar distributions. From Bell *et al.* (2001).

sites (with respect to a given environmental factor) in all blocks then the among-block variance will be small; thus, $-\log(\sigma^2_{\text{among,range}}/\sigma^2_{\text{among,region}})$ will express the accuracy of local adaptation. In practice, however, survey data is still not consistently distinguishable from the output of neutral models, largely because many species with narrow distributions occupy different conditions in different blocks. The failure to observe a clear signal of local adaptation may be attributable to the general principle outlined in §8.2: the genetic correlation falls as environmental variance increases. The variance of average conditions in 1-ha quadrats of forest may not be sufficient to generate substantial local adaptation, which is readily detected only when more pronounced ecological transitions, such as those associated with serpentine soils, are present.

8.5.2 Reciprocal transplant experiments

Individuals belonging to the same species often have different forms in different localities, and it was long a matter of contention whether these were heritable, or attributable to the direct action of the environment. This issue can best be resolved by the transplant experiment, introduced by Kerner in Austria in the 1870s, and later practised by Bonnier in France, Turesson in Sweden, Clausen in the United States, and many others. The simplest type of experiment is to bring plants from different localities into a *common garden*, where the differences they continue to express must be primarily heritable rather than environmental. Such experiments established that many of the distinctive forms of the same species found in different kinds of environment (*ecotypes*) are genetically different from one another. It was natural to speculate that each form has become adapted to the habitat in which it is found, through selection that varies from place to place. The possibility of local adaptation can be investigated by setting up several gardens, each representing a particular set of conditions; Kerner, for example, used alpine and lowland stations, and Clausen and his colleagues grew plants at a series of sites at different elevations. It is then possible to compare the growth of plants in their native locality with their growth elsewhere. In most cases, plants grow poorly in unfamiliar surroundings: lowland species rarely flower in the Alps, and lowland populations of widely distributed species of *Potentilla* or *Achillea* cannot survive at high altitudes, where there are flourishing resident populations of the same species. The most powerful version of such transplant garden experiments would involve following the changing frequencies of different types in populations established at different stations, but I have not found any studies of native vegetation comparable with the composite cross trials of barley. However, a nearly equivalent technique is to move plants between localities, comparing the growth of resident plants, themselves dug up and replanted, with that of individuals from elsewhere. Such reciprocal transplant experiments are relatively easy to perform, and allow local adaptation to be studied at any spatial scale. Individuals are collected from a number of sites, and the output of the experiment is then a matrix whose rows are the sites from which they were collected and whose columns are the sites at which they were grown. The leading diagonal cells then give the growth of residents, enabling two kinds of comparison to be made. The first is the variation among assay sites (columns) for a given source. The resident strain is expected to grow best at its home site, provided that dispersal is unrestricted. The second comparison is the variation among source sites for a given assay site. The resident strain is again expected to be superior, provided that selection is effective. It is this comparison, therefore, which is the appropriate test of local adaptation (see Kawecki and Ebert 2004).

The outcome of experiments involving clearly differentiated ecotypes or closely related species has been reviewed by Schluter (2000, Table 5.1). In most cases (34/42 studies) residents were superior to incomers at their home site. Even at very small spatial scales, local adaptation may still be detectable, if the sites are sufficiently different. In one striking study, for example, ramets of *Ranunculus* were transplanted between adjacent grassland and woodland sites (Lovett Doust 1981). The residents consistently produced more stolons and leaves. Interestingly, the effect was asymmetrical, woodland incomers to the grassland being much more severely handicapped than grassland incomers to the woodland: for total ramet production, for example, the relative performance of incomers was 0.22 in the grassland but 0.93 in the woodland. This may have been because the grassland population may have evolved from the woodland population when the wood was cleared, hardly more than a decade previously. The experiment would then have censused both current selection and the historical effects of past selection, the grassland population still being adapted in some degree to woodland conditions. Studying recently cleared woodland is perhaps uncomfortably close to studying abandoned mines; still, the intensity of selection against incomers is impressive. The selection coefficients involved in similar studies were discussed in §7.2.

I have collated information on similar experiments involving morphologically undifferentiated populations (Table 8.1). These give a different picture: 14 studies found a fairly clear and consistent advantage for residents, but 18 did not, with 2 reporting variable results. There are some striking instances of cryptic local adaptation over very small distances with no obvious environmental transition. Waser and Price (1991) found that seeds of *Delphinium* planted within 1 m of their maternal parent gave more vigorous plants than seeds obtained from 50 m away. The home site advantage was consistent with *outbreeding depression*: crosses between individuals growing 10 m apart in the field produce more seed and more vigorous offspring than crosses between close neighbours 1 m apart, and more than crosses between total strangers 100 m apart (Price and Waser 1979). These cases are exceptional; most of the positive results in the table involve large distances (100–3000 km) or clear ecological transitions (e.g. saline soil). It is true that some of the studies listed in Table 8.1 involved very few sites or individuals and would have had very limited power to detect local adaptation. The predominantly negative tone of the experimental results, however, is consistent with the principle that the cross-environmental genetic correlation becomes negative only when genetic or environmental differences are sufficiently large (§8.2).

The *Bacillus* isolates used to illustrate the ecogenetic landscape furnished an extensive reciprocal transplant experiment, with two replicate clones from each of 100 sites within a 1-ha plot being each cultured at every site (that is, the soil–water medium corresponding to every site). Overall, there was no evidence of a home site advantage. Most isolates, however, grew very poorly at all sites, and most sites did not support growth. If these strains and sites are excluded, the remaining more vigorous isolates showed clear evidence of local adaptation: the fitness of the resident was on average about 50% greater than the mean fitness of other isolates. This advantage declined

Table 8.1 Reciprocal transplant experiments with plants comparing morphologically undifferentiated material

Genus	Context	Locs	k_m	Material	Character	?	Reference
Agrostis	Various habitats	5	100	Tillers	Growth, phenology	No	Rapson and Wilson (1988)
Anoda	Ruderal/ agricultural	2		Field seeds	Fruit yield	No	Rendón and Núñez-Farfán (2001)
Anthoxanthum	Xeric/mesic	2	0.2	Field seeds	Seedling survival	No	Platenkamp (1991)
Arabidopsis	Shaded/unshaded	2	10	Field seeds	Phenology, fecundity	No	Callahan and Pigliucci (2002)
Aristida	Dry sandhills	7	400	Field seeds	Seedling survival	No	Gordon and Rice (1998)
Bouteloa	Deep/shallow soil	2	0.7	Ramets	Survival, fecundity	No[a]	Miller and Fowler (1993)
Bromus	Various habitats	7	100	Field seeds	Net reproduction	No[b]	Rice and Mack (1991)
Carex	Temp/Arctic wetland	5	2000	Tillers	Survival, growth	Yes	Chapin and Chapin (1981)
Carlina	Grassland	12	300	Field seeds	Germination, survival	No	Jakobsson and Dinnetz (2005)
Chamaecristis	Agricultural sites	3	1000	Families	Seed yield	No[c]	Galloway and Fenster (2000)
Delphinium	Meadows	3	0.05	Field seeds	Fitness	Yes	Waser and Price (1985)
Dryas	Fellfield/snowbed	2	0.1	Adults	Survival	Yes[d]	McGraw (1987)
Fraxinus	Woodland sites	8	250	Field seeds	Seedling growth	No	Boshier and Stewart (2005)
Hordeum	Soil salinity	3	50	Cuttings, seeds	Flower yield	Yes	Wang and Redmann (2005)
Hydrocotyle	High/low sand dunes	4		Ramets	Survival	No	Knight and Miller (2004)
Impatiens	Floodplain/hillside	2	0.1	Families	Seed yield	Yes	Bennington and McGraw (1995b)
Impatiens	Open/forest	2	0.1	Inbred lines	Seed yield	Yes[e]	Donohue *et al.* (2001)
Impatiens	Forest edge/interior	2	0.03	Selfed seed	Survival, fecundity	Yes	Schemske (1984)
Impatiens	Forest interior	n/a	0.01	Families	Flower yield	Yes	Schmitt and Gamble (1990)
Impatiens	Forest interior	2	1	Seedlings	Fruit yield	No	Schoen *et al.* (1986)
Impatiens	Forest interior	n/a	0.05	Selfed seed	Fruit yield	No	Wettberg *et al.* (2005)
Lupinus	Grassland/ scrub	3		Field seeds	Fruit yield	No	Helenurm (1998)
Lupinus	Dune/grassland		1	Families	Seed yield	Yes	Kittelson and Maron (2001)
Mimulus	Elevational series	12	100	Families	Fitness	No[f]	Angert and Schemske (2005)
Phlox	Contrasting sites	8	500	Field seeds	Net reproduction	Yes	Schmidt and Levin (1985)
Plantago	Various habitats	5		Field seeds	Seed yield	Varied	Lotz (1990)
Plantago	Agricultural	3	100	Field seeds	Seed yield	Yes	van Tienderen and van der Toorn (1991)
Polemonium	Summit/krummholz	2	–	Families	Survival	Yes	Galen (1996)
Polymnia	Mesic/xeric	4	35	Seedlings	Size, phenology	No	Bender *et al.* (2002)
Potamogeton	Various sites	3	4000	Cuttings	Growth	No	Santamaria *et al.* (2003)
Senecio	Ruderal/agricultural	6	25	Field seeds	Survival, growth	No	Leiss and Müller-Schärer (2001)
Zostera	Marine littoral	2	2	Cuttings	Biomass	Varied[g]	Hammerli and Reusch (2002)
3 spp plants	Various sites	8	3000	Field seed	Survival, fecundity	Yes	Joshi *et al.* (2001)
3 spp shrubs	Tropical forest	2	10	Seedlings	Survival, growth	No	Fetcher *et al.* (2000)

Locs, number of localities; k_m, maximum distance between transplant sites; ?, was the resident more fit at its home site than incomers?
[a]except survival at one site; [b]except for two most extreme sites; [c]some home edge at distances of >1000 km; [d]very few individuals; [e]at low density only; [f]home edge for 2 species, but not for populations within species; [g]home edge in one site, not in other.

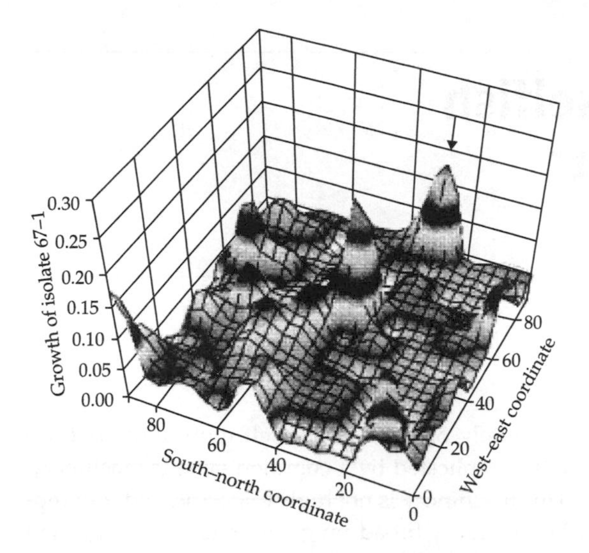

Figure 8.22 Local adaptation in a soil bacterium. The figure shows the growth of one isolate at all sites; it was isolated from the site marked with an arrow. From Belotte *et al.* (2003).

exponentially with distance away from the home site (cf. §7.2). The degree to which the community is fitted to local conditions can be visualized by plotting the growth of an isolate on a map of the hectare plot: the vigorous isolates stand on hilltops or ridges, showing that they have located the site where their fitness is greatest (Figure 8.22). In this case, selection and dispersal seem to have produced a close fit between the community and the landscape it occupies. This may be attributable partly to the unexpectedly high environmental and genetic variation expressed by the community, and partly to the extent of the experiment and the ease and repeatability of growth measurements.

Niche separation should lead to a close and complete fit of the community to the environment it lives in, as exemplified by the *Bacillus* experiment. The ambiguous results of reciprocal transfer experiments and the rather low precision of local adaptation suggest that this fit is in fact far from perfect. Now that the world has been turned into a large (although poorly designed) reciprocal transplant experiment by modern travel and commerce, this point is easy to appreciate. If communities were optimally adapted, successful invasions would be very rare, and when they occurred would result in the extinction of the resident species they displaced. Invasions seem to be commonplace, however, and do not usually cause the extinction of similar species. An unusual opportunity to quantify this impression was provided by the construction of the Suez Canal, linking the Pacific and Caribbean drainages of Central America (Smith *et al.* 2004a). These were occupied by different freshwater fish communities that had evolved independently. Since the Canal was completed in 1914, about 50% of species have crossed from one side to the other and have become established in their new territory. So far as can be ascertained, no species have become extinct as a consequence. In this case at least, each resident community was neither optimally nor completely adapted to its environment.

CHAPTER 9

Autoselection: selfish genetic elements

Parasites such as helminths or rust fungi are organisms that make a living by exploiting the growth system of their host, diverting nutrients obtained or elaborated by the host to their own growth. All organisms have systems for reproduction that are based ultimately on the replication of nucleic acids; and just as the growth system of an organism can be parasitized by other organisms, so the replication system of genes can be parasitized by other genes. We have already seen that a self-replicating system cannot be perfectly precise; nor can it be perfectly specific, replicating its own sequence and no other sequence whatsoever. This creates an opportunity for elements that are not autonomously self-replicating, but which rather utilize the common replication system of the genome. I have already described a simple case of this sort in connection with phage Qβ. The virus itself is a conventional parasite that uses the host cell as a source of raw materials. These can be assembled into new viral genomes only through the viral replicase, a diffusible protein encoded by the viral genome. The replicase, however, will also direct the replication of incomplete viral genomes that do not themselves encode the replicase. Such incomplete viruses (often called *defective interfering particles*) are parasites of the viral replication system that often appear in the later stages of viral infection. The Qβ experiments, indeed, for the most part document the evolution and diversification of such defective interfering particles.

This principle applies with greater force to the genomes of cellular organisms, that comprise a congeries of genes, most of which are incapable, separately or together, of self-replication, but are rather replicated by a common genetic machinery. This machinery is not highly specific, and will replicate a fairly broad range of genes; indeed, it will often replicate genes from very distantly related organisms, which is why genetic engineering is feasible. Replicators that are able to utilize this machinery will therefore persist, even if they are not expressed in the phenotype so as to benefit the genome, or the organism, as a whole. Because they utilize the machinery of replication without contributing to its maintenance, they may be called *selfish* or *parasitic* elements. Their persistence, or spread, is not determined by the degree of adaptedness to external conditions that they confer, but solely by the extent of their ability to be replicated along with the rest of the genome. The process by which such elements evolve can thus be termed *autoselection*.

In asexual organisms there will be little if any tendency for elements to be selected because they are able to replicate more rapidly than others in the same genome. If there is any cost of maintaining superfluous copies of an element in the genome, lineages that bear more over-replicating elements will proliferate less rapidly than lineages that bear fewer; and so the frequency of selfish elements within lineages will be kept low through selection among lineages. Nevertheless, redundant elements, which no longer contribute to the phenotype except as a burden, will continue to be replicated by machinery that is incapable of distinguishing between genes on the basis of their phenotypic effects. Pseudogenes that have long since lost their usefulness will persist for many generations, slowly degenerating as neutral mutations accumulate unchecked. The bits and pieces of genetic gibberish produced by recombinational excision and other processes will likewise continue to be replicated, until they are finally lost by the same sort of accident that created them. No doubt such elements are somewhat deleterious, insofar as energy and materials that could more profitably be deployed elsewhere are used to maintain them, but their effect on the cellular economy is probably very small. Incomplete and defective genes are quite abundant, especially in large genomes where their marginal effect will be less, and constitute the largest category of selfish elements in asexual organisms. They are rudimentary or vestigial structures, akin to the pelvic girdle skeleton of whales and snakes, or the eyes of cave-dwelling shrimps and fish. Vestigial structures do not furnish very good evidence of selection, except that it is very weak in the later stages of the loss of a redundant feature, but they do give conclusive evidence of ancestry. The abundance of vestigial structures in the genome shows how mere passive replication can prevent for long periods the loss of elements no longer functional, and thus emphatically suggests that the replication system might be actively exploited by elements that evolve for no other reason

Eukaryotic genomes, indeed, often house a whole menagerie of active and inactive parasitic elements and their remains (Zeyl and Bell 1995) (Figure 9.1). The evolutionary biology of genomic parasites has been authoritatively interpreted and reviewed by Burt and Trivers (2005), and their book should be consulted for a much more extensive account than it is possible to give here. They distinguish three categories of element in terms of the mechanism responsible for their spread:

- Infective or over-replicating elements such as plasmids and transposable elements replicate more often than the host genome and thereby increase in abundance.
- Interfering elements such as gamete-killers are transmitted to a disproportionate fraction of progeny by destroying competing alleles.
- Gonotactic elements such as B chromosomes achieve the same goal by routing themselves into the germline.

9.1 Infection

Any attempt to explain some interesting feature of an organism is likely to be dominated by the *phenotype paradigm*, that is, the presumption that it has evolved through natural selection because it increases the adaptedness of individuals. For elements within the genome, however, the phenotype paradigm may break down because they can evolve through autoselection without regard to adaptedness. The struggle to understand some kind of element within the genome is therefore usually a struggle between an adaptive interpretation in terms of its benefit to the individual and a non-adaptive interpretation that views it as a strictly selfish genetic parasite. Bacterial plasmids, for example, can be interpreted as effective vectors for beneficial mutations or as parasites that infect new lineages after cell fusion. The transposable elements of eukaryotes can likewise be viewed as engineers of gene regulation or as parasites whose over-replication by transposition is likely to degrade adaptedness. Either point of view may be correct, depending on circumstance. A rare element may be strongly selected for infectiveness despite causing considerable damage to the individual, whereas once it has become common opportunities for infection are few and it may become modified so as to benefit the individual.

9.1.1 Bacterial plasmids

Plasmids are quasi-autonomous genomes residing in bacterial cells, and may be large and powerful creatures with an impressive array of genes (Figure 9.2). In the chemostat, bacterial cells in all stages of growth occur at any given point in time, and the dynamics of the plasmid-bearing fraction of the population can be represented by differential equations in continuous time (Stewart and Levin 1977; Simonsen 1991). The life cycle of an individual plasmid-bearing cell involves a sequence of growth, plasmid loss through segregation, and infection through conjugation. The first of the three recursions representing these processes is the gain or loss of growth s caused by the plasmid: $p' = p(1 + s)/(1 + sp)$. The second is the probability of plasmid loss l through segregation: $p'' = p' - lp'$, where l is

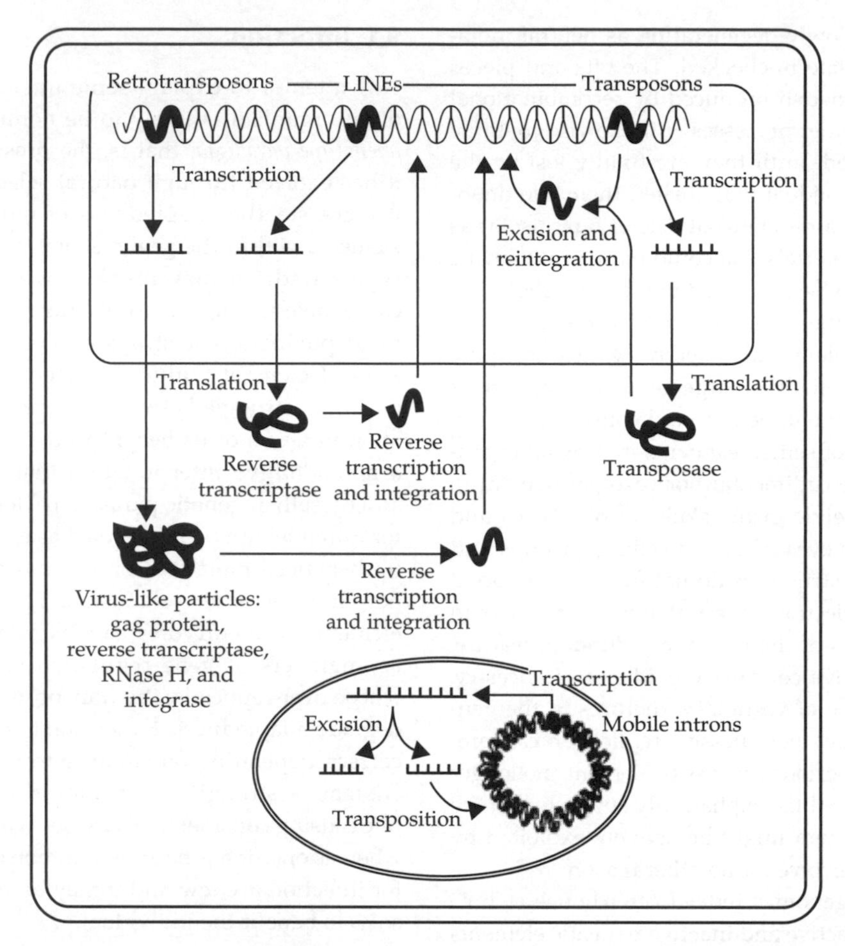

Figure 9.1 The nuclear and plastid genomes of eukaryotes harbour a host of parasitic genetic elements. Transposons encode their own excision and re-insertion, so they move from site to site in the genome. Retrotransposons are transcribed as RNA and the transcript is then inserted as DNA at another site. In structure and life cycle they are similar to retroviruses, except that they do not leave the cell. LINEs (long interspersed nuclear elements) are broadly similar to retrotransposons, although they lack any detailed genetic similarity to retroviruses. Mobile introns are found in the genomes of chloroplasts and mitochondria; a copy spliced from an RNA transcript can be inserted into the homologous site. From Zeyl & Bell (1995)

one-half the probability that a P+ cell will give rise to P+ and P– daughters. The third is the probability v that a P+ cell will encounter, conjugate with and successfully transfer a copy of the plasmid to a P– recipient: $p''' = p''(1 + vq'')$. Chaining these recursions gives an expression for Δp, which shows as expected that the condition for a rare plasmid to invade a largely plasmid-free population is $\Delta p > 0 \mid p \to 0$ if $(1 + s)(1 - l)(1 + v) > 1$, i.e. the plasmid will invade if infection occurs more rapidly than plasmid loss combined with lineage loss when the plasmid reduces growth. If s is modest, an invading plasmid reaches an equilibrium frequency of

$$p^* \approx (X - 1)/[(1 - l)(X - Y) + 2s]$$

where

$$X = (1 + s)(1 - l)(1 + v) \text{ and}$$
$$Y = (1 + s)(1 - l) + s(1 + vl).$$

For a plasmid that has no effect on growth at all ($s = 0$), $p^* = X/v(1 - l)^2$. A plasmid with a slight negative effect on growth will fail to spread so long as

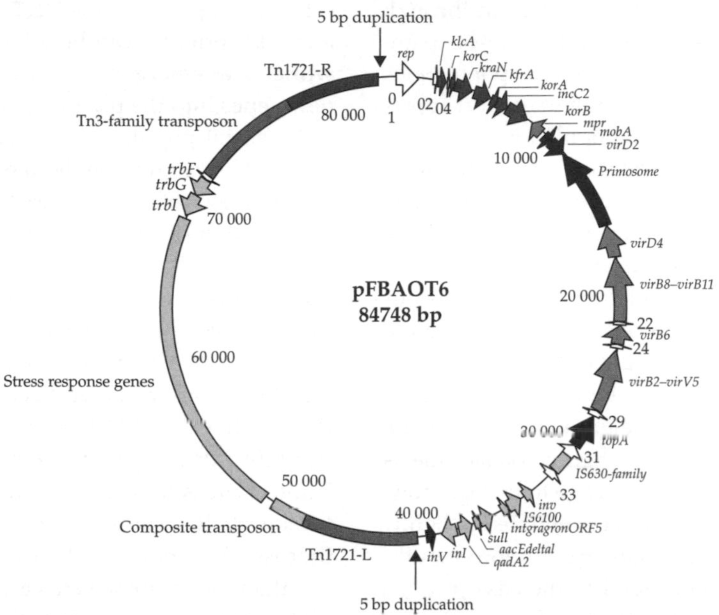

Figure 9.2 A wild plasmid. This large plasmid, which consists of nearly 100 coding sequences, was isolated from *Aeromonas* in hospital effluent. The first one-third of the plasmid (clockwise from 12 o'clock) contains genes responsible for replication, maintenance, and transfer. The remainder of the genome includes resistance genes, an integron, and several kinds of transposable element. See Rhodes *et al.* (2004). Source: NERC Centre for Ecology and Hydrology at *www.ceh.ac.uk*.

$v < |s|$. If the transfer rate is higher there is a narrow band of values of l within which the plasmid invades but is restricted to a low or intermediate frequency, but if $v > |s| + l$ about then the plasmid spreads to very high frequency. In most cases, therefore, the plasmid is either absent or nearly fixed at equilibrium (Figure 9.3)

Experiments that are designed to estimate kinetic parameters such as the specific transfer rate γ by detecting the appearance of transconjugants are necessarily selection experiments that directly demonstrate the invasion of the recipient population by the plasmid. If no invasion occurs then no estimate can be made. Licht *et al.* (1999) introduced a transfer-depressed version of the R1 plasmid of *E. coli* into chemostat populations, for example, and watched it sweep rapidly through them until after a few days almost all cells bore the plasmid. Experiments such as these show that horizontal transfer is rapid relative to segregational and competitive loss when a population is invaded by infective plasmids. Natural isolates may be much less infectious, because efficient transfer will

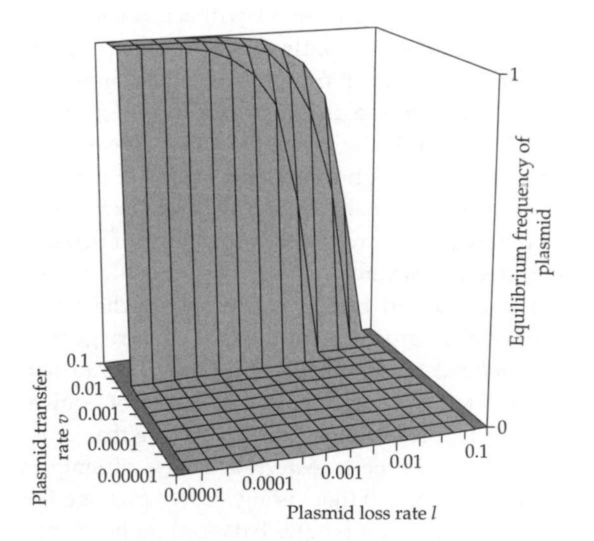

Figure 9.3 Equilibrium frequency of a bacterial plasmid in relation to rates of transfer and loss. The plasmid causes a slight loss of growth, s = 0.01.

become neutral or deleterious once a plasmid has become fixed in a population. Rates of transfer may be lower in natural environments where spatial structure reduces the rate at which uninfected cells are encountered. Nevertheless, infection through horizontal transfer is the straightforward explanation of how plasmids invade bacterial populations and therefore of how they are maintained at appreciable frequency in the metapopulation.

The alternative interpretation of the ubiquity of plasmids is that they contribute to adaptation, especially in patchy or variable environments. Chemostat populations will move towards mutation–selection balance where any given deleterious mutation is rare but each cell carries some load of mutations, which may be light or heavy. A plasmid will initially occur on a random background comprising some number of deleterious mutations; it is unlikely to arise in a very lightly loaded line. A plasmid that shuttles from lineage to lineage, however, will gradually become concentrated in lightly loaded lineages because they grow more rapidly. Its fitness will therefore tend to increase through time. If it happens to bear a beneficial mutation, then the mean fitness of the population is likely to increase in consequence.

If the plasmid bears a beneficial mutation then the enhanced growth of the cell and the horizontal transmission of the plasmid will act in concert to drive the mutation rapidly through the population. Consequently, if a particular beneficial mutation occurs both on the chromosome and on a conjugative plasmid, the plasmid-borne version will spread to very high frequency while the chromosomal version remains rare. The chromosomal version may persist and eventually spread if plasmid carriage has severely deleterious side-effects, but if the enhanced fitness conferred by the mutation is much greater than the cost of carriage then a beneficial mutation situated on a plasmid will be selected over a chromosomal gene of equivalent effect. Bergstrom *et al.* (2000) argue that genes will eventually be transferred from a plasmid to the chromosome if there is any cost of carriage, but moderate costs are readily balanced by horizontal transmission. Suppose there were a chromosomal locus C with wild-type allele C1 and beneficial mutation C2, and likewise a plasmid locus P with equivalent wild-type and beneficial alleles P1 and P2. Granted that C2P1 is more fit than C1P2 because of the cost of carriage, it does not necessarily follow that C2P1 will be fixed, because C1P2 × C2P1 conjugation yields C1P2 and C2P2 descendants. 'Eventually' can be a long time, and meanwhile new emergencies will continually recruit new genes into the plasmid genome

A natural population of bacteria will grow in a range of sites, and may become locally adapted to the conditions of growth in each. Thus, type A might grow well in site *A* where it has long been resident, type B in site *B* and so forth. If the types are specifically adapted to the sites where they are found then $w(\mathrm{A},A) > w(\mathrm{A},B)$ and $w(\mathrm{B},B) > w(\mathrm{B},A)$ where $w(\mathrm{I},J)$ means the fitness of type I at site *J*. Suppose that a mutation arises in A that is beneficial in both *A* and *B* with fitnesses $w(\mathrm{A}^*,A) = w(\mathrm{A},A) + \Delta w$ and $w(\mathrm{A}^*,B) = w(\mathrm{A},B) + \Delta w$. It spreads rapidly through the population at site A because $w(\mathrm{A}^*,A) > w(\mathrm{A},A)$ but fails to invade site B because $w(\mathrm{A}^*,B) < w(\mathrm{B},B)$. This will be the case when the effect of a beneficial mutation Δw is less than the combined effects of alleles responsible for the degree of local adaptation $w(\mathrm{B},B) - w(\mathrm{A},B)$. If the mutation is borne on a plasmid, however, it will be transferred to the B background when a cell from *A* infects a cell from *B*, and can then spread through the type B population.

In a fluctuating environment a plasmid may persist because it is able to shuttle from the current wild type to a new beneficial mutation in the process of spreading. Suppose that the environment is currently in state *A* and the population has become dominated by the well-adapted type A which may (P+) or may not (P–) bear a plasmid. Conditions then change to *B* and a mutant type B begins to increase in a P– background. This type is fated to become fixed, driving all A alleles and the plasmid to extinction. Before this happens, however, the plasmid may escape its fate by transfer to a B cell. This is unlikely to be important if the plasmid is common, since the mutation is then likely to occur in a P+ cell, and unlikely to be effective if the plasmid has long been fixed, because it may then have loss its infectivity. It may, however, effectively prevent the complete loss of the plasmid in a population whose conditions of growth change rather rapidly.

9.1.2 The 2-micron plasmid of yeast

Most strains of yeast, *Saccharomyces*, contain a few dozen copies of a circular DNA molecule about

6 kb in length, the so-called 2-μm circle or 2-μm plasmid. Although it is an unusual creature—free plasmids are rarely found in eukaryotic cells—we understand its maintenance and spread more clearly than that of conventional bacterial plasmids (Futcher *et al.* 1988) (Figure 9.4). In strictly asexual populations of yeast, the 2-μm circle is slowly lost through two processes. The first is vegetative segregation. Because there is no mechanism that ensures an equal partition of copies between the mother cell and its budded daughter, some cells will receive more copies and others fewer. In each generation, a small proportion of cells, descending from parents who themselves possessed few copies, will receive no copies of the plasmid at all. The lineages they found are permanently cured of the plasmid, which cannot re-infect them as long as the population remains asexual. There is therefore a continual increase in the frequency of *cir*^{0} cells, although in practice the average copy number is so high that this is only a weak force. Selection against *cir*$^+$ lineages is probably more important. The plasmid does not encode any vegetative function, and *cir*^{0} strains are apparently normal in all respects. When isogenic *cir*$^+$ and *cir*^{0} strains are mixed, the frequency of *cir*$^+$ falls by about 1% per generation, presumably because of the metabolic burden that 50 or 60 copies of the plasmid imposes on the host cell. Thus, the plasmid cannot spread in asexual

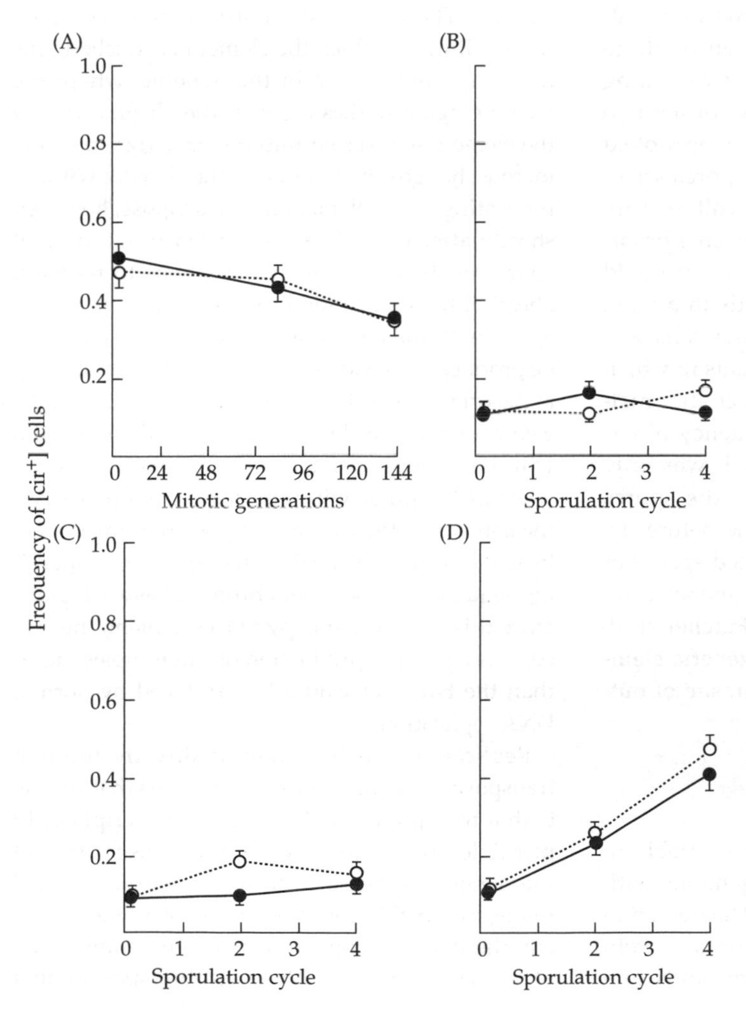

Figure 9.4 The 2-micron plasmid of yeast. These figures show how the frequency of the 2-micron plasmid changes in experimental populations of yeast, from Futcher *et al.* (1988). The two lines in each figure are two different strains. A. In asexual mitotic populations the frequency of the plasmid-bearing cir$^+$ strains declines slowly. B. If cultures pass through successive sexual cycles, with gamete fusion occurring within the ascus, the plasmid does not increase appreciably in frequency. C. The same result is obtained if unfused vegetative cells are killed with ether during the sexual cycle. D. When the ascus is disrupted before gamete fusion, enforcing a high rate of outcrossing, the plasmid spreads rapidly through the population.

populations, and even if it is initially present at high frequency will be lost within a few hundred generations. It can be maintained permanently only in populations where cells go through a sexual cycle from time to time. When *cir*$^+$ and *cir*^{0} cells mate, the plasmid is distributed to all four meiotic products, so that the sexual progeny of *cir*$^+$ and *cir*^{0} parents are all *cir*$^+$. The lineages descending from the four meiospores all have a normal complement of plasmids, showing that the plasmid takes advantage of the opportunity for over-replication provided by the incomplete genome replication of the sexual cycle. If the plasmid is rare, *cir*$^+$ cells will double in frequency whenever the population as a whole goes through a sexual cycle. It must be noted, however, that this process demands, not merely sex, but outcrossing: mating between two *cir*$^+$ cells will not contribute to the spread of the plasmid. Yeast has two mating types, or genders, that define whether or not two haploid cells can mate. After meiosis in the diploid cell produced by sexual fusion, two spores are of one mating type and two of the other; all four are either *cir*$^+$ (if at least one parent was *cir*$^+$) or *cir*^{0} (if neither parent was *cir*$^+$). These cells are held within a thick-walled ascus like seeds in a fruit, and if the ascus is not disrupted sister cells will mate before being released. Experiments in which the ascus was allowed to remain intact after mating showed no increase in the frequency of *cir*$^+$ during six successive sexual cycles. It was only when the ascus was disrupted, and the spores from different asci allowed to mingle before the next round of mating, that the expected spread of the plasmid, roughly doubling in frequency after every sexual cycle, was observed (Futcher *et al.* 1988). This is a clear example of a genetic element that spreads infectiously as a parasite of outcrossed sexual lineages.

9.1.3 Transposable elements

Several kinds of genetic element are capable of moving from place to place in the genome, with or without replicating as they do so. The operation of a transposable element is based on a protein called a transposase or integrase that binds to a particular nucleotide sequence and cuts the adjacent DNA. In principle, the simplest element would consist of a gene encoding such a protein with a recognition site at each end. When the gene was expressed the enzyme would function as a transposase, excising the gene, its flanking recognition sites, and the bound protein. This complex would then diffuse in the nucleoplasm before settling in some different region of the genome, where the protein, acting as an integrase, would cut the DNA and insert the element. There are several kinds of element that are nearly as simple in structure and behaviour: these are DNA transposons, consisting of a single gene bounded by inverted repeats. The gene encodes a transposase which binds to sites in the repeats and cuts the chromatid to release the element. The protein molecules at either end then form a dimer to which the element is attached; this moves to another site in the genome, where the enzyme again makes a cut in the chromatid and the element is inserted into the gap. Transposition merely changes the location of the element without replicating it. An element may transpose, however, shortly after it has been replicated in the normal course of the cell cycle. If it happens to re-insert ahead of the replication fork then it will be copied again, with the consequence that three copies will be produced instead of two (the element remaining at the original site, the element that moved, and the element copied at the new site). One daughter cell will thus inherit a single copy, the ancestral state, whereas the other will inherit two. In other cases the gap left by the excised copy is not immediately ligated but instead repaired using the corresponding sequence on the sister chromatid as a template. Since this is the other copy of the element, the outcome is again the production of three copies rather than the two that would be produced by normal DNA replication

Replication can be achieved directly through transposition in more complex elements that encode both a transposase and a reverse transcriptase. In principle, this would involve the transcription of the element, reverse transcription to DNA, and reintegration: this generates two copies of the element, the parental copy and its reverse-transcribed offspring. In practice, RNA retrotransposons that follow this general theme are more complex because of a developmental constraint: any given RNA molecule can act as a template for DNA synthesis or for protein synthesis, but not for both. If it is translated it produces the transposase and reverse transcriptase, plus two structural proteins that form a capsule containing the enzymes together with two untranslated RNA copies. Reverse transcription takes place within this capsule, which afterwards disintegrates to allow the DNA transcripts to be inserted into the genome. The entire element thus comprises several genes, flanked by long terminal repeats (LTRs) that bear the sites recognized by the encapsulation genes.

There may be several or many copies of a given element in the genome. Copy number often varies among individuals, and may increase within an asexual lineage through new transposition. It does not increase indefinitely, but is rather limited either by selection against clones with more transposon insertions than average, or by active regulation of copy number, by the host genome or by the genome of the transposon itself. The evolutionary genetics of the transposable elements has been reviewed by Charlesworth and Charlesworth (1983), Charlesworth (1987), Charlesworth and Langley (1989) and Charlesworth *et al.* (1994). Transposons are akin to linear plasmids that have become incorporated with the host genome. Like plasmids, they are usually transmitted to all sexual progeny, through replicative transposition in the zygote. Bacterial transposons, like plasmids, often carry beneficial genes encoding functions useful to the organism, such as antibiotic resistance. Unlike plasmids, they often cause serious damage in eukaryotes, by inserting copies into functional genes, making it impossible to transcribe them accurately. Despite this, it is often argued that they are directly beneficial to the host individual or its lineage.

Transposons as regulatory elements

Transposons might contribute to adaptedness in much the same way as ordinary genes, perhaps by acting as developmental regulators or switches. They would then be selected like any other gene. Despite the preference of some transposons for inserting into the upstream regulatory region of genes, the notion that they are part of a subtle system of gene control does not seem very compelling. If they were, there is no clear reason for an element to encode its own transposase. Copy number and location vary among individuals but are substantially the same in different tissues of the same individual; if they behaved so as to regulate development, we would expect the opposite pattern. Transposons may be inactive in somatic cells, or even excluded from somatic cell lines, like B chromosomes. Moreover, individuals that lack transposons seem perfectly normal. The only evidence in favour of this interpretation is that bursts of transposition in certain stocks of *Drosophila* are accompanied by a substantial increase in fitness (Pasyukova *et al.* 1986). However, these stocks have been deliberately selected for *low* fitness under natural selection; they are very sick flies, perhaps because of transposon insertions. It is scarcely surprising that the disease is alleviated when the pathogen releases its hold.

Transposons as mutator elements

A much more popular proposal has been that the mutations induced by transposons are beneficial in the long term, by permitting a more rapid response to selection. Long-term selection is likely to be weak, and will rarely be effective when there is countervailing selection among individuals in the short term. Moreover, the mutations induced by transposition, involving massive disruption of gene sequences, seem unlikely to be beneficial in any circumstances (although if the transposon subsequently excises itself, the reconstituted gene may have changed only slightly). Nevertheless, there is some evidence that populations with active transposons evolve more rapidly.

Most of the experiments of this kind have involved artificial selection on bristle number in *Drosophila*. The transposons involved are the P-elements, which are known to be powerful mutagens. Crossing a P-bearing male with a female lacking P-elements causes a burst of transposition in the offspring, and consequently a great deal of transposon-induced mutation (Figure 9.5). This dysgenic effect persists for several generations,

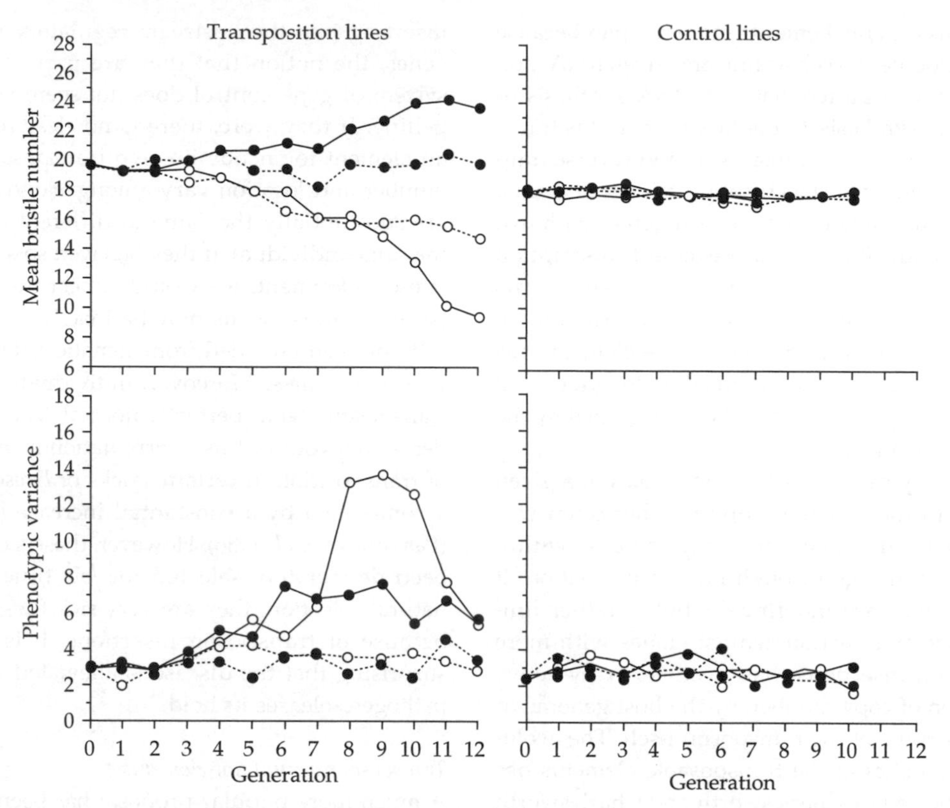

Figure 9.5 Transposons as mutators. These graphs show how P-element transposition affects the response to artificial selection for bristle number in *Drosophila*, from Torkamenzahi *et al.* (1992). The two graphs on the right refer to inbred lines lacking P-elements, the two on the left to lines transformed by P DNA. The lower pair of graphs give the phenotypic variance and the upper pair the mean of bristle number. The four plots on each graph are upward and downward selection lines founded from two base populations. The transposition lines are the more variable, and in three cases out of four respond to selection, whereas none of the control lines show any response.

until the number of copies of the element has approached its equilibrium value. The reciprocal cross (P-bearing female with non-P-bearing male) causes far less transposition, and the offspring are normal. If lines from dysgenic and non-dysgenic crosses are selected, response is much faster, either upwards or downwards, in the dysgenic lines (Mackay 1985, Pignatelli and Mackay 1989). Moreover, in the dysgenic lines a very rapid initial response seems to approach a limit after a few generations of selection, in contrast to the more gradual and nearly linear response of the non-dysgenic lines. At the same time, phenotypic variance in the dysgenic lines remains much higher than in the non-dysgenic lines, even when the response to selection has become very slow. This suggests that the transposon-induced mutations that cause variation in bristle number have deleterious pleiotropic effects on viability or fertility, so that further advance under artificial selection is prevented by natural selection acting in the opposite direction. Indeed, one gene that was identified as having a large effect in reducing bristle number in one of the downward selection lines from the dysgenic cross causes sterility in females, and cannot be maintained as a homozygote.

An alternative procedure is to compare inbred lines that lack P-elements with the lines developed by crossing them with nearly isogenic lines into which P-elements have been introduced. This has the advantage that the control lines (lacking P-elements) are expected to have very little genetic variance, and thus little or no response to selection in the short term. An experiment of this sort has given a clear demonstration of the effect of transposition on the rate of evolution (Torkamanzehi *et al.* 1992). Genetic variance for bristle number was generated by transposition about 30 times as rapidly as would ordinarily be expected from spontaneous mutation. These lines responded both to upward and to downward selection, whereas the control lines showed no response in either direction. The response in the transposition lines was greater for downward selection, where it was roughly equivalent to the response expected in a heterogeneous outbred population. Again, however, a large part of the response to downward selection was caused by a single mutation, causing the loss of bristles from many parts of the body, which had deleterious pleiotropic effects. Females carrying this mutation had very low fertility, and were rapidly lost from the population when artificial selection was suspended. For related experiments, see Mackay (1985), Morton and Hall (1985), Torkamanzehi *et al.* (1988) and Pignatelli and Mackay (1989).

These experiments show that transposon activity may increase the rate of response to selection by generating mutational variation in fitness. However, the severely deleterious side-effects often associated with transposition mutagenesis seem inconsistent with its playing a major role in adaptive evolution.

Transposons as hitch-hiking elements

Studies of the population biology of transposons in cultures of asexual microbes have shown that they can persist or spread as a consequence of being linked to the mutations they cause. This is obviously similar to the idea that they serve to increase the rate of response to selection. However, hitchhiking involves only selection among lineages within a population, a simpler and more plausible process than selection among populations, those linear which bear transposons replacing those which do not.

The yeast genome usually contains 20–30 copies of the retrotransposon *Ty1*, whose evolutionary biology has been investigated by Wilke and Adams (1992) and Adams and Oeller (1986). When a culture is founded by a strain bearing a single copy of *Ty1*, copy number increases through time, to an average of about 3 copies after about 1000 generations. Sex is never induced in these cultures, and probably never occurs. Moreover, adaptation to glucose-limited chemostats seemed to be associated with changes in copy number and location. It appears that *Ty1* transposition somehow benefits the cell. The most direct way of investigating this possibility is to allow strains that have undergone different numbers of transpositions to compete against one another; if transposition is sometimes beneficial to the cell as a whole, then lineages that have undergone appropriate transpositions will be selected, and the class with no transpositions will disappear from the population. The base population can be constructed by taking an exceptional strain that lacks *Ty1* and transforming it with a plasmid containing an active copy of the transposon under the control of a galactose promoter. In its new host, the transposon can be activated by growth on galactose, when it scatters between 1 and 20 copies of itself through the genome in an initial burst of transposition. If the cells are grown on glucose the transposon is quiescent and remains on the plasmid, and few or no transpositions occur. In this way, strains that have experienced different numbers of transposition events can be made. In pure culture, transposition is on average deleterious: the galactose-grown strains in which about a dozen transpositions have occurred reach asymptotic densities about 8% lower than those of the isogenic glucose-grown strains. It seems reasonable to infer that transposition has reduced metabolic efficiency, presumably by disrupting active genes. Lethal disruptions would not be detected by this assay, of course, so transposition is probably more damaging than these results suggest. Strains in which transposition had occurred are more variable, as one would expect; more surprisingly, although the majority are inferior to those in which no transposition had occurred, a few actually reach a higher final population density, suggesting that the mutations caused by transposition can occasionally increase the fitness of the host cell. This was tested by allowing the strains to compete in mixed culture for about 100 generations. The diversity of these mixed populations decreased through time, and the average copy number was less after selection than before. Nevertheless, *Ty1* was not eliminated through selection; instead, strains that had not undergone transposition were rare or absent by the end of the experiment.

Somewhat similar results have been reported by Chao and McBroom (1985) for the bacterial transposon *IS10*. Strains that bear the transposon usually replace strains without it in chemostat competition experiments (Figure 9.6). Moreover, if the *IS10* strain loses, the element has not transposed; and if it is excised from a successful strain, the strain loses its competitive superiority. Unlike *Ty1*, however, *IS10* inserts into a specific region of the genome in successful strains. This location is not far from a locus previously known to affect fitness in glucose-limited chemostats, and it is possible that *IS10* disrupts a regulatory sequence and causes constitutive expression of some uncharacterized enzyme. A similar explanation may explain the dynamics of *Ty1*, and would be consistent with what we know of the early stages of metabolic adaptation to novel environments (§3.7). The transposon *Tn5* is likewise a compound structure, comprising two *IS50* sequences and an internal region. The *IS50* element increases host fitness even without transposition, possibly by increasing promoter activity (Biel and Hartl 1983, Hartl *et al.* 1983). Modi *et al.* (1992) set up chemostat populations of *E. coli* bearing a plasmid containing the transposon *Tn3*; after about 1000 generations, isolates from the evolved populations showed that the insertion of *Tn3* into the bacterial genome was associated with substantial increases in fitness.

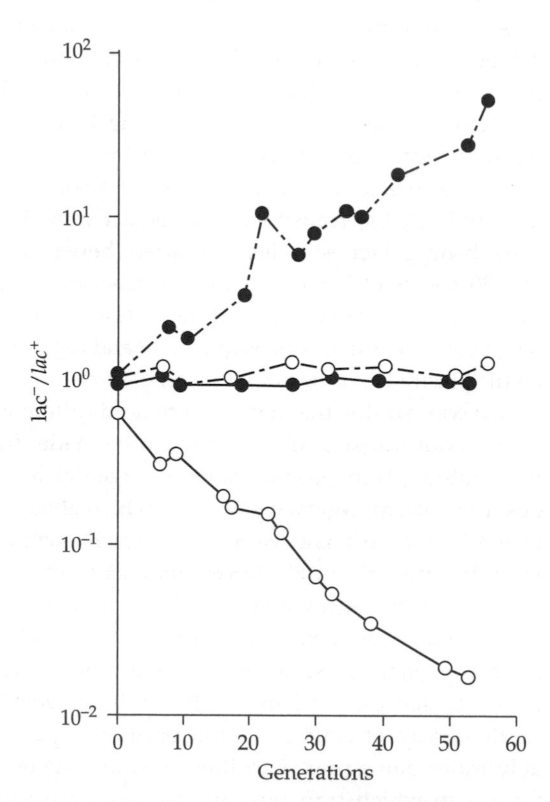

Figure 9.6 Hitch-hiking by a transposon. Chao and McBroom (1985) constructed four strains of *E. coli*, bearing all pairwise combinations of a selectable marker (*lac*⁻ vs *lac*⁺) and the presence or absence of the new *IS10* insertion. The two middle plots show competition between two strains, both of which either do or do not bear *IS10*; there is no appreciable change in frequency, demonstrating the neutrality of the lac marker in glucose-limited conditions of growth. The strain bearing *IS10* increases in frequency, relative to an isogenic strain, whether associated with *lac*⁻ (upper plot) or with *lac*⁺ (lower plot).

These experiments seem to establish that transposons can be maintained in populations through selection of the beneficial mutations which they sometimes cause. They seem entirely consistent with similar experiments on the mutator genes of bacteria. An element, whether a gene or a transposon, may cause a wide range of mutations, most of which are lethal or deleterious, while a few are beneficial in a novel environment. The superior variants produced in this way spread through the population under selection, and as they do so, both the mutant gene and the agent responsible for its mutation increase in frequency together. Mutator genes will behave like this only in asexual populations, where they remain linked to the mutations they cause; transposons normally cause mutations by being inserted into a gene, and are thus tightly bound to the mutant gene whether the population is sexual or asexual. This is a process of hitch-hiking. However, although it will cause the spread of mutation-inducing elements, in appropriate circumstances, it does not follow that the function of transposons is to induce mutations. After all, nobody would argue that the function of polymerase genes is to act as mutators by producing defective enzymes. Despite the clarity of the experimental work, it may be mistaken to interpret transposon structure as being specifically adapted to cause beneficial mutations in the host.

One weakness of the mutator theory of transposons is that it does not explain how a rare transposon can spread. If transposon-bearing cells are sufficiently abundant, a beneficial mutation induced by transposition is likely to appear sooner or later, and as it increases in frequency the transposon must rise too. However, if such cells are initially rare, they are likely to be eliminated, through selection against the deleterious mutations caused by transposition. An unstable equilibrium of this kind has, indeed, been demonstrated for the *IS10* element of *E. coli*. Hitch-hiking may contribute to the dynamics of established transposons, but is unlikely to be responsible for their initial establishment in the population.

Transposons as selfish genetic elements

A quite different interpretation of transposons is that they are genomic parasites that are specialized to exploit the sexual cycle of the host by over-replication, transmitting copies to all four products of meiosis (Doolittle and Sapienza 1980, Orgel and Crick 1980). They should therefore be most abundant in outcrossed sexual populations. In asexual populations they will be eliminated by selection among clones, because the clones that harbour the most transposons will suffer the highest rates of deleterious mutation.

Transposons are not rapidly eliminated from asexual populations, and mass cultures of microbes often seem to retain them, with little change in copy number or location, for long periods of time. Zeyl *et al.* (1994) found no appreciable decline in the abundance of *TOC1* or *Gulliver* (transposons are often more imaginatively named than ordinary genes) in selection lines of *Chlamydomonas* that had been perpetuated vegetatively for about 1000 generations. However, this is not a very powerful test of the hypothesis: when transposons are rather quiescent, with transposition rates of the same order as mutation rates, selection against clones with above-average numbers of transposons will be correspondingly weak.

The crucial issue is whether an active transposon will invade a sexual population, but not a comparable asexual population, when introduced at low frequency. The appropriate experiment is identical in concept with the study of the 2-micron circle of yeast, described above. Zeyl *et al.* (1996) inoculated sexual and asexual populations of yeast with a version of the retrotransposon *Ty3* in which transposition was induced by galactose and repressed by glucose. *Ty3* spread rapidly in sexual lines when transposition was induced in either the haploid (gametic) or diploid (zygotic) phases, with copy number increasing from 1 in the ancestor to an average of 2–3 after 8 sexual cycles. Oddly enough, it failed to spread when transposition was induced in both haploid and diploid phases. Under these conditions the lines are always cultured in galactose, which they are rather poorly adapted to; their fitness increased substantially during the experiment, presumably through the spread of beneficial mutations that would almost certainly arise among the much more abundant transposon-free cells. In asexual lines *Ty3* was usually lost. In two cases, however, a lineage retaining the ancestral insertion site, with no further transposition, spread through the population by virtue of an increased growth rate in galactose; these seem to have been examples of directly beneficial insertions. This experiment validates the interpretation of transposable elements as genetic parasites of sexual eukaryotes, while showing that their dynamics may also be affected by their effects, direct and indirect, on host fitness.

9.1.4 Spread of transposable elements

A new transposable element may spread very swiftly when first introduced into a population. P-elements introduced into laboratory populations of *Drosophila* at first almost double in frequency in each generation and become fixed within 15–20 generations (Good *et al.* 1989) (Figure 9.7). They were introduced into natural populations of *D. melanogaster* from *D. willistoni* (Engels 1992). Strains of *D. melanogaster* isolated before 1960 and since kept secluded are free of P-elements, whereas they infect almost all strains isolated after 1975; thus,

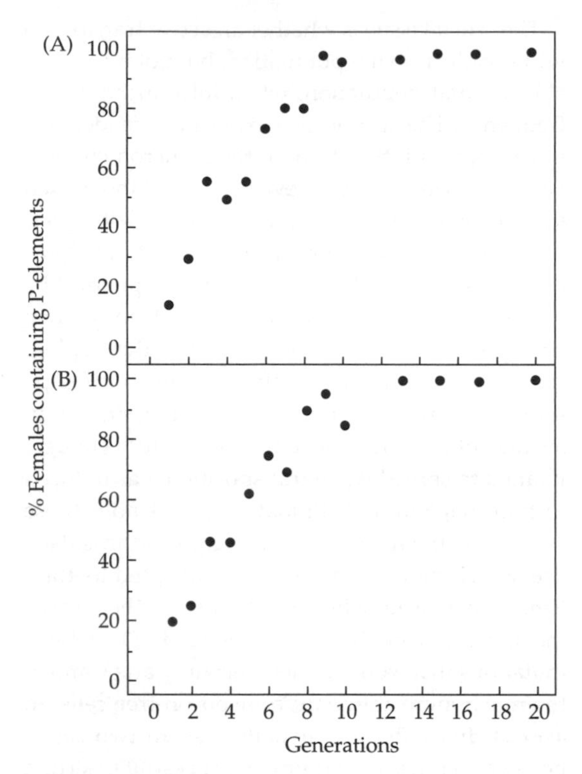

Figure 9.7 Rapid spread of P-elements in experimental populations of *Drosophila*. From Good *et al.* (1989).

these elements spread through the world population of *D. melanogaster* in less than 20 years. It is a remarkable fact that this event occurred in one of the very few organisms in which it would be detected, at precisely the time when its mechanism could be understood. It is difficult to resist inferring that similar events often occur in more obscure circumstances.

Transposable elements may be very active in naive host genomes, with transposition rates of greater than 0.1 (Eggleston *et al.* 1988). This rapidly declines once the element has become abundant because frequent transposition will reduce transmission by damaging already infected hosts, creating selection for more temperate elements and more resistant hosts. The rate of transposition of long-established elements is usually very low, scarcely greater than the mutation rate (Nuzhdin and Mackay 1995). Nevertheless, the replication caused directly or indirectly by transposition will drive a slow but continual increase in the number of copies borne by the genome.

9.1.5 Parasites of transposons

In DNA transposons and LTR retrotransposons the transposase acts in *trans* and can therefore be utilized by defective elements that do not themselves encode a transposase. These defective copies often replicate more rapidly than the intact element because they are shorter, and are consequently very common in eukaryotic genomes. In the limit, they may consist of little more than the repeated sequences that flank the central coding region of the active element. Thousands of more or less incomplete elements may accumulate, to the point where they comprise a large fraction of the genome. Our own genome, for example, contains hundreds of thousands of defective and completely inactive copies of many different kinds of DNA transposons (Lander *et al.* 2001). The proliferation of defective elements impairs the replication of intact elements, by competing for transposase, and may eventually drive them to extinction (Charlesworth and Langley 1989, Lohe *et al.* 1995). Nevertheless, actively transposing elements are not uncommon. This implies that the infection of naive hosts is a fairly frequent event, as the recent irruption of P elements in *D. melanogaster* suggests (Burt and Trivers 2005). On this view, transposable elements have a phoenix-like life cycle on an evolutionary timescale: a brief period of activity causing rapid invasion that is followed by a long period of quiescence and occasionally renewed by horizontal transfer.

In LINE (long interspersed nuclear element) retroposons the transposase acts in *cis*. They produce an mRNA molecule that is translated into a large protein responsible for DNA binding and reverse transcription. This protein binds to the mRNA from which it was translated, and then cuts the host DNA and inserts a reverse-transcribed copy of the attached RNA. This process is often faulty and results in the insertion of defective or incomplete elements, but these are unlikely ever to replicate because they have no access to the reverse transcriptase bound to intact transcripts. Consequently, defective copies do not proliferate as readily, and active elements have a much longer lifespan (Burt and Trivers 2005).

9.1.6 Selfish mitochondria

A small-colony type often appears in isogenic yeast cultures. These 'petite' colonies bear mitochondrial defects that abolish respiration, so the cells must grow through fermentation alone (Ephrussi *et al.* 1955). They arise at surprisingly high frequency (up to 10^{-2} per division), presumably because the phenotype can be caused by a broad range of mutations. They spread because defective mitochondria often replicate faster than normal, perhaps because the mutant mitochondrial genome contains an elevated number of origins of replication (MacAlpine *et al.* 2001). This gives them a within-cell advantage that can more than compensate for their debility in between-cell competition. They are nevertheless rare in natural populations, partly because yeast are highly inbred, and partly because the parental types rapidly segregate out, thus restoring between-cell competition (see Hurst and Hamilton 1992). Petites are a simple example of a widespread tendency for selfish mitochondria and other organelles to spread through selection at one level (such as within-cell competition) while being opposed by selection at another level (such as between-individual competition). The associated phenomena, such as cytoplasmic male sterility in plants, are the subject of a large literature, reviewed by Burt and Trivers (2005).

9.1.7 Population genetic engineering

Homing endonuclease genes (HEGs) encode a site-specific endonuclease that recognizes a unique sequence on the homologous chromosome and cuts it, after which the element is copied into the gap it has created by the cell's repair system (Figure 9.8). The recognition sequence is disrupted by the insertion, so the element does not cut its own chromosome. The host gene is also disrupted, of course, but HEGs are usually associated with self-splicing introns that remove the HEG sequence from the

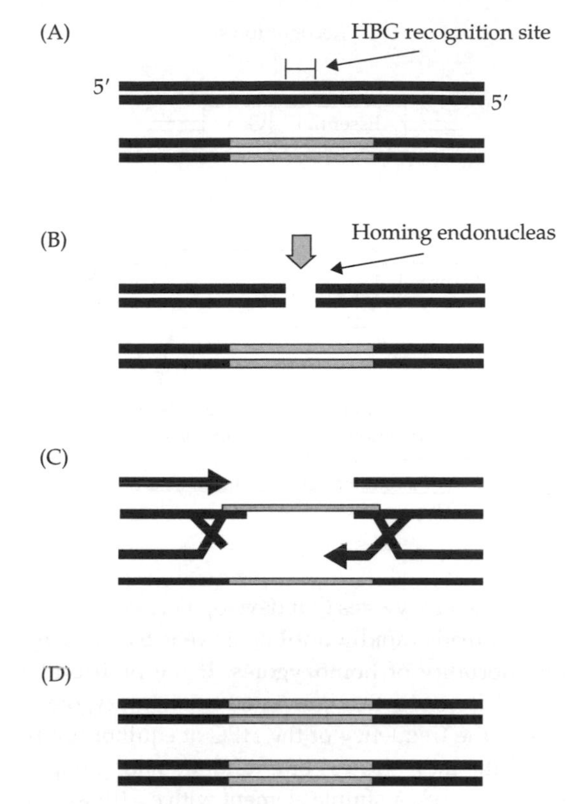

Figure 9.8 Replication of homing endonuclease gene (HEG). A. Genome is heterozygous for HEG. B. Endonuclease is transcribed and cuts homolous region at recognition site. C. Break is repaired using the HEG as template. D. Genome is now homozygous for HEG. From Goddard *et al.* (2001).

transcript and thereby permit normal expression. The outcome is that heterozygotes are converted into homozygotes, and the HEG invades sexual lineages very rapidly (Goddard *et al.* 2001; reviewed by Burt and Trivers 2005). They are widespread among fungi and algae, and occur in the organelle genomes of animals and plants. Burt (2003) has made the audacious claim that HEGs can be engineered to spread rapidly in any outcrossed species, and might therefore be used to exterminate pests such as the mosquito vectors of malaria. He envisages a construct consisting of a HEG with no associated self-splicing intron that would induce a recessive lethal mutation in a gene controlled by a meiosis-specific promoter (Figure 9.9). At low frequency most individuals carrying the construct

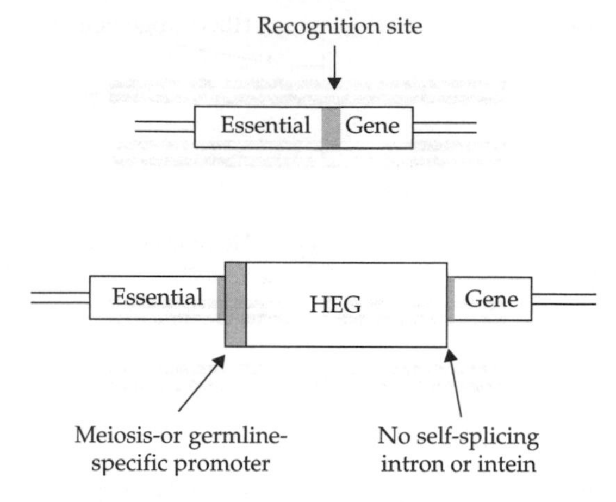

Figure 9.9 A HEG construct for population genetic engineering. From Burt (2003).

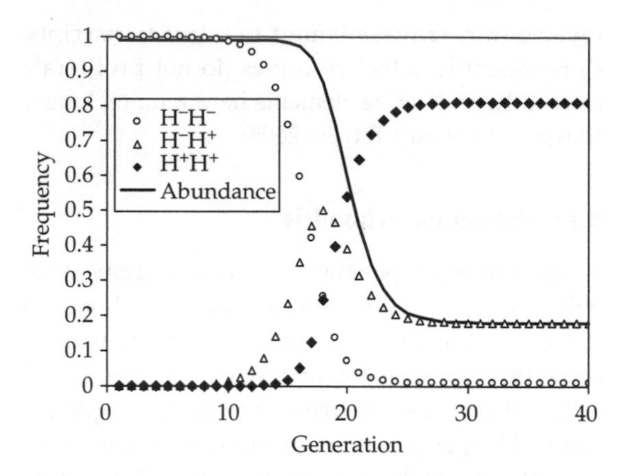

Figure 9.10 Population genetic engineering. A recessive lethal HEG capable of converting 90% of heterozygotes to homozygotes is introduced into a population at a frequency of 10^{-5}. The population is regulated by density-dependent juvenile survival, using a Beverton–Holt model with parameter 0.1.

will be heterozygotes that develop normally, so the HEG spreads rapidly until its drive is balanced by the mortality of homozygotes. If the probability that a heterozygote is converted to a homozygote is c, then the frequency of the HEG at equilibrium is c and the mean fitness of a random-mating population is $1 - c^2$. A simple element with $c = 0.9$ would then be killing 80% of host zygotes only 20 generations or so after its introduction at low frequency (Figure 9.10). Even more effective elements could be constructed by targeting several genes, or by sterilizing females rather than killing males and females indiscriminately. The host population has little opportunity to evolve resistance, and could in any case be frustrated by simultaneously releasing variant constructs. If there were any second thoughts about the programme, however, it could be rescinded by releasing engineered hosts with altered recognition sites that conferred resistance to the HEG. Artificial HEGs are currently being developed and have great promise as agents capable of alleviating some of the worst scourges of mankind.

9.2 Interference

Spiteful genetic elements increase their transmission by destroying their competitors. A Killer element K occurring in a Kk heterozygote is transmitted to all surviving gametes by destroying those that do not bear it. This is an effective strategy only if competition is local. If all members of a population compete together, then each K element displaces only a single k competitor and the selection coefficient will be about equal to the frequency of K, so that it is unlikely to increase when rare (Wade and Beeman 1994). On the other hand, competition that is restricted to the gametes or progeny produced by an individual may cause the displacement of all k elements in the group. If the loss of these elements does not reduce overall success, because male gametes are present in excess or the remaining progeny survive twice as well, then $s = 0.5$ and K will spread very rapidly.

The fundamental barrier to the spread of Killer elements is that they are likely to destroy themselves as readily as their competitors. It will be effective only in combination with some Resistance element R that gives protection against the Killer agent. The K–R complex, however, will be broken up by recombination in K–R × k–r matings, so that it will spread, even in ideal circumstances, only if the rate of recombination between K and R is less than 1/3 (Charlesworth and Hartl 1978). The crucial factor in the spread of spiteful genetic elements is thus the evolution of linkage between K and R loci.

9.2.1 The poison–antidote system

A plasmid that is replicated along with host DNA runs one serious risk: since there is no means of ensuring regular segregation, some daughter cells will by chance receive no copies of the plasmid. These asexual lineages are completely cured, because they are very unlikely to be re-infected by horizontal transmission. These cured lineages will continually increase until the plasmid becomes extinct, so the only plasmids we observe are those that have evolved some means of preventing the accumulation of cured lineages. Some plasmids bear beneficial genes (encoding antibiotic resistance, for example) and persist because of the greater fitness they confer on their host. Others are pure parasites that spread by interfering with plasmid-free hosts through a system in which K produces a poison and R its antidote. Since there is little or no recombination in bacteria, K and R occurring in the same genome are completely linked. Plasmid R1 is very stably maintained in *E. coli* through a poison–antidote system (Gerdes 1988; Figure 9.11). It includes a locus that encodes host-killing (*hok*) and suppression-of-host-killing (*sok*) mRNAs. The *hok* mRNA is translated into a small protein that collapses the proton gradient

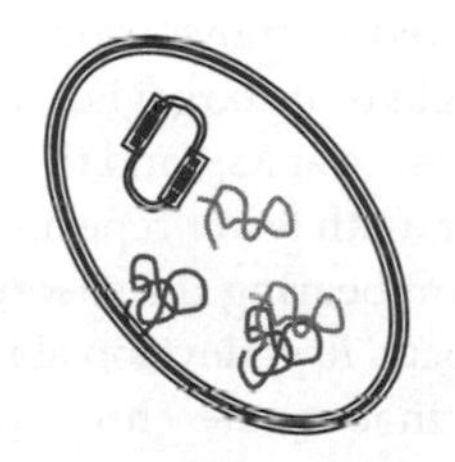

Cells infected by R1 survive because ***hok*** mRNA is bound by ***sok*** mRNA and thus remains untranslated.

There is no special machinery to distribute plasmid copies evenly between daughter cells at cell division. The plasmidis are therefore sometimes lost simply by chance.

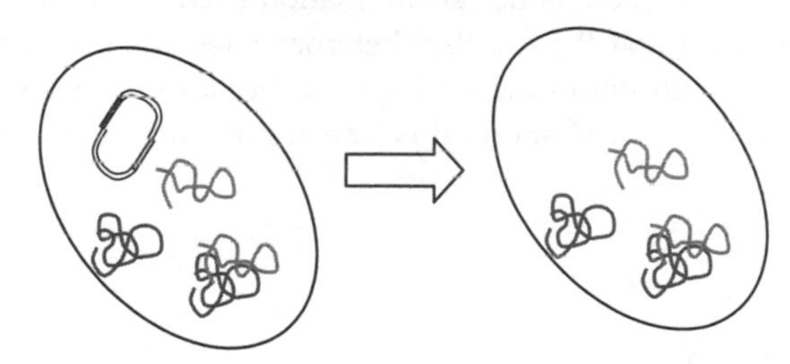

But although the plasmid may have been lost, some plasmid mRNA will inevitably remain ...

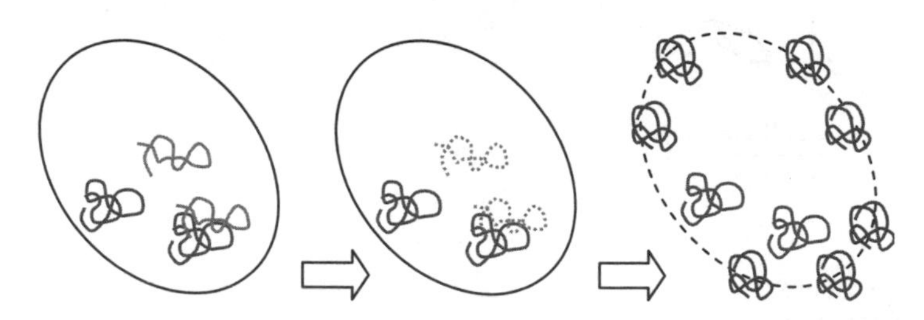

sok mRNA is much less stable than ***hok***

hok is therefore expressed, killing the host cell.

Figure 9.11 Cartoon of the toxin–antitoxin system maintaining plasmid R1.

across the cell membrane, halting respiration and killing the cell. The *sok* mRNA is not translated, but binds to *hok* mRNA and prevents it from being translated. The sok promoter is the more powerful, and so *hok* is effectively unexpressed and the cell is not greatly harmed by carrying the plasmid. Alas, if it should lose it. Although the daughter cell may not receive a copy of the plasmid, it will certainly receive some of the transcribed mRNAs in its cytoplasm. The *sok* mRNA is much less stable than *hok* mRNA, so that in a cell cured of its plasmid *hok* will be expressed and the cell is killed by the *absence* of its parasite (for details of the mechanism, see Gerdes *et al.* 1997). Poison–antidote systems are widespread in bacteria (reviewed by Hayes 2003), but for unknown reasons they are very rare in eukaryotes.

9.2.2 Gamete killers

In sexual organisms gametes, and especially male gametes, compete for fusion partners. A Killer that eliminates k sperm will double its fitness, provided that mating is monogamous and there is an excess of sperm. It must still occur in combination with a Resistant element, which will evolve when selection against recombinants favours tighter linkage of K and R. A modifier of recombination that is linked to K and R will be selected to reduce the rate of recombination between them, whereas if it is unlinked it will be selected to increase recombination because K acts against the interests of unlinked genes. The most straightforward mechanism for creating a non-recombining K-R complex is to include both within an inversion, which is the cytological basis of the classical examples of male gamete killing.

The best-known example is the *Segregation-distorter* of *Drosophila* (Figure 9.12). It consists of a cluster of tightly linked genes, the most important of which are *Sd*, the gene responsible for biased transmission, and *Rsp* ('responder'), which modulates, and can oppose, the effect of *Sd*. The killer allele *Sd* has a partial duplication of *RanGAP*, whose product is involved in transporting proteins from cytoplasm to nucleus. Its target is a highly repeated non-coding sequence at *Rsp*, and the resistant allele is simply shorter, with fewer repeats. In Sd/Sd^+ heterozygotes, sperm bearing the susceptible allele of the responder locus Rsp^S develop abnormally, with a syndrome of inadequate chromatin condensation, the failure of spermatids to become individual cells, and defective sperm maturation. Sperm bearing the insensitive allele Rsp^i develop normally. If there is no recombination between the two loci in $Sd\ Rsp^i/Sd\ Rsp^S$ heterozygotes, then the $Sd\ Rsp^i$ chromosome will spread, because by destroying $Sd^+\ Rsp^S$ sperm it is transmitted to 95% or more of

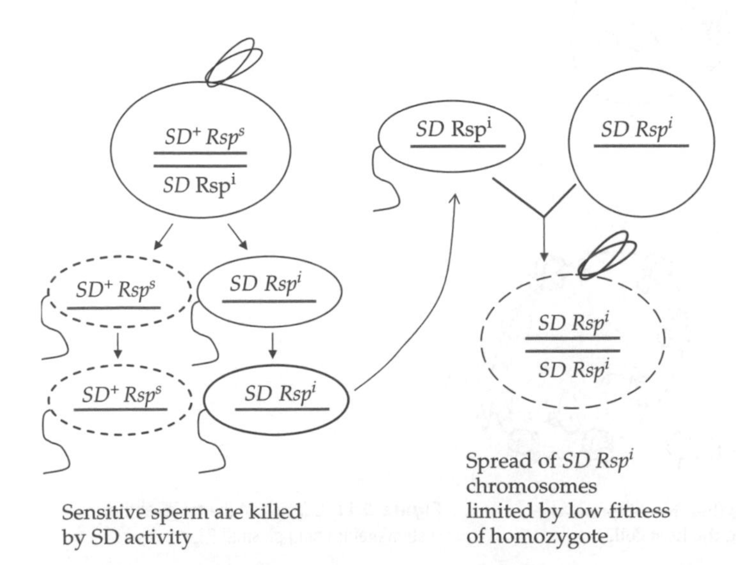

Figure 9.12 Cartoon of Segregation-distorter (Sd) in *Drosophila*.

the progeny. Thus, if *Sd* and *Rsp*^{i} are linked within the same inversion, and therefore recombine very infrequently, autoselection will cause an increase in segregation distortion. *Sd* chromosomes occur in most natural populations of *Drosophila*, but at frequencies of only a few per cent. Their spread is checked by countervailing natural selection. The suppression of recombination near the *Sd* locus relaxes natural selection on all genes within the inversion and thereby promotes the accumulation of deleterious recessive mutations under Muller's ratchet (§12.1.2). When two flies bearing *Sd* mate, the *Sd/Sd* homozygotes they produce are therefore likely to be inviable or sterile. Furthermore, males in which half the sperm are killed may simply have reduced virility; this is a more serious possibility in *Drosphila*, where the sperm are extremely large and few in number, than it would be in most other animals.

Segregation-distorter is an extreme example of epistasis that requires mutations at two loci in order to be effective. A similar difficulty applies to all poison–antidote and killer–resistant systems, and seems to imply that they are very unlikely to evolve. According to Burt and Trivers (2005) the ancestor had short alleles at both *Sd* and *Rsp*. The amplification of the *Rsp* repeats may have been favoured by natural selection, as deleting them reduces fitness. Strains would then quickly come to vary in copy number through unequal crossing-over. The partial *Sd* duplication then arose as a rare variant on a short-*Rsp* chromosome. It would subsequently spread by autoselection, in combination with *Rsp*^{i}, until at moderate frequency an inversion including both arose to constitute the mature K-R system. Any subsequent modifier that enhanced killing would be selected if linked to *Sd*, so the inversion containing the K-R complex will always tend to expand along the chromosome, and in fact the entire *Segregation-distorter* element includes several such modifiers.

A very similar gamete-killing system is well-known in mice, where a series of distorter genes acting on a responder occur in a region where inversions suppress crossing-over. About 25% of wild mice are heterozygous for a *t* haplotype. Males that are heterozygous for the *t* complex transmit it to 90% or more of their offspring; as with SD, however, the homozygotes are severely handicapped. Haploid spores may also be targeted. Meiosis in the filamentous fungus *Neurospora* leads to the production of an ascus bearing eight haploid spores; these germinate to form a haploid but multinucleate mycelium. Spore-killer *Sk* genes kill the four spores from the non-*Sk* parent. *Sk* genes do not kill themselves: a cross between allelic killers yields eight viable spores. This self-protection seems to be attributable to some diffusible substance, since sensitive nuclei are rescued in giant multinucleate ascospores that also contain *Sk* nuclei. It is not known what limits the spread of *Sk*, although it is possible that *Sk* mycelia are vegetatively inferior.

9.2.3 Modifiers of meiotic drive

The *Segregation-distorter* element *Sd* of *Drosophila* is harmful to most of the genome because of the sterility or inviability of SD/SD homozygotes. Consequently, genes unlinked to *Sd* will tend to evolve so as to suppress its effects. Conversely, genes linked to SD would be selected to overcome suppression, because they benefit by being transmitted along with SD. There is therefore an antagonism between different kinds of element within the same genome; a similar antagonism will arise in the case of other autoselected elements, such as B chromosomes or transposons. In many cases, suppression is caused either by alleles at the Responder locus or by dominant genes of large effect on other chromosomes, while enhancement of SD is caused by genes on the same chromosome arm coupled to *Sd* by inversions. The evolution of suppression has been studied experimentally by Lyttle (1979), using stocks in which the second chromosome, bearing SD, is linked to the Y chromosome by a translocation. This causes extreme sex-ratio distortion, because males transmit only the driven Y, so that all their progeny are male. The usual sources of suppression do not evolve in such populations, because major genes of large effect are absent from the base population, while insensitive Responder alleles cannot be transferred to the SD chromosome, because it is linked to Y by translocation, and crossing-over does not occur in male *Drosophila*. Instead, the experimental populations evolved over about 90 generations to suppress distortion through the selection of recessive genes of small effect on the third and fourth chromosomes. It seems likely that many of these genes arose by mutation during the course of the experiment, and therefore that different lines tended to accumulate different sources of suppression. At the same time, genetic enhancers of drive were detected on the compound Y and second chromosome. Autoselected elements may thus inaugurate a complex co-evolutionary succession of antagonistic elements within the same genome.

9.2.4 Meiotic drive on sex chromosomes

The equal numbers of male and female offspring produced by organisms with chromosomal sex determination is a classical instance of Mendelian segregation. The response of the sex ratio to natural selection depends on how males and females interact so as to determine the fitness of the lineage as a whole, and is discussed in §12.3. The sex ratio will also be affected by the autoselection of segregation distorters. Grossly unbalanced sex ratios may be merely the consequence of segregation distortion, or they may be caused by the distorting element so as to enhance its transmission. When genes that bias transmission are located on a sex chromosome, they will cause dramatic changes in the sex ratio of progeny. In many cases, broods may consist entirely of daughters or entirely of sons.

The *sex-ratio* gene SR of *Drosophila* is an X-linked segregation distorter. It acts by destroying Y-bearing spermatids during spermatogenesis; consequently, the offspring of SR males are all daughters. Its spread is opposed by natural selection, primarily because females homozygous for SR have only about half the viability of wild-type flies. There is also a long-term effect of SR, however, since if it were ever fixed there would be no males in the next generation, and the population would die out. It is possible that this plays a part in curtailing the spread of SR chromosomes. A similar distorter, D, is known from the mosquito *Aedes*, but is Y-linked: the offspring of D males are all sons. The D gene causes breaks in X chromatids. As D spreads through a population, the proportion of males increases, and thus the rate of reproduction falls, leading to the decline and eventually the extinction of the population. This has led to proposals for the use of D to control (that is, to eradicate) mosquito populations. The simplest scheme is to release very large numbers of D-bearing males, so that in the following year females (which alone take a blood meal) are scarce. More ingeniously still, it is possible to construct chromosomes that bear both D and a conditional lethal, for example a mutation that is lethal only when it is very cold, or when low concentrations of insecticide are present. This chromosome increases to high frequency in the population, which is then highly vulnerable to a cold snap or chemical treatment.

There is no natural Y-linked distorter in *Drosophila*, but Lyttle (1977) manufactured one by translocating a part of the second chromosome containing the *Segregation-distorter* locus to the Y-chromosome. In its new location, it is selected more effectively than when borne on an autosome, mainly because it is permanently heterozygous. In experimental populations, it behaves in much the same way as an allele with a twofold to fourfold advantage in a haploid population, spreading rapidly when introduced at moderate frequency and becoming nearly fixed within a dozen or so generations. Because males bearing the translocation produce only sons, the frequency of females drops as the SD males become more abundant. This eventually brings about, first the decline, and then the extinction of the population. The system cannot be used straightforwardly to control natural populations, but it is an impressive laboratory demonstration of the effects of sex chromosome drive.

9.3 Gonotaxis

In multicellular organisms the germline is at some point separated from the soma, so that any subsequent mutations in somatic cell lineages will not be transmitted to progeny. This separation may be strictly enforced early in development, as in nematodes and most arthropods and chordates, or it may take effect only after the main features of the adult body plan have been established, as in most molluscs and annelids. In seaweeds, fungi, and land plants, and in animals such as sponges and cnidarians, there is no distinct germline, except insofar as differentiated somatic cells cannot de-differentiate and then develop as gametes. The separation of a sterile soma creates selection for somatic lineages that are able to cross the barrier and re-enter the germline, but in practice this seems to be very difficult to do and very few examples have been reported (see Buss 1987, p. 74). A special case of germ–soma differentiation is the derivation of the vegetative macronucleus from the generative micronucleus after conjugation in ciliates: selfish elements such as transposons that do not contribute to adaptedness are stripped out during the development of the macronucleus. A second opportunity for selfish elements is provided by female meiosis, which in many multicellular organisms gives rise to a single functional product, the other three nuclei being eliminated as polar bodies. This creates selection for elements that can modify or anticipate the outcome of meiosis so as to be transmitted with the successful meiotic pronucleus. The classical example is the 'knobs' of a variant of maize chromosome 10, which act as parasitic centromeres, or neocentromeres (see Dawe and Hiatt 2004). Maize produces a linear tetrad in which the nucleus at one end develops into the megaspore, and by attaching to the meiotic spindle the knobs ensure that they are diverted to the two ends of the tetrad and thereby retained in the germline. This is an exceptional case, but draws attention to the puzzling observation that centromeric sequences evolve exceptionally fast, although the spindle apparatus itself scarcely differs among eukaryotes, as do associated histones, while the non-centromeric histones from which they are derived are highly conserved (Henikoff and Malik 2002). These centromeric sequences, indeed, evolve faster than any other genes except those associated with mating and defence, while the centromeric histones show clear evidence of directional selection in both plants and animals (Talbert *et al.* 2004). One possibility is that the centromeric sequences are under continual selection to use the spindle to navigate to the generative nucleus, giving rise to meiotic drive, whereas the centromeric histones are under selection to frustrate them and restore Mendelian segregation (Henikoff *et al.* 2001).

9.3.1 B chromosomes

The most widespread gonotactic elements are B chromosomes, which occur in about 10% of species of multicellular organisms (Jones 1995). They behave in the main like normal chromosomes, but are clearly not required for normal function because they are lacking in many or most individuals, and in many cases are mildly deleterious (Jones and Rees 1982). They are presumably derived from normal chromosomes, but have evolved as independent selfish replicators because they are directed preferentially into the germline (synthetic review in Burt and Trivers 2005). They were first recognized as genomic parasites by Ostergren (1945). In most cases this drive is based on mitotic non-disjunction, with one daughter cell receiving two copies of the B chromosome and the other none (reviewed by Jones 1991). One common mechanism is directed non-disjunction into the generative nucleus of the pollen grain, but meiotic drive and over-replication of B-containing cells in reproductive tissue have also been described. All such mechanisms cause B chromosomes to spread in outcrossed species, where they are enabled continually to re-infect uninfected lineages, but are ineffective in inbred or asexual species, where they will be eliminated by among-lineage selection through their deleterious effect on growth and reproduction (Burt and Trivers 1998). They are in fact found predominantly in outcrossed species of plant (Jones and Rees 1982). One of the most remarkable properties of B chromosomes is their effect on crossing-over in the normal complement, which is altered in about half of known instances. In most cases both the average number of chiasmata and the between-cell variance are increased, although the reverse may also occur. This may be a by-product of selection for increased chiasma formation by the B chromosome, to prevent univalent formation (Carlson 1994), or the effect of selection on the normal complement to generate variant progeny resistant to parasitism by B chromosomes (Bell and Burt 1990). Burt and Trivers (2005) suggest that B chromosomes typically evolve in a cyclical fashion, invading populations through aggressive drive, later becoming neutralized by host counter-adaptation, whereupon they become benign, degenerate and are lost; a new aggressive B chromosome then appears, or evolves from the previous type, and re-initiates the process. Detailed studies of the unusually numerous and diverse B chromosomes of the grasshopper *Eyprepocnemis* give some support to this view (see Camacho *et al.* 2003): new B chromosomes are more virulent, and old, benign B chromosomes recover virulence when introduced into naive populations.

9.3.2 Infectious agents that control sexual development

SR and *D* are nuclear genes whose effect on progeny sex ratio is incidental, and in fact tends to cause countervailing natural selection. More interesting are elements outside the nucleus that not only cause changes in the gender or sexuality of offspring, but that are autoselected because of these changes. Most of them are bacteria, or less often eukaryotic microbes, that are vertically transmitted in egg cytoplasm, and are therefore selected to induce their host to develop as a female.

Several species of the small wasp *Trichogamma* are wholly asexual, producing unreduced eggs by mitosis. Stouthamer *et al.* (1990) made the remarkable discovery that these species will produce sexual males and females if treated with antibiotics. If the treatment is continued over several generations, the lineage becomes obligately sexual, like most hymenopterans. It seems almost certain that parthenogenetic development is induced by a bacterium transmitted in the egg. In a sexual species its host might be a male, and it can therefore enhance its own transmission by suppressing its host's sexuality and forcing it to develop as an asexual egg-producing individual. Several cases are now known in which sexuality can be induced in parthenogenetic organisms by treatment with antibiotics, or prolonged exposure to high temperature, and the suppression of sex by intracellular microbes may be quite widespread.

Other agents cause the individuals that bear them to develop as sexual females. These have been described from a wide range of organisms. One classic case, described many decades ago by Vandel (1938) and studied in depth by Juchault and Legrand (1989), concerns the isopod *Armadillidium vulgare*. The nuclear sex-determination system is female heterogamety: uninfected XY individuals develop as females. However, XX individuals that are genetically male develop as females if they are infected with a rickettsia-like bacterium. This is transmitted through egg cytoplasm, so that the broods of infected XX neo-females consist largely or entirely of daughters. The condition can be induced by tissue transplantation, and cured by high temperature or antibiotics. Some cases of feminization, however, cannot be cured. They are thought to be caused by a transposable element transferred from the bacterial genome to a host X-chromosome, which is thus converted into a functional Y. It is subsequently inherited in a more or less Mendelian fashion. These baroque genetics are typical of populations where sex determination is influenced by infectious elements, and indeed the details of the system, involving not only the behaviour of the microbe but also the response of its host, are much more complicated than this brief account suggests.

Infectious agents transmitted through the female line can be autoselected if instead of feminizing their host they kill its male offspring. This will increase their rate of transmission if maternal resources that would otherwise support the growth of male offspring are redirected to increase the number or quality of daughters. Such agents may act in female offspring to kill their male sibs. Alternatively, they may act in the males, committing suicide by killing their hosts, but in doing so benefiting copies of the same agent in the sisters of their hosts; this would be an example of kin selection (§10.4). The best-known case is the *sex-ratio* condition of *Drosophila*, which is caused by a spiroplasma that kills male offspring very early in embryogenesis. (This is not related to the SD system described above.) The organism can be cultured outside the host, and used to infect new hosts. When infected flies are introduced into laboratory populations the agent may increase in frequency, but its spread is usually checked, and at equilibrium a nearly equal sex ratio is restored.

The nature of the selection opposing the spread of the spiroplasma has not been identified, although at high population density infected females may be less fecund.

Werren *et al.* (1987) and Beukeboom and Werren (1992) described an unusual sex-ratio distorter in *Nasonia*. This small wasp has the usual hymenopteran genetic system, in which fertilized diploid eggs develop as females, whereas unfertilized haploid eggs develop as males. It harbours several distorters, including Son-killer, a maternally inherited bacterium that is lethal to the male eggs of infected females, and Maternal Sex Ratio, a feminizing cytoplasmic agent that also causes the production of all-female broods. Paternal Sex Ratio causes the production of all-male broods. It is not a microbe, but a B chromosome that is transmitted solely through sperm. It acts by destroying all paternal chromosomes, except itself, in the zygote. This causes fertilized eggs to become haploid and thus to develop into males. These males then transmit the B chromosome to all their progeny, whereas a diploid female would transmit it to only half her progeny. Since it is transmitted only through fertilized eggs which would normally develop into female offspring its dynamics depend on the fraction of eggs that is fertilized; if that fraction is f, then PSR should move asymptotically towards a frequency of $(2-f)/f$. Thus, if all eggs are fertilized, it will become fixed; if fewer than half are fertilized, it will be lost. Figure 9.13 shows the spread of PSR when it is introduced at low frequency into a large cage population of *Nasonia*. The dotted line is the predicted response to selection, given the rate of fertilization observed in the population; with a fertilization rate of about 0.8, the predicted frequency of PSR at equilibrium is 0.75, reasonably close to the observed value of 0.68. The frequency of the element in natural populations will depend in

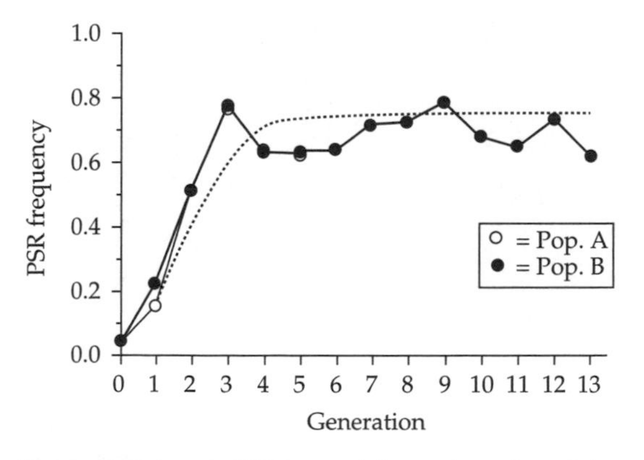

Figure 9.13 Spread of PSR in panmictic experimental populations of *Nasonia*. The dashed line is the expected response. From Beukeboom and Werren (1992).

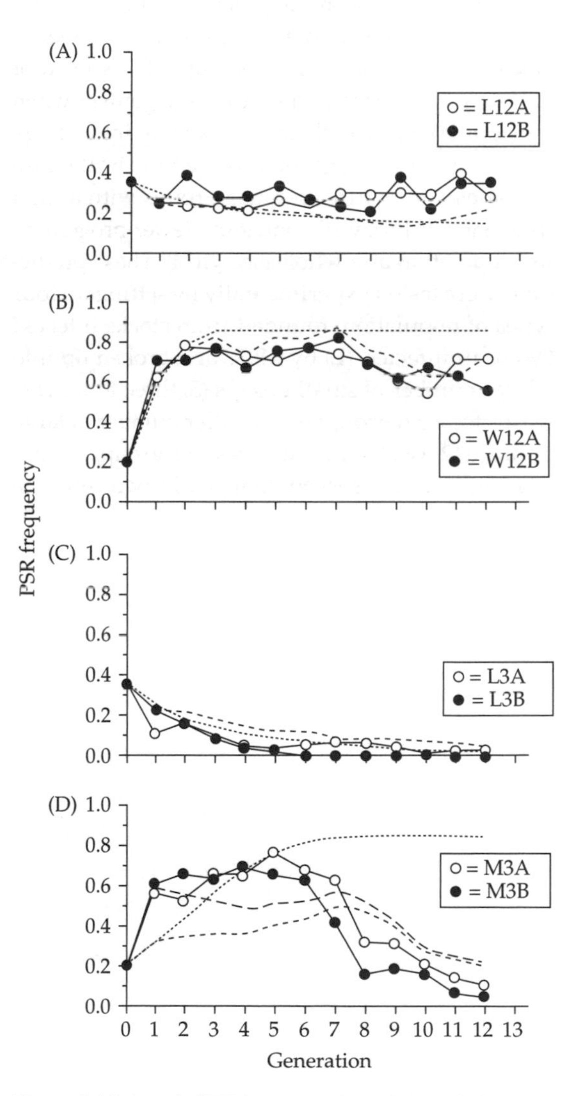

Figure 9.14 Spread of PSR in structured populations. A. Large, no MSR; B. Large, with MSR; C. Small, no MSR; D. Small, with MSR. From Beukeboom and Werren (1992).

part on this powerful autoselection, and in part on two other processes. The first is the rate of fertilization, which will itself depend on the frequency of PSR. Moreover, it will vary with the presence of other genetic elements, such as Maternal Sex Ratio (MSR): its presence will catalyse the spread of PSR because it increases the supply of fertilizable females (see §10.3). Secondly, as PSR spreads it has the same effect as a Y-acting segregation distorter: the frequency of males increases and the productivity of the population is likely to fall. Thus, natural selection among groups is antagonistic to autoselection within groups. The spread of PSR may be constrained by this antagonism when the population as a whole is broken up into a large number of breeding groups (as seems to be the case for *Nasonia* in nature) because groups with a high frequency of PSR will contribute fewer progeny to the population as a whole (see §10.4). These predictions were tested experimentally by setting up four types of population, founded from stocks infected (M) or uninfected (L) by MSR, and broken up into a large number of small groups (3, three foundress wasps for each group) or a smaller number of large groups (12, twelve foundresses per group). Figure 9.14 shows the observed changes in two replicate populations of each type, compared with the predicted response. The prediction is reasonably good for the L12 and L3 treatments: the element is maintained in the population with larger and fewer groups, although as expected its frequency is much lower than in the large undivided cage population, and is strongly reduced or eliminated when there are more, smaller groups. The difference between L and M populations is also correctly predicted, PSR frequency being higher in populations infected with MSR. However, although the observed response is reasonably close to prediction in M12, it is clearly far astray in M3. This was probably because fertilization rates in the experimental populations diverged from those in control populations, on which the predictions were based, and the broken lines show attempts to update the predicted response using the actual fertilization rates in the experimental populations.

Elements such as plasmid R1 and PSR operate in very different ways, but both illustrate the surprising fact that individual adaptation can be severely compromised by selection acting at other levels. The total destruction of the genome with which an element is transmitted is a forceful example of the Darwinian logic of autoselection.

CHAPTER 10

Social selection

The processes of selection that I have described so far are atomistic. The characteristics that are selected are assumed to be expressed by isolated individuals, whose reproduction varies according to the relationship between their phenotype and the state of the external environment. The virtue of this approach is that it makes a clear distinction between the population that is undergoing selection and the environment in which it is selected. It is then possible to investigate selection and predict the course of evolution by studying isolated individuals, or by treating individuals as though they were isolated, and inferring the dynamics of populations from the aggregate properties of their members. However, this simple distinction between organism and environment, though clear, may be misleading. Individuals seldom, if ever, live wholly apart from all other organisms, passively transforming physical resources into offspring; rather, they live embedded in a community of other organisms, whose activities will affect their own. The physical environment, indeed, provides little more than the context for a much richer biotic environment, in which other individuals may be rivals, partners, dangers, or opportunities. These various interactions among individuals give rise to the social selection of attributes whose effect on fitness varies according to the composition of the community. The effect of social selection depends on the type of social interactions that are involved.

In the first place, reproduction will generally be affected by the density of the population. Types that have high fitness as solitary individuals will not necessarily retain their superiority when many individuals are growing together. Even when these neighbours are similar, in the sense of having similar effects on one another's growth, the relative fitness of different types may vary with population density. This gives rise to density-dependent selection. Secondly, individuals of different kinds do not necessarily have similar effects on one another's growth; they may be either more or less severely affected by the presence of their own kind than by that of others. Reproduction will then depend on the composition of the population, giving rise to frequency-dependent selection. Finally, when reproduction depends on the density or composition of the population, groups of individuals will have properties that are not displayed by isolated individuals of the same kinds. These social properties will be selected, irrespective of whether the individuals concerned belong to the same lineage. Social interactions among individuals may thereby lead to competition between social groups, a process of group selection.

This chapter is concerned only with social interactions among individuals belonging to the same species. Social interactions are not by any means restricted to individuals of the same species, or even to individuals following the same general way of life. On the contrary, quite unrelated organisms that are parasites, predators, or partners are a major component of the biotic environment. The selection mutually imposed by dissimilar species leads to a process of co-evolution, which involves different principles and is dealt with in the next chapter.

10.1 Selection within a single uniform population: density-dependent selection

The biotic environment is not merely an additional source of selection. The social selection that it creates is different in kind to selection that is mediated by purely physical factors. The most important difference between biotic and physical factors is not the difference between living and inorganic agents, but rather the difference between depletable and non-depletable resources.

The simplest examples of selection involve non-depletable resources, using the word in a very broad sense: the evolution of tolerance to heavy metals, adaptation to high temperature, the response to selection for high bristle number. In all of these cases, the modification of the population through selection has no effect on the environmental agent responsible for selection. When *Agrostis* evolves metal tolerance, the metal content of the soil does not change; *E. coli* is powerless to alter the temperature of the chemostat; and the experimenter is unmoved by the bristliness of the flies. In such cases, adaptation can be understood by imagining each individual to be separately exposed to selection in its own small vial, being evaluated against some fixed standard, and reproducing more or less successfully as a result. Many environmental factors are non-depletable in this sense. The temperature of a hot spring, the partial pressure of oxygen in the modern atmosphere, the pH of calcareous soil, the depth of the ocean floor; all of these will be unaffected by whether or not a population succeeds in adapting to them.

Other kinds of resources are different, because they can potentially be used up faster than they can be supplied. When a few bacteria are inoculated into a vessel containing lactose as a carbon source, they will at first find themselves in a world of plenty, and will proceed to increase rapidly in numbers. The type with the highest rate of increase will, of course, increase in frequency as well as in abundance. However, a larger population has a larger appetite for lactose, and as the population increases the availability of lactose will fall. The same principle of selection will continue to hold: the type with the greatest rate of increase will spread. However, the type that proliferates the most rapidly when lactose is present in excess of the capacity of the population to metabolize it is not necessarily the type that will be the most successful when the population has grown and lactose is scarce. When a resource is depletable, the environment will change through time as a consequence of the adaptedness of the population, and the direction of selection may change as a result.

10.1.1 Density regulation

As long as the supply of resources is sufficient to support growth, the population will tend to increase exponentially, according to the simple law relating the future population N_{t+1} to the present population N_t as $N_{t+1} = e^r N$, where r is the exponential rate of increase. If individuals are rare and widely scattered then the rate of increase r is as large as it can be, given the rate of supply of resources; call this value r_{max}. As the population grows, the environment will deteriorate. It will become increasingly impoverished, because the ration of depletable resources will fall, and it will become increasingly polluted, because the concentration of toxic metabolites will increase. For either of these reasons, or for both, the rate of increase will decline, until at some point $r = 0$. The population is at equilibrium at this point, because any further addition of individuals will cause r to become negative, so that the population declines, whereas any decrease in numbers will restore a positive value of r, so that the population increases. This process of density regulation can be represented mathematically in any number of ways, but the two simplest depend on whether one is thinking in terms of serial transfer or continuous culture

Fish in a river, sedges in a fen, and bacteria in a laboratory chemostat are all more or less close approximations to continuous culture, where population growth is supported by a constant input of resources. A small inoculum will grow in numbers and consume these resources until the rate of consumption balances the rate of supply: the population is then in equilibrium. The rate of increase of the population will depend on the concentration of the limiting medium in the culture vessel, S, which will be determined in part by the rate of inflow and in part by the population itself. By analogy with the Michaelis–Menton equation describing the kinetics of enzyme action, the per-capita rate of increase can be represented as $r = [u_{max} \; S/(S + K_s)] - D$.

In this more complicated formulation, u_{max} is the maximal rate of division at very low population density; K_s is the so-called half-saturation constant, the concentration of substrate at which the rate of division is equal to half its maximal value u_{max}; and D is the dilution rate of the chemostat. The maximal rate of increase is $r_{max} = u_{max} - D$. This formulation tends to be used by experimentalists who work with microbes in chemostats. Alternatively, a more empirical approach would be to estimate how growth rate falls as density increases. From any number of possible assumptions about how the increasing population density affects the growth rate, the simplest is that birth rates fall and death rates rise linearly with the logarithm of population density. The per-capita growth rate at any population N is then $r = r_{max}(1 - N/K)$, where K is the maximum number of individuals that can be supported indefinitely in a given environment. This formulation is the so-called logistic equation that has been used extensively by ecologists.

A seedling in a canopy gap, a fungal spore that lands on a fallen fruit, or a bacterium in a laboratory vial grow as batch cultures, consuming a finite initial stock of resources. Batch cultures never reach an equilibrium, but instead cycle endlessly from glut to dearth. A small inoculum proliferates until the available resources have been consumed. The number or mass of organisms at this point is the yield or limiting density; it is often called the carrying capacity, because it resembles the asymptotic value of logistic growth, but this is misleading because the population will not remain at this value but rather will decrease as individuals begin to die of starvation. The limiting nutrients are now almost completely exhausted, although there may be some recycling from dead cells. Nevertheless, cells may remain viable for long periods of time, often becoming characteristically specialized for a dormant or semi-dormant lifestyle. *E. coli* cultures enter stationary phase after about a day, when they are still viable but have stopped dividing. After 3 days, most of the cells are dead. A week later, however, the fraction of viable cells increases. Moreover, these cells are dividing; this can be proved by treating the cultures with an antibiotic that inhibits cell division but not cell growth, so that long filaments form in growing cultures. If a small quantity of these aged cultures is inoculated into a 1-day-old stationary culture, the cells from the aged culture rapidly spread, eliminating the younger cells. This is not caused by any inhibitory effect of the aged culture medium itself, because young cells resuspended in a cell-free filtrate of an aged culture remain viable. Moreover, the superiority of the aged cells is maintained through several successive cycles of growth. Mutation in the stationary phase has led the selection of genotypes able to proliferate after the growth of the original strain has long since completely ceased. One of the mutations responsible for continued viability is in *rpoS* which induced in the stationary phase (Zambrano *et al.* 1993). Viable cells from aged cultures bear a frameshift mutation in *rpoS* that causes the final four amino acid residues at the 3′ end to be replaced by 39 new residues. When this mutant gene is transduced into young cells, it confers the ability to spread in stationary phase. Moreover, if the mutant cultures are themselves aged, further, unlinked, mutations occur that enhance the ability to grow in stationary phase, and replace the original mutant strain in mixed cultures. Selection is not, then, confined to young, rapidly growing populations; the characteristic process of sequential substitution also occurs in old and numerically static cultures, causing the evolution of strains adapted to crowded conditions.

Both kinetic and logistic formulations of density regulation lead to a partition of the rate of increase r at a given population density N into two components. The first expresses the maximal rate of increase, attained when the population is so sparse that it is not regulated by its own density. This introduces no new principle: if density can be neglected, types with greater values of r_{max} will spread. The second expresses the way in which population growth rate is regulated through population density, or through the decrease in the availability of resources caused by increasing population density. This introduces the new problem of how selection will act on K or K_s so as to maximize the realized rate of increase r.

10.1.2 Density-dependent fitness

If all genotypes respond in the same way to population density, then one will have the greatest rate of increase at all densities, and will become fixed, no matter what the density, or how density changes through time. Variation in density affects the outcome of selection only if it affects relative fitness, with some genotypes being superior at low density and others at high density. We can, then, treat density as if it were a physical factor, and investigate the magnitude of the genotype–density interaction in experiments where a series of genotypes is scored at a number of different densities

Lewontin (1955) reared from 1 to 40 larvae of 19 different inbred lines of *Drosophila* in standard culture vials. Survival was generally rather low at the highest densities, either because the larvae were starving or because they were being poisoned by their own waste products. More surprisingly, solitary larvae also fared poorly in many cases, their survival being lower than that of larvae growing up with one or a few companions: social interactions may be beneficial. The pattern of response to density, however, varied among the lines; some showed poor survival at low density, surviving best with 4 or 8 larvae per vial, whereas in other lines the solitary larvae survived well, and survival decreased regularly as density increased. This variation can be summarized as a genetic correlation of survival at different densities, in the same way that such genetic correlations can be calculated for other kinds of environmental variable. Lewontin presented his data as a table, each row being an inbred line and each column a density, so that the entry in each cell is the fraction of the larvae of a given strain surviving at a given density. The intraclass correlation coefficient is quite low, at $t_G = 0.30$ about, but because the observations are unreplicated, this reflects stochastic variation from vial to vial, as well as genotype–environment interaction. The interclass genetic correlation is probably more informative; if we take all pairwise combinations of densities the average genetic correlation of larval survival is $r_G = +0.39$, with a standard error of 0.06. The most interesting comparison is that between the density that was, on average, optimal (4 larvae per vial) and that which produced the lowest survival (40 larvae per vial): this correlation is close to zero. In short, genotypes seem to respond to larval density in this experiment much as they generally do to other features of the environment: there is substantial genotype–environment interaction, reflecting variation in relative fitness over environments, and if the environments are sufficiently different the genotypes are uncorrelated.

Agronomists have two reasons to be concerned about density regulation, and in particular about how cultivars differ in their response to density. In the first place, new cultivars are often selected and tested as spaced plants, or in short rows, where their promise may not reflect their performance in commercial conditions, at higher densities in much larger plots. Secondly, the farmer wishes to plant as economically as possible. Increasing the planting rate will increase yield only up to a certain point; beyond that point, yield remains nearly constant, because more means worse: the mature plants are more numerous but individually less fruitful. Different cultivars, with different responses to density, may have different optimal planting rates. Lang *et al.* (1956) planted 9 corn hybrids at densities from 4000 to 24 000 plants per acre (0.4 ha), and then used a low, moderate, or high level of nitrate fertilizer. As the density of plants increased, the average weight of the cobs fell, and an increasing fraction of stems were barren. Total yield thus increased with planting rate to a certain point, but then fell. The density at which increased planting no longer causes increased yield depends on the amount of nitrate applied (corn has a very high demand for nitrogen): when little nitrate is applied, yield is maximized by planting about 12 000 plants per acre, but with heavy nitrate application the maximum yield is got from planting 20 000 plants per acre. This response, however, varies among varieties. The hybrid that yielded most at 24 000 plants per acre was only a mediocre performer at low density, whereas the hybrid that was the poorest at high density was the second best in the group at the lowest density of 4000 plants per acre. There is again an indication that the change in relative fitness with density is such that, if performance is compared at two extreme densities, genotypes may be almost uncorrelated (Figure 10.1).

10.1.3 The principle of frugality

It is evident that the maximal rate of increase at very low population density will not necessarily

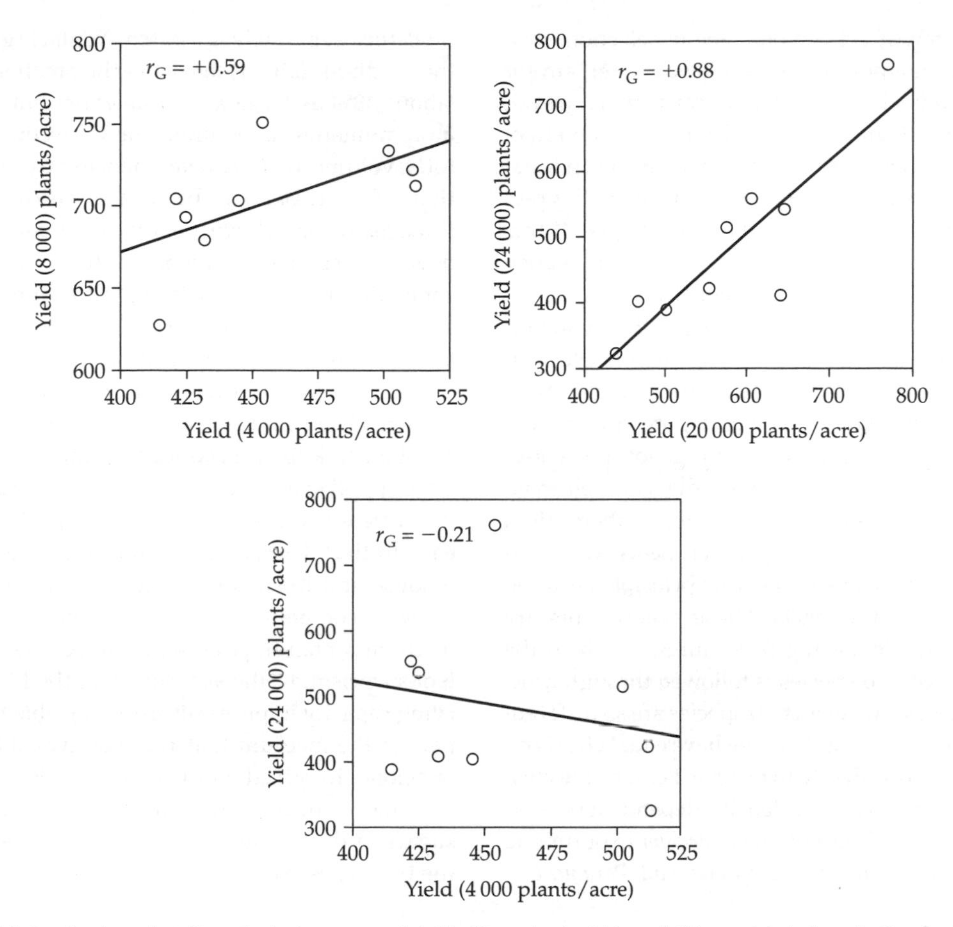

Figure 10.1 Density-dependent fitness in single-cross corn hybrids. Fitness (grain yield per plot) is correlated when planting densities are similar, either low or high (upper diagrams) but uncorrelated when very different planting densities are compared. From Lang *et al.* (1956).

predict the outcome of selection when populations continue to grow to the point where they begin to become severely inhibited by their own density. It should again be emphasized that the types with the greatest rates of increase will spread through selection at any density; but the realized rate of increase is specific to a given density, and the outcome of selection will depend on K or K_s as well as on r_{max}.

A clone inoculated into a fixed volume of medium grows rapidly at first, with $r = r_{max}$, and then more slowly, until finally $r = 0$ and growth ceases. At this point the population has reached its carrying capacity, with $N = K$. The individual ration of any resource is thus proportional to $1/K$. If two clones are inoculated simultaneously, they will begin growing at different rates, the type with the greater value of r_{max} increasing in frequency. After some time, the combined population of the two clones will attain the carrying capacity of the clone with the smaller value of K. This clone will therefore cease growing. The other clone continues to grow, further reducing the ration for the clone with lower K and thus forcing it to decline in frequency. Eventually only the clone with the greater value of K will remain. This simple argument is perfectly general, and applies to any number of competing clones or genotypes: in a density-regulated population, the type with the greatest value of K will become fixed. Another way of expressing this is that the successful clone will be that which can live on the lowest ration. Genetic variance in

r_{max} does not affect the outcome of selection, provided that the population is allowed to remain for long periods close to carrying capacity. However, it is required that all types have the same effect on one another; that is, that any single individual depresses the per-capita rate of growth of all types, including its own type, to the same degree. If this is not the case, selection will depend on frequency rather than (or as well as) on density.

The point of this simple theory is that it should be possible to predict the outcome of selection from the behaviour of pure cultures: a genotype that grows to high density as a pure culture will supplant in mixed culture any genotype whose carrying capacity is lower, regardless of their maximal rates of increase, and regardless of their initial frequencies. A great many experiments were conducted by ecologists to test this principle between the 1930s and the 1960s. These experiments are basically simple sorting procedures, in which the frequency of two species is followed through time. In most cases, however, two species are so different that the requirement that they have equal effects on one another is unlikely to be true. Perhaps the closest approach to a purely density-dependent process was G.F. Gause's famous *Paramecium* experiment. He cultured *Paramecium aurelia* and *Paramecium caudatum* separately in batch, replacing 10% of the medium daily. *P. aurelia* is the smaller animal (about 40% as large as *P. caudatum*) but is much more numerous at equilibrium. Consequently, the total volume of *P. aurelia* consistently exceeded that of *P. caudatum*. In mixed cultures, only *P. aurelia* persisted, with *P. caudatum* being driven down to very low frequencies after 2–3 weeks of competition. This is consistent with simple theory, although it is not a very critical test, because *P aurelia* also had the greater maximal rate of increase. The reason for the superiority of *P. aurelia* is not known; Gause himself suggested that it was more resistant to a toxin produced by the bacteria that were supplied as food. Similar experiments involving different variants of the same species are not easy to find. Perhaps the point is too elementary. Smouse and Kosuda (1977) grew a *lac⁻* mutant of *E. coli* in competition with the normal *lac⁺* type, in medium containing lactose, arabinose and glucose. Not surprisingly, the *lac⁺* clone had the higher carrying capacity in pure culture, being able to utilize part of the medium that was not available to the *lac⁻* clone. In mixed culture, the *lac⁺* clone quickly became fixed, regardless of the concentration of sugars in the medium or the initial frequencies of the two types (Figure 10.2).

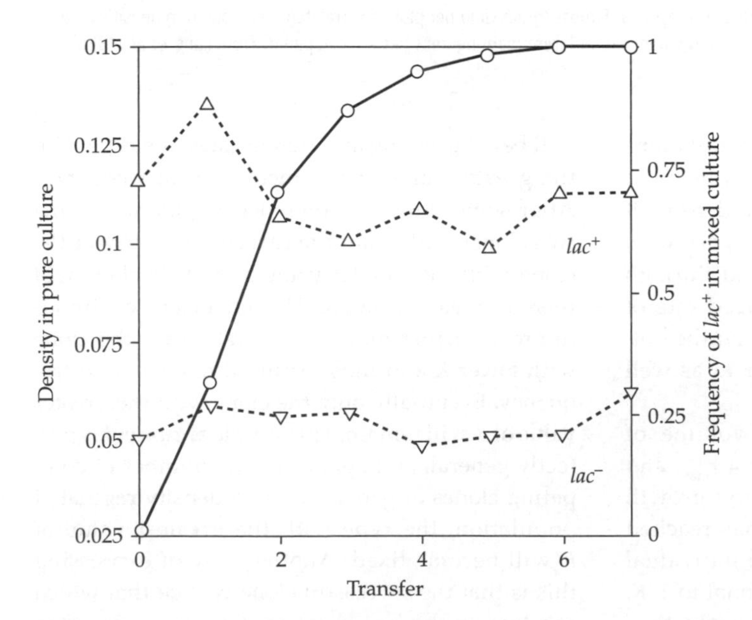

Figure 10.2 Elimination of *lac⁻* by *lac⁺* genotypes of *E. coli* in medium containing lactose. From Smouse and Kosuda (1977).

The analysis of continuous cultures is slightly more complex. It has often been suggested that the type with the lower K_s, and therefore the greater affinity for the substrate, should prevail. However, there is no reason to suppose that the population corresponding to the substrate concentration K_s has any special significance. The correct procedure is to calculate the substrate concentration representing an equilibrium at which the population no longer increases and the substrate itself is no longer depleted. This critical concentration is $S^* = D(K_s/r_{max})$. The dilution rate D is set by the experimenter, but types may vary in K_s or r_{max}, or both, and thereby in the critical concentration S^* at which they can no longer increase. The rule is similar to that for logistic growth: the type with the lowest value of S^* can subsist on the smallest ration, and will therefore exclude all other types. The same conclusion can be reached more rigorously by calculating the total flux J through the limiting metabolic pathway as a function of the kinetic parameters for each substrate in the pathway separately. Provided that the concentration of the limiting substrate at the head of the pathway is low, as it always will be in the chemostat, the relative fitness of a strain will be equal to its flux relative to that of a competitor. The ratio of S^* is equal to the ratio of J for any two strains, so that the ratio of their growth rates is $r_1/r_2 = J_1/J_2 = (K_{s2}/u_{max2})/(K_{s1}/u_{max1}) = S_2^*/S_1^*$. Hence, the principle of frugality follows from standard fitness concepts based on growth rates and can be justified mechanistically from enzyme kinetics (Lunzer *et al.* 2002). It should therefore be possible to predict the dynamics of mixed cultures from a knowledge of the kinetic properties of the component types grown in isolation.

10.1.4 Resource competition in continuous culture

Ecologists have often tested the principle of frugality in species mixtures. Diatoms are unicellular photosynthetic protists that are very abundant in the open water of lakes. They construct a box-like test of silica, and are often limited by the supply of silicate in nature. Tilman (1977) grew *Synedra* and *Asterionella* in silicate-limited chemostats at a dilution rate of $D = 0.11$ per day. By estimating the r_{max} and K_s of pure cultures, the critical concentrations of silica were calculated to be $S^* = 1.0$ μM for *Synedra* and 2.8 μM for *Asterionella*. The actual concentrations of silica in the chemostats when the cultures had reached stationary phase were in fact lower than this (0.4 μM for *Synedra* and 1.0 μM for *Asterionella*), but in the same proportion. It can therefore be predicted straightforwardly that *Synedra* will displace *Asterionella* from mixed cultures, regardless of their initial frequencies, which is exactly what happened: mixtures of the two types reduced the concentration of silica in the medium to about 0.4 μM, at which *Synedra* can maintain itself but *Asterionella* cannot (Figure 10.3). The criterion that is used to predict this outcome is quite general; the particular outcome that is predicted, however, is specific to the conditions in which the experiment was run, because when these are altered r_{max} and K_s are likely to change. In fact, the superiority of *Synedra* holds only above 20 °C, for the growth medium, light level, and dilution rate used in these experiments; below 20 °C *Asterionella* has the lower value of S^*, and, as expected, excludes *Synedra* from mixed cultures

Hansen and Hubbell (1980) used bacterial systems to investigate competitive exclusion in chemostats (Figure 10.4). Strains of *E. coli* and *Pseudomonas* that are unable to synthesize tryptophan have similar r_{max} but very different K_s in tryptophan-limited chemostats. *E. coli* has the smaller K_s and therefore the smaller S^*, and rapidly excludes *Pseudomonas* from mixed cultures. Selection in this strain of *E. coli* for resistance to nalidixic acid and streptomycin gave rise to two lines with distinctive resistance phenotypes that had very similar K_s in tryptophan-limited chemostats, but quite different r_{max}. The selection line that had the greater r_{max} necessarily had the lower S^*, and eliminated the other line from mixed cultures. The successful line happened to be the one that was sensitive to nalidixic acid. When nalidixic acid was added to the growth medium, both r_{max} and K_s of this strain were reduced, whereas the resistant strain was not appreciably affected. Nalidixic acid could thus be used to titrate the growth medium so that the two lines had very nearly the same S^*, although their

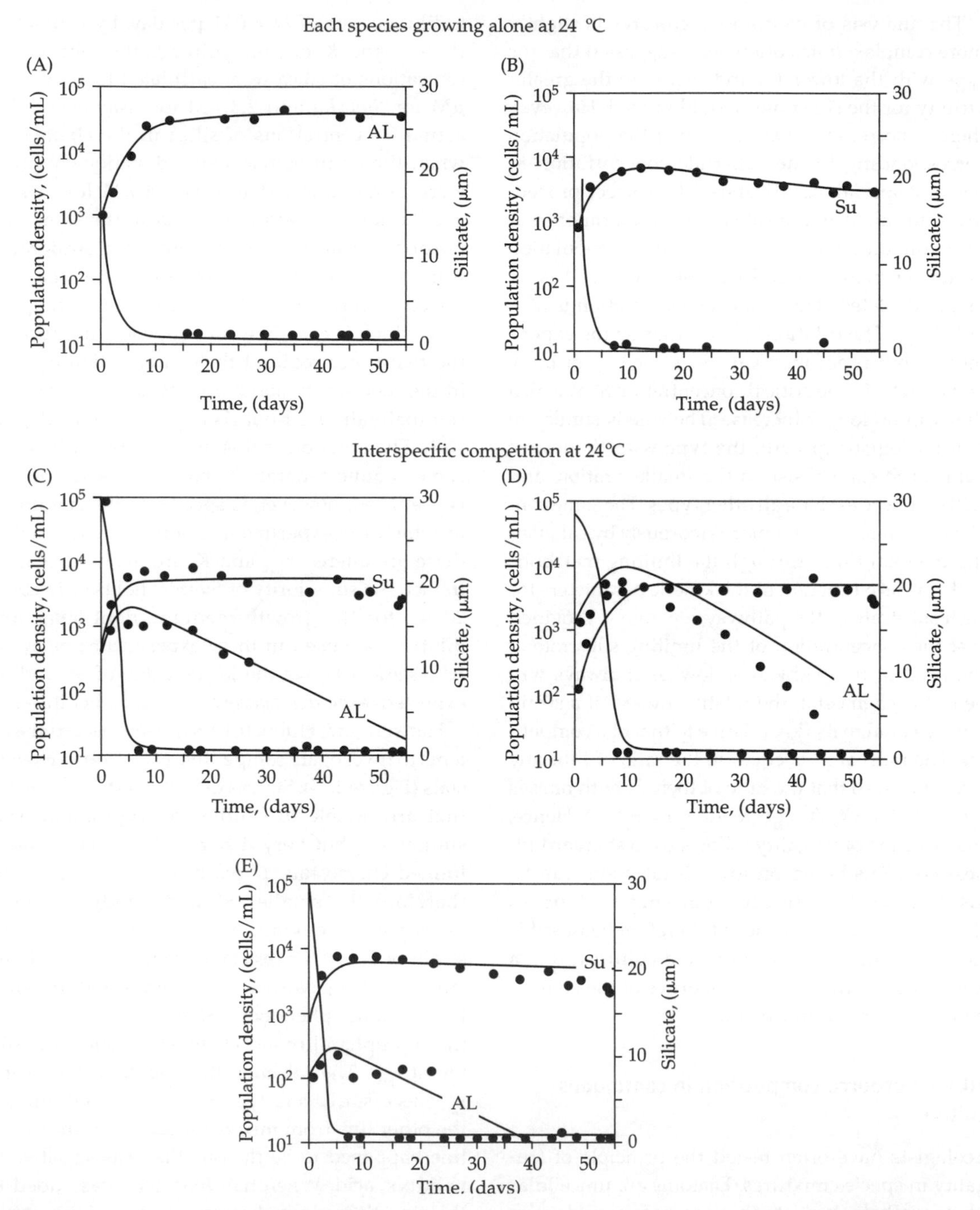

Figure 10.3 Competition between diatoms for silicate. The upper two graphs show the dynamics of population density and resource abundance in pure cultures of (A) *Asterionella* and (B) *Synedra*. Graphs C–E show that *Asterionella* is eliminated through competition with *Synedra* in mixed cultures, regardless of initial frequency, as predicted from the resource dynamics of the pure cultures. From Tilman (1981).

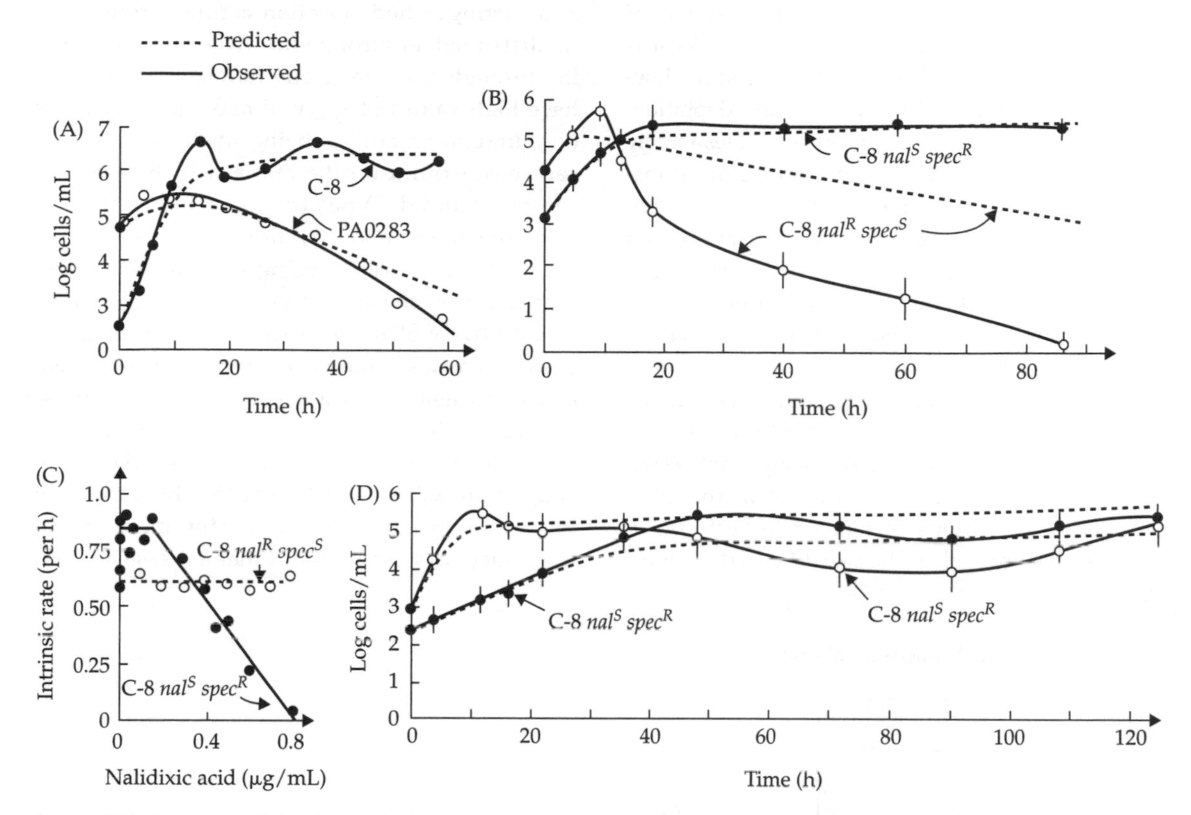

Figure 10.4 Coexistence of *E. coli* strains in chemostats titrated with nalidixic acid. A. *E. coli* replaces *Pseudomonas* in tryptophan-limited chemostats. B. The $nal^S spec^R$ strain of *E. coli* has the greater r_{max}, and eliminates the $nal^R spec^S$ strain from mixed cultures.C. The r_{max} of the nal^S strain is reduced by nalidixic acid, such that at some concentation the nal^S and nal^R lines have the same r_{max}. D. At this critical concentration of nalidixic acid, the nal^S and nal^R lines coexist indefinitely. From Hansen and Hubbell (1980).

K_s and r_{max} were widely different. When mixtures were cultured in these conditions, both types persisted in the chemostat, at roughly equal frequencies, for at least 100 hours.

The process of sorting in density-regulated populations is thus quite straightforward, so long as the competing strains have similar effects on one another. Density-dependent selection will favour the most frugal type able to grow on the smallest ration, because it will reduce the concentration of the limiting nutrient below the point at which other types are able to grow. This is a familiar economic principle: in the same way, unrestricted competition among workers at a particular trade in a saturated economy will drive down the rate of wages to the lowest level at which subsistence is possible.

10.1.5 *r-K* selection

If a population is continually rarefied, so that reproduction is essentially unchecked by density, selection will favour the types that have the highest maximal rates of increase. On the other hand, in populations that are perennially close to carrying capacity, the types that can subsist on the smallest ration will be the most successful. The reproductive characteristics that evolve will thus depend on the demographic history of a population, which will in turn depend on the pattern of ecological change that it experiences. If the rate of supply of resources is nearly constant through time, as it is in a chemostat, the ability to reproduce at the lowest possible ration will be the crucial adaptation. If the rate of supply of resources fluctuates widely in time, as it does when batch cultures are transferred frequently, population density will often be small relative to resource concentration, and the rate of unrestricted growth will alone determine fitness. We therefore expect different kinds of adaptation in stable and in disturbed environments (Figure 10.5).

A theory of this kind was first advanced by Theodosius Dobzhansky, to contrast the expected outcome of selection in the relatively stable economies of tropical regions with that in the more changeable conditions of temperate latitudes. It was later elaborated by Robert MacArthur, who, being an ecologist rather than a microbiologist, thought in terms of logistic population growth, and distinguished selection acting through r_{max} in disturbed environments from selection acting through K in stable environments. Types that have high values of r_{max} will make profligate use of abundant resources, being unchecked by any severe constraint on the rate at which resources can be acquired. Types that have high values of K, on the other hand, will use scarce resources as efficiently as possible, being able to subsist and reproduce on the lowest possible ration. This is an intuitively attractive dichotomy: when fuel is cheap, vehicles tend to be large, over-powered, and inefficient, but if it becomes expensive these types are driven out of the market by smaller cars with better fuel economy. It seems unlikely that adaptation will be indifferent to whether or not rates of uptake are a severe constraint on rates of reproduction, so adaptations that increase r_{max} will

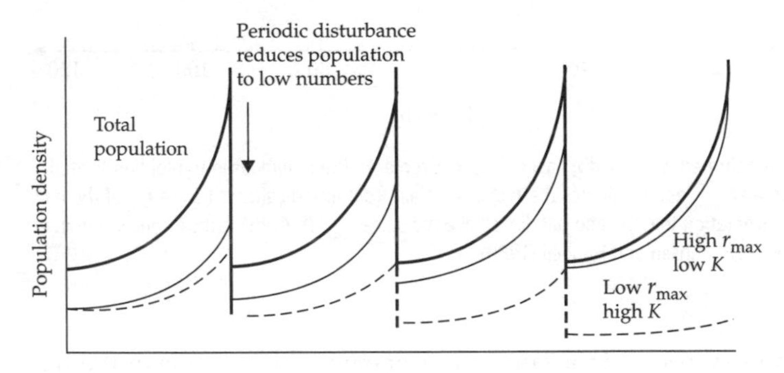

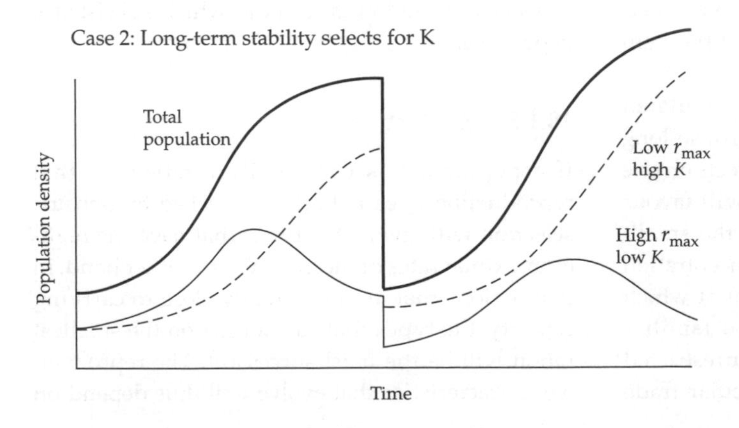

Figure 10.5 Selection for r and K. If the environment is disturbed frequently, the population as a whole being abruptly reduced in numbers, the population is kept in a perpetual state of nearly exponential growth, and the type with highest r_{max} will prevail. If the environment is seldom disturbed, the population is density-regulated most of the time, and total population density may then exceed the carrying capacity of the strain with lower K, which will thus be eliminated.

generally be antagonistic to those that increase *K*. This is the basic concept underlying the theory of *r*-*K* selection, leading to the conclusion that selection in density-independent and density-regulated populations will lead to characteristically different kinds of adaptations.

The term '*r*-*K* selection' is, of course, a misnomer. Selection will always favour types with greater realized rates of increase *r*, and the contrast is properly between r_{max} and *K*. Moreover, the dichotomy between populations that are unconstrained by resource supply and those that are perpetually starving, while useful to organize ideas, is too sharply drawn: most populations will spend most of the time at densities intermediate between zero and carrying capacity. Nevertheless, the realized rate of increase at densities that are kept low relative to resource supply may be more sensitive to changes in r_{max} than to changes in *K*, whereas in populations that reproduce for much of the time at densities close to carrying capacity the reverse is likely to be true.

As an account of how selection will act differently on components of the realized rate of increase in different circumstances, the theory, as I have so far summarized it, is plausible and straightforward. However, it is by no means straightforward to estimate the parameters of logistic growth in density-regulated populations in the field. An alternative approach is to argue that selection on r_{max} will lead to a correlated response of other components of fitness that are easier to measure. In particular, density-independent rates of increase may be elevated by rapid development and the production early in life of a large number of small offspring. In density-regulated populations, on the other hand, selection for high values of *K* is likely to favour long-lived, slowly developing types that produce a few large offspring. The extensive development of *r*-*K* selection theory during the 1960s and 1970s thus became increasingly concerned with the interpretation of life histories, and in particular with the interpretation of age-specific schedules of reproduction in different ecological circumstances. It therefore came into conflict with the parallel development of theories based on costs of reproduction (§5.2). These cost-based theories could be formulated more rigorously and tested more easily, and as they became widely understood during the 1980s the theory of *r*-*K* selection, as an interpretation of life histories, was largely abandoned. Indeed, the two most authoritative recent accounts of the evolution of life histories, by Roff (1992) and Stearns (1992), relegate *r*-*K* selection to a few paragraphs largely concerned with exposing its deficiencies. In my view, this rejection is premature. The central idea that the profligate and the efficient use of resources are antagonistic is at least as clear and compelling as the notion that present and future reproduction are antagonistic. The two approaches are different: *r*-*K* selection generally neglects age structure, just as cost-based theories neglect resource supply. However, they are not mutually exclusive. It is true that selection acting on schedules of reproduction may be independent of density; for example, earlier reproduction will be favoured (other things being equal) whether or not the population is density-regulated. However, this does not mean that the partial effect on fitness of a change in the age at first reproduction, relative to the partial effect of a comparable change in some other feature of the life history, will be insensitive to resource supply or population density. The costs of reproduction on which the balance of selective forces acting on the life history depends derive eventually from a shortage of resources, and will be modulated by population density. The converse is also true: the effects of resource supply on life-history characters will depend on the relationships between them. Moreover, these relationships may themselves change with population density. It is has often been found in crop trials that the correlation structure of components of yield changes with planting density: total yield may be almost independent of a particular component of yield at low density, yet highly correlated with it when the plants are more crowded. Density-dependent selection and age-specific selection should be regarded as complementary rather than competitive interpretations of life history.

An experiment by Taylor and Condra (1980) illustrates the difficulty of testing *r*-*K* selection theory independently of age-specific effects. They set up *K*-selection lines of freshly collected *Drosophila* stocks in population cages where food vials were renewed on a 4-week cycle, causing obvious and severe crowding of the larvae. They felt it to be impracticable to set up comparable r_{max}-selection lines by rarefaction, and instead allowed only the first 100 flies emerging to lay eggs for 2 days on a relatively large amount of medium, so that the larvae were uncrowded. After about 10 months of selection, the life histories of the selection lines had diverged, when tested in a common environment. The r_{max} lines developed more rapidly, and began oviposition about 1 day younger than females from the *K* lines. This seemed consistent with theory, if early reproduction is crucial to increasing r_{max}. However, there was little difference in the number of eggs produced at this time, and the *K*-line females actually produced more eggs later in life and lived longer. The earlier reproduction of the r_{max} lines and the greater fecundity of the *K* lines roughly balanced, so that both had about the same rate of increase overall. These results are difficult to understand in terms of density-dependent selection, but the design of the experiment introduced age-specific as well as density-dependent effects. By taking the first females to emerge and allowing them to lay eggs for only 2 days, the experimenters selected strongly for increased reproduction early in life. We know from other experiments that this is likely to have the pleiotropic effect of reducing expected fecundity in later life. Thus, any density-dependent effects that occurred were overshadowed by the response to age-specific selection.

Some experiments have deliberately incorporated age-specific density effects. Birch (1955), while visiting Dobzhansky's laboratory, set up an experiment to investigate the reasons for the seasonal fluctuation in the frequency of inversion types in some *Drosophila* populations. In population cages under normal conditions of culture, a mixture of Standard and Chiricahua karyotypes reaches an equilibrium of about 70% Standard, because heterozygotes are the fittest genotype, and Standard homozygotes are superior to Chiricahua homozygotes. If both adults and larvae are thinned, the relative fitness of the homozygotes reverses, and at equilibrium about 70% of the chromosomes are Chiricahua; thus, selection on the two types is density-dependent. It probably acted through differences in mating or fecundity, rather than mortality. Thinning the larvae, but crowding the adults, gave the same result; but if the larvae were crowded and the adults thinned, the Standard type was again in a majority at equilibrium. Thus, Chiricahua is selected, not merely at low population density, but specifically at a low density of larvae. Barclay and Gregory (1981) measured the effects of larval and adult crowding on the life history. Their very complicated experiments, which I shall summarize very briefly, involved removing either adults or larvae, or both, from the cultures at frequent intervals for the 6 months of selection, before assaying the selection lines and the controls (in which adults and larvae were both crowded) in a common environment. When both adults and larvae were removed, maintaining low densities of both, the lines evolved a shorter lifespan and greater early fecundity, which is consistent with density-dependent selection. When only one of the two stages was removed, the experiments gave erratic results that are difficult to interpret in terms of either density-dependent or age-specific selection.

r-K selection experiments

The experiments that I have described above seem inconclusive; they seem to involve density-dependent effects, but because of limits of material or time, or because age-specific effects were predominant, they do not clearly demonstrate an antagonism between the components of realized rates of increase in density-regulated populations. Very few well-designed long-term experiments have been attempted. Among the most elegant are those reported by Luckinbill (1984), who selected *E. coli* populations in batch culture with glucose as a carbon source (Figure 10.6). *K*-selected lines were allowed to grow into stationary phase before being transferred; lines selected for r_{max} were transferred more frequently, before the supply of glucose was exhausted. The outcome of selection was unambiguous. The single-mutant selection lines invariably showed the appropriate direct response, the r_{max} lines having a higher r_{max} than the parental strain, and the *K* lines a greater *K*. When tested

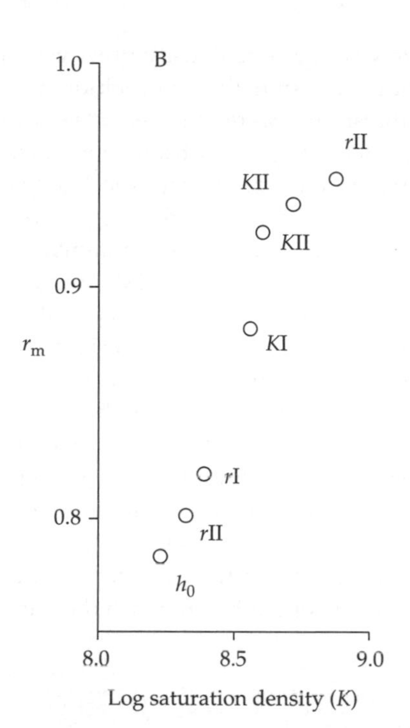

Figure 10.6 *r-K* selection in *E. coli*. The figure shows the characteristics of novel mutants from *r*-selected (transferred at short intervals in log phase) and *K*-selected (transferred at longer intervals in stationary phase) lines of *E.coli*, when tested in pure culture. The base population is h_0; the *r*-selected lines are rI, rII, and rIII; the *K*-selected lines are KI, KII, and KIII. Note the strongly positive genetic correlation of *r* and *K*. In mixed cultures of *r*-selected and *K*-selected lines, pure-culture performance reliably predicts the outcome of selection: any of the *K* lines eliminates rI or rII, but rIII eliminates any of the *K* lines. From Luckinbill (1984).

in the environment of selection, the selection lines displaced the parental strain. However, there was no evidence of any antagonism between r_{max} and *K*; in every case, the selection lines were superior to the parental strain in both r_{max} and *K*. Moreover, when the lines were mixed and allowed to compete for about 30 generations, the outcome of selection was independent of environment: when an r_{max} strain and a *K* strain competed, the same strain always won, regardless of whether transfers were frequent or infrequent. This was not an artefact of the particular system of batch culture used in the experiment. A second experiment that used semi-continuous culture in an anaerobic fermenting vessel gave the same result. Luckinbill concluded that his experiments gave no support to the theory that selection for r_{max} and selection for *K* gave rise to antagonistic adaptations

One flaw in this design is that changes in frequency occurring early in culture growth may be frozen, so to speak, into the composition of the culture at the end of the growth cycle. To avoid this problem, Velicer and Lenski (1999) propagated wild isolates of *E. coli* in batch and continuous culture, arguing that cells in chemostats are always at the edge of starvation whereas those in batch periodically experience the renewal of a superabundance of resources. This is a complex experiment in which two wild isolates were first selected on succinate for 75 days then transferred to 2,4-D as sole carbon source, with some lines switching from batch to chemostat culture and vice versa. In the first phase the lines adapted to the novel environment (s = 0.75, but with much variation among lines, sd 1.08) while the indirect response (of chemostat growth to batch culture, or vice versa) was smaller but positive (s = 0.23, sd 0.36). In the second phase, lines continuing in the same selective regime showed continued increase in fitness (s = 0.65 relative to the phase spores, sd 0.38) with a much smaller indirect response (s = 0.08, sd 0.39), whereas lines whose regime was switched showed a weaker direct response (s = 0.31, sd 0.43) and a negative indirect response (s = −0.17, sd 0.23). (These figures apply to one founding isolate, the other being affected by contamination.) These highly variable results are difficult to interpret confidently, but it is clear that lines were adapting to their novel environment under strong selection in both phases, and the outcome can then be explained by supposing that the first advantageous mutations to appear are simply those that are unequivocally superior in this environment, with greater r_{max} and greater *K* than the unselected initial population.

Viral lines can be propagated either at high or low multiplicity of infection (MOI). At low MOI most cells are initially uninfected and selection favours rapid horizontal spread, whereas at high MOI there is less opportunity for new infections and selection will favour the most thorough exploitation of the host cell. Lines of phage $\varphi6$ (Turner and Chao 1998) and vesicular stomatitis virus (VSV) (Bordería and Elena 2002) propagated at high MOI became adapted to these conditions, but the cost of adaptation varied. In $\varphi 6$ the high-MOI lines performed poorly at low MOI, whereas the low-MOI lines were not impaired at high MOI. In VSV, on the other hand, the low-MOI lines were clearly inferior at high MOI, whereas the two were equivalent at low MOI (Figure 10.7). In foot-and-mouth disease virus low-MOI lines became specifically adapted whereas high-MOI lines did not (Sevilla *et al.* 1998). Virus selection experiments are the clearest quantitative studies of *r*-*K* selection, and show that adaptation to high or low density may (or may not) occur, and that the indirect response is sometimes (but not always) a loss of fitness in the contrasting environment.

The most extensive experimental study of density-dependent selection is that carried out by Mueller and colleagues (Mueller and Ayala (1981), Bierbaum *et al.* (1989), Mueller (1991) and especially Mueller *et al.* (1991); also Joshi and Mueller (1988) on larval feeding rate, and Mueller and Sweet (1986) on pupation height). There is another large-scale experiment in the same general area by Bakker (1969). Mueller's experimental material was derived from a set of strains, homozygous for the second chromosome, derived from recently captured wild flies. Growth at different densities was positively correlated among these strains: those that grew well at one density grew well at other densities, although this correlation was small when the strains were cultured at very different densities (cf. Lang's corn trial, above). These sets were crossed, and the progeny selected divergently for r_{max} and *K*. Lines selected for r_{max} were rarefied by transferring a predetermined number of adults to fresh vials every week, thus keeping larval density quite low; in the *K*-selected lines, all surviving adults were transferred every 4 weeks, so that densities were much higher. After eight generations, the lines were assayed, after having been cultured in a common environment for two generations. At low density the r_{max} lines had the greater rates of growth, although only by a small margin; at high density, the *K* lines had appreciably faster growth (Figure 10.8). This is the expected result of sorting a diverse population in which there is initially

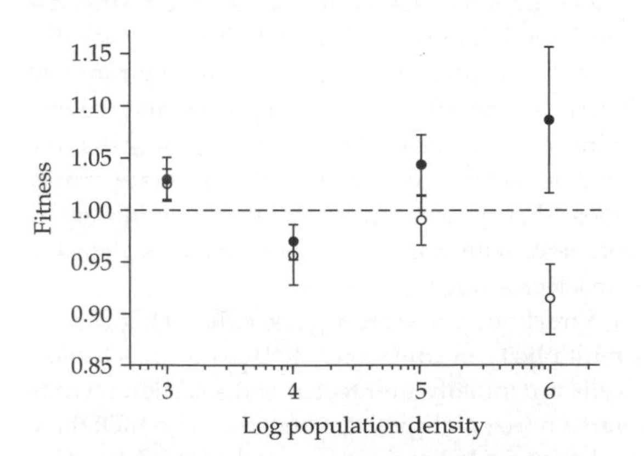

Figure 10.7 *r*-*K* selection in VSV. Competitive fitness of lines selected at high density (multiplicity of infection) (solid circles) and low density (open circles) over a range of densities. From Borderia and Elena (2002).

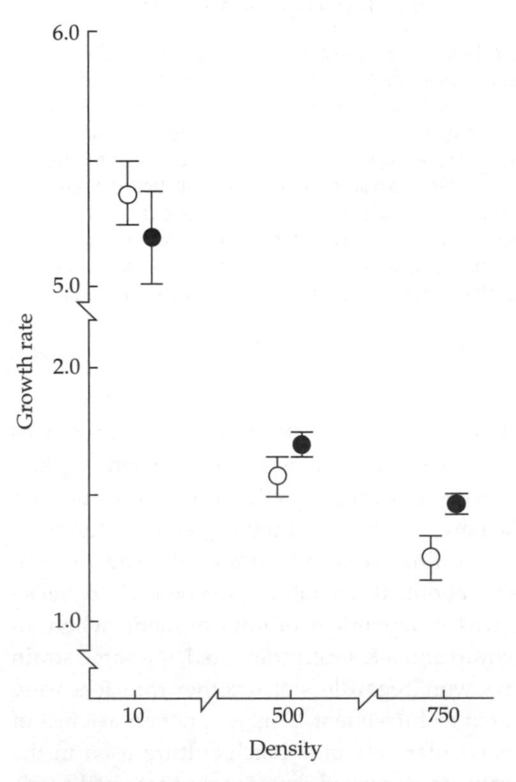

Figure 10.8 *r*-*K* selection in *Drosophila*. Mean growth rate at low and high density in *r*-selected (open circles) and *K*-selected (filled circles) lines. From Mueller and Ayala (1981) by permission of PNAS.

little correlation between r_{max} and K at very low and very high densities. The life history of the flies was studied later in the experiment, after about 20–40 generations of selection. There seemed to be no difference between the lines in adult survival or fecundity at any density, but larval characteristics had been modified: at high larval densities, the K-selected larvae survived better, although at low densities there seemed to be little difference between the lines. The K-selected larvae also grew into somewhat larger adults. The differences in population growth-rate were thus generated largely through larval adaptations. The nature of these adaptations was investigated subsequently, after more than 100 generations of selection. Resource competition among *Drosophila* larvae under laboratory conditions seems to be a fairly simple process: they pump the medium through themselves, and those that pump faster will grow faster and be more likely to survive. Larvae from the K lines pump faster: they move their mouthparts in and out about 15% more quickly than larvae from the r_{max} lines. They also behave differently at the end of larval life. *Drosophila* larvae may pupate directly on the surface of the food medium, or may first crawl some way up the wall of the vial. In uncrowded cultures, it is safe to pupate on the medium, and unnecessary to spend energy crawling up the vial. In crowded cultures, however, the medium becomes as wet and churned as a farmyard in November, and pupae on the surface of the medium are likely to drown. Even those that crawl a little way above the surface are likely to be dislodged by more active larvae crawling past them. In the r_{max} lines, the average pupation height was about 5 mm, and 40% of the larvae pupated directly on the surface of the medium. In the K lines, the larvae crawled nearly twice as high to pupate, and only 15% remained on the surface. Density-dependent selection seems to have modified larval feeding rates and pupation behaviour to increase larval survival and adult size at high larval density.

These experiments demonstrate that characteristic adaptations evolve at high density. They do not directly demonstrate any antagonism between low-density and high-density adaptation, because there seemed to be little response to selection at low density, and in any case the selection lines were not compared with the base population. A form of back-selection was used to resolve this point. After about 200 generations of selection, the r_{max} lines were divided, some remaining at low density, whereas others were transferred to the high-density conditions of the K lines. After 25 generations, the growth-rate and adult productivity of the new K lines had dropped, relative to the lines retained at low density. The low-density conditions therefore seem to require adaptations that are lost when selection is relaxed or changed, although it is not known what these are.

I think that the body of r-K selection experiments is oddly inconclusive. The general tendency for profligate resource use (for example, fermentation) to be antagonistic to more conservative strategies (for example, respiration) seems intuitively obvious and mechanistically sound; and yet selection experiments have so far failed to provide the clinching proof. I shall cling to the theory until a more decisive refutation has been produced.

10.2 Selection within a single diverse population: frequency-dependent selection

Genotypes may sometimes have the same effect on one another, as I have so far assumed; but this must be a special case. It is more likely that some genotypes will have stronger and others weaker effects on their neighbours, or on the individuals with whom they are temporarily interacting. Growth and reproduction will then depend on what kind of neighbours an individual lives alongside; neighbours are one of the features of a particular site, or way of life, and can be treated as an environmental factor like nutrient supply or temperature. The bordered hill plot is a classical design in agronomy for testing the performance of cultivars that provides a concrete example of the effect of neighbours. A single Self plant is grown in the middle of a small clump of Neighbours, to simulate the commercial situation, in which it would be growing within a field of other plants. For example, a 3×3 arrangement provides a single central test plant with a border of eight others. This can be used to compare the effect of a border of similar plants with that of a border of different plants. For example, the arrangements

S	S	S		N	N	N
S	S	S		N	S	N
S	S	S		N	N	N

can be used to identify the environmental effect of similar and dissimilar neighbours on the performance of the central test plant S. Smith *et al.* (1970) used this design to investigate the effect of competition in mixed culture on the seed yield of five oat cultivars (Figure 10.9). They found that the yield of the test plant was substantially affected by the identity of its neighbours; some neighbours enhanced yield, relative to growth in pure culture, whereas others depressed it. One cultivar in particular—Rodney—grew much better in the company of other cultivars than it did when its neighbours were other Rodney plants. At the same time, other cultivars were suppressed by Rodney neighbours. In this sense, Rodney is a strong competitor, perhaps because it was the tallest of the cultivars tested, and may have been able to shade its neighbours. It is not necessary, of course, for neighbours to occupy fixed positions in space. The experiment in which Lewontin measured larval viability as a function of the density of pure cultures (above) was supplemented by a second experiment in which the same inbred lines were reared as mixtures. The mixtures comprised equal numbers of the test strain and of a strain carrying the sex-linked mutation *white*. If genotypes have the same effect on one another, then the viability of the *white* larvae will be the same at a given total density, regardless of the genotype with which it is paired. However, this was not the case: at any density, different strains had different effects on the viability of the *white* larvae. From the point of view of *white*, therefore, different neighbours represent consistently different environments.

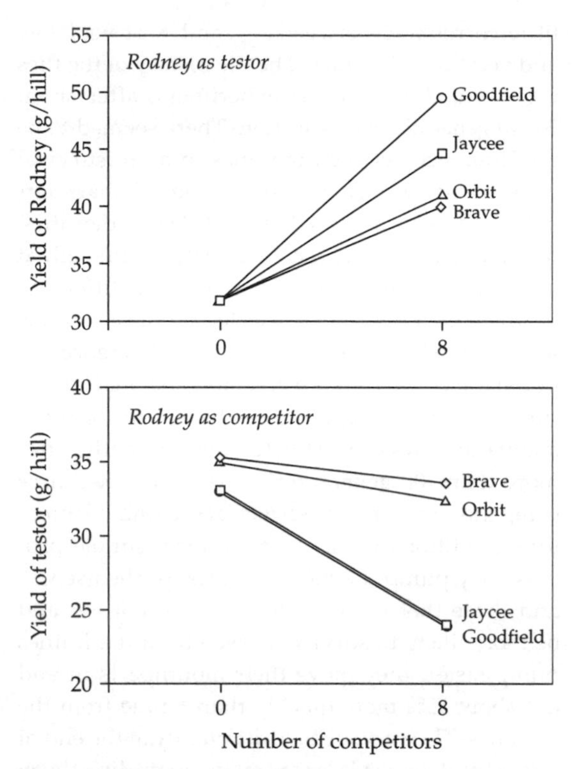

Figure 10.9 The effect of different kinds of neighbour. These two graphs illustrate some of the interactions among oat varieties described by Smith *et al.* (1970). The upper graph shows the yield of Rodney when grown with different kinds of neighbour on the same hill plot. There were four treatments, involving the growth of the central plant in a 3 × 3 hill with zero, two, four, or eight neighbours of a different variety (thus, the zero level is eight neighbours of the same variety); In all four cases, a Rodney plant yields more when presented with different neighbours than when grown with other Rodney individuals. The lower graph shows the effect of Rodney on the yield of other varieties. There are many similar experiments with a variety of crops, for example soybean (Hinson and Hanson 1962), potatoes (Doney *et al.* 1965), maize (Kannenberg and Hunter 1972), and peanuts (Beg *et al.* 1975).

10.2.1 G × *G*

Lewontin's experiment can be viewed from another perspective by comparing the viability of strains in pure culture with their viability when cultured with *white*. As an average over all strains, viability was about the same at a given total density, whether or not white were present. However, the strains responded in different ways: in some, viability was enhanced by the presence of *white* (rather than the same number of flies of the same strain), whereas in others it was reduced. This is a special case of genotype–environment (G × E) interaction. The two environments are the presence of other larvae of the same strain, and the presence of *white* larvae. Genotypes respond differently to these environments, so that performance in one is poorly correlated with performance in the other. Because the environments concerned are genotypes, we can symbolize the interaction, not as $G \times E$, but rather as $G \times G$, the G being the genotype of the neighbour acting as the environment. In Lewontin's experiment, the *white* strain was used as a neighbour for all the test strains. A still more interesting design is to use every strain both as self and as neighbour, so that all the mutual relationships among the strains can be described. Roy (1961) grew three rice cultivars in pairwise combinations in small hill plots. The cultivars do not have euphonious names, and can be referred to simply as 2A, BK, and RP. There were four plants on each plot, two of one sort and two of another, and Roy measured the seed yield of each of them. His results were thus the yield of a given type, when grown in combination with a given neighbour:

		Neighbour		
		2A	BK	RP
	2A	570	574	743
Self	BK	717	553	822
	RP	76	148	321

There are two sorts of effect here. In the first place, there are stronger and weaker competitors. RP is clearly a weak competitor: the other cultivars perform well when they have RP as a neighbour, and RP itself performs poorly when its neighbours are 2A or BK. This is a straightforward environmental effect. However, it is impossible to characterize 2A and BK as being either strong or weak competitors. BK yields better when its neighbours are 2A than it does when grown in pure culture; but 2A in turn yields better when its neighbours are BK than it does when grown as a pure stand. This constitutes $G \times G$: it is impossible to define the relative fitnesses of 2A and BK, unless the neighbours with which they are growing are specified.

10.2.2 Frequency-dependent fitness

Whenever there are unequal competitive interactions among genotypes, the fitness of a genotype must depend on the frequency of different kinds of neighbour. In the simplest case, neighbours have a straightforward environmental effect, as strong or weak competitors. The fitness of a genotype is then reduced when the frequency of stronger competitors is increased. If there are only two types in the population, this implies that the fitness of either type is directly proportional to its frequency. This effect occurs in Roy's hill-plots of rice. Because RP is a weak competitor, its fitness will depend on the frequency of stronger competitors such as 2A and BK: the more frequent they are, the lower will be the relative fitness of RP. However, RP has a lower fitness than the other two at any frequency, and its fitness simply decreases as the other two types become more abundant. The relationship between 2A and BK is different: either may have the greater fitness, depending on the frequency of the other. I have discussed his results in terms of the yield obtained by a given type; however, they can be expressed in a different way, in terms of the effect of a given type on its neighbours' yield. In a 2A environment—that is, at a site where most neighbours are 2A plants—BK produces more seed than 2A does itself. However, the converse is also true: in a BK environment, 2A has the greater fitness. The relative fitness of either type will depend on its frequency in the population. The occurrence of $G \times G$ thus leads to selection that changes in direction, and not merely in magnitude, according to the frequency of competing types.

Some enzyme loci in *Drosophila* have two (or more) alleles that reach characteristic frequencies in cage populations that are not deliberately selected; that is, replicate populations with different initial gene frequencies will usually evolve towards the same frequencies. In the interesting cases, these frequencies are much too large or much too rapidly attained for the maintenance of diversity to be explicable in terms of random processes acting on functionally equivalent genes. These have attracted much attention, because they might reveal a mechanism of selection capable of explaining the allelic diversity typical of many loci in natural populations. Some of the most extensive work was done by Kojima (1971) shortly after the introduction of protein gel electrophoresis as a routine procedure into population genetics. It had until then been widely accepted that a substantial part of genetic variation was maintained through heterozygote advantage. Once the extent of allelic diversity in natural populations was uncovered, it was immediately realized that it was unlikely to be maintained by heterosis, since no population could sustain the cost of selection incurred by the elimination of so many homozygotes. The focus of Kojima's work was thus to show that frequency-dependent selection was an acceptable alternative to heterosis as a Darwinian explanation of allelic diversity. He worked principally on an esterase locus, *Est-6*, and the alcohol dehydrogenase locus *Adh*, both of which have two alleles that coexist more or less indefinitely in cage populations, with the rarer allele at a frequency of about 0.25. There are two ways of finding out how such alleles are maintained. The first is to use the results of the selection experiment itself, by using a particular hypothesis of selection to predict the dynamics of the approach to equilibrium. The main difficulty with this approach is that when genotype frequencies are the only information available, fitness will always appear to be frequency-dependent. Even if the locus is heterotic, the allele or the homozygote that is present in excess will fall in frequency, so that its fitness, as estimated by the ratio of frequencies in successive generations, will be low so long as its frequency remains high. It is possible to correct for this effect, but the differences between populations evolving under heterosis and those evolving under frequency-dependent selection are then rather small, unless gene frequencies are extreme. It is also regrettably true that most of the simplifying assumptions that one must make in order to calculate the expected generational changes produce a spurious impression of frequency-dependent selection if the population does not in fact conform to them. The initial claims for frequency-dependent selection at *Est-6* were, for these reasons, not completely convincing. The second approach is to measure fitness components directly in separate tests where the genotypes are deliberately reared together at different frequencies. Although this seems straightforward, it can also be problematic. For example, the overall viability can be estimated by using homozygous lines and their hybrids to set up a population of pre-mated females that will give rise to progeny genotypes in known proportions, and then scoring the genotypes of adults in the next generation. Experiments of this sort showed that the viability of either homozygote appeared to be greater when it was less abundant. However, the procedure works only if the genotypes are equally fecund; if the heterozygote is the most fecund, it will produce additional homozygous progeny of each kind in equal numbers, but the effect of this will be to increase the frequency of the rarer homozygote by a larger proportion, giving a misleading impression that viability is frequency-dependent. In fact, viability, but not fecundity, was found to be rather strongly frequency-dependent. Although Kojima's experiments have been criticized, they seem to me to show fairly satisfactorily that the evolution of a characteristic allele frequency at these two loci is driven primarily by frequency-dependent selection.

Morgan (1976) set up lines homozygous for the fast F and slow S alleles at the *Est-6* locus from large cage populations of *Drosophila melanogaster*. The populations had been maintained in the laboratory for about 18 months before the experiments were conducted. Four replicate cultures of 200 larvae each were established at frequencies of the F allele ranging from 0.15 to 0.85. The numbers of FF and SS flies emerging from these cultures was counted, and I have calculated the relative fitness of FF as the ratio of the fraction of FF to the fraction of SS surviving (Figure 10.10). The relative viability of FF declines with the frequency of FF in the initial population of larvae, such that FF larvae are superior to SS at frequencies less than about 0.4, but inferior at higher frequencies. Other reports of frequency-dependent selection at enzyme loci include Kojima and Yarborough (1967), Yarborough and Kojima (1967), Kojima and Tobari (1969), Huang *et al.* (1971), and Yamazaki (1971). Comparable experiments with karyotypic variation in *Drosophila* have been conducted by Levene *et al.* (1954) and Nassar *et al.* (1973); there are also similar experiments by Anxolabehère (1971) on a variety of loci, and by Bundegaard and Christiansen (1972) on two marked fourth-chromosome stocks. Harding *et al.* (1966) described the frequency-dependent advantage of

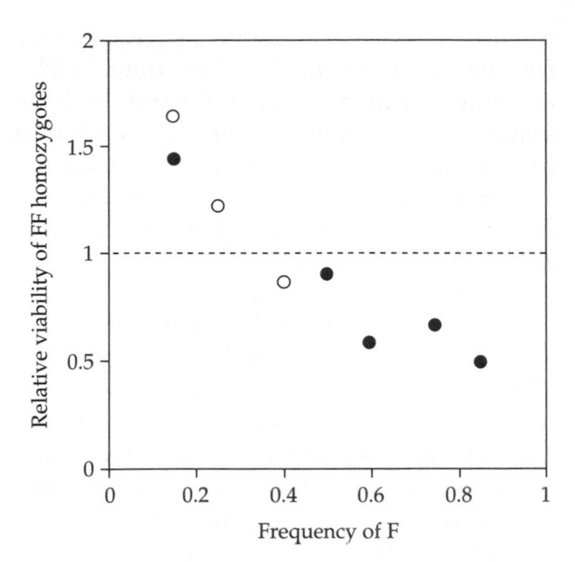

Figure 10.10 Frequency-dependent selection at *Est-6* in *Drosophila*. From Morgan (1976).

heterozygotes at a seed-colour locus in lima beans. The field has been reviewed by Kojima (1971) and by Ayala and Campbell (1976).

10.2.3 The ubiquity of frequency dependence

The sources of frequency-dependent selection were investigated by ecological geneticists led by Bryan Clarke, who emphasized the tendency of predators and pathogens to prefer the most abundant kind of cryptic prey, which he termed 'apostatic selection' (reviewed by Clarke 1979). One simple experiment consisted of exposing coloured pastry baits at different frequencies to predation by wild birds, which took a disproportionate number of the more common type (Allen and Clarke 1968). There can be very large swings in fitness: when green and brown baits were presented at a ratio of 9:1, house sparrows took 167 green and 3 brown baits (s = 5.2 for brown), whereas at a ratio of 1:9 they took 78 green and 761 brown baits (s = −0.08). These experiments are reviewed by Allen (1988). There is also some evidence that common types of host are disproportionately infected by parasites (§11.3). When prey are protected by aposematic coloration, on the other hand, conspicuous types are likely to be at a disadvantage when they are rare ('anti-apostatic selection') (see Lindstrom *et al.* 2001). This can also give rise to large shifts in fitness: for example, Mallet and Barton (1989) released marked *Heliconius* morphs at both ends of a hybrid zone and found that the non-resident type was more likely to be eaten. Overall selection was about $s = -0.5$, which is equivalent to $s = -0.17$ at each of the three loci controlling the polymorphism. Consequently, antagonistic biological interactions may often give rise to strong negative or positive frequency-dependent selection

A second reason for regarding frequency-dependent selection as a general mechanism for protecting diversity is that it is a natural and very general consequence of social interactions, including resource competition. In a complex environment any specialist strain will deplete the resource on which it depends, and as it does so its fitness will fall; thus, it will have high fitness when first introduced invading a novel environment, and is therefore likely to spread, after which its fitness will fall, so that is unlikely to become fixed. The protection of diversity in complex environments has been investigated in detail by community ecologists and population geneticists from somewhat different perspectives. Community ecologists have attempted to explain the diversity of autotrophic algae in the upper waters of lakes in terms of competition for essential resources such as nitrogen or silicon (Tilman 1981), whereas population geneticists have been concerned with heterotrophic bacteria competing for substitutable resources such as sugars or amino acids (Stewart and Levin 1973). In structured environments selection is frequency-dependent even though fitnesses on each substrate are fixed, provided that density regulation is local (see §8.4). This follows from the fixed productivity of each substrate. Consider a simple Levene model in which allele A1 is superior on substrate 1. If there are N initial colonists then the number of survivors on substrate 1 is $N\,(p_{111}s_1 + p_{221}t_1)$ with appropriate signs for the selection coefficients. If this substrate contributes a fraction K_1 of all newborns, the probability that an individual will enter the dispersal pool to form the next generation is $K_1/[N(p_{111}s_1 + p_{221}t_1)]$, and is thus negatively correlated with its initial frequency. This comes about because the local spread of a rare type is fully transmitted to the next generation when its preferred substrate provides a refuge contributing a guaranteed number of offspring to the community. Note that the conditions for preserving a dominant allele in a structured environment with free dispersal are the same as those for preserving a haploid strain in a complex environment (§8.4.2). The close correspondence between the two situations shows that frequency dependence is likely to be ubiquitous in natural communities, which almost always inhabit complex or structured environments.

When two types are limited by a common essential resource, density-dependent selection will cause the single type that can subsist on the lowest ration to become fixed. When there are two depletable resources, frequency-dependent selection may sustain both types. The simplest situation involves resources that are essential and non-substitutable; that is, a certain supply of both is essential for growth; growth is not increased by supply in excess of this minimum; and no supply of one, however great, can compensate for a shortfall in the other. If one type can subsist on a lower ration of both resources, then it will become fixed. However, the requirements of the two types may differ. One type will then be able to subsist on a lower ration of one resource, whereas the second type requires less of the other resource. This implies that the two types will consume the two resources in different proportions. The growth of both types will drive down resource supply to the point where either type is barely able to subsist on the resource for which the other type has the lower requirement. At this point, the two types are limited by different resources: each will be limited by the resource of which it requires the greater ration. This is an equilibrium because neither type is able to increase. This mixture will be stable, preserving diversity, provided that either type represses its own growth more severely than it represses the growth of the other type. This is equivalent to saying that both types must consume proportionately more of the resource that is limiting their own growth. The condition for the indefinite coexistence of two types in a mixture is thus that both should consume proportionately more of the resource for which they require the greater ration. This argument was developed by Tilman (1981), who tested it with mixtures of different species of diatoms in semi-continuous cultures maintained for about 50 generations. At a certain death rate (i.e. outflow rate) in pure culture the minimal subsistence levels of silicate and phosphate for *Asterionella* were 1.9 μM silicate and 0.01 μM phosphate. The corresponding levels for *Cyclotella* were 0.6 μM silicate and 0.2 μM phosphate. The growth of a mixed population will drive silicate down to 1.9 μuM (at which point *Asterionella* is limited by silicate, but not by phosphate), and phosphate down to 0.2 μM (at which point *Cyclotella* is limited by phosphate, but not by silicate). At this point, *Asterionella* consumes proportionately more silicate (1.9/0.01 = 190 for *Asterionella*, whereas 0.6/0.2 = 3 for *Cyclotella*). It therefore represses its own growth, through depletion of silicate, more severely than it represses the growth of *Cyclotella*, through depletion of phosphate. The converse applies to *Cyclotella*: both types are inhibiting their own growth more effectively than they are inhibiting the growth of their competitor. These predictions can be tested by manipulating rates of nutrient supply. If the rate of supply of phosphate relative to that of silicate exceeds the ratio in which the two are consumed by *Cyclotella*, the species limited by phosphate at equilibrium, then *Cyclotella* is released from its limitation by phosphate, while *Asterionella* remains limited by silicate. Consequently, *Cyclotella* will spread to fixation. Conversely, *Asterionella* will become fixed if the supply ratio of silicate exceeds its consumption ratio by *Asterionella*. The results of competition experiments with different supply rates of the two resources agreed well with the predictions from pure-culture dynamics.

Specialists with relatively narrow ranges of substrate preference evolve in mixtures of substitutable resources (§8.3). When one strain is rare its overall fitness is the arithmetic mean of its fitness on each substrate separately, weighted by the abundance of the substrate in the mixture entering the system. The relative fitness of the common strain is then the reciprocal of this quantity, that is, the harmonic mean fitness. The condition that two strains will coexist is therefore that one should have the greater arithmetic mean and the other the greater harmonic mean fitness. This condition will be satisfied within some range of substrate ratios, and diversity is preserved if the actual ratio supplied falls within this range. This approach has been rigorously investigated by Lunzer *et al.* (2002) using two *E. coli* strains carrying different constitutive versions of the *lac* operon that have different activities on lactulose and methylgalactoside (§8A.2). The growth of the strains on each substrate separately successfully predicted the range of mixtures within which they were able to coexist (Figure 10.11), thus forging a chain of reasoning from enzyme kinetics to allozyme frequency.

Natural communities encounter not only resources that support growth but also toxins such as antibiotics that inhibit growth. Resistant strains often detoxify the local environment by removing or neutralizing the toxin, and thereby create an opportunity for susceptible strains to spread. There must be a cost of adaptation (such as plasmid carriage) for diversity to be maintained in a resource–toxin system. The susceptible strain then has the higher growth rate and can invade when the resistant type is common and toxin concentration is low. As it spreads the concentration of toxin will rise and the relative fitness of the susceptible strain will fall, until its higher growth rate is balanced by intoxication. If one type can subsist on a lower ration of the resource and a higher dose of the inhibitor, then it will become fixed. However, one type may be able to subsist on a lower ration of the resource while being repressed by a lower dose of the inhibitor. The two will then coexist if the type that is resource-limited at equilibrium produces proportionately more inhibitor. The dynamics have been worked out in detail by Lenski and Hattingh (1986). The classic experiment involving a situation like this is Gause's study of yeast mixtures (Gause 1934). He used the common budding yeast *Saccharomyces* and a smaller yeast that he refers to (incorrectly) as '*Schizosaccharomyces*'. The limiting resource was sugar; the inhibitor was the ethanol produced by fermentation. In pure culture, *Saccharomyces* has the higher r_{max} and the higher K. Nevertheless, both species persist in mixed cultures for at least 20–30

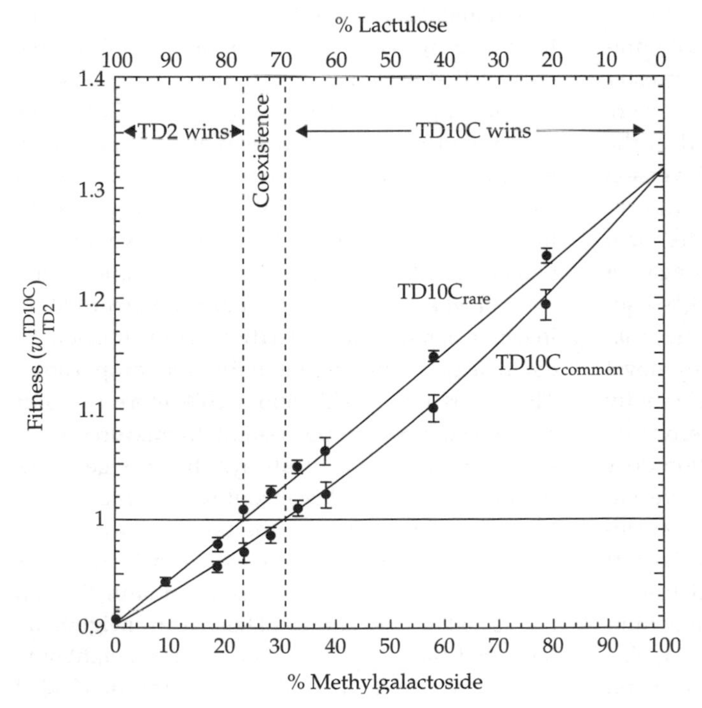

Figure 10.11 Coexistence in complex environments. Fitness of *E coli* strain TD10C relative to strain TD2 when rare and when abundant in relation to the composition of a mixture of two galactosides. The upper line (fitness when rare) is predicted to be the weighted arithmetic mean of fitness on each galactoside separately, the lower line (fitness when common) the weighted harmonic mean. The zone of coexistence defines the range of mixtures within which neither strain has an unequivocal advantage. From Lunzer *et al.* (2002).

generations, though '*Schizosaccharomyces*' is the less abundant. Gause explained the maintenance of diversity in mixed cultures by the greater inhibitory effect of '*Schizosaccharomyces*', which produces about twice as much ethanol per unit volume as does *Saccharomyces*. *E. coli* is normally inhibited by 2.5 μg mL^{-1} of chloramphenicol, but a conjugative plasmid provides resistance at the expense of slower growth. Susceptible strains spread despite the presence of inhibitory concentrations of chloramphenicol if they are introduced at low frequency into chemostats stocked with the resistant strain, as theory predicts (Lenski and Hattingh 1986).

These very successful experiments suggest that the diversity of natural microbial communities may be maintained principally through differential resource utilization, and more broadly that niche separation provides a general mechanism for supporting high levels of functional diversity. Even if frequency dependence were universal, however, it would not necessarily be generally effective. The conditions for stable, protected polymorphism are quite restrictive unless selection is strong and dispersal weak, typically requiring a rather narrow window of intermediate productivities or resource ratios. There is still a gap between the conditions of most experiments, involving only a small number of strains and substrates, and the much greater complexity of natural environments. Where more complex situations have been investigated in the laboratory, they seem to conform to a weak version of niche separation (see §8.4). *Pseudomonas* cultures in rich complete medium evolve without frequent selective sweeps, and competing isolates against one another shows that their relative fitness usually falls with initial frequency (MacLean *et al.* 2005). The variance of fitness generated by novel mutations increases about three times as rapidly in complex environments as in simple environments (Barrett and Bell 2006). This variation does not comprise a large number of narrowly specialized types, but instead seems to reflect the evolution of overlapping imperfect generalists (Barrett *et al.* 2005). If a population initially consisting of two specialists is allowed to evolve for a few hundred generations, diversity might be lost through a rare beneficial mutation arising in one of them and destabilizing the polymorphism; this seldom happens, and specialization either persists or evolves *de novo* if the population were initially monomorphic (Dykhuizen and Dean 2004). The secure foundation provided by studies of simple systems with one or two limiting resources will in future be used to construct experiments and theories that will more closely represent the complexity of natural environments and communities.

10.3 Social behaviour

Crops such as fruit trees and cabbages are normally grown as rather widely spaced individuals that are harvested separately. They are therefore selected as individuals, typically for large size; the modern plants (or at least the harvested parts) are much larger than their wild progenitors. Annual seed crops such as wheat, oats, and barley are harvested as communities, the object being to obtain as large as possible a yield from a given area of land, without regard to the production of individuals. Nevertheless, although they are harvested as communities they have often been selected as individuals. Particularly tall and luxuriant plants with large and abundant seed tend to be preferred as stocks for breeding. Donald and Hamblin (1976) have argued that selecting seed crops on the basis of individual performance is inconsistent. Tall plants with broad horizontal leaves that tiller or branch profusely are superior competitors that will yield well because they are unlikely to be repressed by less vigorous neighbours. For example, Khalifa and Qualset (1974) made a series of mechanical mixtures of wheat in which a tall cultivar (about 120 cm in height at maturity) and a dwarf cultivar (about 80 cm in height) were present in different proportions. The dwarf type yields about 20% more per unit area, when grown in pure stand. In mixtures, however, the tall cultivar is always the stronger competitor, presumably because it shades the shorter type. In mixtures where the tall type was abundant, the dwarf plants had fewer kernels per spike; when the dwarf variety was abundant, the tall plants had heavier kernels. Thus, dwarf neighbours benefited the tall plants, whereas tall neighbours injured the dwarf plants. Consequently, the yield of the dwarf plants, relative to their yield in pure culture, fell as their frequency decreased. The development of rice cultivars provides another example. Traditional rice cultivars are rather tall, leafy plants with high seed yield as individuals. They have now been largely replaced by semi-dwarf varieties that produce a heavier crop on a given area of land. Jennings and de Jesus (1968) planted two tall and three semi-dwarf varieties in equal proportions on experimental plots in the Philippines. Within four generations, the high-yielding semi-dwarf varieties had been almost eliminated from re-sown populations, which became dominated by the lower-yielding of the two tall varieties. In subsequent experiments, heterogeneous mixtures of plants from crosses between tall and semi-dwarf varieties became dominated by taller plants. Tall plants are often superior competitors because they are likely to repress the growth of their neighbours. When grown as pure stands they may not be especially productive in terms of yield per unit area, however, precisely because they compete intensely among themselves. Selection for total yield per unit area is more likely to favour weak competitors, whose modest yield as individuals is accompanied by less adverse effects on their neighbours. Thus, artificial selection for yield in pure culture tends to reduce fitness in mixed populations, and vice versa.

The outcome of competition in a mixture is predictable from performance in pure culture when fitness is governed by the flux through a metabolic pathway (e.g. Dykhuizen and Dean 1990). In this situation strains interact indirectly through 'scramble' competition, each reducing the growth of the other by depleting resources. In other situations strains may interact directly by interfering with one another's growth: one may benefit the other by facilitating its growth, or one may harm the other by suppressing its growth. These interactions do not necessarily maintain diversity and their outcome may or may not be predictable from growth in pure culture. For example, Suneson (1949) described the behaviour of four varieties of barley—Vaughn, Atlas, Hero, and Club Mariout—in California. When they are grown in the usual way, as pure cultures several acres in extent, their yields per unit area are ranked as: Vaughn > Hero > Atlas > Club Mariout. These yields vary from year to year, and from place to place; but certainly Vaughn was superior, exceeding Atlas and Club Mariout for yield in 12 out of 15 years. Suneson grew the four as an annually re-sown mixture between 1933 and 1948, without conscious selection. Barley is almost entirely self-fertilized, so the varieties remained distinct, and the experiment involves only simple sorting. The experiment was begun with equal frequencies of all four varieties. After 16 years, Atlas had risen to a frequency of nearly 90%; Club Mariout at first declined, but appeared to stabilize at about 10% of the population; Hero and Vaughn had almost entirely disappeared, dropping to frequencies of less than 1% (Figure 10.12). Clearly, pure-culture yield would have been a very poor guide to the evolution of the mixed population; in this case, indeed, yield in pure culture and success in mixture were negatively correlated.

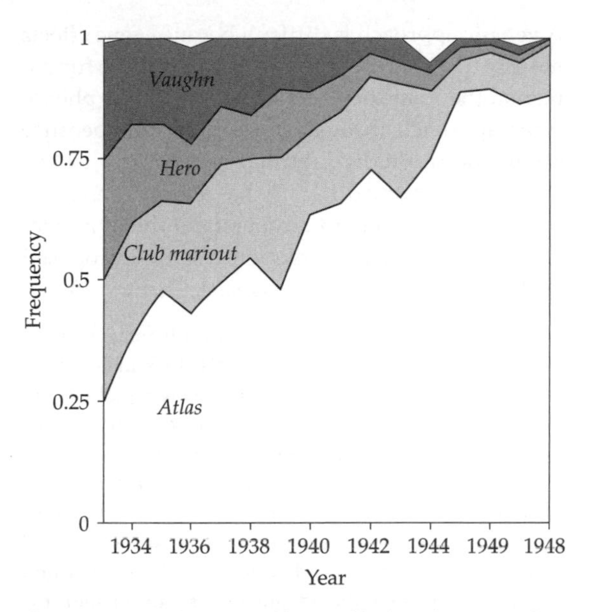

Figure 10.12 Productivity and fitness of barley cultivars. Vaughn is the most productive in pure stand but is rapidly eliminated from mixtures. From Suneson (1949).

10.3.1 The phenotypic theory of aggression and exploitation

In situations like this, classical theory is undermined because its basic assumptions are not met.

A genetic approach is difficult because gene effects necessarily change through time and the fundamental theorem therefore does not hold. A phenotypic approach through optimality fails because no unconditionally optimal phenotype exists. Precisely the same difficulty has always been faced by economists trying to understand the dynamics of the marketplace. A solution was first proposed by Adam Smith (*Wealth of Nations*, Chapter 10):

> The whole of the advantages and disadvantages of the different employments of labour and stock must, in the same neighbourhood, be either perfectly equal or continually tending to equality. If in the same neighbourhood, there was any employment either more or less advantageous than the rest, so many people would crowd into it in the one case, and so many would desert it in the other, that its advantages would soon return to the level of other employments.... Every man's interest would prompt him to seek the advantageous, and to shun the disadvantageous employment.

This approach was later codified as game theory by John von Neumann and others to explain how human conflicts are resolved. The central concept of game analyses is that individuals will come to behave in such a fashion that no alternative behaviour will profit them more, given that their opponents are following the same rule. It has been only partially successful as a theory of human social behaviour, largely because it assumes that the outcome of a conflict can be expressed on a linear scale of utilities, and that players always behave rationally in their own interests so as to obtain the greatest possible utility. It scarcely needs pointing out that people often behave irrationally, and that the different utilities at stake in a conflict may be incommensurable. However, it is much more plausible to assert that in a population whose members have long interacted with one another in similar circumstances, Darwinian fitness provides a single linear scale of utility, and that the population will have become modified through selection acting strictly on the differences in fitness caused by different behaviours. We can then substitute an evolutionary prediction for the outcome of social conflict: social behaviour in a population will evolve towards an unbeatable strategy (Hamilton 1967), that is, a state such that no type expressing a behaviour different from that prevailing in the population can spread. The population must then be at an equilibrium that can be called the *evolutionarily stable strategy*, or, to introduce one of the few useful acronyms in evolutionary biology, the ESS (Maynard Smith 1982).

Evolutionary game theory has been used primarily to understand social and sexual interactions. The main puzzle of social behaviour is why individuals should be helpful and trustworthy, which seems contrary to their Darwinian interests. A very broad range of acts fall in this category, which could loosely be called *helpfulness*. I shall discuss them in order of increasing stringency: accommodating rather than attacking or exploiting partners (sharing); contributing public goods that others use (subsidizing); trading honestly despite risking loss to cheats (cooperation); and sacrificing personal fitness for the gain of others (altruism). There are many other sensible ways of arranging the material (e.g. Sachs *et al.* 2004).

The interaction between aggressive and peaceable individuals was the subject of the original application of evolutionary game theory to social behaviour (Maynard Smith and Price 1973). In a very simplified form it is universally known as the Hawk–Dove game, although more extended forms include a variety of other actors. Imagine that individuals compete in pairs for some resource whose value can be represented as V: this value is fundamentally an increment in fitness, although of course it will be physically and immediately apparent as calories, space, access to water, or some such benefit. When two Doves meet, they behave in such a way that the resource is shared between them. When a Dove meets a Hawk it immediately retreats, so that the Hawk gets all the resource and the Dove gets none. When two Hawks meet, they fight for possession of the resource. The victor gets all the resource and the loser none; moreover, the loser suffers damage that is equivalent to losing C units of resource. We can then represent these interactions in terms of a pay-off matrix representing the effect on fitness of an interaction with a given type of neighbour. The force of this assumption is that Hawks and Doves are equally successful in acquiring uncontested resources, and thus have some equivalent baseline fitness; the difference in fitness between the two types arises solely through social encounters in which they must compete for additional shares of resource. When two Doves meet, they share the resource, each obtaining $E_{DD} = V/2$. When two Hawks meet, each has an equal chance of obtaining the resource of value V, or sustaining an injury equivalent to a loss in value C; the expected value of the contest to either individual is thus $E_{HH} = (V - C)/2$. A Hawk that meets a Dove is always successful, so that $E_{HD} = V$, and conversely $E_{DH} = 0$. The pay-off matrix then looks like this:

		When playing against:	
		Dove	Hawk
Payoff to:	Dove	½V	0
	Hawk	V	½($V - C$)

A behaviour is an ESS if in a population in which almost all individuals display that behaviour no other behaviour causes a greater increment in fitness to the individual displaying it. Another way of expressing the same principle is to say that a behaviour represents an ESS if, when competing with the type displaying that behaviour, individuals with the same behaviour gain more than individuals with any different behaviour. For any two strategies X, Y this implies that either (1) $E_{XX} > E_{YX}$, or (2) $E_{XX} = E_{YX}$ and $E_{XY} > E_{YY}$ (Maynard Smith and Price 1973). In the Hawk–Dove game this requires that $E_{HH} > E_{DH}$ (Hawk is an ESS) or $E_{DD} > E_{HD}$ (Dove is an ESS). As the game has been defined, Dove cannot be an ESS, because if nearly all the population were Doves, a Dove would obtain $V/2$ from each contest, whereas a rare Hawk—all of whose contests would be with Doves—would obtain V. Thus, $E_{HD} > E_{DD}$ for the game, and Hawks would therefore tend to spread in a population of Doves. Doves, however, will not spread in a population of Hawks if they gain less from an encounter with a Hawk than a Hawk itself gains from an encounter with another Hawk. Hawk is thus an ESS, provided that $E_{HH} > E_{DH}$, that is, $(V - C)/2 > 0$. This will be true if $V > C$, that is, if the chance of gaining the resource by fighting for it is worth the risk of being injured. If $C > V$, then neither Hawk nor Dove is an ESS. The ESS is instead a mixture of the two behaviours, in proportions such that either behaviour would be more profitable if it were expressed less frequently. The composition of this mixture follows the rule that the pay-off of any of the pure strategies included in the mixed strategy will be the same when played against the mixed strategy (Bishop and Cannings 1978). This is intuitively clear: if any of the constituent pure strategies was more successful than the others, then its contribution to the mixed strategy should increase; since this cannot be true at equilibrium, equality of pay-offs must define the ESS. The frequency of Hawk in this stable mixture is then $(E_{HD} - E_{DD})/[(E_{HD} - E_{DD}) + (E_{DH} - E_{HH})] = V/C$. The ESS is thus a mixture if $E_{HD} > E_{DD}$ and $E_{DH} > E_{DD}$: that is, if either behaviour is more profitable when rare, because neighbours suppress similar individuals more severely.

Ruzzante and Doyle (1991) manipulated feeding interactions in the medaka (a small fish, *Oryzias*), by presenting floating food within a small cork ring (high interaction) or broadcasting it over the surface of the pond (low interaction). In either case the food was present in excess. They selected for faster growth in two successive generations by choosing the largest offspring from each brood, because of the large environmental variance among broods, with selection intensities of 0.2–0.4. The correlated response to artificial selection for rapid growth was less aggressive behaviour in the high-interaction lines but not in the low-interaction lines. This suggests that Doves prosper in contests for food. This interpretation is heavily dependent on a single outlying line, however, and the parallel with the Hawk–Dove game is far from exact.

When individuals search for food in groups interactions may occur between several individuals simultaneously. The Producers search actively, whereas the Scroungers loiter about until food has been discovered by a Producer, whereupon they all rush to share it. Before they can get there, the Producer has consumed the 'finder's share' V_{finder}, but must share the rest with the Scroungers. This is the Producer–Scrounger game of Barnard and Sibly (1981). In this case the Scroungers exploit the Producers and lose by interacting with other Scroungers, although the gains and losses are less dramatic than in the Hawk–Dove game. It is clear that Scrounger is not an ESS, as a pure population would die of inanition. The rate of consumption by each Producer is proportional to $V_{\text{finder}} + (V - V_{\text{finder}})/(1 + qN)$, whereas that of a Scrounger is proportional to $pN(V - V_{\text{finder}})/(1 + qN)$, where $p = 1 - q$ is the frequency of Producers in the local group of N individuals. Hence, Scroungers can invade if $1 - (V_{\text{finder}}/V) > 1/N$, and if so will reach an ESS frequency of $[1 - (V_{\text{finder}}/V)] - (1/N)$ (Vickery *et al.* 1991). Alternatively, the Producer might just give up in disgust when the Scroungers arrive, in which case the ESS frequency of Scroungers is $[1 - (V_{\text{finder}}/V)] [(N + 1)/N] \approx 1 - (V_{\text{finder}}/V)$. These equations are interesting because they suggest that the frequency of Scroungers may often be quite high. The game is not as well known as the Hawk–Dove game, but may be more widespread: careful observation of a flock of pigeons or sparrows will often reveal a few scroungers.

Whether or not aggressive or exploitative behaviour evolves in a population depends on how severely these behaviours injure or impede the individuals that display them when they become abundant. The argument is entirely phenotypic, and is rather insensitive to assumptions about genetics. Each strategy may be governed by an allele in a haploid population, in which case the ESS is the frequency of individuals bearing this allele in the population at equilibrium. Alternatively, every individual may display aggressive or pacific behaviour at random in any given encounter: in this case, the ESS is the frequency with which any given individual is aggressive when the population is at equilibrium. In part because of its freedom from any particular genetic model, evolutionary game theory can be applied very broadly to any situation involving social interactions, and its initial introduction led to an efflorescence of theoretical and comparative studies. Experimental work, however, has concentrated on demonstrating adaptedness, usually by showing how the behaviour of individuals changes appropriately when their environment is manipulated, and it is only very recently that selection experiments with microbes that trace the evolution of the ESS have been attempted (see Pfeiffer and Schuster 2005).

10.3.2 Cross-feeding

A shell vial with glucose minimal medium is perhaps the simplest of environments and should not be capable of supporting more than a single strain. This is often true: there was no detectable genetic variance in the long-term *E. coli* lines (beyond that expected from recurrent deleterious mutation) after 2000 generations (Lenski *et al.* 1991). But there are many observations of diversity appearing in the simplest of conditions, and a range of different types appeared even in the *E. coli* lines after 10 000 generations or so. Helling *et al.* (1987) found that nearly neutral genotypes of *E. coli* conferring resistance to phage T5 fluctuated in frequency in long-term glucose-limited chemostats, flagging the spread of novel favourable mutations in the usual way. Less expectedly, they also found that variants giving small colonies on agar also appeared in all cultures that were maintained for more than about 100 generations. These variants, which bore all the markers of the initial inoculum, and therefore could not be dismissed as recurrent contaminants, never took over the population, but instead persisted indefinitely at intermediate frequencies. Further experiments showed that the small-colony types processed glucose much more rapidly than the large-colony types. One might expect that they would quickly eliminate their competitors. However, the large-colony types can grow in medium that has been exhausted by the small-colony types. These large-colony types, which frequently appear in long-term cultures, evolve as scavengers that are able to grow on substances excreted by the rapidly growing but inefficient small-colony types (Figure 10.13). The genetic and biochemical mechanisms responsible for cross-feeding were elucidated by Rosenzweig *et al.* (1994). The small-colony strain CV103 has evolved a greater rate of glucose uptake and replaced ancestor JA122 because it drives down glucose concentration to a much lower level. Acetate is excreted by *E. coli* growing by fermentation in aerobic conditions, and is normally utilized once glucose has been exhausted. CV103 excretes increased quantities of acetate, perhaps because extracellular acetate contributes to glucose uptake by permitting

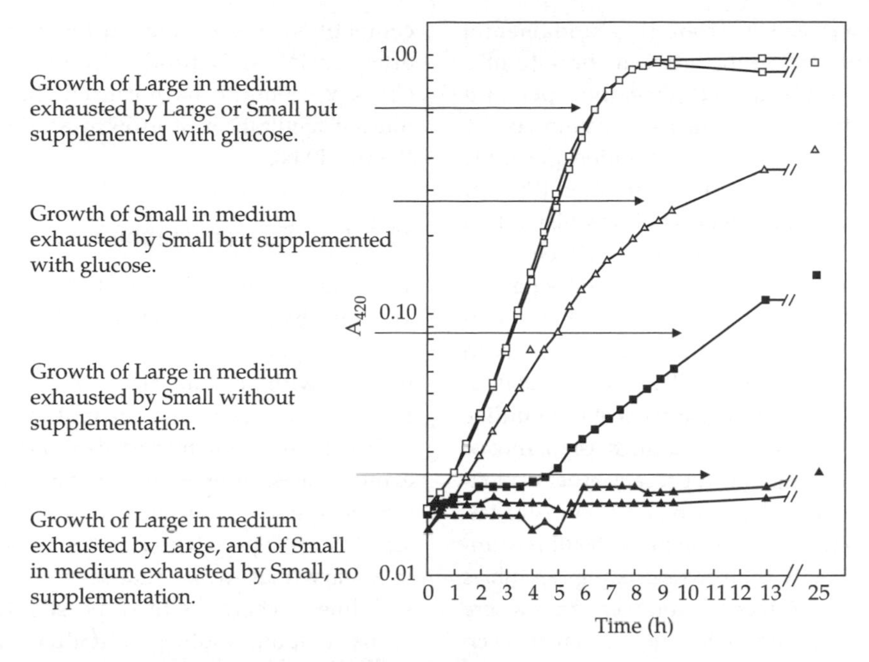

Figure 10.13 Cross-feeding in *E. coli* chemostats. From Helling *et al.* (1987).

higher rates of phosphorylation. This creates an opportunity for CV101, which has a high rate of acetate uptake. There is thus a stable polymorphism of CV103 and CV101 that can be titrated by adding acetate. CV116 is a second small-colony strain that grows better on glucose than CV101 but not as well as CV103. It has increased uptake of glycerol, which is excreted by both CV101 and CV103. CV101 persists in mixed culture because it can utilize the acetate that CV116 secretes. CV116 also coexists with CV103 despite its inferior glucose uptake through its enhanced ability to utilize glycerol; this polymorphism is density-dependent because CV116 cannot persist when population density is too low to generate insufficient glycerol. All three types can coexist when only glucose is supplied, with CV103 acting as a glucose specialist, CV101 as an acetate specialist and CV116 as a generalist. Acetate cross-feeding evolves repeatably in *E. coli* lines maintained on glucose minimal medium through the over-expression of acetyl CoA synthetase caused by regulatory mutations in the *acs* gene attributable to base substitution or transposon insertion (Treves *et al.* 1998). A similar cross-feeding polymorphism was described by Rozen and Lenski (2000) in long-term lines of *E. coli*, where the cross-feeders appeared after about 6000 generations and rapidly increased in frequency to 80–90% of the population before declining and then fluctuating irregularly for unknown reasons during the ensuing 15 000 generations.

It is striking how quite novel kinds of social relationship may evolve after several hundred generations of culture, even in physically constant and uniform conditions, and in doing so support diversity in a system where only uniformity might have been anticipated. This is probably only a small sample of the diversity that can be supported by a single substrate, because the variants were recognized only by the fortuitous difference in colony morphology. In more realistic situations where many substrates are available the diversity that can be maintained through these cross-feeding interactions is presumably greater still. Thus, in the long term the competitive exclusion principle is undermined by the evolution of novel social interactions that may involve selfishness, exploitation and cooperation.

Cross-feeding arises from the fundamental thermodynamic trade-off between the rate of a reaction and its yield: ATP generation per unit time in heterotrophic organisms is maximized by fermentation, whereas ATP generation per mole of substrate is maximized by respiration (Pfeiffer *et al.* 2001). Hence, populations of organisms that utilize some external nutrient source will be swept by fermenters at high nutrient supply rates despite the fact that they can maintain only a low population density. Once the nutrient supply has been depleted, they must either switch to respiration, or be invaded by respiring organisms able to utilize incompletely oxidized products such as ethanol or acetate. Cross-feeding may represent either exploitation or cooperation, depending on circumstances. If the respirers deny the fermenters the opportunity of diauxic growth then cross-feeding resembles exploitation in a Producer–Scrounger game where the fermenter is limited to the finder's share in each molecule of glucose and leaves once this share has been obtained. On the other hand, if the respirers alleviate end-product inhibition by consuming the metabolites of the fermenters then the relationship is mutalistic, and cooperation is favoured because both partners gain directly from the interaction.

The same principles presumably apply to multicellular organisms, although there is little experimental evidence. Barley Composite Cross V was created by intercrossing 31 disparate varieties, and afterwards propagated in bulk without conscious selection. The population was thus a highly heterogeneous mixture of many different self-fertilizing lineages. Allard and Adams (1969) extracted genotypes from this population at Davis 18 generations after its foundation, and measured their response to one another as neighbours in 3×3 hill plots. In most cases, the seed yield of the central plant was greater when its neighbours were dissimilar than when they were members of the same selfed line. Moreover, neighbours of a given genotype tended to repress the growth of the same genotype more strongly than they repressed other genotypes. This was in contrast to experiments with arbitrary pairs of cultivars, where they repeated Suneson's work with more or less the same result: some cultivars, such as Vaughn, were unequivocally inferior competitors that usually suffer in mixture with other varieties. Natural selection in Composite Cross V seems to have caused the evolution of mutual facilitation of growth among genotypes (Figure 10.14).

10.3.3 Selfish cooperation

These very simple ideas explain why populations are not always dominated by the most aggressive or exploitative kinds of individual, but not why individuals routinely assist others or how communities organized for mutual benefit, such as lichens or plant–mycorrhizal associations, can evolve. Hawk, after all, may not always be an ESS, but Dove never is. It has been difficult to understand how cooperative social behaviour can evolve in a Darwinian world that seems to reward only selfishness. There are three possibilities. The first is that helping another individual may return a direct benefit to the actor, in terms of increased fitness. Although this does not involve any new principle, evolutionary game theory has been

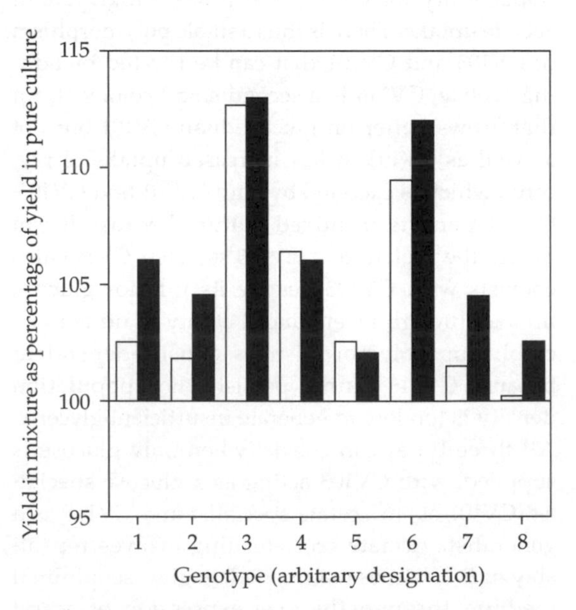

Figure 10.14 Mutual facilitation in barley Composite Cross V after 18 generations of natural selection. Open bars: all neighbours belong to a single different genotype, solid bars: neighbours belong to eight different genotypes. From Allard and Adams (1969).

useful in understanding how helpful behaviour can follow from purely selfish motives. There are useful reviews by Nowak and Sigmund (2004) and Doebeli and Hauert (2005). The second possibility is that the actor suffers a net cost, but confers a benefit on other individuals who bear copies of the gene responsible for the behaviour. This is *kin selection*, developed in its modern form by Hamilton (1964a, b) and discussed at length by Frank (1998). It is identical with a process called *group selection* by Wilson (1980) and others. The third possibility is that the actor suffers a net cost, but confers a benefit on its neighbours. If they are unrelated to the actor (are no more likely to bear a copy of the gene responsible for the behaviour than random individuals from the metapopulation) this is classical group selection, as argued by Wynne-Edwards (1962, 1986). Kin selection and group selection are discussed in §10.4. The very large literature on the evolution of social behaviour and cooperation has been reviewed recently by Sachs *et al.* (2004) and West *et al.* (2006). At this point it should be stated for the sake of clarity that altruistic behaviour which reduces the direct fitness of the actor can evolve only through kin selection.

The most elementary form of cooperation is to share the costs of exploiting a resource. If two cars are stranded on either side of a snowdrift, the drivers will benefit by sharing the cost C of digging a way through, so that each can proceed on their way and hence realize a benefit V. On the other hand, if one driver sits in his car it still pays the other to dig, provided that the risk of heart attack is ignored, so that $V > C$. The pay-off matrix for the Snowdrift game is:

		When playing against:	
		Dig	Sit
Payoff to:	Dig	$V - ½ C$	$V - C$
	Sit	V	0

This is a variant of the Hawk–Dove game in which it is better to help an unhelpful partner than to refrain from acting. The outcome is always a mixed ESS with Sit at a frequency of $C/(2V - C)$: the greater the cost, provided it does not exceed the value of the resource, the greater the advantage of letting the other fellow do the work.

Digging a path through the snowbank is an example of a particular kind of public good, one that can be enjoyed by the actor himself as well as by all other members of the local community travelling on that road. Substances secreted by microbes that function outside the cell are public goods of this kind, including extracellular enzymes, siderophores and adhesive substances. For example, yeast secretes an invertase that hydrolyses sucrose into glucose moieties that are then taken up across the cell membrane. Loss-of-function mutations are favoured when they are rare, because they take advantage of the invertase secreted by neighbours without themselves bearing the cost of production. Consequently, the fitness of these mutants on sucrose–agar plates increases with population density: at low density the mutant is less fit because isolated cells cannot metabolize sucrose, whereas at high density it is more fit because it is able to exploit its neighbours (Greig and Travisano 2004). The mats formed by wrinkly spreaders (§8.1.1) in static cultures of *Pseudomonas fluorescens* are vulnerable to invasion by the ancestral type, which gains the benefit of access to oxygen within paying the cost of secreting the adhesive cellulose-like polymer that glues the mat together (Rainey and Rainey 2003).

Pseudomonas fluorescens is named for its fluorescent siderophores, iron-scavenging proteins that are secreted by the cell and subsequently recovered. They give colonies a characteristic yellow-green glow, so that loss-of-function mutations are easily recognized because they form white colonies. Cultures of intact cells are readily invaded by non-secreting mutants. Suppose we inoculate several vials, mixing their contents at the end of growth and using this mixture to inoculate the next set of vials. The fate of non-secretors will depend on the balance between competition within vials, favouring the non-secretors, and competition among vials, favouring the secretors. The balance can be tipped in favour of the non-secretors by transferring a fixed number of cells from each vial to the inoculation pool, eliminating among-vial competition and thereby procuring the rapid spread of non-secretors. These predictions were verified by Griffin *et al.* (2004). All of these seem to be Snowdrift situations where a behaviour that is beneficial to the local group and costly to the individual persists despite being exploited because on balance it increases the fitness of the actor: it pays to produce siderophores, even though most will be used by other individuals. This permits the spread of non-cooperative 'defectors', while cooperation is maintained for purely selfish reasons.

10.3.4 Prisoners' dilemmas

A more stringent concept of cooperation is that defectors should punish cooperators more than other defectors. In this case, cooperation cannot increase the fitness of the actor directly unless it is reciprocated (Trivers 1971). This is the famous Prisoner's Dilemma, which I shall illustrate as a trading game, since this may be a more common context in biology and economics. Two parties each possess a good that the other requires; one, say, grows corn and the other makes ploughs. They arrange to meet at some rendezvous, each bearing a large box, and then exchange boxes and leave. If both are honest, they will cooperate: the agriculturalist will receive a box containing a plough, and the industrialist will receive a box containing corn; both will benefit. But what prevents dishonesty? After all, either may be tempted to defect, that is, to bring an empty box, obtaining what they require—they hope—without supplying anything in return. If both succumb to temptation, then neither will benefit, and the social arrangement breaks down. The equivalent of Dove, then, is an honest trader who always cooperates with his neighbour; the equivalent of Hawk is the dishonest trader who defects by providing an empty box. The pay-off matrix looks like this:

		When playing against:	
		Cooperate	Defect
Payoff to:	Cooperate	a = award—for honesty	b = booby prize—for cooperating with a dishonest neighbour
	Defect	c = cheating—gets something for nothing	d = deadlock—of two dishonest traders

It will be cheaper to bring an empty box, so we can assume that $c > a$ and $d > b$. On the other hand, if both traders are honest, each gains more than if both are dishonest: $a > d$. The game is therefore defined by $c > a > d > b$. The difference from the Snowdrift game is the final inequality, which implies that if your partner defects you should also defect. The solution is obvious from the preceding section: the only ESS is the Hawk-like behaviour of always defecting. That is the dilemma.

Either trader would benefit if they behaved honestly, provided that the other reciprocated. If they meet only once, no such arrangement will be honoured, because neither can know whether or not their neighbour will defect. However, if they were to meet repeatedly, their past behaviour may indicate what they are likely to do in the future. The expected future behaviour of one's neighbour could then serve to guide one's own. To be sure, the calculation is a delicate one. If one's neighbour has a fixed behaviour—always cooperate, or always defect, or any alternation of cooperating and defecting—then it is obviously profitable to defect in every encounter. It is only if one's neighbour's behaviour depends in some appropriate way on one's own response that any sustained cooperation will be possible. The 'appropriate way', however, is not easy to define. Complete dishonesty is clearly an ESS: if all one's neighbours always defect, it is folly to cooperate. This does not rule out the possibility that some less adamantly uncooperative behaviour might also be an ESS. To identify this behaviour, however, required the unlikely collaboration between a political economist, Robert Axelrod, and an evolutionary biologist, W.D. Hamilton, and involved one of the most extraordinary selection experiments ever devised (Axelrod 1984).

Axelrod approached the problem empirically by inviting people who might reasonably be supposed to be experts—game theoreticians in economics and sociology, for example—to specify the best possible social behaviour, by defining the response to one's neighbour in each of a long series of encounters. There was no limit on the complexity of such behaviours, which were expressed as computer programs of any length. Whether one cooperated or defected in a given encounter could involve remembering the outcome of every previous encounter with one's neighbour, weighting more and less recent outcomes appropriately, inferring their likely future behaviour, projecting its consequences for one's own behaviour, expressing the result as a probability distribution, and then drawing a random number to decide how to act. Or, of course, you could just always defect. The failure of previous attempts to solve the problem was confirmed by the fact that all 14 programs submitted were different. They were entered into a tournament, along the lines of a hill-plot crop trial. Each program was matched with a partner for a long series of encounters, accumulating a score from their outcome. After all the pairwise combinations had been tried, the program with the best average score was declared the winner. Rather surprisingly, this was the shortest program in the tournament, devised by Anatol Rapoport, a philosopher at the University of Toronto. Its recipe for social behaviour was very simple, involving none of the laboured ingenuity of other entries: always cooperate at the first meeting, and thereafter do what your neighbour did at the previous meeting. Rapoport called this pattern of behaviour 'Tit-for-tat'. It is friendly on first encounter, promptly and invariably punishes any dishonesty, and when honest trading is resumed immediately forgives—and forgets. However attractive these straightforward virtues might seem, however, it was not easy to explain why they should be so successful. Axelrod therefore organized a second tournament, in which more than 60 contestants, profiting from the outcome of the first, invented yet more ingenious methods of exploiting one's neighbour. Rapoport, with really superb aplomb, simply submitted Tit-for-tat again, unrevised. It won.

For an evolutionary biologist, this is not quite conclusive: it is perfectly conceivable that a behaviour might have a high average score over pairwise contests, and yet fail to spread in a heterogeneous population. The programs were next entered in a selection experiment. The programs constituted a population in which each was at first equally frequent, there being equal numbers of copies of all of them. Each copy was paired with a random partner, and its score over a long series of encounters used as a measure of fitness to adjust the frequency of copies of that program in the population entering the next cycle. As the experiment progressed, different programs waxed and waned. Aggressive programs able to exploit foolishly cooperative neighbours at first began to spread rapidly; but as they became more abundant, they became more likely to meet one another, and fail together in deadlock. After 1000 generations, Tit-for-tat was the clear winner, not only more abundant but still spreading more rapidly than any of its competitors.

With the benefit of hindsight, it was subsequently possible to prove that Tit-for-tat (or some very similar strategy) is an ESS, provided that neighbours interact for long enough (see Nowak and Sigmund 1992). It nonetheless remains as surprising as it is instructive that such an agreeably social pattern of behaviour should be so successful. After all, Tit-for-tat never wins. At best, it will draw (if its neighbour always cooperates); more often (if its neighbour ever defects, even once) it will lose, though not by much. It is successful because more aggressive behaviours end up by punishing themselves. By rewarding similar patterns of behaviour, Tit-for-tat gradually creates a social environment in which it prospers. Moreover, it does not require a capacious memory or advanced mental powers, but merely the ability to follow a very simple rule that can evolve as readily in bacteria as in apes. The mutual profitability of reciprocating past favours will lead to the evolution of genuinely cooperative social behaviour between partners who interact repeatedly, without involving any process more onerous than straightforward Darwinian selection.

The phage evolving at high MOI in the experiment by Turner and Chao (1999) turned out to be defective in supplying the diffusible resources required for successful virus replication and encapsidation inside the cell. Its fitness is therefore low when cells are co-infected by two (or more) defective phage, but high when it co-infects with an intact phage. Hence, its fitness fell as its frequency increased, and once it had become fixed mean fitness was lower than that of the ancestor. By competing the ancestor φ6 against an evolved clone φH2 in all pairwise combinations the payoff matrix at high MOI could be estimated (Turner and Chao 1999):

		When playing against: φ6 (Cooperate)	φH2 (Defect)
Payoff to:	φ6 (Cooperate)	$a = 1$	$b = 0.65$
	φH2 (Defect)	$c = 1.99$	$d = 0.83$

These payoffs have $c > a > d > b$ and thus conform to the Prisoner's Dilemma. Thus, *K*-selected viral genotypes adapted to high MOI can evolve either to exploit host cells more thoroughly, and thereby become more virulent, or to economize on their contribution to co-infection, and thereby become less virulent. Most wild virus are intact, presumably because they are usually in low-MOI regimes where cells are seldom infected by more than a single virus.

Myxococcus xanthus is a predatory social bacterium where roving groups of cells lyse their prey and take up the nutrients that are released. When starved for amino acids they actively aggregate into larger groups of about 10^5 individuals which form a fruiting body within which dispersive spores are produced and matured. During the formation of the fruiting body the majority of cells are lysed, dying in order to benefit the few that will be propagated as spores; this is an example of kin selection, which is discussed below. If the organism is maintained in stirred liquid culture it proliferates exclusively as free cells. The growth rate increases as a direct response to selection, whereas social motility, fruiting-body formation, and sporulation ability are all severely degraded, showing that these social behaviours are costly (Velicer *et al.* 1998). Asocial mutants are complemented by mixture with wild-type cells, and in many cases are over-represented among spores (Velicer *et al.* 2000). Their success is attributable to the death of the cooperative wild-type cells forming the fruiting body, who consequently suffer more in the presence of the asocial defectors than in normal colonies. Similar developmental mutants are known from other aggregative microbes, especially the eukaryotic amoeba *Dictyostelium* (Filosa 1962). These pay-offs seem to conform to the Prisoner's Dilemma, but whether reciprocal altruism plays a major role in the evolution of non-human social behaviour is doubtful. Costly cooperative behaviour involving individuals of the same species, including microbial sociality, almost always evolves through kin selection, as discussed below.

10.3.5 Intransitive social interactions

Most familiar games are transitive, in the sense that players can be ranked unequivocally in terms of their ability: if A beats B, and B beats C, then A will beat C. It is generally assumed that populations behave in the same way, with the type that is on average most successful in pairwise encounters spreading through a mixed population under selection. Furthermore, the same rule can be applied to successive gene substitutions. A population placed in a novel environment will adapt through a succession of substitutions, and if genotypes are extracted from the population at intervals and afterwards made to compete in mixed cultures, those that evolved later will eliminate those that evolved earlier (see §3.5). However, games, and perhaps evolutionary processes, are not necessarily transitive. The standard example is the Rock–Scissors–Paper game played by children. Rock (a clenched fist) blunts Scissors (two spread fingers), and Scissors cut Paper (the open hand); however, Paper wraps Rock, so the profitability of the three behaviours cannot be expressed on a single scale. If evolutionary processes are likewise intransitive, their analysis will be greatly complicated. In socialized environments, the concept of adaptedness must be replaced by a stricter concept of fitness; but if social relations are intransitive then there is no fixed scale on which fitness can be expressed.

Intransitivity represents a higher-order interaction among genotypes, which might be symbolized as G×G×G: the outcome of competition between two types depends on whether a third is present. Because neighbours may modify their environment, while the outcome of competition between two types is often sensitive to the state of the environment, it is quite possible that intransitivity is widespread. The empirical evidence is not yet sufficient to know whether this is really the case. The simplest approach is to measure the outcome of selection in pairwise mixtures of a large number of types. The degree of intransitivity can be expressed by considering all the possible combinations of three types. The pairwise interactions of these three types are either transitive or intransitive; the frequency of transitive triplets is then a measure of the overall transitivity of the social matrix. A large experiment of this kind was reported by Goodman (1979), who set up all the pairwise combinations of 19 strains of *Drosophila*, representing 15 different species. The pairs were set up with 5 females of each kind, the proliferating population being transferred from vial to vial for 9 weeks, when it was censused. The type that was then in a majority can be regarded as the winner; in many cases, one type was almost or completely eliminated. Assessed in this way, the social matrix was strikingly transitive. The strains could be arranged in an almost perfect linear sequence, with the common laboratory species *melanogaster*, *simulans*, and *pseudoobscura* at the head, and the less familiar *paulistorum*, *pallidipennis*, *funebris* and others at the foot of the ladder. About 95% of triplets were transitive, and the few exceptions almost all concern two species of similar competitive ability, both of which were still present in substantial numbers at the end of the experiment.

The outcome of selection in pairwise mixtures leads naturally to selection experiments involving three or more strains: if social relations are transitive, then the outcome of selection in mixtures of three or more types can be predicted from the outcome of pairwise competition. Richmond *et al.* (1975) investigated this prediction in the triplet of *Drosophila pseudoobscura*, *D. willistoni*, and *D. nebulosa*. In three of four cases, *pseudoobscura* eliminated *nebulosa* from mixtures, the exception being highly unstable, so *pseudoobscura* > *nebulosa*. *Pseudoobscura* and *willistoni* usually coexisted for at least 200–300 days, although in one case *willistoni* was eliminated; thus, *pseudoobscura* ≥ *willistoni*. *Willistoni* and *nebulosa* also usually continued to coexist, but in one case *nebulosa* was eliminated, so *willistoni* ≥ *nebulosa*. These pairwise experiments thus suggest that *pseudoobscura* is the strongest competitor and *nebulosa* the weakest. When the mixture of all three species was tested, *nebulosa* was eliminated in all six replicates, and *willistoni* from three; *pseudoobscura* persisted in all. The actual social hierarchy was thus *pseudoobscura* > *willistoni* > *nebulosa*, which is consistent with the results of the pairwise trials.

Experiments such as these seem to show that as a general rule the social hierarchy of arbitrary mixtures of related species is transitive, or nearly so. I have myself investigated competitive fitness among about a dozen unicellular green algae with characteristic colony morphology isolated from soil and found a completely transitive hierarchy, although the experiment was never written up because the outcome seemed so mundane. One fascinating exception, however, is provided by the toxin–antitoxin system (§9.2.1) based on colicins that many bacteria possess. Colicinogenic types possess a plasmid encoding the colicin, a lysis protein that lyses the cell, and an immunity protein that confers resistance to the colicin. When nutrients are scarce, the lysis gene is expressed in a small fraction of the colicinogenic cells, killing them and releasing the colicin. Other colicinogenic cells are immune, but non-colicinogenic neighbours are killed, releasing nutrients while at the same time relieving pressure on exogenous nutrient supply. Resistance to the colicin can evolve through mutations in membrane receptors, creating three categories of cells: colicinogenic (C), susceptible (S), and resistant (R). Colicin production reduces growth through the cost of plasmid carriage, and resistance through the partial disruption of membrane function. Hence, the ranking of growth rate is S > R > C. This leads to intransitive fitnesses when cells compete locally: C kills S; R then displaces C because it grows faster; S displaces R because it grows faster still; and then C re-invades. This system was worked out by Kerr *et al.* (2002). When the three types are grown in stirred liquid culture (or mixed at every transfer after growth on plates), R rapidly eliminates the other two types. When local spatial structure is preserved by replica-plating, however, all three types are preserved and patches of each type 'chase' one another around the plate (Figure 10.15).

10.3.6 Time-lagged social interactions

The basic dynamic consequence of intransitive interactions can be appreciated by imagining the

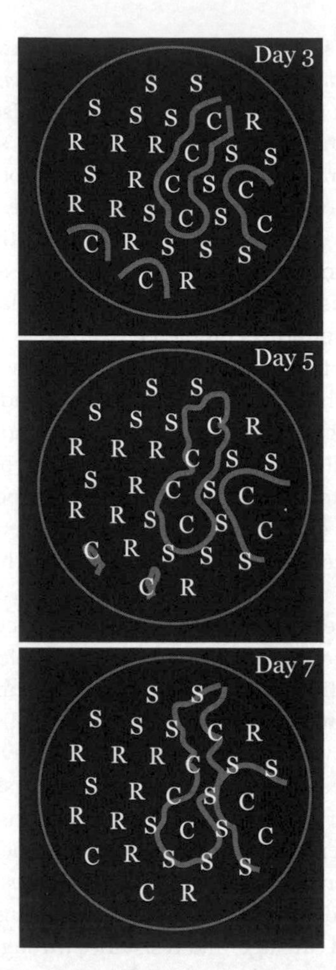

Figure 10.15 Intransitive fitness. The rock–scissors–paper game played by colicinogenic bacteria on replica-transferred plates. Letters give the initial location of genotypes spotted onto the agar surface. Lines show borders between C and R and C and S. From Kerr *et al.* (2002).

outcome of the Rock–Scissors–Paper game. It is probably easy to understand that when individuals are free to change their behaviour, the best that can be done in any given encounter is to choose one of the three possibilities at random, each with probability 1/3. Children soon learn this. But what if Rock, Scissors, and Paper are three different genotypes, with fixed behaviour resulting in the intransitive social relations that their names suggest? To settle draws, I shall assume that if two similar individuals meet they behave in a Dove-like manner and share whatever resource is being contested. Suppose that one type—Paper, say—is initially somewhat more abundant than the others, which are about equally frequent. The premium given to draws means that Paper has a slightly greater fitness, since more of its encounters will be with Paper neighbours, and it will increase in frequency. As it does so, its fitness will increase—because there are more draws—and it will eventually become very abundant, displacing the other two types. But at some point, a largely Paper population will be invaded by Scissors, since they will win almost all their encounters. A population consisting largely of Scissors is no more stable, however, because it will be invaded by Rocks. Not only is there no equilibrium, there is no end point: the population will cycle endlessly between its three possible states (see Maynard Smith 1978)

A population under selection in a novel physical environment will increase asymptotically in adaptedness, with the interval between successive beneficial mutations lengthening through time. On the other hand, if selection is caused largely by intransitive social interactions among genotypes, the population may never reach any definite end point, fitness will not tend to increase, and genotypes will replace one another in an endless sequence, with a more or less constant interval between the spread of successive clones. The conventional view is that evolution through selection is elicited by exogenous changes in the physical environment, and if these were to cease any directional evolution would soon run down, with the fixation of the best-adapted type. The alternative view is that social selection based on largely intransitive interactions will drive an endless process of change, even if the physical environment remains precisely the same. Whether this alternative view has much substance is not yet known: there have simply not been enough careful long-term experiments. Lenski and Bennett's bacterial populations (see §3.4) seem to evolve more and more slowly through time, but in Paquin and Adams' yeast populations the interval between successive substitutions did not lengthen appreciably during their experiment.

When social relations are intransitive, any genotype will by increasing in frequency depreciate its own social environment in the future. The spread of Scissors through a Paper population is checked, not by self-inhibition (draws are profitable), nor by any increase in the ability of its opponent (Scissors always wins against Paper), but because its very abundance creates a social environment in which Rock is a superior competitor: its present success ensures its long-term ruin. The three types all experience negative frequency-dependent selection, but the negative effect of increasing frequency on fitness is postponed. The effect of this time-lag is to destabilize the genetic equilibrium that is normally associated with negative frequency-dependent selection. When a given increase in frequency will only cause a reduction of fitness several generations hence, the frequency of a type will overshoot its equilibrium value before being restrained by selection. Instead of settling down to a stable mixture, therefore, the different types endlessly cycle or fluctuate in frequency.

A population of *Drosophila* growing in a vial will soon exhaust the food available for the larvae, and must be transferred to a fresh vial. However, if the larvae were killed or removed before the food was used up, the vial could be restocked with new larvae. It would not be surprising to find that these new larvae did not grow as well on medium that had been conditioned by the growth of a previous cohort of larvae, if only because the food supply will have been depleted to some extent; but it would be interesting to know whether the effect of conditioning were in any way specific, with different types having more or less severe effects on succeeding generations. Weisbrot (1966) grew various strains of *Drosophila* as larvae for 3 days before killing them and restocking the vials with the same or with different strains. The survival of a Californian strain of *melanogaster* was actually enhanced in medium conditioned by *pseudoobscura*, whereas *pseudoobscura* was strongly repressed by the previous activity of *melanogaster*. This Californian wild type also repressed other strains of *melanogaster*. The social effects of species or strains may thus be propagated through time, with one generation of larvae affecting the next, in a way that varies among genotypes. The effect did not seem to be caused by the depletion of yeast or by the presence of dead larvae in the medium, so the accumulation of metabolites seems the most likely explanation. In Weisbrot's experiment, the four *melanogaster* strains tested were generally repressed less by the previous activity of their own strain than they were by that of other strains. This would lead to a positive frequency-dependent effect on fitness, and the rapid fixation of one type. However, this is probably not a general phenomenon: the main lesson of this and similar experiments is that they show how specific social effects can persist through time.

The cause of time-lagged negative frequency-dependent selection is the projection of self-inhibition through time. The belief that any single type, continually re-sown, will sour its own environment is the basis of crop rotation, a very ancient practice (the Latin poet Virgil mentions it) whose modern literature is rather scanty. Ripley (1941) published a series of experiments to show that cereal–cereal rotations generally increased seed yield. Most of these were done in agronomic conditions, but one particularly simple experiment involved growing corn, rye, and oats in two-gallon earthenware jars. After the crop had been harvested, the jar was re-sown, either with the same cereal or with a different one. It is interesting to compare his results with Roy's hill-plots of rice cultivars. When the yield of a particular temporal sequence is expressed in terms of the yield of a sequential monoculture (the same cereal sown in successive seasons), the social matrix looks like this:

		Neighbour		
		Corn	Rye	Oats
	Corn	1	1.27	1.21
Self	Rye	1.29	1	1.59
	Oats	2.11	1.98	1

In Roy's experiment, neighbours were grown on the same plot at the same time; in Ripley's experiment, neighbours were grown on the same plot but in the preceding season. There is clearly a tendency for self-inhibition to extend through time: each cereal yields more when its plot was previously occupied by a different type.

The classic crop-rotation experiment is the Broadbalk Wheat Experiment, originally designed to investigate the performance of continuous wheat monocultures under different soil treatments, and pursued, though with several changes in design, at the Rothamsted station since 1843. The design followed in 1968–1978 included a third-year fallow and a sequence of potatoes—beans—wheat, and allows continuous wheat to be compared with two sorts of rotation. There were also different levels of nitrogen application. The mean yield of wheat (over all levels of nitrogen) were 4.53 tons/ha for continuous wheat; 4.69 tons/ha for the second wheat crop after fallow; 4.81 tons/ha for the first wheat crop after fallow; and 5.34 tons/ha for wheat after potatoes and beans. Continuous culture of the same crop thus depresses seed yield, in the sense that yield is about 20% greater when the two preceding crops were different species of plant. The difference is most pronounced—amounting to almost a doubling of yield—at the lowest level of nitrogen application (Figure 10.16). One possible explanation for this is that the different crops have complementary resource requirements and that their effects on relative levels of soil nutrients persist from one growing season to the next, being reduced or removed by large inputs of extraneous nutrients.

The Rock–Scissors–Paper game is admittedly fanciful, and the rotation of different species of crops rather remote from conventional issues in evolutionary biology. Nevertheless, the occurrence of intransitive social relations and the propagation of social effects through time raise the possibility of a continuously dynamic process of selection. This becomes more substantial when we consider how selection within one population will affect the evolution of completely different kinds of organisms in the surrounding community, which is the subject of Chapter 11.

10.4 Kin selection and group selection

Cooperation shades into altruism when an act unequivocally harms an individual (by reducing its fitness) in the course of helping others. This is a very widespread phenomenon: the development of any but the simplest multicellular creature involves the death or sterilization of many of the cells involved. The germline may be segregated early

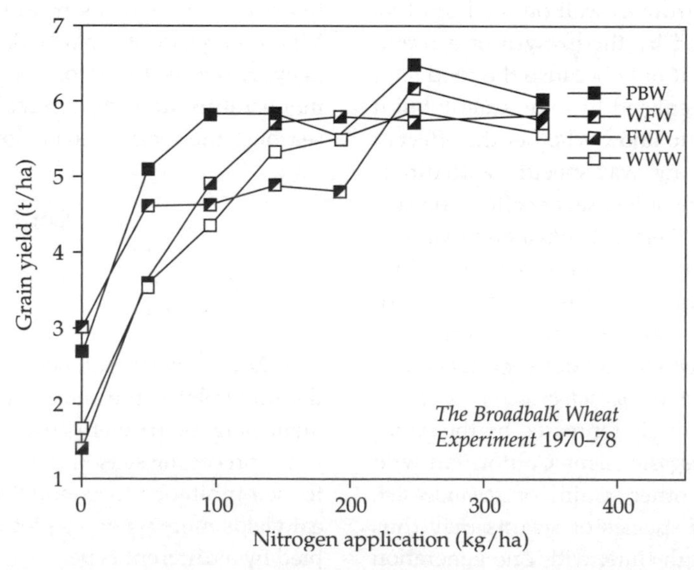

Figure 10.16 Time-lagged fitness effects. This diagram shows recent results from the Broadbalk Wheat Experiment, from Dyke *et al.* (1983). The four rotation treatments are WWW, continuous wheat; FWW, second wheat crop after fallow; WFW, first wheat crop after fallow; PBW, wheat crop following potatoes and beans. Continuous wheat yield is calculated as the average of the two sections on which this sequence was followed. Yields at a given level of nitrogen application are unweighted means over all treatments (involving other elements) representing this level of nitrogen. The results are for three rotations in 1970–1978. For other crop-rotation experiments, see Ripley (1941; historical review) and Wooding (1981).

in development (as in *Volvox* or vertebrates), or it may be continually recruited from a population of stem cells (as in polyps or plants), but in either case a cell that has become differentiated for a particular somatic function very rarely gives rise to germ cells. Moreover, some developmental processes, such as the separation of the digits in tetrapods, require the death of some of the cells involved. At a coarser scale, the zooids of some colonial hydrozoans and ectoprocts develop as purely somatic creatures, serving the reproductive zooids of the colony; in highly differentiated colonies, such as siphonophores, a zooid may become little more than a bract or scale that protects its neighbours. Vassiliki Koufopanou and I obtained some experimental evidence for the utility of somatic tissue in *Volvox*, where a dozen or so much large germ cells are set aside early in development. The germ cells develop into new colonies that are eventually released from the parental colony, leaving it as a hulk of somatic cells that soon die. We macerated colonies to isolate germ cells, which are capable of growing as unicells in nutrient medium, in the absence of a soma. However, the specific growth rate of these isolated germ cells was less than that of the intact colonies (Koufopanou and Bell 1993). The rate of proliferation of the clone as a whole is thus increased when most of its members are unable to reproduce, but serve instead to accelerate the reproduction of a small set of clone-mates.

10.4.1 Kin selection

When the cells produced by division grow independently, the position is less clear. Having similar phenotypes, they are likely to compete intensely among one another, so that the more abundant they become the less successful each individual is likely to be. The most obvious response to social selection is to avoid one another. There is, however, a second possibility. If members of the same clone are often associated with one another, they will be selected to reduce the degree to which they repress the growth of their neighbours. The characteristics that evolve in this way may go beyond simple cooperation. A type that reproduces very slowly, or even does not reproduce at all, may be selected, provided that it enhances the growth of its neighbours. The reason is almost self-evident. Imagine a genotype that represses the growth of the individual in which it is expressed, but enhances the growth of its neighbours. If the positive effect on its neighbours exceeds in sum its negative effect on the individual concerned, the reproduction of the local group as a whole will be increased. If in addition its neighbours are members of the same clone, they will bear the same genotype, and the number of copies of this genotype will increase, relative to the number of copies of an alternative genotype that neither repressed the growth of the individual nor enhanced the growth of its neighbours. A genotype may increase in frequency even if it is lethal, provided that it has the pleiotropic effect of adding more than a single individual to the combined reproduction of its clone-mates.

I have previously explained how the concept of fitness refers to a lineage, and expresses its capacity to proliferate. This capacity was regarded as being completely reducible to the reproduction of independent individuals, so that one could speak loosely of the fitness of an individual. When we drop the assumption that individuals reproduce independently of one another, and recognize instead that social relations may exist among members of a clone, we do not dilute in any way the basic principle that the future representation of that clonal lineage in a wider population of such lineages will depend solely on its rate of proliferation. We must, however, retreat from the view that the rate of proliferation of a clone is reducible to individual reproduction, and must therefore modify the concept of individual fitness. The equivalent concept in the situation where social relations exist between members of the same clone extends beyond the individual expressing a particular behaviour to include the effects of its behaviour on its clone-mates. The individual has a certain fitness by virtue of the reproduction that it is able to accomplish alone, corresponding to the concept of individual fitness in an asocial context, but we must also take into account the effect of its behaviour on the combined reproduction of other members of the same clone. It is this *inclusive fitness* that determines the overall rate of proliferation of the clone, and thus its future representation in the population:

inclusive fitness =
  reproduction achieved by individual, independently of any aid given by relatives
  + reproduction achieved by clone-mates through assistance given by individual.

A trait may be selected, even though it reduces the reproductive capacity of the individuals that express it, provided that it increases inclusive fitness through enhancing the reproduction of other individuals of the same type. Selection that favours nepotistic behaviour that aids relatives is called *kin selection*.

The same principle applies to sexual relatives. A gene that directs cooperative behaviour will spread even if it harms its bearer, by benefiting related individuals that bear a copy of the same gene. In outcrossed sexual lineages, however, there is the complication that only the gene itself is transmitted clonally, the genome being diluted and recombined in every generation with genes from other lines of descent. Consequently, the probability that two individuals in a sexual lineage both possess copies of the same gene varies with their co-ancestry. For example, the probability that two full sibs have both inherited a copy of a nuclear gene for altruistic behaviour present in one of their parents is 1/2. In more remote relatives, this probability falls off geometrically: for an aunt and a nephew it will be 1/4, for two first cousins 1/8, and so forth. This probability is the coefficient of relatedness, r. Altruistic behaviour will be favoured by kin selection only if it is directed towards individuals who bear a copy of the gene that directs it, so among sexual kin there is a substantial risk that it will be misdirected, and benefit instead individuals bearing a different allele. In sexual populations, therefore, the benefit b of a nepotic act, in terms of its effect on the rate of increase of the genes that direct the act, must be discounted by the relatedness r of the recipient. Kin selection will favour such acts only if the cost c to the individual expressing them is sufficiently small, such that $rb > c$. This is the basis of J.B.S. Haldane's famous remark, to the effect that he could reasonably be expected to lay down his life to save eight cousins. The modern theory of kin selection based on inclusive fitness was developed by Hamilton (1964a, b), and the condition $rb > c$ is often called *Hamilton's rule*. It has given rise to a very large and at times confusing literature that I cannot hope to do justice to here; it can be sampled by reading the synthetic review of Lehmann and Keller (2006) and the ensuing commentaries.

One general point that may be worth making for the benefit of non-specialists is that relatedness is not in itself necessary for the evolution of altruism, but serves only to indicate the probability that the recipient individual bears a copy of the allele governing the altruistic behaviour expressed by the actor. This is readily understood through a thought experiment in which two starving haploid cells meet and one, by virtue of bearing an allele directing altruistic behaviour, consents to be consumed by the other, whose chances of surviving are thereby increased many-fold. This behaviour will be favoured only if the recipient also bears a copy of the altruism allele. In one case, actor and recipient both descend immediately from the same mother cell and the allele is identical by descent. In the other case, actor and recipient are unrelated but both independently acquired the allele by mutation, so it is identical in state only. It is obvious that the evolutionary consequences of the altruistic act are the same in both cases: the same increase in the expected number of copies of the allele that will be transmitted. Hence, it is possible to derive Hamilton's rule from the secondary theorem of natural selection (§7.3.7) (Queller 1985, Frank 1998). The actor has breeding value (genetic value) G for the altruistic trait. The phenotypic values of actor and recipient are P and P^*, so the fitness of the actor can be expressed as w = constant + $\beta_{wP.P^*}$ + $\beta_{wP^*.P}$, where the β are partial regression coefficients of fitness on phenotype (§7.3). Substituting into the Price formulation $\bar{w}\Delta G = \mathrm{Cov}(w,G)$ yields $\Delta G = \beta_{wP.P^*}\,\mathrm{Cov}(G,P) + \beta_{wP^*.P}\,\mathrm{Cov}(G,P^*)$. This is positive provided that $\beta_{wP.P^*} + [\mathrm{Cov}(G,P^*)/\mathrm{Cov}(G,P)]\,\beta_{wP^*.P} > 0$. The cost is the partial regression of the fitness of the actor on the altruism it displays, whereas the benefit is the partial regression of the fitness of the recipient on its own phenotype, and the covariance ratio is a measure of relatedness expressing the degree to which altruistic genotypes in recipients are associated with the altruistic phenotypes in actors. The equations can be modulated to provide for non-additivity and other effects, providing a general theoretical basis for kin selection.

10.4.2 Kin proximity and kin choice

If altruistic acts are directed towards random members of the population the genetic association is zero and consequently altruism cannot evolve. In sessile organisms with poor powers of dispersal, such as many plants and soil microbes, neighbours are likely to be more closely related, on average, than random members of the wider metapopulation, so helping neighbours is likely to be favoured by kin selection. This effect is tempered by the effect of local competition: when relatives are aggregated they will tend to compete with one another, so the benefit donated by the actor is depreciated by competition between the recipient and other relatives of the actor (Queller 1992, Frank 1998). Griffin *et al.* (2004) monitored the production of siderophores (iron-scavenging molecules) in *Pseudomonas* microcosms where both relatedness and local competition could be manipulated (§10.3.3). Local competition alone suppressed the spread of the siderophore-producing strain. A high degree of relatedness was imposed by inoculating each vial with either producer or non-producer strains, whereas low relatedness was imposed by inoculating vials with a mixture of the two. Siderophore production is always favoured by high relatedness, but its effect is tempered by competition: high relatedness allowed siderophore producers to spread rapidly when local competition was weak, but had less effect when local competition was strong. This is not altruism in the strict sense, since siderophore production is probably directly beneficial even in the presence of cheats, but the agreement with theory is reassuring.

Alternatively, individuals could choose to direct altruistic behaviour preferentially towards relatives, or directly towards other altruistic individuals. Kin choice is common in mycelial fungi, where it is based on multiallelic somatic incompatibility loci. Hyphae bearing the same alleles at all loci will usually anastomose, creating a common cytoplasm and sharing all resources. If they differ at any locus they are unlikely to fuse, and may instead form a barrage zone of dead, vacuolated hyphae separating the two mycelia. A similar system based on a single, highly polymorphic incompatibility locus is found in tunicates, where one partner often partly or completely resorbs the other, or at least displaces its gametes from the combined germline (Pancer *et al.* 1995). The ability to invade the germline or soma of the partner is hierarchical and heritable (Stoner *et al.* 1999). This would evolve through kin selection if the reproduction of the chimera exceeded that of the two isolated zooids, which does not seem to have been established.

10.4.3 Spite

The opposite of altruism is unprovoked attacks that reduce the fitness of the recipient. Spiteful behaviour can evolve even if it is risky, provided that it is directed preferentially towards individuals who are less closely related than the average of the population. This is a very onerous condition that ensures that spite is rare in nature. It has nevertheless attracted more attention from experimenters than altruism, either because it is easier to study or just more interesting. The production of colicins by bacteria is an example of spiteful behaviour evolving through kin selection: colicinogenic cells that lyse under stress benefit other colicinogenic cells by killing their competitors. Toxin producers have a local advantage insofar as killing non-producers necessarily increases their own frequency. They will also have a global advantage if, as generally occurs, they gain access to more resources by killing their potential competitors, thereby producing more individuals to colonize other sites. Given a cost of adaptation in terms of slower growth of the toxin producer, it will spread only if it is sufficiently abundant to produce enough toxin to overcome the intrinsic advantage of the non-producers. Toxin/antidote systems thus have an unstable equilibrium, regardless of the details of how they work (for a simple formal model, see Levin 1988). If the toxin producer is rare it is unable to invade and the population will consist entirely of non-producers;

if its density exceeds a certain value it will spread to fixation. Social conflict thus engenders positive frequency-dependent selection that enforces uniformity rather than supporting diversity. The hurdle that the producers must surmount is rather high, however, say about 1%, so as they will hardly ever be able to invade if introduced by recurrent mutation or as rare immigrants. This difficulty arises from the homogeneous conditions of growth assumed by simple models, with a uniform diffusion of toxin acting in *trans*; it is less severe in structured environments with high local concentrations of toxin here and there. The crucial role of spatial structure in the evolution of social conflict was investigated by Chao and Levin (1981). In a metapopulation, empty sites will occasionally be colonized by producers. These toxic sites cannot be invaded by non-producers, and will provide a source of colonists for sites that are subsequently vacated. On the surface of a thinly spread agar plate, each site is the place where a cell happens to settle and grows to produce a colony. A producer colony will establish a toxic zone around itself that inhibits the growth of neighbouring non-producers and thereby enables it to grow to a larger size, provided that the plate is not so thinly spread that colonies are too widely separated to interact at all. Consequently, when cultures are maintained on soft agar colicin-producing strains are able to spread from any initial frequency, however low (Chao and Levin 1981)

Some plants also produce substances that inhibit the growth of neighbours different from themselves; this phenomenon is known as *allelopathy*. Whether or not specific growth inhibition often occurs has been controversial, primarily because of the difficulty of interpreting experiments that use macerated or dying plant tissues. However, a careful experiment by Newman and Rovira (1975) uncovered allelopathic interactions between most of the pasture herbs that they investigated. They grew eight common species in separate pots, collecting the liquid that percolated through the growth medium and applying this leachate to other individuals of the same eight species. By measuring the growth of the recipient plants, they could construct the 8×8 matrix defining the effect of the leachate of each of the donors on each of the recipients. On average, this leachate repressed growth; the substances excreted by plants therefore have a general inhibitory effect on growth. There was no overall tendency for plants to be less strongly repressed by leachate from their own species. However, some species were less strongly and others more strongly repressed by self-leachate. Species such as *Anthoxanthum* and *Cynosurus* were less strongly repressed by leachate from their own species than by leachate from other species; on the other hand, species such as *Plantago* and *Hypochoeris* were actually more strongly repressed by their own species. This seems to be related to their growth habit in natural populations. *Anthoxanthum* and *Cynosurus* form more or less continuous swards that can dominate permanent grassland, whereas *Plantago* and *Hypochoeris* are found as isolated individuals or small groups, never as pure stands. Thus, although there is no general tendency for self-inhibition to be weaker than inhibition by different types, there does seem to be a tendency for self-inhibition to be weaker in species that usually grow with neighbours of the same species. Whether a similar pattern is displayed by different genotypes of the same species is not known. Nor is it possible to infer cause and effect from these data: plants with relatively weak self-inhibition may for this reason grow as pure stands, or plants that grow as pure stands may evolve weak self-inhibition. There is a fertile and so far neglected field for evolutionary experiments here. Natural experiments are provided by the invasion of exotic plants producing novel allelopathic substances; surviving residents have elevated levels of resistance (Callaway *et al.* 2005).

10.4.4 Group selection in structured populations

The social behaviour of individuals defines the society that they constitute. This society has itself some of the attributes of an individual: it may grow, divide, and die. These attributes may vary with the behaviour of its members: cooperative individuals are more likely to form a stable society than those who generally attempt to exploit or suppress their neighbours. This will lead to a political process of selection, in which more vigorous, stable, and rapidly proliferating societies will tend to supplant neighbouring societies in which social behaviour is less effective in enhancing the fitness of the group as a whole. Indeed, this political selection, through which one kind of society is replaced by another, has been very widely regarded as being the chief process by which social behaviour evolves.

The notion that characters evolve through the selection of more or less extensive groups of individuals, because such characters affect the fitness of the population or the species, or even the local community or the ecosystem, has a long history in evolutionary biology, and in related disciplines such as ecology and animal behaviour. The classical model of group selection is due to Sewall Wright (1945; see also Boorman and Levitt 1973, Levin and Kilmer 1975). Small, isolated populations will diverge through genetic drift (§4.4): roughly speaking, if population size N and the rate of migration among populations m are sufficiently low, such that $Nm < 1$, populations will tend to become fixed for one or another type. The variance that is generated in this way is available for selection among populations. Those with a high proportion of helpful individuals may well flourish, sending out migrants that can occupy vacant sites, or providing a continual stream of helpful immigrants for predominantly selfish populations. Altruistic behaviour may thus spread infectiously through the whole assemblage of populations. This view was crystallized in a famous treatise on social behaviour published in 1962 by V.C. Wynne-Edwards. To put it very briefly, Wynne-Edwards argued that a society in which every individual was adapted to harvesting resources as rapidly as possible would soon exhaust the local resource supply, and all would then starve. The efficacy of selection acting at the level of individuals would thus encompass the ruin of the population as a whole. On the other hand, if individuals were to regulate their behaviour, refraining from growth and reproduction when resources were scarce, the local resource supply would be self-sustaining, and the population could survive indefinitely. This self-denial would evolve because prudent and frugal societies would eventually replace their spendthrift neighbours. Arguments of this kind, less explicitly and comprehensively developed, were commonplace at the time, and can still be found in elementary textbooks of biology. They have a very grave weakness. Granted that societies where individuals conserve resources by refraining from reproduction will arise, they will continually be invaded by types that reproduce at the greatest possible rate, even when resources are scarce. Selection among groups is therefore opposed by the selection of individuals within groups. The latter is expected to be the more effective, because individuals are more numerous and reproduce more rapidly than societies, and can therefore be selected more intensely. These and other objections were marshalled by George Williams (1966), in a counterblast to Wynne-Edwards's thesis so convincing that the interpretation of social behaviour as evolving through the selection of groups was utterly discredited.

There is nothing illogical, however, in supposing that selection among groups may be effective, supposing that there is genetic variation for social behaviour among groups. Although the cooperative behaviour Tit-for-tat is an ESS, a single Tit-for-tat mutant would not spread in a population where every other individual always defected. Tit-for-tat could spread in the selection experiment only because it was originally present at an appreciable frequency, and therefore occasionally encountered itself, or a program with similar tendencies. The situation simulated by the experiment, indeed, was a population divided into random groups of two individuals. In any given group, Tit-for-tat was nearly neutral: it did no better than its partner, and if paired by an uncooperative program would do a little worse. It spread because of the greater productivity of cooperative pairs. The same principle applies more broadly. Imagine any behaviour that is cooperative, in the sense that it benefits others (by trading honestly, for example), without appreciably harming the individual expressing it. Within a group of any size, the trait is neutral, because the benefit is conferred on cooperative and uncooperative individuals alike, and will not tend to spread. Within the population at large, however, the trait will be selected, because those groups that have a higher frequency of cooperative individuals reproduce more successfully. At the level of the population, of course, cooperative behaviour spreads because cooperative individuals have the greater relative fitness. Nevertheless, this would not be apparent from a study of the consequences of cooperation at the level of the local group in which the behaviour is actually expressed. A character may be selected in a structured population because it increases absolute fitness, although it has no effect, or a negative effect, on relative fitness within the local group. A model of this sort has been developed extensively by Wilson (1975, 1980). Individuals bearing the A1 allele experience an increment of fitness c while donating an increment b to each of the $(N - 1)$ other individuals in the local group, so they have fitness $\lambda + c + N(p_i - 1/N)b$ where p_i is their frequency in the ith local group. Selfish individuals bearing A2 have fitness $\lambda + Np_ib$. Groups containing altruistic individuals produce more offspring than selfish groups, and given free dispersal will therefore spread through the metapopulation. If the behaviour is costly ($c < 0$) then it can spread only if $\sigma^2_p > (1/N)pq$, that is, that the variance of gene frequency over groups should be greater than binomial. This means that when A1 individuals are somewhat aggregated they will tend to interact with one another more often than expected by chance, and will thereby reap the benefit of their mutually altruistic behaviour. Altruism is then selected provided that $(1 - N\sigma^2_p/pq)\, b - c > 0$. This is reminiscent of Hamilton's rule, and indeed 'trait-group' selection is essentially equivalent to kin selection arising because relatives are selected for helping behaviour because they encounter one another more frequently than expected by chance (Grafen 1985, Queller 1992). This immediately suggests how helping behaviour might evolve: if there is a genetic basis for choosing different kinds of site, either for feeding or for breeding, then any mutation causing helpful behaviour will arise in linkage disequilibrium with habitat preference and consequently is likely to spread.

10.4.5 Productivity and diversity

The primary group property is productivity, which should be maximized by group selection. The type that spreads in mixed cultures is not necessarily the type that is the most productive in pure stand, however, so the outcome of selection among individuals may differ from the outcome of selection among groups. A population is unlikely to become fixed for a helpful allele, since when this reaches high frequency it will inevitably be vulnerable to invasion by selfish mutants. Group selection and individual selection will lead to different outcomes only if the ESS mixture does not maximize productivity. A population is mixed at equilibrium if either type is more fit than the other, or others, when rare, so that it tends to spread when its neighbours are predominantly of the other type; in terms of the *abcd* model, the ESS is mixed if $c > a$ and $b > d$ (§10.3). The overall productivity of a mixture, however, is not necessarily maximized at the ESS. A mixture is more productive than either pure stand if either type is on average more productive when its neighbours are different than it is as a pure stand. The conditions for this to be the case are thus $(b + c) > 2a$ and $(b + c) > 2d$. When both conditions are satisfied, the *most productive population* (MPP) is a mixture in which the frequency of type 1 is

$$f_{\text{MPP}} = \tfrac{1}{2}[(b - d) + (c - d)]/[(c - a) + (b - d)].$$

The ESS and the MPP are obviously closely related, inasmuch as both involve mixtures if there is less competition (or more cooperation) between unlike than between like types. However, these mixtures do not have the same composition; they are related as

$$f_{\text{MPP}} = \tfrac{1}{2} f_{\text{ESS}}[1 + (b - d)/(c - d)].$$

The ESS is the outcome of pure individual selection in a single homogeneous population; the MPP can be thought of as the outcome of pure group selection in a spatially structured population. One important practical consequence of this distinction is that agronomists cannot rely on individual mass selection—harvesting and re-sowing a population without conscious selection—to produce a mixture with maximal yield. This conclusion is potentially important, because if the MPP is often a mixture it suggests that agronomic trials might be directed towards selecting the best mixture rather than the best single variety.

The simplest experiment is to measure the productivity of equal mixtures of several arbitrary types in all pairwise combinations. This procedure is equivalent to the diallel cross of conventional genetics, but generates combinations of individuals rather than combinations of genes. Such mixtures may be superior in one of two respects. In the first place, they may be more productive than the average of their components in pure culture. This demonstrates some degree of social interaction between the components of the mixture, but the MPP is nonetheless a pure stand of the more productive variety. However, in extensive trials the more productive variety is not known in advance, and sowing a mixture may be less risky. Secondly, the mixture may be more productive than the better component. Mixtures that constitute an MPP are said to be transgressive, and if they could be identified and selected would be unequivocally superior in agronomic practice. There is an extensive literature on the yield of mixtures of lines or cultivars in crop plants, and rather than cite individual cases I shall refer to two extensive surveys. Trenbath (1974) reviewed crops that are harvested for their vegetative parts, covering over 300 trials from 15 studies. Mixtures out-yielded the average of their components in about 60% of these trials. About a quarter of all mixtures were transgressive, the mixture yielding more than the better component, although about half as many were transgressive in the opposite direction, the mixture being inferior to the poorer component. There is thus a rather pronounced tendency for mixtures to be more productive than expected. However, the results are erratic and highly variable; there are few, if any, cases in which a specified mixture has been shown by repeated trials to be consistently over-yielding. I have carried out a similar survey of seed yield, collating nearly 200 trials from 20 studies (Figure 10.17). It leads to similar conclusions. Mixtures were more productive than the average of their components in nearly 70% of all cases, but out-yielded the better component in only about 20% of cases. The superiority of mixtures seems to be very general—it was reported in 19 of the 20 studies—but it is also very slight: on average; mixtures yielded only about 20% more than the mean of their components, and yielded nearly 5% less than the better component.

The natural extension of work on equal binary mixtures has been to ask whether a more pronounced increase of yield can be got from mixtures with several or many components. Agronomic experiments of this sort have given indecisive results that are difficult to summarize briefly; certainly there is no regular and predictable increase in yield with diversity. It seems that different species or cultivars of crop plants may interact quite strongly with one another, but these interactions are as likely to be antagonistic as to be cooperative. When the plants are grown close together, different mixtures vary widely, some being over-yielding and others under-yielding; as spacing increases, both antagonistic and cooperative tendencies weaken, the mixtures become more alike, and eventually any mixture effect disappears. In recent years similar experiments have been carried out on species of wild plants (e.g. Hector *et al.* 1999). The results have again been indecisive. Mixtures of types that have clearly different strategies of resource use can be consistently over-yielding, the clearest example being legumes and grasses, already known from agronomic trials. In other cases any increase in yield seems to be slight and it has been difficult to separate the effect of complementary resource utilization from that of selection for high-yielding species (see Loreau and Hector 2001). The results that I have obtained with mixed cultures of *Chlamydomonas* are consistent with general agronomic experience (Bell 1990a, b). When the types are different species, pairwise mixtures reliably exceed the average of their components, and this excess increases steadily with diversity. There was little evidence, however, that these mixtures, however diverse, consistently exceed the productivity of their best component. T. Bell *et al.* (2005) showed that community respiration increased with diversity in bacterial microcosms re-assembled from a natural community. This experiment separated the effect of diversity from the effect of composition and provides the best available evidence that diversity in itself contributes to community productivity.

Although the mixture that evolves through frequency-dependent selection does not generally have the most productive composition, the

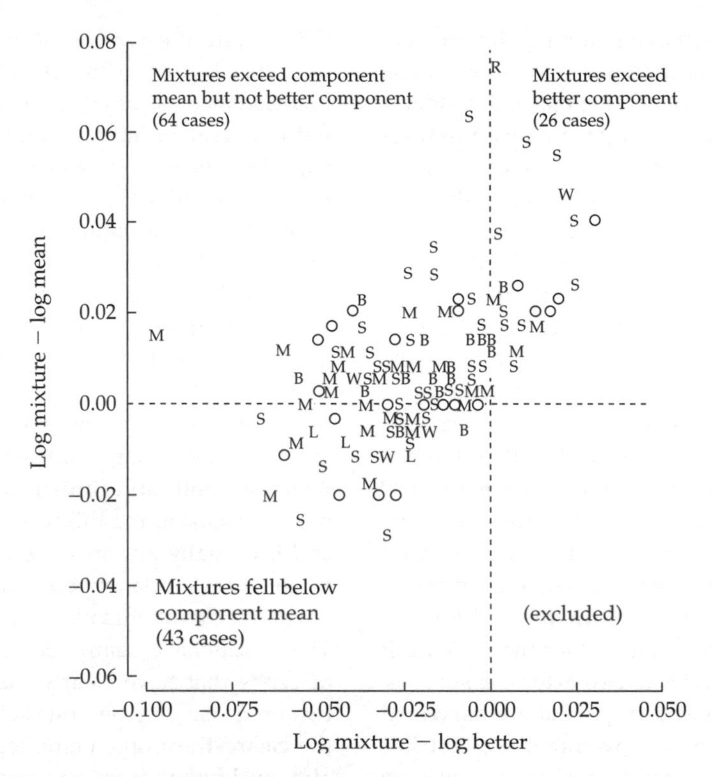

Figure 10.17 Seed yield of varietal mixtures of grain crops. This diagram summarizes my own review of seed yield in equal binary mixtures of cereal cultivars. I collated 169 experiments from 21 studies published in *Crop Science*, involving seven species. Yields vary greatly with the crop, of course; however, the log ratio of mixture and monoculture yield is Normally distributed, and does not vary among species of crop. The *y*-axis if this graph is (log mixture yield – log mean yield of components in pure culture), while the *x*-axis is (log mixture yield – log yield of better component). The symbols identify different crops: B barley, L lima bean, M maize, O oats, R rye, S soybean, W wheat. Points falling in the upper right quadrant are 27 cases in which the mixture outyields the better component; the 64 cases in the upper left quadrant are those in which the mixture exceeds the mean of the components, but not the better component; in 43 cases, in the lower left quadrant, the mixture is inferior to the component mean. Overall, mixtures yielded more than the mean of their components, but the difference was on average only 1.8%. Moreover, they were generally not as productive as the better monoculture, the mixture yield being on average only 95.6% as great as that of the better component.

conditions that the ESS and the MPP be mixtures rather than pure stands are similar, in that both require that b and c should be large relative to a and d. It is reasonable to expect that the components of a mixture whose composition has evolved through individual selection should be more productive in combination than sets of genotypes chosen at random. There are some indications that this may be the case: Allard and Adams (1969) found that lines from Composite Cross V (described in §10.3.2) showed a much stronger tendency for elevated yield in mixture after 18 generations of mass selection. Allard (1963) has reported a somewhat similar experiment with lima beans propagated for up to seven generations. He compared seed yield in five different kinds of population:

- the two parental lines P_1 and P_2 (both, of course, previously selected independently for high yield), with mean P
- the equal mechanical mixture of these lines, M_{12}
- a mechanical mixture of equal amounts of seed from 30 F_4 families, M_{F4}
- the mean of the same F_4 families grown in pure culture, F_4
- an unselected F_7 bulk, B_{F7}

The general result was that yields were ranked in this way:

$$B_{F7} = P_1 > M_{F4} > P > M_{12} > F_4 > P_2.$$

Because $M_{F4} > F_4$ even though $M_{12} < P$, the greater diversity of the F_4 mixture may be responsible for its superiority. However, $B > M_{F4}$ suggests a further increase in yield attributable to individual selection acting during bulk propagation, although it is possible that the type resembling the higher-yielding parent had increased in frequency through selection in the bulk. A more effective way of producing high-yielding mixtures might be to select directly on mixture composition.

As a very broad generalization, these results suggest that the properties of mixtures are dependent on their genetic scale. Different species may interact strongly, sometimes constituting consistently over-yielding mixtures. Different varieties of the same species may show the same effect, although on average it is very weak; in mixtures of random genotypes from the same family, any more subtle social effects are erased as one genotype comes to dominate the mixture. If this is the case, it will be possible to select over-yielding mixtures only when their components have sufficiently distinct ecological properties.

10.4.6 Artificial group selection

Despite a considerable amount of interest in social evolution and in the agronomic properties of mixtures, however, there seems to have been no attempt to prolong the artificial group selection of mixture composition beyond the single generation of a crop trial. The only applied research programme I am aware of set out to modify the social behaviour of domestic fowl in broiler houses. Birds that are kept together in the same cage will often peck one another, causing injury and reducing production. Cannibalistic pecking can be reduced by selection, with a realized heritability of 0.65 (Craig and Muir 1996). Since production and aggression are positively correlated, individual selection for production will tend to increase damage when birds are housed together, reducing the production of the group as a whole (Craig *et al.* 1975). The usual remedies (clipping the beak or turning down the lights) seem to harm the birds, and artificial group selection of production has been suggested as an alternative. The procedure used was actually kin selection, since the units were single-sire families. Six generations of selection doubled longevity and increased the rate of lay, so that the lifetime production of individual birds increased from 5.3 kg to 13.3 kg (reviewed by Muir and Craig 1998).

The most extensive laboratory programme has involved the 'population phenotype' of the flour beetle *Tribolium*. In most cases, this phenotype is population number, reflecting the traditional concern of group-selection theorists with the evolution of density regulation (Wade 1977). A population is founded by a small number of adult beetles put into a glass vial containing the nutrient medium, a mixture of flour and yeast. The characteristics of the population that develops in this vial can then be selected: that is, the most extreme vials can be chosen to propagate the line, just as in most experiments the most extreme individuals are chosen. This process of selecting vials rather than individuals amounts to artificial group selection, and the main purpose of the experiments is to find out whether it is effective in modifying the population phenotype. Other papers in this series concern the correlated response of competitive ability (Wade 1980), the genetic and demographic basis of the response (McCauley and Wade 1980), the effect of different rates of migration (Wade 1982), the response to relaxation of group selection (Wade 1984) and the interaction of group size and migration (Wade and McCauley 1984). A parallel experiment was reported by Craig (1982). Group selection in two-species communities of *Tribolium* has been studied on similar lines by Goodnight (1990a, b). This research programme has been reviewed by Goodnight and Stevens (1997).

The experiments typically involve a large number (of the order of 100) of small populations. The populations really are small, usually being founded by only 16 adult beetles, which give rise after a few weeks to 100–200 descendants. When a small proportion of the populations is chosen at random to reconstitute the population array in each cycle, there is no consistent selection for any particular phenotype, so the mean productivity—the number of beetles alive in each vial at the end of the cycle—is not expected to change. However, each population will drift independently, when each is a separate unit that propagates itself from cycle to cycle, and the variance among populations will increase through time. By choosing a few populations at random to reconstitute the whole array, a process of populational drift is superimposed on the drift within each population. Just as the effect of drift within a population is to reduce the genetic variance among individuals by fixing one or another lineage or allele, the effect of populational drift will be to reduce the genetic variance among populations within the array. This will not necessarily happen if each population is undergoing a similar process of individual selection: this will tend to make populations more similar, and the random selection of populations will retard their convergence. If we neglect individual selection, the divergence of populations will be promoted by the random sampling of individuals within each population, and restrained by the random sampling of populations within the array. The rate of divergence will be increased by low population number and a low rate of population extinction. Broadly speaking, these expectations are borne out by the experiments, although small populations founded by 6 or 12 adults, after diverging rapidly during the first few cycles, were no more variable after 10 or 11 cycles than larger populations founded by 24 or 48 adults. After 10 cycles of random selection, there was about a fourfold variation in productivity among populations. These differences were heritable, in the sense that daughter populations tended to resemble one another, and resembled the parental population from which their founders were drawn. The genetic variance of population productivity arising in this way is then the basis for selection among populations.

The random selection of populations can be contrasted with two other modes of selection. The first is no selection at the group level, with all populations being propagated in the same way. There will be selection among individuals within each population, of course, but this need not necessarily lead to any change in mean productivity. The second is directional group selection, for increased or decreased productivity, by propagating the whole array from the most extreme vials. Randomly selected and unselected populations diverged through time, as expected; indeed, the randomly selected populations diverged more, suggesting that unselected populations were experiencing strong individual selection of some sort. This variance was harvested by selection among vials: after nine cycles, the upward selection line was producing on average nearly 200 adults per vial, and the downward line only 20 (Figure 10.18). The cause of the divergence was that the upward line had retained more or less the same productivity as the base population, whereas the productivity of the downward line had declined steeply. There was also, however, a substantial decline in the productivity of the randomly selected and unselected lines. The most straightforward explanation of these results is that productivity was generally reduced in these small populations by inbreeding depression, which could be countered by group selection for high productivity, or exacerbated by group selection for low productivity. Both the divergence of randomly selected populations and the response to selection are reduced if the selected populations are supplemented by a few migrants from the discarded populations. Nevertheless, even when 25% of the individuals transferred in each cycle were migrants, there was still a detectable response to

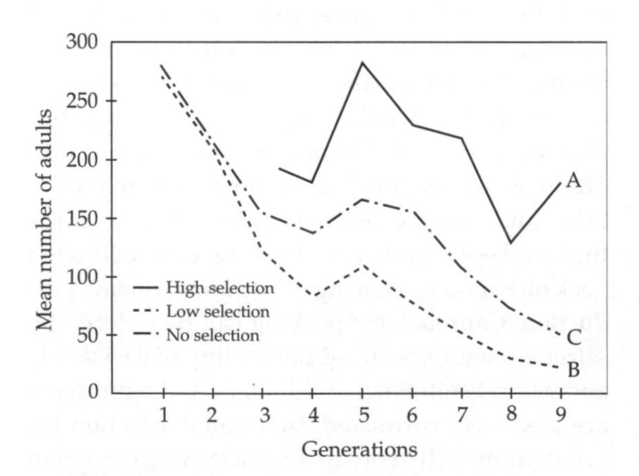

Figure 10.18 Response to upward and downward selection for population number in *Tribolium*. From Wade (1977).

directional group selection. Getty (1999) has questioned the interpretation of these experiments on the grounds that randomly selecting offspring from the group is equivalent to individual natural selection, since the most fecund individuals will contribute more.

Populations were selected for productivity without regard to the particular individual and social behaviours underlying the population phenotype, so that Wade's experiments do not directly address the evolution of nepotism or altruistic behaviour. Nevertheless, they have the usual force of experiments in artificial selection: if conditions are such that small populations with particular characteristics are favoured, these characteristics can be modified by selection among groups, and the social behaviour of individuals may evolve as a correlated response. When these behaviours were investigated, it was found that they had often evolved differently in populations selected in the same direction. Low productivity, for example, can evolve through group selection by reduced fecundity early in life, or by increased rates of cannibalism. When the upward and downward selection lines were reared as mixtures with another species of *Tribolium*, the downward lines were often considerably more successful. Group selection for low productivity had thus enhanced interspecific competitive ability, selecting for aggressive types that performed well in mixtures. Selection for high productivity had the converse effect of enhancing social behaviour that could be exploited by another species. The beetles thus responded to artificial group selection rather like crop plants selected for high yield per unit area as pure stands.

10.4.7 Cultural evolution

Social behaviour may be directly encoded by genes and expressed constitutively, like siderophore production, or in response to an environmental signal, such as social swarming by myxobacteria or slime moulds. The social lives of bugs and beetles, although more interesting than might at first appear, can then be explained in straightforward terms and investigated in microcosm experiments, whereas the much more complex societies of other organisms, and especially humans, seem to require explanations that go beyond simple kin selection and group selection. These explanations are required by the appearance of culture, the set of interacting social behaviours governing the lives of animals that live together in societies. In cultural animals, social behaviour is acquired by learning during the lifetime of an individual; that is, only the general ability to learn is inherited, whereas the specific skill that is learned is not.

Simple skills can be transmitted culturally in various ways, for example when offspring forage in the same sites as their parents and therefore encounter similar problems, which each individual solves independently. Many examples of cultural transmission have been documented in vertebrates (e.g. Lefebvre and Palameta 1988) and are probably found in one form or another in all organisms. They are very seldom combined and modified, however, to form behaviours that an individual could not discover for itself. The central problem of anthropology is thus to explain why the rudiments of culture, which can be found in a thousand species, have given rise to indefinitely complex societies based on a learned division of labour in only a single instance. Boyd and Richerson (1988) observe that the cultural evolution of skills that transcend individual learning requires a cumulative process of observational learning: individuals imitate a skilful act (that is, learn to do it from seeing it done), occasionally modify it, and are in turn imitated by individuals in the following generation. If more skilful individuals are more likely to act as models (because they are more likely to survive, or because learners prefer them) then skill will cumulate. Boyd and Richerson view imitation, or observational leaning, as a specific trait, not closely associated with high intelligence, expressed only by birds (with respect to song dialects), apes, and people. They explain the rarity of imitation, relative to individual learning, through a simple model in which unskilled individuals have fitness w_0 and skilled individuals $w_0 + b$, representing a benefit to possessing the skill so long as the environment does not change so that the skill becomes useless. The skill is acquired through individual learning with probability p at a cost c_i, so the fitness of individuals restricted to individual learning is $w_i = w_0 + pb - c_i$. It can also be acquired through observing N other individuals and imitating the behaviour of skilled individuals, if there are any. The cost of acquiring a specific skill by observation is c_s, which may be less than c_i, but there is a fixed cost c_0 of the general ability to observe and imitate. The fitness of individuals with social learning is then

$$\gamma(w_0 + pb - c_i) + (1 - \gamma)\,[w_0 + q(D - c_s) + (1 - q)(\gamma D - c_i)] - c_0,$$

where γ is the probability of environmental change from one generation to the next, and q is the probability that the N individuals observed include at least one skilled individual. The crucial conditions for the initial spread of observational learning are then that the environment does not change very often and that there are enough skilled individuals present for imitation to be effective. If individual learning is seldom successful, therefore, observational learning cannot spread because there are too few models, whereas if individual learning is usually effective observational learning does not spread because it offers no net advantage (Figure 10.19). Hence, the conditions for the invasion of imitators are rather restrictive, although once they have gained a foothold they will rapidly spread rapidly because they themselves provide an increasing number of models.

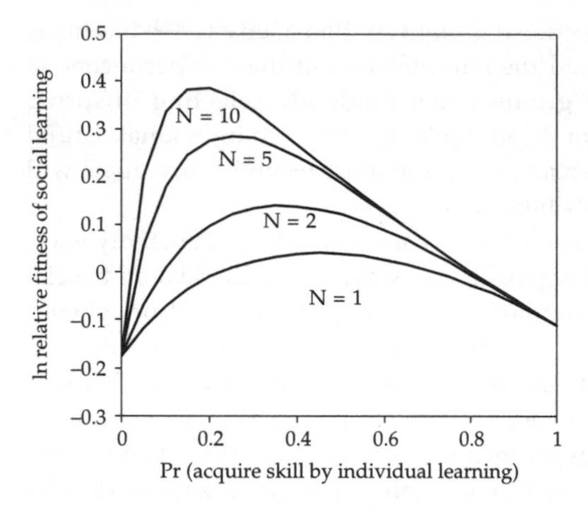

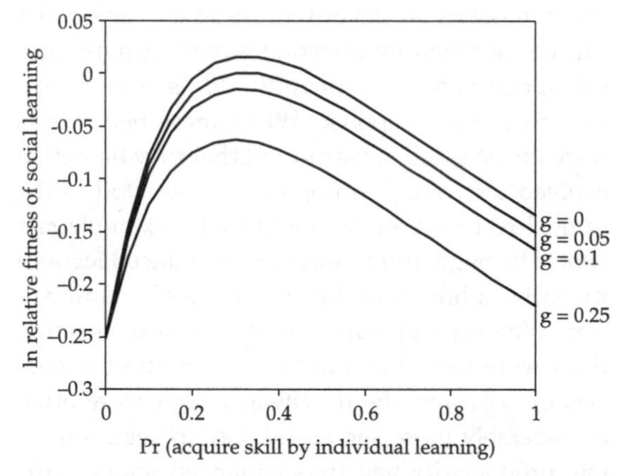

Figure 10.19 Social learning. A. Advantage of social learning in relation to the probability of acquiring a skill without imitation and the number of individuals observed by imitators. Model from Boyd and Richerson (1988) with $D = 1$, $C_i = 0.1$, $C_s = 0.05$, $K = 0.2$, $g = 0.1$. B. Advantage of social learning in relation to the probability of acquiring a skill without imitation and the rate of environmental change. Model from Boyd and Richerson (1988) with $D = 0.3$, $C_i = 0.1$, $C_s = 0.05$, $K = 0.2$, $N = 5$.

More complex models of social learning include instruction by parents or by unrelated individuals. Human offspring tend to resemble their parents both in useful features, such as hunting techniques, and in useless or harmful features, such as belief in superstitions (Cavalli-Sforza *et al.* 1982, Hewlett and Cavalli-Sforza 1986). Instructive learning requires at least two loci, one for transmitting and the other for receiving information, besides the phenotypic responses of the actors, and an extensive formal theory of cultural evolution based on this situation has been developed (Cavalli-Sforza and Feldman 1981). The only selection experiments seem to be microsociety games where the evolution of social behaviour in small groups is traced as they change in membership through time. Baum *et al.* (2004) gave groups of four volunteers the choice of two kinds of problem of equal difficulty that could usually be solved in roughly x minutes. Solving one brought a low monetary reward, but was followed immediately by the opportunity to solve another, whereas the other kind, although much more rewarding, was followed by a longer or shorter time-out of y minutes. Group members had to agree which to attempt, and once it was solved received the appropriate reward. After one social generation of 12 minutes, the longest-serving member was replaced with a newcomer who was taught by the remaining members how to play the game. This was then repeated for a dozen generations so that both the overall strategy and the teaching tradition could be followed through time. The best strategy clearly depends on the solution time in relation to the time-out. Most groups preferred the high-reward option to begin with, but if $x \ll y$ all groups converged on the low-reward option that maximizes total pay-off after 5 generations or so, whereas if $x \approx y$ there were broad fluctuations in choice without any clear trend. Most instructions were correct, although erroneous information was given more often when the situation was more difficult to evaluate (longer time-out); coercion (instruction without explanation) was very rare. These traditions seemed to be expressed at a more or less constant rate through time, and therefore did not themselves evolve, although there was some tendency for coercion to diminish.

What is evolving in situations like this? It is clearly not the participants, a random sequence of individuals, but rather the ideas that are transmitted from one generation to the next. Where these ideas have concrete realizations, such as tools, a phylogeny can be constructed to show how more complex and effective devices evolve from and replace simpler and less effective precursors. In the simplest cases, such as the transition from simple pandanus-leaf strips to tapered tools made by New Caledonian crows for extracting insects from crevices, this can be fairly straightforward (Hunt 2003). It is less easily applied to human technologies because radically new designs can be produced by imagination; nevertheless, the most sophisticated devices may incorporate functionless features descending from distant precursors, the classical example being the arrangement of the keyboard of the laptop computer on which I am typing this sentence. Basalla (1988) gives a synthetic account of the cumulative evolution of technology, and the case for a Darwinian theory of culture is outlined by Mesoudi *et al.* (2004). The most radical interpretation of cultural evolution is that it is the concepts themselves that evolve, with our brains serving merely as the medium through which these memes (Dawkins 1976, Blackmore 1999) are transmitted. A meme is any mental construct, or the product of any mental process, such as a song, that can be imitated. This is perfectly consistent (it seems to me) with the general formulation of selection given by Price (1972), which could serve as a formal basis for the theory of memetics. The great attraction of the theory is that it has the potential to provide a direct explanation of irrational beliefs and self-harmful acts that gene-based accounts must view as the non-adaptive by-products of behaviour that increases inclusive fitness, or did so in the past. At present, unfortunately, the field is riven by polemic, as sociobiology used to be, and no regular scientific research programme, nor tradition of experimentation, has yet emerged.

CHAPTER 11

Co-evolution

Scotch broom (*Cytisus scoparius*) is a low shrub that grows on heaths, especially on disturbed soils. At Silwood Park in southern England it is a rather rare plant that is found in acid grassland dominated by *Agrostis*, *Holcus* and other grasses, herbs, and shrubs, where it has been the subject of detailed ecological investigations (Waloff 1968, Memmott *et al.* 2000). Below ground it has two important mutualistic associations, with nitrogen-fixing rhizobial bacteria in root nodules and with vesicular–arbuscular mycorrhizal fungi. Above ground it is consumed by about 20 species of insects in several orders, which in turn are consumed by about 60 predators and about the same number of parasitoids. Exclusion experiments show that these insects severely reduce plant growth (Waloff and Richards 1977). The plant is thus bound within a web of material and energy flow between all the other plants, the animals, fungi, and microbes that live together on the same heath. Some will influence it directly; a few, such as the broom beetle, *Phytodecta olivacea*, will have a powerful and specific effect on its growth and reproduction. Others will influence it indirectly, by parasitizing its herbivores, or, more broadly still, by trapping the parasitoids of the herbivores of competing species of plant. Almost all organisms are likewise embedded within food webs which confront them at every turn with rivals, partners and enemies.

The interaction between two species is conventionally classified by the direction of material and energy flow between one and the other. It may be negative for both parties (competition), or positive for both (mutualism), or positive for one and negative for the other (predation or parasitism). It is the fervent hope of any reductionist approach to evolutionary ecology that selection on any target species can be related, as a good approximation, to its direct pairwise interactions with the other species in its community (John Maynard Smith, in a reference I have mislaid). I shall first describe pairwise interactions between ecologically equivalent species, like broom and ferns, whose interaction is entirely negative, and then between ecologically non-equivalent species, whose interaction is positive for at least one of the two, like broom and beetles, or broom and *Rhizobium*. I shall finish by attempting to visualize how selection acts within a trophically structured community of indefinite complexity, that is, how broom and its rivals, partners, and enemies evolve together as components of a single, ecologically connected ecosystem.

11.1 Rivals

11.1.1 The social environment

The distinction between adaptedness and fitness implies that neighbours mutually modify one another's environment. This will lead directly to selection that causes changes in the frequency of lineages in the population. However, it may lead indirectly to selection for altered or improved competitive abilities. That is to say, a lineage might evolve so as to become adapted to neighbouring or interacting individuals of other lineages, just as it might become adapted to any other sort of environmental modification. There are three sorts of experiment that might address this issue. In the first kind, a population is cultured repeatedly in the presence of some stock genotype. In this case, the neighbour to whose presence the population is adapting is always the same, being drawn in every generation from some uniform line, maintained separately. This is essentially the same as adapting to some feature of the physical environment, the only novelty being to discover whether different genotypes do in fact modify the environment in substantially different ways. A rare grass in a meadow, for example, will be grazed along with the rest of the herbage by a herd of cattle. The relative fitness of different types of the grass may vary according to the intensity of grazing, whereas the cattle are indifferent to whether such a minor item in their diet occurs or not. From the point of view of a species that is perennially rare, generalized grazers are a fixed component of the environment, akin to frost or drought. Their effects may vary through time, but they vary independently of any evolutionary modification of the rare prey. This situation can be represented in the laboratory by culturing a population generation after generation with neighbours that are freshly recruited in every generation from a stock that is itself maintained as a pure culture. The experimental population consists of a number of lines that can readily be distinguished by some marker, and which do not interbreed, or are kept from interbreeding. In an outcrossing population such lines would not remain distinct, and could not evolve as units. Competitive ability can be modified only among different species, or among different lineages of asexual or self-fertilizing organisms. Although the lines must be stable, they must nevertheless also be sufficiently variable to respond to the selection caused by their neighbours. They must therefore be populations, and not merely different mutants. The interesting question is one of scale: how different must these interacting populations be in order to elicit specific adaptation? It would not be surprising if experimental populations of flies evolved in response to geckos on the walls of the cage, or nematodes in the food medium (although I do not know of any such experiments). It would be more surprising if similar neighbours of the same species generally modified the environment so distinctively that specific adaptations evolved as a response.

A favourite system for studying the evolution of social interactions between related species is the sibling pair *Drosophola melanogaster* and *Drosophila simulans*, which are morphologically almost indistinguishable, but do not interbreed. Futuyma (1970) reared 10 generations of a highly heterogeneous *melanogaster* population in the company of an equal number of *simulans* recruited in every generation from a stock that was marked, for ease of identification, with a mutation causing dark body colour. Control *melanogaster* populations were maintained at the same time as pure cultures. After 10 generations, flies from the control or the selected *melanogaster* lines were mixed with equal numbers of stock *simulans*. The offspring of these flies grew up together, and eventually emerged as adults. If social selection had been effective, the proportion of *melanogaster* emerging from the mixed cultures should be greater for the selection lines than for the controls. In fact there appeared to be little if any difference. In one of the replicate selection lines, an excess of *melanogaster* emerged; but in most lines there was no appreciable difference, and in two cases the control lines actually yielded more *melanogaster* than the corresponding selection lines. When Futuyma tested the individual strains (kept in mass culture during the selection experiments) from which his base population had been constructed, he could find no evidence that they differed in fitness when cultured with equal numbers of *simulans*. Selection for social adaptation to a closely related species was therefore ineffective: the base population may have contained no genetic variance for specific social behaviour, and none emerged through sorting.

11.1.2 Mutual modification

The second kind of experiment is to perpetuate a mixed population for several generations, without conscious selection, and then to compare the competitive relationships among its component lines with those among the equivalent lines in the base population, or in concurrent pure cultures. This design allows lineages to become mutually adjusted to one another, provided that the population remains diverse for long enough for selection to be effective. This can readily be enforced, by transferring fixed numbers of types in each generation, regardless of population frequencies. The experimental population thus consists of a definite number of types (in practice, two) that are deliberately maintained in the same proportions. When two species are both moderately abundant, each will form part of the other's environment. They will then behave quite differently from physical features of the environment, because each will respond to the other's presence, and in responding will change the social environment that the other experiences. This may result in specific mutual modification through a dynamic process of social selection. The same species-pair of *Drosophila melanogaster* and *Drosophila simulans* has been used in this kind of experiment by Barker (1973) and van Delden (1970). The basic procedure is to set up pure and mixed cultures from laboratory strains of the two species that can be distinguished because they carry genes conferring different eye colour.

Barker transferred 10 pairs of *melanogaster* and 10 of *simulans* in every generation in his mixed cultures, and 20 pairs in each of the control pure cultures. The flies could be extracted from the lines at intervals and tested to see whether the competitive ability of the selection lines had changed relative to that of the controls, as judged by the proportions of the two species emerging from mixed cultures. The experiment was continued for about 60 generations, but was bedevilled by all the difficulties of long-term studies: the author moved his laboratory from Chicago to Sydney during the experiment, the culture medium was changed twice, some of the mixed cultures had to be discontinued because there were not enough flies from one of the two species, and at least one *melanogaster* line was contaminated by immigrants. Perhaps as a result of these misadventures, the behaviour of the pure-culture stocks changed through time, and the behaviour of the selection lines is difficult to interpret. It does seem that there was little if any change in social behaviour for the first 30 generations or so. After 40 generations, there was some indication that one of the *simulans* lines had evolved somewhat different social properties. These seemed to involve the production of greater numbers of *simulans*, without a corresponding reduction in the production of *melanogaster*. It is possible that in this one line the two species had diverged somewhat so as to compete less intensely than they did at first. Whatever changes took place, however, were certainly slight and difficult to identify with certainty.

Van Delden's experiment was even longer, being carried out to about 80 generations. It combines the two designs I have discussed so far. In a restocked selection line, the *simulans* were supplied in every generation from the control pure culture, whereas in a co-evolved line they were transferred from the previous generation of the mixed culture. There appeared to be no response to selection after 10 generations, but quite pronounced effects began to emerge after about 60 generations. In either kind of selection line, the production of both *melanogaster* and *simulans* increased. The effect in *melanogaster* was more marked for the line in which it co-evolved with *simulans*. However, the production of pure cultures set up from the selection lines also increased, relative to that of the controls. Van Delden's results are unfortunately described too tersely for one to be certain of what happened, but it seems as though there was a change in general social behaviour (as shown by the increased productivity of the co-evolved selection lines in pure culture) and a more specific response (as shown by the improved performance of *melanogaster* from the restocked line when tested against stock *simulans*). In both cases, the response seemed to involve some sort of mutual facilitation, similar to that which may have occurred in Barker's experiment, with flies of one species increasing in abundance without causing a corresponding loss to the other species.

Flour beetles in the genus *Tribolium* were used in several classic ecological experiments on interspecific competition by Thomas Park at Chicago in the 1930s and 1940s. Dawson (1972) subsequently used *Tribolium castaneum* and *Tribolium confusum* to investigate the evolution of competitive ability. This was a conscious attempt to repeat the *Drosophila* experiments, using the same design to culture co-evolving mixtures over 5 or 10 generations. The beetles showed the same range of social behaviour as the flies—an unequivocally stronger competitor in one case, frequency-dependent fitnesses in another—but there was no tendency at all for this behaviour to change through time.

11.1.3 Social co-evolution

The final kind of experiment is to propagate the population by bulk transfer, if it is expected that its diversity will be maintained through natural selection. The experimental population is in this case an indefinite mixture, whose composition is not directly controlled. Selection can act in two ways in such populations. It may, as before, modify the properties of the various types within the population. It may also, however, modify their frequencies, thereby causing a change in the average behaviour of the population, regardless of whether any types have been individually modified. Pairs of related species can sometimes be maintained as self-perpetuating mixtures in which the two species coexist in roughly constant proportions for long periods of time. In some cases, these proportions have been seen to shift abruptly, so that one species, formerly rare, becomes prevalent. For example, Pimentel *et al.* (1975) cultured the housefly *Musca* and the blowfly *Phaenicia* together. One always eventually eliminated the other, but species that seemed to be prevailing was sometimes rapidly replaced by its competitor. In one experiment, for example, *Phaenicia* struggled along for about 40 weeks at low frequency, and eventually became very rare. At this point, however, it began to increase in numbers, and 20 weeks later had completely eliminated *Musca* from the community. This may have been caused by social selection within the *Phaenicia* population. While it was rare, social selection would have favoured individuals that succeeded in competition with *Musca*, rather than those successful in intraspecific competition, and a variant appearing in about the fortieth generation was so well-adapted to competing against *Musca* that it drove it to extinction. Other explanations are possible. A subtle shift in laboratory environment or handling procedures might have tipped the balance decisively against the type that was formerly predominant. This possibility can be investigated by comparing the supposedly evolved population with the base population, or with a population maintained in pure culture, in mixtures with the other species. This is convincing only if the environment in which the lines are tested is the same as that before the shift took place, and this may be difficult to establish. In Pimentel's experiment, the selection lines were deliberately supplemented from time to time by unselected flies taken from wild population, with the intention of reducing the level of inbreeding, and the successful strain of *Phaenicia* may simply have been introduced in this way. Reversals in social dominance are thus often difficult to interpret as unequivocal evidence for social evolution in one of the competing species, although the circumstantial evidence is sometimes quite strong.

Strains of the same species are even more similar than related species, and selection for altered competitive abilities is likely to be correspondingly weak and non-specific. Nevertheless, a well-known experiment by Seaton and Antonovics (1967) seemed to demonstrate rather strong and specific selection of this kind. They studied competition between wild-type flies and a stock carrying the wing mutation *dumpy*. The experiment was founded with progeny from crosses within the wild-type and *dumpy* stocks. These were cultured together until they eclosed as adult flies. While still virgin, these flies were removed and the two stocks, distinguishable from the *dumpy* marker, separated. In this way, two sexually isolated populations could be maintained in competition with one another. In every generation, the selected flies from one of the two stocks were tested against unselected flies of the other stocks, reared as pure cultures. After only three or four generations of selection, there was a remarkable increase in the specific competitive ability of the selected flies, demonstrated by an increase in the proportion of flies from the selected strain emerging from mixtures with the unselected stock of the other strain. This was caused by an increase in the number of the selected flies, rather than by a decrease in the number coming from the unselected stock. Hence, it seemed that the main effect of selection in mixture was a reduction in the intensity of competition between the strains.

Sokal and Sullivan (1963) conducted a similar experiment, using strains of the housefly, *Musca*. After about 10 generations of selection, however, neither of the two strains they used seemed to have responded to social selection in mixed culture. Bryant and Turner (1972) argued that one reason for this discrepancy might be that Seaton and Antonovics had given *dumpy* a head start in mixed cultures by putting its eggs into the vials a couple of days before adding wild-type eggs; *dumpy* is so sickly that unless it is given this head start it is quickly eliminated from the mixture. They therefore set up another experiment giving a similar head start to a rather weak green-eyed strain of *Musca*. This time, the survival of the mutant strain increased markedly in mixed culture within four generations, whereas it remained about the same in pure culture. The wild-type strain that it was competing with did not change. The improvement in the mutant strain was traced to an acceleration of hatching caused by selection against late-hatching larvae in the mixtures, although it was not clear why the same improvement should not also occur in the pure cultures. At all events, this second experiment seemed to confirm Seaton and Antonovics' original result.

This result has not been quantitatively confirmed in subsequent experiments. For example, Sulzbach (1980) created two highly heterogeneous strains by matings among different sets of unrelated isolates from widely separated localities, one set being marked by the eye-colour mutants *vermilion* and *brown*. After about 20 generations of selection in mixture, there were no signs of any substantial modification of competitive ability, although some replicate lines may have evolved weak and rather erratic tendencies to perform better in the presence of their neighbour. Like Bryant, Sulzbach reasoned that the head start given to a weak competitor might be the critical factor in getting a rapid improvement in competitive ability, so he tried the effect of giving a 2-day start to the brown-eyed strain for the first 8 generations of selection (Sulzbach and Emlen 1979). In one replicate—but not in other lines—survival of the brown-eyed flies increased sharply during the early generations of selection. It seems doubtful whether this improvement was caused by social selection; the line was originally weaker than the control line, for unknown reasons, and merely regained the same level of performance as the control.

The longest social selection experiment was reported by Pruzan-Hotchkiss *et al.* (1980), who maintained a wild type and a mutant in pure culture and in restocked mixture for 8 years. Both strains were rather unusual: the wild-type line was founded by a single fertilized female, and the mutant was a multiply marked compound autosome construct. They were used in the experiment because the hybrids are inviable, so that no special procedures are necessary to keep the two strains isolated. They seemed to show no specific adaptation to each other's company. Unfortunately, the crucial test of rearing flies from the mixture in the company of neighbours with a history of mixed or pure culture was not attempted, but it was established that flies from the mixture performed as well in pure culture as they did in mixture. It seems that 200 generations was not sufficient for these two strains to evolve any specific response to one another.

There is a very strong tendency to publish experiments that work. The phrase itself is revealing: an experiment that works provides striking support for an hypothesis. Everyone is aware that experiments should be attempts to falsify hypotheses, but everyone nevertheless organizes their experimental work in the hope that it will support some hypothesis they favour. After all, it is easy to produce negative results through laziness or lack of skill, whereas a clear demonstration of some new or hitherto obscure phenomenon is a great deal more difficult to accomplish. The proportion of negative results in experiments designed to detect the evolution of specific social interactions between different strains therefore carries some weight. It is most unusual for selection experiments to repeatedly fail to procure a response, and the most likely explanation for this plethora of negative or marginal results is that social selection of the kind envisaged in the experiments simply does not occur very often. Even the positive results are not very convincing—ironically, because they are too strikingly positive. They all seem to involve mutant strains that are feeble in pure culture, and it seems likely these instances of rapid improvement following selection in mixture merely represent the immediate effects of selection on populations whose mean fitness is low, either because the base population carries unconditionally deleterious mutations, or because it is at first poorly adapted to the environment in which the experiment is conducted.

The rationale for these experiments is that strains of the same species are likely to have very similar ecological properties, and will therefore compete intensely. This may be so; but when the base populations are very similar, or highly heterogeneous, or both, the difference between 'self' and neighbour is slight or variable, and there will rarely be strong or consistent selection for a *specific* response to the competing strain. In the limit, the two strains are identical, and only a general density-dependent response is possible. The lack of a specific response has been attributed to epistasis, the social interaction between related strains involving combinations of genes that cannot be selected effectively in sexual populations. This may be true, but it seems equally likely that genes causing a specific response towards similar neighbours are just very rare. Moreover, the prevalence of one species or another in mixed-species cultures is usually strongly affected by the physical environment, and very slight changes in culture conditions may cause the frequency of related strains of the same species to change markedly from one trial to the next. The character that is measured in these experiments—percentage emergence of one type—is thus likely to show a high degree of genotype–environment (G × E) interaction, to have a correspondingly low heritability, and therefore to respond only very slowly to selection. The generally negative outcome of the experiments is not uninformative, however: arbitrary pairs of closely related species, or strains of the same species, seems to represent a scale at which social adjustments evolve only very feebly or slowly. This is perfectly consistent with the widely held view that in sexual organisms the species is a natural ecological unit constituting the minimal level of distinctiveness that competition can effectively enforce.

11.2 Partners

The social relations I have described in previous sections have concerned similar kinds of organisms: closely related species, or genotypes within the same species. Entirely different kinds of organisms, however, often form very close associations, either as partners or as enemies. The selection that is generated by such associations has been studied most intensively in agriculture, where the fungal and bacterial communities that live on crop plants are of enormous economic importance.

Crop trials that investigate how relative fitness varies in different physical environments are usually conducted so that other organisms, especially competitors, herbivores, and pathogens, have a minimal effect on the results. If genotypes vary in their ability to withstand competition or attack, the routine application of herbicides and pesticides to the experimental plots will cause the amount of environmental and G × E variance to be underestimated. There is no doubt that this is the case. Briefly, two main kinds of experiment have been attempted. The first involves measuring the impact of disease on cultivars grown in a range of more or less infested environments. The literature shows that disease scores vary in much the same way as other agronomic characters, the relative health of genotypes varying in different localities and years. The second kind of experiment is to measure the yield of cultivars that are either exposed to or protected from pathogens. This can be done by manipulating the environment, so that the plants are grown either in infested or in uninfested conditions; alternatively, it can be done by manipulating the genotype, so that lines known to be resistant or susceptible to a particular pathogen are grown over a series of more or less infested environments. An example of an experiment that combines the two approaches was reported by Kayasthar and Heyne (1978), who constructed five pairs of nearly isogenic lines of wheat, one member of each pair being resistant and the other susceptible to the soil-borne mosaic virus of wheat. Some of the plots in which the wheat was sown were known to be clear of the virus, whereas others were infested. The genotype-environment matrix for mean yield looked like this:

		Environment	
		Uninfested	Infested
Genotype	Resistant	2672	2222
	Susceptible	2705	1737

This neatly illustrates the usual results of such experiments. There is a pronounced environmental effect, yield being reduced by disease in the infested plots. There is also a distinctive pattern of G × E: the resistant plants produce more seed on the infested plots, as expected, but the susceptible plants have the higher yield in the absence of disease.

The interaction between the wheat and the virus is not only substantial, it is also highly specific, being caused largely by a single gene for resistance to this disease in the wheat. Such genes are not uncommon; moreover, they are often matched by specific genes in the pathogen that determine whether or not it is virulent on a given strain of host. In contrast to the rather vague genetic basis for social relations among strains of the same species, there is often a simple Mendelian basis for the relationships between resistant or susceptible host plants and the virulent or avirulent fungi and bacteria that live on them. These relationships can be explored in trials where lines or cultivars of the crop are exposed to defined clonal isolates of the pathogen. The analysis of this genotype–genotype matrix introduces a new and important complication.

We might measure the seed yield of each plant strain when exposed to each virus strain. This would enable us to express how the relative fitness of plants varied according to the viral isolate they were exposed to. However, it is only for economic reasons that we are primarily interested in the plant; from a biological standpoint, we will be just as interested in how the relative fitness of viral strains varies according to the strain of plant they grow on. We would then measure the productivity of the virus, although this is never done in agronomic trials, except as the intensity of disease symptoms expressed by the plant. There are thus two social matrices:

	Pathogen		
		P1	P2
Host	H1	a	b
	H2	c	d

	Host		
		H1	H2
Pathogen	P1	*a*	*b*
	P2	*c*	*d*

Here, the abcd represent the fitness of a host type when exposed to a particular pathogen type (or the effect on host fitness of that type), whereas the *abcd* represent the fitness of a pathogen type when exposed to a particular host type. These matrices are irreducible: they cannot be expressed, as all previous cases have been, as a single matrix. The reason is that the abcd cannot be interpreted as fitnesses (or effects on fitness) relative to the *abcd*. Thus, if a is large and *c* small, we cannot infer anything about the outcome of selection in either population. It is in this sense that the two components of the system, such as a plant host and a fungal pathogen, are not ecologically equivalent. When we are dealing with ecologically non-equivalent organisms, we can no longer describe the social evolution of a single population or set of populations, but must instead think in terms of the co-evolution of two distinct entities. This consideration extends, of course, to any number of host and pathogen genotypes, and, what is more important, to any number of interacting but ecologically non-equivalent organisms.

The social matrices of non-equivalent organisms are coupled, so that selection in one will drive evolution in the other. The dynamics of these coupled systems are quite complex, but for asexual organisms, or alleles of a single gene, we can make a basic distinction between two types of system (Figure 11.1). In the first place, the two matrices may be similar in form, in the sense that they can be arranged so that the greatest fitness in either is found in the same cell, or along the same diagonal. For example, suppose that the two matrices can be arranged so that a > d > b > c and *a* > *d* > *b* > *c*. We can symbolize the two types of one species as H1 and H2, and those of the other species as P1 and P2. The joint population may have any composition to begin with; say that it is dominated by H1 and P2, and so can be represented as H1P2. Because one population is dominated by H1, the frequency of P1 will rise in the other, since *a* > *c*, leading to a H1P1 population. This is stable against the spread of either H2 (because a > c) or P2 (because *a* > *c*). A similar argument shows that if the population were initially H2P1, it would evolve too a different stable state, H2P2. The greater the frequency of H1, the greater the fitness of P1, and vice versa; the greater the frequency of H2, the greater the fitness of P2, and vice versa. This type of system

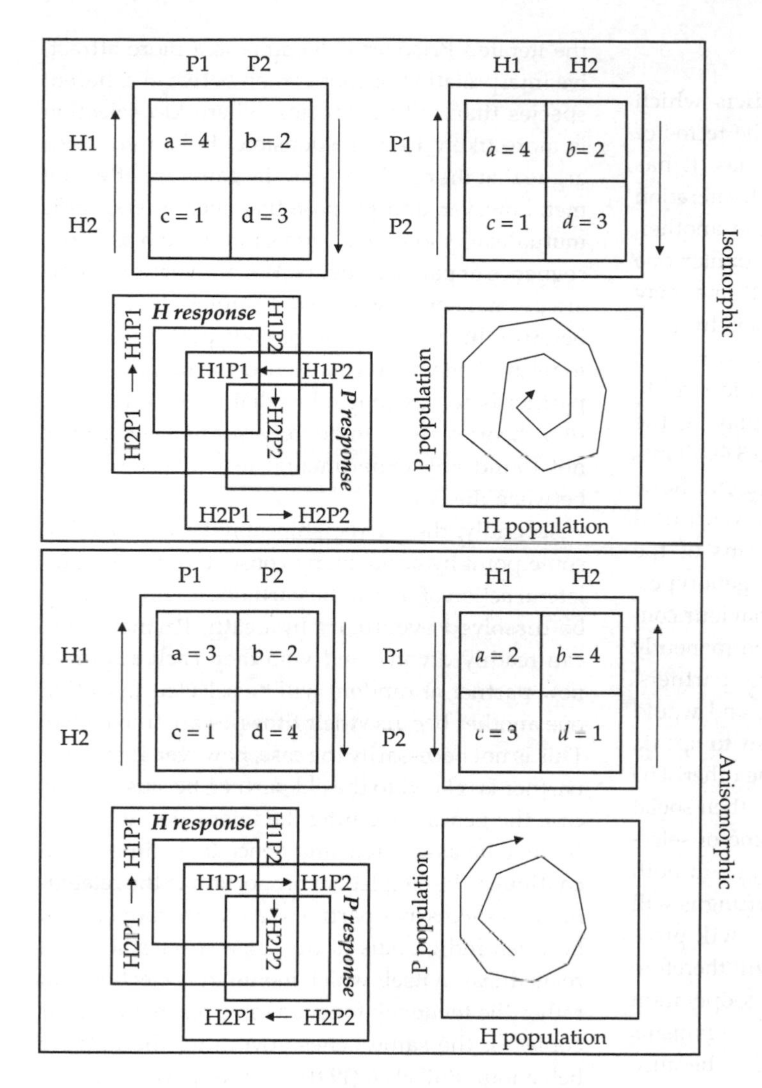

Figure 11.1 Isomorphic and anisomorphic social matrices. Superimposing the matrices (lower left) so that given H-P combinations occupy the same positions, makes the coupled responses clear. These can be expressed as a phase diagram (lower right) which shows genotype frequencies converging to a point (from isomorphic matrices, upper) or diverging to a cycle (anisomorphic matrices, lower).

thus generates positive frequency-dependent selection, leading to the fixation of a single type in both populations. Because the two matrices have the same form, they can be said to be isomorphic. The condition for isomorphism is that $(a - c)(a - c) > 0$, or $(b - d)(b - d) > 0$, or both. The second possibility is that the two matrices are unlike, in which case the dynamics of the coupled system are very different. Anisomorphic matrices have both $(a - c)(a - c) < 0$ and $(b - d)(b - d) < 0$. For example, $d > a > b > c$ and $b > c > a > d$ are anisomorphic. If the joint population is initially H1P2, as before, then it will evolve towards H2P2 (because $d > b$); which will in turn evolve towards H2P1 (because $b > d$), subsequently to H1P1 (because $a > c$), and finally to H1P2 again (because $c > a$). The system is therefore never at rest, but continually cycles H1P1–H1P2–H2P2–H2P1–H1P1–....

The form of the social matrices determines how two interacting species will evolve. If the matrices are isomorphic the outcome may be a fundamentally stable partnership between partners who each gain from the other's presence. If the matrices are anisomorphic the outcome is a potentially unstable struggle between enemies whose interests are incompatible.

11.2.1 The transmission hypothesis

There is a general theory that predicts which kind of interaction will evolve from the terms of the social contract between two species. It has two components that jointly express the iteration of interactions between one lineage and another. In the first place, individuals may encounter one another only once, or, at the other extreme, may be bound together indissolubly all their lives. A long series of encounters between the same two ecologically equivalent individuals would tend to select cooperative partners through playing the iterated prisoner's dilemma game (§10.3.4). Plants and fungi are not ecologically equivalent, however: the cooperative behaviour of a mycorrhizal symbiont cannot be explained in terms of the interactions between different fungal genotypes. Nevertheless, the logic of tit-for-tat behaviour continues to hold. Plant and fungus are permanently associated, and are unlikely to change partners. Each has something valuable to trade, and would suffer if its partner defected; each, so to speak, constitutes a renewable resource for the other. The symmetry of their interests implies that their social matrices are isomorphic, so the outcome of selection will be the fixation of a single type in both populations. However, both plant and fungus will prosper if they cooperate, and neither will prosper if both defect. Cooperative types will therefore spread in both populations. Indeed, cooperation is more likely to evolve than if the two organisms were in direct reproductive competition, because quantitative symmetries are irrelevant: a plant will benefit from cooperating with its mycorrhizal associate if it thereby increases its fitness, if only very slightly, even if as a consequence it provides an enormous benefit to the fungus. On the other hand, it might pay either to defect if the terms of trade change, so that a partner can more profitably be treated as a non-renewable resource: a green hydra growing in the dark, for example, draws no benefit from its photosynthetic symbionts, and is likely to digest them. It will also pay to defect if the association is coming to an end; if your partner is sick and likely to die, it is better to realize the capital, rather than have the investment fail. Hence, the iterated Prisoner's Dilemma is a more attractive interpretation of cooperation between different species than within species, where kin selection is more likely to be responsible. Bull *et al.* (1991) argue that the conditions for the game are often not met, however, and propose that such interspecific mutualisms should instead be interpreted as a consequence of partner fidelity. When two individuals are permanently associated it cannot pay to defect because this would damage your partner's ability to return benefits to you. Hence, the fitness of each partner becomes coupled to that of the other. The two theories seem to be very similar and I have not found any experimental test distinguishing between the two.

Secondly, the partnership may be dissolved at some point by accident, by consent, or by the unilateral action of one of the partners; it will certainly be dissolved eventually by death. Partners who can readily divorce and who then each acquire a new partner at random will be selected to exploit one another because their fitnesses are uncoupled. This is not necessarily the case, however, if the new partner is related to the old partner, because in this case the genes governing the interaction will continue to be associated, and hence their fitness will continue to be coupled, in proportion to the relatedness of successive partners. As with cooperation between individuals of the same species, it is not relatedness in itself which modulates selection, but rather the tendency for successive partners to bear copies of the same genes governing the helping behaviour. Bull *et al.* (1991) call this 'partner choice' to explain how helpful behaviour can evolve when interactions are short-lived, as in cleaning symbioses: the host observes the behaviour of potential partners and chooses the most dedicated cleaners; the cleaners avoid host individuals who succumb to temptation and eat their cleaner. The same principle applies to the relatives of the partners, and in particular to their offspring. If the offspring of either species encounters new hosts at random then building trust has to begin anew in every generation. Short-term encounters with a succession of unrelated partners, or a single partner, leading to the production of offspring which in turn encounter partners unrelated to those of their parents, will lead to the exploitation of one partner by the other. Conversely, the transmission of offspring to related partners, or to partners which for any reason are more likely than average to bear similar alleles for genes governing helpfulness, will favour helping and lead to cooperation. The offspring of partners may be packaged together, as the fungal and algal components of lichen often are, so that the offspring resume the partnership of their parents. In the extreme, the partners are indissolubly united during life and transmitted in the same germplasm, like mitochondria, and their relationship becomes not merely mutualistic but obligatory. This is the basis for believing that horizontally transmitted symbionts are likely to evolve as pathogens or other kinds of enemy whereas vertically transmitted symbionts will evolve a high degree of mutualism.

11.2.2 A novel protist–bacterium partnership

Jeon and Jeon (1976) observed a remarkable instance of the evolution of a novel partnership in the laboratory. Their cultures of *Amoeba* became accidentally infested by an unknown bacterium able to multiply inside the host cells. The bacterium was initially pathogenic, killing most of the amoebas it infected. The bacteria released by the lysis of the dead host would then infect other amoebas. A few hosts survived this infection, though retaining a residual population of bacteria in their cytoplasm. When these amoebas divided, both daughters inherited an intracellular population of bacteria, so the bacteria were necessarily vertically transmitted with their host. The evolution of a specific relationship between the amoebas and the bacteria could be demonstrated by reciprocal transplant experiments: bacteria extracted from the perennially infected host lines were pathogenic in naive hosts, and unevolved bacteria were pathogenic to evolved hosts whose resident bacteria had been cleared out by antibiotics. Moreover, in some lines this relationship became obligate after about 100 host generations—the amoebas were no longer able to grow successfully without the bacteria. This shows how it is possible to reconstruct the initial stages of the evolution of highly modified bacterial endosymbionts such as chloroplasts and mitochondria in the laboratory. Such burgeoning relationships can be found in a range of extant protists, such as the bacterial community of *Pelomyxa*, or the photosynthetic entities of *Glaucocystophora*.

11.2.3 Experimental evolution of cooperation in a bacterium-phage system

The best evidence for the evolution of cooperation through the selection of vertically transmitted combinations of lineages comes from an elegant experiment by Bull *et al.* (1991), who used a filamentous phage that infects *E. coli*. Filamentous phages are relatively benign parasites whose progeny can pass out through the host cell envelope without disrupting it. The bacterium thus continues to divide—distributing phage to both daughters—although at a diminished rate, because of the metabolic burden of maintaining 100 or so replicating copies of the phage. Infected cells are immune to further infection. This is important, because if the phage has no monopoly on a host, it will have less interest in maintaining host viability. The phage used by Bull and his colleagues had additional DNA inserted into the region between its two origins of replication; this DNA encoded antibiotic resistance, and was used as a selectable marker to ensure that all bacteria carried copies of the original phage or one of its evolved derivatives. Selection for 'cooperative' phage—phage genotypes that interfered less with the reproduction of their hosts—was imposed simply by serial transfer in medium containing antibiotics. This ensured a high degree of fidelity between partners (as Bull expresses it), because all surviving cells carried the phage and were thus immune from further infection; thus, genotypes able to infect new hosts would gain no advantage, whereas types that enhanced the reproduction of their hosts (relative to the usual deleterious effect of infection) would increase in frequency through the increased frequency of the host lineages that bore them. Selection was applied in the opposite direction by collecting free phage from the culture, and inoculating them into a fresh culture of uninfected bacteria. This enforced low partner fidelity by separating phage and host lineages, selecting for infectious phage that should evolve so as to maximize the production of infectious progeny, regardless of their deleterious effect on host reproduction. The high-fidelity and low-fidelity lines were propagated for 15 growth cycles, each comprising about 10 bacterial generations.

The hypothesis that cooperation is favoured by the permanent association of lineages makes several predictions about the consequences of mixing cells (and the phage they contain) from the high-fidelity and low-fidelity lines. If the mixture is simply inoculated into fresh medium, the high-fidelity phage should increase in frequency, by virtue of the higher rate of growth of bacteria whose phage are cooperative. The same thing should happen if the culture also includes a large number of other bacteria that are resistant to the phage; but if there is a large proportion of uninfected susceptible bacteria in the culture, then the low-fidelity phage, with their greater infective ability, will be the more successful, and might even spread. All of these predictions were confirmed. Either the bacteria or the phage, or both, had evolved so that bacterial lines maintained in close association with their resident phage were no longer harmed as severely by infection. By alternating episodes of vertical or horizontal transmission it can be shown that the level of virulence increases with the probability of horizontal transmission (Messenger *et al.* 1999).

The simplicity of the phage genome (7000 nucleotides specifying 10 genes) made it feasible to investigate in detail the genetic changes associated with the evolution of cooperative behaviour. These turned out to be of great interest, because different changes occurred in different selection lines. In one case, there were loss-of-function mutations that created non-infectious phage; these were infrequent because most mutations that knock out phage genes are lethal to the host cell. In other cases, the phage genome became integrated with the host genome. This presumably enhanced host reproduction by reducing the number of phage copies from 100 or so down to 2 or 3. Finally, several lines evolved highly defective phage-like plasmids from which almost all the phage genome had been eliminated: these consisted of the two phage origins of replication, the antibiotic resistance factor, and a short length of DNA presumably derived from the original phage genome. These were incapable of autonomous self-replication, depending on a functional integrated copy of the phage. Almost as notable as the divers adaptations that did evolve were those that did not. For example, replication-deficient phage did not appear, although they might have been expected, perhaps because there is no mechanism to regulate the distribution of phage between daughter cells, so that lines with few phage copies would continually segregate lineages that lacked phage altogether. Social evolution in this system is thus highly contingent, leading to similar phenotypes by quite different genetic routes.

11.2.4 Co-evolution of bacteria and plasmids

Plasmids are more highly integrated than phage with their bacterial hosts. Conjugative and mobilizable plasmids can spread through autoselection despite their deleterious effect on the growth of their hosts (§9.1.1). Some plasmids encode functions beneficial (in certain environments) to their host, such as antibiotic resistance, and will be selected for this reason. Others, however, confer no distinctive phenotype, cannot be transferred from cell to cell, and are mildly deleterious to their host. In such cases, a more cooperative relationship should evolve. Bouma and Lenski (1988) studied a 4-kb plasmid of *E. coli* that encodes multiple antibiotic resistance but reduces bacterial growth-rate in the absence of antibiotics. A cell line into which the plasmid had been inserted was cultured for about 500 generations in the presence of antibiotics; this ensured the continued association of bacterial and plasmid lineages, because the plasmid cannot be transmitted from cell to cell, and any cells that had lost the plasmid would be killed by the antibiotic. At the end of the experiment, the mutual adaptation of bacterial and plasmid genomes was measured by constructing all combinations of evolved and unevolved bacteria and plasmids. This showed that the bacterial genome had been modified through selection for types in which the deleterious effect of the plasmid was reduced; no modification of the plasmid could be detected. More surprisingly, the evolved bacterial strain outcompeted the original plasmid-free strain, even in growth medium containing no antibiotics. This new cooperative relationship had been established through a unilateral modification of the host genome, because the same advantage was realized by the evolved bacterial strain when transformed with the original, unmodified plasmid.

A similar experiment was conducted by Modi and Adams (1991), who cultured a population founded by a single plasmid-infected clone for about 800 generations in a chemostat (Figure 11.2). They did not select against cured cells, which were generated by unequal segregation and increased in frequency throughout the experiment because of their higher rate of growth. The decline in the frequency of infected cells was very erratic: mutations occurring either in the bacterial or in the plasmid genome that enhance bacterial growth-rate occasionally spread through the infected sector of the population, causing it to increase briefly in frequency. The fitness of the evolved population—under the conditions of glucose-limited continuous culture—was greatly increased through the periodic selection of the beneficial mutations whose passage was marked in this way; the fitness of the original strain, with its original plasmid, was only one-tenth to one-half that of the evolved strain, with its evolved plasmid, when the two are put into competition with one another. By comparing cured lines from the original and the evolved strains, it could be shown that much of this advance was caused by the modification of the bacterial genome, as would be expected. However, exchanging plasmids between lines showed that it was also attributable in part to changes in the plasmid, and to altered interactions between bacterial and plasmid genomes. The adaptation of plasmid-infected lineages to a novel environment thus involves some degree of co-evolution towards a more cooperative relationship. The changes in the plasmid genome that contributed to this are not known precisely, but resulted in fewer copies per cell, presumably through reduced rates of replication.

11.2.5 Quora and consortia

The extent of evolved partnerships among different kinds of organisms and the way in which they influence the structure of communities remain poorly understood. An interesting line of approach is suggested by the experiments of Senior *et al.* (1976) on the community of microbes that develops in soils treated with the herbicide Dalapon, a chlorinated propionic acid. When soil samples are cultured in chemostats with Dalapon as the only source of carbon and energy, they quickly and consistently give rise to a characteristic community of seven different microbes (Figure 11.3). Three of these are able to grow as pure cultures in Dalapon medium: a species of *Pseudomonas,* an unidentified Gram-negative bacterium, and the filamentous fungus *Trichoderma*. The other four members of the community were *Pseudomonas putida,* an unidentified pseudomonad, a *Flavobacterium,* and a budding yeast. They were unable to grow on Dalapon as pure cultures, but formed a conspicuous part of the Dalapon community: they presumably fed on metabolites excreted by the primary Dalapon users. The composition of the community changes somewhat with conditions of culture: when the dilution rate rises above 0.2 per hour the yeast drops out, and if it exceeds 0.45 per hour *Pseudomonas* displaces the Gram-negative bacterium as the most abundant primary user. Nevertheless, the community is remarkably stable, showing no qualitative change in composition over several thousand generations of continuous culture. This represents a sorting experiment that selects a group of six or seven interacting species from the teeming microbial community of the soil. Unfortunately, the nature of their interactions has not yet been ascertained. The primary users are clearly supporting the secondary users, but whether the secondary community enhances the growth of the primary users by removing metabolic wastes, and what the relationships among the species in either community are, remain unknown. Interestingly, the experiments also showed that the community can continue to evolve: a strain of *Pseudomonas putida* with a modified dehalogenase able to hydrolyse Dalapon to pyruvic acid appeared and spread after a few hundred generations, adding a new component to the community of primary users.

Whole model communities have also been the target of selection. Swenson *et al.* (2000a, b) added

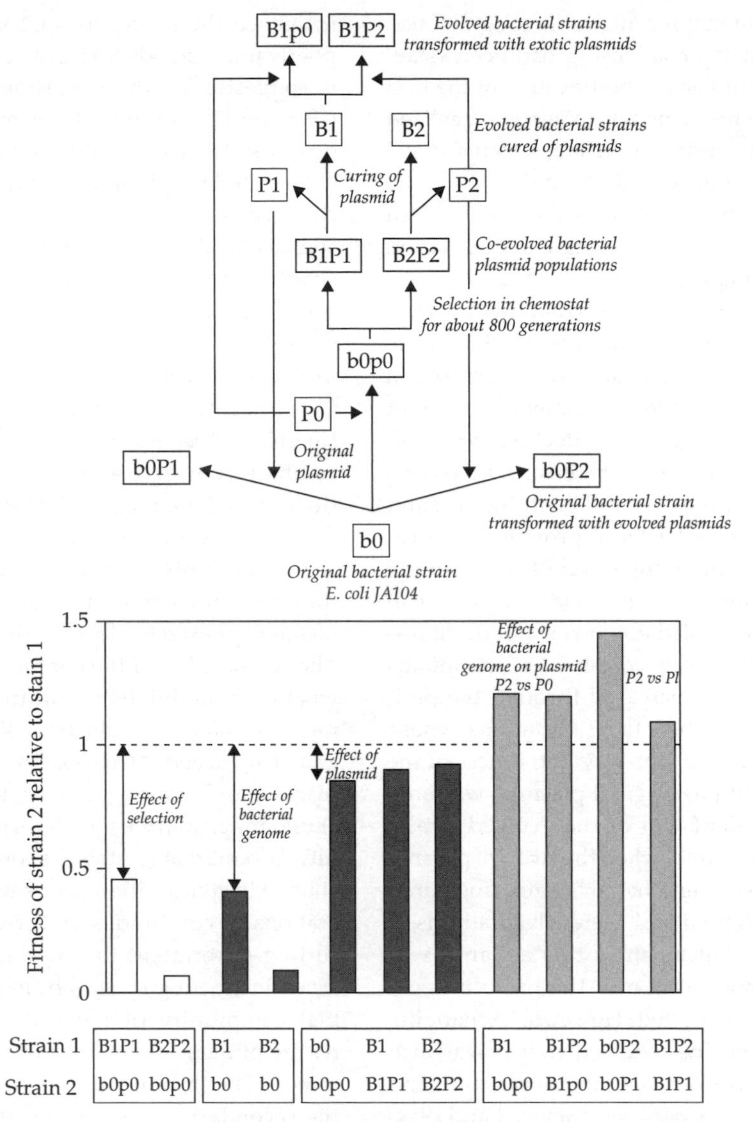

Figure 11.2 Experimental evolution of a bacteria-plasmid system. The upper diagram shows how the experiment by Modi and Adams (1991) was conducted. The original bacterial and plasmid strains are designated b0 and p0 respectively, and the evolved strains B1, B2 and P1,P2. When different combinations of bacteria and plasmids were constructed, the main results shown in the lower diagram were as follows. (i) Selection causes adaptation to chemostat conditions (e.g. B1P1 vs b0p0). (ii) This adaptation is largely attributable to changes in the bacterial genome (e.g. B1 vs b0). (iii) The plasmid is deleterious, but its effect on the bacterium is less in the evolved lines (e.g. B1P1 vs B1, compared with b0p0 vs b0). (iv) The effect of the plasmid depends on the bacterial genome with which it is associated (the effect of p0 vs P2 is the same in b0 and B1, but P2 is more deleterious in b0 than in B1).

samples of forest soil to cultures of *Arabidopsis* and selected the soils that gave the greatest plant growth. The selected soils were then re-sown with stock seed (so that the plants could not evolve) and the selection repeated. For about 40 generations there were large and erratic fluctuations in plant yield, caused by agents such as fungal pathogens, which are difficult to interpret in terms of group

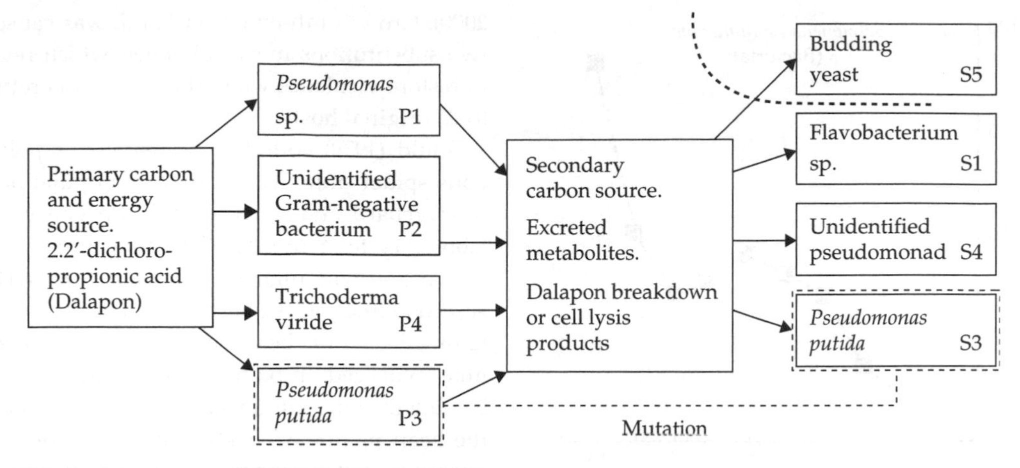

Figure 11.3 The Dalapon community. From Senior *et al.* (1976).

selection or any other process. A similar but simpler experiment selected replicate samples of pond water for the ability to degrade 3-chloroaniline. In this case there may have been some response, although there were too few replicates to be sure. Since in both cases (soil and water) the environmental samples were taken from the same site, it is not clear where the variation among communities would have come from. The explanation proffered, citing the mysterious properties of complex systems, is not fully satisfying, but it is possible that relatively rare components of the original community became extinct stochastically early in the history of the cultures. More highly replicated experiments with greater initial variability might well be successful, and so provide a valuable technique for enhancing the effectiveness of bioremediation.

Quorum sensing, microbial consortia, and the properties of biofilms are currently very fashionable research topics. To what extent natural communities of microbes represent evolved mutualistic entities, rather than simply growing in the same place, is not yet clearly established. Some of the reports in the literature have been rather unguardedly enthusiastic, and occasionally seem to be hinting at superorganisms. More precise measurements of fitness will be required before a primarily mutualistic interpretation of microbial communities is tenable.

11.3 Enemies

11.3.1 Serial passage

The evolution of virulence can be studied by infecting individuals with pathogens, harvesting the pathogens after growth and then re-inoculating new host individuals taken from the same non-evolving ancestral stock. This continued serial passage is a special kind of selection experiment that has often been used in medical research, for example in the development of vaccines. The outcome of serial passage experiments has been reviewed from an evolutionary point of view by Ebert (1998), who found that serial transfer in a novel host is almost invariably accompanied by an increase in virulence, which often increases very rapidly to high levels. Ten passages of *Salmonella typhimurium* through naive mice, for example, increased mortality from less than 10% to more than 90% (Figure 11.4). The host's body provides a huge stock of resources which the pathogen is at first poorly adapted to exploit; types that exploit it more effectively will be selected, and increased virulence—damage to the host—is the consequence of their spread. Very high levels of virulence can evolve because the parasites are transmitted by the experimenter and are thus indifferent to the health of their host.

Effective transmission between hosts may constrain growth within the host, for example because it requires protective capsules or the production of

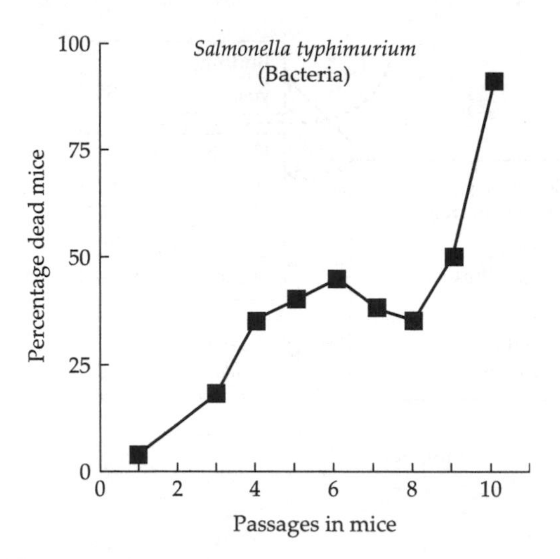

Figure 11.4 Mortality after passages of *Salmonella typhimurium* through naive mice. From Ebert (1998)

specialized infective stages. This will temper the advantage of high growth rates because it may be advantageous to switch to the transmission mode at low density, rather than continuing to reproduce, if the host is likely to die. In serial passage experiments these stages are unnecessary and will tend to be lost as a correlated response to selection for increased growth. Like Qβ in culture tubes, serially passaged pathogens may lose the ability to persist as natural infections.

11.3.2 Genetic specificity

Serially passaged viruses and microbes also almost invariably lose the ability to grow in their ancestral host (Ebert 1998 Table 1), proving that the infective ability of a pathogen is rather highly specific to a particular host species. The outcome from the point of view of the ancestral host is then an attenuated pathogen (such as the Sabin polio vaccine) whose host range has shifted. The attenuation of a canine parvovirus grown in dog kidney cells for over 100 passages was associated with the fixation of 16 mutations (Badgett *et al.* 2002). The correlated response may not be symmetrical: culturing phage ϕ174 on *Salmonella* reduced its ability to grow on *E. coli*, its natural host, whereas adaptation to *E. coli* did not affect growth on *Salmonella* (Crill *et al.* 2000). Growth inhibition on *E. coli* was caused by two substitutions in capsid genes, which reverted to restore virulence when the phage was returned to its original host.

Gould (1979) collected *Tetranychus*, a polyphagous spider mite, from a natural population in a pear orchard, and reared them on lima beans in the laboratory for 8 months. The mites grow well on lima beans, with high fecundity and low mortality. A second line was then established in a cabinet containing both lima beans and cucumber plants. The mites will feed on both the beans and the cucumber, but juvenile mites usually die on cucumber, so the bean plants are destroyed more rapidly. The remaining mites must then feed on cucumber alone until transferred to a new cabinet. After nearly 2 years of selection, the mites had evolved much higher survival rates on cucumber. The evolution of new sources of virulence can therefore occur quite rapidly, within less than 20 generations. The evolved populations were also tested on exotic hosts that are normally resistant to the mites. The correlated response to selection on cucumber included an increased survival rate on potato and tobacco, although not on plantain. Adaptation to one antagonist was thus not completely specific, and may have manifold consequences for interaction with others. Fry (1990) selected *Tetranychus* for 10 generations on tomato and lima bean and found that the bean-adapted lines became less well-adapted to tomato, whereas tomato-adapted lines showed no loss of fitness on bean plants. Agrawal (2000) found that adaptation of *Tetranychus* to a low-quality host (cotton) did not lead to any loss of fitness on a high-quality host (cucumber), although the ability to grow on the low-quality host quickly attenuated after transfer to the high-quality host.

Hosts may also be selected for resistance to particular pathogens, although this kind of experiment is not as often attempted. *Drosophila* lines selected for resistance to particular species of parasitoid wasps evolved an increased ability to encapsulate their eggs, but the correlated response to other species was inconsistent (Fellowes *et al.* 1999). Passage of the nematode *Howardula* on species of *Drosophila* failed to detect any degree of specific adaptation (Jaenike and Dombeck 1998).

Likewise, selection of snails for resistance to *Schistosoma* showed no specific response, although only two strains of host and parasite were used (Webster and Woolhouse 1998).

11.3.3 Anisomorphic social matrices

The anisomorphic social matrices of Figure 11.1 assume that the phenotypes of host and pathogen were:

		Pathogen	
		P1	P2
Host	H1	resistant/ avirulent	susceptible/ virulent
	H2	susceptible/ virulent	resistant/ avirulent

—or the converse. This is an example of matching-alleles specificity: allele H1 of the host is resistant to allele P1 of the pathogen, and likewise H2 resistant to P2. This kind of specificity will tend to generate perpetual oscillations H1P1 → H1P2 → H2P2 → H2P1 → H1P1. The alternative is either:

		Pathogen	
		P1	P2
Host	H1	susceptible/ virulent	susceptible/ virulent
	H2	resistant/ avirulent	susceptible/ virulent

or:

		Pathogen	
		P1	P2
Host	H1	susceptible/ virulent	resistant/ avirulent
	H2	resistant/ avirulent	resistant/ avirulent

This constitutes gene-for-gene specificity. In the left-hand scheme, the virulent allele P2 infects all hosts, and consequently the system moves H1P1 → H2P1 → H2P2 and then stops, with the resistant allele fixed in the host population and the virulence allele fixed in the pathogen population, a permanent genetic stand-off (Parker 1994). If there is a cost of resistance ($b > d$), however, then H2P2 → H1P2, and if there is also a cost of virulence ($a > c$) then H1P2 → H1P1. The same conclusions apply to the right-hand matrix, with a host resistant to all pathogens. Gene-for-gene interaction is expected to drive oscillations of gene frequency only if there are substantial costs of both resistance and virulence.

Matching-allele mechanisms resemble somatic incompatibility systems in fungi and animals (discussed below) but there is little evidence that they are involved in plant or animal disease. Thompson and Burdon (1992) reviewed 40 plant–pathogen interactions and found that all were consistent with the gene-for-gene mechanism, with detailed genetic support in about a dozen cases. Parker (1996) came to the same conclusion and pointed out that resistance was typically mediated by pairs of alleles at several loci. There is relatively little work on animals; specific interactions have been reported (for example, Carius *et al.* 2001), but it is not clear how general they are, and the genetic details are poorly understood. T-series phage infect bacteria through a specific interaction between the phage tail fibres and a receptor on the bacterial cell surface (reviewed by Bohannan and Lenski 2000). Bacteria evolve resistance by modifying or deleting the receptor. If the receptor is modified then the phage can evolve a new specificity for the modified receptor, and in principle this establishes a matching-allele mechanism. In practice an invulnerable bacterial strain that lacks the receptor appears sooner or later, so that the system reverts to the lower gene-for-gene mechanism sketched above.

11.3.4 The cost of virulence and resistance

Antagonistic co-evolution is constrained in the usual way by costs of adaptation, arising either through functional interference or through mutational degradation (§8.3.4) (reviewed by Coustau *et al.* 2000). The enfeeblement of resistant genotypes in uninfested environments has been studied carefully in two contexts, the development of crops resistant to disease (discussed below) and the relationship between bacteria and phage (reviewed by Bohannan and Lenski 2000). The cost of phage resistance in bacteria has been documented by Lenski (1988a, b) in *E. coli* and its lytic phage T4, which consumes all available cellular resources and then kills the cell when the progeny phage are released. *E. coli* has two means of preventing phage from entering the cell. The first is to develop a mucoid capsule that prevents phage from gaining access to the cell envelope. Lenski studied a second mechanism of resistance, involving mutations that alter the lipopolysaccharide core of the cell envelope. There are three loci, or closely linked sets of loci, that encode the structure of the core. A series of independent isolates resistant to T4 have mutations at one or another of these loci: there are thus several different routes to resistance. All were associated with reduced fitness when grown in competition with susceptible bacteria, in media where no phage were present, demonstrating a primary cost of resistance. The main point of Lenski's experiment was to investigate how this cost might be modified through selection. There are two possibilities. In the first place, the different isolates did not all express the same cost of resistance. There were two groups, probably representing mutations at different loci. One was greatly enfeebled in the absence of phage, with a fitness, relative to susceptible strains, of only about 0.55; the other was less severely handicapped. with a relative fitness of about 0.85. Provided that these fitnesses are transitive, selection will favour the isolates whose resistance was less costly, thereby reducing the average cost of resistance. The second possibility is that the cost might be further reduced by mutations at other loci that compensated for the physiological defects associated with resistance. This possibility was investigated by growing pure cultures of susceptible and resistant strains by serial transfer in batch culture for about 400 generations. Selection during this period increased the fitness of the susceptible strains by about 10%, the result of adaptation to the novel conditions of culture; the fitness of the evolved susceptible strains relative to that of the initial clones was thus about 1.10. The relative fitness of the resistant strains, however, increased during the same period from 0.66 to 1.03. These evolved resistant strains had thus restored fitness to a level comparable with that of the base population, and were little inferior to the evolved susceptible strains. This advance was not caused or accompanied by a reversion to sensitivity; the evolved lines were all still fully resistant to T4. It must instead have been caused by selection for genes that ameliorated the initially deleterious side-effects of the resistant phenotype.

If this experiment were tried with types that secrete a mucoid capsule to hinder infection by the virus, it would be anticipated that selection in a permissive environment, with no virus present, would cause an increase in fitness through selection for the loss of capsule synthesis. The surprising feature of Lenski's experiment is that relative fitness increased without any concomitant loss of resistance. However, mutations that modify the cell envelope have no substantial economic effects on the cell; it is probably no more expensive to make a defective envelope than it is to make a normal one. Selection can therefore favour mutations that compensate in some way for this defect, without affecting the resistance that it confers. Moreover, because resistance is in this case associated with the partial loss of normal function, it will not be degraded through time in permissive environments by mutation accumulation, in the way that a gain-of-function mutation would be. Thus, selection may exacerbate or ameliorate the side-effects of disease resistance, depending on the genetic and economic consequences of a particular resistance mechanism.

The persistence of susceptibility in natural populations of plants suggests that disease resistance is generally costly. A broad review of resistance to herbicides, pathogens, and herbivores in crop plants, however, showed that it was costly in only about 50% of cases (Bergelson and Purrington 1996). There are several reasons for not taking this surprising conclusion at face value (as the authors acknowledge). First, induced defences are usually found to be costly (Cipollini *et al.* 2003). Secondly, the expression of a cost may vary with conditions of growth. Thirdly, the phenotype associated with a genetic source of resistance can be evaluated accurately only when genetic background is controlled, for example by backcrossing. Bergelson *et al.* (1996) transformed *Arabidopsis* with an acetolactase synthase allele conferring resistance to the herbicide chlorsulfuron, backcrossed the transformants for two generations to the standard Columbia strain, selfed these plants and isolated resistant and null homozygotes. Lifetime seed production of the resistant transgenics was 34% less than that of comparable susceptible plants, revealing a major cost levied by this resistant allele. The costs of herbicide resistance, however, may generally be greater than those associated with resistance to natural enemies. The best-studied resistance locus is *Mlo* of barley, which represses the apoptotic response to attack by pathogens such as mildew (reviewed by Brown 2002). The loss-of-function allele *mlo* is resistant to mildew but as a side effect suffers spontaneous necrotic lesions on the leaf, which reduce grain yield by about 5%. It is also more susceptible to other fungal pathogens, which proliferate in the dead tissue, so that strains resistant to mildew are often highly susceptible to blast. Recent reviews have re-instated the traditional view, that resistance to pathogens or herbivores usually exacts a cost, often a substantial one, which varies with physical and biotic conditions of growth (see Strauss *et al.* 2002). *Chlorella* cultures grazed by the rotifer *Brachionus* evolved unpalatability at the expense of reduced competitive ability in the absence of the predator (Yoshida *et al.* 2003).

Crops that are grown in novel environments where an otherwise endemic pathogen does not occur should, therefore, often lose their resistance, although curiously enough there seem to be few careful quantitative studies of this phenomenon, and no deliberate selection experiments, despite its obvious importance. On the other hand, there is a larger literature on the cost of virulence in pathogens. Leonard (1969) collected sexual spores of oat stem rust from barberry (the alternate host), and then recycled asexual generations on single oat cultivars for eight generations, scoring pathogenicity to a different set of test cultivars in each generation by inoculating them with spores from the experimental plants. He found that virulence against the test cultivars decreased through time, presumably through selection against unnecessary virulence in populations evolving on a single host cultivar. The selection coefficients involved were quite large (of the order of 10%), so the evolution of virulence directed specifically against one cultivar may result in the rapid loss of virulence towards others.

The cost of resistance limits the degree to which hosts can become adapted to the local community of pathogens; likewise, the cost of virulence limits the degree of adaptedness of the pathogen. This is why most host individuals are susceptible to most pathogen genotypes, and most pathogen individuals avirulent on most host genotypes. The range of adaptation of the scald fungus, *Rhynchosporium*, was estimated by inoculating a test series of 14 barley cultivars. This defined 75 races by the different combinations of cultivars they could attack; only one of the 75 was virulent on all 14 hosts. The cultivars in turn varied in the extent of their resistance; the most susceptible resisted 21, the most resistant 56 of the fungal genotypes. Now, suppose that a heterogeneous barley population such as Composite Cross II is propagated in the presence of the native population of *Rhynchosporium;* what breadth of adaptation will evolve? Fungal strains that have a narrow host range are rather abundant, and will select for resistance among the genotypes susceptible to them. Strains with a very broad host range are rare (because multiple virulence is costly to maintain), and the weak selection they impose is unlikely to cause hosts with a broad range of pathogen resistance to evolve (because multiple resistance is costly). Broadly speaking, the barley population behaved as expected: resistance to simple races of fungi, with narrow host ranges (as gauged by the 14 test cultivars), evolved rapidly, whereas plants remained susceptible to fungal strains with broad host ranges (Figure 11.5).

11.3.5 Evolution of virulence

The classical interpretation of virulence is that it is a primitive feature of host–pathogen interactions. When a naive host population first encounters a pathogen it has no evolved defences and rapidly succumbs. This is consistent with serial passage experiments and with the lethality of introduced pathogens such as *Ophiostoma* (Dutch elm disease) and *Cryphonectria* (chestnut blight). Unless dead hosts transmit pathogens, this will favour less-virulent strains of the pathogen that farm their hosts and thereby increase the number of propagules they generate to infect new hosts. In highly

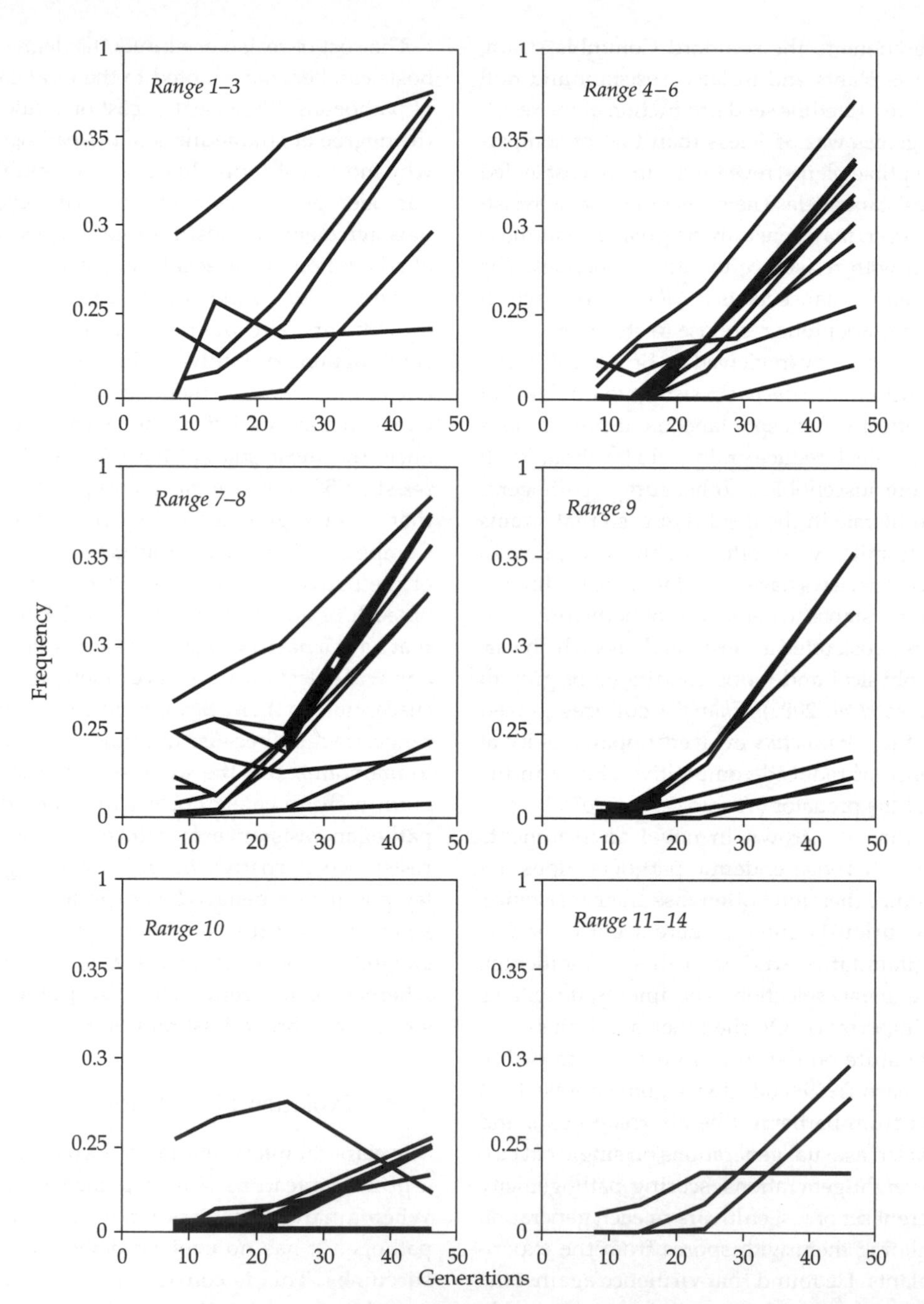

Figure 11.5 Evolution of resistance in barley Composite Cross II. The *y*-axis of each graph is the frequency of individuals in the barley population that are resistant to strains of the scald fungus, *Rhynchosporium secalis*, whose breadth of adaptation is assessed as the number of a series of 14 test cultivars they are able to exploit. The *x*-axis is the number of generations of propagation, the experiment being scored in generations 8, 14, 24, and 46. The simplest strains can attack only a few cultivars (range 1–3 or 4–6, meaning virulent on 1–3 or 4–6 of the test cultivars). A higher level of resistance to almost all these strains evolves in the barley population, with resistance in many cases increasing from very low frequencies to dominate the population. The more complex strains can attack many of the test cultivars. Resistance to these strains evolves much more slowly, and in many cases does not evolve at all; in the highest category (range 11–14) resistance was demonstrated for only three of the nine strains. From Webster *et al.* (1986).

evolved relationships, pathogens may proliferate within hosts while causing only mild symptoms of disease. This school of thought was overturned at the 'Enlightenment' by the theoretical demonstration that, given reasonable assumptions, pathogens will often evolve intermediate rather than minimal levels of virulence (Anderson and May 1982; for historical background see Levin 1996). Suppose that a few pathogen individuals are introduced into a population of N susceptible hosts, each infecting a different host individual. If α is the rate of host mortality caused by the pathogen, m the rate of mortality from other sources, and v the rate of clearance of the pathogen by the host, the time available for the pathogen to transmit propagules is $T = 1/(\alpha + m + v)$. During this period, propagules are produced at a rate λ and each propagule has probability B of successfully infecting any given individual in the host population, so that the number of successful propagules produced by a single pathogen is λBNT. Other things being equal, this will favour pathogen strains that reproduce rapidly or maintain dense populations within the host (λ large) and cause highly infectious (B large) incurable (v small) diseases with mild symptoms (α small), as the classical hypothesis predicts. It seems reasonable to suppose, however, that the growth of a pathogen is inherently damaging to its host, that is, that λ and α are positively correlated, in which case λ and T are negatively correlated. A benign pathogen will be poorly adapted because it proliferates slowly, whereas a malign pathogen will be equally poorly adapted because it proliferates for only a short period of time; the optimal strategy will often be an intermediate rate of growth and thus an intermediate level of virulence (Levin and Pimentel 1981, Antia *et al.* 1994).

The microsporidian *Glugoides* is an intracellular parasite of the cladoceran *Daphnia* whose spores are acquired by feeding and after reproduction within the host pass out from the gut to infect new individuals. It can have a severe effect on its host, and Ebert (1994) found that a larger spore load was associated with a greater reduction of host reproduction. He also cultured neonates with infected females and found that infection of the neonate increases with the spore load of the female. Similarly, there is a positive genetic correlation between parasite load and virulence in *Plasmodium* infections of mice, and transmission to mosquitoes increases with parasite load (Mackinnon and Read 1999a). Soon after the introduction of highly virulent myxoma virus into rabbit populations the level of virulence declines, but it becomes even more attenuated by serial passage experiments, suggesting that intermediate levels of virulence are optimal in the wild (Fenner and Ratcliffe 1965). Laboratory experiments showed that fleas transferred the virus from only 12% of rabbits inoculated with fully virulent strains (because they died in a few days) and from only 8% of those inoculated with the most attenuated strains, whereas the infection rate was 42% from rabbits carrying moderately virulent strains (Mead-Briggs and Vaughan 1975).

When two or more strains of a pathogen infect the same host individual they will compete for host resources, and other things being equal the most rapidly growing strain will be the most successful. We have seen that parasite strains that severely damage their hosts may limit their own transmission, for example if transmission depends on host motility or longevity. The profit from a high growth rate is realized by the fast-growing strain itself, however, whereas the penalty is paid by all strains within the host. Hence, higher levels of virulence should evolve when hosts are infected simultaneously by several strains (Nowak and May 1994). When vesicular stomatitis virus (VSV) is passaged in hamster kidney cells for about 100 generations, single infection caused a 5% increase in fitness whereas co-infection caused 15–30% increase in fitness (Miralles *et al.* 2000, Carrillo *et al.* 2007). Mackinnon and Read (1999b) selected *Plasmodium* lines for increased and reduced virulence in mice, but found that the low lines increased in virulence about as much as the high lines. They concluded that within-host natural selection for rapid growth, created by mutation in a novel host species, was powerful enough to over-ride countervailing artificial selection.

The effect of co-infection will be mitigated by selection among hosts. Cooper *et al.* (2002) selected the polyhedrosis virus of gypsy moth (*Lymantria dispar*) for early and late transmission. Early transmission favours rapid growth, and after nine passages the early-transmission lines were more virulent, as expected. The late-transmission lines produced more progeny in total, however, so within-host selection would be opposed by among-host selection. Selecting small plaques of VSV for transmission causes a fall in mean fitness and virulence (Miralles *et al.* 1997). The evolution of virulence under individual (within-host) selection may thus be restrained by group (among-host) selection, and the classical view is in part rescued.

11.3.6 Epidemics

One consequence of pathogen evolution is that genetically uniform host populations are vulnerable to epidemic disease. The likelihood of an epidemic depends on the size and the rate of turnover of the host population and its pathogens. Any uniform population of hosts will cause the selection of the pathogen genotypes best able to exploit the particular host type. If the host population is very small, this selection will be inappreciable; any tendency for a specific virulence to be selected will be opposed by its cost. But if the host population represents a large resource, relative to other available hosts, then selection among the pathogens for specific virulence will be very strong. If the pathogen population is in turn very small, or reproduces very slowly, this selection will again be ineffective; if specific virulence were to evolve in the pathogen, specific resistance would arise more rapidly and spread more quickly in the host. However, if the pathogen population is very large, and reproduces very rapidly, then virulent types will arise quickly, and spread through the pathogen population before the host population is able to respond to selection for the new source of resistance now required. The result is a devastating epidemic that may destroy the host population completely.

The scenario of a very large uniform host population, and an even larger and more rapidly cycling pathogen population, is supplied by many crop plants and some breeds of livestock. Cereal crops, in particular, are often grown as populations of hundreds of millions of individuals of the same cultivar. One of the main preoccupations of agronomists is the essentially evolutionary problem of how to prevent epidemics in such extensive pure stands; I shall illustrate some of the issues involved with barley and its fungal pathogen *Erisyphe*, powdery mildew. Powdery mildew was reported as a minor pest in the nineteenth century, but the first epidemics occurred in the early years of the twentieth century. Its rise coincided with the changeover from genetically heterogeneous landraces of barley to the extensive monocultures of single varieties that characterizes modern agriculture. By the 1930s, many different sources of virulence had been identified in the fungus, and almost all the commercial varieties then available were susceptible to one or another of them. One variety, however, possessed a source of resistance to all known strains of *Erisyphe* that turned out to be attributable to a single major gene, and was therefore readily transferred to other high-yielding cultivars. These first began to be planted extensively in the late 1940s; 10 years later, they occupied hundreds of thousands of hectares, including 70% of the barley area in Germany. By this time, however, all were susceptible to some strains of mildew, and shortly afterwards they were withdrawn from commercial use. This pattern of a highly resistant variety selecting virulent strains of pathogen when planted over very large areas has become familiar in barley and other grain crops as the single most damaging side effect of monoculture farming. The conventional response has been to engage in an arms race with the pathogen by selecting for resistance to current pathogen strains. However, the resources of the pathogen are so great—10 generations for every barley generation, and a million spores for every barley plant—that sequential selection for resistance on the experimental farm may be too slow to counter selection for virulence in the commercial fields. An alternative approach is to check the evolution of virulence by incorporating several different sources of resistance into the barley population.

Epidemics seldom occur in natural populations. The ancestral populations of *Hordeum spontaneum* from which cultivated barley was selected still grow wild in the Middle East, in the company of other grasses and their herbivores and pathogens, including *Erisyphe*. A large proportion of plants in these populations is resistant to heterogeneous bulk inoculates of *Erisyphe*, and must bear many different sources of resistance, encoded by different genes. On the other hand, an equally large proportion is susceptible to at least some of the *Erisyphe* strains in the inoculum. The host population is thus a mixture of resistant and susceptible plants, although whether this mixture is stable, or fluctuates in composition over time, is not known. Likewise, the populations of *Erisyphe* infesting these wild grasses have a wide host range that matches the broad resistance shown by their hosts. In these conditions, disease is endemic, and most plants—perhaps nearly all—show some symptoms of disease, but epidemics do not seem to occur.

The interdependence of host and pathogen variation can be studied in cultivated populations. The cultivars Hassan, Midas, and Wing incorporate three different sources of resistance to powdery mildew, each of which opposes a specific source of virulence in the fungus. If the cultivars are grown as pure stands, the pathogen populations that infest them are dominated by the corresponding simple races, each with a single source of virulence. In binary mixtures of Hassan and Wing, mildew strains able to grow on both components are abundant early in the season, although later they tend to be replaced by the two simple races, each able to grow on only one component of the mixture. In the three-way mixture, strains of *Erisyphe* with all three sources of virulence become more abundant. The host range of the mildew thus increases with the diversity of the barley. The effect is rather slight, however: strains with all three sources of virulence, for example, increase in frequency from about 5% in pure stands to about 12% in the three-way mixture (Chin and Wolfe 1984).

These observations suggest that genetically diverse populations of crop plants could be used to mitigate the effects of disease and reduce the incidence of epidemics. There are three ways in which this could be done. The most obvious is to incorporate several different sources of resistance into a single line. The main problem with this approach is the cost of resistance: multiply resistant varieties are likely to have low yields in most years, which is why wild populations contain a substantial proportion of susceptible plants. A more sophisticated alternative, to which there is no close analogue in natural populations, is to introduce different sources of resistance into nearly isogenic lines of a high-yielding strain. The result is a multiline, in which individuals vary in specific resistance, but are nearly identical in yield when virulent races of the pathogen are absent. The main drawback of developing multilines is that by the time the backcrossing programme has been completed, the high-yielding recurrent parent has been overtaken by even higher-yielding varieties. The third strategy overcomes this problem by simply using a mechanical mixture of current high-yielding selections into which several single sources of resistance have been introduced by a single generation of crossing.

Mixtures reduce the rate of spread of disease chiefly through the lower density of plants susceptible to particular strains of pathogen, so that fewer pathogen spores from any given lesion are likely to reach a susceptible host. Plants that are resistant to a particular strain act like flypaper, trapping fungal spores and thereby reducing the effective rate of spore dispersal. For example, in mixed stands of barley and wheat, the wheat is of course completely resistant to powdery mildew; the rate at which disease spreads in the barley is directly proportional to the percentage of barley in the mixture. Wolfe and Barrett (1981) conducted a large-scale trial of 47 mixtures of 25 barley cultivars representing 12 resistance phenotypes. They were unusually productive: 39 exceeded their component means, and 26 exceeded the highest-yielding component. The effect of mixture varied with the intensity of mildew infestation. In the least infested sites, mixtures yielded about 3% above their component means, which is consistent with general experience (§10.4). In heavily infested sites, however, the mixtures exceeded their component means by nearly 10%. Mixtures can thus be effective in suppressing disease and preventing epidemics. Moreover, the identification of specific sources of resistance provides a rational basis for the artificial selection of productive mixtures—though this has not yet been attempted. However, there have been two objections to the widespread utilization of varietal mixtures. The first is merely the legal difficulty of registering mixtures for commercial use. The second is the argument that presenting the pathogen population with several different sources of resistance simultaneously night result in the evolution of a 'super-race' capable of overcoming them all. Whether or not this is a very likely outcome of selection has been controversial. However, from an evolutionary standpoint, it is clearly preferable to challenge the pathogen population with several sources of resistance simultaneously than to deploy them sequentially.

11.3.7 Arms cycles and arms races

The correlated response to selection imposed by other organisms is a special case of the cost of adaptation to any kind of novel environment. In most cases, the evolution of specific resistance by hosts and virulence by parasites will be associated with lower relative fitness in environments where the particular pathogen or host involved does not occur. The social matrices defining the relationship between host and pathogen genotypes assuming gene-for-gene interaction and costly resistance and virulence can then be represented like this:

		Pathogen	
		Virulent	Avirulent
Host	Resistant	a	b
	Susceptible	c	d

		Host	
		Resistant	Susceptible
Pathogen	Virulent	a	b
	Avirulent	c	d

These matrices represent the interaction of a particular pair of host and pathogen genotypes, the 'virulent' pathogen type and the 'resistant' host type being defined relative to one another. The 'avirulent' and 'susceptible' categories, on the other hand, may be very heterogeneous mixtures of genotypes displaying varying degrees of resistance and virulence towards one another. A cost of resistance implies that $\mathrm{c} < \mathrm{a}$ and $\mathrm{b} < \mathrm{d}$; a cost of virulence implies that $a < c$ and $d < b$. This might be called the canonical form of the anisomorphic matrices that govern the co-evolution of antagonistic organisms. It will tend to cause sustained oscillations of genotype frequency in both populations.

In the late 1960s barley cultivars carrying a new gene *Mla12* for resistance to powdery mildew began to be used on a large scale, occupying about a quarter of the acreage by the end of the decade. At this point they began to lose their effectiveness as the corresponding virulence genes *Va12* spread in the fungal population, and were largely withdrawn from cultivation. Nevertheless, varieties bearing *Mla12* were still being used in breeding programmes, where they were necessarily selected in the presence of virulent populations of mildew. The outcome was the re-emergence of *Mla12*, combined in new ways with other sources of resistance. This new range of cultivars were planted extensively during the 1970s, occupying about a third of the acreage by 1977. However, their resistance to mildew was eroded year by year, until by the end of the decade it had been overtaken by the evolution of virulent strains of mildew. Naturally, these cultivars were in turn largely abandoned, and a third range developed. The coupled evolution of virulence in the pathogen and resistance in the host thus drives a cyclical process in which the response of fungal strains carrying *Va12* virulence first checks and then reverses the spread of barley varieties carrying *Mla12* resistance (Figure 11.6).

In a simple system of this sort, history continually repeats itself. However, there is no need to think that history ever precisely repeats itself, except in models. The expression of virulence or resistance is often epistatic, depending not only on the state of other loci with similar effects, but also on genes that normally play no part in host-pathogen interactions. At every turn of the cycle, continued directional selection on virulence in the pathogen (or resistance in the host) is likely to be a contingent process whose outcome depends in part on interacting genes that appear during the course of selection. Moreover, to the extent that pathogen virulence is a specific response to the particular mechanism of host resistance, and vice versa, the contingent outcome of selection in the pathogen will drive a contingent response by the host. At the level of overall virulence or resistance, host–pathogen systems may converge to regular oscillations, but

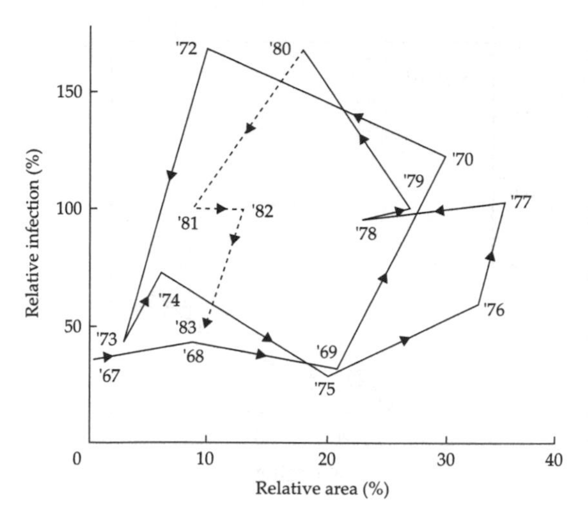

Figure 11.6 Coupled oscillations of barley cultivars carrying *Mla12* resistance and mildew strains carrying *Va12* virulence. From Wolfe (1987).

the underlying genetic and physiological mechanisms of resistance and virulence may nonetheless diverge rapidly. Indeed, this is what happens in the barley–mildew system, where *Mla12* is deliberately combined with other genes in order to overcome the existing sources of fungal virulence.

The barley–mildew system seems to be an isolated case, and I do not know of any other clear example of coupled genetic oscillations. The Composite Cross barley populations set up 40 or 50 years ago in California have been monitored at intervals for various agronomic characters, including disease resistance. They are attacked by a variety of pathogenic fungi, including *Rhynchosporium*, which causes scald (see §11.3.5). Both host and pathogen are variable: different barley genotypes are resistant to different races of the fungus. Most of the varieties used to set up the initial hybrid populations were highly susceptible to most or all races of scald. After 45 generations of propagation, however, the frequency of resistance to most races had increased substantially (Figure 11.7). Moreover, many plants bore genes at different loci conferring resistance to several races of scald. An originally rather susceptible population, then, that is exposed to pathogens under more or less natural conditions evolves higher levels of disease resistance, through selection sorting both the initial variation and the variation arising subsequently from recombination. The plants in turn drive evolution in the pathogen populations. Composite Crosses II and V are closed populations that have been grown on isolated plots for many years, and each has evolved a characteristic community of pathogens. The *Rhynchosporium* populations are genetically quite diverse, as assessed by enzyme electrophoresis, with three or four common genotypes making up the bulk of the population, and a great many less common types. The same two genotypes together made up about 60% of the fungal population on both barley populations. However, there was no concordance among the remaining genotypes, some of which occurred at frequencies of 10–20% on one population but were rare or absent on the other. Selection on host populations that differ in genetic composition thus gives rise to pathogen populations that differ in composition. The virulence of the *Rhynchosporium* genotypes was assessed by inoculating a standard series of barley cultivars. The most virulent genotypes, attacking the widest range of test cultivars, were more frequent in Composite Cross II, which is the more resistant of the two barley populations. Thus, the interaction between pathogen and host caused the coupled evolution of resistance and virulence, although they seem to accumulate rather than cycle.

Host–pathogen co-evolution will be detected only in long-term studies of systems whose genetics are accessible, and this may in large part explain the rarity of good examples in the literature. However, it is also possible that the abcd model of social matrices cannot be applied straightforwardly to most natural populations because of the assumption that the component populations each comprise a small set of defined genotypes, whose evolution occurs through a process of sorting. It is entirely possible that there is more often a continued selection of novel genotypes arising through mutation. The evolution of the system will still be driven by time-lagged frequency-dependent selection, but instead of the endless alternation of the same range of types, there will instead be an endless succession of different types, the outcome of an arms race between a population and its antagonists.

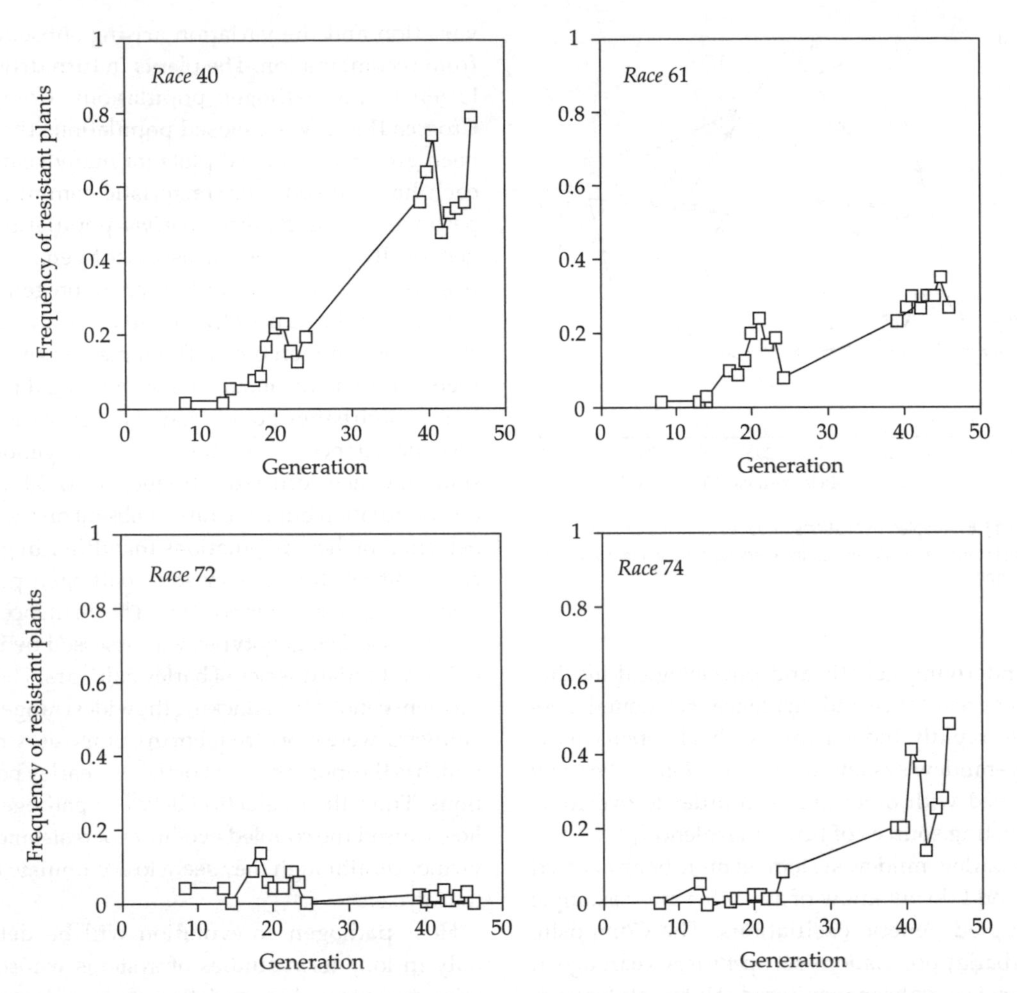

Figure 11.7 Evolution of resistance to races (strains) of scald, *Rhynchosporium secalis*, in barley Composite Cross II over 45 generations. From Webster *et al.* (1986).

11.3.8 Phage wars

Bacteria produce restriction enzymes that cut DNA molecules wherever a particular sequence of a half-dozen or so bases occurs; different enzymes recognize different sequences. They also, necessarily, produce modification enzymes that, in most circumstances, prevent their own DNA from being cut, usually by methylating a particular sequence of bases; different enzymes modify different sequences. The combination of a restriction enzyme and an appropriate modification enzyme constitutes a restriction-modification system that will destroy foreign DNA, such as invading viral genomes, without endangering the bacterium itself. It is not quite a foolproof system; occasionally, the viral DNA will be modified by the host enzyme, and the virus will then be able to reproduce. Its progeny inherit the modified DNA, and this virulent genotype can then spread through—and destroy—the bacterial population. Consequently, there should always be selection for new restriction-modification types that will provide immunity against phage so long as they remain rare. This would account for the diversity of bacterial restriction-modification systems. Some experiments by Korona and Levin

(1993), however, cast doubt on this simple interpretation of restriction-modification as a generally effective source of resistance to pathogens. They introduced an R+M+ strain bearing a functional system into liquid cultures of an isogenic R–M– *E. coli* population lacking restriction-modification function. In the absence of phage, R+M+ is nearly neutral, and continues to persist at low frequency. When phage is present, the R+M+ strain increases in frequency, especially if there are two or more types of phage in the culture. However, this advantage is transient; within a few generations, resistant genotypes with altered cell membranes that prevent phage adsorption appear, and subsequently spread through the bacterial population, so that after one or two transfers the population is almost completely resistant. Because these mutations are likely to occur in the more abundant R–M– genotype, R+M+ cannot spread when introduced at low frequency. However, experiments on solid medium, where each viable cell gives rise to a discrete colony, had a different outcome. The growing colony will eventually make contact with a phage particle, which initially infects cells at the margin and then proliferates inwards to destroy the whole colony. In R–M– colonies, there are too few cells for resistant mutants to arise before the colony is destroyed. R+M+ colonies, on the other hand, are usually able to digest the invading phage before they acquire the appropriate modification, preventing the infection from taking hold, provided that phage density is not too high. Even at high density, restriction-modification slows down the pace of the initial attack, giving time for envelope-based resistance to evolve. The course of the arms race in bacteria-phage systems is thus unexpectedly sensitive to bacterial ecology.

The co-evolution of phage and bacteria is usually brought to a halt by the appearance of invulnerable bacteria which have lost the crucial cell-surface binding site. This is a costly adaptation that may reduce bacterial fitness in the absence of phage by up to 50% (Lenski 1988a), allowing susceptible strains to survive when the prevalence of resistance has reduced phage density (Lenski 1988b). Long-term chemostat communities may thus consist of both susceptible and resistant bacteria in equilibrium with the remaining phage. The long-term coexistence of *Pseudomonas fluorescens* and a DNA phage, however, appeared to involve continued co-evolution rather than a static equilibrium (Buckling and Rainey 2002). Bacteria sampled at any time during the history of a culture remained resistant to the ancestral phage, but the proportion resistant to the phage population of the near future (estimated by plating stored material) declined precipitately on several occasions before recovering to normal levels. The resistance of the bacteria and the infectivity of the phage both increased by a factor of 4–5 over the course of 50 transfers. The genetic basis of the interaction is not known, but this is the best experimental evidence for a sustained process of co-evolution during which both host and pathogen increase in fitness.

11.3.9 Perpetual evolution

Parasites and pathogens will evolve so as to be best able to exploit the most abundant types of host, whose fitness will decline as a result. A similar generalization holds for the pathogens themselves: hosts will be selected for resistance to the most abundant types of pathogen, whose fitness will likewise decline. In either case, the average individual has relatively low fitness. From the point of view of the most abundant lineages in the population, the environment continually tends to get worse. Perhaps this is unjustifiably pessimistic; after all, when partners evolve so as to cooperate more closely, the environment will appear to both of them to be improving. However, once a partnership has evolved, the genotypes responsible will be fixed in the population, with perhaps some slight further accommodation being made from time to time. Mutualistic relationships, especially highly integrated ones, do not change much over time. Most of the change that organisms experience is therefore contributed by antagonistic relationships, which are continually tending to change for the worse.

In one of the most famous metaphors of ecology, G.E. Hutchinson spoke of the ecological theatre and the evolutionary play. The environment is thought of as a fixed frame of reference, within which the evolutionary action is contained. This may be the case for geological or meteorological processes that may be adapted to but cannot be deflected. However, it does not apply to neighbours—to the host of rivals, partners, and enemies that are the most important features of the environment. Social selection will readily cause a population to become adapted to its neighbours; but will with equal facility enable its neighbours to respond, leading to closer cooperation or to a ceaseless struggle. Once the importance of biotic features of the environment is recognized, the distinction between the theatre and the play is blurred; once the co-evolutionary response of antagonistic neighbours is seen as the most important source of continuing selection, the distinction all but disappears. There is no fixed frame of reference, but only local and temporary structures that are continually recast by the action of the play itself. Indeed, the imagery of a play enacted on an unresponsive and unchanging stage is one that might appeal more to an ecologist than to an evolutionist. A better analogy might be a street market, crowded, noisy, and intensely competitive. The merchants, in family groups that might extend over four generations, use any device to sell at the highest price their wares will command. They may even conspire with one another to maintain the price—but only if they know their neighbours well. Their customers attempt to buy at the lowest price they can persuade the merchants to accept. A merchant will occasionally discover some new trick of selling, and grow fat for a while, but when it is seized on by others it will eventually become known to the customers, and lose its effectiveness. There is ceaseless manoeuvring, constant bargaining, an endless succession of local triumphs and failures, all organized by the common theme of buying cheap and selling dear. But although there is a common theme, there is no plot. The action does not move in a predestined way to a foreseeable end, but instead unfolds as an historical sequence of events, shaped by simple forces, but unpredictable in detail. The market is affected by physical events, of course—it may slacken when it rains, will be disrupted by an earthquake, may be abandoned altogether if the sea reclaims the sandspit it is built on. But the evolution of the market, the rise and fall of lineages of merchants and customers, the development of techniques for buying and selling, cannot be understood in terms of geology or meteorology; the springs of action are provided by the actors themselves.

The shift from the analogy of a play, scripted and confined, to the analogy of a market, represents the shift from viewing the environment as an unresponsive external constraint, causing the selection of certain phenotypes, to appreciating that adaptation is seldom final or conclusive, but instead tends to procure its own overthrow by selecting compensatory adaptations among neighbours. General Darwinism is not only a theory of how populations respond to the environment in which they live; it is also a theory of the environment itself. The lack of fit between the population and its social environment is not only restored by selection, but is also caused by selection. I think that this is an important idea because it puts an arrow on ecological time. In principle, it could be tested by maintaining large clonal populations, perhaps of sedges or duckweed, both in the field and in the greenhouse. The field populations are continually replenished from the surplus reproduction of the greenhouse stocks. In the field, the clones will offer new resources to the local community of herbivores and pathogens, who will evolve strains capable of exploiting them. The host population cannot evolve resistance, because the plants are continually being replaced from greenhouse stocks; the rate of reproduction in these populations will then simply decline through time as individuals become increasingly vulnerable to the increasingly sophisticated adaptation of local antagonists. No such process will occur in the greenhouse, where the plants can be protected from pests; thus, the fitness of a given clone growing in the field will continually decline, relative to the fitness of the same clone maintained at the same time in the greenhouse.

No such experiment has yet been attempted. An alternative approach is to compare the distributions of resistance and virulence in space or time, with all the uncertainties of attempting to infer process from pattern. A naive prediction from a matching-alleles model is that common host strains will be more heavily infected than rare strains, because selection will favour pathogen strains able to attack them. Conversely, common pathogen strains will be less virulent than rare strains, because selection will favour host strains able to resist them. Within a single isolated population this does not follow, however, if host and parasite have the same generation time, in which case an abundant genotype will be under-infected half the time (when it is increasing in frequency) and over-infected half the time (when it is declining). Common genotypes in the host population will be consistently over-infected only if the host has a longer generation time than the parasite (Nee 1989). In metapopulations the comparison is further obscured by dispersal, which offers hosts an opportunity to escape from virulent pathogens, and likewise pathogens an opportunity to escape from resistant hosts. If pathogens have the higher dispersal rate this gives them an evolutionary advantage similar to that of shorter generation time, and they will tend to be locally adapted. Patterns in time and space are thus most likely to occur when evolution is asymmetrical, with the species that grows and disperses faster showing local adaptation whereas the other is consistently maladapted. In gene-for-gene models local adaptation may occur if resistance and virulence are costly, but otherwise most populations will be similar, probably consisting mainly of resistant hosts and virulent pathogens. Finally, any particular pathogen may infest many species of host, and each species of host may be attacked by many pathogens, in which case strong and specific interactions may fail to evoke substantial adaptation because each is only a minor component of overall mortality.

It is hardly to be expected, therefore, that surveys will yield unequivocal evidence of co-evolution. Nevertheless, striking patterns have sometimes been reported. *Synchytrium decipiens* is a chytrid whose sole host is the annual vine *Amphicarpum bracteum*. Infection by a single zoospore results in the liberation of sporangia containing several thousand zoospores about a month later, so the generation time of the fungus is much shorter than that of its host. Plants from one site were uniformly susceptible to pathogen isolates from that site, whereas plants from other sites were partially or completely resistant (Parker 1985); the pattern of infection, with selfed families being completely resistant or completely susceptible to experimental inoculations, suggests a gene-for-gene interaction.

Larger-scale studies have produced less clear-cut results. *Melampsora lini* is a rust fungus attacking the perennial herb *Linum marginale* (wild flax), probably with gene-for-gene interaction. The fungus is extremely variable, with about $N/3$ races, defined by infection of a standard set of differentially susceptible flax lines, in each sample of N isolates from a flax population. The plants were markedly less variable, with only one or two resistance phenotypes, determined by infection with a standard set of rust races. Some rust strains were virulent on all plants, many plant populations were susceptible to most or all of the standard rust races, and almost all populations contained at least some fully susceptible individuals. Local adaptation of the pathogen would lead to a positive correlation among populations between the frequency of a race and the frequency of susceptible plants, but no general association could be detected (Jarosz and Burdon 1991). A major epidemic was associated with marked shifts in the resistance structure of the plants, as expected (Figure 11.8). This could not be explained by the virulence structure of the fungus, however, since the most common race was virulent on all plants and the second most common was avirulent on the two plant phenotypes that were wiped out by the epidemic (Burdon and Thompson 1995). Faced with vague and conflicting patterns, Thompson and Burdon (1992) concluded that natural selection for resistance was not a powerful force within plant populations, which were more strongly influenced by drift and dispersal.

Likewise, Little (2002) concluded that specific adaptation to parasites has seldom if ever been observed despite the severe effects of many pathogens and widespread occurrence of genetic variation for virulence and resistance. I think that this scepticism is overstated. In the first place, a substantial majority of published studies demonstrate local adaptation of parasites to hosts. They were reviewed by Kaltz and Shykoff (1998), whose survey included 16 studies of plants or invertebrates

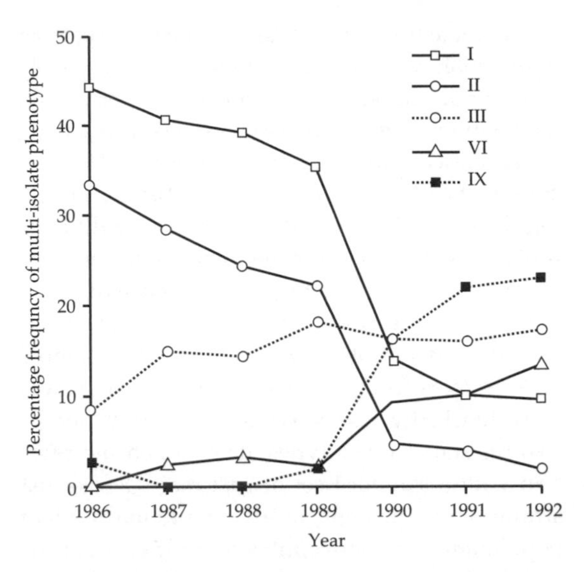

Figure 11.8 Rapid change in frequency of resistance phenotypes of *Linum* during an epidemic of *Melampsora lini*. From Burdon and Thompson (1995).

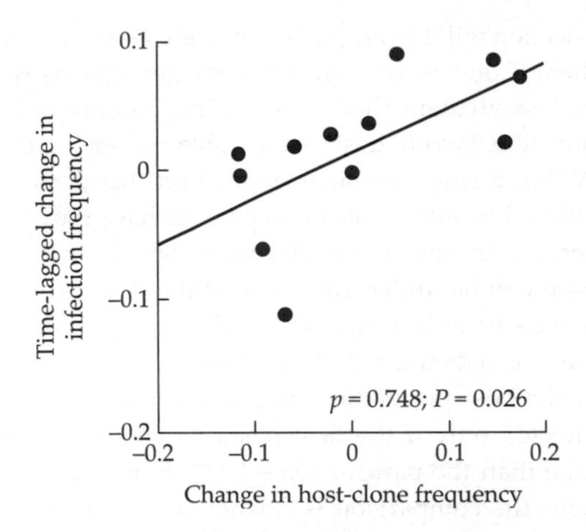

Figure 11.9 Time-lagged response of infection by *Microphallus* to host genotype frequency in *Potamopyrgus*. The graph related the change in infection of clones of the snail between years $y + 1$ and $y + 2$ to changes in its frequency between years y and $y + 1$. From Dybdahl and Lively (1998).

that found consistent local adaptation, 2 that found maladaptation, and 15 that found only weak or variable trends. Secondly, horizontal surveys may fail to reflect the operation of selection, for the reasons given above. The crucial observation is not the correlation between parasite and host, either within or among populations, but rather coupled genetic change through time. Long-term studies that would give direct evidence of host–parasite coupling are rare, however. The most detailed work on wild animal populations has concerned the apomictic snail *Potamopyrgus antipodarium* and the trematode *Microphallus*. A geographical survey showed that the most common snail clone was over-infected in one population whereas it was under-infected in two others (Dybdahl and Lively 1995). When longer time series were available, however, it was possible to show that clones became more vulnerable after they had become common, and that chronically rare clones were consistently under-infected (Dybdahl and Lively 1998) (Figure 11.9). Serial passage experiments, laboratory microcosms, agricultural systems and transplant experiments in the field have all shown that selection in response to enemies is frequent, strong, and specific. It will drive rapid adaptive change, however, only in simple systems where interaction with a single pathogen or host is a major force of mortality. A species embedded in a more complex ecosystem may face a multitude of enemies whose conflicting effects are difficult to discern from field surveys.

11.4 Ecosystems

The demographic interaction between predator and prey, or between parasite and host, is fundamentally unstable: the population of an effective predator will increase as the prey are consumed and then crash as the predators starve. If the prey recover swiftly there may be sustained oscillations, but predators in particular will always be in danger of stochastic extinction in the troughs of the cycle, so we expect that the composition of trophically structured communities will fluctuate wildly through time and will often be simplified by the extinction of predators. This is readily verified in microcosms, for example by the classical experiments with *Didinium* preying on *Paramecium* by Gause (1934). At the same time, there may be specific genetic interactions between pathogens and hosts that can generate strong selection for virulence and resistance. This is also readily verified in serial passage experiments and microcosm community studies. Nevertheless, natural communities retain their trophic structure over long periods of time, while the evidence for persistent large-amplitude genetic cycles in natural systems is scanty. Why are species diversity and genetic structure not destabilized by strong interactions among enemies? One possibility that has long been understood by ecologists is that the physical environment is more complex than a culture vial, and in particular that it provides refuges that protect part of each prey or host population. (Gause 1934, Huffaker 1958). A second possibility that would affect both ecological and evolutionary dynamics is that the biotic environment is much more complex than a laboratory community, so that most species interact with many enemies. Each individual must be able to resist a wide variety of enemies, and consequently we might expect to find that defence is governed by many genes, each responding to a particular range of enemies. The selection acting on these defence genes might lead either to diversification or to rapid change. In the first place, most parasites or predators will destroy only a small fraction of the focal species and will therefore impose only weak or infrequent selection. Consequently, species that are embedded in complex communities will seldom be exposed to a single predominant source of mortality and selection. Selection will tend to favour rare types able to resist prevalent genotypes and species, so defence genes are likely to be highly polymorphic. The second possibility is that from time to time a particular genotype or species of pathogen will become highly destructive, perhaps for reasons unconnected with the abundance or adaptation of a given host species. This will impose repeated episodes of strong directional selection at some defence loci, which will drive rapid divergence of related species if they are attacked by different suites of pathogens. Whether selection is frequency-dependent (balancing) or directional, however, the agents of selection may be difficult to identify because change in any one species may affect others either directly or indirectly. If broom becomes more abundant because rabbit grazing is relaxed by a myxomatosis epidemic, for example, both a moth that feeds on it and its hymenopteran parasitoids might increase, thereby endangering a second moth that feeds on a different plant but shares parasitoids. Trophic chains such as that connecting the myxoma virus with the moth raise two important issues. The first is how species adapt to a complex and ever-changing spectrum of competitors and enemies. The second is how the community itself, and in particular the food web that describes the trophic interactions among species, evolves over time, either through the sorting of species or as the result of adaptive evolution within species. These difficult problems cannot be addressed by a simple model, but I shall try to show how a digital ecosystem based on simple principles can help us to understand the evolutionary dynamics of complex systems, before describing system-wide indicators of co-evolution.

11.4.1 Uqbar

One of the most severe limitations of most current community models is that the range of diversity and the pattern of interaction are specified in advance and cannot change, although there has been considerable interest recently in modelling communities whose attributes can evolve (e.g. Loeuille and Loreau 2005, Stauffer *et al.* 2005). To gain some insight into selection within complex evolvable systems I have developed an individual-based community model that operates through simple rules rather than through equations, and which is capable of effectively indefinite variation and evolution (Bell 2007). It simulates trophic interactions through imaginary creatures bearing structures represented by sequences of binary numbers (Figure 11.10). When two creatures encounter one another, one may consume the other if their sequences are complementary, that is, if the sequence borne by one has the opposite pattern of bits to that borne by the other (such as 00101 and 11010) (Figure 11.11). Individuals may bear one or many of these sequences, called social interaction loci (SIL), which may be of any length. If they bear more than one SIL, they may interact only through homologous SIL (interaction in *cis*) or through any pair of complementary SIL (interaction in *trans*). Complementarity at several SIL increases the probability that the prey is captured. There are two kinds of creature: those that

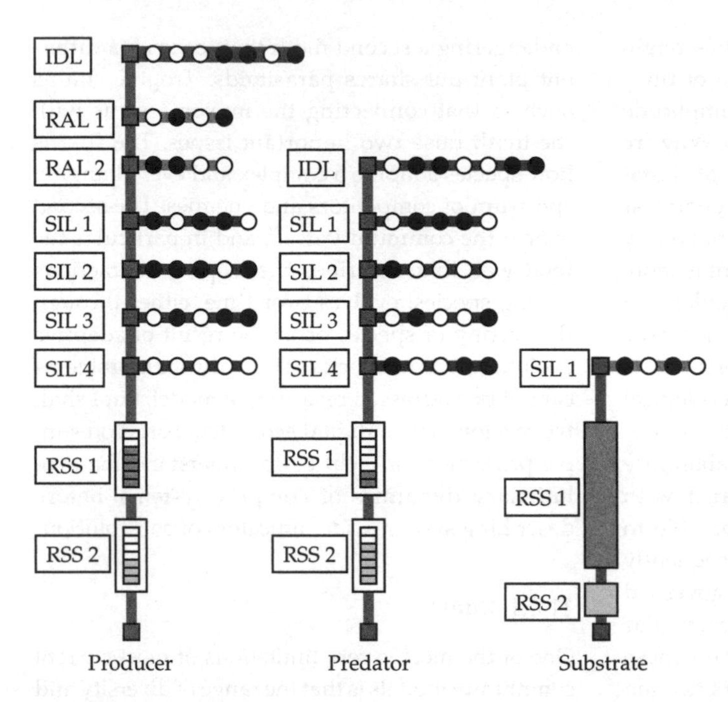

Figure 11.10 An evolvable ecosystem model. The structure of the three kinds of entity in Uqbar, showing the identity locus (IDL), resource acquisition loci (RAL), social interaction loci (SIL) and resource storage sites (RSS). The circles along IDL, RAL and SIL are examples of the bit string (open = 0, filled = 1) that defines the genotype at a given locus. The shading of the RSS for Producer and Predator denotes the quantity of each resource possessed by an individual. The size of the RSS for the substrate indicates the quantity of that resource that the substrate contains.

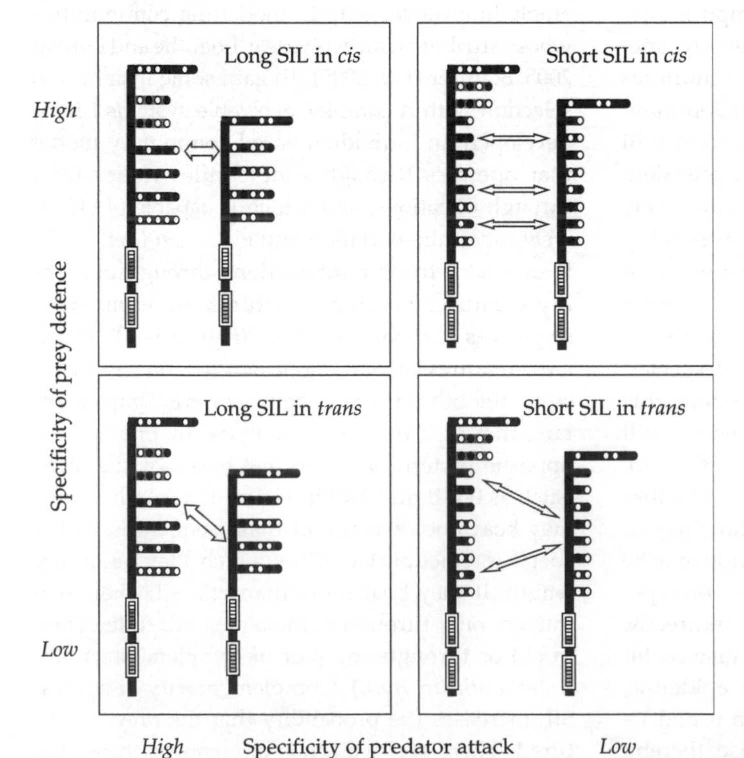

Figure 11.11 An evolvable ecosystem model. Consumption of Producer (left) by Predator (right).

consume only other creatures are called Predators, whereas those that consume substrates by an analogous set of resource acquisition loci (RAL) are called Producers. Individuals grow because the consumption of a substrate or another organism transfers the different kinds of resource they contain to the resource storage sites (RSS). These resources are continually depleted by activity through a maintenance cost that depends on the resource-gathering potential of an individual: thus, the maintenance of a Producer is proportional to the number of RAL it bears, and that of a Predator on the number of SIL. If its resource status falls below some minimal level the creature dies. When the resources held by an individual exceed some threshold it reproduces by copying its code (the values held in the SIL and RAL, plus the identity locus IDL that specifies which species it belongs to) to another site in the ecosystem, along with a ration of each kind of resource. When an offspring is produced by copying the parental code, errors occur with probabilities assigned by the programmer. These result in deletions (one or more SIL or RAL completely removed), duplications (a second copy of an existing SIL or RAL added) or point mutations (one or more bits flipped in a SIL or RAL). This creates a very large range of possible genotypes of which the community will usually constitute only a very small sample. Consequently, the community is free to evolve and may eventually consist of creatures and patterns of interaction none of which were originally present. I have called the system Uqbar, after the imaginary parallel world whose fragmentary records were described by Jorge Luis Borges.

Trophic interactions depend on the length and number of the SIL. Long SIL represent specialized prey, in the sense that the longer the sequence the less likely is a predator to possess an exactly complementary sequence. Moreover, if the prey population is currently susceptible, it is more likely to evolve resistance with long SIL because longer SIL are more likely to undergo mutation. Given a susceptible prey population, the probability of a resistance mutation appearing in a given SIL of length L is Lu per genome per replication, where u is the mutation rate per bit, whereas the probability of a mutation restoring virulence is u. This reflects the fundamental asymmetry between prey and predators: any mutation in the relevant SIL of a susceptible prey genotype will produce resistance, but only the precisely corresponding mutation in the predator will restore virulence. On the other hand, many SIL represent specialized predators, in the sense that having more SIL makes it more likely that a predator will find a complementary sequence in the prey it encounters. It also makes the evolution of virulence more likely, once a point mutation for resistance has become common in the prey, because with N SIL similar in state the probability of a mutation restoring virulence will be Nu. In practice, interaction in *cis* or *trans* is used to represent variation in the number of SIL, *cis* being equivalent to few and *trans* to many SIL, because this allows the physical genome size (total number of bits in SIL) to be kept constant. In short, long SIL represent specialized prey which readily evolve resistance, whereas many SIL, or interaction in *trans*, represent specialized predators that readily evolve virulence. In this way, although the SIL are clearly a highly formalized representation of the mechanisms involved in predator–prey interaction, they are nevertheless capable of being configured so as to imitate a broad range of ecological and evolutionary scenarios.

11.4.2 Evolution of trophic structure through sorting

When an isolated community is allowed to proliferate, poorly adapted species starve or are consumed, and become extinct. This process of sorting dissects out a simpler persistent community from the original inoculum. This community maintains trophic structure only if the rate of consumption by Predators is sufficient to pay for their cost of maintenance. The rate of consumption is governed by the relationship between the degree of complementarity and the probability of prey capture. Provided that the probability of capturing the prey when it is encountered is high enough, Predators paying some level of maintenance are able to persist. If the interaction is too weak, the Predators die of starvation. If the interaction is very strong, however, Predator abundance fluctuates with large amplitude and Predators often become extinct when rare through demographic stochasticity. A low level of immigration from an external pool of species rescues all taxa from permanent extinction and may support trophic complexity even when predator–prey interaction is weak. The general result is that the pattern of connection in webs under immigration and sorting is wholly determined by the microscopic rules of engagement. In Long-Cis systems with high specificity the complementarity of any two random types is likely to be zero, and connectance is correspondingly low. Many Producers have no predator, and most vulnerable Producers are consumed only by a single predator. Most Predators consume only one or a few types of prey, and few Predators are capable of consuming other Predators. The food web is thus dominated by very short, largely unconnected food chains, and resembles a lawn. In Long-Trans and Short-Cis systems complementarity and connectance are considerably greater. Isolated Predators that lack a predator are scarce, food chains lengthen, and trophic diagrams take on a bushy or 'webby' appearance. It is often difficult or impossible to assign a type to a single trophic level unambiguously; indeed, recursive loops involving non-transitive trophic relationships occur quite frequently. In Short-Trans systems connectance is complete, although complementarity and thus the strength of interaction may vary widely among predator–prey couples. The food web resembles a tree, with a single long food chain. Among Predators, interactions are completely hierarchical, meaning that any predator can consume all (or almost all) the Predators at lower levels. The whole-system trophic structure is thus determined by the microscopic rules of engagement (Figure 11.12).

11.4.3 Evolutionarily stable webs

The outcome of sorting may be a simpler persistent community, but it is not necessarily the only such community, or the most stable; some or all of the components of a more stable system may have been lost stochastically early in the sorting process. An evolutionarily stable (ES) web is a stable community that cannot be invaded by any other member of a defined initial pool of taxa. Whether or not the community acquires a more or less fixed composition depends on the diversity of the inoculum. With a low-diversity inoculum (several tens of taxa) an ES web usually emerges within a few thousand cycles. With a high-diversity inoculum (several hundreds of taxa) a stable community of roughly constant composition has never been observed. The Producer community does indeed change little in composition through time. There is a perpetual succession of Predator taxa, however, that never repeats itself, as far as I have been able to ascertain. This observation applies to all systems and does not seem to depend on the specificity of predator–prey interactions.

11.4.4 Evolved webs

Mutation prolongs the existence of trophically complex communities through continued co-evolution. This is a very prominent effect in Long-Cis systems with very high specificity, where communities quickly become dominated by resistant Producers and so lose trophic complexity. With no mutation there is an irreversible decline in Predator diversity in closed systems, culminating in complete extinction. When Predators are allowed to mutate they can recover virulence, with the result that trophically complex communities persist for much longer. The magnitude of the effect depends on the flux of mutants and thus on community size. In large systems, point mutation seems to immortalize Predator communities that would otherwise disappear within a few hundred cycles. The underlying dynamics can be investigated by tracking the abundance of particular genotypes, or categories of genotypes. The Producer community is relatively stable, changing slowly in composition over periods of thousands of cycles. The Predator community, by contrast, fluctuates violently in composition over periods of tens to hundreds of cycles. Most Predator taxa, most of the time, are rare specialists barely able to persist through occasional encounters with equally rare vulnerable prey. Many Producer taxa, on the other hand, are abundant, and are invulnerable to any existing Predator. From time to time, a mutation confers on the offspring of a Predator the ability to consume one of these abundant Producers. This novel

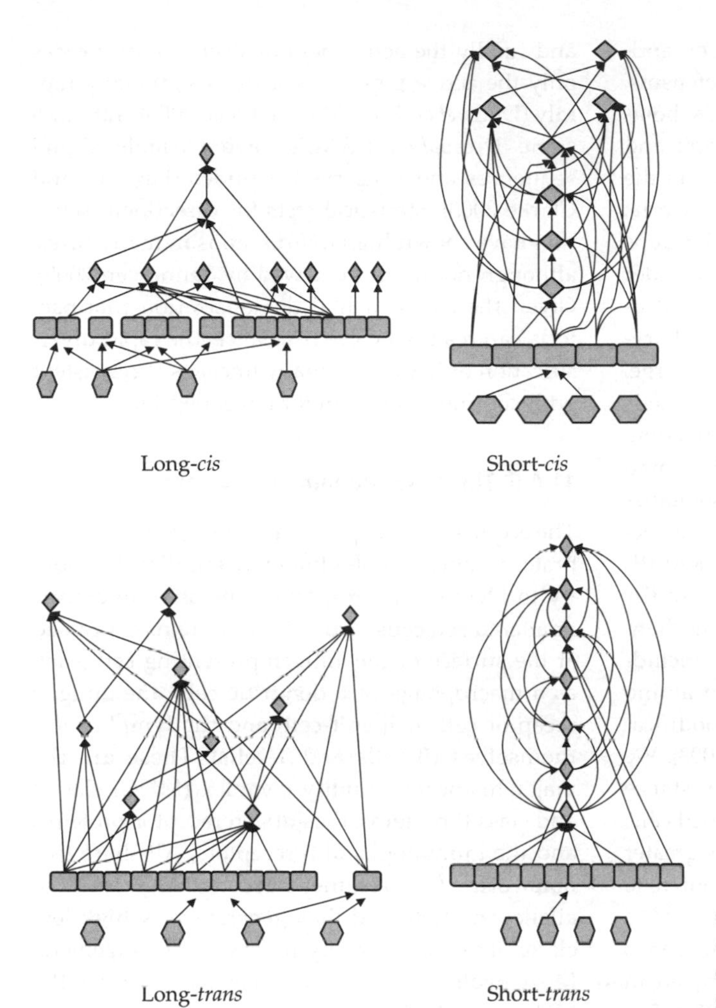

Figure 11.12 An evolvable ecosystem model. Cartoons of trophic structures generated by the rules of engagement. Basal hexagons are resource packets; oblongs are Producers; diamonds are Predators.

Predator proliferates rapidly, wiping out its prey as it does so. It therefore crashes quite soon, but during its brief period of rapid expansion it will have given rise to a cloud of new mutants, some of which will persist and may in turn give rise to the next virulent type. At the level of the community as a whole, this process cannot be perceived; the overall abundance of Predators and Producers does not fluctuate strongly through time. In systems with less highly specific interactions the fluctuations of Predators are less extreme, but the Predator community is always more dynamic than the Producers. The main reason for this is that the fitness of Producers depends in part on the non-evolving composition of available substrates.

The Uqbar communities thus display high levels of diversity and continual flux on a timescale of tens to hundreds of generations, as might be expected from a matching-alleles type of model. Whether this accurately mirrors the dynamics of natural communities is not known, because extensive long-term genetic surveys have not yet been reported. The properties of immune systems, however, suggest that it may not be far out.

11.4.5 The innate immune system

The primary defence of eukaryotes against bacterial infection is the release of small cationic peptides that are highly toxic to bacteria (see Zasloff

2002) (§7.5.3). These ribosomally synthesized antimicrobial peptides (RAMPs) include the defensins, indolicidins, and cathelicidins in mammals; bombinin, magainin, and buforin in frogs; cecropin and melittin in insects; the thionins in plants; and the bacteriocins, epidermidins, and nisin in bacteria. The negatively charged phospholipid head groups on the outer surface of bacterial membranes render them highly vulnerable to electrostatic and hydrophobic interactions with RAMPs, whereas eukaryotic cell membranes, with little or no net charge, are almost immune. RAMPs are abundant in cells and tissues responsible for controlling infection, such as leucocytes and the Paneth cells of the small intestine, and kill bacteria very rapidly by solubilizing the cell membrane. It has been argued that bacteria will be unable to evolve resistance to RAMPs because they target a fundamental feature of the bacterial cell, but this is incorrect. Several mechanisms of resistance have been described, including the substitution of positively charged amino acids into the outer membrane and the modification of efflux pumps (Bell and Gouyon 2003). We selected *E. coli* and *Pseudomonas* lines for resistance to magainin, a synthetic RAMP, and obtained cells resistant to concentrations a hundred times greater than those fatal to the ancestor after about 1000 generations of selection (Perron *et al.* 2006).

RAMPs are encoded by about eight genes in humans (see Ganz 2003) and about seven in *Drosophila* (see Hoffmann and Reichhart 2002). The genes are often duplicated and may exist in several or many copies that are then free to diverge, so that related species often have markedly different RAMPs that may also differ in copy number and site of expression. This suggests that a species' repertoire of RAMPs is tailored to its unique spectrum of pathogens. For example, frogs express 10–20 RAMPs in their skin, whose sequences vary markedly among species; there is often no clear homology between the RAMPs produced by different species, even within the same class, so that the 5000 or so species of frog may be producing up to about a million different RAMPs (Nicolas *et al.* 2003). RAMPs such as defensins are encoded as molecules with three segments: an N-terminal signal sequence, an anionic intervening sequence, and finally the active peptide itself. In most cases only the active peptide evolves exceptionally rapidly (Duda *et al.* 2002, Maxwell *et al.* 2003), although some *Drosophila* RAMPs have variable signal sequences and conserved peptides (Lazzaro and Clarke 2003). Statistical tests for directional selection have shown high d*N*/d*S* ratios in many cases, although not in all (reviewed by Tennessen 2005). Given the experimental demonstration that bacteria can readily evolve resistance, the rapid diversification of RAMPs in many lineages is consistent with continual selection for novel peptides.

11.4.6 The acquired immune system

The acquired, or adaptive, immune system of vertebrates is more complex but more familiar. The ability to identify and respond to invasive microbes, or altered self cells, depends on proteins expressed at the surface of the antigen-presenting cell (such as a macrophage or a dendritic cell), the antigen receptor cell (helper T-cell) and the lymphocytes themselves (B-cells and T-cells). These are the major histocompatibility complex (MHC) class I and class II proteins, the antigen receptor proteins, and the immunoglobulin receptors. All share two noteworthy features: they have very high levels of allelic variation, and they generate very high levels of antigenic diversity by DNA rearrangement. Most evolutionary studies have concerned the MHC, and the large literature has been reviewed by Bernatchez and Landry (2003), Garrigan and Hedrick (2003), and Piertney and Oliver (2006).

The class I and II MHC proteins are encoded by gene families that have arisen through duplication and remained together as closely linked blocks of genes. Members of each family are thought to be generated and to be lost or inactivated rather frequently, in a birth–death process that gives rise to MHC with different numbers of functional genes and a high proportion of pseudogenes (see Nei and Rooney 2005). MHC is the most polymorphic gene cluster in humans (where it is called HLA) and other vertebrates; HLA genes have up to at least 500 alleles, and nucleotide diversity is roughly 100 times greater than the genomic average (reviews by Kelley *et al.* 2005, Pancer and Cooper 2006, Cooper and Alder 2006). MHC genotype is often associated with disease resistance, with heterozygotes being more resistant than homozygotes (Doherty and Zinkernagel 1975). The best-known examples are HIV and hepatitis B infection in humans, but a range of studies in fish, birds, and mammals uniformly report an advantage of heterozygotes (see Wegner *et al.* 2004 Table 1). This might be attributable to overdominance, but is more likely to reflect a general advantage of particular rare alleles (Clarke and Kirby 1966), which will generally be found as heterozygotes. For example, intense selection for resistance in Atlantic salmon to the virulent bacterium *Aeromonas salmonicida* is associated with a particular allele of the MHC class IIB gene (Langefors *et al.* 2001, Lohm *et al.* 2002). Infection by nematodes and its effect on juvenile survival in a population of feral sheep was mediated by particular MHC class II alleles rather than by heterozygosity (Paterson *et al.* 1998). Other cases are reviewed by Bernatchez and Landry 2003).

Studies of genotype frequencies within and between populations seem to support the past action of selection at MHC. The few studies of wild populations that have measured allele frequencies report that they are more even than expected through drift alone. The high level of linkage disequilibrium in the MHC region is consistent with selection for particular combinations of alleles. Genetic variance among populations of fish tends to be greater than expected under drift, but the data are scanty and studies of mammals show little if any effect. In human populations, the extent of MHC polymorphism is correlated with viral diversity (Prugnolle *et al.* 2005). The peptide-binding region of HLA genes has an excess of non-synonymous substitutions whereas the non-peptide-binding region does not (Figure 11.13). A review of 48 studies found $dN/dS > 1$ in all cases but one, suggesting systematic directional selection (Bernatchez and Landry 2003). Trans-species polymorphisms were detected by 40/42 studies, as expected if variation is protected by frequency-dependent selection (Bernatchez and Landry 2003). There is also a great deal of variation between species, however. Class I MHC genes show no sequence orthology between orders, or even in some cases between genera in the same family (Kelley *et al.* 2005). The number of genes may also differ greatly between closely related species. Similar patterns have been found in other defensive systems, such as plant chitinases (Bishop *et al.* 2000).

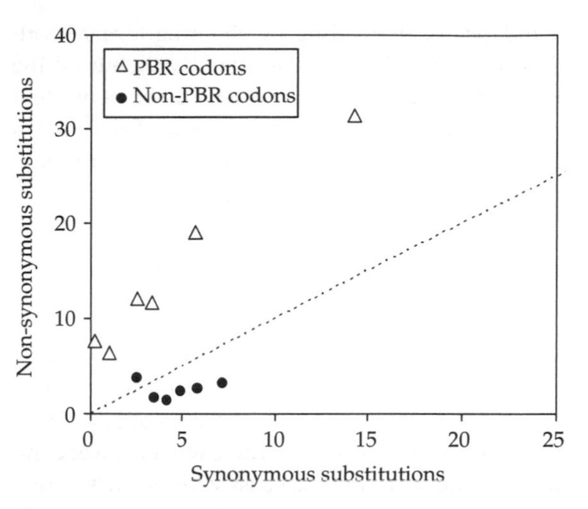

Figure 11.13 Rapid evolution of immune-system genes. Number of synonymous and non-synonymous substitutions in the peptide-binding region (PBR) and non-peptide-binding region of six human HLA (MHC) genes. From Garrigan and Hedrick (2003).

The innate and acquired immune systems both bear witness to the pervasive selection pressures applied by short-lived pathogens. The ability to recognize billions of antigens and to produce the appropriate antibodies shows that multicellular individuals are continually being attacked by a wide range of pathogenic microbes and viruses. Moreover, parasites may evade the immune response by similar means, such as the ability of *Plasmodium* (malarial parasite) to alter the antigenic properties of infected red blood cells. Indeed, Endo *et al.* (1996) computed dN/dS for nearly 4000 homologous sequences and found that 9 of the 17 gene groups where directional selection seemed to be acting encoded surface antigens of pathogens. The extreme polymorphism of genes encoding immunoglobulins and other defence-related proteins points to selection for rare alleles likely to confer resistance to common pathogen species or genotypes. The pronounced differences between species show that selection is often exceptionally rapid. Other kinds of selection may also act on these genes, for example avoidance of inbreeding by choosing a mate with dissimilar MHC (Penn and Potts 1999), but all the lines of evidence I have reviewed are consistent with the frequent occurrence of powerful selection driven by pathogens and capable of causing both diversification and rapid change.

11.4.7 Selection at the ecosystem level

The animals and plants that consume or are consumed by one another form a food chain that is linked to other food chains by common interactions to form a food web. If predators or pathogens are sufficiently specialized there will be fewer common interactions and the ecosystem as a whole will comprise several isolated food webs that ultimately depend on the same supply of resources but do not overlap in species composition. Each isolated food web replicates (because its component species replicate) and may tend to grow or to shrink, that is, to comprise a greater fraction of community biomass and to process a greater fraction of community resources. In complex ecosystems of this kind, there will be selection among isolated food webs that may alter the properties of the ecosystem as a whole, and Uqbar can be used to investigate how it might be expected to act.

In the simple systems described above all individuals have SIL of the same length. Individuals having SIL of different length cannot interact, so when the length of SIL is allowed to vary the system initially consists of a number of subsystems, isolated food webs each with its own rules of engagement. All Producers in the system compete for the same pool of substrates, but Predators can consume only those individuals belonging to the same food web. Subsystems may differ in any way, including length and mode of operation of SIL, interaction strength and maintenance cost. When alternative subsystems all comprise Producers alone, then the most frugal subsystem (that capable of reducing substrate concentration to the lowest value) replaces all others. This is merely an extension of the well-known result for competing species (§10.1.3). The situation is more complex when the food web as a whole consists of several source webs with similar or dissimilar rules of engagement. These subsystems will compete because they draw on a common pool of resources, and it might be imagined that the more frugal or the more stable subsystems would be favoured by selection, and by replacing all others would enhance ecosystem properties such as average productivity, variation in productivity, trophic transfer efficiency, constancy of composition, or robustness to perturbation. This does not seem to be generally true: the outcome of competition between trophically complex subsystems is usually coexistence rather than replacement. The reason for this is that as frugal Producers become very abundant their Predators increase, thus raising the death rate of Producers and reducing their abundance. Hence, selection among subsystems is negatively frequency-dependent, and the most frugal subsystem will replace all others only if the difference in maintenance costs is sufficiently high. Subsystems that differ only in interaction strength will coexist, provided that each is capable of persisting in pure culture in similar conditions. These persistent subsystems may become more or less abundant: when rules of engagement are similar, those with weaker interactions tend to become the more abundant. This is because the Predators of subsystems with strong interactions drive down Producer numbers and thereby restrict their own abundance. When the rules of engagement differ among subsystems, I have found no general rule specifying the replacement of one by another, or the prevalence of a particular kind of web. Systems that are able to persist as isolated systems in pure culture are likely to coexist as mixtures, given comparable conditions of growth. In any particular case, component webs with different rules of engagement may differ in frequency, but I have not been able to find any general relationship between the pattern of specialization and relative abundance. There is no tendency in Uqbar, therefore, for one web topology to replace all others, and the ecosystem will usually consist of component webs with a variety of different topologies.

11.4.8 Evolution and whole-system properties

It has sometimes been held that complex ecological systems will tend to become more stable or productive through time, for example because weaker interactions are favoured (see Paine 1992, McCann 2000) or because interactions tend to become compartmentalized (Krause *et al.* 2003). Montoya and Solé (2003) found that random Lotka–Volterra systems resembled real food webs after sorting only if maximum interaction strength was low. It might be imagined that at some level selection will favour more stable over less stable sets of interacting species in any given part of the overall food web, thus tending to stabilize the system as a whole through the evolution of weak interactions. This does not appear to be the case in Uqbar. When the rules of engagement of Uqbar are fixed, selection among individuals will favour more effective predators and more resistant prey, and the overall strength of interactions may either increase or decrease as a consequence. For any particular predator–prey interaction, selection will always favour predator genotypes with stronger interactions and prey genotypes with weaker interactions. When deletion and duplication are allowed, for example, selection acts to optimize expenditure on SIL (or in Producers on RAL). In any move, the cost is a fixed metabolic expenditure, whereas the benefit depends on the probability of encountering prey, the probability of capturing an individual once it has been encountered, and the nutritive value of the prey. The optimal number of SIL can then be calculated for any given capture function in the standard way, and when deletion and duplication are permitted communities evolve towards this optimal value from above or below within a few thousand cycles. Thus, simple systems comprising a single food web do not naturally tend to become more stable, because selection among individuals will not generally produce this outcome. Where there are several isolated systems with different rules of engagement, however, selection may operate among these trophically isolated communities, as I have described above. In general, those with weaker interactions between predator and prey tend to become the more abundant, and in this case the overall strength of interaction may fall, relative to that of a simpler system.

A rival account of ecosystem evolution is given by the Gaia hypothesis (Lovelock 1979), according to which the whole-Earth ecosystem has evolved towards maximal stability and productivity through fundamentally mutualistic interactions at all trophic levels. So far as I can understand the argument, this is the outcome of selection in a population of size one, as there are no competing planetary ecosystems. The hypothesis has been given respectful coverage by the most highly respectable journals (e.g. Lovelock 2003) despite being based on a fallacy evident to any evolutionary biologist (see Williams 1992a). Together with the continued survival of the theory of divine creation, support for the Gaia hypothesis provides the most eloquent testimony of the tenuous grasp of Darwinism, even among professionals in other branches of science. As one of many possible general-purpose models of evolvable ecosystems, Uqbar fails to show any consistent tendency for whole-system properties to be optimized as the consequence of the ceaseless struggle among the myriad rivals, partners, and enemies of the ecosystem.

CHAPTER 12

Sexual selection

Natural selection is caused by differential reproduction, and in an asexual world this concept would be sufficient to account for adaptedness. However, in many organisms—perhaps, fundamentally, in all eukaryotes—the vegetative processes of growth and reproduction are occasionally interrupted by sexual episodes. Although the nature of sexuality was worked out before the end of the nineteenth century, misconceptions about the process continue to cloud thinking about evolution. This is largely attributable to the fact that sex and reproduction are intimately connected in large and familiar animals, especially in vertebrates, and to a lesser extent in flowering plants. Respectable textbooks of biology still refer to 'sexual reproduction' (I probably have somewhere), as though, in Michael Ghiselin's phrase, sex were a kind of reproduction. To understand what sex is, and why it introduces new kinds of evolutionary principle, it is helpful to put dogs and daisies on one side, and think instead of ciliates, seaweeds, foraminiferans, or yeasts; the bulk of living diversity, in fact. Ciliates may as well serve as an example, although I shall suppress the details of their rather imaginative genetic and developmental systems. They are large predatory unicells that are abundant in freshwater and marine environments. The vegetative cells are diploid. After growing to a sufficient size, they reproduce exclusively by binary fission, dividing by mitosis to form two nearly equal daughters. In certain conditions (usually when they are crowded or starving) a pair of cells will fuse; the nucleus of each cell goes through meiosis to form two surviving haploid nuclei, one of which remains in the parental cell, while the other migrates into its partner, there to fuse with the corresponding product of meiosis. Diploidy having been restored, the two cells separate and swim away. The complementary processes of reduction and fusion together constitute the sexual cycle. The important point is that sex and reproduction are completely distinct. They are, indeed, opposed: growth and reproduction generate two similar cells from one parent, whereas sex changes the nature but not the number of individuals.

All unicellular and many multicellular eukaryotes follow the same pattern, which is obscured only when new individuals develop from fertilized eggs exclusively. The eukaryotic life cycle thus comprises two quite different processes. The spore—any uncommitted cell—may develop either vegetatively or sexually. Vegetative development involves growth, with or without development as a multicellular individual, followed by reproduction that regenerates similar spores. Sexual development involves differentiation as a gamete, followed by fusion with another gamete, after which the spore is regenerated by meiosis. The complete life cycle of sexual organisms involves two vegetative cycles linked by a sexual cycle (Figure 12.1), although one or the other vegetative cycle is often reduced to a rudiment (Figure 12.2). Natural selection, acting through vegetative development, is thus incomplete as a description of evolution in eukaryotes. Selection acting on the sexual cycle differs from natural selection in that it acts through sexual fusion, rather than through vegetative fission.

12.1 Evolution of sex

12.1.1 Calkins's experiment

So long as everyday experience confirmed that sex is necessary for reproduction there was no need to explain why sex should be so general among animals

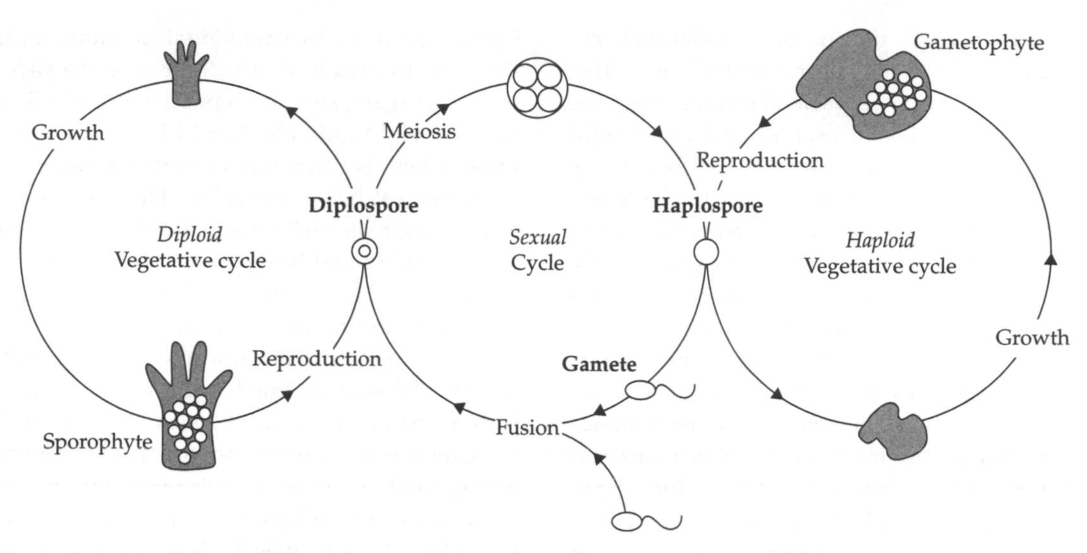

Figure 12.1 The sexual life cycle of eukaryotes.

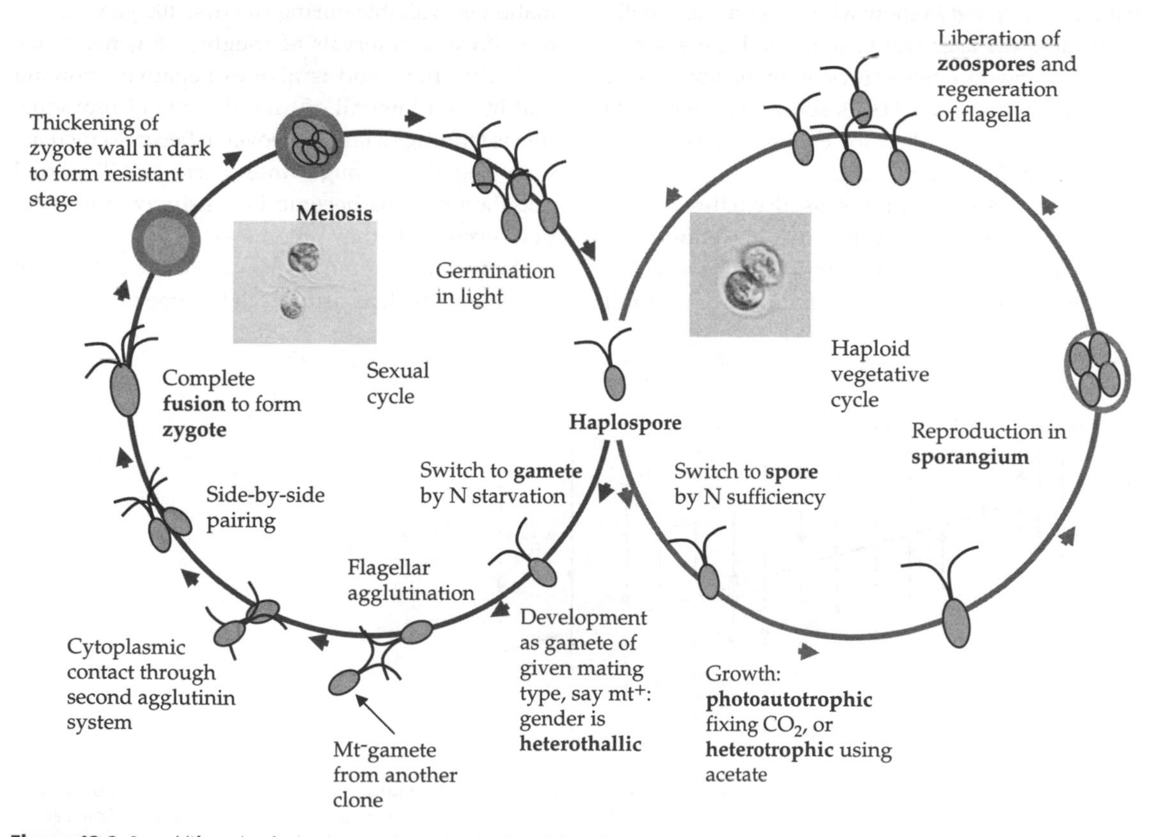

Figure 12.2 Sexual life cycle of a haplontic eukaryotic microbe, *Chlamydomonas*.

and plants. Towards the end of the eighteenth century, however, curious naturalists had found that hydras and aphids are capable of reproducing without any obvious sexual process, and nineteenth-century biologists subsequently described many kinds of asexual propagation in animals, plants, and protists. But if sex is not necessary for reproduction, what is it necessary for? The answer given by the French-Algerian biologist Emile Maupas was that asexual propagation gradually dilutes or degrades some vital principle that can be fully restored only by a sexual episode: sex halts a process of racial senescence and rejuvenates the lineage. This idea chimed with notions that human races and civilizations have life cycles comparable to those of individuals, passing from vigorous youth to maturity and then to senescence and decay before being reborn in another form. It became very influential, and inspired a fascinating series of experiments performed by several groups of biologists between about 1880 and 1920 that were designed to show whether or not asexually propagated lineages were immortal. I came across these by accident while browsing in the library and eventually wrote a short book about them (Bell 1988) that should be consulted for documentation of the brief account I shall give here.

Most of the experiments used ciliates, which reproduce quite rapidly but are big enough to manipulate easily. A single individual is isolated on a microscope slide and left for a day or so, until it has divided; the two individuals produced by fission may in turn have divided before the slide is inspected again, so a small population of between two and eight individuals will be found. One of these is then chosen at random and moved to a new slide, the rest being discarded. This technique of isolate culture is the limiting case of serial transfer, a single individual being transferred in each line at the end of the growth cycle. In most cases, it was impossible to propagate isolate cultures indefinitely: over the course of a few hundred generations the rate of fission declined, and the line eventually became extinct. I reviewed 26 experiments of this kind, involving about 80 lines maintained for up to 4000 generations. Until recently these were the longest-running of all selection experiments, the record being held by L.L. Woodruff (1926), who propagated lines of *Paramecium* for 19 years, representing about 11 000 generations. Figure 12.3 summarizes the trend in fission rate in all lines for which information is available during the first 400 generations of culture, at intervals of roughly 25 generations. Note that the trend is always negative, showing that in each interval of time the rate of reproduction is tending to fall; moreover, it becomes increasingly negative through time. These small asexual populations thus become increasingly enfeebled, until eventually they become extinct.

Horace Calkins ran a series of isolate cultures of *Uroleptus*, which as usual each became extinct after

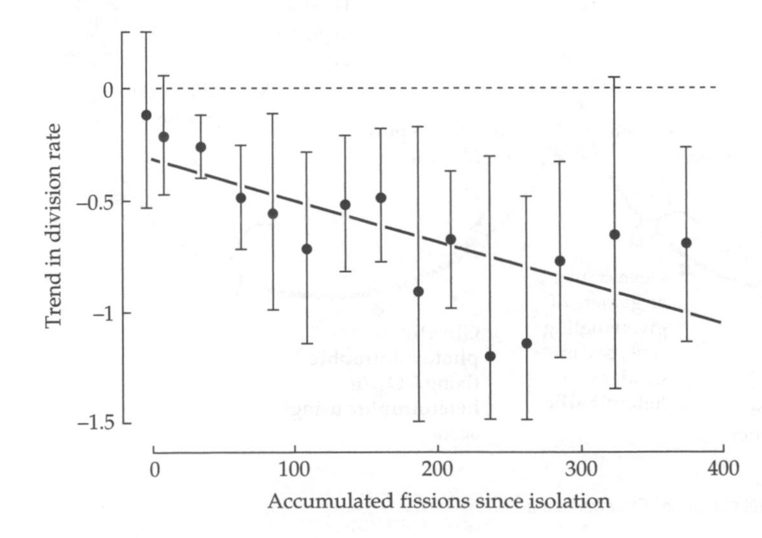

Figure 12.3 The accelerated decline of rate of division in isolate cultures of ciliates. From Bell (1988).

a few hundred generations. At the same time, however, he extracted sexual progeny at intervals from each line to found a new isolate culture, and found that although every asexually propagated isolate line was mortal, the sexually propagated sequence of lines was potentially immortal (Calkins 1919, 1920) (Figure 12.4). This beautiful experiment proved that Maupas had been right: sex is necessary to restore the vigour that is somehow exhausted by long-continued asexual propagation. These ciliate experiments were extensively discussed in the genetics texts of the time, but they were difficult to reconcile with the burgeoning research on inheritance in *Drosophila*, maize, and mice, and by the end of the 1930s they had disappeared from view. There was always a slight odour of vitalism about the whole affair. It had at least provided an indisputable function for sex, even if nobody could understand how it worked.

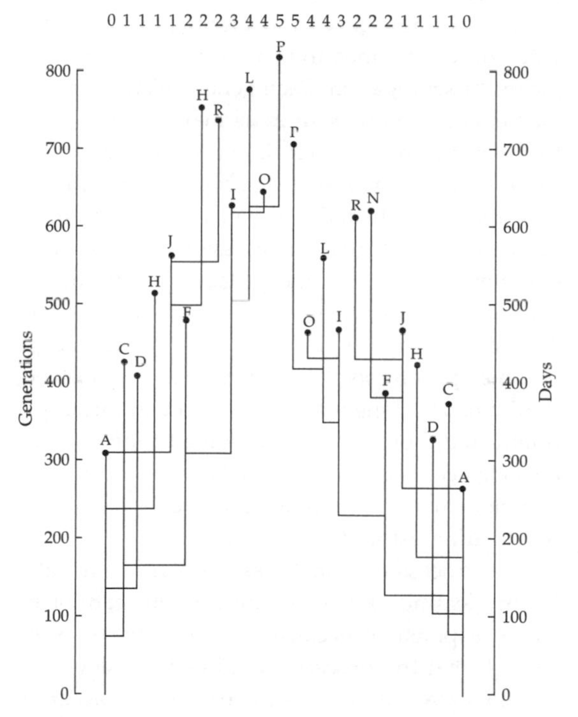

Figure 12.4 Calkin's experiment. The vertical lines indicate the lifespan of an asexual culture, the solid circle marking the point at which it becomes extinct. The horizontal lines indicate the sexual origin of a new culture. Time is given in generations on the left, and as calendar time on the right. Although each asexual culture declines quite rapidly, becoming extinct within a few hundred generations, the sexual lineage as a whole can be propagated indefinitely. From Calkins (1919, 1920).

The most straightforward interpretation of sexuality is that it is essential for normal vegetative function. Many of the genes that govern sexual processes, especially recombination, are found in bacteria, where they function in DNA repair, and it has been suggested that meiotic recombination is preserved in contemporary populations because it is essential for repairing double-strand breaks (see Bernstein *et al.* 1985). This seems unlikely, because species that lack recombination or crossing-over have normal development, although the cooption of existing repair systems has certainly shaped the evolution of the sexual cycle. It is always possible to argue that sex is so deeply implicated with reproduction in organisms such as mammals and birds that parthenogenetic variants never arise. This is not a very convincing assertion either; if there are parthenogenetic beetles, fish, and lilies, it does not seem likely that there is any physiological obstacle to the development of complex multicellular creatures without sex. In any event, sex is only an occasional interruption for the large number of organisms whose life cycle consists for the most part of repeated episodes of asexual proliferation, and they could certainly abandon sex entirely. Most people who have thought about the problem have therefore concluded that the sexual cycle is preserved indirectly through its effect on adaptation (§4.3). This is a rather onerous argument, however, because at first glance sex seems more likely to reduce fitness than to increase it. In the first place, random fusion followed by recombination will tend to break up epistatic combinations of alleles, causing an immediate loss of adaptedness. Any modifier that increases the rate of recombination should therefore be lost from the population. The immediate effect of sex in *Chlamydomonas* lines subject to strong selection is to reduce fitness, usually by a few per cent (Colegrave *et al.* 2002, Kaltz and Bell 2002). The progeny from crosses among or within selection lines of (haploid) *Chlamydomonas* maintained for 100 sexual cycles and 1000 vegetative cycles in benign conditions are consistently inferior to their parents (Renaut *et al.* 2006). Asexually derived spores increase in frequency immediately after inducing germination in yeast (Greig *et al.* 1998). Secondly, sex usually reduces the short-term rate of growth because the complicated processes of cell fusion, chromosome rearrangement, and meiosis consume time and material that could otherwise be used for growth. Sexual microbes will readily go through several vegetative cycles each day in culture, whereas the sexual cycle requires at least a day or two and thereby greatly reduces the rate of proliferation. This drawback is often mitigated in natural populations by inducing the sexual cycle when conditions are unfavourable for growth, for example at the end of the season or when the population is very crowded (Bell 1982). Finally, in multicellular organisms that have highly differentiated male and female gametes, sex retards vegetative proliferation because only one gender is reproductive (Williams 1975, Maynard Smith 1978). Imagine a population that consists of (say) 50 males and 50 females; the argument works just as well for outcrossed hermaphrodites, but is slightly more complicated. If each inseminated female produces 10 offspring, then the population as a whole produces a total of 500 new individuals. But a comparable population that consisted of 100 parthenogenetic adults, whose eggs did not need to be fertilized, would produce 1000 new individuals. An asexual lineage that produces unreduced spores by mitosis will therefore proliferate twice as fast as a sexual lineage that produces reduced gametes by meiosis. Consequently, any element that suppresses meiosis, and allows offspring to develop without fertilization, will tend to spread through a sexual population; so long as it is rare, indeed, it will double in frequency in every generation. This simple conclusion will hold whenever the male contribution to the offspring is physiologically negligible. These simple arguments lead to the conclusion that the sexual cycle, and especially outcrossed sexual systems with differentiated male and female gametes, should not exist. Sexuality is a very general and perhaps primitive feature of eukaryotes, however, and the difficulty of reconciling general observation with simple theory has given rise to a large theoretical and comparative literature (see reviews by Barton and Charlesworth 1998, Otto and Lenormand 2002).

12.1.2 Muller's ratchet

One cause of senescent decline in isolate cultures of ciliates is their unusual cytogenetic machinery. A single diploid nucleus would be inadequate for the levels of protein synthesis required by these large cells. Instead it gives rise to a much larger nucleus containing many copies of each gene that is transmitted to asexual progeny but broken up and subsequently reconstituted anew in sexual progeny, thus establishing a dichotomy between a germline micronucleus and a somatic macronucleus. The micronucleus has conventional chromosomes that segregate regularly in mitosis and meiosis, whereas the macronucleus is a soup of sequences down to the size of single genes and simply pinches in two when the cell divides and has no spindle or regular means of segregation. Each genetic element in the macronucleus is present in as many copies in the first generation after meiosis and sexual fusion, but in subsequent generations the copy number fluctuates stochastically and the element may eventually be lost. Once lost it cannot be restored, so all subsequent members of the lineage lack the element and cannot perform whatever functions it encoded. As more and more elements are lost the lineage becomes progressively and irreversibly enfeebled, until it can no longer survive. The micronucleus is unaffected, however, and provided that a sexual episode occurs before the line becomes extinct the lost elements are supplied and the vigour of the line is fully restored.

This process of somatic assortment contributes to the decline of isolate cultures but cannot be alone responsible because copy number is often so high that the process would be very slow. The second source of decay is the irreversible accumulation of deleterious mutations. Deleterious mutations inherited from the founding micronucleus or arising during vegetative growth will eventually become homozygous through assortment in the macronucleus. High copy number is no protection because the number of mutations that arise must be directly proportional to the number of copies.

This is probably the main factor responsible for the senescence of the ciliate lines. At the same time, mutations will also accumulate in the micronucleus; indeed, they will accumulate rapidly because all are neutral during vegetative growth. Thus, the ciliate isolate cultures are examples of mutation-accumulation experiments (§2.2) in which Muller's ratchet (§3.3.3) will ensure a steadily increasing load of mutations and a corresponding fall in vigour. Sex renews the macronuclear genome by reconstituting it from a fresh micronuclear template. It also restores the integrity of the micronucleus by recombination and fusion. Since most deleterious mutations will be heterozygous, recombination will produce some lightly loaded gametic pronuclei, and since mutations will have occurred independently in different lines the products of mating between lines will have normal vigour provided the mutations are recessive. Sex usually results in a large proportion of feeble or lethal exconjugants when induced in old clones, because the micronuclear load is heavier, and when clones are self-fertilized, because the products are likely to be homozygous. The ciliate experiments, therefore, suggest that outcrossed sex might evolve because it halts the ratchet and thereby immortalizes sexual lineages, as Calkins's experiment showed.

Isolate culture and mutation–accumulation experiments provide other examples of fitness decay in small asexual populations. Fitness should be restored by recombination. Chao (1990) propagated 20 lineages of the segmented bacteriophage $\phi 6$ by enforcing a bottleneck of a single particle in each growth cycle. After 40 cycles fitness relative to the ancestor varied between 1.06 and 0.22 with an average of 0.78, showing how mutations, most of them deleterious, had been trapped by drift. When a cell is infected by virus from different lineages recombination between the segments halts the ratchet and restores full viability (Chao *et al.* 1992).

Isolate cultures suffer most severely from Muller's ratchet, but it will turn in populations of any size. Meiosis in male *Drosphila* happens to be achiasmate: autosomes in a male experience recombination only when transmitted to a daughter. Rice (1994) prevented recombination altogether in a lineage of autosomes by mating males with females bearing marked autosomes. The heterozygous male progeny of these crosses must have inherited the unmarked copy of each autosome from their father, and by propagating lines in this way the paternally transmitted autosomes are prevented from recombining. Although 48 males were used each generation, the vigour of these lines fell to about half that of controls within 35 generations. The classic example of a ratcheted genome is the Y chromosome, which originates as an autosome bearing a major sex-determining gene and then degenerates once it no longer recombines with the X. Any region of the genome where recombination is reduced or prevented, such as inversions, centromeric genes, and mitochondria, is liable to accumulate deleterious mutations.

The expected distribution of mutational load is Poisson, and the frequency of the least-loaded class in a population of N individuals is thus $n_0 = N \exp(-U/s)$ (Haigh 1978). If n_0 is small then this class will occasionally become extinct, after which the entire distribution will move back by one class. One click of the ratchet thus involves two phases: an establishment phase of length t_{est} generations in which the next-least-loaded class with n_1 individuals becomes the least-loaded class with n_0 individuals, and an extinction phase of length t_{ext} generations in which the least-loaded class with an expected number of n_0 individuals fluctuates stochastically to zero. The establishment phase proceeds nearly deterministically and depends only on selection and mutation: $t_{est} \approx -\ln (U/s) / \ln(1-s)$ (Haigh 1978). The extinction phase depends on n_0 and thus on N; if the least-loaded class is treated as an isolated subpopulation, which will be approximately true if it is rare, then a good rule of thumb is $t_{ext} \approx 10n_0$. More exact solutions involve U and s separately as well as U/s (Stephan *et al.* 1993) although t_{ext} seems to increase proportionately with n_0s (Gordo and Charlesworth 2000). Mutations of very small effect are likely to be very common and the least-loaded class will then comprise very few individuals: the establishment phase will be much longer than the extinction phase, and the interval between clicks may be very long. Conversely, the least-loaded class will comprise the majority of the population for rare mutations of large effect, so the extinction time will be much longer than the establishment time and the total time may again be very long. For mutations of moderate effect and moderate frequency, however, the ratchet may turn rather rapidly. To take Mukai's estimates of $U = 0.4$ and $s = 0.025$ with an average dominance of 40% for *Drosophila* (Mukai 1964, Mukai *et al.* 1972; §2.3.2) as an example, the establishment phase will be about 100 generations while the extinction phase will depend on population size. In a population of 100 000 individuals or fewer the least-loaded class will usually comprise a single individual and will swiftly become extinct, so the ratchet will turn every 100 generations or so; 100 clicks of the ratchet will reduce fitness to less than 50% of its original value, which will drive a microbe reproducing by binary fission to extinction in about 10 000 generations. In a larger population of 10^{10} individuals the extinction phase is roughly 10 000 generations and the population will be endangered after about 10^6 generations.

The ratchet is halted by recombination, which tends to restore the deterministic Poisson distribution of load. The minimal level required depends on n_0: higher rates of recombination are necessary when population size is small or mutation rate is high. Modest amounts of recombination, generated by sexual episodes at long intervals, will be sufficient in microbes with population sizes of 10^{10} individuals or more. In much larger populations of 10^{15} individuals or more the ratchet turns slowly in geological time and sex is unnecessary. Small populations, on the other hand, will require obligate sexuality and a high level of meiotic crossing-over, and very small sexual populations of 10^5 individuals or fewer, in which the least-loaded line usually comprises a single individual, may be doomed to extinction regardless of the amount of recombination that occurs. This indicates a wide range of population size in which a sexual population will maintain adaptedness whereas an equivalent asexual population will decay and eventually disappear. If sexual and asexual populations within this range are competing with one another the sexual populations should eventually prevail, although this might take a very long time.

Unlike most theories of sex, Muller's ratchet is readily investigated and has been repeatedly validated by experiment. Its severity depends chiefly on population size, so it explains why bacteria remain asexual. It is also consistent with the brief lives of most species of large asexual eukaryotes. There are ancient lineages that seem to be entirely asexual, the best-known being bdelloid rotifers, but these are exceptional: most asexual taxa are young, showing that they quickly become extinct. Parthenogenetic vertebrates, for example, are mostly Postglacial forms that have arisen through hybridization. The theory has two main limitations. The first is that during the long process of asexual decay the environment is assumed to remain constant in all important respects, so that a given mutation is always deleterious. If selection coefficients fluctuate through time, sometimes taking negative and sometimes positive values (§7.3), the effectively unloaded class will be continually re-established by selection. The second is that the short-term advantages of asexuality imply that sexual populations should often be invaded and displaced by new asexual lineages. To understand why that does not often happen, the short-term advantages of sex need to be identified.

12.1.3 Artificial selection for recombination

The sexual cycle will evolve in the short term if there is genetic variation for features such as the rate of recombination. A number of genes in various organisms are known to affect the rate of recombination, either in some defined region between two loci, or across the genome as a whole. These genes are identified by making a set of crosses and then looking for families with unusually high or low numbers of recombinant offspring. Even if a large number of offspring can be scored from each family, this is a very coarse procedure that will detect only genes with extreme effects, usually those that abolish recombination altogether. Such genes are often loss-of-function mutations that may tell us very little about how rates of recombination evolve. However, it is impracticable to set up a screen that would detect small differences in the proportion of recombinants, because these differences will be obscured by sampling error unless the number of offspring scored from each family is extremely large. The point at issue is whether recombination rates can evolve through the selection of individuals whose differences are attributable to genes of slight effect at many loci. Both the evolutionary question and the genetic question are best answered by a selection experiment. If recombination rates cannot be altered through selection, then there is no heritable variation for the character in the experimental population. If the rates shift abruptly under selection, rapidly attaining a new value and thereafter failing to respond to further selection, then recombination is controlled by one or a few genes with large effects on the character, of the sort that would be detected by conventional screens. If recombination rates change rather smoothly and continuously over time, then the selection line may be accumulating alleles at different loci with individually small effects on recombination, of the sort that would be missed by a single-generation screen.

Most experiments of this kind have used *Drosophila* (Detlefson and Roberts 1921, Parsons 1958, Acton 1961, Mukherjee 1961, Chinnici 1971, Kidwell 1972a, 1972b, Abdullah and Charlesworth 1974, Charlesworth and Charlesworth 1985) although there are a few in other organisms: *Neurospora* (Calef 1957); lima bean, *Phaseolus* (Allard 1963); flour beetle, *Tribolium* (Dewees 1970); and silkworm, *Bombyx* (Turner 1979, Ebinuma and Yoshitake 1981, Ebinuma 1987). They involve crossing individuals bearing different mutant genes with easily visible effects on the phenotype. The mutations are at loci on the same chromosome, so that the combinations of characters seen in the progeny are generated by crossing-over rather than merely by the random assortment of chromosomes. There are two ways in which the combinations of characters affected by these marker genes can be used to select for higher or lower rates of recombination. The first is through straightforward individual selection: individuals bearing recombinant genotypes are identified, and used to propagate a high-recombination line. To propagate a corresponding low-recombination line, one would deliberately select non-recombinant individuals that retained the parental combination of genes. There is one difficulty with this otherwise simple procedure: if a gene in one of the parents changes the rate of recombination between the marker loci, it may also affect recombination between itself and the marker loci, so the recombinant offspring may not bear the gene responsible for their condition. An alternative procedure is to scrutinize each family separately, and select individuals from those families with the highest or lowest rates of recombination. This will be more likely to trap genes responsible for altered rates of recombination, but it is also much more laborious.

Kidwell (1972a) used family selection to alter recombination between Glued *Gl* and Stubble *Sb*, two genes on the third chromosome of *Drosophila*. The mutant alleles at both loci are dominant, and their phenotypes are easily scored. In the base population used to initiate the experiment, an average of about 15% of progeny were recombinant. This rate was estimated by crossing *Gl Sb* / + + females, where the + indicates the wild-type allele, to wild-type males, and counting the number of *Gl* + / + + and + *Sb* / + + progeny produced. Two selection lines were set up, one for increased and the other for decreased recombination. In both lines, 5 females were crossed to 5 wild-type males in all combinations to yield 25 families in each generation. The rate of recombination was estimated from the first progeny to be produced; these were then discarded, and the subsequent progeny produced by the 5 females with the highest or lowest used to propagate the line. Selection for reduced recombination was ineffective, but in the upward selection line recombination was doubled, from about 15% to about 30%. Most of this increase occurred in the first 12 generations of selection; although selection was continued for another 8 generations, there was little further response. Attempts to increase recombination beyond 30% seemed to be unsuccessful because the fertility of the selected flies declined markedly once recombination exceeded about 20%; thus, the tendency of artificial selection to increase recombination was countered by natural selection for lower rates of recombination. Charlesworth and Charlesworth (1985) carried out a similar experiment using individual rather than family selection.

In the upward selection line, *Gl* + / + *Sb* females were mated with wild-type males, and the recombinant *Gl Sb* / + + female progeny selected. These were crossed again with wild-type males in the next generation; this generated *Gl* + / + + females, which could then be crossed with + *Sb* / + + males to regenerate *Gl* + / + *Sb* females, and the cycle repeated. Downward selection was simpler, with *Gl* + / + *Sb* males and females being mated and the non-recombinant *Gl* + / + *Sb* progeny selected. This procedure was almost entirely ineffective. Only 1 of 16 lines showed any appreciable response to selection, an increase to about 22% recombination. This was presumably because, as anticipated, individual selection is less effective than family selection. One must be cautious, then, in extrapolating from the outcome of artificial selection to evolution in natural populations: natural selection, after all, will usually act through the phenotypes of recombinant individuals, rather than through the average recombination of families. Chinnici (1971) used a combination of individual selection and family selection to increase and decrease recombination between two loci on the X chromosome. Both upward and downward lines responded smoothly and more or less continuously, so that after about 30 generations of selection recombination had fallen from about 15% to about 8% in the downward line, and had increased to about 22% in the upward line. In experiments with *Drosophila*, it is always possible that gross chromosomal rearrangements may be responsible for changes in recombination rate, and like mutations of large effect these are not very relevant to the evolution of rates in natural populations. However, in Chinnici's experiment, as well as the other two I have described, no such rearrangements could be detected cytologically, so the response to selection was probably caused by the accumulation of several or many alleles with individually small effects on recombination in the selection lines. Chinnici was in addition able to show that recombination in neighbouring regions of the chromosome had not changed, so that crossing-over had been specifically altered in the selected region, rather than being redistributed to or from nearby regions. The general conclusion from these experiments is thus that recombination can be modified through selection, and that by inference the maintenance of high levels of recombination in sexual populations requires a powerful source of short-term selection.

12.1.4 Sex and the rate of adaptation

At about the same time that the ciliate experiments were documenting the decay of asexual isolate cultures, August Weismann was pioneering the genetical interpretation of natural selection. To explain the prevalence of sexuality, he suggested that 'the communication of fresh ids [genes] to the germplasm implies an augmentation of the variational tendencies, and thus an increase of the power of adaptation.' (Weismann 1889, in translation). This was radically different from the usual explanation of how attributes evolve through their benefit to individuals. Sex does not benefit the individual, or even its offspring; instead, it benefits the lineage by enhancing its ability to adapt and thereby ensuring its success in the distant future. Weismann's idea and its modern versions are succinctly explained by Burt (2000). Since the environment continually deteriorates (§2.3), sexual populations will eventually displace their asexual competitors. It is also possible that alleles causing a greater tendency to enter the sexual cycle, or a higher rate of recombination, will spread within a population by hitch-hiking with the advantageous combinations of genes they create. Such alleles are likely to be separated from their effects, however, by virtue of the very recombination they induce, so selection on the modifiers of recombination is likely to be very weak, although artificial selection is effective (Parsons 1958, Charlesworth and Charlesworth 1985).

In either case, sex will accelerate adaptation only if it increases SV_A, the standardized additive genetic variance of fitness. This in turn requires that fitness effects are negatively correlated among loci. If the population displays positive linkage disequilibrium, with an excess of + + + + + + + and − − − − − − − chromosomes then sexual offspring will be intermediate and the variance of fitness is reduced; recombination is deleterious because it impedes progress towards the fixation of the + + + + + + +

type. With negative linkage disequilibrium there is an excess of + − + − + − + − chromosomes and recombination is favoured because it increases genetic variance and facilitates selection of the fittest type. This situation can be created in several ways: by synergistic epistasis, by disruptive sorting, or by chance. It has been persuasively argued that chance is likely to be the more important agent (Otto and Barton 2001, Barton and Otto 2005). The immediate loss of fitness often observed after a sexual episode suggests that epistasis may also be influential. In the first place, suppose that genes interact in such a way that a few + alleles cause a large increase in fitness, relative to having none, whereas the addition of more + alleles causes only a small further increase in fitness. Thus, the fitness of a genotype increases less steeply than multiplicatively as the number of + alleles it bears increases, rising asymptotically as it approaches the composition of the types favoured in the new environment. This kind of epistasis will lead to negative linkage disequilibrium, because types with intermediate numbers of + alleles will have much higher fitness than those with few + alleles, and will therefore increase swiftly in frequency, whereas extreme types with many + alleles are not much fitter than the intermediate types, and will spread more slowly. Secondly, negative linkage disequilibrium will also arise naturally through disruptive sorting. Mutation or environmental change will lead to genetic variance among alleles at any two (or more) loci. The effects of these alleles may be initially uncorrelated, but genotypes with low-fitness alleles at both loci will be selectively removed, creating a negative correlation among loci and thus negative linkage disequilibrium. Finally, negative linkage disequilibrium may arise by chance. It will tend to be generated by drift in small populations because combinations of similar numbers of + and − alleles are nearly equivalent in fitness and thus only slowly resolved by selection (Hill and Robertson 1968). It will often be generated by novel beneficial mutations, since these are likely to arise in different lineages to give − + − − − − − − and − − − − − − + − chromosomes. This is the original Fisher–Muller hypothesis, that sex combines independently arising mutations into the same lineage (Fisher 1930, Muller 1932) and thereby accelerates evolution by the assembly of beneficial mutations or the clearance of deleterious mutations (§4.3.7). One interesting version of this hypothesis is that sex can liberate beneficial mutations from the inferior backgrounds in which they arise (Peck 1994).

A few experiments have compared the response of sexual and asexual populations to stressful conditions of growth (§4.3) and they all support the Weismann theory. Sexual populations of both *Chlamydomonas* and yeast evolve more rapidly under stressful conditions (Colegrave *et al.* 2002, Kaltz and Bell 2002, Goddard *et al.* 2005) but not under benign conditions (Goddard *et al.* 2005, Renaut *et al.* 2006), and the advantage of sex is greater in large populations, as expected (Colegrave 2002). Birdsell and Wills (1996) made mixtures of diploid sexual and asexual yeast strains in complete medium: the sexual bore both *a* and *α* alleles at the mating-type locus *MAT* whereas the asexual strains bore two copies of one or the other. The strains were constructed by crossing two dissimilar haploid strains to create a heterozygous genome or by crossing within the same strain to make a homozygous genome. The sexual strain rapidly excluded its asexual competitor in all cases, even when both strains were homozygous. This experiment shows a consistent and substantial advantage for sexuality, but it is not clear what it might be: perhaps meiosis is mutagenic, or heterozygosity at *MAT* affects gene expression. Greig *et al.* (1998) made asexual yeast strains by deleting *IME1* and thus abolishing meiosis, then constructed homozygous and heterozygous diploids and cultured them at high temperature. Sex is expected to affect the response to selection only in the heterozygous lines. When these strains were mixed the asexual strains at first increase in frequency, as has been found in *Chlamydomonas*; the fitness of sexual heterozygotes was particularly low, presumably through recombinational load. Before the cultures could become fixed, however, the sexual strains rallied and subsequently eliminated the asexuals in 10/12 cases for the heterozygous lines and 6/12 for the homozygous lines. This dynamics of these experiments (Figure 12.5) supports a short-term advantage for sex generated through increasing the genetic variance of fitness.

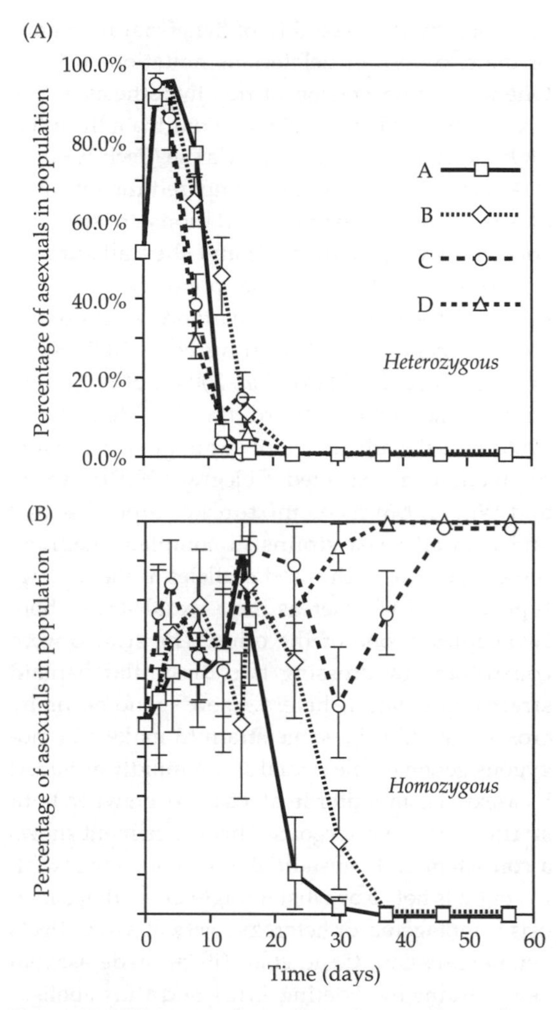

Figure 12.5 Short-term selection for sex in yeast. Change of frequency of asexually derived cells in four replicate lines with heterozygous (upper diagram) or homozygous (lower diagram) ancestors. From Greig *et al.* (1998).

12.1.5 Correlated response to directional selection

Muller's ratchet is a special case of mutation clearance in small populations, where recombination can re-establish the unloaded category by virtue of the variance in load it creates. There is little experimental evidence for mutation clearance in large populations, but several instances of mutation assembly have been described (§4.3.10). The crucial experiment is whether recombination increases as an indirect response to strong directional selection. Flexon and Rodell (1982) investigated the response of recombination to selection for DDT resistance in *Drosophila*. They set up three selection lines by placing strips of filter paper soaked in a solution of DDT on the surface of the food medium, gradually increasing the concentration of DDT throughout the experiment. After about 20 generations of selection, DDT resistance had increased markedly, the flies being able to tolerate DDT at more than 10 times the concentration that was initially fatal. The rates of recombination between marker loci on the three major autosomes were then measured. On one chromosome, they were no different from rates in wild-type flies or unselected control lines. On the other two, however, they had increased substantially. Moreover, the two chromosomes where recombination had increased were the two that were primarily responsible for the increased resistance to DDT. This promising experiment does not seem to have been followed up, although there are some experiments by Russian workers that I have not read but which are described by Burt (2000). Gorodetskii et al. (1990) subjected selection lines to diurnal temperature fluctuations of increasing intensity, with the temperature eventually falling to 18°C and rising to 33 °C every day (*Drosophila* populations are usually maintained in the laboratory at a constant 25 °C). There was the expected direct response to selection: the selection lines evolved greater viability and fecundity after being cultured in these conditions for about 50 generations. There was also a correlated response of recombination, which increased substantially in several regions of the autosomes, and perhaps also in a restricted region of the X chromosome. In some regions of the autosomes, recombination rates were twice as great in the selection lines as in the controls. Korol and Iliadi (1994) selected for geotaxis in *Drosophila* and found increased recombination on most chromosomes after 50 generations. More generally, domestication seems to be associated with elevated levels of recombination in mammals (Burt and Bell 1987). Thus, there seems to be good evidence for the expected indirect response. (It is, in fact, unexpectedly strong—nearly as strong as the direct response to selection for enhanced recombination.) In laboratory conditions this effect must be caused by modifiers of recombination, and might be much greater if sexual and asexual populations were to compete.

12.1.6 The Red Queen

Weissmannian models have been extensively analysed by population geneticists because an indefinite number of alleles are imagined to be straightforwardly deleterious or beneficial and consequently theory can be developed along traditional Fisherian lines. There is little curiosity about the drivers of change in this literature: the only empirical project that has been pursued with any energy is the attempt to estimate the genomic rate of deleterious mutation. The environmental change required to fuel continued mutation assembly or liberation is assumed to occur, but with little effort to elucidate the rate of change or to identify the factors involved. An alternative theory is that environmental change causes different combinations of alleles to be favoured from time to time. Since gene effects change through time the Fundamental Theorem cannot be applied, and much of the development of the theory is based on numerical analysis or simulation. Models of this sort were pioneered by George Williams (1975) and John Maynard Smith (1978) during the renaissance of interest in the evolution of sex in the 1970s. The basic scenario can be outlined as follows (see Maynard Smith 1978 Chapter 6). Suppose that there is a series of environmental factors α, β, γ, etc., each of which takes one of two states, denoted by upper and lower-case letters at any particular time. Thus, the current state of a site might be denoted AbcDEf, while another site has the state abCdEf, and so forth. A single locus with two alleles corresponds to each factor, each allele conferring high fitness in one of the two states: thus, the optimal genotype in the first site would be *AbcDEf*, in the second site *abCdEf*, and so forth. From time to time some or all of the sites change state, so the genotypes adapted to these sites, which have become frequent in the population, are now maladapted. New adaptive combinations can readily be generated by recombination, so it might be thought that a sexual population would be able to keep pace with a changing environment more readily. The flaw in this idea is that recombination is just as likely to create inferior combinations, so there is no net advantage of sex. Free recombination will tend to bring the population towards linkage equilibrium, but an asexual population will move towards the same composition, given multiplicative fitness interactions among loci. The key insight was provided by Maynard Smith (1978): sex facilitates adaptation to a changing environment only when the sign of linkage disequilibrium alternates over time. For example, suppose that the optimal genotypes were *ABCDEF* and *abcdef* in the past, but are now *AbCdEf* and *aBcDeF*, and will again be *ABCDEF* and *abcdef* in the future. Recombination will generate an excess of these newly adaptive genotypes, provided that individuals from different sites mate with one another, and a sexual population will then be always better adapted than an asexual competitor. Selection will drive changes in the sign of linkage disequilibrium, however, only if the correlation between adaptively important features of the environment often changes sign, and it is difficult to see why this should happen. Hence the emphasis that this line of argument has placed on identifying the agents of environmental change.

The difficulty is that the theory requires the population to be engaged in a cyclical game with the environment. The only way in which such cyclical dynamics can naturally arise is when the environment is another organism, when mutually antagonistic species such as parasites and hosts may become locked into a perpetual evolutionary struggle in which each must continually counter the adaptation of the other. This struggle can drive persistent cycles in genotype frequency (§10.3), and in some circumstances it will favour sex and recombination. This theory was developed by Jaenike (1978), Hamilton (1980), Hutson and Law (1981) and Bell (1982). A simple situation might be that resistance and virulence are governed by two loci with additive effects. The virulence of a parasite is greatest when its phenotype matches that of its host, whereas the resistance of a host is greatest when its phenotype differs from that of an invading parasite. Thus, if the virulence of a parasite with genotype *AB* is v on an *ab* host it would be $2v$ on an *AB* host, and if the resistance of an *AB* host towards an *ab* parasite is r it would be $r/2$ towards an *AB* parasite. The repulsion genotypes *Ab* and *aB* are intermediate in both resistance and virulence. These social matrices drive strong fluctuation in genotype frequency, with each population cycling $ab \rightarrow aB + Ab \rightarrow AB \rightarrow Ab + aB \rightarrow ab$, etc. Selection for recombinants when the population consists mostly of the repulsion genotypes leads to the rapid spread of a rare allele at a third locus causing recombination (Bell and Maynard Smith 1987). This elegant theory explains much of the comparative biology of sex: why it is generally more prevalent in the tropics than at high latitudes, for example, or in the sea rather than in fresh water, because biological interactions will be more varied, intense, and prolonged in more permanent and diverse communities (see Levin 1975, Bell 1982). It also chimes pleasingly with the fact that the most effective defence against invading pathogens, the adaptive immune system, operates through recombination, as do the efforts of pathogens such as *Plasmodium* to evade this defence. It has several shortcomings, however. In the first place, other kinds of genetic interaction are not as favourable to sex. If the genes governing resistance and virulence specifically antagonize one another then *Ab* and *aB* genotypes will have dissimilar properties. This will again drive strong fluctuations in genotype frequency, but the population is usually almost fixed for one genotype or another, and there is no strong or consistent selection for recombination. Secondly, although cyclical alternations of genotype frequency have been documented in parasite–host systems it is not clear that the genetic interactions involved will favour recombination, nor that the genetic dynamics of natural populations are generally cyclical. It is at least as plausible that each new source of virulence will be countered by a novel source of resistance, in which case mutually antagonistic species will indeed provide a perpetual source of continued selection, but sex will derive its advantage from mutation assembly or liberation.

12.1.7 Sib competition

The fundamental feature of the mass-action theories described above is that sexual progeny are different from their parents. An alternative view is that sexual progeny differ among themselves. If competition occurs principally among sibs, the greater diversity of a sexual brood could enhance its success either through replacement or through complementation.

The first possibility is that selection will be more effective among sexually generated than among clonal sibs because of their greater genetic variance. In the extreme case, only the single fittest individual survives: this is the lottery model of George Williams (1975) who likened an asexual parent to someone who buys tickets for a lottery with one prize, and finds they all have the same number. In a patchy, heterogeneous environment where sexually and asexually produced individuals compete, the winner in most sites is likely to be a sexual individual with an extreme genotype. Maynard Smith (1976, 1978) investigated formal models of this kind in which each site receives S seeds from each of P parents. Fitness is governed by five loci with alleles *A*, *a*, etc. that correspond to five attributes A, a, etc. of the sites, as in the section above; a single individual survives at each site, chosen at random from those with the fittest genotype. If $S = 1$ then there is no advantage to recombination because sibs do not compete; if $P = 1$ there is likewise no advantage because sexual and asexual broods do not compete. Thus, S expresses the strength of within-family selection and P the strength of between-family selection, their product SP setting the limit to the total intensity of selection. If SP is sufficiently large, recombination can have a substantial advantage. This is because each asexual clone tends to fall in abundance, because its sibships seldom supply the fittest genotype, and eventually becomes rare. It is then likely to be lost altogether by sampling error, and in the absence of mutation it cannot be restored; genotypes in the sexual population, on the other hand, can be restored by recombination if they happen to become extinct.

Models of this sort have fallen from favour, I think, because they seem to refer to particular kinds of environment or ways of life. There has been a revival of interest in the effect of population structure on Weismannian models, however

(Agrawal and Chasnov 2001, Waxman and Peck 1999, Martin *et al.* 2005), because linkage disequilibrium will be generated stochastically in the small population at each site. Migrants arriving at some focal site will tend to carry different combinations of alleles, reflecting past selection in their natal sites. When they mate together, the negative genetic correlation between sites will increase the genetic variance among their progeny (Lenormand and Otto 2000). Consequently these progeny will include new combinations of alleles, some of which may be highly adapted to the focal site and will then increase in frequency. Dispersal should therefore favour outcrossing, and moreover outcrossing should occur during dispersal as the result of mating between individuals from different sites. Conversely, if local adaptation is based on a particular combination of alleles then mating within a site should involve close inbreeding because this will minimize the effect of genetic recombination.

The second possibility is that sexual sibships might be more successful than clones because of the general tendency of mixtures to outyield the average of the components. The argument can be understood through a simple economic analogy. In a small, closed community, some professions will be more profitable than others; the dentist (say) is likely to earn more than the shop assistant. Nevertheless, parents raising a family in the community would be ill-advised to have all their children trained as dentists, because they would then compete among themselves, and their average income would fall. It would clearly be preferable to enter them into a variety of different professions. The diversification of offspring genotypes through outcrossing might serve the same function of reducing competition among sibs, and thus increasing the success of the brood as a whole. I called this theory the Tangled Bank, after an expression in the final paragraph of the *Origin of Species*, referring to the diversity of natural communities; it was developed by Ghiselin (1974) and Bell (1982). The basic experimental evidence for the theory is that mixtures often produce more than the mean of their components (§10.4). The effect is usually slight, however, and there is little evidence that complementation rather than replacement is usually responsible. When the same experiment was tried with sibs from crosses within a species, pairwise mixtures were roughly as productive as the better of the two components (Bell 1991b). In this case, however, there is no doubt that the effect of mixture is caused simply by the displacement of the less productive genotype, because the genetic variance of growth-rate is greatly attenuated by a single cycle of growth. Doney *et al.* (1965) measured the total vegetative yield of plots receiving 10 potato seeds, where each group of 10 was either from the same clone or from the same outcrossed family. There was no overall difference between the two groups.

Outcrossed sexual sibships in which diverse sources of resistance are randomly combined may be less vulnerable to disease epidemics. Whether or not this would represent a substantial advantage for outcrossing seems again to be a question of scale, depending on whether or not sexual sibships present a sufficiently broad resistance spectrum to inhibit the spread of disease when growing as neighbours. There is little experimental work that directly addresses the issue. Schmitt and Antonovics (1986), working on the intensively studied *Anthoxanthum* population at Duke University, set up a hill-plot design in which the central plant was surrounded by four neighbours, which were either full sibs of the central plant or unrelated plants from random seed; other plants grew singly. The experiment was fortuitously attacked by aphids, with infestations approaching 50%. The single plants suffered the most damage, showing that a fringe of neighbours may dilute an infestation. Plants surrounded by sibs or by unrelated plants survived equally well in groups that were not attacked by aphids, but when aphids were present the plants with unrelated neighbours survived better. This suggests that disease is less debilitating in mixtures of dissimilar individuals. Unfortunately, clonal groups were not used in the experiment, so it is not known whether the more limited diversity of the sib groups might have had any effect on the severity of damage by aphids.

The current consensus among theoretical biologists, I believe, is that Muller's ratchet sets a limit to the longevity of asexual lineages while

Weismannian mutation assembly or liberation explain the short-term success of sexual lineages. It may be, however, that these theories have found favour because they are more directly related to the standard models of population genetics and because experiments can be done with single species in standard microcosm conditions. The ecological processes that maintain sexuality in natural populations have not yet been convincingly identified.

12.1.8 Somewhat sexual

There is a wide range of processes in eukaryotes and bacteria that entail recombination without a typical sexual cycle. The parasexual cycle of fungi and oomycetes involves the fusion of two haploid nuclei within the same hypha to form a diploid nucleus that subsequently regains haploidy by chromosome loss. The two fusion nuclei are usually sisters, since somatic incompatibility prevents mycelial fusion and the formation of heterokaryons. Recombination of novel mutations occurs when different copies of a chromosome from the fusion nuclei are lost, or through mitotic recombination in the diploid nucleus. Schoustra *et al.* (2005) selected for vegetative growth rate in *Aspergillus nidulans* by transferring mycelium from the growing front of rapidly growing colonies for 3000 cell cycles. The ancestor carried a mutation conferring resistance to the fungicide fludioxonil but reducing vegetative growth rate in the absence of the fungicide by about 50%. Haploid and diploid lines showed little or no response to selection, but haploid segregants from diploid lines expressed much higher growth rates on permissive medium. This appeared to be attributable to recessive mutations arising in the diploid phase that were recombined during re-haploidization and interacted to compensate for the cost of resistance.

Sex is the most complicated mechanism for genetic exchange. There are several simpler systems, such as bacterial transformation and conjugation, which have somewhat similar consequences, insofar as two genes from different lineages are brought into the same lineage. All are clearly adaptations, but they are much more difficult to explain than morphological or physiological adaptations because they alter the rules of the game, so to speak, rather than the positions of the pieces. Imagine an hypothetical genetic element κ that is able to move by some unspecified mechanism from the genome where it currently resides to another random genome. How could it spread through a population? There seem to be at least four possibilities.

In the first place, it might supply some useful or even essential feature to its new lineage. This is a natural interpretation of integrons that confer the ability to utilize a novel substrate, for example, or conjugative plasmids that bear antibiotic resistance genes (§4.2.2).

Secondly, it might be passively swept along by the genetic flux of the population. If κ happens to arise in the generating lineage its future is assured, albeit its propensity to move will be a handicap. If it arises in any other lineage, it will persist only if it moves into the generating lineage. Given that it can move at all, however, this is not unlikely, because the generating lineage for the population in the near future is likely to be very abundant. Most of the cells in the generating lineage will themselves fail to leave descendants in the more distant future, so κ must again move if it is to remain associated with the generating lineage. In a population at equilibrium between selection and deleterious mutation the generating lineage is almost certainly one of the least-loaded lineages, whereas κ will probably arise in a genome with moderate or high load. When it moves, it will sometimes move into a more lightly loaded genome. This is not a member of the generating lineage and is doomed to extinction, but before this can happen κ will have moved again. As time goes on the chance that a random move will encounter a still less heavily loaded lineage becomes less, because such lineages become steadily less frequent, but the number of moves that κ can make will increase, because it occupies a more fit lineage. In this way it will become steadily more concentrated in highly fit lineages until it enters the generating lineage. It will then be steadily lost from this lineage (because any move takes it to a doomed lineage) but will nonetheless persist provided that the lineage proliferates rapidly enough. In order to move up the ranks in this way, κ must transfer itself to a lineage with one fewer mutation before all the least-loaded cells in which it currently resides have acquired one more mutation.

Thirdly, an element that can replicate as well as move will be autoselected as a genomic parasite. A broad variety of such elements infect the genetic systems of eukaryotes (§9.1).

Finally, an element may not only replicate and move but also carry unrelated genetic material with it. In this case the genes it transports are also moved from one lineage into another, creating new genetic combinations that may be favoured by selection. Elements such as the F plasmid of *E. coli* may therefore spread as the indirect consequence of the recombination they cause, although it seems more likely that they function primarily as parasites and their ability to transport host DNA is an accidental or even deleterious side effect.

Some plasmid genes, such as those governing recombination, also participate in the sexual cycle of eukaryotes. Whether bacterial conjugation is directly ancestral to sex is far from clear, although gender and fusion in eukaryotic microbes often seems to involve mobile elements. It is conceivable that mating-type genes arose independently as transposable elements able to encode fusion, behaving in a manner similar to the conjugative plasmids of bacteria (§4.2.2) (Hickey 1982). Indeed, I have suggested that most of the characteristic features of the eukaryote genome involve domesticated selfish genetic elements that originally drove the sexual cycle or were parasitic on it (Bell 1994). Telomeric sequences at the tips of chromosomes, for example, are restocked by transposition. Gender would then initially evolve through the autoselection of parasitic genetic elements. This view is not widely accepted. The origins of linear chromosomes, segregation, meiosis and other components of the sexual machinery have often been discussed (see Hurst and Randerson 2000) but remain obscure.

12.2 The alternation of generations

12.2.1 Vegetative theories of the life cycle

The life cycle of sexual eukaryotes comprises two growth cycles, representing the proliferation of haploid and diploid individuals, linked by the sexual cycle. Haploid individuals (gametophytes) produce haploid gametes. These fuse to form a zygote that develops as a diploid individual (sporophyte). The sexual germ cells undergo meiosis to form haploid spores that close the cycle by developing as haploid individuals. Because sexual fusion must be balanced by a reduction division, this alternation of haploid and diploid generations is a necessary and fundamental feature of sexual lineages. However, the relative importance of haploid and diploid growth varies enormously. In some organisms, such as charophytes and many unicellular and filamentous algae and fungi, the zygote is the only diploid structure in the life cycle, and only haploid individuals grow and reproduce. Conversely, in groups such as animals and flowering plants it is the diploid generation that is responsible for growth; in animals, indeed, the gametes are the only haploid structures in the life cycle. Other organisms, including many seaweeds, grow both as haploid and as diploid individuals. Most attempts to explain the evolution of the life cycle refer to the different effects of natural selection in haploid and diploid populations growing vegetatively (reviews by Valero *et al.* 1992, Mable and Otto 2001).

The simplest difference between haploids and diploids is gene dosage. The consequences of this difference are well known in yeast: diploid cells are bigger, contain twice as much DNA, RNA, and protein, and express higher specific activities for a range of enzymes. This need not lead to any difference in fitness: if the rate of work is twice as high in diploid cells, the amount of machinery needed to sustain it is twice as great. When the volume of a cell is doubled without change in shape, however, its surface area does not increase proportionately, so that the ratio of surface area to volume decreases. If growth is limited by some externally supplied nutrient, diploid cells may have a lower rate of growth because the rate of flux through the cell surface is less per unit volume (Lewis 1985). One weakness of this interpretation is that it defines an advantage for haploidy, but not for diploidy. It might be that the greater rate of transcription in diploid cells is advantageous when nutrients are present in excess, by permitting the evolution of rapid but wasteful resource utilization. In dilute media, haploids would grow faster because of their greater relative surface area, and also because the genome itself would not represent as great a drain on limited resources. Thus, the interaction of ploidy with environment would favour diploid growth when resources are plentiful, but haploid growth when they are scarce. Adams and Hansche (1974) grew isogenic haploid and diploid strains of yeast in different media and found that there appeared to be little if any difference in growth rate related to ploidy when all nutrients were supplied in excess. (This was surprising, because one of the commonplaces of yeast genetics is that diploids almost always outgrow haploids in normal laboratory conditions.) When organic phosphate was limiting to growth, the haploid strain had a selective advantage of about 7%, and displaced the diploid in competition trials. Yeast cells cannot take up organic phosphate directly, but first hydrolyse it with an extracellular acid phosphatase. It seems likely, then, that selection for haploids in this experiment is attributable to their greater relative uptake rate. Mable (2001) found that haploid yeast grow faster than isogenic diploids in rich medium and equally well in minimal medium. Destombe *et al.* (1993) found that the gametophyte of *Gracilaria verrucosa* (Rhodophyta) grew better than the isomorphic tetrasporophyte in normal seawater but less well in supplemented seawater. Mable and Otto (2001) cite inconsistent results from other organisms, however. In short, the experimental evidence does not support a metabolic interpretation of ploidy.

The most popular theory of ploidy is genetic redundancy: diploidy provides a protection against deleterious recessive mutations by supplying a spare copy of every gene. Korona (1999) accumulated deleterious mutations in haploid mutator (mismatch-repair-deficient) lines of yeast until growth had been reduced by about 25%. This handicap was expressed by homozygous diploids made from the haploids, whereas heterozygotes were much less affected. Mable and Otto (2001) found that diploid strains suffered less growth reduction than haploids after treatment with a chemical mutagen. Thus, if a population with haploid growth were to switch abruptly to diploid growth, its mean fitness would increase because many of the deleterious mutations maintained in the population would be shielded by normal alleles. Diploid growth is then unconditionally advantageous; the theory does not explain why so many creatures are haplontic. This advantage is transient, because new deleterious mutations would then proceed to accumulate in the diploid population. Indeed, the mutational load at equilibrium will be greater in the diploid population, because the selection acting against deleterious mutations is weakened when they are complemented by dominant alleles. However, if the population were now to revert abruptly to haploidy, all the deleterious recessive mutations that had accumulated would now be expressed, causing a substantial reduction in mean fitness. Sliwa *et al.* (2004) accumulated mutations in diploid strains of yeast for several hundred generations until growth had fallen by about 10%. When these were sporulated the haploid progeny were severely handicapped, with cell density reduced by a factor of 1000 (Figure 12.6). The evolution of diploidy is then virtually irreversible. This is not the same as historical contingency, where the course of evolution is essentially irreversible because it is highly improbable that selection will cause a population to retrace precisely the same pathway of genetic change. There is nevertheless no objection in principle to it doing so: contingency operates precisely because the many different pathways to the same outcome are all more or less equally navigable. Diploidy may be irreversible because it blocks the path behind it.

Vegetative ploidy will affect the rate at which populations respond to natural selection because the rate of appearance of beneficial mutations should be about twice as great in diploid populations. Paquin and Adams (1983) surveyed successive selective sweeps in chemostat cultures of yeast. The rate of fixation of beneficial mutations in these experiments was estimated to be 3.6×10^{-12} per cell per generation for the haploid cultures, but 5.7×10^{-12} for the diploids (cf. §4.3.2). The diploid populations were thus evolving about 60% faster than the haploids, at least on a per-cell basis (Figure 12.7). This is consistent with the occurrence

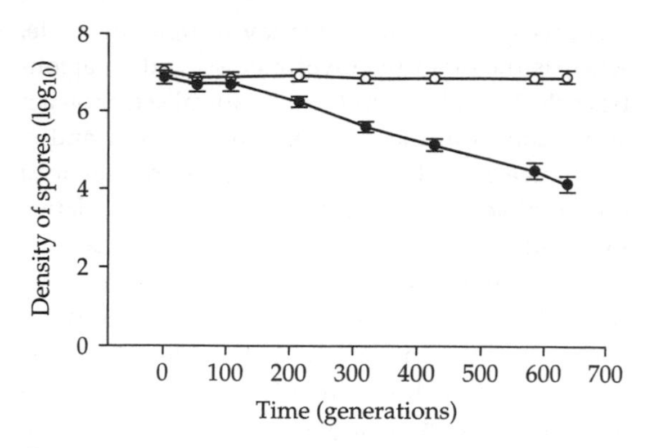

Figure 12.6 Loss of fitness on reverting to haploidy after mutation accumulation in diploidy. From Sliwa *et al.* (2004).

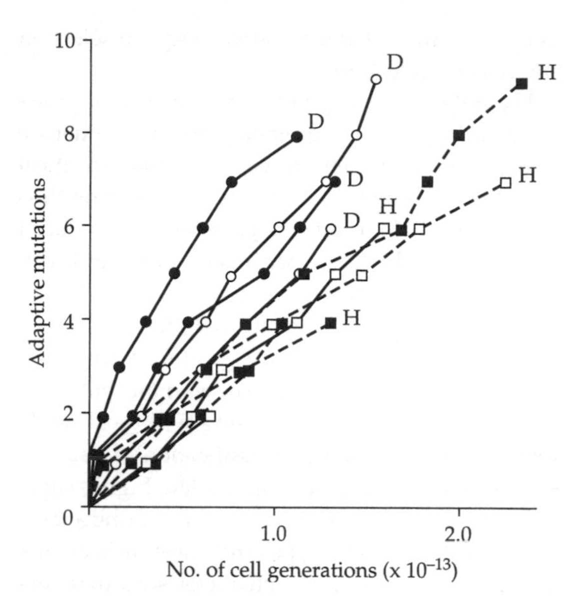

Figure 12.7 Rate of evolution in haploid and diploid yeast lines. Number of beneficial mutations fixed was estimated from observations of selective sweeps. Haploid and diploid lines are labelled H and D. From Paquin and Adams (1983).

of partly dominant mutations at about twice the rate in diploids. However, it is offset by the greater abundance of the haploids: because they are smaller, they were about 40% more dense than the diploids. The overall rate of fixation per population was thus only slightly greater for the diploids. The interpretation of this experiment has been clouded because the frequency of sweeps seems too high, and it is possible that the populations were polymorphic. Moreover, diploids will not necessarily adapt more rapidly. If a diverse population is sorted by selection, a haploid population will respond more quickly than a comparable diploid population, because every gene is expressed in haploids, whereas the variance of fitness in diploids is reduced by dominance. It is only if novel mutations are the primary source of adaptation to a novel environment that a diploid population will evolve more rapidly because it contains twice as many targets for mutation. Hence, diploidy reduces the waiting time whereas haploidy reduces the fixation time (Orr and Otto 1994). The effect of ploidy on the rate of response to selection will then depend on the dynamic regime: diploid populations should evolve faster (by a factor of about 2h) at low mutation supply rate and haploids should evolve faster (by a factor of about h) at high mutation supply rate. Zeyl *et al.* (2003) grew isogenic yeast lines in minimal medium and found that large (10^7 cells) haploid populations evolved about 70% faster than diploids during the first 2000 generations of culture, whereas there was little or no difference in small (10^4 cells) populations.

Diploidy might also permit more extensive adaptation because a mutation that represents the acquisition of a new activity, at the expense of losing its original activity, can be selected, because the original activity is still expressed by the unmutated allele. In the experiment of Zeyl *et al.* (2003) there was no difference between haploid and diploid strains in large populations after 5000 generations, however, indicating that they had equivalent capacities for the ultimate level of adaptation.

12.2.2 Sexual theory of the life cycle

Theories that attempt to explain sexual life cycles solely in terms of natural selection acting on vegetative characteristics seem to me likely to be incomplete. An alternative point of view is that the life cycle comprises two cycles of vegetative growth linked by the sexual cycle, and interprets the relative extent of haploid and diploid growth as representing a balance between sexual selection and natural selection.

The only clear-cut genetic consequence of growing as a haploid or a diploid concerns the nature of the gametes. When growth is haploid, identical gametes are produced by mitosis; when growth is exclusively diploid, diverse gametes are produced by meiosis (Bell 1982). In organisms with highly differentiated male and female gametes, sexual competition among male gametes will be very severe, because so many more are produced than can possibly succeed in fusing. Any variant that has a slightly greater chance of fusing will be strongly selected, so the meiotic diversification of gametes will be worthwhile if it creates a few highly superior gametes, even at the expense of creating a large proportion of inferior types that have little chance of success. Moreover, the characteristics that contribute to the sexual success of a male gamete are unlikely to be the same as those that enhance vegetative performance. The opposed effects of sexual selection and natural selection will exacerbate this antagonistic pleiotropy over the life cycle as a whole, because alleles that enhance sexual ability, even at the expense of vegetative ability, will be favoured during the sexual cycle, whereas those that enhance vegetative performance, even at the expense of sexual ability, will be favoured during the vegetative cycle. Sexual selection will therefore continue to act on variation in gametic performance, and this will indirectly select for diploid growth. The situation is quite different if gametes of different gender are nearly the same size, because most will succeed in fusing, and sexual selection will be weak. Moreover, because the gametes are not strongly differentiated, they may be selected for nearly the same qualities as vegetative cells; in many microbes, especially, gametes and vegetative spores are very similar in structure and behaviour. Sexual selection and natural selection will then act predominantly in the same direction, and will jointly favour a single type that is successful both as a gamete and as a spore. This will indirectly select for a haploid growth cycle that proliferates uniform gametes mitotically.

The life cycle could be controlled by a gene acting in the zygote to determine that the first division should be mitotic or meiotic. One allele (M) gives mitosis, and thereby a diplontic cycle, whereas the other (m) gives meiosis and a haplontic cycle. Sexual selection and natural selection act at a second unlinked locus with alleles A and a. Vegetative growth of a (in the haploid phase) or aa (in the diploid phase) individuals is $1 - v$ relative to A and AA individuals, whereas the sexual success of A gametes is $1 - s$ relative to a gametes. This fitness locus will go to fixation for one or the other allele, and in doing so it induces a change in the frequency of alleles at the life-cycle locus. This is because selection creates a correlation between m, the allele that leads to haplonty, and the allele with higher vegetative growth. Linkage disequilibrium between the two loci is expressed by $D = f_{MA} f_{ma} - f_{Ma} f_{mA}$, and if $v > 0$ then $D < 0$, with an excess of mA and Ma chromosomes. If $s > v/2$ the advantage of the Ma chromosome in gametes more than balances its disadvantage in vegetative individuals, assuming no dominance. Diplonty will then increase if $D(s - v/2) < 0$ (Bell 1994). Thus, when A and a alleles have opposed effects on vegetative and gametic function diplonty will be favoured, and conversely if they have similar effects haplonty will be favoured.

Unicellular eukaryotes with unspecialized gametes resembling vegetative cells are usually haplontic. Multicellular eukaryotes with specialized gametes are usually diplontic, and within groups such as phaeophytes, chlorophytes, and land plants there is a tendency for types with more highly differentiated gametes to be diplontic. The sexual theory of the life cycle is thus broadly consistent with comparative patterns, but there has not yet been any experimental work in this area.

12.2.3 Sporophyte and gametophyte

In organisms with both vegetative cycles haploid individuals (gametophytes) and diploid individuals (sporophytes) may have completely different structure, so that the life cycle involves an alternation of morphologically different types. This is Hofmeister–Strasburger alternation, named for two nineteenth-century biologists principally responsible for elucidating it. It must be clearly distinguished from the Steenstrup alternation of phases (named for another nineteenth-century biologist) in groups such as hydrozoans and trematodes, which does not involve a change of ploidy and has no sexual consequences. A purely vegetative theory of gametophytes and sporophytes is possible: multicellular haploid and diploid phases can coexist if they are ecologically distinct, occupying niches with separate density regulation (Stebbins and Hill 1980, Jenkins 1993, Hughes and Otto 1999). For isomorphic phases with equal demographic rates, Thornber and Gaines (2004) show that the frequency of gametophytes at equilibrium is $2/(2 + \sqrt{2}) \approx 0.6$ and compare this prediction with data from field surveys. Organisms like ferns and seaweeds, however, are typically heteromorphic: the sporophyte is usually a large erect structure extending into the air or water column whereas the gametophyte is a small filamentous or thalloid creature living in the sediment. Bower (1908) explained this dichotomy through the different ecologies of spores and gametes. Spores require resources in order to grow, so they are likely to benefit from dispersal to unoccupied sites, which can best be achieved by release from a large upright sporophyte. Gametes require fusion partners and so benefit from aggregation, which is most effectively achieved by scattering small prostrate individuals close to neighbours. This gives a satisfactory interpretation of life cycles in groups such as phaeophytes (Bell 1997b). There are many complications (such as the existence of haploid and diploid sporophytes), however, that I have no space to discuss and that still await theoretical and experimental investigation.

Where gametophyte and sporophyte are different in form they will respond to different sources of selection. In flowering plants haploid growth is very restricted, and the male gametophyte comprises only three haploid nuclei. Mulcahy *et al.* (1978) suggested that large populations of pollen grains could supply a very effective selective screen for characters that are subsequently expressed in the sporophyte. This is an attractive idea, because a substantial fraction of the genome is expressed in the gametophyte, including many loci that are also expressed in the sporophyte. (It would not work in animals, where the properties of sperm seem to be controlled by the diploid genome of the gamont, rather than by the haploid genome of the gamete.) Moreover, it is sometimes possible to grow haploid plants from germinated pollen grains. The technique has not been used extensively so far, largely because of technical difficulties in recovering the selected pollen and using it to fertilize female plants. It will be effective only if there is a positive genetic correlation between gametophytic and sporophytic expression, so that sporophytic performance advances as a correlated response to gametophytic selection. It may be doubted that this will be generally true. There is some evidence for a positive correlation, especially with regard to tolerance of stresses such as high temperature, salinity, or toxins. However, a positive correlation is to be expected when plants are exposed to novel conditions of growth, and may be reversed under continued selection. Moreover, one might also expect that selection of gametophytes for vegetative performance would induce a deterioration in their sexual performance. A negative genetic correlation between gametophytic and sporophytic success will impede adaptation to either way of life and may preserve genetic variation at loci which are expressed in both phases.

12.3 Gender

Selection of any kind acts through the differential proliferation of lineages; how it acts depends on the structure of these lineages, that is, on the way in which alternative genes are associated during development and transmission. This intertwining of genealogies is expressed through the concepts of gender and species. The two concepts are complementary, and between them define the limits to sexual fusion. Gametes are unable to fuse if they belong to the same gender or to different species; they may fuse only if they have different gender and belong to the same species. Genes, or other entities, may compete only if they belong to a common category, within which one may displace another. Natural selection is the consequence of competition within the species; sexual selection is the consequence of competition within a gender. This is a clear and useful point of view for thinking about most familiar animals and plants. For other organisms the situation is less straightforward, as the concept of gender becomes extended, and the concept of species breaks down.

12.3.1 Many genders

Genes that control gender enforce a simple rule: recognize and reject self. Beyond that, few generalizations are possible. Sexual populations of many green algae and ascomycete fungi are bipolar, with two mating types. In the simplest cases, such as *Chlamydomonas* and *Neurospora*, each mating type is controlled by a single locus: each individual bears a single copy of the gene, which is necessary and sufficient for the expression of mating type, and individuals do not bear copies of both mating-type genes. They are thus heterothallic: all the members of a given clone express the same mating type. One of the most intriguing recent discoveries of molecular genetics is the extraordinary nature of the mating-type genes themselves. They occupy homologous regions of the genome (and therefore segregate in Mendelian ratios at meiosis), but they are not allelic. In all the bipolar fungi that have been investigated so far, they are of different size, and incorporate coding regions that are in different positions, may be transcribed in different directions, and encode substantially different products. In *Chlamydomonas* the mating-type locus is an extensive tract within which recombination is suppressed that bears several genes and other genetic elements, some clearly related to sexual activity and others not, which differ between plus and minus mating types (Ferris *et al.* 2002). They are so different that it is difficult to believe that they have diverged from a single common ancestor. Such genes are said to be idiomorphic. Some microbes, such as yeast, are bipolar but homothallic with genders termed a and α. Yeast mating-type genes are idiomorphic and transposable. Each cell bears a copy of one of the two mating-type genes at each of the silent cassette loci *HMRa* and *HML*α, one of which has been copied to the *MAT* locus responsible for mating-type expression. When a daughter cell is formed by budding, the resident copy is removed from the expression locus via a double-strand break made by an endonuclease encoded by the *HO* locus and replaced by a copy from the other cassette, resulting in a switch of mating type. The cell therefore has a single sexual identity, but the clone expresses both. Heterothallic clones are easily produced by loss-of-function mutations in *HO*. Other microbes, including basidiomycetes and ciliates, have multipolar heterothallic systems (review in Brown and Casselton 2001). A common arrangement is for two loci to control sexual compatibility, with each locus having a series of alleles. In the basidiomycete *Coprinus cinereus* the *A* locus encodes transcription factors similar to the yeast MAT proteins whereas the *B* locus encodes two pheromones and a pheromone receptor. There are 3 regions within each locus, each with a series of alleles, yielding a total of about 12 000 mating types. There are conspicuous exceptions, however. The basidiomycete *Cryptococcus neoformans* has a non-switching bipolar system with a long recombination-suppressed mating-type region encoding several genes and containing remnants of transposons and other elements, rather like that of *Chlamydomonas* (Lengeler *et al.* 2002). Most metazoans have bipolar gender, but mating-type genes and sex-determination systems vary widely. Most of the genes involved in determining gender in the model organisms *Drosophila*, *Caenorhabditis*, and mouse are unrelated, albeit with elements that are phylogenetically conserved (review by Marin and Baker 1998). Tunicates are exceptional in permitting fusion only when the allele at a mating-type locus in sperm is absent from the diploid female parent (see Grosberg 1988). In many plants the gender of morphologically bipolar gametes or gametophytes is overlain by self-incompatibility systems which allow fusion only when the allele borne by the pollen at a bipartite S-locus is absent from the diploid pistil (gametophytic self-incompatibility) or in a more complex manner depending on the diploid genomes of male and female parents (sporophytic self-incompatibility) (reviewed by Lawrence 2000, Hiscock and McInnis 2003, Castric and Vekemans 2004, Charlesworth *et al.* 2005). The self-incompatibility systems of tunicates and plants manifest 10–100 genders.

This diversity suggests that gender may evolve rapidly. Probes specific to *mid*, which specifies mating type in *Chlamydomonas reinhardtii*, fail to hybridize to any related species, except one species in which the corresponding gene bears a large insertion. The *fus1* gene, which mediates successful gamete fusion, had no detectable homology with any other species (Ferris *et al.* 1997). In hemiascomycete yeasts the evolution of the mating system of *Saccharomyces sensu stricto* has involved two main events, the acquisition first of silent mating-type cassettes and subsequently of endonuclease-mediated gender switching (Butler *et al.* 2004). Distantly related species lack both, while the very distantly related *Schizosaccharomyces pombe* has a gender-switching system directed by a completely different molecular mechanism. The gamete surface proteins involved in signal transduction and fusion often evolve especially rapidly. They are often remarkably divergent in sexual microbes such as ciliates and *Chlamydomonas*. In *Drosophila*, accessory gland proteins transferred in the seminal fluid are about twice as variable as normal somatic proteins (Civetta and Singh 1995). Sperm of *Haliotis* (abalone) release a lysin from the acrosome that binds to a receptor on the egg surface, causing a hole to appear through which the sperm enters; lysins evolve many times faster than other genes (Metz *et al.* 1998). The proteins governing self-incompatibility in flowering plants can differ by up to 50% of their amino acids (Richman and Kohn 2000). These and other cases reviewed by Vacquier (1998) and Swanson and Vacquier (2002) provide good comparative evidence for rapid and sustained evolution of genes that govern mating and gamete fusion.

In short, gamete fusion is an exceptionally idiosyncratic process that varies widely among related organisms and evolves rapidly. In this respect it resembles immune systems, which are likewise concerned with self-recognition, and differs sharply from meiosis, the other pole of the sexual cycle, which is highly conserved. The agent of change is undoubtedly sexual selection.

12.3.2 Male and female

Microbes are almost always isogametic, the gametes having the same gross morphologicy irrespective of gender. Multicellular organisms (except fungi) are almost always oogametic, with two distinct classes of large and small gametes. This is mirrored on a small scale within the Volvocales, where gamete dimorphism is related to the grade of somatic complexity: unicellular species are almost all isogametic whereas large multicellular species have relatively small motile sperm and large immotile eggs. The evolution of specialized male and female gametes is thought to be caused by the opposition of sexual selection for numerous gametes and natural selection for successful zygotes (Parker *et al.* 1972, Bell 1978b, Charlesworth 1978, Maynard Smith 1978; review by Bulmer and Parker 2002). If gametes encounter one another largely by chance, the number of fusions involving a particular type of gamete will be proportional to its abundance. Producing more gametes from a more or less fixed mass of material implies that the gametes will be smaller. Sexual selection thus favours genes that direct the production of numerous small gametes. Males producing a larger number of smaller sperm fathered more offspring in a cricket, for example (Gage and Morrow 2003). Natural selection pulls in the opposite direction. The competitive ability of vegetative organisms is generally enhanced by large size. A large zygote is more likely to give rise to a successful vegetative lineage; this is especially true for multicellular organisms whose early embryonic development is fuelled by zygotic reserves (see Levitan 2000). Natural selection thus favours large gametes that form large zygotes when they fuse. The opposed tendencies of sexual and natural selection may cause gametes to evolve towards an optimal intermediate size or it may lead to disruptive selection on gamete size, depending on the form of the relation between gamete size and zygote survival. Larger gametes will give rise to larger zygotes, which are expected to be more successful. If zygote survival is a decreasing function of gamete size, stabilizing selection favours an isogamete of intermediate size, whereas if it is an increasing function disruptive selection may lead to the elimination of gametes of intermediate size in favour of microgametes and macrogametes. One reasonable model is that gametes have some minimal size δ and that zygote survival for gametes of size s_m is a sigmoidal function of size with an inflection point at some multiple $k\delta$ of the minimal gamete size. The size m of the locally stable isogamete is the size at which s_m/m is tangent to the zygote survival curve. If the point of inflection of the survival curve is far to the left (k small) this isogamete is likely to be locally and globally stable, but if it moves far enough to the right (k large) an isogametic population can be invaded by microgamete-producers, resulting in anisogamy (Bell 1982). A proof that the ESS is anisogamy when the survival curve is shifted to larger values of size is given by Bulmer and Parker (2002). Hence, markedly dimorphic gametes are likely to evolve in multicellular organisms.

In a population containing gametes of different sizes, natural selection will act against the zygotes produced by fusion between two small gametes. Small gametes will therefore be selected to fuse preferentially with larger gametes. This will lead to a correlation between gamete size and gamete gender, small gametes being of one gender and large gametes another. The simplest model consists of two loci with two alleles: alleles at the G locus control gamete size while the M locus is mating type. Germ cells in gametophytes bearing the g allele divide one or more times to form microgametes, whereas the G allele directs maturation without division to form macrogametes; hence g gametophytes produce n times as many gametes. A zygote formed by fusion of a microgamete with a macrogamete survives with probability s_{Gg} and a zygote from a fusion between two microgametes with probability s_{gg}, relative to the largest zygotes formed by fusion between two macrogametes. Rare macrogamete-producers can invade a population of microgamete-producers if the enhanced survival of their zygotes compensates for their lower gamete production, i.e. if $s_{Gg} > ns_{gg}$. Conversely, microgamete-producers can invade when rare if $s_{Gg} > 1/n$. These conditions are equivalent to those given in the previous paragraph. Gamete size becomes correlated with gender in an anisogametic population only if linkage disequilibrium develops between G and M. This requires not only that some initial disequilibrium should exist but also that G is linked to M, since a single generation of random mating restores linkage equilibrium when one of the loci concerned is mating type, and no correlation can then evolve. This implies that when the conditions for anisogamy are satisfied the male–female distinction will follow only if linkage with mating type is sufficiently close (Charlesworth 1978). This critical value is quite large when gametes are not strongly differentiated, although it becomes less restrictive when n is large and s_{gg}/s_{Gg} small. In the limit, when zygotes formed from two microgametes are inviable, the critical value is $\frac{1}{2}\,[(ns_{Gg} - 1)/(2ns_{Gg} - 1)]$, or $\approx \frac{1}{4}$ if those from microgamete–macrogamete fusion are fully viable. If gamete size is determined by the sporophyte genotype, however, male–female gender will not evolve with comparable values of gamete production and zygote viability unless G is within 1 cM of mating type. Thus, we expect that alleles governing the number of divisions in gametogenesis are likely to have large effects and will be linked to mating type.

The same process of specialization will extend to other properties of gametes, especially their motility. Small gametes competing for fusion partners will often evolve or maintain a high degree of motility, because the rate of encounter between gametes will be proportional to their relative velocities. Large gametes whose fitness is determined largely by the cytoplasmic resources they donate to the zygote will be selected in the opposite direction, because being more or less assured of fusion by the superabundance of small gametes of different gender they can conserve resources by evolving or maintaining immobility. This greatly reduces macrogamete–macrogamete fusion and thereby strengthens linkage disequilibrium. The specialized oogametic male–female system characteristic of metazoans and other large multicellular organisms has now been established. Although the process is usually modelled by allowing more divisions in microgametogenesis, this may not reflect the actual course of evolution in many groups. Microgametes resemble zoospores in phaeophytes and some chlorophytes, where the female gamete appears to be the derived type (Bell 1997a).

12.3.3 The gender of individuals and parts

Gender is an attribute of gametes, but gender can also be assigned to individuals on the basis of the gametes they bear. In outcrossed organisms, there is no automatic advantage in being an hermaphrodite, and hermaphroditism will be favoured only if total reproduction is greater when both types of gamete are produced. The fitness of a rare type that produces Q_f successful macrogametes (fertilized eggs) and Q_m successful microgametes is proportional to $(Q_f/\hat{Q}_f + Q_m/\hat{Q}_m)$ where $\hat{Q}_f$ and $\hat{Q}_m$ are the population averages (see Charnov 1982 for a descriptive proof). Consequently, if a shift in allocation causes a change in fitness through female function ΔQ_f and a corresponding change through male function ΔQ_m it will be favoured if $\Delta Q_f/\hat{Q}_f + \Delta Q_m/\hat{Q}_m > 0$. Selection favours a division of sexual labour within an individual in the same circumstances that it favours a similar division between individuals within a brood, or a division of vegetative labour between different kinds of cell: labour should be divided when similar structures interfere with one another, causing diminishing returns on further investment. Animals must make a substantial minimum investment in oviducts, shell glands, uterus, and the like before they can function properly as females; males require a similar list of equipment. Thus, to deploy a single ovum with any appreciable chance of success a male might have to deploy a complete female system, creating a small positive ΔQ_f at the expense of a large negative ΔQ_m. These large fixed costs thus favour the separation of the sexes.

Gender may also be assigned to parts according to their contribution to differentially facilitating the transmission of male or female gametes. We could argue, for example, that as the mass of a flower increases from C to $C + \Delta C$, the fraction of ovules fertilized increases from O to $O + \Delta O$ and the fraction of pollen dispersed from P to $P + \Delta P$. For any given flower mass the gender of the flower can be expressed in terms of its relative effectiveness in dispersing pollen as $\alpha = P/(P + O)$. As floral mass increases to $C + \Delta C$, floral gender will become $\alpha' = (P + \Delta P)/(P + \Delta P + O + \Delta O)$, so the increment of gender will be $\alpha' - \alpha$. Summing all these increments up to the actual size of the flower gives a definition of floral gender as $\int (O\,dP - P\,dO)/(P + O)^2$ plus a constant of integration representing the gender as C approaches zero. This will be zero if P does not change with C, unity when O does not change with C, and ½ when P and O are influenced identically by changes in C. Consequently, the gender of hermaphroditic organisms can be quantified by estimating the gender of parts, although this is far from easy in practice. I think that flowers have largely male gender in animal-pollinated plants, because larger flowers will attract more pollinators; the fraction of pollen dispersed is a steadily increasing function of the number of visits, whereas all or most ovules may be fertilized by a single visit (Bell 1985). The most important implication of this view, should it be correct, is that sexual selection will occur very generally among plants. Secondary sexual structures, such as the anthers and styles crowded into the middle of a flower, may also show functional interference; if a given mass of tissue promotes sexual performance more effectively when it is divided between anthers and styles, the flower itself, not only the individual, should be hermaphroditic. In organisms that have a unitary rather than a disseminated system of sexual structures, the reverse is more likely to be the case.

12.3.4 Equality under Fisher

In haplontic microbes with fixed mating types the zygote is necessarily heterozygous at the mating-type locus. In bipolar systems this implies that the frequency of either type is reset to equality in every sexual generation regardless of any disparity in frequency caused through natural selection by the differential vegetative proliferation of the mating types, provided that unfused gametes die. The same principle applies to diploid multicellular organisms with heteromorphic sex chromosomes, where we expect males and females to be nearly equally frequent at birth. Thus, the genetic control of mating type often ensures that the genders are equally frequent, and no evolutionary rationale is required. The sex ratio, indeed, is one of the few quantitative characters that do not readily respond to artificial selection in mice and *Drosophila* (Falconer 1955).

The sex ratio is more labile in more complex genetic systems, or when a simple Mendelizing system is perturbed (for example by meiotic drive), or when parents can influence the gender of their offspring (for example when gender depends on ploidy or the conditions of incubation). In such cases the proportion of male and female individuals in the population is governed by the sexual advantage of the minority gender (Fisher 1930; see Bull and Charnov 1988). If females are more abundant than males, a male will on average mate with more than one female, and will therefore produce more offspring than the average female. A gene that causes more sons than daughters to be produced will be selected; the reason is not that individuals bearing the gene will produce more children—the total number of sons and daughters may be fixed—but rather that they will produce more grandchildren, through their sons. The spread of a son-producing gene will increase the proportion of males in the population. If they become more abundant than females, the converse argument applies, and genes that cause the over-production of daughters will be favoured. The outcome will be equal numbers of males and females, provided that they are equally costly to produce; if an individual of either gender is more costly to produce, in units of fitness, it will be correspondingly less frequent at equilibrium. Negative frequency-dependent selection is therefore a very general feature of the evolution of gender that follows immediately from the equal genetic contributions of male and female parents to offspring. When this assumption fails, the principle fails (see Karlin and Lessard 1986 for theoretical treatments of a range of systems). It does not apply to sex chromosomes or mitochondria, for example, so meiotic drive on X or Y can spread, distorting the population sex ratio as it does (§9.2.4). It may not apply in multipolar systems, where selection is frequency-dependent but the representation of mating types is affected by vegetative success and mating dynamics (Iwasa and Sasaki 1983). The multiple mating types of ciliates such as *Paramecium* and *Tetrahymena* seem nevertheless to be nearly equally frequent in natural populations (Kosaka 1991, Doerder *et al.* 1995).

Selection experiments designed to test the basic Fisherian principle have seldom been attempted because of the strong genetic constraints on offspring sex ratio in most situations. Carvalho *et al.* (1998) propagated lines of *Drosophila* that carried the X-linked meiotic drive gene *sex-ratio* (§9.2.4) and were therefore strongly female-biased, but which were also segregating for autosomal suppressors of drive. Sex ratio shifted slowly in the expected direction, from less than 20% to more than 30%, in 50 generations. Gender in the fish *Menidia* is influenced by temperature; broods raised at low temperatures usually develop as females, whereas those raised at high temperatures develop as males. However, there also appear to be some genetic effects that modify the effect of temperature. Conover *et al.* (1992) established laboratory populations at different temperatures, so that their sex ratios were initially highly skewed in one direction or the other. In almost all cases the minority sex increased in frequency in the following generation. After a few generations of selection, the sex ratio in all populations had approached equality, where it remained for the duration of the experiment (Figure 12.8). Moreover, when selection lines were tested at a different temperature, sex ratios often became skewed again; selection for equal sex ratios was thus specific to the environment in which the selection was practised. The rapid response to selection was presumably fed by autosomal genes of large effect, which would behave like sex chromosomes. Gender in the fish *Xiphophorus* is controlled by a three-factor system, with three female genotypes (WX, WY, XX) and two male (XY, YY). Mixed populations founded with highly skewed sex ratios attained a nearly equal sex ratio in two generations, as expected from the segregation of the three sex factors (Basolo 1994).

12.3.5 Gender allocation

Nevertheless, a large excess of either sons or daughters is sometimes observed. In some cases this is caused by selfish genetic elements that manipulate the gender of their host to further their own transmission (§9.2). When this can be ruled out, selection must for some reason favour greater investment in gametes or offspring of one gender than of the other. The field of gender allocation, or sex allocation, has given rise to a large theoretical literature devoted to predicting the optimal investment in male and female function, generally by calculating

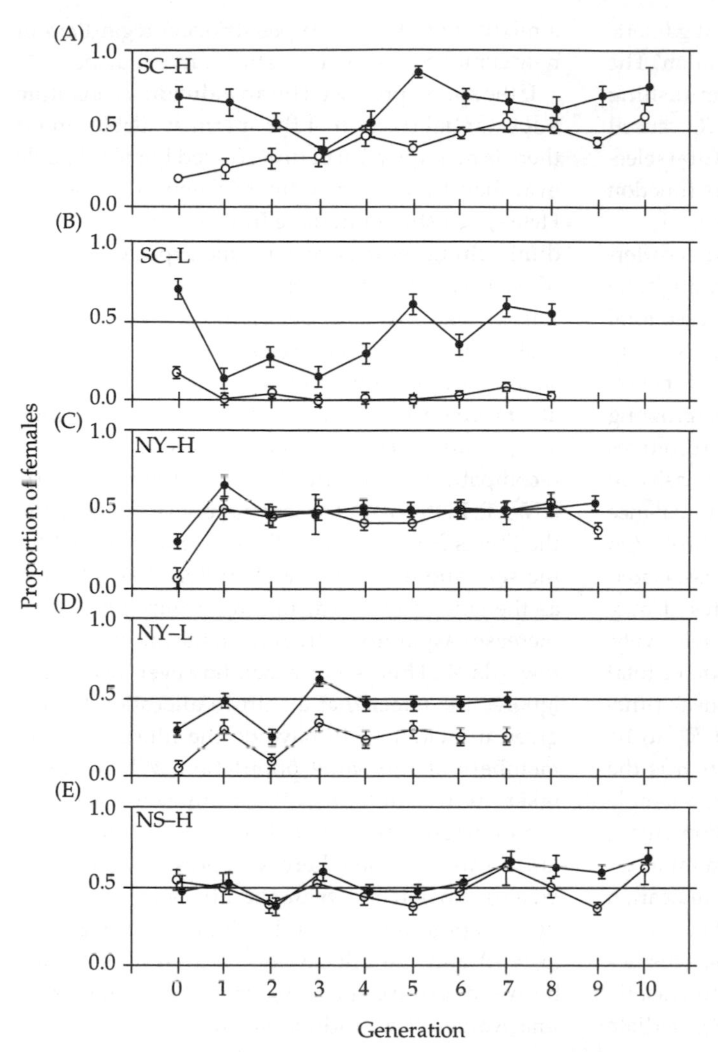

Figure 12.8 Response of environmentally determined sex ratio to selection. This diagram shows the outcome of propagating experimental populations of *Menidia* at different temperatures, from Conover *et al.* (1992; see also Conover and Van Voorhees 1990). Fish were collected from localities in South Carolina (SC), New York (NY) and Nova Scotia (NS), and subsequently raised at a constant temperature of either 28 °C (high temperature, H) or 17 °C (low temperature, L) during the sensitive period of larval development during which gender is determined. The solid lines show sex-ratios during ten generations of propagation, tested at the same temperature as the selection environment. The broken lines show the sex ratio when tested at the other temperature.

the ESS, which appears as a concept before the term was coined in treatments such as those by Shaw and Mohler (1953) and Hamilton (1967).The basic theory was developed and reviewed by Charnov (1982), who based his account on the fitness set, the plot of male on female components of fitness. Here I give a brief theoretical account of selection acting on gender allocation that is developed in a different way, in order to identify the mechanism of selection more explicitly, but that leads to equivalent conclusions. I assume that we are dealing with an outcrossed hermaphrodite in which gender allocation is governed by unspecified autosomal genes.

Since every individual has exactly one mother and one father, the fitness of an individual which allocates a fraction α of its total production G of gametes is $w = G(1 - \alpha)Q_f/\Phi + \alpha Q_m/\varphi$ where Q_f is the fraction of ova or ovules that are fertilized and Φ is the mass of an ovum, and Q_m and φ are the corresponding quantities for sperm or pollen. If we define $R_f = Q_f/\Phi$ as the unit return on investment in female function and $R_m = Q_m/\varphi$ as the unit return on investment in male function, then $w = G[(1 - \bar{\alpha}) R_f + \alpha R_m]$. It follows that there are three sources of selection on gender allocation leading to a balance between investment in male and female function.

The first is natural selection through total gametic output G, which I shall call 'overcompensation'. The second is sexual selection through diminishing returns to allocation to female function R_f, caused by local mate competition. The third is natural selection through diminishing returns to male function R_m, caused by local resource competition.

If male and female gametes are somewhat different in composition then their production might be limited by different resources, in which case total production would be maximized by producing both of them. Likewise, if male and female function is expressed at different times during the growing season (like flowers and fruits), while resources are accumulated continuously but are expensive to store, then there will some optimal overall balance of investment in male and female structures. On the other hand, if there are fixed costs associated with deploying male and female gametes, it may be optimal to produce one or the other exclusively. A combination of all these factors might make total gamete production G a complex curvilinear function of allocation α. Supposing R_f and R_m to be unaffected, $w = G[1 - \bar{\alpha}(2 - 1/\bar{\alpha})]$, where $\bar{\alpha}$ is the average allocation in the population. In a simple case, the plot of G on α, the compensation curve, might have a single intermediate maximum at α^*. The non-invasible pure culture then has allocation defined by $\tilde{\alpha} = [1/(2-1/\alpha)] - (G/G')$, where G' is the derivative of G with respect to α at $\bar{\alpha}$. This lies between $\alpha = ½$ and $\alpha = \alpha^*$, and is locally and globally stable. Conversely, if G has a single intermediate minimum then selection will lead to a mixture of pure males and pure females. More generally, if the compensation curve is quadratic, the ESS is either a single hermaphrodite or a mixture of pure males and females for almost all combinations of parameter values, and ESS hermaphrodites almost always have $0.2 < \alpha < 0.8$ and are seldom extremely biased. A compensation curve of arbitrary shape can readily lead to mixtures of hermaphrodites and females (gynodioecy) or hermaphrodites and males (androdioecy). In gynodioecious populations the allocation of the hermaphrodite lies between α^* and 1 if G for pure females exceeds G for any hermaphrodite, and otherwise lies between ½ and α^*; the converse holds for androdioecy. The ESS is never a mixture of all three types (trioecy) regardless of how complex the compensation curve may be.

If the sperm produced by an individual constitute a substantial fraction of the sperm available, and if there is no large pool of unfertilized but fertilizable ova, then they must compete among themselves. Hence, investment in male function will meet with diminishing returns because the expected success of each sperm falls with the number produced. This situation will arise when the population is broken up during the breeding season into many small groups each comprising only N individuals. Disregarding any effect on G, fitness is then $w = (1 - \alpha) + N(1 - \bar{\alpha})\{\alpha/[\alpha + (N - 1)\bar{\alpha}]\}$ given self-incompatibility, and the ESS is then $\tilde{\alpha} = (N - 1)/(2N - 1)$. In the case of strictly random gamete fusion the ESS is $\tilde{\alpha} = ½[N/(N + 1)]$, as first worked out for the sex ratio of wasps by Hamilton (1967). Hence, as the size of the local mating group increases $\tilde{a}$ increases asymptotically from some minimal value towards ½. This is not exact, however, because it ignores the effect that a shift in allocation by any given individual will have on the fitness of other members of the same population. When this is taken into account, $\tilde{\alpha} = ½$ for any single isolated group, irrespective of N. The Fisherian allocation fails, however, when there is dispersal, because this creates tension between the allocation favoured by selection between individuals within groups ($\alpha = ½$) and the allocation favoured by selection between groups ($\alpha = 0$, or rather some minimal value). If all individuals disperse at random and immigrants are received from M other local groups, the ESS allocation is $\tilde{\alpha} = H\{1 + 1/[N + M(N + 1)]\}$, where H is the Hamiltonian allocation. More generally, if m is the fraction of zygotes produced in each group which remain within that group rather than joining the dispersal pool, the ESS allocation is $\tilde{\alpha} = ½\,[1/(1 + X)]$, with $X = (1/N)[(1 - m)M/(1 + mM)(1 + M)]$. This is nearly Hamiltonian, provided that M is large and m fairly small. The theory of sex allocation in structured populations was developed by Wilson and Colwell (1981), and extended to continuously distributed populations by Charlesworth and Toro (1982).

Allocation to macrogametes will meet with diminishing returns if they compete for some limiting resource that becomes progressively depleted as the number of macrogametes produced increases. The simplest case is sib competition: after fertilization, sibs are reared together and compete for some limiting resource such as space within a brood pouch or nourishment provided by the parent. The ESS then depends on how survival is related to sib number, although it will generally represent an hermaphrodite that spends more on male than on female function. More generally, individuals may compete in small groups of $(N + 1)$ individuals with a fraction $(1 - m)$ of fertilized ovules and $(1 - \beta)$ of pollen being dispersed to other groups. The survival of offspring depends on the total number of propagules germinating at the site, which for simplicity I take to be negative exponential with slope k. There are now five components of fitness: surviving fertilized ovules which remain in the same site, fertilized ovules dispersed to other sites, pollen fertilizing ovules in the same site and remaining there, pollen fertilizing ovules which are then dispersed, and pollen fertilizing ovules in other sites. If the number of other sites is large, the ESS allocation is a root of the quadratic equation

$$0 = -X\tilde{\alpha}^2 + (X - Y - 1)\tilde{\alpha} + Y,$$

with $X = km^2(1 + h)$, $Z = 1 - h^2/N$, and $h = \beta N/(N + 1 - \beta)$. As expected, ESS male allocation increases with seed residency (m) and the severity of local resource competition (k). Pollen residency (β) has little effect on allocation unless N is small, when local mate competition participates in setting the ESS.

The theory of gender allocation has provided an elegant interpretation of patterns of male and female allocation that has been supported by many comparative studies (reviewed by Charnov 1982). I have given only a brief sketch of the theory because I have not found any selection experiments designed to test its predictions. Decisive tests of the theory in hermaphroditic plants have been hampered by the difficulty of measuring both allocation α and fitness gain R; the attempts to do so have been reviewed by Campbell (2000). There are several experimental studies in insects, however, showing that individuals can alter progeny sex ratio appropriately as circumstances change (e.g. Werren 1980).

12.3.6 Homothallism

In homothallic microbes self is not rejected and fusion occurs between clonal gametes, either because gender switches from mother to daughter or because no gender distinction exists. Homothallism occurs in most or all major groups of protists, although I can find no estimates of its frequency. Several genera of ascomycetes contain both homothallic and heterothallic species. In many cases homothallism is associated with loss of one of the two mating type loci in bipolar species, but in *Neurospora terricola* both mating-type loci are present in the same mycelium (Glass *et al.* 1990). Homothallism arises from heterothallism in *Cochliobolus* by an unequal cross-over that effectively fuses the two mating-type idiomorphs (Turgeon 1998). *Chlamydomonas* contains both heterothallic and homothallic species.

Intense sexual selection applied to experimental populations of *C. reinhardtii*, a heterothallic species, resulted in the spread of homothallic strains in most selection lines (Bell 2005). The phenotype was very labile, disappearing during vegetative propagation but being restored by heterothallic mating. Its behaviour is reminiscent of a known mutant of *C. reinhardtii* in which a copy of the *mid* mating-type gene has been transposed to an autosome. When this copy is transferred to a mt+ background by mating, it creates a genotype that mates as minus. A clone of this genotype will be rendered unstable by excision of the unknown element (designated *ele*) responsible for transposition. Complete excision of *ele* and the associated copy of the *mid* gene restores a stable plus line. Incomplete excision that removes part of *ele* but retains *mid* yields a stable minus line. A growing culture will in this way come to contain both plus and minus gametes, and will therefore express homothallic mating. Crosses between plus and minus strains isolated from the culture will show an apparently regular segregation of mating type. Homothallic mating, between a minus gamete bearing *ele* and a plus gamete from which *ele* has been excised, will often yield zygotes homozygous for *ele* through replicative transposition during meiosis. The progeny will be a mixture of gametic phenotypes: plus, if excision occurs early in culture growth; minus, if no excision occurs; and homothallic if excision occurs during colony growth to produce a mixture of gametes bearing and lacking *ele*. This hypothesis has not yet been confirmed by molecular analysis, but the ease of obtaining homothallic lines from a heterothallic ancestor demonstrates the evolutionary lability of gender.

Homothallic mating in haploid microbes has few direct genetic consequences, since the zygote is a doubled haploid and meiosis restores the original haploid genomes. The most obvious advantage of homothallism is the greater probability of mating, since any other gamete is an acceptable partner. This may be important in natural populations where encounters with conspecifics are infrequent. Homothallism also protects the population against sexual sterilization by selective sweeps. In the *Chlamydomonas* experiment, asexual lines soon became fixed for a single mating type. This is very unlikely to have been the result of drift or sampling, as the number of cells transferred was in excess of 10^5. It was not caused by an inherent competitive advantage of one mating type, because that mating type would have been fixed in all lines. It is most plausibly attributable to a selective sweep by some mutation conferring greater fitness in the conditions of culture, arising on an arbitrary mating-type background and thereby carrying this mating type with it as it spread through the population. The fixation of one mating type, of course, precludes mating in a heterothallic population. This process of sexual sterilization, which seems to occur so rapidly and predictably in laboratory cultures, might be an important constraint on the evolution of heterothallic sexual systems in natural populations.

12.3.7 Self-fertilization

In self-compatible diplontic organisms self-fertilization involves the fusion of male and female gametes produced by the same diploid parent. In microbes several or many vegetative generations necessarily intervene between successive sexual episodes. Homothallic fusion in a diplont such as yeast makes selection more effective by inflating the genetic variance of fitness, and thereby allows recessive deleterious mutations to be eliminated more rapidly, or the most beneficial combination of beneficial alleles to be substituted more rapidly. Whether or not this will be favourably selected depends on the dominance of the mutations and the number of intervening mitoses, but I can find no theoretical account of the process. Knop (2006) has shown that the products of intratetrad mating in yeast quickly eliminate vegetative competitors. The extreme longevity of Woodruff's isolate lines of *Paramecium* (Woodruff 1926) was associated with autogamy and may have been attributable in part to the occasional purging of deleterious mutations from the micronuclear genome.

At least 30% of hermaphroditic plants and animals are self-compatible and produce offspring by selfing (see Jarne and Charlesworth 1993). Like homothallism, self-fertilization provides reproductive assurance when sex is obligate and mating partners rare. Other things being equal, however, selfing should replace outcrossing because an allele increasing the rate of selfing spreads automatically, even at high density. Suppose in a population of outcrossed hermaphrodites a rare variant arises that makes the same allocation to male function but that causes all its ovules to be self-fertilized at negligible expense in terms of pollen. If all the ovules of outcrossed plants are fertilized then selfed and outcrossed plants yield an equal number of fertilized zygotes, as the selfed plants produce as many ovules and nearly as much outcrossed pollen as do the obligately outcrossed plants. Each selfed zygote includes two haploid genomes derived from the parent, however, whereas each outcrossed zygote contains only a single genome from each parent. A selfed parent is therefore transmitting genes through its ovules twice as fast as an outcrossed parent: it will transmit in all $[2(1 - \alpha) + \alpha]$ times as many haploid genomes. A rare allele that causes selfing will thus spread $[2(1 - \alpha) + \alpha]$ times as fast as a prevalent allele causing obligate outcrossing (Fisher 1930, Nagylaki 1976). Furthermore, selfing preserves adaptive combinations of alleles that would be disrupted by outcrossing, provided that the optimal genotype is an homozygote. This principle is routinely used by animal and plant breeders, who inbreed selected lines until the desired combination of characters is transmitted.

The main drawback of selfing is that in multicellular organisms the expression of deleterious recessive mutations in progeny constitutes inbreeding depression, which reduces the mean fitness of progeny and thereby counteracts the automatic advantage of selfing. The modern theory of selfing and inbreeding depression was developed principally by the Charlesworths (Charlesworth 1980, Charlesworth and Charlesworth 1978, 1979,) and others (e.g. Lloyd 1979, Wells 1979, Lande and Schemske 1984) in the 1970s. Supposing the mean relative fitness of selfed progeny to be $1 - s$ then it follows that an allele causing selfing will be fixed if $s < 0.5$ and lost if $s > 0.5$, there being no stable intermediate frequency. The precise threshold of s may differ if reproductive assurance or the genetic mechanism of self-incompatibility is taken into account, but in any case the condition tends to be self-reinforcing because low inbreeding depression will favour selfing, which tends to reduce the mean load and thereby further reduce inbreeding depression, whereas the load will tend to increase under outcrossing and thus penalize selfing more severely. This explains why most species of plant are almost exclusively outcrossed or almost exclusively selfed (Vogler and Kalisz 2001). Populations of many organisms retain both selfing and outcrossing, however, the cleistogamous and chasmogamous flowers of *Impatiens* providing a familiar visual reminder. If these intermediate cases represent optima, then the severity of inbreeding depression must increase with the frequency of selfed zygotes (Maynard Smith 1978, Charnov 1987, Jarne and Charlesworth 1993). One possible mechanism is that the relative fitness of selfed zygotes will fall with the frequency of selfing. Supposing that parents are well-adapted to their home sites, that selfed offspring more closely resemble their parents, and that selfed offspring are consistently inferior at random germination sites, then selfed seed should fall close to the parental plant whereas outcrossed seed should be widely dispersed. Granted that this correlation has evolved (as in the cleistogams and chasmogams of *Impatiens*) then as the production of selfed seed increases the local density of selfed sibs will necessarily increase, limiting their combined success. An intermediate level of selfing may then result from a combination of local adaptation and linkage between fertilization and dispersal. In yeast the ascus wall is partly digested during passage through the gut of *Drosophila*, and consequently gametes voided by the flies can readily outcross (Reuter *et al.* 2007), so that asci which remain at the parental site will self-fertilize whereas those dispersed by flies are likely to outcross. A second possibility is that the performance of outcrossed gametes should rise as selfing becomes more common, because for any given level of male function the ratio of outcrossing pollen to available ovules will be greater when more ovules are selfed. This may prevent the fixation of selfing if its main benefit is reproductive assurance, with the success of outcrossing being strongly restricted by fertilization.

The level of self-fertilization modulates selection on gender allocation. When selfing is permitted, the number of haploid genomes transmitted by some rare variant will be twice the number of selfed ovules plus the number of crossed ovules plus the number of successful pollen. Calculating the ESS from this sum shows that a pure female type will invade a population of hermaphrodites with male allocation α_i and fraction of selfed ovules Y_i if $(1 - \alpha_i) < ½(1 - sY_i)$, which is less restrictive than the corresponding condition for outcrossed populations, $(1 - \alpha_i) < ½$, and permits the coexistence of a pure female with a female-biased hermaphrodite. A population of partially selfed hermaphrodites will be invaded by males if $\alpha_i < (1 - Y_i)/(2(1 - sY_i)$, a more restrictive condition implying that partly selfed hermaphrodites are more likely to coexist with females than with males (Lewis 1942 and many subsequent authors). This is not because pure females cannot inbreed (it holds even if $s = 0$) but rather because the supply of fertilizable ovules and thus the unit success of outcrossing pollen is reduced by selfing.

12.4 Beauty and the Beast

Sexual selection is ineradicably associated with peacocks. In order to understand how it is different from natural selection, it is necessary to return to microbes. During the vegetative cycle, individuals developing from spores compete to acquire resources. The surviving individuals are larger, and reproduce by transforming captured resources into spores. Types that can do this more rapidly increase in frequency through natural selection. The organisms involved may be obligately outcrossing, so that progeny can only be produced by a combination of two parents, but the theory would require only the most trivial amendment to accommodate this curious restriction on reproduction. It remains a vegetative theory, grounded on the concept of differential growth. Suppose, however, that the spore differentiates into a gamete, switching development from the vegetative cycle to the sexual cycle. It will not then compete for resources with other gametes in order to grow and reproduce more rapidly. Gametes do not grow or reproduce at all. They do nevertheless give rise to other spores. When two gametes fuse, the fusion product is a diploid spore, the zygote. This may develop directly, if vegetative growth occurs in the diploid phase, or it may first go through a meiosis to yield haploid spores, if growth occurs in the haploid phase. In either case, gametes that succeed in fusing give rise to a new generation of spores that are capable of vegetative growth. Gametes that do not fuse have one of two fates: they may re-differentiate into spores and re-enter the vegetative cycle, or they may die. In either case, they are unable to proceed further in the sexual cycle. Just as the vegetative cycle is completed by the fission of a single individual, so the sexual cycle is completed by the fusion of two gametes.

The subject of competition in the sexual cycle is, then, not growth, but fusion. Gametes compete for fusion partners. The availability of fusion partners is often limited, although it need not necessarily be, and it is more severely limited in some circumstances than in others. Competition among gametes for fusion partners gives rise to sexual selection. Types that are more apt to fuse increase in frequency because they are incorporated disproportionately into the spores that will enter the succeeding vegetative cycle of growth and reproduction.

In many microbes, the distinction between sexual selection and natural selection is perfectly clear, because cells that are morphologically alike have contrasting fates, either fusing in pairs to form a single spore, or growing to divide into two equal spores. In other organisms this distinction is blurred because gamete fusion is mediated by the fusion or copulation of individuals that bear gametes. These gamete-bearing individuals are called gamonts. In most fungi and ciliates, for example, it is gamonts that fuse, prior to the fusion of the gametes they bear. In multicellular animals and plants, gametes may fuse inside or outside the bodies of gamonts, and it is very often the behaviour of the gamonts—such as peacocks—that determines the success of gametes, rather than the behaviour of the gametes themselves. In these cases, sexual selection modifies the properties of gamonts. Nevertheless, it does so by causing changes in the frequencies of different types in the population of competing gametes, through their success in completing the sexual cycle. Natural selection in these cases acts through the relative productivity of gametes; sexual selection through the relative success of the gametes that are produced. The life cycles of animals and plants are often so highly modified from the simpler cycles of microbes that the distinction between sex and reproduction, and thus between sexual selection and natural selection, is easily misapprehended. The underlying principle, however, remains the same.

12.4.1 Experimental sexual selection

The simplest experiment is to find out whether, or to what extent, gamete fitness can be increased in a sexual microbe (Bell 2005). I set up lines of *Chlamydomonas* that were periodically induced to mate by nitrogen deprivation and afterwards exposed to chloroform vapour to kill all cells except zygotes. These lines were first allowed 2 hours to mate, but this period was reduced over time to a few minutes, until eventually the step was omitted altogether because the cultures were mating spontaneously. At this point the need for nitrogen deprivation to initiate gamete development had been lost, and a number of new phenotypes began

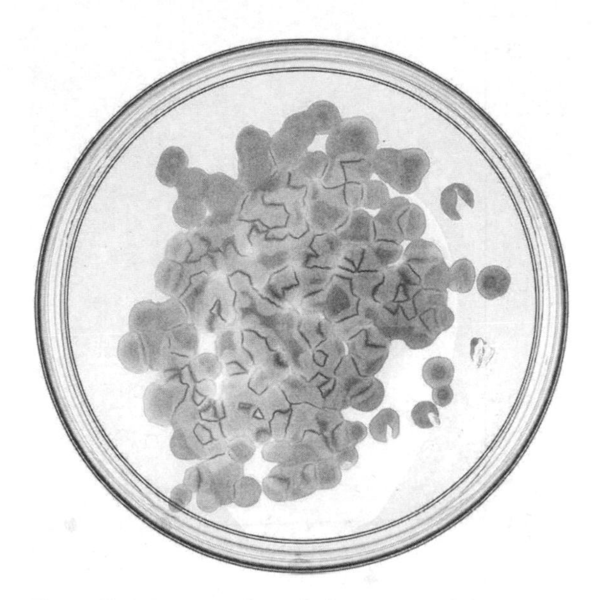

Figure 12.9 Experimental sexual selection. A novel phenotype: dark bands of zygotes form where colonies of different mating type meet. From Bell (2005).

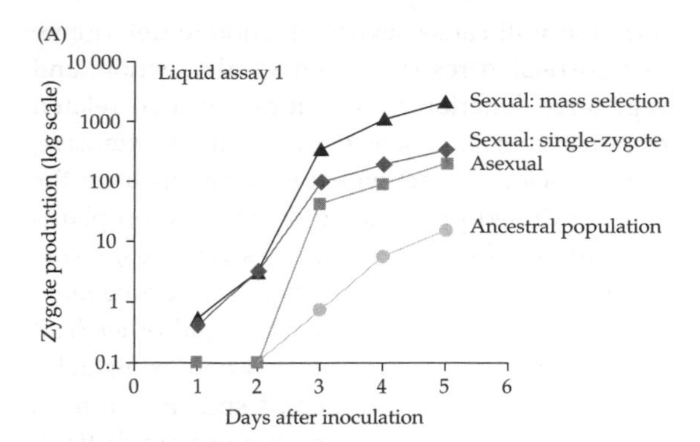

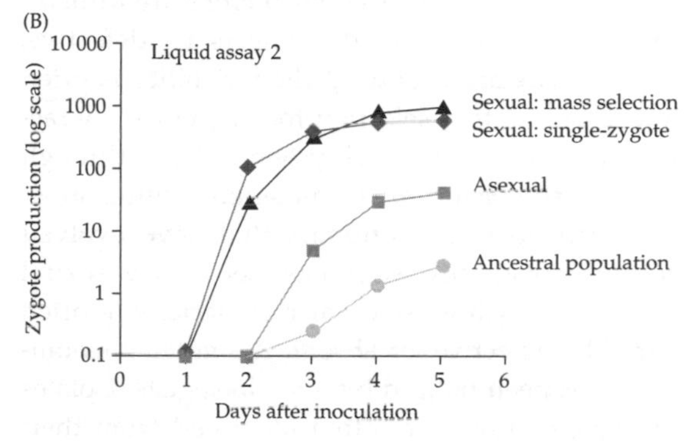

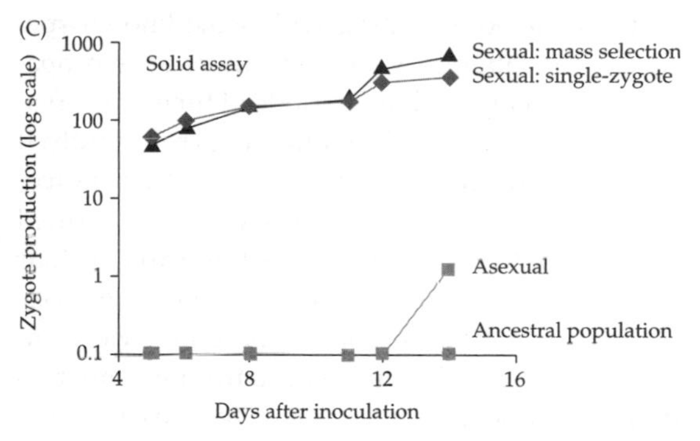

Figure 12.10 Experimental sexual selection. Enhancement of rate of mating by sexual selection. From Bell (2005).

to appear. In liquid cultures, conspicuous strings and clusters formed after 3–4 days of growth, while in solid cultures dark bands appeared at the boundary between colonies of different mating type (Figure 12.9). Both the clusters and the bands consisted of zygotes formed by mating early in culture growth. This behaviour would normally be heavily penalized, because zygotes do not reproduce, but ensured transmission to the next cycle, whereas failure to mate in time would be lethal. After about 100 sexual cycles, and about 1000 intervening vegetative cycles, the rate of zygote production was at least 100 times greater in the sexual lines than in asexual controls or in the ancestor (Figure 12.10). Hence, this experiment showed that sexual selection will induce both qualitative and quantitative changes in the sexual cycle, as effectively as natural selection will modify growth and reproduction in the vegetative cycle.

12.4.2 Antagonism of sexual selection and natural selection

The overall fitness of any type with a complete life cycle has thus two components: its vegetative success through growth, and its sexual success through fusion. If the two were always proportionate, there would be no special problem of sexual selection: more vigorous types would both reproduce more profusely and fuse more readily. A special theory of sexual selection is required only if vegetative and sexual success are antagonistic. The antagonism of sexual selection and natural selection will cause sexual function to deteriorate as a correlated response to natural selection, and vegetative function to deteriorate as a correlated response to sexual selection. The most extensive, if unconscious, selection experiment has been the long-continued propagation of certain crop plants without sex. Root crops such as potatoes, sweet potatoes, and yams, for example, are propagated as tubers, or tuberous roots; citrus and other fruit trees are grown as buds grafted onto rootstocks; many berries are propagated as suckers, runners, or cuttings. In most cases, such crops are difficult to induce to flower, or the flowers are defective, and if seeds are produced their viability is often low. It seems that selection for vegetative characters has caused the sexual system to decay, through antagonistic pleiotropy or mutation accumulation. In parthenogenetic animals that have evolved recently from sexual stocks, secondary sexual structures such as seminal receptacles are often variable, defective, or absent. A similar phenomenon has been noticed by microbiologists. Isolates of algae or fungi are often identified from their sexual behaviour, mating with some known species in the laboratory, and are afterwards propagated asexually for long periods of time. It is often then discovered that the strain mates reluctantly, if at all. Sexual function can often be fully restored, provided that a few zygotes can be obtained from the old asexual culture, within three or four sexual generations of intense selection for zygotes. In another *Chlamydomonas* experiment we compared mating ability and vegetative reproduction in sexual and asexual lines. The asexual lines evolved the higher rates of increase, but their mating ability declined; after about a hundred asexual generations, indeed, their zygote production fell to only a few per cent of its original value (da Silva and Bell 1992) (Figure 12.11). This confirmed the casual observation that sexual ability declines in asexual cultures. The sexual lines mated rapidly and vigorously, as one would expect in a situation where failure to do so was lethal, but grew less rapidly as vegetative cultures. This seems to be the clearest experimental evidence that sexual selection and natural selection are fundamentally antagonistic.

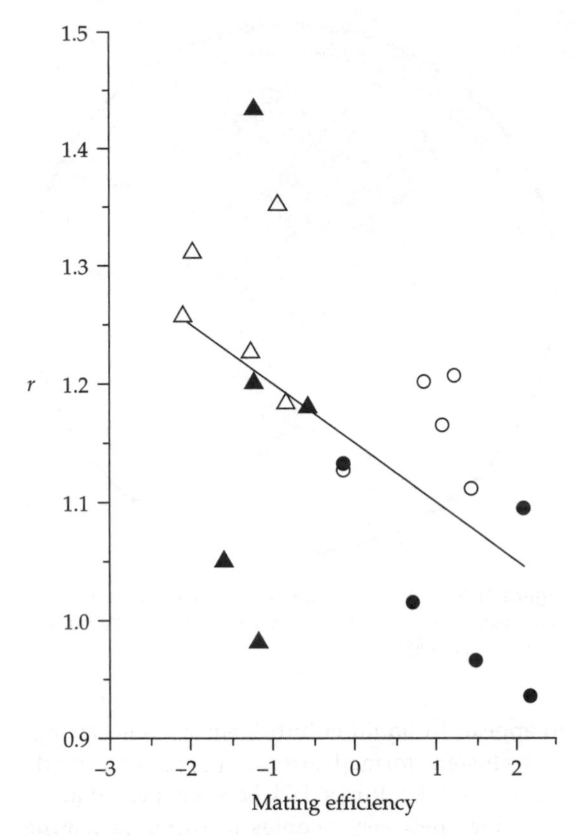

Figure 12.11 Antagonism of sexual selection and natural selection. Maximal rate of increase and mating efficiency in asexual lines (triangles), obligately sexual lines (open circles) and lines selected for rapid mating (filled circles). From da Silva and Bell (1992).

12.4.3 Gamete competition

The gametes of multicellular organisms are essentially sexual microbes that compete for fusion partners, and if they are shed into water and fuse at random the microgamete is usually a typical spermatozoon and the macrogamete an ovum. If sperm are shed into a confined space, however, such as the female reproductive tract, they may compete directly with one another, so that the success of a spermatogonium depends on the quality as well as the quantity of microgametes it yields. In this case, large sperm of atypical structure may evolve to navigate the female duct more rapidly or to obstruct the passage of rivals. In flowering plants, for example, haploid growth is very restricted, and the male gametophyte comprises only three haploid nuclei. A substantial fraction of the genome is expressed in the gametophyte, and sexual selection of gametophytes should enhance their own sexual abilities. Thus, pollen grains can be selected for their ability to fertilize by placing many pollen grains on the stigma, creating intense competition among them, and harvesting the seeds produced. The results of such experiments have so far been equivocal. Several authors have reported success, but a careful study of pollination in the wild radish yielded no response to selection (Snow and Mazer 1988). One explanation of the variable experimental results is that fertilization success depends primarily, not on the male gametophytic genome, but rather on the interaction between the male gametophyte and female sporophytic tissue.

In animals, mature spermatids usually undergo alterations of chromatin structure that prevent gene expression, but early spermatids are transcriptionally active. The archetypal animal spermatozoon is a minute, highly motile creature, produced by the hundred million and engaged in scramble competition for fertilization (Parker 1982). There is a great deal of variation in sperm size and morphology, however, which happens to be exemplified by the two major invertebrate models: *Caenorhabitis* has amoeboid sperm that crawl up the female reproductive tract, while the more conventionally shaped sperm of *Drosophila* may be almost as long as the fly, if the tail is included. In *Caenorhabditis*, larger sperm are more expensive to make but crawl more rapidly and fertilize more eggs (LaMunyon and Ward 1998). Sperm competition is normally minimal because the worms are usually self-fertilizing hermaphrodites, but can be induced by constructing populations consisting of males and females (male-sterile hermaphrodites). After 60 generations sperm volume had increased by about 20% in lines where sperm competition was intense (LaMunyon and Ward 2002) (Figure 12.12). So far as I know, this is the only experimental demonstration that the characteristics of male gametes can be altered through sexual selection, although there is comparative evidence that sperm are more competitive in species where they are produced in greater abundance (e.g. Gomendio *et al.* 2006).

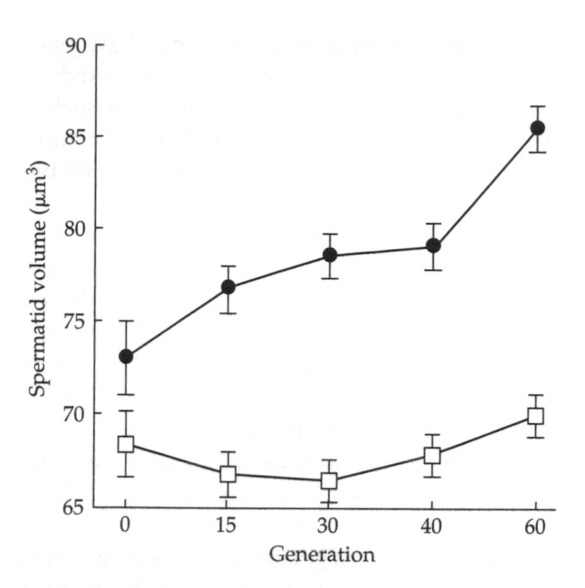

Figure 12.12 Experimental evolution of sperm size in *Caenorhabditis.* Filled circles, lines with intense sperm competition; open squares, lines with no sperm competition. From LaMunyon and Ward (2002).

12.4.4 Bateman's principle

In organisms such as seaweeds or sponges that shed their gametes into the sea the sexual success of an individual, whether as male or female, depends mainly, or solely, on the bulk of the gametes it produces, and consequently the appearance and behaviour of individuals is little affected by the gender of the gametes they bear. In other organisms fertilization involves intimate contact between two or a few individuals and sexual success will depend on the ability of the soma to facilitate the transmission of male or female gametes by courtship and mating. In this case individuals or parts of individuals will be divergently modified through sexual selection. This holds whether individuals bear gametes of one gender only or of both: the notion that sexual selection does not occur in hermaphrodites such as plants with perfect flowers is mistaken. It also holds whether mating is direct (for example, through copulation) or indirect (for example, through pollination). There are many cases, especially among organisms with separate sexes, where these modifications reach bizarre proportions. It was, indeed, the difficulty of accounting for characters such as the plumage of male peacocks in terms of natural selection that led Darwin to develop a special theory of sexual selection.

Sexual selection is caused by sexual competition, in which each gender is a resource for the other. The limiting resource is the gender that is the less abundant. Gametes of the gender that is present in excess must compete for access to the limiting gender, and will thereby be modified through sexual selection. In multicellular organisms, gamete success is often greatly enhanced because the gametes are brought into contact by specialized structures or behaviours displayed by vegetative individuals. These vegetative individuals will then compete, and are liable to be secondarily modified through sexual selection. The same principle applies: individuals bearing gametes of the gender present in excess will compete for access to individuals bearing gametes of the minority gender. Male gametes are usually much more numerous than female gametes, and therefore males will usually compete for access to females, and are more likely to respond to sexual selection. The principle extends to hermaphrodites: hermaphroditic individuals acting as males will compete for access to those acting as females, even though the same individual may act as male and female at different times, or even, when copulation is reciprocal, at the same time. The extreme modification of structures and behaviours associated with mating is therefore often confined to one gender (this is, indeed, the main circumstantial evidence that they evolve through sexual selection), and this gender is usually, although not always, the male. The principle that governs this process is that sexual selection modifies the type whose gametes are present in excess.

This principle has been expressed in different ways by a variety of biologists, reflecting the context of the problem they set out to investigate and the organism they were studying. The classical restatement was made by A.J. Bateman (1948) as the result of experiments in which he used genetic markers to estimate the number of progeny produced by individual male and female *Drosophila* when several flies of each gender were cultured together and allowed to mate freely. Both male and female flies may copulate several times, and the average number of copulations must be the same for males and females, but the number of copulations has a different effect on the number of offspring they produce. Females require one copulation in order to produce a batch of fertilized eggs, but their output of eggs was not increased by additional copulations. The number of offspring fathered by a male, on the other hand, increased in proportion to the number of females he mated with. Almost all females produced offspring; the number each produced would depend primarily, not on sexual success, because they are almost certain to be courted and inseminated, but on the quantity of resources they have been able to garner and turn into eggs. Variation among females in resource-gathering ability will evolve through natural selection. A substantial proportion (nearly a quarter) of males, on the other hand, failed to copulate successfully, and thus fathered no offspring at all. The remainder copulated more or less frequently, and gave rise to varying numbers of descendants as a result. Reproductive success thus varied more among males than among females. The greater variance of male reproductive success (strictly speaking, sexual success) is often called Bateman's Principle. It is the outcome of the more intense sexual competition among males, arising from the greater abundance of male gametes. This need not necessarily have any evolutionary consequences. If we were to choose pairs of flies at random, allowing them to mate before returning them to the population and repeating the procedure, the variance of the number of copulations would be the same for males and females, but the variance of reproductive success would nevertheless be greater for the males than for the females. Because the flies were chosen at random, the fact that some male flies were much more successful than others would cause no consistent change in male attributes. It is only if the greater variance among males is caused in part by heritable variation in sexual ability that the response to sexual selection will be greater among males than among females.

Another way of expressing the general principle of sexual competition is that the response to sexual selection will be greater in the sex that invests less in the progeny (Trivers 1974). Male gametes being more numerous, they are also smaller. Although their genetic contribution to the zygote is the same as that of the female gamete, their cytoplasmic contribution is much less. In some cases, however, the male may provide resources that can be used by the developing embryo: a nuptial meal attached to the spermatophore in some insects, brooding of the eggs by sea-horses, or feeding the growing nestlings in many birds. When the parental investment of the male approaches or exceeds that of the female, male gametes, however abundant, may be no more readily available than female gametes. Females must then compete among themselves for access to males, and as a result may be strongly modified by sexual selection.

12.4.5 Artificial sexual selection

There have been many experimental demonstrations that artificial selection can alter the vegetative characteristics of populations. I presume that it would be just as easy to modify sexual characteristics, but there have been very few attempts to study sexual selection in the same way. The only familiar example is provided by the florist. The form, coloration, and scent of flowers evolve through sexual selection for an adequate level of pollen import, to ensure the fertilization of the plant's own ovules, and, more importantly, for a maximal rate of pollen export, to fertilize the ovules of other plants. These characters have also been deliberately selected, yielding cultivars to decorate the garden or the table. In most cases rather little has been accomplished by selection, beyond a general increase in the size of the bloom, and sometimes brighter or different coloration. One of the few experimental attempts to manipulate the form of flowers by selection was made by Stanton and Young (1994). In natural populations of wild radish petal size increases with pollen and nectar production, but not with ovule or seed number, as one would expect from the predominantly male role of flowers in serving pollen export rather than pollen import. Artificial selection for the greatest and smallest ratios of petal size to pollen production for only two generations removed the correlation between secondary sex allocation to petals and the primary allocation to pollen, suggesting that in natural populations this correlation is maintained by selection.

Inseminated female flies usually discourage courting males, and other things being equal males that succeed in mating quickly will have a considerable sexual advantage. Manning (1961) selected for mating speed by choosing the 10 fastest and the 10 slowest pairs from samples of 50, establishing 2 replicate fast lines and 2 slow ones. After about 20 generations, the fast and slow lines had diverged substantially from the unselected controls (Figure 12.13). Half the unselected flies had mated after about 5 minutes; in the fast lines this time had been reduced to less than 2 minutes, whereas in the slow lines it had increased to more than 15 minutes. Most of the fast flies, indeed, had mated before the first of the slow flies had done so. The base population, a large cage population that had been maintained in the laboratory for several years, must have contained a large amount of genetic variance for mating behaviour that was readily harvested by selection. The realized heritability in the first few generations of selection was about 30%, after which the response was much slower. Crossing the lines showed that the differences in mating speed were heritable. They were transmitted both by males and by females, although crosses between the lines suggested that the males had been more affected by selection. A subsequent experiment in which only one sex was selected (Manning 1963) showed that selection for slow mating was much more effective in males than in females, although selection among males only was not nearly as effective as the selection of pairs practised in the first experiment. The behavioural changes that caused the response of mating speed to selection were most interesting. When pairs were selected, there was a change in general levels of activity, but in an unexpected direction: the fast maters were relatively inactive flies. The reason is that rapidly moving females cannot be successfully courted. In the slow-mating lines the flies would move around for several minutes when introduced

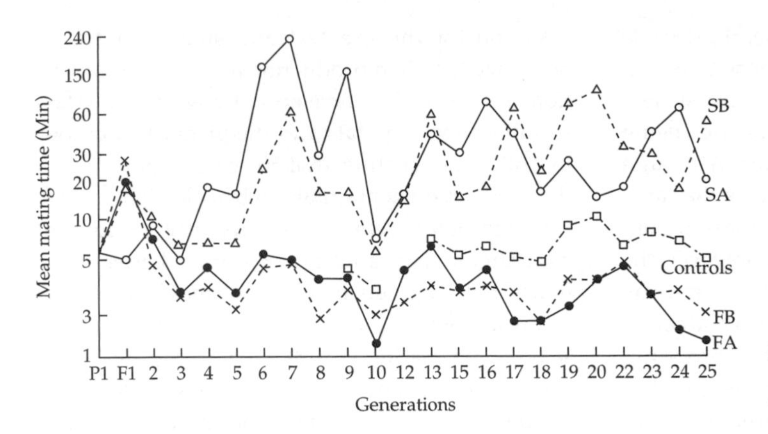

Figure 12.13 Selection for mating speed in *Drosophila*. FA, FB are the fast lines, SA and SB the slow lines. From Manning (1961).

into the mating chamber, and would rarely court or mate during the period. The flies from the fast-mating lines, on the other hand, crawled only a short distance before beginning courtship. However, this outcome depends on an interaction between the sexes. When flies are selected through the speed with which they mate with an unselected control stock, as in the second experiment, the situation is different. Generally speaking, the first females to mate are inactive, slow-moving individuals, but the first males to mate are active flies that are the first to discover the females. Selection for slow mating applied to males alone therefore led to a decrease in male activity: many of the slow males remained almost motionless when introduced into the mating chamber, and could not mate until a female happened to walk by within range. Interestingly, the same phenotype was produced by back-selecting the fast lines in the pair-selection experiment—extremely inactive flies that mated very slowly because they were unable to find partners. The behavioural causes of mating success thus differed from experiment to experiment. In particular, slow mating can evolve through either very high or very low levels of general activity, depending primarily on whether selection is applied to one sex or to both. For comparable experiments, see Manning (1968), Manning and Hirsch (1971) and Spiess and Stankevych (1973). Gromko *et al.* (1991) reported a response to divergent selection for copulation duration in *Drosophila*.

12.4.6 Sexual contests

The sexual specialization of male and female individuals parallels that of the male and female gametes they bear. At one extreme, male and female offspring are scattered into the water column and the adult males compete simply by seeking females. When females mature a bulky crop of large gametes, they are likely to become sedentary and specialized through natural selection for gathering resources. A large male with an equally bulky testis would be carrying around far more male gametes than could be used in a single mating, and would be likely to find and inseminate fewer females than would a large number of small males amounting to the same total mass. Producing numerous small sons will therefore be favoured by sexual selection in most circumstances (Ghiselin 1974). In most invertebrates, not to mention a host of other organisms, from filamentous algae to ferns, males are smaller than females. In some cases, this reduction in size is carried to extremes that are more bizarre than antlers or wattles. A male rotifer, for example, is little more than a testis equipped with a locomotor organ. The dwarf males of some barnacles and angler-fishes settle and live parasitically on, or in, the females they discover, and are in some cases so highly reduced as adults that the females in which they embedded themselves were originally described as hermaphrodites.

When mating occurs within a confined space males may struggle among themselves for the physical control of access to females, and this direct competition may favour large rather than small size, just as copulation may favour large sperm. Male red deer are much larger than the females, bear antlers that grow only during the rut, and engage one another in grim combat where the prize is a harem of hinds and the penalty may be a lingering death. Such contests among males provide a vivid enactment of sexual competition, and even in less theatrical cases the evolution of male structures and behaviours through sexual selection is universally accepted. Moreover, combat ability has on occasion been deliberately selected, especially in fighting cocks and Siamese fighting fish, although I have not, unfortunately, come across any careful analysis of these unintentional selection experiments. The reverse experiment is to remove males with highly developed weaponry such as horns and antlers, which is accomplished by trophy hunters. The response should be the evolution of less exaggerated structures, but I do not know whether this has ever been verified.

12.4.7 Male × female interaction

The divergent specialization of male and female gametes or individuals implies that an allele that is beneficial when expressed in one gender may be deleterious when expressed in the other. This has been called 'intralocus sexual conflict', which imposes a cost on one gender if the allele deleterious in that gender becomes fixed. It is somewhat similar to G × E, because each gender is akin to a site in which one of the alleles is favoured. There is extensive male × female interaction for fitness in *Drosophila*, which is largely attributable to negative genetic correlation between male and female reproductive success (Chippindale *et al.* 2001). In the simplest diploid case, an autosomal allele A_m is favoured in males and A_f in females; opposite homozygotes have fitness 1 and $1 - s_m$ in males, and $1 - s_f$ and 1 in females, the heterozygotes being intermediate. The A_f allele beneficial in females is fixed if $s_f < s_m/(1 - s_m)$ and that beneficial in males if $s_m < s_f/(1 - s_f)$ (see Kidwell *et al.* 1977). If neither condition is satisfied, there is a stable equilibrium and consequently both males and females carry suboptimal alleles. For sex-linked genes the conditions for polymorphism are less stringent, because they will be predominantly transmitted from parents to offspring of the same gender. Since alleles that are beneficial in both males and females are likely to be fixed, whereas those deleterious in both will be eliminated, much of the variation within either gender might be attributable to sexually antagonistic genes. Rice (1992) imitated the introduction of a novel sex-determining gene in *Drosophila* by breeding from females heterozygous for eye-colour markers which males chosen for breeding did not bear. Hence, sexually antagonistic alleles linked to the marker are transmitted predominantly to daughters, and those beneficial to females should increase. The sex ratio of experimental lines became steadily more female-biased, due to the increased vigour of female flies. At the same time the eye-colour genes, after about 30 generations of selection, reduced the fitness of males to which they were transferred. Rice (1998) later used clone-generator lines of *Drosophila* to transmit entire haploid genomes intact from father to son over many generations without passing through a female body. These lines were thus shielded from counter-selection in females and should accumulate alleles with beneficial effects limited to males. After about 40 generations the experimental males mated more rapidly than controls, both with virgin and with non-virgin females. Moreover, when the evolved genome was transferred to females it reduced their rate of development, although other components of fitness were not severely affected. These experiments provide good evidence for the segregation of substantial numbers of sexually antagonistic alleles capable of impeding the evolution of sexual dimorphism. I would expect such effects to be attenuated through selection among females for sex limitation of expression. This would be effective if the antagonistic effects were merely incidental by-products of the male behaviour, although not if they were themselves the source of the advantage to the male.

12.4.8 Sexual choice

Males may also compete by soliciting the attention of females, who may choose one male over another for one of two reasons.

- The first is that sexual and vegetative performance are positively correlated, so that sexually successful individuals give rise to vegetatively superior progeny. Natural selection and sexual selection then act in the same direction. In most cases, this implies that females prefer to mate with the more vigorous males, or at least with males that will yield more vigorous progeny.
- The second possibility is that sexual and vegetative performance are negatively correlated, so that sexually successful individuals have relatively low viability or fecundity. Natural selection and sexual selection then act in different directions. In most cases, this implies that males compete directly among themselves for access to females.

The essential features of the situation are captured by a 'MOP' model in which alleles at three loci control mating type, ornamentation and preference. The simplest version has two alleles at each locus in a haplont: *M*/*m* determines female or male gender, *O*/*o* gives fancy or plain males, and *P*/*p* gives choosy or random-mating females. The fancy males (*O*) are helped or handicapped by their sexual ornament so that their survival is $(1 - s)$ relative to plain males (*o*). The mating behaviour of females is governed by a parameter λ that expresses their mating preference, such that the fraction of matings of choosy females (*P*) with fancy males is $x \exp[\lambda(1 - x)]$ where x is the frequency of fancy males: random-mating females (*p*) have $\lambda = 0$, choosy females $0 < \lambda < 1$. Any simple function yields similar results: Kirkpatrick (1982) gives a detailed account of a similar model using $x\lambda/[(1 - x) + x\lambda]$. Choosy females pay a cost for exercising choice, such that the number of matings, or the probability of mating, is $(1 - c)$ relative to random-mating females. Given a population consisting largely of *op* individuals, with a few *Op* and *oP* individuals, we want to know whether the *OP* type will spread.

The first possibility is that fancy males have higher survival ($s < 0$). This is likely to favour choice because females will be selected to mate preferentially with males whose progeny are likely to be vegetatively superior. Partridge (1980) compared the offspring of females that were allowed to mate freely with those of females that were constrained to mate with a single partner chosen at random by the experimenter. The offspring of the females that were given the opportunity to choose their partners were on average about 4% more viable than those from the arranged marriages. In the MOP model the *O* allele increases in frequency and as it does so confers an increasing benefit on choosy females, so that *P* spreads at the same time. Once *O* has become fixed mate choice is neutral and the population is a mixture of *OP* and *Op* individuals. The final frequency of *OP* is greater when the benefit of the male ornament is less (s negative and small) because the longer passage time of *O* provides more time for *P* to be selected. Thus, if females show strong preference ($\lambda = 0.5$) for a weakly selected ($s = -0.01$) male ornament the final frequency of *OP* is over 97%, whereas stronger natural selection ($s = -0.1$) reduces this to a few per cent, depending on initial frequencies. If choice is costly, however, *Op* individuals are more fit than *OP* once the ornament has become fixed; in this case, *OP* initially spreads but is afterwards eliminated, leaving the population fixed for *Op*. This is the main shortcoming of the theory: a beneficial male trait will tend to become fixed, at which point female choice is pointless or harmful and therefore will not be maintained. Active female choice implies that genetic variation for male ornamentation is perpetually renewed. The various theories of sexual selection through female choice have been developed largely in response to this objection. They fall into two categories, depending on whether the lack of fit between population and environment is caused by genetic deterioration or by environmental change.

Recurrent deleterious mutation will maintain genetic variance for vegetative fitness, and females may avoid males whose appearance or behaviour signals a high mutational load. Some major mutations in *Drosophila* are indeed associated with poor male mating success. Inbred flies from an outbred population suffer inbreeding depression, and may be avoided by females. Courtship in *Drosophila* involves a dance in which the partners, face to face, perform a series of rapid shuffles from side to side. Maynard Smith (1956) found that inbred males were unable to keep up the rapid tempo of the dance, and suggested that females rejected dancing partners whose inferior athletic abilities might indicate a heritable defect. Zahavi (1975) even suggested that females should choose to mate with males whose phenotype is clearly maladapted, on the grounds that, having survived despite their handicap, they must be exceptionally vigorous in other respects. If the handicap is not heritable—a leg broken by accident and subsequently healed, for example—then it is possible that handicapped males might sire unusually well-endowed offspring, although it is not clear that females use accidental injuries of this sort as criteria in choosing mates. If the handicap is heritable, such as long tail plumes that impede flight, the superior genes that allow males to survive will be transmitted to daughters, where they can be expressed without the handicap. Whether this compensates for the production of handicapped sons, however, is highly debatable.

Females might discriminate against poorly adapted males in a novel environment. These might be immigrants in a patchy environment, or males adapted to the previous state of a changing environment. If the environment is changing through time, female discrimination will continue to be selected only if genetic variation continues to be maintained among the males. This led Hamilton and Zuk (1982) to argue that disease resistance is an appropriate criterion for female choice, because alternative resistance genes will cycle in frequency through time in response to the delayed frequency-dependent selection induced by pathogens and parasites (§11.3). Male displays are then viewed as medical examinations that allow females to scrutinize potential sexual partners for infection. This attractive idea continues to be controversial.

12.4.9 Sexual ornaments

Ornaments are not an infallible guide to mate quality because males may falsely advertise their genetic worth. Females should therefore discriminate among possible mates only on the basis of characteristics that cannot readily be misrepresented, such as body size. It is not clear that many interesting sexual modifications, such as bright patches of colour or conspicuous crests, sacs, or wattles, are unequivocal signals of heritable vegetative superiority. In some cases, indeed, it seems clear that secondary sexual characters have been exaggerated to the point where they must impair vegetative fitness. It is equally clear that such characters can evolve despite natural selection, if females prefer to mate with bizarre males. A series of experiments over the last decade or so have established that in strongly dimorphic species females do indeed prefer the extremes of male showiness. Andersson (1982) demonstrated this kind of preference in a wild population of African widow-birds. Male widow-birds have enormously long tail feathers that appear to be used in courtship displays but are aerodynamically unsound; the females have more conventional tails. By cutting the tail feathers, it is possible to create males with very short tails, with normal long tails (by re-attaching the cut section with glue), or with extremely long tails (by gluing a long piece on to a tail that had not been shortened much). The males with very short tails attracted fewer females than the normal birds, and those with extremely long tails attracted more females than did males of normal appearance. Highly exaggerated structures may, then, evolve because they are favoured by sexual selection, despite their disadvantage under natural selection, as Darwin realized long ago. It is the aesthetics that is more puzzling: why should females possess such apparently inappropriate tastes?

The answer given by Fisher (1930) is that females will prefer to mate with males that are attractive to other females because their sons will be sexually successful. Given $s > 0$ the spread of *OP* genotypes generating choosy females and fancy males requires that the sexual advantage conferred by the ornament exceeds the handicap it confers, if we ignore the cost of choice for the time being. This advantage depends on the choice parameter λ and the frequency of choosy females y. Roughly speaking, *OP* will spread if $\lambda y > s$. Consequently, if *P* is initially rare the population will become fixed for o while continuing to segregate at the choice locus because P and p are almost neutral; any small cost will lead to the fixation of op. Once the initial frequency of P exceeds a threshold of about s/λ the O allele will spread, generating positive linkage disequilibrium between M and P and between O and P. If P is initially very common the population will eventually become fixed for P while continuing to segregate for O/o, the frequency of OP being about $1 - s/\lambda$. Hence, exaggerated sexual ornaments may evolve because mate choice necessarily implies that sexual selection will generally be positively frequency-dependent. The P allele quickly comes into linkage disequilibrium with mating type and spreads in oP individuals. At this time Op declines rapidly, depressing the frequency of OP and creating a correlation between O and P. When P has become sufficiently common the decline of OP is reversed and it spreads rapidly as the p allele is eliminated. At equilibrium the population is dimorphic for the male ornament because the fraction of matings that choosy females devote to fancy males depends on the frequency of fancy males, so a male polymorphism is maintained by the opposition of a fixed viability handicap with a frequency-dependent mating advantage. It is only if $s \ll \lambda$ that the population consists almost entirely of OP individuals. Recurrent revertible mutation preserves linkage disequilibrium between M and P to give partial sex linkage of both preference and ornament, and between O and P to give a positive genetic correlation between ornament and preference. If choice is costly the same dynamics are seen, except that the threshold frequency of choice is raised. Thus, the mechanics of selection favouring harmful male ornaments is fairly straightforward, but the theory leads to two difficulties. The first is that plain males always persist at equilibrium, so that in more realistic scenarios a good deal of genetic variance in male ornamentation is to be expected. The second and more serious difficulty is that mating preference must be very strong to overcome a moderate male handicap, and it is not obvious why this should be.

The simplest answer is that females are sometimes made like that. Some quite unrelated area of activity, such as gathering food or building nests, happens to make a certain kind of structure or a certain sound or movement more apparent or more attractive to females. A variant male that develops a long tail or croaks at a certain pitch is favoured as a mate because of the pre-existing aesthetic or sensory bias of the female, and such variants may spread because of their sexual advantage, even if they reduce male viability. This was Darwin's original notion, which has received some experimental support (see below).

The alternative is that female preference will evolve. R.A. Fisher (1930) proposed that female preference might be based originally on some trait that is favoured through natural selection, for any of the reasons mentioned above. Females that are better able to choose the more fit males will produce more fit offspring, and selection will thereby enhance the discrimination that females exercise among males. Males in which the character is more clearly and strongly developed will be more likely to be chosen as mates, and will thereby receive a double advantage, having greater viability and greater sexual success. As the character becomes more exaggerated its sexual effect will increase, but it will become less well adapted for its vegetative function. Sexual selection and natural selection now pull in opposite directions. By this time, however, most females have inherited a tendency to use this character as a criterion in choosing mates. Those who choose males in which it is exaggerated beyond usefulness will continue to be favoured, because their sons will be more likely to obtain mates, and their sexual prowess will more than outweigh their loss of viability. The success of the offspring of such matings will create a correlation between the tendency, expressed in males, to develop an exaggerated version of the character, and the tendency, expressed in females, to prefer to mate with such males. The process now feeds upon itself: a more pronounced female preference favours a greater exaggeration of the male character, which in turn favours the most discriminating females. Fisher called this a 'runaway' process, which will be brought to a halt only when natural selection against the modified males becomes overwhelmingly strong. This truly beautiful theory explains the extreme contingency of bizarre sexual modification. Characters that evolve through natural selection are often highly convergent. The mottled plumage of ground-nesting birds, the countershading of pelagic fish, the stripe passing through the eye in frogs: such characters evolve because they render animals less conspicuous, and they evolve, in more or less the same form, in many independent lines of descent. Conspicuous warning or frightening patterns such as fake eyespots or bands of highly contrasted colours have likewise evolved independently on many occasions. Sexual modifications, on the other hand, are usually peculiar to a small clade, and involve different structures, or different kinds of modification, in nearly every case. This is to be expected, if the ultimate exaggeration of a male character depends on the prior establishment of one particular female preference from the very large number that might have arisen.

In terms of a simple MOP model a highly exaggerated ornament can evolve through the successive substitution of O alleles each with a somewhat stronger expression than its predecessor, such that s remains small because the fitness of each new mutation is only slightly less that of the resident allele. This process could eventually give rise to fancy males with much lower survival than the original male type, although the population would then be vulnerable to invasion by the original type unless female preference were very strong. A quantitative model of runaway sexual selection was developed by Lande (1981), who treated the evolution of a female preference as the correlated response to natural selection on the male ornament. He concluded that as any given handicap s is balanced by a corresponding preference λ the relationship between them defines a line of neutral equilibria. The population will evolve towards this line and reach a stable equilibrium if the line is steeper than the regression of the genetic values of preference on ornament. Otherwise it will diverge indefinitely and with gathering speed, leading to extreme preferences for exaggerated ornaments. With recurrent mutation there will be positive genetic correlation between ornament and preference at equilibrium, as in the MOP model.

The runaway process has never been observed experimentally. Indeed, the Fisherian process of runaway sexual selection has a somewhat similar history to the 'shifting balance' of Sewall Wright: both have fascinated generations of theoreticians and generated models by the cartload (Table 1 of Mead and Arnold 2004 lists about a dozen variants of Lande's model alone) while seldom receiving any decisive empirical test. There is a good deal of evidence, however, for the evolution of male ornaments as a balance between sexual and natural selection. One particularly attractive system with a large literature is provided by guppies (*Poecilia*), small freshwater fish in which males bearing spots of bright colours actively court the drab females. The conspicuous male coloration has long been interpreted as representing a balance between sexual and natural selection; the earlier literature has been reviewed by Endler (1983). The colour and position of spots is heritable and often sex-linked (see Lindholm and Breden 2002). In *P. parae*, for example, five colour morphs correspond to a series of alleles on the Y chromosome (Lindholm *et al.* 2004), and much of the colour variation in the classic aquarium guppy *P. reticulata* may be Y-linked (Brooks and Endler 2001). The intensity of coloration may depend on nutrition, however, as colours such as orange require carotenoids from algae, and may thereby indicate the health and vigour of the male. More brightly coloured males, especially those with intense orange spots, are more attractive to females, who spend more time in their vicinity (Brooks and Endler 2001). Female guppies are more attracted to orange objects in non-sexual contexts, so the underlying female preference may have evolved prior to the male ornament (Rodd *et al.* 2002). The more attractive males in turn mate more frequently (Brooks 2000, Brooks and Endler 2001). They pay a heavy price, however. Even in laboratory conditions of growth the sons of the most attractive fathers are about 30% less likely to survive to maturity than those of the least attractive fathers. In the aquarium, predatory cichlids attack and consume more brightly coloured males first (Godin and McDonough 2003). In the field the more conspicuous males are also more vulnerable to predators, so that in natural populations spot number and size are inversely related to the intensity of predation (Endler 1978). When guppies from a high-predation locality were transplanted to a comparable site that lacked predators, the number and size of spots increased within four or five generations (Endler 1980). The number of mature daughters produced by males in the laboratory was not correlated with their colouration because of its contrary effects on survival and attractiveness (Brooks 2000). Artificial selection for orange colouration was rapidly effective, and females from lines selected for more orange males showed stronger preference for orange (Houde 1994), confirming a genetic correlation between ornament and preference (Figure 12.14). Hall *et al.* (2004) failed to replicate this result after three generations of selection, but since they obtained no direct response to selection for attractiveness or preference their base population may not have contained much genetic variation. In short, the experimental work on guppies, of which this is a very brief abstract, provides a broad confirmation of simple models of how male ornaments and female choice respond to opposed sexual and natural selection.

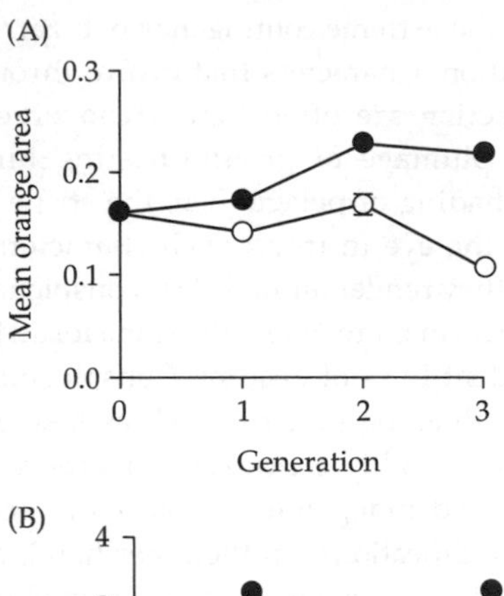

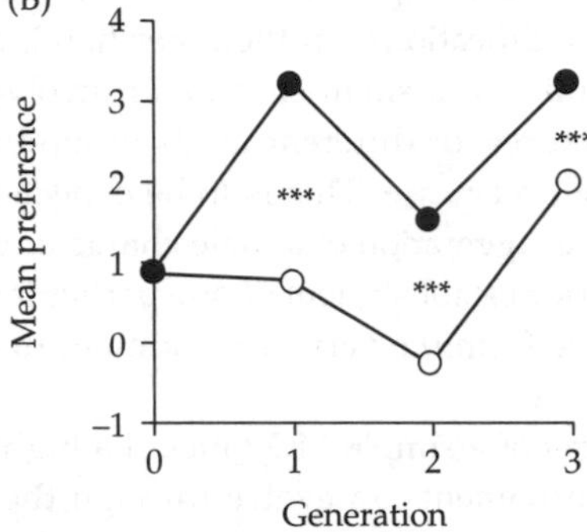

Figure 12.14 Correlated response to artificial sexual selection in guppies. A. Direct response of male coloration (extent of orange). B. Correlated response of female preference (for orange). One of four replicate lines are controls. From Houde (1994).

12.4.10 Battles of the sexes

Females that exercise a choice may pay a cost for doing so, for example an opportunity cost in terms of missed opportunities to mate. This affects the dynamics of selection because choice is no longer neutral, and may lead to a single stable equilibrium in place of a line of neutral equilibria, although other outcomes are possible (see Houle and Kondrashov 2001). More generally, females are often physically damaged during courtship and mating because males may benefit from rapid or frequent mating even if this reduces the average success of each brood. This may be plainly evident, as when females are bullied or raped, but may also be more subtle and widespread. For example, sperm and seminal fluid transferred during copulation in *Drosophila* contains substances that inhibit re-mating, and hence maximize paternity, but also have the effect of shortening lifespan (Chapman *et al.* 1995); the agent chiefly responsible is a small protein bound to the sperm tail (Liu and Kubli 2003). Because exploitation by the male and resistance by the female will involve different loci, this type of sexual antagonism has been called 'interlocus' sexual conflict. It is inevitable in anisogametic species, except in special cases such as lifelong monogamy, and may often cause the antagonistic co-evolution of harmful behaviour by males and resistance by females. It has recently become the subject of much speculation and is reviewed by Chapman *et al.* (1995), Cordero and Eberhard (2003), Arnquist and Rowe (2005), and Parker (2006).

Parker (1979) introduced a simple model in which an allele expressed in males gains B times as many matings but damages the female so that only $(1 - C)$ times as many progeny are produced per mating. The harmful male will benefit if $B(1 - C) > 1$. The position of the female is ambiguous because she may benefit indirectly by producing harmful sons. Each female will mate once at most with a harmful male when the harmful allele is rare, and $(m - 1)$ times with normal harmless males; a fraction p of the broods produced by harmless males bear the cost of a previous mating by their female partners with a harmful male. The inclusive fitness of females comprises three components: grandchildren through matings with harmless males, grandchildren via harmful sons from matings with harmful males, and grandchildren via daughters from these matings. Adding these up, he harmful allele will increase the fitness of females if $B(1 - C) > 1 + uC[1 + p(m - 1)]/(1 - C)$, where u expresses the mode of inheritance of the harmful allele. While the harmful allele is rare it is reasonable to suppose that p is very small, in which case the condition approaches $B(1 - C) > 1 + u[C/(1 - C)]$. As expected, the condition is always more difficult to satisfy for females than for males; it is least stringent for Y-linked genes ($u = 2$), more stringent for autosomal genes ($u = 4$) and most stringent for X-linked genes ($u = 8$) (André and Morrow 2003). Hence, harmful alleles may spread through their benefit to males despite reducing the fitness of the females with whom these males mate.

Rice (1996, 1998) detected the accumulation of conflict genes in his clone-generator lines, where they were shielded from the evolution of resistance by females. Unselected females housed with these males had considerably greater mortality in the first 12 days of adult life and were more often sterile. Females in turn can evolve resistance to male behaviour (Wigby and Chapman 2004). Several authors have raised the possibility that an arms race between males and females would produce idiosyncratic change and thereby sexual divergence of conspecific populations. Population crosses might then result in lower levels of female morbidity because foreign males are likely to be sexually maladapted, and there is some evidence for this in *Drosophila* (Long *et al.* 2006). In the long run this would result in sexual isolation, the subject of the next chapter.

Richard Dawkins has memorably remarked that the bodies of multicellular organisms are little more than lumbering robots that serve to nourish and transport the germline. In asexual organisms this is a reminder that evolutionary change is mediated solely through changes in gene frequency among germ cells, microbes that proliferate within bodies and which succeed in proportion to the ability of these bodies to produce and sustain them. In sexual organisms it has a somewhat different connotation, because the germ cells are not merely microbes, but sexual microbes that succeed not only in proportion to their number but also in proportion to their ability to find fusion partners. Natural selection then operates through the quantity of gametic material produced, whereas it is the quality of the gametes drives sexual selection. In many circumstances, natural selection may be a subordinate principle, operating only to modulate slightly the strife among gametes. It is often thought that sexual selection, running counter to natural selection, obstructs the process of adaptation. This is a misconception: in sexual organisms, sexual success *is* adaptation, and although bodies may be recruited in the process, its principle feature is the apotheosis of the gamete.

CHAPTER 13

Speciation

One of the chief tasks of biology is to classify organisms into discrete categories called 'species'. This is relatively easy for large organisms such as fish or ferns, despite some uncertainties (such as hybridization) in particular cases. It has been much more problematic for organisms such as fungi and microbes that often lack the familiar attributes of most animals and plants such as individuality and sex. This has led to a very lengthy discussion of the nature of species and the proposal of a wide range of 'species concepts', some intended to be universal whereas others are tailored for particular circumstances. Concepts are not testable hypotheses, however, and this lengthy discussion has added little if anything to our understanding of evolution. I shall acknowledge the lumpiness of nature by using some neutral term such as 'taxon' or 'cluster' while reserving the term 'species' for sexual organisms, where it has a clear meaning. Gender and species are the two main attributes of sexual organisms: gametes can fuse successfully if they are of like species and unlike gender. I shall moreover reserve the term 'sex' for the complete gametic fusion and subsequent meiosis typical of eukaryotes. Other kinds of process may have similar consequences—for example, transformation in bacteria or genome segmentation in virus can lead to genetic recombination—but they are not homologous to eukaryotic sexuality. This may seem unduly restrictive, but I hope it will help to avoid some of the confusion notoriously associated with the topic of speciation.

The approach that I shall take, then, will be to recognize the distinctive nature of the eukaryote sexual cycle and to treat species formation as one of its consequences. This is essentially the original biological species concept of Mayr (1940), except that I would prefer to speak of a sexual (rather than biological) species concept based on sexual (rather than reproductive) isolation. Moreover, I shall use a strong version of the concept: speciation requires that populations belong to different species if they seldom interbreed when mingled together, and on the rare occasions that they do the hybrids are sickly or sterile. The reason is that speciation is only interesting in its own right if it is distinctively different from divergent specialization. Clearly, there are no species in primitively asexual organisms such as bacteria, where adaptive radiation leads to a potentially indefinite divergence of lineages, although in practice ecological constraints may lead to the maintenance of finite number of similarly specialized types. Secondarily asexual groups of animals and plants do not form species either, although they may retain the characteristics of their parental species for a long time. In either case, recognizable types are often designated by Linnaean binomials for the sake of convenience; this does not imply that they have the same general characteristics or were formed in the same way as true species. Strictly homothallic or self-fertilizing organisms would not form species, but in practice most will outcross occasionally and a modest level of genetic exchange will ensure coherence.

13.1 Speciation and diversification

Species are (more or less) sexually isolated and (more or less) ecologically specialized. They may arise abruptly as the consequence of major mutations that enforce inbreeding because outcrossed progeny are inviable. Alteration of chromosome number or structure may have this effect, by obstructing meiosis or generating aneuploid progeny, and speciation via polyploidy is well known among plants. Conversely, the fertile hybrids produced by two well-defined species may be unable to mate successfully with either parent and thus constitute a new species. Hybrid speciation has been described from several groups of plants such as sunflowers (Rieseberg 2001). In most cases, however, one species gives rise to another only after a lengthy process of divergence, at the end of which the two species are sexually and ecologically distinct. Broadly speaking, there are two principal theories of how this could happen, corresponding to the two fundamental events of the sexual cycle. The first is that species formation is the direct response to divergent natural selection; the indirect response is sexual incompatibility, and genetic isolation is maintained primarily by the failure of the zygote to complete meiosis successfully ('postzygotic' isolation). I shall refer to this as 'adaptive speciation'. The second is that species formation is the direct response to divergent sexual selection; the indirect response is ecological specialization, and isolation is maintained primarily by the failure of gametes to fuse successfully ('prezygotic' isolation). I shall call this 'sexual speciation'. This distinction is similar to but does not correspond precisely with 'ecological' and 'sexual' speciation as distinguished by Schluter (2000) and Rundle and Nosil (2005), who would include any case where sexual selection was influenced by external conditions (such as the visual or chemical environment) in the category of ecological speciation. I have preferred to make a primary distinction between natural and sexual selection because this reflects the diversifying and integrating tendencies whose balance governs the rate of speciation.

Several authors have remarked on the sharp divide between the theory of adaptation and the theory of speciation (for example, Barton and Partridge 2000, Kirkpatrick and Ravigné 2002). Adaptation has clearly defined population-genetic mechanisms based on broadly applicable principles (however questionable these may sometimes be) whereas speciation has become 'balkanized' into a host of models narrowly crafted for particular circumstances without providing a synthetic general theory. I have found the minimalist model of Kirkpatrick and Ravigné (2002) useful for focusing concepts. There are two haploid loci at which equally frequent alleles *A*, *a* and *B*, *b* are maintained by an unspecified source of balancing selection. Natural selection favours *AB* and *ab* (fitness 1) over *Ab* and *aB* (fitness $1 - s$), and sexual selection operates through assortative mating such that the frequency of mating between two genotypes with frequencies f_1, f_2 is $f_1 f_2 (1 - \alpha)^d$ where α is the strength of assortative mating and d is the squared difference in gene dosage, upper-case alleles counting 1 and lower-case 0. The criterion for sexual isolation is then linkage disequilibrium $D = f_{AB} f_{ab} - f_{Ab} f_{aB}$, with $D = 0$ for a single random-mating population and $D = ¼$ for two non-interbreeding subpopulations constituting different species. The condition for speciation involves s, α and the initial state of D. Starting from linkage equilibrium, $s \neq 0$ is a necessary condition for divergence, and in the absence of any assortative mating substantial linkage disequilibrium is created by strong selection. This is never complete or nearly complete, however, whereas assortative mating can lead to nearly complete isolation even when selection is weak (Figure 13.1); for $s = 0.01$, for example, D is 96% of its maximal value if $\alpha > 2/3$. The reason is very simple: recombination breaks up combinations of genes favoured by natural selection whereas assortative mating preserves them. Unless selection is overwhelmingly strong, therefore, sexual isolation must precede divergence and sexual selection is likely to be the prime mover of speciation.

13.1.1 Allopatric and sympatric divergence

The process of diversification in sexual and asexual populations has been clarified by Barraclough *et al.* (2003). A set of ecologically equivalent clones constituting a population of N individuals will have a branching phylogeny described by the neutral coalescent descending from the most recent common ancestor N generations previously. They will be distributed among genetic clusters of random size, and if they are phenotypically distinguishable they will form phenotypic clusters that could be used as the basis for classification.

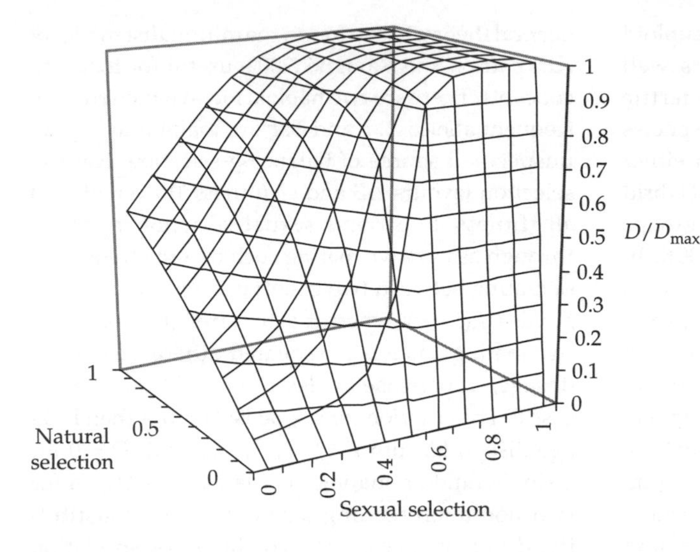

Figure 13.1 Kirkpatrick and Ravigne's toy model of speciation. 'Natural selection' is the selection coefficient s; 'sexual selection' is the assortative mating coefficient α. The vertical axis is the extent of linkage disequilibrium, relative to the maximum value: D/D_{max} = 1 represents complete sexual isolation.

In an outbreeding sexual population the alleles of any given gene will likewise be related by a branching process, but the topology of descent will be different for unlinked genes and therefore a phenotypic character governed by several genes will form only a single cluster. Now suppose there are two distinct sites providing different conditions of growth, so that divergent selection favours two functionally differentiated genotypes. In an asexual population selection may maintain two major clusters, provided that population size is regulated independently at each site, so that fitness is frequency-dependent (§10.2.2). Each of these major clusters in turn consists of the series of minor clusters generated by the neutral coalescent process. In a sexual population the same pattern will apply to a single gene responsible for adaptation, but adaptation to a particular site or way of life is seldom based on a single gene, but is rather likely to depend on a constellation of genes at many loci. It is therefore hindered by mating with differently adapted individuals, because recombination will break up a successful combination of genes to produce inferior progeny. This is simply the converse of the view that recombination enables sexual lineages to adapt quickly to continually changing circumstances. Sex allows mutations that have arisen independently in different lines of descent to be combined into the same line of descent. But the other side of the coin is that sex causes genes that are combined in the same line of descent to be distributed among different lines of descent.

If two sites are very distant, or separated by impassable barriers, then sexual isolation is enforced and the sexual population overall will fall into two clusters, within each of which there is no further substructuring of multilocus characters because of recombination. Should the barriers be removed, both asexual and sexual populations will return to a single coalescent for any given gene, but crossing will rapidly efface any substructuring for multilocus genotypes in the sexual population (Felsenstein 1981). The preservation of distinct phenotypic clusters thus requires the prior evolution in allopatry of sexual isolation, either through drift or divergent sexual selection. The strength of selection needed to overcome the dissipative tendency of recombination explains the lumpiness of obligately sexual clades. A minor specialization that confers some slight advantage will scarcely ever become established as a distinct lineage, because the advantage of mating assortatively will be too slight to counteract the effect of recombination in an outcrossing population. Sex thereby reduces the mean fitness, or level of adaptation, of the population. It is only major modifications that entail large adaptive differences that can engender selection powerful enough to drive the evolution of sexual preference, and outcrossing species therefore tend to be widely separated and highly distinctive. In asexual or self-fertilizing organisms this constraint does not apply, and instead of a few clearly distinct species there is often instead an extensive assemblage of many similar lineages. Many sections of *Taraxacum* (dandelions), for example, consist of dozens of consistently but very slightly demarcated 'microspecies'. In the limit there may be an almost continuous distribution of variation, as in *Alchemilla* (everlastings). This resembles a single sexual species, but may be more extensive because genotypes specialized for growth in rare sites can persist, whereas they would be swamped by immigration and crossing in a sexual population. In short, we expect sexual species to form fewer, more different, and more variable clusters than asexual taxa. They will be fewer because recombination will often erase differentiation, more distant because stronger selection is necessary to overcome recombination, and more variable because divergent selection will slow down the loss of variation.

This view implies that sister species will usually diverge in allopatry, as famously argued by Mayr (1963). When they subsequently come into contact again, ecological differentiation will be enhanced by character displacement and sexual isolation by reinforcement. The case for purely sympatric speciation has been argued repeatedly (for a recent defence, see Via (2001)) but its theoretical difficulties are difficult to surmount except in special cases. It is entirely plausible, on the other hand, that nascent species occupying the same tract of land or water should breed at different times or in different places, and I shall mention some examples of such microallopatric separation below.

13.1.2 Asexual diversification

Conversely, strictly asexual organisms should fall into many narrowly specialized clusters. This seems to be contradicted by the paucity of named taxa of bacteria: about 5000–6000, the same as frogs and only 2–3 times as many as single genera of plants such as *Piper* or *Carex*. Molecular trawls of seawater and soil have shown that the real number must be much greater, however, since many sequences recovered from amplifying bulk DNA cannot be assigned to any named taxon. A quantitative estimate of diversity can be obtained from DNA reassociation kinetics: when DNA that has melted at high temperature is cooled, homologous single strands reassociate at a rate that depends on its complexity. Torsvik *et al.* (1990) isolated DNA from 30 g of forest soil containing about 1.5×10^{10} cells g^{-1} and estimated its complexity to be equivalent to 2.7×10^{10} bp. Given the average genome size of a soil bacterium to be 6.8×10^6 bp, this suggests that the number of taxa in the sample was $(2.7 \times 10^{10})/(6.8 \times 10^6) = 4000$, almost as many as have been named in the whole world. Dykhuizen (1998) argued that this was an underestimate. First, taxa need not be completely different; if 30% sequence divergence is regarded as sufficient to distinguish different taxa, then the reassociation kinetics suggest that 40 000 taxa contributed. Secondly, taxa are not equally abundant: if there are 25 rare taxa to each common taxon then besides 20 000 common taxa the community contained 500 000 rare taxa, each represented by an average of somewhat more than 10 000 individuals. These are staggering numbers for a cupful of soil, and if they are accepted the number of bacterial taxa in the world can only be guessed at: estimates based on apparently reasonable assumptions suggest 10^{10}–10^{14} different types (Bell 2008). Dykhuizen attributes this vast diversity to divergent selection for resource specialization, which would be consistent with the interpretation I have given above.

I am not aware of any direct comparison of divergence in comparable sexual and asexual clades, but detailed genetic and phenotypic descriptions of the paradigmatic group of asexual metazoans, the bdelloid rotifers, are now becoming available. There are fewer species (*c*.360) than in the related sexual monogonont rotifers (*c*.900). Species names based on morphological criteria seem to be stable, and levels of synonomy are actually less in bdelloids than in monogononts, suggesting that bdelloid species are more distinctive and easily recognized (Holman 1987). Birky *et al.* (2005) sequenced the mitochondrial *cox1* gene from six nominate genera of bdelloids and derived an unusual phylogeny in which single strains or small groups of related strains descend from the basal node as an unresolved polytomy. This seems contrary to the bacterial situation but consistent with the natural history of the group: a relatively small number of cosmopolitan taxa that are found wherever a suitable habitat (e.g. *Sphagnum* bog) is available. The bdelloids do not seem distinctively different from a sexual group.

13.1.3 The poverty of the protists

The number of described species in the major eukaryote clades is shown in Figure 13.2. There are hundreds of thousands of species of animals and plants, the largest and most complex organisms. There are tens of thousands of species of simpler multicellular organisms such as fungi and algae. Most microbial clades, on the other hand, have only a few hundred species: only diatoms, foraminiferans, 'radiolarians', ciliates, and dinoflagellates have diversified into a few thousand species, about the same as much narrower groups such as grasses, orchids, or sponges. The common ancestor of eukaryotes was undoubtedly unicellular, and large complex organisms appear only late in the fossil record. Modern protists are extremely abundant, occupy every environment, and follow every way of life available to eukaryotes. Nevertheless, despite their ancient origins and current ubiquity only about 50 000 species have been described, about the same as the number of snails. This is unlikely to be attributable to their lack of easily recognizable morphological characters, because many groups (such as coccolithophores, testate amoebas, or all the groups mentioned above) have very distinctive and variable appearances under the microscope. It may be attributable in part to a shortage of taxonomic work, but the same is likely to be true for fungi. There are two sharply divergent interpretations of the unexpected poverty of protists.

The first is that there really are relatively few species, most of which can be distinguished morphologically and many of which have very broad geographical distributions (Finlay 2002, Finlay *et al.* 2004). There are only about 3000 species of ciliates, for example, and as they are well studied and morphologically distinctive there may be few more remaining to be discovered. Microbes are readily dispersed as dormant or resistant spores by wind and water, so many are cosmopolitan and can be found at suitable sites anywhere in the world. A litre of fresh water taken anywhere in the world, for example, is likely to contain the ciliate *Tetrahymena pyriformis*; a litre of seawater from the photic zone of any ocean will probably contain the coccolithophore *Emiliana huxleyi*. The chrysomonad *Paraphysomonas* occurs in all parts of the world, and yet 32 of the 41 named species were found in 25 μl of sediment from a single Cumbrian pond (Finlay and Clarke 1999). Consequently, microbes lack biogeography and species number will show little if any tendency to increase with the area or volume sampled (Finlay 2002). A strong justification of the cosmopolitan distribution of microbes is that they are resistant to extinction by virtue of their large populations. The world population of organisms between 10^{-10} and 10^{-3}g in mass, most of which are protists, is about 4×10^{27} individuals, so the average total abundance of a species is about 10^{23} individuals. Protected against stochastic extinction by their abundance and against environmental stochasticity by their broad distribution, they are almost immortal. Conversely, speciation rates are correspondingly low because populations are seldom allopatric.

The alternative view is that the uniformity of cosmopolitan microbes is more apparent than real. In the first place, it is now clear that extremely small picoeukaryotes, little larger than bacteria, are extremely abundant and since they often lack features such as flagella and scales their diversity, like that of bacteria, may be greatly underestimated by morphological surveys. Secondly, large numbers of undescribed taxa have been identified by molecular surveys. Major clades of unknown or poorly known marine heterotrophs and deep-sea picoeukaryotes have been detected (Moreira and Lopez-Garcia 2002, Behnke *et al.* 2006) and even in fresh water a large fraction of sequences could not be attributed to named species (Slapeta *et al.* 2005). No DNA reassociation experiments have yet been described, however, so no quantitative assessment of diversity can be attempted. Finally, morphological uniformity may conceal genetic diversity and geographical

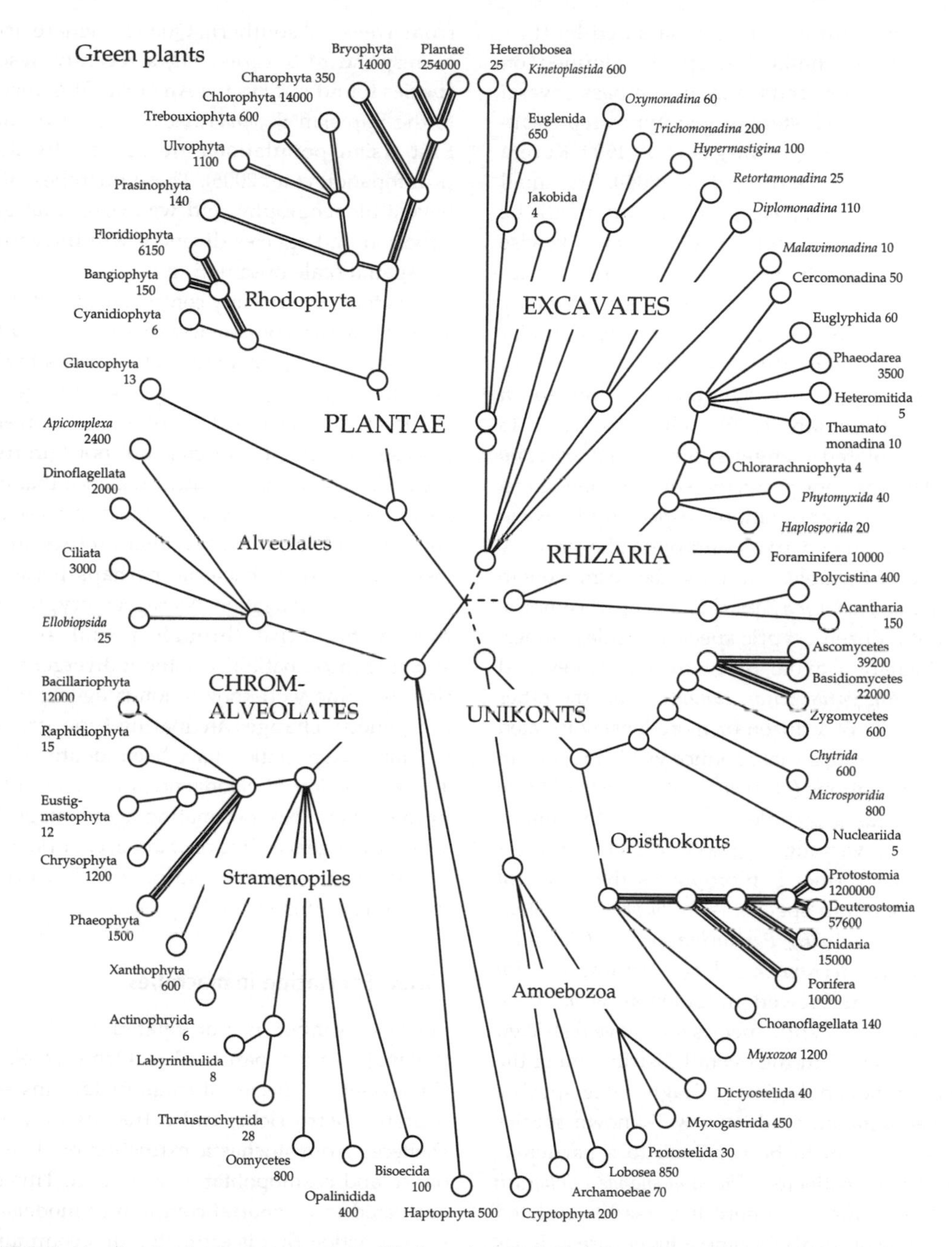

Figure 13.2 Species diversity of major eukaryote clades. Thin line: microbes; double line: variable; triple line: multicellular. Parasites in italics. Adapted from Keeling *et al.* (2005). The discussion in the text is not sensitive to the topology of the tree, except insofar as unicellularity is primitive and multicellularity has been independently derived several times.

structure. Foraminiferans are classified by their intricate coiled shells, but species defined on morphological grounds may encompass several distantly related clusters of genotypes representing cryptic species (de Vargas *et al.* 1999, Kucera and Darling 2002, Darling *et al.* 2004)). The small prasinophyte *Micromonas pusilla* occurs in all the oceans of the world and genetic surveys likewise show that it consists of a series of sibling species (Slapeta *et al.* 2006). One source of hidden diversity is the tendency for cosmopolitan protists of similar appearance to constitute swarms of cryptic or sibling species called 'syngens' by Sonneborn (1957; see Coleman 2005), who described 14 sexually isolated syngens within *Paramecium aurelia*. They are not perfectly isolated: mating can occur in some combinations, especially between very closely related pairs, although the progeny are sterile or inviable. Similarly, the cosmopolitan ciliate *Tetrahymena pyriformis* is a syngen comprising about a dozen cryptic species, besides numerous obligately asexual derivatives (Nanney and McCoy 1976). *Paramecium caudatum*, on the other hand, consists of a dozen or more loosely isolated mating type groups that produce viable progeny in many combinations (Hori *et al.* 2006). Each of these ciliates is thus a complex of ecologically similar lineages with varying degrees of direct or indirect sexual contact. Among chlorophytes, the abundant and cosmopolitan species *Gonium pectorale* (Stein and McCauley 1976), *Pandorina morum* (Coleman 1977), and *Closterium ehrenbergii* (Ichimura 1985) are syngenic (reviewed in Coleman *et al.* 1994, Ichimura 1996). Cryptic species may have restricted distributions within the overall distribution of the syngen or morphologically diagnosable species. Moreover, detailed studies of well-known species often show them to be confined to a particular region. I have collected *Chlamydomonas reinhardtii* in southern Quebec, where it is readily isolated from the soil of cornfields and adjacent areas. It has been found at several other sites in eastern North America, but a careful survey of *Chlamydomonas* isolates from around the world failed to find it in any other region (Pröschold *et al.* 2005). I have also collected *Saccharomyces sensu stricto* (the group that includes baker's yeast, *Saccharomyces cerevisiae*) from trees in southern Quebec, where isolates correspond to *S. cariocanus*, a recently described species found only in the Americas. It is very close to the Eurasian *S. paradoxus*, where European and East Asian populations are genetically distinct (Koufopanou *et al.* 2006). These microbes, at least, have a biogeography, and we expect that genetic variation and species diversity will increase with the spatial scale of sampling.

This topic is currently controversial, but the facts so far known are consistent with two generalizations. The first is that protist diversity exceeds the list of described species, but not by a very large factor: protists with very broad distributions often encompass several cryptic species, but not hundreds of them. Secondly, there is little evidence that cryptic species are ecologically specialized. If these results continue to hold, it is likely that protists are much less diverse than bacteria, perhaps because sex obstructs specialization. Moreover, cryptic species presumably arise through sexual speciation, evolving incompatibility without divergent adaptation. In some well-known non-syngenic microbes the genetic changes in mating-type genes that accompany speciation have been identified (§12.3). The association of cladogenesis with abrupt shifts in the genetic basis of mating again suggests that sexual divergence is the primary event in speciation. The field is developing so rapidly, however, that firm conclusions would be premature.

13.1.4 Speciation in macrobes

Species number is not related to size among the major metazoan clades (Orme *et al.* 2002). Consequently species of small metazoans such as rotifers, gastrotrichs, and ostracods may also be sheltered from stochastic extinction by their abundance and cosmopolitan distribution. This can be evaluated from neutral community models where the extinction flux is estimated in a community of fixed total size that is set up with the population size N of each species being a Poisson variable. These founding species then disappear in a manner similar to the loss of least-loaded line in asexual populations under Muller's ratchet (§12.1.2), and the scaling law is similar. For relatively small communities of 10^3–10^6 individuals I find from numerical work that the half-life T_{50} of species depends on the mean population size N as $T_{50} = aN^z$ with $z = 1$ and $a \approx 5$–6 almost independently of total community size. About 1–2×10^6 species of macroscopic eukaryotes ($W > 1$ mg) have been described, and perhaps as many as 10^7 exist. From size spectra, the world population of multicellular organisms is about 4×10^{20} individuals (§2.2.10), so mean population size is 4×10^{13} individuals. If the numerical results hold for much larger numbers, the half-life of the macroscopic biota is about 2×10^{14} generations. The average generation time G is unknown; using $G = aW^{0.25}$ (Peters 1983) and taking the value of 1.5 days for marine autotrophic bacteria cited by Whitman *et al.* (1998) gives an average for macroscopic organisms of 0.85 y, or roughly annual. The number of species becoming extinct each year is then only $(5 \times 10^6)/(2 \times 10^{14}) \ll 1$ even if estimates are out by two or three orders of magnitude. Hence, small multicellular organisms such as rotifers are unlikely to become extinct through demographic stochasticity; when they are lost, it must be through environmental change.

The average lifespan of species of marine animals was estimated by Raup (1991) from fossil data at about 4 my. Since diversity tends to increase over time the average interval between speciation events in a lineage should be somewhat less; Sepkoski (1999) gives a value of 3.3 my. The well-preserved and intensively studied series of fossil horses in North America gives a similar estimate of 3.1 my (Hulbert 1993). Other mammals have speciation intervals of 1–3 my (Avise *et al.* 1998) and marine gastropods of 2–4 my (Jablonski 1986). Rates of diversification at different taxonomic levels in animals have been reviewed by Sepkoski (1998). Among flowering plants, estimates range from about 1 my for herbs to 4 my for shrubs and 8 my for trees (Levin and Wilson 1976). These estimates are based on fossil data and take extinction into account. Where this is not possible, speciation rates can be calculated for a given group from its phylogenetic tree, although the speciation interval will be overestimated because it is based on extant taxa only (methods reviewed by Barraclough and Nee 2001). The average values from Table 12.1 of Coyne and Orr (2004) are 12.8 my for marine invertebrates, 8.8 my for terrestrial arthropods (excluding Hawaiian taxa), 4.3 my for vertebrates (excluding lake and island endemics), and 7.5 my for plants. The stability of these averages is rather surprising, and the variation is not clearly related, as one might have expected, to generation time. Some clades diversify much more slowly, however (Avise *et al.* 1998). Cycads and conifers have speciation intervals of 25–35 my(Levin and Wilson 1976). Plants very similar to the modern *Gingko biloba* lived over 100 my ago, and modern populations of the notostracan branchiopod *Triops* (tadpole shrimp) are morphologically almost identical to their Triassic ancestors (Longhurst 1955). Conversely, there are well-known examples of rapid speciation, with speciation intervals of 0.1–0.5 my estimated for African lake cichlids and Hawaiian silverswords. Broadly speaking, however, the speciation interval in most clades seems to be 1–10 my with no strongly marked correlation with body size, generation time, habitat, or taxon.

Sexual isolation between related species of *Drosophila* increases with genetic distance D as $1 - \exp(-kD)$, with $k \approx 10$ (Coyne and Orr 1997). The measure of overall sexual isolation used in this study took a minimal value of 0.9 for closely related species living in sympatry, equivalent to $D \approx 0.2$. Assuming that distance increases linearly with time and that $D = 1$ corresponds to about 5 my, the time required for speciation is about 1 my, which is comparable with estimates from fossils. There were pronounced differences, however, between sympatric and allopatric species pairs. They showed similar levels of meiotic (postzygotic) isolation, whereas gametic (prezygotic) isolation evolved much more rapidly in sympatry. Indeed, almost all sympatric pairs failed to mate regardless of genetic divergence, suggesting that the time required for speciation was as little as 0.2 my in sympatry following secondary contact (Figure 13.3). Mate choice in *Drosophila* is based primarily on response to cuticular hydrocarbon pheromones, which can evolve rapidly in the laboratory when related allopatric species are cultured together (Higgie *et al.* 2000). The number of genes responsible for morphological differences between

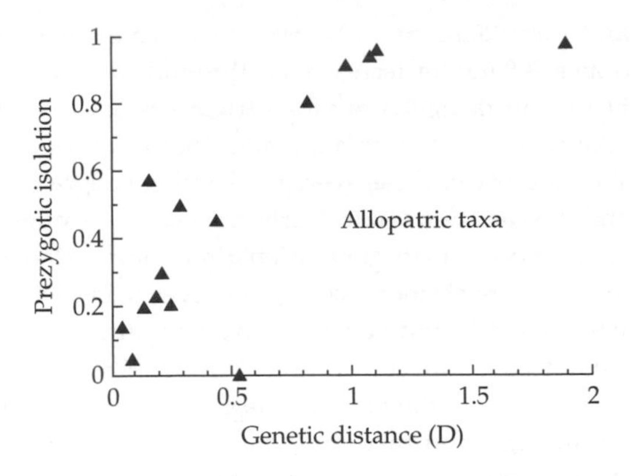

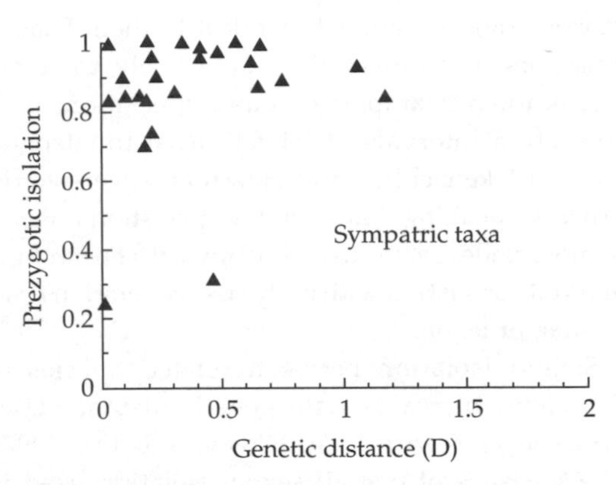

Figure 13.3 Sexual isolation in relation to genetic distance for allopatric and sympatric species. From Coyne and Orr (1997).

closely related species varies from one to many (reviewed by Orr 2001). At one extreme is the *YUP* locus responsible for the deposition of carotenoids in the petals of *Mimulus* (monkey flower). Moving alleles between *M. lewisii* and *M. cardinalis* by repeated backcrossing produces, in either species, pink flowers attractive to bumblebees or orange-red flowers attractive to humming-birds (Bradshaw and Schemske 2003). The integrity of these two species appears to be a clear case of sexual speciation driven by a shift in pollinators caused by a single mutation. Other cases involve several or many genes, but differences will accumulate through time and it is not known how many are responsible for the initial separation of sister species. Funk *et al.* (2006) found that the degree of sexual isolation is consistently (although weakly) correlated with the degree of ecological divergence in a range of plants and animals, and interpret this as evidence for the general prevalence of ecological speciation. It is not clear, however, that this pattern would not also arise if enhanced sexual isolation permitted more extreme ecological divergence.

Schluter (2001) has reviewed the evidence for ecological speciation and Panhuis *et al.* (2001) for sexual speciation, recalling that these terms do not precisely correspond to adaptive and sexual speciation as used here. Any brief review such as this risks selecting cases to support a particular view, and besides it is quite plausible that both adaptive and sexual speciation are common in animals and plants. A comprehensive review of the field, including detailed discussion of species concepts, isolation mechanisms and modes of speciation was published by Coyne and Orr (2004). This chapter is limited to some brief notes linking it with previous chapters on adaptive radiation and sexual selection, with an account of the relevant experimental work on mating behaviour.

13.2 Experimental speciation

13.2.1 Sexual divergence by drift

Populations adapting to different conditions of growth may diverge sexually through drift or selection. The first possibility is that a mutation altering mating behaviour might drift to fixation in a small population which would thereby become sexually isolated from its ancestor. The main difficulty with this idea is that any such mutation would be likely to reduce mating success. If it occurred in only one sex then the other would remain fully fertile with the ancestor, so genetic continuity would be restored on contact. For isolation to occur it would be necessary for selection to favour mutations in the other sex that increased its fertility by enhancing compatibility. This seems a cumbersome idea, although something of the sort may well be involved in the origin of syngens in sexual microbes. Carson (1968) envisaged a series of populations colonizing disturbed sites, in which they repeatedly increased in numbers, because of adaptation or lack of competition, before abruptly declining to very low numbers, because the environment has changed or because they have exhausted local resources. The population is thus repeatedly sampled from a very few individuals, who might by chance be sexually idiosyncratic. After a number of cycles of increase and decrease, some populations, while still outcrossing, might have lost the ability to cross with members of other populations.

A number of selection experiments have attempted to identify a process of this sort. Powell (1978) assembled a highly diverse population by crossing isofemale lines of *Drosophila* collected from several localities. This was allowed to increase for two or three generations to about 10 000 individuals. Eight pairs of flies were then chosen to found new populations, each of which again increased in numbers before being reduced to a single mated pair. After four such population cycles, the mating preferences of flies from different lines were compared with those of flies from control lines that had been maintained more or less constant in size. Flies from these control lines showed no particular preference for mates from the same line. The experimental lines, on the other hand, often showed departures from random mating when flies from different lines were mixed together. When a substantial preference was observed, it always involved flies from the same line preferring to mate with one another. About a third of all the lines behaved like this. The strongest preferences persisted for some time after the end of the experiment, when the flies had been cultured at more or less constant density for about 10 generations.

At least three other experiments, with similar protocols, have found a similar but less pronounced tendency for mating preferences to evolve. The most extensive was run by a team led by Galiana *et al.* (1993), who used large numbers of replicate lines, set up controls for inbreeding and other complications, and used between one and nine pairs to found new populations after each crash. After four or more population cycles, about 10% of crosses between lines showed some evidence of assortative mating. This seemed more likely to happen when the number of founders was very small. However, the mating preferences shown by these lines were rather weak. In short, severe fluctuations in population number do lead to the appearance of mating preferences, provided that the residual populations are extremely small. It is not important that the effect occurs in only a minority, perhaps a small minority, of experimental lines: nobody expects speciation to occur in every isolated group. However, it is cause for concern that the mating preferences that have been observed are, for the most part, rather weak. They certainly do not even approach sexual isolation in any case.

The severe bottlenecks in population number that the experimental populations passed through would have caused high levels of inbreeding, and the chance fixation of alleles at many loci. It is unlikely, however, that inbreeding itself caused the mating preferences that were observed. Powell (1978) also tested random inbred lines, maintained without a periodic reduction in numbers, and failed to find any consistent deviation from random mating. Indeed, crosses between inbred lines of *Drosophila* usually produce superior F1 progeny, and selection would rapidly disperse any preference that had arisen. It is rather the combination of inbreeding with rapid population expansion that seems to be responsible for the divergence of mating preferences. The argument usually made is that the increase in variation during population expansion coupled with the opportunity for intense selection during the subsequent crash can lead to a 'genetic revolution', and the establishment of novel genotypes unique to each line. I have not been able to understand clearly what constitutes a genetic revolution, but if one were to occur I would expect it to lead to hybrid breakdown, rather than to mating preferences. Instead, crosses among the experimental lines give normal offspring in both the experiments described above. The experiments might instead be interpreted in terms of sexual selection. If the original population were segregating for genes that affected the choice of males by females, the single pair chosen to perpetuate the line might be highly unrepresentative. The particular female preference established by chance would then induce selection, during the subsequent expansion, for the preferred male type. Fisherian sexual selection on female preference would then become more intense, and lines that had be chance acquired somewhat different preferences would begin to diverge rapidly. The preferences exhibited during these experiments were relatively modest, and might soon be dissipated if the lines were allowed to interbreed freely; their relevance to general theory is that they suggest how Fisherian sexual selection may participate in the rapid evolution of sexual isolation.

13.2.2 Disruptive natural selection

The second possibility is that mating preferences are selected indirectly, as a correlated response to disruptive selection. The basic concept is that adaptive divergence through natural selection creates sexual selection for assortative mating as a consequence of the superiority of the offspring from matings between similar parents. In this kind of experiment, individuals with extreme phenotypes are chosen as parents to generate disruptive selection and the selected adults are mated as a group, with no control being exercised over their behaviour. This mimics disruptive selection in sympatry. Hybrid offspring may be produced throughout the experiment and are penalized only because they are likely to have intermediate phenotypes that will be discarded by the experimenter.

Over 40 years ago, a remarkable experiment was published by Thoday and Gibson (1962). In each generation, they extracted 80 females and 80 males from the experimental population, and selected the 8 females and 8 males with the most sternopleural bristles, and an equal number with the fewest. The 32 selected flies were then mingled together in the same vial and allowed to mate. The males were then discarded, and the females moved to rearing vials to lay their eggs, keeping the high and low females separate. The variance of bristle number increased through time, so that by the twelfth generation of selection the frequency distribution was bimodal. One mode consisted entirely of the offspring of the low-selected females, the other of the offspring of the high-selected females. However, when high and low individuals were deliberately crossed, their offspring were intermediate. Since few or no intermediate individuals were at this point present in the experimental population, it seems likely that flies with many bristles and those with few had become separated into two sexually isolated groups. This process occurs, as the authors remarked, 'with astonishing rapidity'.

Alas, it has not recurred. The dramatic result of the experiment spurred another dozen or more laboratories to repeat it, but they were uniformly unsuccessful (see Scharloo *et al.* 1967a, b, Chabora 1968, Barker and Cummins 1969a, b, Spiess and Wilke 1984). Indeed, the one generalization that may safely be made from these experiments is precisely the opposite of the original conclusion: sexual isolation does not generally evolve as a correlated response to disruptive selection of moderate intensity. This is not in itself surprising, in view of the effect that recombination is likely to have in uncoupling mating preference from bristle number; what remains surprising is the original result. Barker and Cummins (1969a) ran a more elaborate version of the experiment, using the same original base population. They obtained a comparable direct response to selection for bristle number only when using much more intense selection on a larger number of flies in each generation. In these circumstances—but not with weaker selection, or smaller samples—some degree of sexual isolation did evolve, at least in the earliest generations. In this case, however, selection was so intense that after the third generation all matings were between high flies or between low flies, so that the procedure was equivalent to discarding all the hybrid offspring. This will not explain Thoday and Gibson's result, because in their experiment selection was relatively weak. It is possible that the very small number of individuals selected created by chance a correlation between bristle number and mating preference. This does not appear to have been present in the base population—the evidence is equivocal—but could have arisen in subsequent generations. This would account for the rapid divergence of the high and low lines, and would be consistent with Powell's boom-and-bust experiments. Whatever the truth of this may be, the important point is that the many attempts to repeat Thoday and Gibson's experiment have demonstrated that mating preferences do not usually evolve as a correlated response to disruptive selection of moderate intensity on an arbitrary character. This confirms that recombination effectively obstructs adaptive speciation. The exceptions are experiments where selection on geotaxis in houseflies resulted in substantial assortative mating even when 30–50% gene flow was allowed (Soans *et al.* 1974, Hurd and Eisenberg 1975).

13.2.3 Evolution of isolating mechanisms

In order to start up a coupled process of natural and sexual selection, it is therefore necessary, early in the history of diverging populations, to protect them from recombination. The most obvious way in which this might happen in nature is if they are in different places, with little exchange of migrants between them. Rice and Salt (1990) ran a particularly colourful experiment to find out how fast habitat choice could evolve, and how complete would be the sexual isolation it caused. Young flies entered a maze that presented them with a sequence of different kinds of choice. The first chamber had a light and a dark exit; passing through either, they could then move upwards or downwards; finally, they could enter chambers scented with either acetaldehyde or ethanol. They thus sorted themselves among eight vials, depending on their phototactic, geotactic, and chemotactic behaviour (Figure 13.4). They mated and laid eggs in these vials, and the eggs were collected either at the beginning, in the middle, or at the end of the oviposition period. The combination of spatial and temporal criteria thus provided a total of 24 different habitats, or ways of life. All but two were lethal. Flies were reared only from eggs laid early in the light, downward, acetaldehyde habitat, or laid late in the dark, upward, ethanol habitat. These were deemed to represent two types of fruit that would support larval growth; other choices led to inappropriate oviposition sites, where the larvae could not survive. The larvae from the two habitable vials were reared separately, and after eclosion the young flies were re-introduced to the maze for the next generation of selection. After about 30 generations of selection, the separation between flies utilizing the two habitable vials was almost complete: those born in one of these vials almost invariably returned to the same vial as adults (Figure 13.5). Because mating occurred almost exclusively within the vials, rather than at intermediate points in the maze, habitat choice enforced sexual isolation, so that by the end of the experiment the population comprised two sexually independent strains.

Similar experiments have been done unintentionally whenever crop plants have been introduced into a new region. The clearest example of isolation caused by the evolution of habitat choice is the tephritid fly *Rhagoletis,* studied by Bush (1969). Its native host plant in North America is hawthorn. When orchards of exotic fruits were planted in its range during the nineteenth century, some strains were able to exploit them. It was reported from apple in 1864, and has since established itself on pear and cherry. The adults have a strong tendency to mate on the host they fed on as larvae. Moreover, the rate of larval development depends on the host species—cultivated fruits provide a richer pasture—so the flies emerge from different hosts at different times of the year. About 100 generations of selection has produced distinct strains specializing on different host species that are sexually isolated by virtue of their evolved habitat preferences. A second example is provided by ecologically distinct strains of fish in postglacial lakes that spawn at different sites within the same lake (§13.3.2). Strains that mate in different places or at different times are necessarily sexually isolated, because their members rarely meet. It seems reasonable to suppose that they will proceed to evolve distinctive mating structures or behaviours, so that members of different strains will not mate with one another, should they ever come into contact. However, this has not yet been observed. The *Drosophila* selection lines from Rice and Salt's experiment mate randomly when mingled in the same vial, and host-specific strains of *Rhagoletis* and ecotypes of fish readily mate among one another in the laboratory. There is as yet no experimental evidence that mating preferences evolve as a correlated response to selection for habitat preferences.

Local adaptation can also be conserved when recombination is prevented by asexuality,

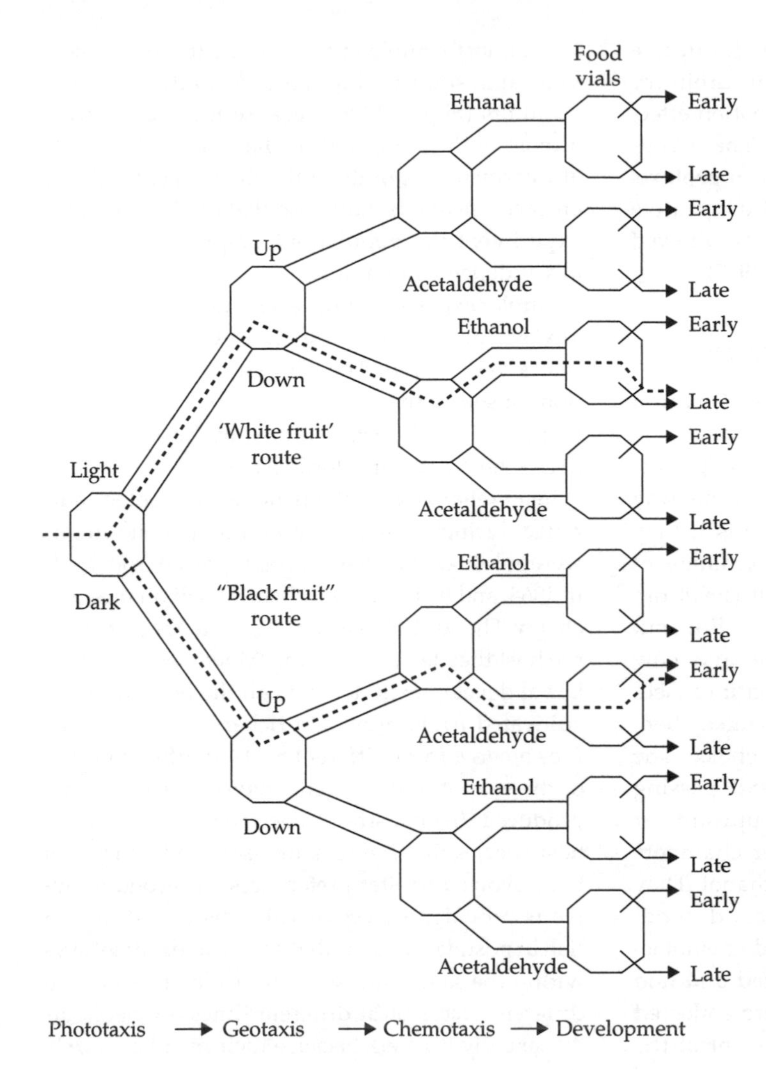

Figure 13.4 Sexual isolation through habitat choice. Only flies choosing the 'white fruit' (light, downwards, ethanol, late) or 'black fruit' (dark, upwards, acetaldehyde, early) sites survived. Flies making different decisions in the maze were identified by an ingenious phenocopy technique involving adding a supplement to the food medium in the vials representing one habitat that suppressed the expression of an eye-colour mutation, providing a permanent but non-heritable marker. From Rice and Salt (1990).

self-fertilization, or polyploidy. Polyploid lineages are established in a single generation, either by the failure of meiosis after fusion, or as the result of a vegetative doubling. The diploid gametes produced by a tetraploid parent will produce triploid plants on crossing with the original strain, and these are likely to be inviable or sexually sterile. A gene that directs parthenogenesis has a similar effect and moreover will tend to spread because by suppressing recombination it causes its own transmission to be associated with that of locally adapted genotypes. A shift from a sexual to an asexual life cycle may evolve only rarely in some groups, from the paucity of genetic variation for so radical a change. Self-fertilization, on the other hand, evolves very readily in hermaphroditic organisms such as flowering plants. The local adaptation of grasses such as *Agrostis* and *Anthoxanthum* to the polluted soil of former mines is continually diluted by pollen from non-tolerant plants on the adjacent pastures (§7.5.2). Antonovics (1968) found that metal tolerance was associated with self-fertility: all non-tolerant plants produced very few or no seeds in isolation, whereas some tolerant plants were highly self-fertile (Figure 13.6). Moreover, average self-fertility fell abruptly at the mine boundary,

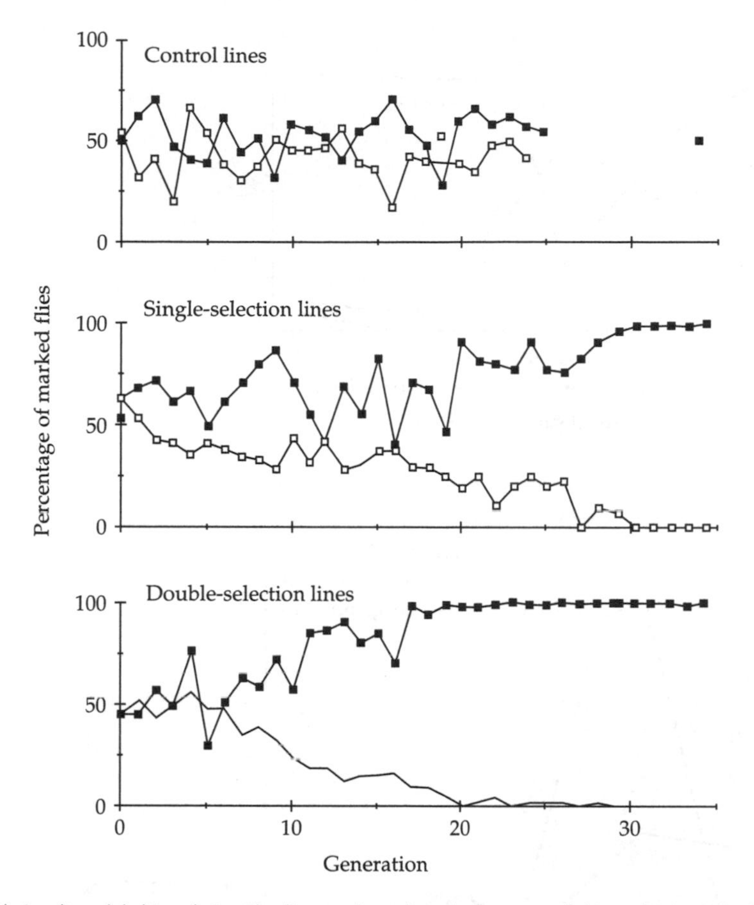

Figure 13.5 Sexual isolation through habitat choice. The diagrams here show the frequency of this marker in each of the two preferred habitats; an absolute preference for one of the two habitats would be indicated by a frequency of unity in one habitat and zero in the other. There are three types of line. The upper panel shows unselected control lines; the middle panel shows single-selection lines in which only flies moving to one of the two preferred types of habitat were allowed to breed; the lower panel shows double-selection lines in which flies were retained not only if they moved to one of the preferred habitats, but also if their parents had displayed the same preference. From Rice and Salt (1990).

in parallel with the decline in metal tolerance. Selection against the fertilization of pasture plants by pollen from the mine is much weaker, and there was no indication that selfing was selected on the pasture. McNeilly and Bradshaw (1968) found that mine populations also flower about a week earlier than pasture populations; this difference was retained in common-garden experiments, and may have arisen through selection for sexual isolation under outcrossing. The difference was especially marked at the mine boundary, or on small mines. Similarly, Snaydon and Davies (1976) found that plants collected near the boundary of plots in the Rothamsted Park Grass Experiment (§7.4.1) flowered several days earlier than their neighbours on either side.

13.2.4 Divergent natural selection in complete isolation

The experimental issue is then whether completely isolated divergent selection lines eventually exhibit some degree of sexual isolation. There are about a dozen experiments of this sort; none

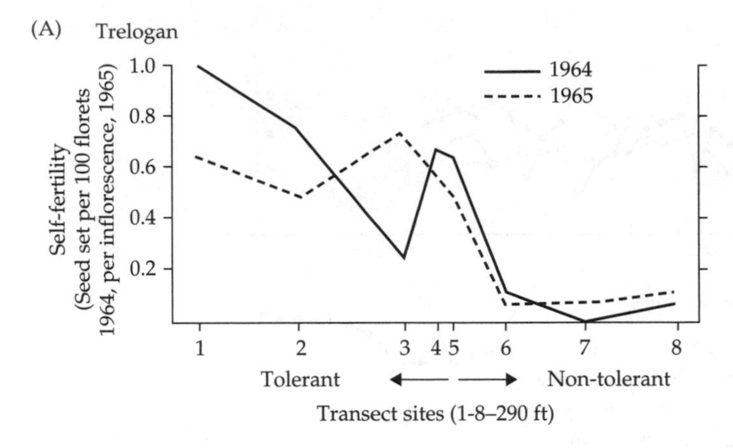

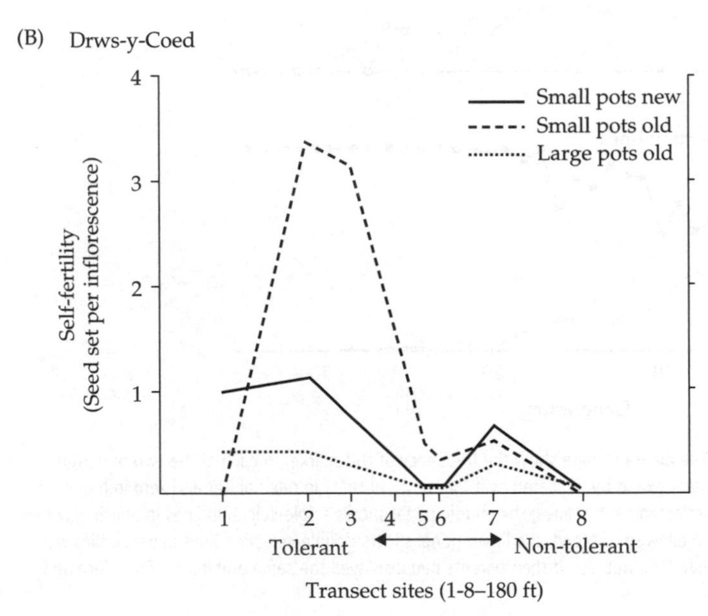

Figure 13.6 Sexual isolation through self-fertilization. These diagrams show the self-fertility of plants sampled across the boundary of a lead mine at Trelogan and a copper mine at Drws-y-Coed, from Antonovics (1968). They should be compared with the pattern of change in metal tolerance along similar transects (Figure 7.25).

have involved the evolution of complete sexual isolation, but several have reported some degree of assortative mating. The simplest experiments are those in which there is no contact between selection lines for some considerable period of time, after which their sexual compatibility is assessed. Barker and Cummins (1969a, b) selected for high and low bristle number in *Drosophila* in the usual way and assessed sexual compatibility after about 30 generations by mingling flies from the 2 lines and scoring their mating behaviour. The morphological selection seemed to have caused changes in mating behaviour; in particular, flies from the High line mated much more quickly than flies from the Low line. This meant that most pairs that formed within a few minutes of the flies being introduced to one another comprised a High male and a High female, the Low pairs being rare. This was a consequence of mating speed, however, not of any preference of one type over the other. After a few hours, all types of pairs were equally frequent. Divergent selection had caused no tendency for flies selected in the same direction to prefer to mate with one another.

De Oliveira and Cordeiro (1980) selected *Drosophila* on excessively acid and basic food media, as well as on medium of normal composition. After about 100 generations of selection the lines had diverged considerably, with the number of offspring surviving to adulthood being much greater on the medium on which their parents had been selected. This adaptation was most pronounced for the lines on acid medium, which produced about five times as many offspring as the basic lines, when tested on acid media; the basic lines produced about twice as many offspring as the acid lines, when tested on basic media. When the lines were crossed, the hybrids were notably inferior to their adapted parent on the acid medium, but there was little difference on the basic medium. Some degree of hybrid inferiority had thus arisen, although this cannot have affected the mating behaviour of the flies, the selection lines being maintained quite separately. Mating behaviour was assessed by putting males from one line into a vial containing females from the same line and from one of the other two lines. There was an overall tendency for matings between flies from the same selection treatment to occur in excess. The design of the assay seems to imply that this was caused by male preferences; however, there was a pronounced bias only when females from the acid lines were available as mates, so female choice may have contributed to the expression of the male preference. The biases that were displayed were rather weak and erratic, and did not seem to increase during the course of the experiment.

Robertson (1966) added EDTA to the growth medium to select for resistant lines. Crosses between the selection lines and the unselected controls showed varying degrees of hybrid inferiority. This was largely attributable to genes on the third chromosome; when third chromosomes from the control population were introduced into the selection lines they caused sterility or death. However, the lines mated randomly, and attempts to select for assortative mating were unsuccessful.

One advantage of using physiological rather than morphological characters is that natural selection can be used in place of artificial selection, greatly increasing the size of the selected sample. Behaviour is even more awkward to study than morphology, because selection may require a lengthy scrutiny of each individual. This difficulty has been circumvented for geotaxis and phototaxis by the construction of ingenious mazes, designed so that the flies sort themselves out into different compartments according to their behaviour, making large-scale selection experiments feasible. Del Solar (1966) scored mating behaviour in lines that had previously been selected for phototaxis and geotaxis, positive and negative. A single generation of selection produced a very weak tendency for flies from the same line to mate with one another; this tendency was much stronger after 5 and 11 generations of selection, and was maintained after 3 generations of reduced selection (Figure 13.7). Hurd and Eisenberg (1975) selected for positive and negative geotaxis in houseflies and observed strong isolation (albeit in very limited trials) after 16 generations.

Several experimenters have found that the modification of behaviour through selection is soon accompanied by the appearance of preferential mating. This is presumably because flies that move in opposite directions are unlikely to mate, so the same gene or genes are responsible for fitness under artificial selection and for assortative mating and therefore cannot be broken up by recombination. This is only true for directional movement, however. Van Dijken and Scharloo (1979a, b) selected for high and low general activity using the 'locometer', a series of conical compartments that allow passage in only one direction, so that the most active flies accumulate in the furthest compartments. When the mating behaviour of the flies was examined after 30 or 40 generations of selection,it was found to be highly non-random; but this was almost entirely attributable to the fact that the males selected for high activity were quicker off the mark; in trials where one kind of female was mingled with both kinds of male, the high-activity males mated much more quickly, regardless of which kind of female was offered. Like selection for phototaxis or geotaxis, selection for general activity led indirectly to the modification of sexual behaviour, but in this case one selection line acquired an unconditional sexual advantage.

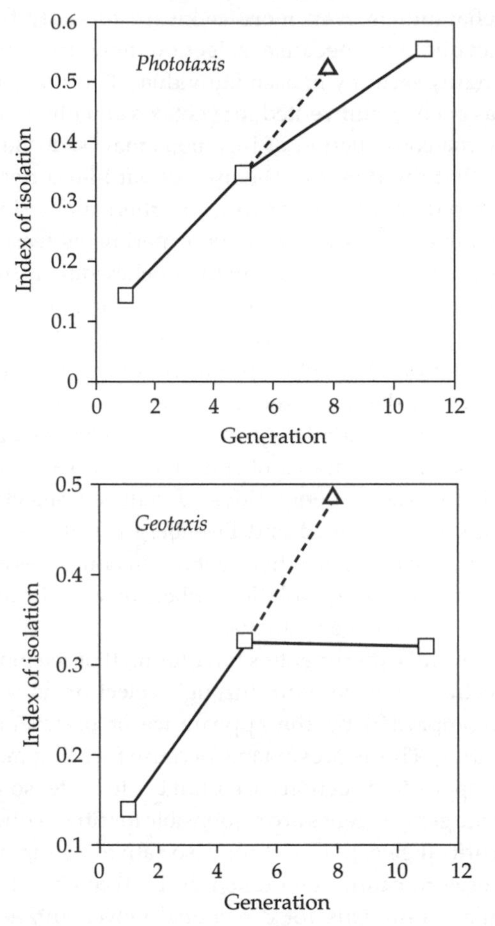

Figure 13.7 Sexual isolation as a correlated response to directional selection. These diagrams shows the outcome of mating trials between lines of *Drosophila* selected for positive and negative phototaxis (upper diagram) or for positive and negative geotaxis (lower diagram), from del Solar (1966). The index of isolation is an increasing measure of the frequency of matings between individuals from lineages selected in the same direction. The broken line and triangle represent lines in which selection was relaxed for three generations; it is somewhat surprising that isolation seems to increase at the same rate, or an even greater rate, as in the selection lines.

13.2.5 Reinforcement

These experiments demonstrate that some degree of isolation may evolve between lines that are kept completely separate. They suggest that hybrid inferiority is likely to arise through selection on physiological characters, and, rather more strongly, that mating preferences may evolve as a correlated response to selection on behavioural characters. Neither generalization is based on a very extensive body of experiments, but both are what common sense might lead us to expect. Dobzhansky (1937) argued that the hybrid inferiority arising while the lines are separate might cause the evolution of mating preferences when they are mingled. Any tendency towards mating within lines will be reinforced by selection against the inferior offspring produced by mating between lines. The difficulty with this attractive idea is, of course, that the association between genes for specific adaptedness and genes for preferential mating will be disrupted by recombination. The experimental evidence for reinforcement remains sketchy. The experiments in artificial disruptive selection that I have described above seem to show that preferences rarely evolve in freely mating populations that are initially undifferentiated. They do evolve when hybrids are deliberately eliminated, but there is then no mutual reinforcement of adaptedness and sexual isolation, because the isolation is already complete. There are very few experiments that investigate the reinforcement of partial isolation. One attempt was made by Forbes Robertson (1966), who re-established contact between the lines that he had selected with and without EDTA added to their food, by the simple device of connecting the population cages by glass tubes. Twenty generations later, there was no sign that any substantial mating preferences had evolved.

13.2.6 Artificial selection for sexual isolation

If large populations often contain genetic variance for female mating preferences, it should be possible to modify these preferences through selection. The standard design is to permit males and females of two different strains to mingle freely, subsequently discarding the hybrid progeny (recognized as recombinants for a pair of markers), so that the lines are perpetuated mainly or only through matings between parents of the same strain. Artificial selection is thus applied directly to the markers, and acts indirectly through sexual selection on the mating preferences. Such experiments—almost all with *Drosophila*—usually succeed in modifying mating preferences, creating or enhancing some degree of sexual isolation between the strains.

Crossley (1974) used Ebony and Vestigial mutants extracted from the same laboratory population. The mutants are interesting because their phenotypes are likely to affect mating behaviour. In both cases, males are less successful in mating than wild-type flies. Ebony males have poor vision and are likely to lose sight of their partner; Vestigial males have highly reduced, non-functional wings, and are unable to produce the buzz that is an important part of normal courtship behaviour. In the light, Vestigial females mate randomly, whereas Ebony females, for unknown reasons, prefer Ebony males. In the dark, the Vestigial males cannot buzz, and cannot see their partners either: Ebony males are thus preferred by both types of female. These initial preferences can be modified through selection, by discarding recombinant wild-type progeny. About 40 generations of selection caused a pronounced reduction in the frequency of mating between Ebony and Vestigial, both in the light and in the dark (Figure 13.8). Sexual selection increased the mating abilities of Vestigial males in the dark, and of Ebony males in the light. However, this did not enhance sexual isolation, and may have reduced it. It was instead female behaviour that was responsible for the decrease in mating between the two strains. In the dark, Ebony females repelled Vestigial males, either by wing-flicking (which halts male courtship), or simply by jumping away, so that most of their matings were with Ebony males. Vestigial females did not modify their behaviour in the dark, but in the light both Vestigial and Ebony females actively repelled males of the other strain. The complex courtship behaviour of female *Drosophila* can thus be tuned by sexual selection when powerful artificial selection is enforced against hybrids.

Drosophila pseudoobscura and *D. persimilis* are very similar species that can hybridize in nature. Kessler (1966) selected for both an increased and a decreased degree of isolation between them. To select for Low isolation, females from the Low line of one species were mingled with unselected males of the other species: the first females to mate were chosen to perpetuate the line by removing them and allowing them to re-mate with males that had been selected

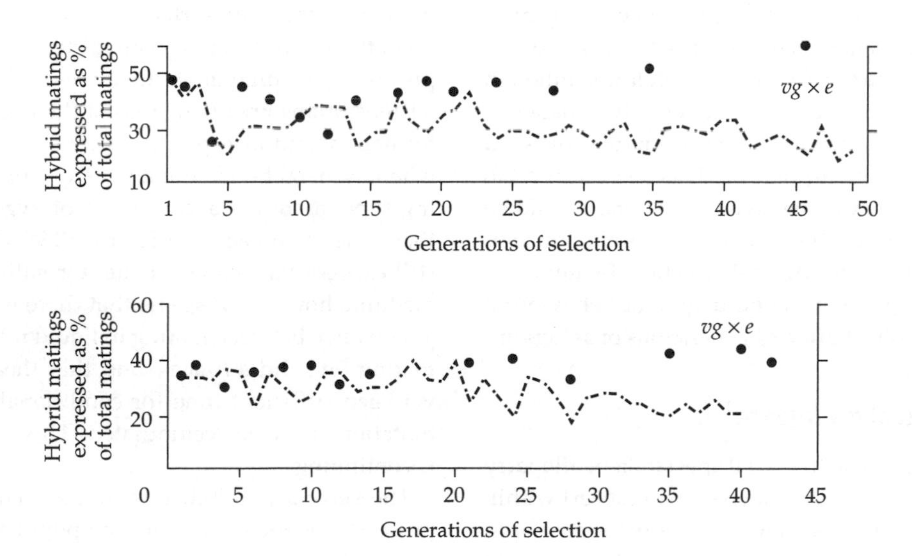

Figure 13.8 Artificial selection for sexual isolation. The line is the frequency of matings between Ebony *e* and Vestigial *vg* individuals in the light (lower) and dark (upper) selection lines; the plotted points are the corresponding estimates for unselected control lines. The decline in heterogamous matings, relative to the control, was caused by an increase in both $e \times e$ and $vg \times vg$ matings in both light and dark conditions.

in a similar manner. To select for High isolation, all the flies from the selection lines that mated with the other species were discarded, the surviving conspecific males and females were reintroduced to one another, and the first pairs to mate were selected. The response to selection was erratic, but after 18 generations the High and Low lines had diverged in both male and female descent. The effect was again largely attributable to the behaviour of the females. *D. persimilis* females are reluctant to mate with *D. pseudoobscura* males, even in the unselected lines, and selection had little effect on this behaviour; if their partners are *D. pseudoobscura* males, there is almost no contact between them. The behaviour of *D. pseudoobscura* females was more malleable: in particular, the Low females became much more receptive to *D. persimilis* males. Thus, it is possible either to enhance or to diminish the degree of isolation between two species that have only recently become distinct, through selection in the short term.

Dobzhansky and Pavlovsky (1971) conducted similar experiments with races of *Drosophila paulistorum*. These races are to a large extent sexually isolated, both because they are reluctant to mate with one another and because male hybrids are sterile. One isolate, originally classified as belonging to the Orinocan race, was found to produce sterile male progeny with Orinocan mates when retested after four or five years in culture, although the flies mated freely. The cause of the change is unknown, although infectious agents of some kind might have been responsible. This line was used to investigate whether behavioural isolation could be superimposed on the newly arisen sexual incompatibility through artificial selection. By eliminating hybrid progeny, some degree of behavioural isolation evolved after 70 generations of selection.

13.2.7 Sexual divergence

The main obstacle to sexual speciation in allopatry is that mating rules must change in concert within each population as they diverge between populations. There are two ways in which this might happen: because sexual and vegetative characters are correlated, or because sexual characters in males and females are correlated.

Divergent sexual selection in itself would lead to ecologically equivalent cryptic species. If different vegetative adaptations are also favoured, however, then two populations will become ecologically specialized while becoming capable of mating effectively only in their home site. Both mates and prey might be detected visually in a brightly lit site, for example, and by smell in a dark site, so that the outcome of long-continued sexual selection would be that individuals from either site would be unable to find either mates or food in the other. Since mating preferences are readily created by artificial sexual selection they likely to evolve through natural selection when conditions that modulate the success of alternative mating behaviours or structures vary among sites. Sexual incompatibility will follow if superior mating ability at one site reduces success at others, or if conditionally neutral mutations affecting mating at other sites tend to accumulate. *Chlamydomonas* initiates mating by producing a gametic autolysin that digests away the cell wall and by expressing an agglutinin on the flagella which causes motile cells of different mating type to become entangled, after which a second agglutinin system mediates complete cell fusion. I happened to observe cells in one culture mating on the agar surface, where the conditions for both growth and fusion are different from those prevailing in liquid medium. Intense selection over 80 generations greatly increased the rate of zygote formation, and also produced a novel phenotype whereby neighbouring colonies of opposite mating type lay down a dark band of zygotes along their line of contact (see Figure 12.9). These lines still express flagella and mate normally in liquid medium, however. It seems that there is no strong antagonism between mating in liquid medium and mating on solid surfaces, and that there has not yet been sufficient time for conditionally neutral mutations to have accumulated. This experiment is continuing.

The second possibility is that sexual selection proceeds independently in each population as the result of antagonistic male–female co-evolution (§12.4). If conflict were mainly responsible for sexual divergence it should not require divergent natural selection; independent lines selected in similar environments should show the same sort of behaviour. There is little indication, however, that replicate selection lines become isolated within periods of up to 100 generations. For example, Kilias *et al.* (1980) collected *Drosophila* from two localities in Greece, and then cultured their descendants either in cold, dry conditions in the dark, or in warm, humid conditions in the light. After 5 years in the laboratory, flies from either locality showed a rather strong preference for partners from the same line, so that, as in many other experiments, divergent selection led to some degree of sexual isolation. However, flies from different localities selected in the same conditions mated freely with one another, showing no preference for partners from their own line. Dodd (1989) reported a similar result for *Drosophila* selected on starch or maltose medium. The sexual preferences displayed by flies cultured in different environments therefore represent a correlated response to selection, rather than purely historical divergence.

The available experimental evidence about the mechanism of speciation has been reviewed by Rice and Hostert (1993) and by Kirkpatrick and Ravigné (2002). I would summarize it as follows. Disruptive selection in sympatry is not accompanied by any substantial degree of sexual isolation, with the possible exception of two small experiments involving behaviour. Sexual isolation is necessary for adaptive divergence to occur, and can readily evolve through selection for mating at specific sites. Divergent natural selection in complete isolation sometimes indirectly creates some degree of preferential mating on secondary contact, the best examples involving modified behaviour whose potential effect on mating is self-evident. Artificial sexual selection is usually effective. There is therefore good experimental evidence for a sequence from primary isolation due to selection for using particular mating sites, to divergent sexual selection imposed by conditions at those sites, which permits effective natural selection for ecological specialization. Although the links in this chain seem reasonably sound, and are perfectly consistent with simple theory, the final outcome of ecologically specialized types that do not mate when mingled has yet to be achieved.

13.2.8 A new species of yeast

Saccharomyces cerevisiae and *S. paradoxus* are two closely related yeasts that can occupy the same sites and mate readily, but 99% of hybrids die. Greig *et al.* (2002) selected viable F1 haploid progeny and selfed them to obtain fertile homozygotes. The F2 generation included viable strains that had high fertility when intercrossed but low fertility when back-crossed to either parent. Fertility dropped in F3 but recovered to very high levels in the F4. The inability of the F2 to mate successfully with either parent was associated with the replacement of some *paradoxus* chromosomes by *cerevisiae* chromosomes, suggesting that successful hybrids require particular mixtures of *paradoxus* and *cerevisiae* genomes. This remarkable experiment is the only successful attempt so far to create a new species in the laboratory, but I am sure that others will follow as speciation is gradually drawn into the research programme in experimental evolution.

13.3 Emerging species

The mechanism of speciation should be most clearly apparent when populations are in the process of diverging to form separate species. The classical examples of rampant speciation in exotic places—cichlids in the Rift Valley lakes, Darwin's finches on the Galapagos islands, *Partula* in New Caledonia—have often been described and I have nothing new to say about them, so I shall instead illustrate the theme with common fish and plants from a landscape more familiar to me.

13.3.1 Sticklebacks

The three-spined stickleback *Gasterosteus aculeatus* is a small, rather bizarre fish widely distributed in the colder waters of the northern hemisphere. It has male parental care: the male constructs a nest from plant fragments, attracts one or more females to lay eggs in it, and subsequently guards the developing offspring. The marine form, living in coastal and brackish water, is heavily armoured with dorsal and pelvic spines capable of locking into place and connected by a series of bony plates that extends to the root of the tail. The spines and plates protect sticklebacks from predators by making them difficult to swallow, protecting them from injury, and facilitating their escape (Reimchen 1989, 2000). These marine populations have repeatedly invaded streams since the last glacial retreat 10 000–15 000 years ago, giving rise to a very large number of independently derived freshwater populations. Freshwater sticklebacks are often different in appearance from their marine ancestors, and since these differences must have evolved within a few thousand years, or perhaps much less, they have been intensively studied as examples of rapid diversification and even speciation (Hagen and Gilbertson 1973, Bell and Foster 1994, McKinnon and Rundle 2002).

The heavy armour of marine sticklebacks is strongly reduced in populations that have adapted to living in streams. Armour development is controlled largely by a single QTL (Colosimo *et al.* 2004) that maps to the *Ectdysoplasin* (*Eda*) gene (Colosimo *et al.* 2005). The product of this gene is a signal molecule that is required for normal scale development in other fish and for the development of ectodermal structures such as hair and teeth in mammals. *Eda* alleles from different populations have common ancestry, suggesting that the invasion of fresh water was accompanied by sorting low-plated alleles from the ancestral marine population. The large pelvic spines of marine sticklebacks are also reduced or completely lost in freshwater populations. Pelvic spine reduction is a Mendelian character involving a single QTL that appears to represent a regulatory mutation in *Pitx1*, a gene whose homologue is necessary for normal hindlimb development in mice (Shapiro *et al.* 2004). Plates and spines are reduced very rapidly in freshwater. Bell (2001) cites several cases in which marine populations have moved into newly created or newly vacated freshwater sites and have evolved greatly reduced armour and a deeper body within 10 generations or so. The agent of selection has not been identified, although deep-bodied, weakly armoured individuals may be more manoeuvrable and thus more proficient in capturing benthic invertebrates. The strong, repeatable natural selection based primarily on a few genes of large effect echoes the outcome of experimental evolution in laboratory microcosms.

The stream populations are free to invade lakes, where they may retain more or less the same morphology or evolve along different lines. On Graham Island off the coast of British Columbia, for example, the stream form is the normal sturdy, mottled brown type whereas the open lake is occupied by long, slim, melanic individuals that are relatively heavily armoured and have a small mouth and many gill-rakers (Thompson *et al.* 1997). These fish are specialized to resist the attacks of trout and predatory birds, while also being specialized to capture planktonic crustaceans. Fish in outlet streams are often more similar to the lake fish than those from inlet streams, showing how the selection responsible for this divergence is partially obstructed by dispersal (Reimchen *et al.* 1985, Hendry *et al.* 2002). In some lakes there are two distinct ecotypes: a deep-bodied type with a long jaw and relatively few gill-rakers feeding on large benthic prey in the littoral zone, and a slimmer limnetic type foraging on plankton in the open water (Lavin and McPhail 1986) (Figure 13.9). These types could represent successive invasions brought about by marine incursions (McPhail 1993), but mtDNA studies have shown that different ecotypes from the same lake generally resemble one another more closely that the same ecotype from different lakes, showing that benthic and limnetic forms have diverged in sympatry (Taylor and McPhail 1999) as the result of ecological character displacement (Schluter and McPhail 1992). The maintenance of both types in the same lake requires a cost of adaptation in the form of a lower rate of growth of each type in the other's site (Figure 13.10). This is reflected in the outcome of competition between sticklebacks of intermediate phenotype, from lakes where radiation has not occurred, and trophic specialists: when the limnetic type is present, individuals of the intermediate type that have more benthic-like phenotypes grow better (Schluter 1994).

The evolution of benthic and limnetic ecotypes has stimulated a great deal of research, as it seems to offer replicated natural experiments in which the agents responsible for speciation can be identified. The ecotypes within each lake are, indeed, referred

Figure 13.9 Benthic and limnetic ecotypes of the stickleback *Gasterosteus aculeatus*. Cleared and stained specimens. Scale bar is 5 mm. From Peichel *et al.* (2001) .

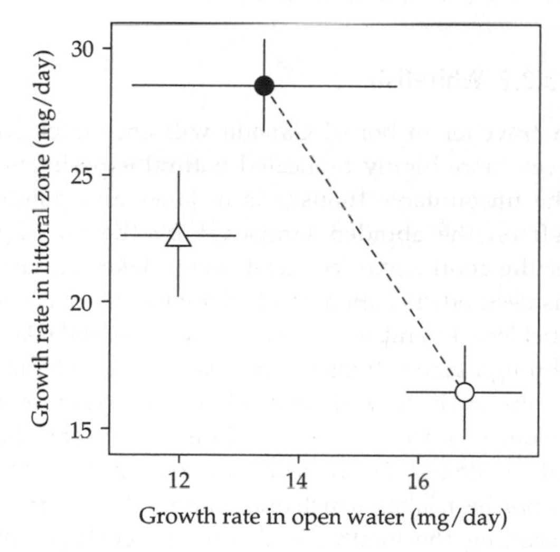

Figure 13.10 Cost of adaptation in benthic and limnetic sticklebacks. Open circle, limnetic; filled circle, benthic; triangle, hybrids. From Hatfield and Schluter (1999).

to as species by workers in the field. The F1 and F2 hybrids are nevertheless fully viable in the laboratory, although their growth rate in field enclosures is somewhat less than that of either parent in the zone to which it is specialized (Hatfield and Schluter (1999). Mate-choice experiments in the laboratory have shown that females prefer to mate with males of the same type (Ridgway and McPhail 1984), apparently as a consequence of differences in size and coloration (Boughman *et al.* 2005). In the field, both types nest in the littoral zone, benthics among macrophytes and limnetics mainly in shallower unvegetated areas. Mate choice in sticklebacks is strongly influenced by the nuptial colouration of the male, which evolved in the common ancestor of the *Pungitius–Culaea–Gasterosteus* clade (McLennan 1996). The ancestral state is a black throat, which has become a vivid red in *G. aculeatus*. This is used both as a threat display towards other males and as an enticement for females: females prefer intensely red-throated males, and males with redder throats receive more eggs (Bakker and Mundwiler 1994). They also pay a price, as red throats are easily detected by visual predators (Whoriskey and Fitzgerald 1985). Red is most conspicuous in the clear water where the limnetic type usually nests, and less conspicuous where the water is stained by organic matter. Males living in clear water display a larger area of red throat, while females from clear water are more sensitive to red. Hence, populations differ both in signal and response in a way that reflects the difference between benthic and limnetic sites. When male and female sticklebacks from different populations are paired in the laboratory, the probability of spawning is greater when the populations are most similar in signal and response, indicating some degree of isolation between benthic and limnetic types (Boughman 2001). Some degree of assortative mating based on body size has also been reported (Nagel and Schluter 1998), the smaller limnetic males preferring limnetic females from their own lake (Albert and Schluter 2004).

The preference for red has not been explained but may be an unmodified ancestral bias ('sensory drive'), since both male and female *Pungitius* (with black-throated males) and male and female *Gasterosteus* (with red-throated males) show more interest in red objects in mundane situations (Smith *et al.* 2004b), and sticklebacks are most sensitive to light with a wavelength of about 600 nm, in the red region (Cronly-Dillon and Sharma 1968). There are three interesting departures from a universal preference for red. The first are the giant black-throated males of the west coast, which are conspicuous in peat-stained water and seem to be preferred by local females (Moodie 1982, Reimchen 1989). The second is the unnamed small white stickleback of Nova Scotia, whose males sport a highly conspicuous white dorsal patch. They breed in the same general area as *G. aculeatus*, but construct nests above the substrate in filamentous algae (Jamieson *et al.* 1992); the males are careless parents and remove the eggs from the nest to strew them over the algae (Blouw 1996). Where their territories are adjacent, *G. aculeatus* females follow white males to their nest but then (perhaps deploring their domestic skills) break off courtship (Jamieson *et al.* 1992). They will not mate together in the laboratory, although artificial insemination is said to yield fully viable offspring. The third exception is the only other named species in the genus, *G. wheatlandi*, which has small bright green males with black-spotted throats. It occupies the same areas in estuaries but does not interbreed with *G. aculeatus*, which indeed displaces *G. wheatlandi* males from their nests (Cleveland 1994). I have not found an account of hybridization between *G. aculeatus* and *G. wheatlandi*.

The diversification of *G. aculeatus* in freshwater is a striking example of rapid evolution and seems to represent the initial phase of speciation. A modest level of assortative mating, comparable to that achieved in *Drosophila* experiments, has arisen through differences in body size and also through differences in the visual environment at the nesting sites that generate a correlation between natural selection through trophic specialization and sexual selection through mate preference. Despite the large number of opportunities and the very rapid response of trophic and morphological characters to selection, however, new species with strong sexual isolation have not evolved. Moreover, morphologically distinct benthic and limnetic populations may rapidly collapse into a single intermediate type (Taylor *et al.* 2006). It has been persuasively argued that the repeated evolution of benthic and limnetic forms provides a strong argument for ecological speciation, and there is no doubt of the merit of the case. Complete speciation, however, seems to require stronger sexual selection through the more extreme modification of nuptial coloration seen in *G. wheatlandi* and the white stickleback of Nova Scotia.

13.3.2 Whitefish

A traveller in boreal Canada will encounter an even more highly replicated natural experiment, the uncountable thousands of lakes and ponds left on the abraded landscape by the melting of the continental ice sheet. Many lakes remain fishless; others support salt-tolerant forms such as sticklebacks and smelt that arrived in coastal lakes through connections to the sea, or cold-tolerant forms such as whitefish, char, and trout that colonized interior lakes after shifts in drainage. The lake whitefish *Coregonus clupeaformis* is abundant in northern lakes, which often support two morphs recalling the benthic and limnetic ecotypes of sticklebacks: the benthic or 'normal' ecotype is larger in size, with fewer and coarser gill-rakers, feeding on larger prey captured on the bottom or on vegetation in the littoral zone, whereas the limnetic or 'dwarf' ecotype is a smaller and more agile type that feeds on plankton and has morphological adaptations for this way of life, such as a larger number of finer gill-rakers. In whitefish the benthic and limnetic ecotypes are similar in shape but the benthic is about 10 times larger (Lu and Bernatchez 1999), and this difference is attributable to 2–8 QTL associated with growth (Rogers and Bernatchez 2005). In whitefish the two forms are thought to represent descendants of distinct races that originally diverged in the Illinoan (Riss) glaciation in the middle Pleistocene, about 150 000 BP, and established secondary contact after the last glacial retreat about 12 000 BP. The closely related European *C. lavaretus* has undergone a similar but even more marked diversification. There are four ecotypes in Lake Femund, southern Norway, which differ in the number of gill-rakers, the length of the lower jaw, and the position of the mouth. They spawn in different locations and often at different times, and the bulk of the within-lake genetic variation, from microsatellite surveys, is between ecotypes (Østbye *et al.* 2005). *Coregonus* has undergone a modest radiation in North America and Europe involving about 20 species or species complexes (Figure 13.11). Some of them are small limnetic forms such as *C. artedii* or *C. sardinella*, and the limnetic ecotype of *C. clupeaformis* is found only in lakes from which they are absent. It seems likely that the ecotypic variation within *C. clupeaformis*, *C. lavaretus* and other species recapitulates this broader adaptive radiation. Detailed mapping studies have found extensive genetic differences between the ecotypes (Rogers *et al.* 2006), so the process of speciation may be further advanced than in *G. aculeatus*, perhaps because of the relatively long time passed in isolation before secondary contact was made after the last glaciation.

Whitefish neither construct nests nor care for their young. Male and female are alike; the eggs are externally fertilized during spawning on shoals in late fall, are deposited on the bottom, and hatch in the following spring. Ecotypes may

Figure 13.11 Whitefish (*Coregonus* species) from Swiss lakes. From the website of Ole Seehausen at *www.fishecology.ch*.

spawn at different times or in different places, but there appears to be much less opportunity for sexual selection than in sticklebacks. Nevertheless, there is good evidence from mtDNA surveys that benthic and limnetic types in Yukon lakes are sexually isolated (Bernatchez *et al.* 1998). Lu and Bernatchez (1999) show that the degree of morphological differentiation of the ecotypes is correlated with genetic distance and interpret this to mean that ecological differentiation strengthens sexual isolation, although the alternative interpretation, that ecological differentiation is facilitated by stronger sexual isolation, seems equally plausible. Moreover, hybrids between the whitefish ecotypes suffer higher embryonic mortality, equivalent to a selection coefficient of about 0.2–0.4 (Lu and Bernatchez 1998) (Figure 13.12). This may be attributable to incompatibility between genes governing development time, resulting in asynchronous emergence of hybrid larvae (Rogers and Bernatchez 2006).

Remarkably, many other fish that have succeeded in colonizing postglacial lakes have undergone parallel radiations. For example, four morphs of Arctic char (*Salvelinus alpinus*) have been described from an isolated volcanic lake in Iceland: a limnetic planktivore, a large and a small benthic type, and a large piscivore (Skÿlason *et al.* 1996). Smelt (*Osmerus mordax*) have evolved anadromous and freshwater forms, with lake populations evolving sympatrically into limnetic and benthic ecotypes (Taylor and Bentzen 1993). North American lakes may support several kinds of benthic and limnetic round whitefish *Prosopium* with different diets that are named as species but may be comparable to ecotypes in *Coregonus* (see Kennedy *et al.* 2006). In short, the occupation of new lakes by the few species of fish capable of expanding into a recently glaciated landscape is followed by the evolution of trophically specialized benthic and limnetic types almost as predictably as wrinkly spreaders evolve in laboratory vials.

13.3.3 Sedges

Almost any terrestrial habitat outside the tropics will contain sedges in the genus *Carex* (Figure 13.13). There are more than 2000 species of sedges, which are so diverse that they often make up 10% or more of all plant species at a site: tropical diversity in a temperate setting. The genus embraces several more or less extensive clades whose relationship is shown in Figure 13.14 for the species found in a subarctic fen and a temperate forest in Québec. The genus dates back about 50 my to the early Tertiary, and the great majority of the hundreds of Canadian

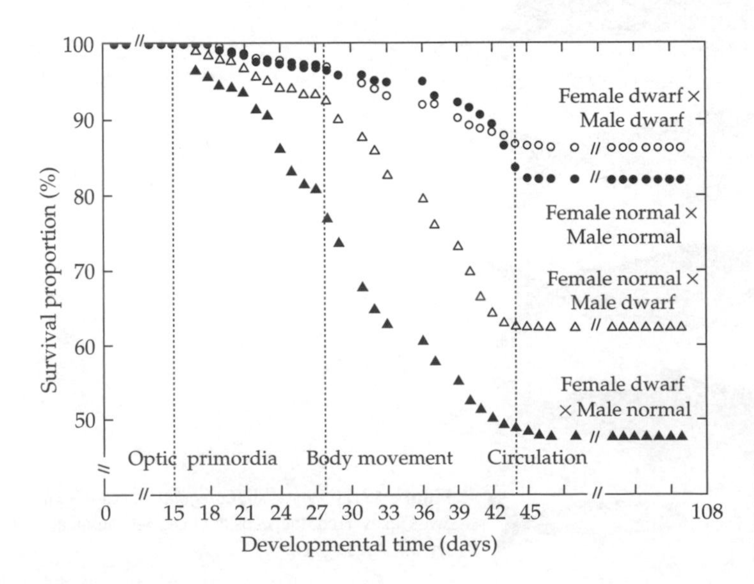

Figure 13.12 Hybrid breakdown in crosses between normal and dwarf whitefish ecotypes. Points are survival during early development. From Lu and Bernatchez (1998).

species must have originated further south and then moved northward following the receding glacial front. Besides being diverse they are also rather similar: a few species can be distinguished at sight by amateurs, but most can be identified only during the spring and early summer by examining the flowers or fruit under a lens. It is obvious at sight that benthic and limnetic sticklebacks and whitefish are divergently specialized, but what underlies the luxuriant radiation of sedges?

The very broad distribution of sedges is attributable in some measure to aerenchyma, a tissue providing a diffusion pathway for air from shoot to root and rhizosphere that enables the plants to grow in stagnant anoxic water and thus to colonize the northern morasses. Aerenchyma can be induced in many if not all species, but is most prominent in wetland species. There is a clear distinction between wetland and forest sedges: the subarctic fen and the temperate forest have only a single species in common. It is clear that sister species are ecologically correlated because fen-dwelling species are phylogenetically aggregated, with more than half the community belonging to a single clade of the subgenus *Vignea* and the rest occurring mostly as isolated small clades elsewhere in the phylogeny. If fen species were distributed randomly among clades the number per clade would follow a multinomial Poisson distribution with variance/mean = 1. Numerical analysis using the observed number of total species per clade yields an average value of variance/mean of about 1 for randomized data whereas the observed value is 3.9. Thus, sister taxa within clades tend to be similar with respect to emergent vs submerged growth. This specialization has evolved independently on several occasions and is associated with superior development of aerenchyma.

Figure 13.13 *Carex plantaginea* on the forest floor. Photo: Marcia Waterway.

Within forest or fen, specialization is not so clearly apparent, and a more quantitative approach is necessary to analyse the less pronounced environmental variation within either kind of site. This requires an extension of the argument that related species will be ecologically similar. Consider two sister species A and B together with the most closely related outgroup species C. Any evolutionary innovation that occurred in the C lineage, or in the lineage leading to the most recent common ancestor of A and B, will make A and B more alike than either is to C. It is also possible, however, that adaptation is associated with speciation, in which case sister species will be negatively correlated. The core concept of ecological phylogenetics, or ecophylogenetics, is thus that the correlation of sister species will depend on the mode of speciation, whereas the neutral theory predicts that sister species will be ecologically uncorrelated. Suppose that some allele affecting fitness becomes fixed in species A, causing it to occupy a range of sites with mean x_A and range from $x_A - y_A$ to $x_A + y_A$ (Figure 13.15). The extent of its overlap (number of sites occupied in common) with a sister species B that is fixed for a different allele is then $O_{AB} = 0$ if $x_B - x_A > y_A + y_B$, $O_{AB} = 2y_B$ if $x_B - x_A < y_A - y_B$, and $O_{AB} = (x_A - x_B) + (y_A + y_B)$ otherwise. The greater its overlap, the more

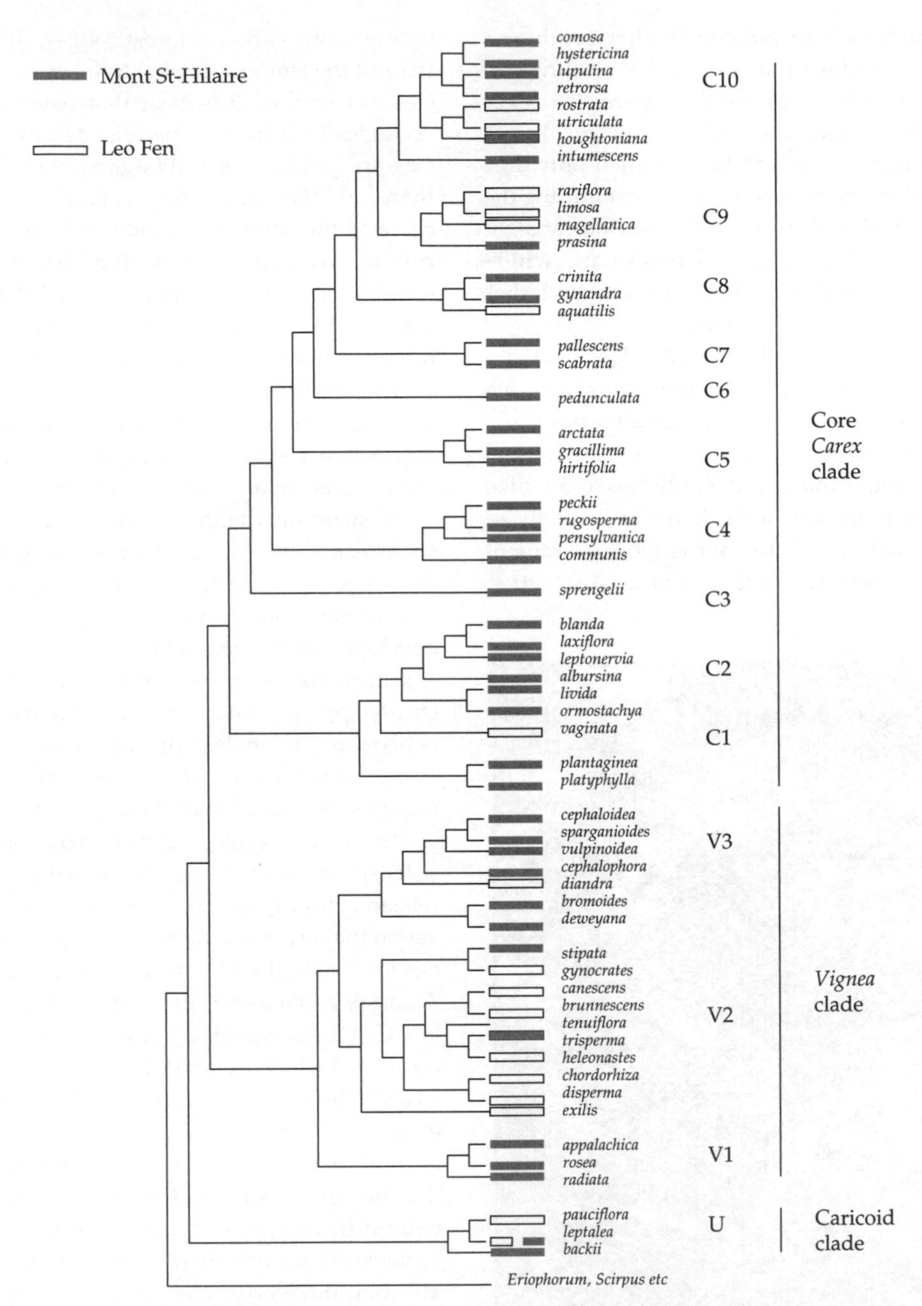

Figure 13.14 Phylogenetic relationships of sedges found in temperate mixed forest and a boreal fen in Quebec. Source: Waterway & Starr (2006).

similar will be the distributions of the two species. The degree of similarity can be expressed by the binary covariance $\mathrm{Cov}_{AB} = n_{00}n_{11} - n_{01}n_{10}$, where n_{10} is the number of sites occupied by species A but not by species B and so forth. Positive covariance indicates a large overlap in range, $\mathrm{Cov}_{AB} = O_{AB} - R_A R_B$, where R_A is the overall range (number of sites occupied) of species A.

Co-occurrence is greater for the two sister species than for the outgroup with the combination of sister species if $Cov_{AB} > Cov_{AB.C}$, that is, $O_{AB}/R - R_A R_B/R^2 > O_{AB.C}/R - R_{AB}R_C/R^2$, or $(o_{AB} - o_{AB.C}) > (r_A r_B - r_{AB} r_C)$ where the lower-case letters represent quantities as fractions of the total sites available. If the species have equal ranges, a sufficient condition is $o_{AB} > o_{AB.C}$. This requires $(x_B - x_A) < (x_C - x_B) + (½y_A + ½y_B - y_C)$. With equal ranges, this requires that the outgroup be less similar to the closer of the two sister species than these two are to themselves. Thus, the co-occurrence of ecologically similar sister species will exceed their joint co-occurrence with an outgroup species. When the ranges are grossly unequal the situation is less clear, but numerical analysis shows that this conclusion holds in the great majority of cases, with the disparity in co-occurrence increasing with the disparity in the means. This argument applies to sister taxa at any level, so that the phylogeny can be decomposed into sister clades of successively higher rank until no outgroup clade is available for comparison. If ecological similarity is expressed chiefly at the level of sister species then the neutral hypothesis is completely rejected. If it is expressed at the level of more inclusive clades then neutrality is rejected at this level, and higher levels, but may still hold within clades, In this way, sister group decomposition an be used to map the limits of neutrality.

There is little evidence of ecological similarity between sister groups at Mont St-Hilaire. A null hypothesis of neutrality was constructed by comparing random sets of species, with approximately the same distribution of number of species per set as the sister clades in the phylogeny. Most observations fell within the neutral zone ($r_{AB} < 0.1$), or failed to exceed the outgroup, and most of the exceptions are at a rather high phyletic level,

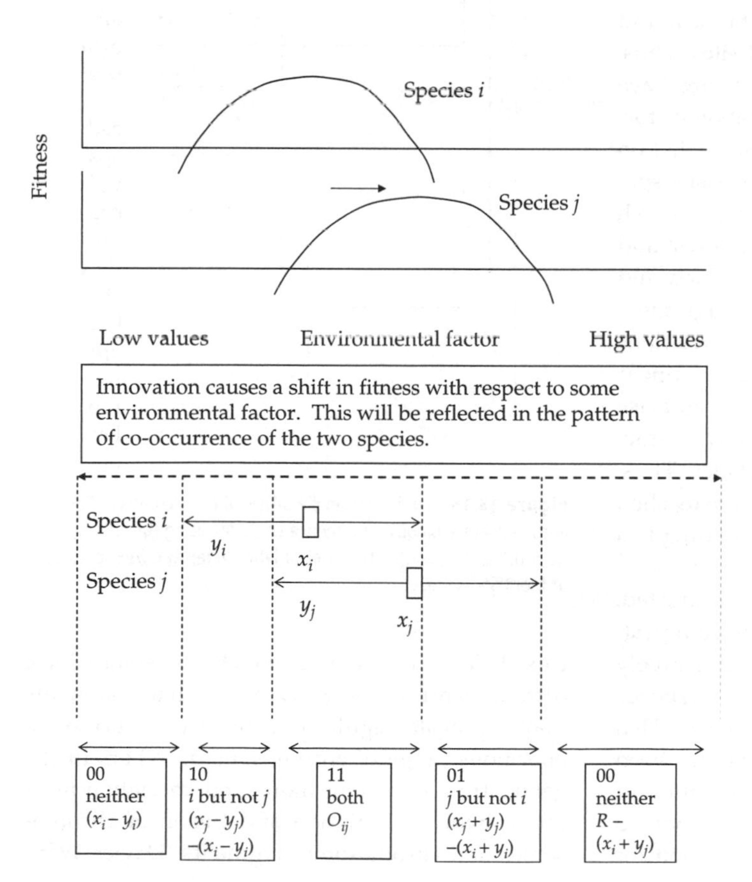

Figure 13.15 Ecological divergence and co-occurrence of sister species.

reflecting diversification among clades (Figure 13.16). Moreover, there is no correlation between ecological similarity and phylogenetic distance between species. This suggests that distantly related species may have independently evolved adaptations with similar effects. The relatives of Leo Fen species, for example, generally grow in wet situations at Mont St-Hilaire. Moreover, clades C1–C4 of subgenus *Carex* include mainly forest species found mainly in dry, well-insolated sites, whereas clades C5–C9 comprise species growing in wet sites such as pond margins and stream channels. The forest group occupies a large fraction of the drier quadrats, which constitute most of the survey area, whereas the wetland group is mainly restricted to the minority of wetter quadrats. Consequently, the correlation between the combined distributions of these two non-monophyletic groups is negative. There are some exceptions to these generalizations, however; in particular, *C. plantaginea* is a member of clade C1 but is found mainly in damp and poorly insolated sites. Thus, lineages specialized for growth in wet sites have evolved several times during the radiation of forest sedges. The DNA sequence data shows that on average there is <1% divergence between sister species, about 1% within clades, about 2% within each set of *Carex* clades, about 3.5% between forest and wetland *Carex*, and about 7.5% between *Carex* and *Vignea* (Waterway and Starr 2006). Although there are some clear examples of ecological correlations between sister species, most correlations appear at a rather high phyletic level and seem to arise chiefly from a forest-wetland dichotomy characteristic of different clades or sets of clades. Thus, the limits to neutrality in this radiation are reached for a divergence of about 2–4%, corresponding to a divergence time of roughly 4–8 my.

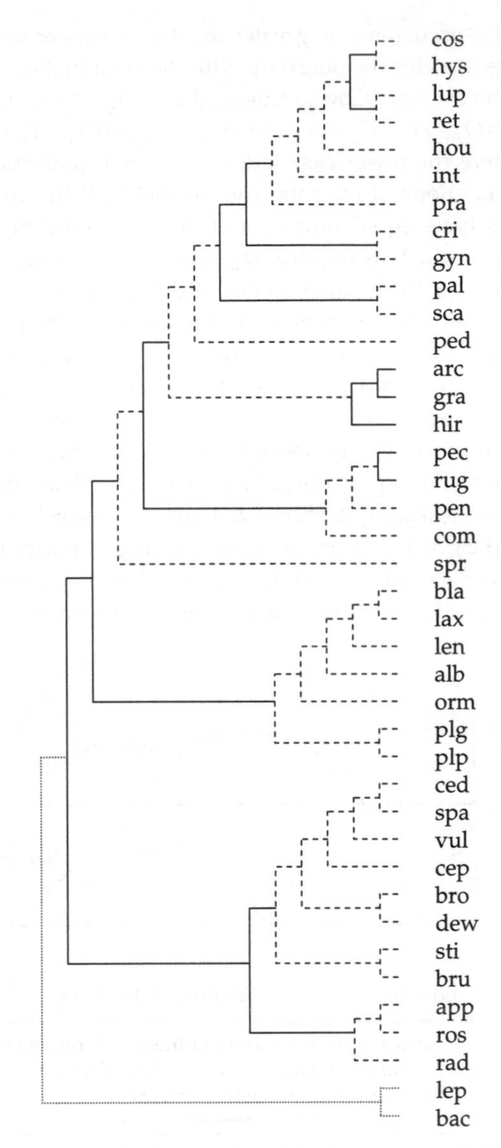

Figure 13.16 Ecophylogenetic analysis of co-occurrence. Solid lines indicate ecological divergence of clades using the criterion $r_{AB} > 0.1$ and $r_{AB} > r_{AB.C}$. Three-letter labels refer to *Carex* species at Mont St-Hilaire.

Sticklebacks and sedges are strongly contrasted. Sticklebacks and other northern fish have repeatedly evolved parallel adaptations but scarcely ever advanced to full speciation. The divergence of forest and wetland sedges has likewise arisen independently on several occasions, and has been accompanied by an uncontrolled proliferation of hundreds of reliably identifiable species with a regular phylogeny. The most plausible reason for this difference is that *Carex* chromosomes have diffuse centromeric activity, so that fragments may segregate regularly at meiosis. There is an enormous range of chromosome numbers in the genus, from 6 to 60 or more, and speciation must often involve the fission and fusion of chromosomes and chromosome fragments (Davies 1956).

Moreover, most species are self-compatible and many may be mostly selfing. Hybrids are rather readily formed, although success is often low and the fertility of the F1 progeny decreases with the difference in chromosome number between the parents (Whitkus 1988). Several species have been diagnosed as stabilized hybrids. *Carex* taxonomy is based almost exclusively on sexual characters such as the structure of the flower and the shape of the perigynium, a bract which encloses the carpel and persists after fertilization to enclose the achene. Hence, new species readily arise abruptly, are protected by self-compatibility and diverge in sexual structures while remaining for the most part vegetatively unmodified until ecological specialization emerges at a higher phyletic level.

CHAPTER 14

Epitome

The experimental study of evolution has shown that the classical account of adaptive evolution, by gradual change involving shifts in the frequencies of many genes of small effect over countless generations, is incomplete and must be supplemented by an account in which much adaptive change in the laboratory, the farm, or the field is driven by strong selection acting on a few loci whose identity and function can be ascertained. In this closing section I have tried to epitomize some of the material presented in the bulk of the text under headings which reflect the introductory section in consisting of declaratory sentences. This is not because I hold that any of them is true beyond doubt, although I think that each is probably sound, but rather to provide a framework for discussion that will help to move our field into the next stage of evolutionary studies.

14.1 Adaptation through selection works, although not always and not perfectly

Microcosm experiments usually (but not always) demonstrate specific adaptation to particular conditions of growth. Artificial selection usually (but not always) shifts a population rapidly beyond the ancestral limits of variation. Natural selection in the field often (but not always) drives populations towards specifically adapted phenotypes, although these are not necessarily the best attainable. Individual-based computer simulations show that selection usually improves but seldom perfects. The impact of these very well-supported observations should not be underestimated by professionals: they show that the core of evolutionary biology is normal reductionist science whose principles can readily be demonstrated to schoolchildren or lay people through observation and experiment.

14.2 Selection can be used to engineer new kinds of organisms and molecules

The ineluctable Baconian argument is that experiments of light must inform experiments of use: the ultimate proof and justification of evolutionary biology is that it leads to new useful products and devices. The classical proof is the improvement of crops and livestock through artificial selection. The modern potential includes the prediction of responses to global change, the development of rational processes of bioremediation, the control of disease by population genetic engineering, and the improvement of enzymes and binding agents by directed evolution. These should be given high priority by researchers and funding agencies for the next generation of evolutionary biologists.

14.3 Microbes and macrobes evolve in different adaptive regimes

Evolutionary dynamics are governed by two scaled mutation rates, the genomic mutation rate U and the mutation supply rate M. Bacteria have small genomes but large populations and hence small U and large M, whereas eukaryotes, and especially large multicellular eukaryotes such as animals and plants, have large U and small M. Microbial populations may be so large that all single mutants are usually present. In this case, selection will initially proceed by sorting, and the rate of adaptation will be limited by passage time in the presence of clonal interference. Lineages bearing exceptionally beneficial mutations will become abundant, until one or more experiences a secondary beneficial mutation and thereby initiates a further round of selection, expansion, and variation. This results in the ultimate fixation of the single best-adapted genotype after a more or less protracted period during which the evolving population has a complex and continually shifting composition. Most multicellular organisms are much less abundant, so some considerable time may elapse after conditions have changed before a beneficial mutation arises and surpasses the threshold for nearly deterministic spread. The rate of adaptation is thus limited by waiting time, with long periods during which the population is genetically uniform (with respect to adaptive variation) elapsing between episodes of selective replacement. Adaptation proceeds by an adaptive walk consisting of successive substitutions of beneficial mutations.

The adaptive regime within which a population evolves is thus characterized by the mutation supply rate, or equivalently by the ratio of passage time to waiting time. For abundant kinds of bacteria these quantities are large ($M \gg 1$ and $t_{pass}/t_{wait} \gg 1$) whereas for most multicellular organisms they are small ($M \ll 1$ and $t_{pass}/t_{wait} \ll 1$), with uncommon bacteria, most eukaryotic microbes, and abundant micrometazoans (such as rotifers and small nematodes) evolving in intermediate regimes.

14.4 Interaction is local; adaptation is global

The simplest environment is a well-stirred laboratory microcosm in which every individual interacts equally with every other, although even liquid microcosms have walls and a surface that provide more or less discrete niches. Natural populations are often rather highly aggregated, so that individuals interact with a few others in small groups, whether these be fish in a shoal or trees on a hill or bacteria in a soil pore. The actions of an individual thus reflect on itself rather than being broadly diffused and diluted. They reflect directly because an individual experiences the conditions it has itself modified. They reflect indirectly because its neighbours may be kin, or if unrelated may become long-term partners. The outcome of each local interaction is then exported to the general population by dispersal. This makes little difference, in principle, to the evolution of purely individualistic traits, but will necessarily modify any adaptation that affects neighbours, and thereby extends the range and complexity of social behaviour. Even bacteria may have interesting social lives that have not yet been fully explored.

14.5 Mutation is usually inconsequential in benign laboratory conditions but deleterious in the field

There are two categories of genes, as defined by the effect of complete deletion in benign laboratory conditions of culture: deleting Fisherian genes is lethal, whereas deleting non-Fisheran genes has little or no average effect on fitness. Most genes are non-Fisherian. There are few or no intermediates between these categories. Thus, organisms have the capacity to buffer complete loss of function at most loci. In natural populations most mutations seem to be deleterious, as there is strong circumstantial evidence for purifying selection, although the agents responsible have not yet been clearly identified. Hence, estimates of the effect of mutation from laboratory studies may be misleading, and estimates made under field conditions are badly needed.

14.6 Adaptive walks are usually short and involve beneficial mutations of large effect

The effect of beneficial mutations on fitness appears to be exponentially distributed, but the bulk of mutations of small effect are lost stochastically or suppressed by clonal interference. Adaptation to novel environments is thus attributable largely to mutations of moderate or large effect at a few loci, and not to mutations of small effect at many loci. In low-M conditions this leads to short adaptive walks of decreasing effect, because those of larger effect are likely to be substituted earlier, and thereby to a curvilinear, saturating increase of fitness or adaptedness over time. The response to artificial selection for a continuously varying character is likewise often associated with a few loci of large effect. There may well be some ascertainment bias, and extensive surveys will usually uncover many loci of small effect, but there is little doubt that mutations of large effect will often modify the target of selection, in which case they will be responsible for most of the response to selection. The genetic basis of adaptation in natural populations is poorly understood, but in several well-characterized situations such as land snails, ground finches, and sticklebacks it again involves one or a few genes of large effect. The bulk of adaptive change, in short, sometimes involves the substitution of mutations of large effect at a few loci, and this may be the general rule in the laboratory and in the field. If so, this will be good news for evolutionary biologists because it holds out the prospect of being able to identify precisely the genes responsible for any given instance of adaptation.

14.7 The response to selection slows down and eventually ceases

The response to natural selection and to artificial selection is often initially rapid but later slows down and eventually reaches a plateau, when no further progress can be made. The half-life of this process is typically a few tens of generations for artificial selection on quantitative characters in multicellular organisms, and a few hundreds of generations for natural selection on adaptation to novel conditions of growth in microbes. This is consistent with the substitution of mutations of diminishing effect, the bulk of adaptation being attributable to the few mutations of large effect that are substituted earliest in the series. The response ceases when no further beneficial mutations are available, although it may be prematurely terminated, for example by countervailing natural selection.

14.8 Selection is commonplace and often strong

The cumulation of beneficial mutations in poorly adapted populations usually restores adaptedness to novel or stressful conditions in the laboratory within a few hundred generations. Selection in the field is also often strong enough to drive appreciable phenotypic change in the short term. It is quite likely that a study is more likely to be attempted when there is some reason to expect selection to be acting, and more likely to be reported when selection is in fact detected. But this is beside the point. Field studies do not show that selection acts strongly on all characters at all times, of course, nor can it possibly do so. They do show that weak selection producing gradual change over long periods of time is not the only and may not be the principal mode of adaptation. It will often happen, no doubt, but even so is likely to be overridden from time to time by episodes of much more intense selection.

14.9 Adaptation often follows broadly similar routes

The evolution of novel metabolic capacities by bacteria presented with a substrate they are poorly adapted to utilize sometimes proceeds in four characteristic stages of exaptation, deregulation, amplification, and modification that I have called the EDAM process. The first stage is simply the use of some enzyme which normally acts on some other substrate, which often happens because enzymes are not perfectly specific. The second stage is a loss-of-function mutation causing constitutive expression of this enzyme, as it is unlikely to be induced by the exotic substrate. The third stage involves increased expression, by gene duplication or promoter mutation or some other means, which will increase fitness if the substrate is limiting. The fourth stage is beneficial mutations in the structural gene itself (or one of the copies of this gene) that enhance its catalytic properties. Once this has been achieved, the new system is able to regain appropriate regulation, although as yet this has seldom been observed. Other kinds of adaptation may follow broadly similar routes, for example the exaptation of a morphological structure for a novel use, followed by developmental changes and modification of the structure itself. The EDAM stereotype does not fit all cases of adaptation, but shows how different instances can often follow broadly comparable evolutionary paths.

14.10 Complex integrated structures are readily generated by selection

When adaptation involves interactions among several or many genes it may not be possible to evolve complex structures by the successive substitution of a series of mutations each of which is individually beneficial. Complex structures such as new metabolic pathways in bacteria or new logical capabilities in algorithms nevertheless evolve rather readily in experimental conditions. This requires that some intermediate phenotypes, at least, are more fit than their ancestors, although this may involve adaptation to different conditions rather than a monotonic increase in adaptation to some given environment. It does not require an unbroken sequence of beneficial mutations provided that there is a diversity of neutral or deleterious mutants segregating at low frequency in the population, since a second mutation in one of these lineages can create a superior epistatic genotype. This has been observed in high-*M* conditions but may be a severe impediment to adaptation in low-*M* populations.

14.11 Adaptation is accelerated by sex and horizontal transfer

The genetic interference that arises when beneficial mutations arise on a deleterious background impedes adaptation but can be resolved by recombination or horizontal transfer. There is reasonably good experimental evidence that recombination is effective, and this may largely account for the prevalence of outcrossed sexuality in natural populations, although some critical experiments remain to be done. The environmental agents responsible have not been conclusively identified, but enemies such as pathogens supply the only known consistent source of strong directional selection (apart from mating itself) and are the most plausible candidate. Horizontal transfer between unrelated lineages is widespread in bacteria and can lead to the immediate acquisition of new metabolic or ecological capabilities. It is excluded from conventional microcosms, however, and understanding its contribution to bacterial adaptation is an urgent priority for new kinds of experiment.

14.12 Replicated adaptation shows themes and variations

Whether the modern biota would be the same, or very similar, if our most recent common ancestor had been slightly different is on the face of it undecidable, but microcosm experiments have begun to reveal the outcome of replicated evolution. In some cases, adaptation involves a stereotyped succession of substitutions whose order and final combination are highly predictable. More generally, replicate selection lines of algorithms, viruses, and bacteria converge in fitness but diverge in other characters, showing that the genetic basis of adaptation comes to differ among independently evolving lines descending from the same ancestor. In some well-studied cases this has been confirmed decisively by identifying and sequencing the genes responsible. Although there may be several outcomes, however, it seems likely that there are not very many completely distinct ways of becoming adapted to given conditions. Although the evidence is scanty as yet, it seems likely that there are relatively few major themes on which adaptation can be based, albeit each theme can have many minor variations. The themes correspond, more or less, to the genes that contribute to the response, which will be few in number according to an oligogenic view of adaptation. The variations correspond, again more or less, to the spectrum of beneficial mutations at each locus, which may be extensive, for example loss-of-function mutations at regulatory loci. The short-term course of evolution, then, is neither predictable nor completely unpredictable, but rather predictable within limits. In the long term uncertainty will grow and the outcome may become unpredictable for all practicable purposes, but no truly long-term experiments have yet been reported in sufficient detail.

14.13 Simple environments support only a single type, except sometimes

If there is a single limiting depletable resource then the principle of competitive exclusion states that a single superior type will exclude all others. The sole remaining type is that able to grow at the lowest concentration of the limiting resource, by the principle of frugality. Both principles are often borne out by experience, although many exceptions have been reported. A simple environment may support several types when the limiting resource is incompletely processed, or when superior resource utilization is balanced by toxin production, or when density varies during culture growth. Variation may be conserved in sexual populations because epistatic combinations of nearly equal fitness are continually disrupted by recombination. These exceptions have been so often reported, indeed, even from the simplest of microcosms, that competitive exclusion must often be tempered in natural populations by social and sexual interactions that tend to perpetuate diversity.

14.14 Environments are fine-grained in space and time

The difference between two sites with respect to the state of some environmental factor increases with their distance apart. This relationship follows a power law, showing that environments vary on all scales, so that organisms will experience the same degree of patchiness regardless of their size. The exponent of the power law is often small, showing that environments are often fine-grained, such that a large proportion of the regional variation occurs locally. Temporal change follows a similar trend: conditions at a given site are likely to be more different after a longer period of time. This relationship is likewise described by weak power laws, showing that many factors are fine-grained in time, with individuals experiencing during their lifetimes a large proportion of the variation that occurs between generations. Both spatial and temporal rules are well documented for simple physical factors such as temperature, and also seem to apply to growth and fitness. Mutation and immigration will thus continually act to erode adaptedness; it seems likely that selection increases fitness by a few per cent in each generation as a consequence.

14.15 Selection fluctuates over periods shorter than the passage time

The pattern of temporal change implies that selection fluctuates in magnitude and occasionally in sign at any timescale. This may explain why natural populations often exhibit substantial genetic variance for characters related to fitness. Temporal change has been much less thoroughly studied than spatial variation, however, and the notion of continually fluctuating strong selection has far to go before being firmly established. It would not in itself protect genetic variation, except in special conditions, because plastic genotypes with high geometric mean fitness should eventually become fixed, but it will effectively retard fixation.

14.16 Relative fitness changes in proportion to environmental variance

If two genotypes differ in fitness they are likely to differ to nearly the same extent if conditions change slightly. If conditions change more drastically then the difference between the types may increase or decrease, or even reverse, so that the fitter type becomes the less fit. In general, the shift in relative fitness will be greater when conditions change more, provided that the factors responsible affect growth. Hence, the genetic correlation of fitness will tend to fall as the environmental variance of growth increases. This implies that the correlation will become negative when conditions are sufficiently different, each type being superior in some conditions and inferior in others. A similar argument can be made for genetic differences. Two genotypes that are closely related, or otherwise very similar, will usually respond in the same way to environmental change. Unrelated genotypes, or those that differ with respect to mutations that have a large effect on growth, may respond more or less, or in the opposite sense, to environmental change. Hence, an environmental change of given magnitude will alter the fitness rank of genotypes that are sufficiently dissimilar.

14.17 Specific adaptation is limited by its cost

Negative genetic correlation may arise simply because the direct response to selection exceeds the indirect response. In populations that are well-adapted to the general conditions of growth, however, specific adaptation to a novel stress will impede growth in the ancestral environment through functional interference. This may arise merely from the chance occurrence of a beneficial mutation in an inferior background, in which case it will be resolved by recombination or compensatory mutation. On the other hand, it may reflect a fundamental antagonism arising from developmental, mechanical, or physiological constraints, in which case it will be expressed immediately and permanently. If selection is long-continued then performance in the ancestral environment will also be impeded by mutational degradation through the accumulation of conditionally deleterious mutations. Both functional interference and mutational degradation have often been reported from experimental populations, and no doubt both contribute in some proportion to the cost of adaptation.

14.18 Costs of adaptation lead to optimal compromises

The cost of adaptation will generally imply that fitness will be maximized by an optimal intermediate character state. Selection will act to minimize both the distance between the optimum and the population mean, and the variance among individuals around the mean. It is not completely effective, and in practice adaptation is neither very accurate nor very precise, possibly because selection fluctuates over time. Optimality models have been more successful when applied to components of fitness and used to predict the timing and intensity of reproduction.

14.19 Complex and structured environments support diverse communities

In structured environments the state of non-depletable factors, such as temperature or pH, or the supply of depletable resources, such as sugars or prey species, varies among discrete sites. Consequently, a range of specialized types may be maintained by selection when the cost of adaptation prevents the fixation of a single universally adapted generalist. The criteria are quite strict, however: soft selection, comparable productivity and restricted dispersal. In a complex environment consisting of a mixture of depletable resources, the criteria are the same (given that dispersal is unrestricted). In both cases diversity is protected by the frequency-dependent selection that arises when local (or substrate-specific) density regulation creates a refuge for each specialized type. Hence, a population living in a structured environment will tend to become specifically adapted to the site it occupies, and many striking examples of local adaptation have been reported. The most widely cited evidence for the prevalence of local adaptation (and the easiest to obtain) is genotype–environment correlation, but this is not necessarily trustworthy. The output of spatially explicit neutral models is essentially indistinguishable from survey data unless selection is strong, dispersal is weak, and the environment is coarse-grained relative to the sampling unit. Hence, distribution patterns can be used to document the precision of local adaptation if there is independent evidence for site-specific selection, but they do not themselves provide convincing evidence for local adaptation. The tendency for sister clades to occur together, outside their region of phyletic divergence, is the most convincing kind of comparative evidence, but careful studies are rare, so the limits of neutrality—the phyletic level at which divergent specialization characteristically appears—are not yet reliably mapped. The most extensive experimental evidence is provided by reciprocal transplantation, but although this has demonstrated the divergent specialization of phenotypically distinct species or ecotypes, the evidence for widespread local adaptation is not entirely convincing. In short, frequency-dependent selection is likely to be ubiquitous in natural systems, although it will often fail to preserve variation, when functional interference is too weak, or immigration is too high, or productivity is too unequal. Highly specialized types are arresting proofs of adaptation, but are remarkable because they are exceptional. Most genotypes will cope reasonably well with a certain range of conditions that constitutes part of the range exploited by the population as a whole, so that a population living in a complex or structured environment consists of overlapping imperfect generalists.

14.20 Sex and disease generate strong directional selection

The biotic environment seems to vary in space and time like the physical environment, although we have much less quantitative information about it. Rivals and enemies have the further property that they are capable of rendering any adaptation rapidly obsolete, so they can in principle generate perpetual directional selection. Apart from epidemics in crop plants the experimental evidence for this view is currently limited, although the development of phage–bacterium microcosms is beginning to extend it. Disease and starvation are very frequent causes of death and debility, however, so antagonists are potentially the most important selective agents acting on natural populations. The other strong interaction that most eukaryotes experience is with mates, or potential mates, or rivals for mates. Sexual selection may be as powerful as natural selection, or more powerful, especially in species where it is obligately associated with reproduction. It is also to some extent self-generating, like co-evolution, insofar as mates are continually selected to overcome current male rivalry and female resistance. The experimental evidence for sexual arms races is intriguing, although not yet conclusive. There is no doubt, however, that genes associated with mating, together with those governing attack and defence against enemies, often evolve exceptionally fast. In many cases, moreover, a small number of genes of large effect participate in sexual or co-evolutionary adaptation. The tendency for natural selection to involve strong selection on a small number of identifiable major genes seems to be emphatically reinforced by sexual selection and co-evolution.

14.21 Selfish genetic elements can spread very rapidly

Even stronger selection may occur when sex and disease meet. Amphimixis necessarily creates an opportunity for infection that is exploited by a zoo of parasitic genetic elements such as transposable elements, homing endonuclease genes, and B chromosomes. An effective parasite can at first nearly double in frequency in every generation, spreading so rapidly that the host cannot counter-adapt in time and is overwhelmed. This process can be engineered to drive the host (and the parasite) extinct, but the host usually survives and once the element is fixed there may be little or no further selection for parasitism and it proceeds slowly to degenerate or, more interestingly, becomes domesticated. Several important genome services in eukaryotes are provided by domesticated selfish elements.

14.22 Helpful and altruistic behaviour is favoured by kin selection and reciprocity

Aggressive behaviour towards potential competitors will normally be favoured and is therefore the rule. It only pays to be helpful when aggression is too risky. Altruistic behaviour that harms the actor cannot evolve unless it is directed preferentially towards close relatives, who are likely to bear the genes responsible for the behaviour. Helping neighbours may be favoured, precisely because in many species they are likely to be relatives. Close collaborations between individuals of different species usually evolve only when there is a long series of interactions that cannot be readily broken off, so that the fitness of one partner entrains the fitness of the other.

14.23 Arms races lead to perpetual evolution

Simple genetic models of interactions between enemies such as a pathogen and its host lead to persistent genetic cycles. These have very beautiful dynamics, but have been frustratingly difficult to find in the field. It seems more likely that challenge and response often leads to a less tightly choreographed process in which adaptation involves new phenotypes at every stage, with little or no recurrence of historical resistance or virulence. The consequence would be perpetual directional selection with a continually changing target. In ecosystems this will be driven by a continually changing spectrum of direct and indirect interactions with enemies. At present, however, we do not have a convincing laboratory model of such processes.

14.24 The life cycle is a balance between natural selection and sexual selection

Arms races provide the renewable source of directional selection that favours sex in natural populations. The current trend in theory is hostile to this interpretation, but it is consistent with what we know about selection in natural populations, and I find it difficult to believe that the complex sexual ballet has evolved merely to increase the rate at which deleterious mutations are eliminated. Sex sets up a fundamental antagonism between natural selection for growth and reproduction among spores (or vegetative individuals) and sexual selection for mating success among gametes (or sexual individuals). I have interpreted the fundamental features of the eukaryote life cycle, such as the alternation of generations and the evolution of males and females, from this perspective.

14.25 Sexual isolation is often the first step in speciation

In sexual organisms offspring are produced by mating between individuals of like species and unlike gender; neither species nor gender occurs in asexual organisms such as bacteria. Speciation may be precipitated either by ecological divergence, followed by the evolution of sexual isolation, or contrariwise by sexual isolation, followed by ecological divergence. My main reason for preferring the latter route as a general rule is that ecological divergence is effectively frustrated by outcrossing, whereas ecological divergence occurs very readily once isolation has been achieved.

14.26 No principle other than natural selection and sexual selection is required to explain adaptation

Selection is an effective mechanism for producing adaptation, and by adding to the list of experimentally verified principles we are able to provide convincing explanations for a growing range of natural phenomena. These explanations involve two central concepts, natural selection and sexual selection, together with their natural extensions, such as kin selection, and the constraints within which they must act. Other kinds of concept have been advanced from time to time to supply rival explanations. These include Lamarckian or directed evolution, self-assembly, intelligent design, holistic or emergent properties of complex systems, the benefit of the species (or the whole biosphere), and macroevolutionary processes operating only at long timescales. So far as I am aware, none of these has ever been shown, by experiment or any other compelling procedure, to contribute substantially to adaptation. Moreover, there is no aspect of adaptation, so far as I am aware, that is incapable of being explained by selection. Natural selection and sexual selection are together sufficient to explain how blind variation is transformed into complex adaptation.

References

Abdullah, N.F., and Charlesworth, B. (1974). Selection for reduced crossing-over in *Drosophila melanogaster*. *Genetics* **76:** 447–451.

Abzhanov, A., Protas, M., Grant, B.R., Grant, P.R., and Tabin, C.J. (2004). *Bmp4* and morphological variation of beaks in Darwin's finches. *Science* **305:** 1462–1465.

Acton, A.B. (1961). An unsuccessful attempt to reduce recombination by selection. *American Naturalist* **95:** 119–120.

Adams, J., and Hansche, P.E. (1974). Population studies in microoganisms. I. Evolution of diploidy in *Saccharomyces cerevisiae*. *Genetics* **76:** 327–338.

Adams, J., and Oeller, P.W. (1986). Structure of evolving populations of *Saccharomyces cerevisiae*: Adaptive changes are frequently associated with sequence alterations involving mobile elements belonging to the Ty family. *Proceedings of the National Academy of Sciences of the USA* **83:** 7124–7127.

Adams, J., Puskasrozsa, S., Simlar, J., and Wilke, C.M. (1992). Adaptation and major chromosomal changes in populations of *Saccharomyces cerevisiae*. *Current Genetics* **22:** 13–19.

Agar, W.E. (1913). Transmission of environmental effects from parent to offspring in *Simocephalus vetulus*. *Philosophical Transactions of the Royal Society of London Series B* **203:** 319–351.

Agar, W.E. (1914). Experiments on inheritance in parthenogenesis. *Philosophical Transactions of the Royal Society of London Series B* **205:** 421–489.

Aggrey, S.E., Ankra-Badu, G.A., and Marks, H.L. (2003). Effect of long-term divergent selection on growth characteristics in Japanese quail. *Poultry Science* **82:** 538–542.

Agrawal, A.A. (2000). Host-range evolution: adaptation and trade-offs in fitness of mites on alternative hosts. *Ecology* **81:** 500–508.

Agrawal, A.F., and Chasnov, J.R. (2001). Recessive mutations and the maintenance of sex in structured populations. *Genetics* **158:** 913–917.

Albert, A.Y.K., and Schluter, D. (2004). Reproductive character displacement of male stickleback mate preference: reinforcement or direct selection? *Evolution* **58:** 1099–1107.

Al-Hiyaly, S.A.K., McNeilly, T., and Bradshaw, A.D. (1988). The effect of zinc contamination from electricity pylons. Contrasting patterns of evolution in five grass species. *New Phytologist* **114:** 183–190.

Allard, R.W. (1963). Evidence for genetic restriction of recombination in the lima bean. *Genetics* **48:** 1389–1395.

Allard, R.W., and Adams, J. (1969). Population studies in predominantly self-pollinating species. XIII. Intergenotypic competition and population structure in barley and wheat. *American Naturalist* **103:** 621–645.

Allen, J.A. (1988). Frequency-dependent selection by predators. *Philosophical Transactions of the Royal Society of London* **319:** 485–503.

Allen, J.A., and Clarke, B. (1968). Evidence for apostatic selection by wild passerines. *Nature* **220:** 501–502.

Andel, J.V. (1998) Intraspecific variability in the context of ecological restoration projects. *Perspectives in Plant Ecology, Evolution and Systematics* **1:** 221–237

Anderson, R.M., and May, R.M. (1982). Coevolution of hosts and parasites. *Parasitology* **85:** 411–426.

Andersson, M. (1982). Female choice selects for extreme tail length in a widowbird. *Nature* **229:** 818–820.

André, J.A., and Morrow, E.H. (2003). The origin of interlocus sexual conflict: is sex-linkage important? *Journal of Evolutionary Biology* **16:** 219–223.

Angert, A.L. and Schemske, D.W. (2005). The evolution of species distributions: reciprocal transplants across the elevation ranges of *Mimulus cardinalis* and *M. lewisii*. *Evolution* **59:** 1671–1684.

Antia, R., Levin, B.R., and May, R.M. (1994). Within-host population dynamics and the evolution and maintenance of microparasite virulence. *American Naturalist* **144:** 457–472.

Antonovics, J. (1968). Evolution in closely adjacent plant populations. V. Evolution of self-fertility. *Heredity* **23:** 219–238.

Antonovics, J. (1976). The nature of limits to natural selection. *Annals of the Missouri Botanical Garden* **63:** 224–247.

Anxolabehère, D. (1971). Sélection larvaire et fréquence génique chez *Drosophila melanogaster. Heredity* **26:** 9–18.

Araus, J.L., Slafer, G.A., Buxo, R., and Romagosa, I. (2003). Productivity in prehistoric agriculture: physiological models for the quantification of cereal yields as an alternative to traditional approaches. *Journal of Archaeological Science* **30:** 681–694.

Arnold, S.J., and Wade, M.J. (1984). On the measurement of natural and sexual selection: theory. *Evolution* **38:** 709–719.

Arnquist, G., and Rowe, L. (2005). *Sexual conflict.* Princeton University Press, Princeton, NJ.

Asthana, S., Schmidt, S., and Sunyaev, S. (2005). A limited role for balancing selection. *Trends in Genetics* **21:** 30–32.

Atchley, W.R., Rutledge, J.J., and Cowley, D.E. (1982). A multivariate statistical account of direct and correlated response to selection in the rat. *Evolution* **36:** 677–698.

Atwood, K.C., Schneider, L.K., and Ryan, F.J (1951a). Periodic selection in *Escherichia coli. Proceedings of the National Academy of Sciences of the USA* **37:** 146–155.

Atwood, K.C., Schneider, L.K., and Ryan, F.J. (1951b). Selective mechanisms in bacteria. *Cold Spring Harbor Symposia on Quantitative Biology* **16:** 345–355.

Avise, J.C., Walker, D., and Johns, G.C. (1998). Speciation durations and Pleistocene effects on vertebrate phylogeography. *Proceedings of the Royal Society of London, Series B* **265:** 1707–1712.

Axelrod, R. (1984). *The evolution of cooperation.* Basic Books, New York.

Ayala, F.J., and Campbell, C.A. (1976). Frequency-dependent selection. *Annual Review of Ecology and Systematics* **5:** 115–138.

Badgett, M.R., Auer, A., Carmichael, L.E., Parrish, C.R., and Bull, J.J. (2002). Evolutionary dynamics of viral attenuation. *Journal of Virology* **76:** 10524–10529.

Baer, C.F.,Shaw, F., Steding, C. *et al.* (2005). Comparative evolutionary genetics of spontaneous mutations affecting fitness in rhabditid nematodes. *Proceedings of the National Academy of Sciences of the USA* **102:** 5785–5790.

Bakker, H. (1974). Effect of selection for relative growth rate and body weight of mice on rate, composition and efficiency of growth. *Mededelingen Landbouwhogeschool, Wageningen* 74–8.

Bakker, K. (1969). Selection for rate of growth and its influence on competitive ability of larvae of *Drosophila melanogaster. Netherlands Journal of Zoology* **19:** 541–595.

Bakker, T.C.M., and Mundwiler, B. (1994). Female mate choice and male red coloration in a nutural three-spined stickleback (*Gasterosteus aculeatus*) population. *Behavioral Ecology* **5:** 74–80.

Barclay, H.J., and Gregory, P.T. (1981). An experimental test of models predicting life-history characteristics. *American Naturalist* **117:** 944–961.

Barker, J.S.F. (1973). Natural selection for coexistence or competitive ability in laboratory populations of *Drosophila. Egyptian Journal of Genetics and Cytology* **2:** 288–315.

Barker, J.S.F. (1988). Population structure. In W.G. Hill and T.G.C. Mackay (eds.), *Evolution and animal breeding,* pp.75–80. CAB International, Wallingford.

Barker, J.S.F., and Cummins, L.J. (1969a). Disruptive selection for sternopleural bristle number in *Drosophila melanogaster. Genetics* **61:** 697–712.

Barker, J.S.F., and Cummins, L.J. (1969b). The effect of selection for sternopleural bristle number on mating behaviour in *Drosophila melanogaster. Genetics* **61:** 713–719.

Barker, J.S.F., and East, P.D. (1980). Evidence for selection following perturbation of allozyme frequencies in a natural population of *Drosophila. Nature* **284:** 166–168.

Barnard, C.J., and Sibly, R.M. (1981). Producers and scroungers: a general model and its application too captive flocks of house sparrows. *Animal Behaviour* **29:** 543–550.

Barnes, B.W. (1968). Stabilizing selection in *Drosophila melanogaster. Heredity* **23:** 433–442.

Barraclough, T.G., and Nee, S. (2001). Phylogenetics and speciation. *Trends in Ecology and Evolution* **16:** 391–399.

Barraclough, T.G., Birky, C.W., and Burt, A. (2003). Diversification in sexual and asexual organisms. *Evolution* **57:** 2166–2172.

Barrett, R.D.H., and Bell, G. (2006). The dynamics of diversification in evolving *Pseudomonas* populations. *Evolution* **60:** 484–490.

Barrett, R.D.H., MacLean, R.C., and Bell, G. (2005) Experimental evolution of *Pseudomonas fluorescens* in simple and complex environments. *American Naturalist* **166:** 470–480.

Barrowclough, G.F. (1980). Gene flow, effective population sizes, and genetic variance components in birds. *Evolution* **34:** 789–798.

Barton, N.H., and Charlesworth, B. (1998). Why sex and recombination. *Science* **281:** 1986–1990.

Barton, N.H., and Keightley, P.D. (2002). Understanding quantitative genetic variation. *Nature Reviews Genetics* **3:** 11–21.

Barton, N.H., and Otto, S.P. (2005). Evolution of recombination due to random drift. *Genetics* **169:** 2353–2370.

Barton, N.H., and Partridge, L. (2000). Limits to natural selection. *BioEssays* **22:** 1075–1084.

Basalla, G. (1988). *The evolution of technology.* Cambridge University Press, Cambridge.

Basolo, A.L. (1994). The dynamics of Fisherian sex-ratio evolution: theoretical and experimental investigations. *American Naturalist* **144:** 473–490.

Bataillon, T. (2000). Estimation of spontaneous genome-wide mutation parameters: whither beneficial mutations? *Heredity* **84:** 497–501.

Bateman, A.J. (1948). Intra-sexual selection in *Drosophila. Heredity* **2:** 349–368.

Bauer, G.J., McCaskill, J.S., and Otten, H. (1989). Travelling waves of in vitro evolving RNA. *Proceedings of the National Academy of Sciences of the USA* **86:** 7937–7941.

Baum, W.M., Richerson, P.J., Efferson, C.M., and Paciotti, P.M. (2004). Cultural evolution in laboratory microsocieties including traditions of rulegiving and rule following. *Evolution and Human Behavior* **25:** 305–326.

Bayliss, C.D., Field, D., and Moxon, E.R. (2001). The simple sequence contingency loci of *Haemophilus influenzae* and *Neisseria meningitides. Journal of Clinical Investigations* **107:** 657–662.

Beaudry, A.A., and Joyce, G.F. (1992). Directed evolution of an RNA enzyme. *Science* **257:** 635–641.

Beg, A., Emery, D.A., and Wynne, J.C. (1975). Estimation and utilization of inter-cultivar competition in peanuts. *Crop Science* **15:** 633–637.

Behnke, A., Bunge, J., Barger, K., Breiner, H.-W., Alla, V., and Stoeck, T. (2006). Microeukaryote community patterns along an O_2/H_2S gradient in a supersulfidic anoxic fjord (Framvaren, Norway). *Applied and Environmental Microbiology* **72:** 3626–3636.

Bell, A.E., and Moore, C.H. (1972). Reciprocal recurrent selection for pupal weight in *Tribolium* in comparison with conventional methods. *Egyptian Journal of Genetics and Cytology* **1:** 92–119.

Bell, A.E., Moore, C.H., and Warren, D.C. (1955). The evaluation of new methods for the improvement of quantitative characters. *Cold Spring Harbor Symposia on Quantitative Biology* **20:** 197–212.

Bell, G. (1978). Further observations on the fate of morphological variation in a population of smooth newt larvae (*Triturus vulgaris*). *Journal of Zoology* **185:** 511–518.

Bell, G. (1982). *The masterpiece of nature.* Croom Helm, London.

Bell, G. (1984a). Evolutionary and non-evolutionary theories of senescence. *American Naturalist* **124:** 600–603.

Bell, G. (1984b). Measuring the cost of reproduction. II. The correlation structure of the life tables of five freshwater invertebrates. *Evolution* **38:** 314–326.

Bell, G. (1985). The origin and evolution of germ cells as illustrated by the Volvocales. In Halvorson, H.O. and Monroy, A. (eds.) *The origin and evolution of sex,* pp. 221–256. Alan R. Liss, New York.

Bell, G. (1988) *Sex and death in protozoa: the history of an obsession.* Cambridge University Press, Cambridge.

Bell, G. (1990a). The ecology and genetics of fitness in Chlamydomonas. I. Genotype-by-environment interaction among pure strains. *Proceedings of the Royal Society of London Series B* **240:** 295–321.

Bell, G. (1990b). The ecology and genetics of fitness in *Chlamydomonas.* II. The properties of mixtures of strains. *Proceedings of the Royal Society of London Series B* **240:** 323–350.

Bell, G. (1991a). The ecology and genetics of fitness in *Chlamydomonas.* III. Genotype-by-environment interaction within strains. *Evolution* **45:** 668–679.

Bell, G. (1991b). The ecology and genetics of fitness in *Chlamydomonas.* IV. The properties of mixtures of genotypes of the same species. *Evolution* **45:**1036–1046.

Bell, G. (1992a). The ecology and genetics of fitness in *Chlamydomonas.* V. The relationship between genetic correlation and environmental variance. *Evolution* **46:** 561–566.

Bell, G. (1992b). The emergence of gender and the nature of species in eukaryotic microbes. *Verhandlung der Deutschen Zoologischen Gesellschaft* **85:** 161–175.

Bell, G. (1992c). Five properties of environments. In P.R. Grant and H.S. Horn (eds.), *Molds, molecules and metazoa,* pp. 33–56. Princeton University Press, Princeton, NJ.

Bell G. (1994). The comparative biology of the alternation of generations. *Lectures on Mathematics in the Life Sciences* **25:** 1–26.

Bell, G. (1997a). Experimental evolution in *Chlamydomonas.* I. Short-term selection in uniform and diverse environments. *Heredity* **78:** 490–497.

Bell, G. (1997b). The evolution of the life cycle of brown seaweeds. *Biological Journal of the Linnean Society* **60:** 21–38.

Bell, G. (2000). The distribution of abundance in neutral communities. *American Naturalist* **155:** 606–617.

Bell, G. (2002). Neutral macroecology. *Science* **293:** 2413–2418.

Bell, G. (2003). The interpretation of biological surveys. *Proceedings of the Royal Society of London Series B* **270:** 2531–2542.

Bell, G. (2005). Experimental sexual selection in *Chlamydomonas. Journal of Evolutionary Biology* **18:** 722–734.

Bell, G. (2008). The poverty of the protists. *British Ecological Society Symposium on Speciation and Ecology* (in press).

Bell, G., and Burt, A. (1990). B-chromosomes: germ-line parasites which induce changes in host recombination. *Parasitology* **100:** S19–26.

Bell, G., and Gouyon, P.-H. (2003). Arming the enemy: the evolution of resistance to self-proteins. *Microbiology* **149:** 1367–1375.

Bell, G., and Koufopanou, V. (1991). The architecture of the life cycle in small organisms. *Philosophical Transactions of the Royal Society of London B* **332:** 81–89.

Bell, G., and Lechowicz, M.J. (1991). The ecology and genetics of fitness in forest plants. I. Environmental heterogeneity measured by explant trials. *Journal of Ecology* **79:** 663–685.

Bell, G., and Maynard Smith, J. (1987). Short-term selection for recombination among mutually antagonistic species. *Nature* **328:** 66–68.

Bell, G., and Mooers, A.O. (1997). Size and complexity among multicellular organisms. *Biological Journal of the Linnean Society* **60:** 345–363.

Bell, G., and Reboud, X. (1997). Experimental evolution in *Chlamydomonas*. II. Genetic variation in strongly contrasted environments. *Heredity* **78:** 498–506.

Bell, G., Lechowicz, M.J., and Schoen, D.J. (1991). The ecology and genetics of fitness in forest plants. III. Environmental variance in natural populations of *Impatiens pallida*. *Journal of Ecology* **79:** 697–713.

Bell, G., Lechowicz, M.J., Appenzeller, A. *et al.* (1993). The spatial structure of the physical environment. *Oecologia* **96:** 114–121.

Bell, G., Lechowicz, M.J., and Waterway, M.J. (2000). Environmental heterogeneity and species diversity of forest sedges. *Journal of Ecology* **88:** 67–87.

Bell, G., Lechowicz, M.J., and Waterway, M.J. (2001). The precision of adaptation in forest plants. In Silvertown, J. and Antonovics, J. (eds.) *Integrating ecology and evolution in a spatial context*, pp 117–138. Blackwell Science, Oxford.

Bell, M.A. (2001). Lateral plate evolution in the threespine stickleback: getting nowhere fast. *Genetica* **112–113:** 445–461.

Bell, M.A., and Foster, S.A. (1994). *The evolutionary biology of the threespine stickleback*. Oxford University Press, Oxford.

Bell, M.A., Baurngartner, J.V., and Olson, E.C. (1985). Patterns of temporal change in single morphological characters of a Miocene stickleback fish. *Paleobiology* **11:** 258–271.

Bell, T., Ager, D., Song, J.-I. *et al.* (2005). Larger islands house more bacterial taxa. *Science* **308:** 1884.

Bell, T., Newman, J.A., Silverman, B.W., Turner, S.L., and Lilley, A.K. (2005b). The contribution of species richness and composition to bacterial services. *Nature* **436:** 1157–1160.

Belotte, D., Curien, J.-B., Maclean, R.C., and Bell, G. (2003). An experimental test of local adaptation in soil bacteria. *Evolution* **57:** 27–36.

Bender, M.H., Baskin, J.M., and Baskin, C.C. (2002). Phenology and common garden and reciprocal transplant studies of *Polymnia canadensis* (Asteraceae), a monocarpic species of the North American temperate deciduous forest. *Plant Ecology* **161:** 15–39.

Bengtsson, J., Baillie, S.R., and Lawton, J. (1997). Community variability increases with time. *Oikos* **78:** 249–256.

Bennett, A.F., and Lenski, R.E. (1993). Evolutionary adaptation to temperature .II. Thermal niches of experimental lines of *Escherichia coli*. *Evolution* **47:** 1–12.

Bennett A.F., and Lenski R.E. (1996). Evolutionary adaptation to temperature .V. Adaptive mechanisms and correlated responses in experimental lines of *Escherichia coli*. *Evolution* **50:** 493–503.

Bennett, A.F., Lenski, R.E., and Mittler, J.E. (1992). Evolutionary adaptation to temperature. I. Fitness responses of *Escherichia coli* to changes in its thermal environment. *Evolution* **46:** 16–30.

Bennington, C.C., and McGraw, J.B. (1995a). Environment-dependence of quantitative genetic parameters in *Impatiens pallida*. *Evolution* **50:** 1083–1097.

Bennington, C.C., and McGraw, J.B. (1995b). Natural selection and ecotypic variation in *Impatiens pallida*. *Ecological Monographs* **65:** 303–323.

Bergelson, J., and Purrington, C.B. (1996). Surveying patterns in the cost of resistance in plants. *American Naturalist* **148:** 536–558.

Bergelson, J., Purrington, C.B., Palm, C.J., and Lopez-Gutiérrez, J.-C. (1996). Costs of resistance: a test using transgenic *Arabidopsis thaliana*. *Proceedings of the Royal Society London B* **263:** 1659–1663.

Bergstrom, C.T., Lipsitch, M., and Levin, B.R. (2000). Natural selection, infectious transfer and the existence conditions for bacterial plasmids. *Genetics* **155:** 1505–1519.

Bergthorsson, U., and Ochman, H. (1999). Chromosomal changes during experimental evolution in laboratory populations of *Escherichia coli*. *Journal of Bacteriology* **181:** 1360–1363.

Bernatchez, L., and Landry, C. (2003). MHC studies in nonmodel vertebrates: what have we learned about natural selection in 15 years? *Journal of Evolutionary Biology* **16:** 363–377.

Bernatchez, L., Chouinard, A., and Lu, G. (1998). Integrating molecular genetics and ecology in studies of adaptive radiation: whitefish, *Coregonus* sp., as a case study. *Biological Journal of the Linnean Society* **68:** 173–194.

Bernstein, H., Byerly, H.C., Hopf, F.A., and Michod, R.E. (1985). Genetic damage, mutation and the evolution of sex. *Genetics* **174:** 2173–2180.

Berry, R.J., and Crothers, J.H. (1970). Stabilizing selection in the dog-whelk (*Nucella lapilus*). *Journal of Zoology* **155:** 5–17.

Beukeboom, L.W., and Werren, J.H. (1992). Population genetics of a parasitic chromosome: experimental analysis of PSR in subdivided populations. *Evolution* **46:** 1257–1268.

Biebricher, C.K. (1983). Darwinian selection of self-replicating RNA molecules. *Evolutionary Biology* **16:** 1–51.

Biebricher, C.K., and Luce, R. (1993). Sequence analysis of RNA species synthesized by $Q\beta$ replicase without template. *Biochemistry* **32:** 321–327.

Biebricher, C.K., and Orgel, L.E. (1973). An RNA that multiplies indefinitely with DNA-dependent RNA polymerase: Selection from a random copolymer. *Proceedings of the National Academy of Sciences of the USA* **70:** 934–938.

Biebricher, C.K., Eigen, M., and Luce, R. (1981a). Kinetic analysis of template-directed and template-free RNA sythesis by $Q\beta$ replicase. *Journal of Molecular Biology* **148:** 391–410.

Biebricher, C.K., Eigen, M., and Luce, R. (1981b). Product analysis of RNA generated de novo by $Q\beta$ replicase. *Journal of Molecular Biology* **148:** 369–380.

Biel, S.W., and Hartl, D.L. (1983). Evolution of transposons: Natural selection for Tn5 in *Escherichia coli* K12. *Genetics* **103:** 581–592.

Bierbaum, T.J., Mueller, L.D., and Ayala, F.J. (1989). Density-dependent evolution of life-history traits in *Drosophila melanogaster*. *Evolution* **43:** 382–392.

Bigler, B.S., Welch, D.W., and Helle, J.H. (1996). A review of size trends among North Pacific salmon (*Oncorhynchus* spp.). *Canadian Journal of Fisheries and Aquatic Science* **53:** 455–465.

Birch, L.C. (1955). Selection in *Drosophila pseudoobscura* in relation to crowding. *Evolution* **9:** 389–399.

Birdsell, J., and Wills, C. (1996). Significant competitive advantage conferred by meiosis and syngamy in the yeast *Saccharomyces cerevisiae*. *Proceedings of the National Academy of Sciences of the USA* **93:** 908–912.

Birky, C.W., Wolf, C., Maughan, H., Herbertson, L., and Henry, E. (2005). Speciation and selection without sex. *Hydrobiologia* **546:** 29–45.

Bishop, D.T., and Cannings, C. (1978). A generalized war of attrition. *Journal of Theoretical Biology* **70:** 85–124.

Bishop, J.A. (1972). An experimental study of the cline of industrial melanism in *Biston betularia* (L.) (Lepidoptera) between urban Liverpool and rural North Wales. *Journal of Animal Ecology* **41:** 209–243.

Bishop, J.G., Dean, A.M., and Mitchell-Olds, T. (2000). Rapid evolution in plant chitinases: molecular targets of selection in plant-pathogen coevolution. *Proceedings of the National Academy of Sciences of the USA* **97:** 5322–5327.

Bjedov, I., Tenaillon, O., Girard, B. *et al.* (2003). Stress-induced mutagenesis in bacteria. *Science* **300:** 1404–1409.

Blackmore, S. (1999). *The meme machine*. Oxford University Press, Oxford.

Blayney, D.P. (2002). *The changing landscape of U.S. milk production*. Electronic report from the Economic Research Service (www.ers.usda.gov/Publications/SB978/).

Bloom, F.R., and McFall, F. (1975). Isolation and characterization of D-serine deaminase constitutive mutants by utilization of D-serine as sole carbon or nitrogen source. *Journal of Bacteriology* **121:** 1078–1084.

Blouw, D.M. (1996). Evolution of offspring desertion in a stickleback fish. *Ecoscience* **3:** 18–24.

Blueweiss, L., Fox, H., Kudzma, V., Nakashima, D., Peters, R. and Sams, S. (1978). Relationship between body size and some life history parameters. *Oecologia* **37:** 257–272.

Boag, P.T., and Grant, P.R. (1981). Intense natural selection in a population of Darwin's Finches (Geospizinae) in the Galapagos. *Science* **214:** 82–85.

Bodnar, A.G., Ouellette, M., Frolkis, M. et al. (1998). Extension of life-span by introduction of telomerase into normal human cells. *Science* **279:** 349–352.

Bohannan, B.J.M., and Lenski, R.E. (2000). The relative importance of competition and predation varies with productivity in a model community. *American Naturalist* **156:** 329–340.

Bohonak, A.J. (1999). Dispersal, gene flow and population structure. *Quarterly Review of Biology* **74:** 21–45.

Bohonak, A.J., and Jenkins, D.G. (2003). Ecological and evolutionary significance of dispersal by freshwater invertebrates. *Ecology Letters* **6:** 783–796.

Bonner, J.T. (1965). *The molecular biology of development*. Clarendon Press, Oxford.

Boorman, S., and Levitt, P.R. (1973). Group selection on the boundary of a stable population. *Theoretical Population Biology* **4:** 85–128.

Boorman, S.A., and Levitt, P.R. (1980). *The genetics of altruism*. Academic Press, New York.

Boraas, M.E., Seale, D.B., and Boxhorn, J.E. (1998). Phagotrophy by a flagellate selects for colonial prey: a possible origin of multicellularity. *Evolutionary Ecology* **12:** 153–164.

Bordería, A.V., and Elena, S. (2002). r- and K-selection in experimental populations of vesicular stomatitis virus. *Infection Genetics and Evolution* **2:** 137–143.

Borisov, V.M. (1978). The selective effect of fishing on the population structure of species with a long life cycle. *Journal of Ichthyology* **18:** 896–904.

Bos, M., and Scharloo, W. (1973a). The effects of disruptive and stabilizing selection on body size in *Drosophila melanogaster*. I. Mean values and variances. *Genetics* **75:** 679–693.

Bos, M., and Scharloo, W. (1973b). The effects of disruptive and stabilizing selection on body size in *Drosophila melanogaster*. II. Analysis of responses in the thorax selection lines. *Genetics* **75:** 695–708.

Bos, M., and Scharloo, W. (1974). The effects of disruptive and stabilizing selection on body size in *Drosophila melanogaster*. III. Genetic analysis of two lines with different reactions to disruptive selection with mating of opposite extremes. *Genetica* **45:** 71–90.

Boshier, D., and Stewart, J. (2005). How local is local? Identifying the scale of adaptive variation in ash (*Fraxinus excelsior* L.): results from the nursery. *Forestry* **78:** 135–143.

Bossart, J.L., and Prowell, D.P. (1998). Genetic estimates of population structure and gene flow: limitations, lessons and new directions. *Trends in Ecology and Evolution* **13:** 202–206.

Bost, B., de Vienne, D., Hospital, F., Moreas, L., and Dillman, C. (2001). Genetic and nongenetic bases for the L-shaped distribution of quantitative trait loci effects. *Genetics* **157:** 1773–1787.

Boughman, J.W. (2001). Divergent sexual selection enhances reproductive isolation in sticklebacks. *Nature* **411:** 944–948.

Boughman, J.W., Rundle, H.D., and Schluter, D. (2005). Parallel evolution of sexual isolation in sticklebacks. *Evolution* **59:** 361–373.

Bouma, J.E., and Lenski, R.E. (1988). Evolution of a bacteria/plasmid association. *Nature* **335:** 351–352.

Bovallius, A., Bucht, B., Roffey, R., and Anas, P. (1978). Three-year investigation of the natural airborne bacterial flora at four localities in Sweden. *Applied and Environmental Microbiology* **35:** 847–852.

Bower, F.O. (1908). *The origin of a land flora*. Macmillan, London.

Boyd, R., and Richerson, P.J. (1988). *Culture and the evolutionary process*. University of Chicago Press, Chicago.

Bradley, D.E., and Williams, P.A. (1982). The TOL plasmid is naturally derepressed for transfer. *Journal of General Microbiology* **128:** 3019–3024.

Bradshaw, A.D. (1952). Populations of *Agrostis tenuis* resistant to lead and zinc poisoning. *Nature* **169:** 1098–1100.

Bradshaw, A.D. (1991). Genostasis and the limits to adaptation. *Philosophical Transactions of the Royal Society of London Series B* **333:** 289–305.

Bradshaw, H.D., and Schemske, D.W. (2003). Allele substitution at a flower colour locus produces a pollinator shift in monkeyflowers. *Nature* **426:** 176–178.

Braun, H.J., Rajaram, S., and van Ginkel, M. (1996). CIMMYT's approach to breeding for wide adaptation. *Euphytica* **92:** 175–183.

Bridge, R.R., and Meredith, W.R. (1983). Comparative performance of obsolete and current cotton cultivars. *Crop Science* **23:** 231–237.

Bridges, B.A. (2001). Hypermutation in bacteria and other cellular systems. *Philosophical Transactions of the Royal Society of London Series B* **356:** 29–39.

Brooks, R. (2000). Negative genetic correlation between male sexual attractiveness and survival. *Nature* **406:** 67–70.

Brooks, R., and Endler, J.A. (2001). Direct and indirect sexual selection and quantitative genetics of male traits in guppies (*Poecilia reticulate*). *Evolution* **55:** 1002–1015.

Brown, A.J., and Casselton, I.A. (2001). Mating in mushrooms: increasing the chances but prolonging the affair. *Trends in Genetics* **17:** 393–400.

Brown, C.J., Todd, K.M., and Rosenzweig, R.F. (1998). Multiple duplications of yeast hexose transport genes in response to selection in a glucose-limited environment. *Molecular Biology and Evolution* **15:** 931–942.

Brown, J.K.M. (2002). Yield penalties of disease resistance in crops. *Current Opinion in Plant Biology* **5:** 339–344.

Brown, W.M., George, M., and Wilson, A.C. (1979). Rapid evolution of animal mitochondrial DNA. *Proceedings of the National Academy of Sciences of the USA* **76:** 1967–1971.

Brown, W.P., and Bell, A.E. (1961). Genetic analysis of a 'plateaued' population of *Drosophila melanogaster*. *Genetics* **46:** 407–425.

Bryant, E.H., and Turner, C.R. (1972). Rapid evolution of competitive ability in larval mixtures of the housefly. *Evolution* **26:** 161–170.

Buckling, A., and Rainey, P.B. (2002). Antagonistic coevolution between a bacterium and a bacteriophage. *Proceedings of the Royal Society of London Series B* **269:** 931–936.

Bull, J.J., and Charnov, E. (1988). How fundamental are Fisherian sex ratios? *Oxford Surveys in Evolutionary Biology* **5:** 96–135.

Bull, J.J., and Wichman, H.A. (2001). Applied evolution. *Annual Review of Ecology and Systematics* **32:** 183–217.

Bull, J.J., Molineux, I.J., and Rice, W.R. (1991). Selection of benevolence in a host-parasite system. *Evolution* **45:** 875–882.

Bull, J.J., Badgett, M.R., Wichman, H.A. *et al.* (1997). Exceptional convergent evolution in a virus. *Genetics* **147:** 1497–1507.

Bull, J.J., Badgett, M.R., and Wichman, H.A. (2000). Big-benefit mutations in a bacteriophage inhibited with heat. *Molecular Biology and Evolution*, **17**(6): 942–950.

Bulmer, M.G. (1972). Multiple niche polymorphism. *American Naturalist* **106:** 254–257.

Bulmer, M.G. (1985). *The mathematical theory of quantitative genetics.* Clarendon Press, Oxford.

Bulmer, M.G., and Parker, G.A. (2002). The evolution of anisogamy: a game-theoretic approach. *Proceedings of the Royal Society of London Series B* **269:** 2381–2388.

Bumpus, H. (1899). The elimination of the unfit as illustrated by the introduced sparrow, *Passer domesticus. Marine Biological Laboratory (Wood's Hole) Biology Lectures,* 1898: 209–228.

Bundegaard, J., and Christiansen, F.B. (1972). Dynamics of polymorphisms. I. Selection components in an experimental population of *Drosophila melanogaster. Genetics* **71:** 439–460.

Bunger, L., and Hill, W.G. (1999). Inbred lines of mice derived from long-term divergent selection on fat content and body weight. *Mammalian Genome* **10:** 645–648.

Bunger, L., Renne, U., Dietl, G., and Kuhla, S. (1998). Long-term selection for protein amount over 70 generations in mice. *Genetics Research,* **72**(2): 93–109.

Bunger, L., Laidlaw, A., Bulfield, G. *et al.* (2001). Inbred lines of mice derived from long-term growth selected lines: unique resources for mapping growth genes. *Mammalian Genome* **12:** 678–686.

Burch, C.L., and Chao, L. (1999). Evolution by small steps and rugged landscapes in the RNA virus ϕ6. *Genetics* **151:** 921–927.

Burch, C.L., and Chao, L. (2000). Evolvability of an RNA virus is determined by its mutational neighbourhood. *Nature* **406:** 625–628.

Burdon, J.J., and Thompson, J.N. (1995). Changed patterns of resistance in a population of *Linum marginale* attacked by the rust pathogen *Melampsora lini. Journal of Ecology* **83:** 199–206.

Burleigh, B.D., Rigby, P.W.J., and Hartley, B.S. (1974). A comparison of wild-type and mutant ribitol dehydrogenases from *Klebsiella aerogenes. Biochemical Journal* **143:** 341–352.

Burley, N. (1986). Comparison of the band-colour preferences of two species of estrilid finches. *Animal Behaviour* **34:** 1732–1741.

Burt, A. (1995). The evolution of fitness. *Evolution* **49:** 1–8.

Burt, A. (2000). Sex, recombination and the efficacy of selection: was Weismann right? *Evolution* **54:** 337–351.

Burt, A. (2003). Site-specific selfish genes as tools for the control and genetic engineering of natural populations. *Proceedings of the Royal Society of London Series B* **270:** 921–928.

Burt, A., and Bell, G. (1987). Mammalian chiasma frequencies as a test of two theories of recombination. Nature 326, 803–805

Burt, A., and Trivers, R. (1998). Genetic conflicts in genomic imprinting. *Proceedings of the Royal Society of London Series B* **265:** 2393–2397.

Burt, A., and Trivers, R. (2005). *Genes in conflict: the biology of selfish genetic elements.* Harvard University Press, Cambridge, MA.

Bush, G.L. (1969). Sympatric host race formation and speciation in frugivorous flies of the genus *Rhagoletis. Evolution* **23:** 237–251.

Buss, L. (1987). *The evolution of individuality.* Princeton University Press, Princeton, NJ.

Butler, G., Kenny, C., Fagan, A., Kurischko, C., Gaillardia, C., and Wolfe, K.H. (2004). Evolution of the MAT locus and its Ho endonuclease in yeast species. *Proceedings of the National Academy of Sciences of the USA* **101:** 1632–1637.

Cain, A.J. (1964) The perfection of animals. *Viewpoints in Biology* **3:** 36–63 (see *Biological Journal of the Linnean Society* **36:** 3–29).

Cain, A.J., and Sheppard, P.M. (1950). Selection in the polymorphic land snail *Cepaea nemoralis. Heredity* **4:** 275–294.

Cain, A.J., and Sheppard, P.M. (1952). The effects of natural selection on body colour in the land snail *Cepaea nemoralis. Heredity* **6:** 217–223.

Cain, A.J., and Sheppard, P.M. (1954). Natural selection in *Cepaea. Genetics* **39:** 89–116.

Cain, A.J., Cook, L.M., and Currey, J.D. (1990). Population size and morph frequency in a long-term study of *Cepaea nemoralis. Proceedings of the Royal Society of London Series B* **240:** 231–250.

Cairns, J. (1975). Mutation, selection and the natural history of cancer. *Nature* **255:** 197–200.

Cairns, J., and Foster, P.L. (1991). Adaptive reversion of a frameshift mutation in *Escherichia coli. Genetics* **128:** 695–701.

Cairns, J., Overbaugh, J., and Miller, S. (1988). The origin of mutants. *Nature* **335:** 142–145.

Calef, E. (1957). Effects on linkage maps of selection of crossovers between closely linked markers. *Heredity* **11:** 265–279.

Calhoon, R.E., and Bohren, B.B. (1974). Genetic gains from reciprocal recurrent and within-line selection for egg production in the fowl. *Theoretical and Applied Genetics* **44:** 364–372.

Calkins, G.F. (1919). *Uroleptus mobilis* Engelm. II. Renewal of vitality through conjugation. *Journal of Experimental Zoology* **29:** 121–156.

Calkins, G.F. (1920). *Uroleptus mobilis* Engelm. III. A study in vitality. *Journal of Experimental Zoology* **31:** 287–305.

Callahan, H.S., and Pigliucci, M. (2002). Shade-induced plasticity and its ecological significance in wild populations of *Arabidopsis thaliana*. Ecology **83:** 1965–1980.

Callaway, R.M., Ridenour, W.M., Laboski, T., Weir, T., and Vivanco, J.M. (2005). Natural selection for resistance to the allelopathic effects of invasive plants. *Journal of Ecology* **93:** 576–583.

Camacho, J.P.M.. Sharbel, T.F., and Beukeboom, L.W. (2003). B-chromosome evolution. *Philosophical Transactions of the Royal Society of London Series B* **355:** 163–178.

Campbell, D.R. (2000). Experimental tests of sex-allocation theory in plants. *Trends in Ecology and Evolution* **15:** 227–232.

Campbell, L.G., and Lafever, H.N. (1980). Effects of locations and years upon relative yields of the soft red winter wheat region. *Crop Science* **20:** 23–28.

Campo, J.L., and Tagarro, P. (1977). Comparison of three selection methods for pupal weight of *Tribolium castaneum*. *Annals of Genetics and Selection in Animals* **9:** 259–268.

Carius, H.J., Little, T.J., and Ebert, D. (2001). Genetic variation in a host-parasite association: potential for coevolution and frequency-dependent selection. *Evolution* **55:** 1136–1145.

Carlson, W.R. (1994). Crossover effects of B chromosomes may be 'selfish'. *Heredity* **72:** 636–638.

Carrillo, F.Y.E., Sanjuan, R Moya, A., and Cuevas, J.M. (2007). The effect of co- and superinfection on the adaptive dynamics of vesicular stomatitis virus. *Infection, Genetics and Evolution* **7:** 69–73.

Carson, H.L. (1968). The population flush and its genetic consequences. In Lewontin, R.C. (ed.) *Population biology and evolution*, Syracuse University Press, Syracuse, NY.

Carvalho, A.B., Sampaio, M.C., Verandas, F.R., and Klaczko, L.B. (1998). An experimental demonstration of Fisher's principle: evolution of sexual proportion by natural selection. *Genetics* **148:** 719–731.

Castle, W.E., and Phillips, J.C. (1914). Piebald rats and selection. An experimental test of selection and of the theory of gametic purity in Mendelian crosses. *Carnegie Institute of Washington Publication* **195:** 1–31.

Castleberry, R.M., Crum, C.W., and Krull, C.F. (1984). Genetic yield improvement of U.S. maize cultivar under varying fertility and climatic environments. *Crop Science* **24:** 33–36.

Castric, V., and Vekemans, X. (2004). Plant self-incompatibility in natural populations: a critical assessment of recent theoretical and empirical advances. *Molecular Ecology* **13:** 2873–2889.

Caswell, H. (1976). Community structure: a neutral model analysis. *Ecological Monographs* **46:** 327–354.

Cavalli-Sforza, L.L., and Feldman, M.W. (1981). *Cultural transmission and evolution*. Princeton University Press, Princeton, NJ.

Cavalli-Sforza, L.L., Feldman, M.W., and Dornbusch, S.M. (1982). Theory and observation in cultural transmission. *Science* **218:** 19–27.

Chabora, A.J. (1968). Disruptive selection for sternopleural chaeta number in various strains of *Drosophila melanogaster*. *American Naturalist* **102:** 525–532.

Chao, L. (1990). Fitness of RNA virus decreased by Muller's Ratchet. *Nature* **348:** 454–455.

Chao, L., and Cox, E.C. (1983). Competition between high and low mutating strains of *Escherichia coli*. *Evolution* **37:** 125–134.

Chao, L., and Levin, B.R. (1981). Structured habitats and the evolution of anticompetitor toxins in bacteria. *Proceedings of the National Academy of Sciences of the USA* **78:** 6324–6328.

Chao, L., and McBroom, S.M. (1985). Evolution of transposable elements: an IS10 insertion increases fitness in *E. coli*. *Molecular Biology and Evolution* **2:** 359–369.

Chao, L., Vargas, C., Spear, B.B., and Cox, E.C. (1983). Transposable elements as mutator genes in evolution. *Nature* **303:** 633–635.

Chao, L., Tran, T., and Matthews, C. (1992). Muller's Ratchet and the advantage of sex in the RNA virus $\phi 6$. *Evolution* **46:** 289–299.

Chapin, F.S., and Chapin, M.C. (1981). Ecotypic differentiation of growth processes in *Carex aquatilis* along latitudinal and local gradients *Ecology* **62:** 1000–1009.

Chapman, T., Liddle, L.F., Kalb, J.M., Wolfner, M.F., and Partridge, L. (1995). Cost of mating in *Drosophila melanogaster* females is mediated by male accessory gland products. *Nature* **373:** 241–244.

Charlesworth B., Charlesworth D., and Morgan M.T. (1990). Genetic loads and estimates of mutation rates in very inbred plant populations. *Nature* **347:** 380–382.

Charlesworth, B. (1978). The population genetics of anisogamy. *Journal of Theoretical Biology* **73:** 347–357.

Charlesworth, B. (1980). *Evolution in age-structured populations*. Cambridge University Press, Cambridge.

Charlesworth, B. (1987). The population biology of transposable elements. *Trends in Ecology and Evolution* **2:** 21–23.

Charlesworth, B., and Charlesworth, D. (1978). A model for the evolution of dioecy and gynodioecy. *American Naturalist* **112:** 975–997.

Charlesworth, B., and Charlesworth, D. (1979). The maintenance and breakdown of distyly. *American Naturalist* **114:** 486–498.

Charlesworth, B., and Charlesworth, D. (1983). The population dynamics of transposable elements. *Genetical Research* **42:** 1–27.

Charlesworth, B., and Charlesworth, D. (1985). Genetic variation in recombination in *Drosophila*. I. Responses to selection and preliminary genetic analysis. *Heredity* **54:** 71–83.

Charlesworth, B., and Hartl, D.L. (1978). Population dynamics of the segregation distorter polymorphism of *Drosophila melanogaster. Genetics* **89:** 171–192.

Charlesworth, B., and Langley, C. (1989). The population genetics of *Drosophila* transposable elements. *Annual Reviews of Genetics* **23:** 251–287.

Charlesworth, B., and Toro, M.A. (1982) Female-biased sex ratios. *Nature* **298:** 494.

Charlesworth, B., Lande, R., and Slatkin, M. (1982). A neo-Darwinian commentary on macroevolution. *Evolution* **36:** 474–498.

Charlesworth, B., Sniegowski, P., and Stephan, W. (1994). The evolutionary dynamics of repetitive DNA in eukaryotes. *Nature* **371:** 215–220.

Charlesworth, D. (2006). Balancing selection and its effects on sequences in nearby genomic regions. *PLoS Genetics* 2: e64 doi: 10.(1371)/journal.pgen.00(2006)4.

Charlesworth, D., and Charlesworth, B. (1978). Population genetics of partial male sterility and the evolution of monoecy and dioecy. *Heredity* **41:** 137–155.

Charlesworth, D., Vekemans, S., Castric, V., and Glmin, S. (2005). Plant self-incompatibility systems: a molecular evolutionary perspective. *New Phytologist* **168:** 61–69.

Charnov, E.L. (1982). *The theory of sex allocation*. Princeton University Press, Princeton, NJ.

Charnov, E.L. (1987). On sex allocation and selfing in higher plants. *Evolutionary Ecology* **1:** 30–36.

Chetverin, A.B., Chetverina, H.V., and Munishkin, A.V. (1991). On the nature of spontaneous RNA synthesis by Qβ replicase. *Journal of Molecular Biology* **222:** 3–9.

Chin, K.M., and Wolfe, M.S. (1984). Selection on *Erisyphe graminis* in pure and mixed stands of barley. *Plant Pathology* **33:** 535–546.

Chinnici, J.P. (1971). Modification of recombination frequency in *Drosophila*. I. Selection for increased and decreased crossing-over. *Genetics* **69:** 71–83.

Chippindale, A.K., Gibson, J.R., and Rice, W.R. (2001). Negative genetic correlation for fitness between sexes reveals ontogenetic conflict in *Drosophila. Proceedings of the National Academy of Sciences of the USA* **98:** 1671–1675.

Christiansen, F.B., and Frydenberg, O. (1977). Selection-mutation balance for two nonallelic recessives producing an inferior double homozygote. *American Journal of Human Genetics* **29:** 195–207.

Chung, C.S., and Chapman, A.B. (1958). Comparisons of the predicted with actual gains in selection of parents of inbred progeny of rats. *Genetics* **43:** 594–600.

Cipollini, D., Purrington, C.B., and Bergelson, J. (2003). Costs of induced responses in plants. *Basic and Applied Ecology* **4:** 79–89.

Civetta, A., and Singh, R.S. (1995). High divergence of reproductive tract proteins and their association with postzygotic reproductive isolation in *Drosophila melanogaster* and *Drosophila virilis* group species. *Journal of Molecular Evolution* **41:** 1085–1095.

Clare, M.J., and Luckinbill, L.S. (1985). The effect of gene-environment interaction on the expression of longevity. *Heredity* **55:** 19–29.

Clark, R.M., Linton, E., Messing, J., and Doebley, J.F. (2004). Pattern of diversity in the genomic region near the maize domestication gene *tb1*. *Proceedings of the National Academy of Sciences of the USA* **101:** 700–707.

Clarke, B. (1979). The evolution of genetic diversity. *Proceedings of the Royal Society of London Series B* **205:** 453–474.

Clarke, B. (2003). The art of innuendo. *Heredity* **90:** 279–280.

Clarke, B., and Kirby, D.R. (1966). Maintenance of histocompatibility polymorphisms. *Nature* **211:** 999–1000.

Clark, G. (1991). Yields per acre in English agriculture, 1250–1860: evidence from labour inputs. *Economic History Review, New Series* **44:** 445–460.

Clarke, J.M., Maynard Smith, J., and Sondhi, K.C. (1961). Asymmetrical response to selection for rate of development in *Drosophila* subobscura. *Genetical Research* **2:** 70–81.

Clarke, P.H. (1984). Amidases of *Pseudomonas aeruginosa*. In R.P. Mortlock (ed.), *Microorganisms as model systems for studying evolution*, pp. 187–232. Plenum Press, New York.

Clausen, J., Keck, D.D., and Hiesey, W.M. (1940). Experimental studies on the nature of species. I. Effect of varied environments on Western North American plants. *Carnegie Institute of Washington Publication*, 520.

Clayton, G., and Robertson, A. (1955). Mutation and quantitative variation. *American Naturalist* **89:** 151–158.

Clayton, G.A., and Robertson, A. (1957). An experimental check on quantitative genetical theory. II. The long-term effects of selection. *Journal of Genetics* **55:** 152–170.

Clayton, G.A., Morris, J.A., and Robertson, A. (1957a). An experimental check on quantitative genetical theory. I. Short-term responses to selection. *Journal of Genetics* **55:** 131–151.

Clayton, G.A., Knight, G.R., Morris, J.A., and Robertson, A. (1957b). An experimental check on quantitative genetic theory. III. Correlated responses. *Journal of Genetics* **55:** 171–180.

Clegg, M.T., Kahler, A.L., and Allard, R.W. (1978). Estimation of life cycle components of selection in an experimental plant population. *Genetics* **89:** 765–792.

Cleveland, A. (1994). Nest site habitat preference and competition in *Gasterosteus aculeatus* and *Gasterosteus wheatlandi*. *Copeia* 1994: 698–704.

Clutton Brock, T.H. (ed.) 1988. *Reproductive success*. University of Chicago Press, Chicago.

Colegrave, N. (2002). Sex releases the speed limit on evolution. *Nature* **420:** 664–666.

Colegrave, N., Kaltz, O., and Bell, G. (2002). The ecology and genetics of fitness in *Chlamydomonas*. VIII. The dynamics of adaptation to novel environments after a single episode of sex. *Evolution* **56:** 14–21.

Coleman, A.W. (1977). Sexual and genetic isolation in the cosmopolitan algal species *Pandorina morum*. *American Journal of Botany* **64:** 361–368.

Coleman, A.W. (2005). *Paramecium aurelia* revisited. *Journal of Eukaryotic Microbiology* **52:** 68–77.

Coleman, A.W., Suarez, A., and Goff, L.J. (1994). Molecular delineation of species and syngens in volvocacean green algae. *Journal of Phycology* **30:** 80–90.

Collins, S., and Bell, G. (2004). Phenotypic consequences of 1000 generations of selection at elevated CO_2 in a green alga. *Nature* **431:**566–569.

Collins, S., and Bell, G. (2006). Evolution of natural algal populations at elevated CO_2. *Ecology Letters* **9:** 129–135.

Collins, S., Sultemeyer, D., and Bell, G. (2006a). Changes in C uptake in populations of *Chlamydomonas reinhardtii* selected at high CO_2. *Plant, Cell and Environment* 29: 1812–1819.

Collins, S., Sultemeyer, D., and Bell, G. (2006b). Rewinding the tape: selection of algae adapted to high CO_2 at current and Pleistocene levels of CO_2. *Evolution* **60:** 1392–1401.

Colosimo, P.F., Peichel, C.L., Nereng, K. *et al.* (2004) The genetic architecture of parallel armor plate reduction in threespine sticklebacks. *PLoS Biology* **2**(5): e109 doi:10.1371/journal.pbio.0020109.

Colosimo, P.F., Hosemann, K.E., Balabhadra, S. *et al.* (2005). Widespread parallel evolution in sticklebacks by repeated fixation of ectodysplasin alleles. *Science* **307:** 1928–1933.

Conover, D.O., and Van Voorhees, D.A. (1990). Evolution of a balanced sex-ratio by frequency-dependent selection in a fish. *Science* **250:** 1556–1558.

Conover, D.O., Van Voorhees, D.A., and Ehtisham, A. (1992). Sex ratio selection and the evolution of environmental sex determination in laboratory populations of *Menidia menidia*. *Evolution* **46:** 1722–1730.

Cook, L.M. (1965). Inheritance of shell size in the snail *Arianta arbustorum*. *Evolution* **19:** 86–94.

Cook, L.M. (2000). Changing views on melanic moths. *Biological Journal of the Linnean Society* **69:** 431–441.

Cook, L.M., Dennis, R.L.H., and Mani, G.S. (1999). Melanic morph frequency in the peppered moth in the Manchester area. *Proceedings of the Royal Society of London Series B* **266:** 293–297.

Cooper, M., and Alder, M. (2006). The evolution of adaptive immune systems. *Cell* **124:** 815–822.

Cooper, T.F., Rozen, D.E., Lenski, R.E. *et al.* (2003). Parallel changes in gene expression after 20,000 generations of evolution in *Escherichia coli*. *Proceedings of the National Academy of Science of the USA* **100**(3): 1072–1077.

Cooper, V.S., and Lenski, R.E. (2000). The population genetics of ecological specialization in evolving *Escherichia coli* populations. *Nature* **407:** 736–739.

Cooper, V.S., Schneider, D., Blot, M., and Lenski, R.E. (2001). Mechanisms causing rapid and parallel losses of ribose catabolism in evolving populations of *Escherichia coli* B. *Journal of Bacteriology* **183:** 2834–2841.

Cooper, V.S., Reiskind, M.H., Miller, J.A. *et al.* (2002). Timing of transmission and the evolution of virulence of an insect virus. *Proceedings of the Royal Society of London Series B* **269:** 1161–1165.

Cordero, C., and Eberhard, W.G. (2003). Female choice of sexually antagonistic male adaptations: a critical review of some current research. *Journal of Evolutionary Biology* **16:** 1–6.

Cotgreave, P. (1993). The relationship between body size and abundance in animals. *Trends in Ecology and Evolution* **8:** 244–248.

Coustau, C., Chevillon, C., and ffrench-Constant, R. (2000). Resistance to xenobiotics and parasites: can we count the cost ? *Trends in Ecology and Evolution* **15:** 378–383.

Cowperthwaite, M.C., Bull, J.J., and Meyers, L.A. (2005). Distributions of beneficial fitness effects in RNA. *Genetics* **170:** 1449–1457.

Cox, E.C. (1976). Bacterial mutator genes and the control of spontaneous mutation. *Annual Reviews of Genetics* **10:** 135–156.

Cox, E.C., and Gibson, T.C. (1974). Selection for high mutation rates in chemostats. *Genetics* **77:** 169–184.

Coyne, J.A., and Orr, H.A. (1997). 'Patterns of speciation in *Drosophila*' revisited. *Evolution* **51:** 295–303.

Coyne, J.A., and Orr, H.A. (2004). *Speciation*. Sinauer Associates, Sunderland, MA.

Craig, D.M. (1982). Group selection versus individual selection: an experimental analysis. *Evolution* **36:** 271–282.

Craig, J.V., and Muir, W.M. (1996). Group selection for adaptation to multiple-hen cages: beak-related mortality, feathering, and body weight responses. *Poultry Science* **75:** 294–302.

Craig, J.V., Jan, M.L., Polley, C.R., Bhagwat, A.L., and Dayton, A.D. (1975). Changes in relative aggressiveness and social dominance associated with selection for early egg production in chickens. *Poultry Science* **54:** 1647–1658.

Crill, W.D., Wichman, H.A., and Bull, J.J. (2000). Evolutionary reversals during viral adaptation to alternating hosts. *Genetics* **154:** 27–37.

Cronly-Dillon, J., and Sharma, S.C. (1968). Effect of season and sex on the photopic spectral sensitivity of the three-spined stickleback. *Journal of Experimental Biology* **49:** 679–687.

Crossley, S.A. (1974). Changes in mating behavior produced by selection for ethological isolation between Ebony and Vestigial mutants of *Drosophila melanogaster. Evolution* **28:** 631–647.

Crow, J.F. (1993). Mutation, mean fitness and genetic load. *Oxford Surveys of Evolutionary Biology* **9:** 3–42.

Crow, J.F., and Kimura, M. (1970) *An introduction to population genetics theory.* Harper & Row, New York.

Crow, J.F., and Nagylaki, T. (1976). The rate of change of a character correlated with fitness. *American Naturalist* **110:** 207–213.

Crow, J.F., and Simmons M.J. (1983). The mutation load in *Drosophila.* In M. Ashburner, H.L. Carson, and J.N. Thompson (eds.) *The genetics and biology of* Drosophila, vol 3C, pp. 1–35. Academic Press, New York.

Cuevas, J.M., Elena, S.F., and Moya, A. (2002). Molecular basis of adaptive convergence in experimental populations of RNA viruses. *Genetics* **162:** 533–542.

Curtis, S.E., and Clegg, M.T. (1984). Molecular evolution of chloroplast DNA sequences. *Molecular Biology and Evolution* **1:** 291–301.

Curtis, T.P., Head, I.M., Lunn, M., Woodcock, S., Schloss, P.D., and Sloan, W.T. (2006). What is the extent of prokaryotic diversity? *Philosophical Transactions of the Royal Society of London Series B* **361:** 2023–2037.

da Silva, J., and Bell, G. (1992). The ecology and genetics of fitness in *Chlamydomonas.* VI. Antagonism between natural selection and sexual selection. *Proceedings of the Royal Society of London Series B* **249:** 227–233.

Dallinger, W.H. (1887). Presidential address. *American Monthly Microscopical Journal* **VIII:** 114.

Damuth, J. (1987). Interspecific allometry of population density in mammals and other animals: the independence of body mass and population energy-use. *Biological Journal of the Linnean Society* **31:** 193–246.

Darling, K.F., Kucera, M., Pudsey, C.J., and Wade, C.M. (2004). Molecular evidence links cryptic diversification in polar planktonic protists to Quaternary climate dynamics. *Proceedings of the National Academy of Sciences of the USA* **101:** 7657–7662.

Darlington, C.D. (1958). *Evolution of genetic systems.* Oliver & Boyd, Edinburgh.

Davies, E.W. (1956). Cytology, evolution and origin of the aneuploid series in the genus *Carex. Hereditas* **42:** 349–366.

Davies, M.S. (1975). Physiological differences among populations of *Anthoxanthum odoratum* L. collected from the Park Grass Experiment, Rothamsted. IV. Responses to potassium and magnesium. *Journal of Applied Ecology* **12:** 953–964.

Davies, M.S., and Snaydon, R.W. (1973a). Physiological differences among populations of *Anthoxanthum odoratum* L. collected from the Park Grass Experiment, Rothamsted. I. Response to calcium. *Journal of Applied Ecology* **10:** 33–45.

Davies, M.S., and Snaydon, R.W. (1973b). Physiological differences among populations of *Anthoxanthum odoratum* L. collected from the Park Grass Experiment, Rothamsted. III. Response to phosphorus. *Journal of Applied Ecology* **10:** 699–707.

Davies, M.S., and Snaydon, R.W. (1976). Rapid population differentiation in a mosaic environment. III. Measures of selection pressures. *Heredity* **36:** 59–66.

Davis, B.K. (1991). Kinetics of rapid RNA evolution in vitro. *Journal of Molecular Evolution* **33:** 343–356.

Dawe, R.K., and Hiatt, E.N. (2004). Plant neocentromeres: fast, focused and driven. *Chromosome Research* **12:** 655–669.

Dawkins, R. (1976). *The selfish gene.* Oxford University Press, Oxford.

Dawson, P.S. (1965). Genetic homeostasis and developmental rate in *Tribolium. Genetics* **51:** 873–885.

Dawson, P.S. (1972). Evolution in mixed populations of *Tribolium. Evolution* **26:** 357–365.

de Jong, G. (1994). The fitness of fitness concepts and the description of natural selection. *Quarterly Review of Biology* **69:** 3–29.

de la Fuente, L.F., and San Primitivo, F. (1985). Selection for large and small litter size of the first three litters in mice. *Génétique, Sélection, Évolution* **17:** 251–264.

de la Pena, M., Elena, S.F., and Moya, A. (2000). Effect of deleterious mutation-accumulation on the fitness of RNA bacteriophage MS2. *Evolution* **54:** 686–691.

de Oliveira, A.K., and Cordeiro, A.R. (1980). Adaptation of *Drosophila willistoni* populations to extreme pH medium. II. Development of incipient reproductive isolation. *Heredity* **44:** 123–130.

de Vargas, C., Norris., R., Zaninetti L., Gibb, S.W., and Pawlowski, J. (1999). Molecular evidence of cryptic speciation in planktonic foraminifers and their relation to oceanic provinces. *Proceedings of the National Academy of Sciences of the USA* **96:** 2864–2868.

de Varigny, H. (1892). *Experimental evolution.* Macmillan, London.

de Visser, J.A.G.M., and Hoekstra, R.F. (1997). Synergistic epistasis between loci affecting fitness: evidence in plants and fungi. *Genetic Research, Cambridge* **71:** 39–49.

de Visser J.A.G.M., Hoekstra, R.F., and Van den Ende, H. (1996). The effect of sex and deleterious mutations on fitness in *Chlamydomonas*. *Proceedings of the Royal Society of London Series B* **263:** 193–200.

de Visser, J., Zeyl, C.W., Gerrish, P.J., Blanchard, J.L., and Lenski, R.E.. (1999). Diminishing returns from mutation supply rate in asexual populations. *Science* **283:** 404–406.

Dekkers, J.C.M., and Hospital, F. (2002). The use of molecular genetics in the improvement of agricultural populations. *Nature Reviews Genetics* **3:** 22–32.

del Solar, E. (1966). Sexual isolation caused by selection for positive and negative phototaxis and geotaxis in *Drosophila pseudoobscura*. *Proceedings of the National Academy of Sciences of the USA* **56:** 484–487.

Délye, C., Menchari, Y., Michel, S., and Darmency, H. (2004) Molecular bases for sensitivity to tubulin-binding herbicides in green foxtail. *Plant Physiology* **136:** 3920–3932.

Dempster, E.R. (1955). Maintenance of genetic heterogeneity. *Cold Spring Harbor Symposia on Quantitative Biology* **20:** 25–32.

Denamur, E., and Matic, I. (2006). Evolution of mutation rates in bacteria. *Molecular Microbiology* **60:** 820–827.

Deng, H.-W., and Lynch, M. (1996) Estimation of deleterious-mutation parameters in natural populations. *Genetics* **144:** 349–360.

Denver, D.R., Morris, K., Lynch, M., Vassilieva, L.L., and Thomas, W.K. (2000). High direct estimate of the mutation rate in the mitochondrial genome of *Caenorhabditis elegans*. *Science* **289:** 2342–2344.

Denver, D.R., Morris, K., Lynch, M., and Thomas, W.K. (2004). High mutation rate and predominance of insertions in the *Caenorhabditis elegans* nuclear genome. *Nature* **430:** 679–682.

Destombe, C., Godin, J., Nocher, M., Richerd, S., and Valero, M. 1993. Differences in response between haploid and diploid isomorphic phases of *Gracilaria verrucosa* (Rhodophyta: Gigartinales) exposed to artificial environmental conditions. *Hydrobiologia* **260–261:** 131–137.

Detlefson, J.A., and Roberts, E. (1921). Studies on crossing-over. I. The effect of selection on crossover values. *Journal of Experimental Zoology* **32:** 333–354.

Dewees, A.A. (1970). Two-way selection for recombination rates in *Tribolium castaneum*. *Genetics* **64** (Suppl.): s16–s17.

di Cesnola, A.P. (1907). A first study of natural selection in *Helix arbustorum* (Helicogena). *Biometrika* **5:** 387–399.

Diamond, J., and Hammond, K. (1992). The matches, achieved by natural selection, between biological capacities and their natural loads. *Cellular and Molecular Life Sciences* **48:** 551–557.

Dickerson, G.E. (1955). Genetic slippage in response to selection for multiple objectives. *Cold Spring Harbor Symposia on Quantitative Biology* **20:** 213–224.

Dobzhansky, T. (1937). *Genetics and the origin of species.* Columbia University Press, New York.

Dobzhansky, T., and Pavlovsky, O. (1971). Experimentally created incipient species of *Drosophila*. *Nature* **230:** 289–292.

Dobzhansky, T., Hunter, A.S., Pavlovsky, O., Spassky, B., and Wallace, B. (1963). Genetics of natural populations. XXXI. Genetics of an isolated marginal population of *D. pseudoobscura*. *Genetics* **48:** 91–103.

Dodd, D.M.B. (1989). Reproductive isolation as a consequence of adaptive divergence in *Drosophila pseudoobscura*. *Evolution* **43:** 1308–1311.

Doebeli, M., and Hauert, C. (2005). Models of cooperation based on the Prisoner's Dilemma and the Snowdrift game. *Ecology Letters* **8:** 748–766.

Doebley, J. (2004). The genetics of maize evolution. *Annual Review of Genetics* **38:** 37–59.

Doerder, F.P., Gates, M.A., Eberhardt, F.P., and Arslanyolu, M. (1995). High frequency of sex and equal frequencies of mating types in natural populations of the ciliate *Tetrahymena thermophila*. *Proceedings of the National Academy of Sciences of the USA* **92:** 8715–8718.

Doherty, P.C., and Zinkernagel, R.M. (1975). Enhanced immunological surveillance in mice heterozygous at the H-2 gene complex. *Nature* **256:** 50–52.

Donald, C.M., and Hamblin, J. (1976). The convergent evolution of annual seed crops in agriculture. *Advances in Agronomy* **36:** 97–143.

Doney, D., Plaisted, R.L., and Peterson, L.C. (1965). Genotypic competition in progeny performance evaluation of potatoes. *Crop Science* **5:** 433–435.

Donnelly, P., and Tavaré, S. (1995). Coalescents and genealogical structure under neutrality. *Annual Review of Genetics* **29:** 401–421.

Donohue, K., Pyle, E.H., Messiqua, D, Heschel, M.S., and Schmitt, J. (2001). Adaptive divergence in plasticity in natural populations of *Impatiens capensis* and its consequences for performance in novel habitats. *Evolution* **55:** 692–702.

Doolittle, W.F., and Sapienza, C. (1980). Selfish genes, the phenotype paradigm and genome evolution. *Nature* **284:** 601–603.

Dowell, C.E. (1980). Growth of bacteriophage ϕX-174 at elevated temperatures. *Journal of General Virology* **49:** 41–50.

Dower, W.J., and Mattheakis, L.C. (2002). In vitro selection as a powerful tool for the applied evolution of proteins and peptides. *Current Opinion in Chemical Biology* 6: 390–398.

Drake, J.W. (1991). A constant rate of spontaneous mutation in DNA-based microbes. *Proceedings of the National Academy of Sciences of the USA* **88:** 7160–7164.

Drake, J.W., Charlesworth, B., Charlesworth, D., and Crow, J.F. (1998). Rates of spontaneous mutation. *Genetics* **148:** 1667–1686.

Drickamer, L.C. (1981). Selection for age of sexual maturity in mice and the consequences for population regulation. *Behavioral and Neural Biology* **31:** 82–89.

Duarte, C.M., Agusti, S., and Peters, R.H. (1987). An upper limit to the abundance of aquatic organisms. *Oecologia* **74:** 272–276.

Duda, T.F., Vanhoye, D., and Nicolas, P. (2002). Roles of diversifying selection and coordinated evolution in the evolution of amphibian antimicrobial peptides. *Molecular Biology and Evolution* **19:** 858–864.

Dudley, J.W., and Lambert, R.J. (2004). 100 generations of selection for oil and protein in corn. *Plant Breeding Reviews* **24:** 79–110.

Dunham, M.J., Badrane, H., Ferea, T. *et al.* (2002). Characteristic genome rearrangements in experimental evolution of *Saccharomyces cerevisiae*. *Proceedings of the National Academy of Sciences of the USA* **99:** 16144–16149.

Dunnington, E.A., and Siegel, P.B. (1996). Long-term divergent selection for eight-week body weight in White Plymouth Rock chickens. *Poultry Science* **75:** 1168–1179.

Dybdahl, M.F., and Lively, C. (1995). Host-parasite interactions: infection of common clones in natural populations of a freshwater snail (*Potamopyrgus antipodarium*). *Proceedings of the Royal Society of London Series B* **260:** 99–103.

Dybdahl, M.F., and Lively, C. (1998). Host-parasite coevolution: evidence for rare advantage and time-lagged selection in a natural population. *Evolution* **52:** 1057–1066.

Dyke, G., George, B.J., Johnsyon, A.E., Poulton, P.R., and Todd, A.D. (1983). The Broadbalk Wheat experiment 1968–1978: Yields and plant nutrients in crops grown continuously and in rotation. *Rothamsted Experimental Station Report for 1982*, Pt. 2: 5–44.

Dykhuizen, D.E. (1978). Selection for tryptophan auxotrophs of *Escherichia coli* in glucose-limited chemostats as a test of the energy conservation hypothesis of evolution. *Evolution* **32:** 125–150.

Dykhuizen, D.E. (1990). Experimental studies of natural selection in bacteria. *Annual Reviews of Ecology and Systematics* **21:** 373–398.

Dykhuizen, D.E. (1998) Santa Rosalia revisited: Why are there so many species of bacteria? *Antonie van Leeuwenhoek* **73:** 25–33.

Dykhuizen, D.E., and Davies, M. (1980). An experimental model: bacterial specialists and generalists competing in chemostats. *Ecology* **61:** 1213–1227.

Dykhuizen, D.E., and Dean, A.M. (1990). Enzyme activity and fitness: evolution in solution. *Trends in Ecology and Evolution* **5:** 257–262.

Dykhuizen, D.E., and Dean, A.M. (2004). Evolution of specialists in an experimental microcosm. *Genetics* **167:** 2015–2026.

Dykhuizen, D.E., and Hartl, D.L. (1978) Transport by the lactose permease of *Escherichia coli* as the basis of lactose killing. *Journal of Bacteriology* **135:** 876–882.

Dykhuizen, D.E., and Hartl, D.L. (1980). Selective neutrality of 6PGD allozymes in E. coli and the effects of genetic background. *Genetics* **96:** 801–817.

Dykhuizen, D.E., and Hartl, D.L. (1981) Evolution of competitive ability in *Escherichia coli*. *Evolution* **35:** 581–594.

Dykhuizen, D.E., and Hartl, D.L. (1983). Functional effects of PGI allozymes in *Escherichia coli*. *Genetics* **105:** 1–18.

Earl, D.J., and Deem, M.W. (2004). Evolvability is a selectable trait. *Proceedings of the National Academy of Sciences of the USA* **101:** 11531–11536.

East, E.M. (1910). A Mendelian interpretation of variation that is apparently continuous. *American Naturalist* **44:** 65–82.

Ebert, D. (1994). Virulence and local adaptation of a horizontally transmitted parasite. *Science* **265:** 1084–1086.

Ebert, D. (1998). Evolution—experimental evolution of parasites. *Science* **282:** 1432–1435.

Ebinuma, H. (1987). Selective recombination system in *Bombyx mori*. I. Chromosome specificity of the modification effect. *Genetics* **117:** 521–531.

Ebinuma, H., and Yoshitake, N. (1981). The genetic system controlling recombination in the silkworm. *Genetics* **99:** 231–245.

Edley, M.T., and Law, R. (1988). Evolution of life histories and yields in experimental populations of *Daphnia magna*. *Biological Journal of the Linnean Society* **34:** 309–326.

Edwards, M.D., Stuber, C.W., and Wendel, J.F. (1987). Molecular-marker-facilitated investigations of quantitative trait loci in maize. I. Numbers, genomic distribution and type of gene action. *Genetics* **116:** 113–125.

Eggleston, W.B., Johnson-Schlitz, D.M., and Engels, W.R. (1988). P-M hybrid dysgenesis does not mobilize other transposable element families in *Drosophila melanogaster*. *Nature* **331:** 368–370.

Ehrman, L., White, M.M., and Wallace, B. (1991). A long-term study involving *Drosophila melanogaster* and toxic media. *Evolutionary Biology* **25:** 175–209.

Eigen, M. (1983). Self-replication and molecular evolution. In D.S. Bendall (ed.), *Evolution from molecules to men*, pp. 105–130. Cambridge University Press, Cambridge.

Eisen, E.J. (1975). Population size and selection intensity effects on long-term selection response in mice. *Genetics* **79:** 305–323.

Eisen, E.J. (1980). Conclusions from long-term selection experiments with mice. *Journal of Animal Breeding and Genetics* **97:** 305–319.

Elena, S.F. (1999). Little evidence for synergism among deleterious mutations in a nonsegmented RNA virus. *Journal of Molecular Evolution* **49:** 703–707.

Elena, S.F., and Lenski, R.E. (1997). Test of synergistic interactions among deleterious mutations in bacteria. *Nature* **390:** 395–398.

Elena, S.F., and Moya, A. (1999). Rate of deleterious mutation and distribution of its effects on fitness in vesicular stomatitis virus. *Journal of Evolutionary Biology* **12:** 1078–1088.

Elena, S.F., Ekunwe, L., Hajela, N., Oden, S.A., and Lenski, R.E. (1998). Distribution of fitness effects caused by random insertion mutations in *Escherichia coli*. *Genetica* **102/103:** 349–358.

Ellstrand, N.C. (1992). Gene flow by pollen: implications for plant conservation genetics. *Oikos* **63:** 77–86.

Emmerson, D.A. (1997). Commercial approaches to genetic selection for growth and feed conversion in domestic poultry. *Poultry Science* **76:** 1121–1125.

Endler, J.A. (1977). *Geographic variation, speciation and clines*. Princeton University Press, Princeton, NJ.

Endler, J.A. (1978). A predator's view of animal color patterns. *Evolutionary Biology* **11:** 319–364.

Endler, J.A. (1980). Natural selection on color patterns in *Poecilia reticulata*. *Evolution* **34:** 76–91.

Endler, J.A. (1983). Natural and sexual selection on color patterns in poeciliid fishes. *Environmental Biology of Fishes* **9:** 173–190.

Endler, J.A. (1986). *Natural selection in the wild*. Princeton University Press, Princeton, NJ.

Endo, T., Ikeo, K., and Gojobori, T. (1996). Large-scale search for genes on which positive selection may operate. *Molecular Biology and Evolution* **13:** 685–690.

Enfield, F.D. (1977). Selection experiments in *Tribolium* designed to look at gene action issues. In E. Pollak, O. Kempthorne, and T.B. Bailey (eds.), *Proceedings of the International Conference on Quantitative Genetics*, pp. 177–190. Iowa State University Press, Ames, IA.

Enfield, F.D., Comstock, R.E., and Braskerud, O. (1966). Selection for pupa weight in *Tribolium castaneum*. I. Parameters in base populations. *Genetics* **54:** 523–533.

Engels, W. R. (1992). The origin of P elements in *Drosophila melanogaster*. *BioEssays* **14:** 681–686.

Englert, D.C., and Bell, A.E. (1970) Selection for time of pupation in *Tribolium castaneum*. *Genetics* **64:** 541–552.

Engström, G., Liljedahl, L.-E., and Björklund, T. (1992). Expression of genetic and environmental variation during ageing. II. Selection for increased lifespan in *Drosophila melanogaster*. *Theoretical and Applied Genetics* **85:** 26–32.

Ephrussi, B., Hottinguer, H., and Roman, H. (1955). Suppressiveness: a new factor in the genetic determinism of the synthesis of respiratory enzymes in yeast. *Proceedings of the National Academy of Sciences of the USA* **41:** 1065–1071.

Erickson, D.L., Fenster, C.B., Stenøien, H.K., and Price, D. (2004). Quantitative trait locus analyses and the study of evolutionary process. *Molecular Ecology* **13:** 2505–2522.

Estes, S., and Arnold, S.J. (2007). Resolving the paradox of stasis: models with stabilizing selection explain evolutionary divergence on all time scales. *American Naturalist* **169:** 227–244.

Etches, R.J. (1998). A holistic view of poultry science from a reductionist perspective. *British Poultry Science* **39:** 5–10.

Evans, L.T. (1993). *Crop evolution, adaptation, and yield*. Cambridge University Press, Cambridge.

Ewens, W.J. (1972). The sampling theory of neutral alleles. *Theoretical Population Biology* **3:** 87–112.

Ewens, W.J. (1989). An interpretation and proof of the Fundamental Theorem of Natural Selection. *Theoretical Population Biology* **36:** 167–180.

Ewing, H.E. (1914). Notes on regression in a pure line of plant lice. *Biological Bulletin* **27:** 164–168.

Eyre-Walker, A., Keightley, P.D., Smith, N.G.C., and Gaffney, D. (2002). Quantifying the slightly deleterious mutation model of molecular evolution. *Molecular Biology and Evolution* **19:** 2142–2149.

Falconer, D.S. (1952). The problem of the environment and selection. *American Naturalist* **86:** 293–298.

Falconer, D.S. (1953). Selection for large and small size in mice. *Journal of Genetics* **51:** 470–501.

Falconer, D.S. (1955). Patterns of response in selection experiments with mice. *Cold Spring Harbor Symposia on Quantitative Biology* **20:** 178–196.

Falconer, D.S. (1957). Selection for phenotypic intermediates in *Drosophila*. *Journal of Genetics* **55:** 551–561.

Falconer, D.S. (1960). Selection of mice for growth on high and low planes of nutrition. *Genetical Research* **1:** 91–113.

Falconer, D.S. (1971). Improvement of litter size in a strain of mice at a selection limit. *Genetical Research* **17:** 215–235.

Falconer, D.S. (1981) *Introduction to quantitative genetics*, 2nd edn. Longman, London.

Falconer, D.S. (1990). Selection in different environments: effects on environmental sensitivity (reaction norm) and on mean performance. *Genetical Research* **56:** 57–70.

Fay, J.C., and Wu, C.-I. (2003). Sequence divergence, functional constraint, and selection in protein evolution. *Annual Review of Genomics and Human Genetics* **4:** 213–235.

Fay, J.C., Wyckoff, G.J., and Wu, C.-I. (2001). Positive and negative selection on the human genome. *Genetics* **158:** 1227–1234.

Feller, W. (1957). *An introduction to probability theory and its applications*. Wiley, New York.

Fellowes, M.D.E., Kraaijeveld, A.R., and Godfray, H.C.J. (1999). Cross-resistance following artificial selection for increased defence against parasitoids in *Drosophila melanogaster*. *Evolution* **53:** 966–972.

Felsenstein, J. (1981). Skepticism towards Santa Rosalia, or why are there so few kinds of animals? *Evolution* **35:** 124–138.

Fenchel, T., and Finlay, B.J. (2004). The ubiquity of small species: patterns of local and global diversity. *BioScience* **54:** 777–784.

Fenner, F., and Ratcliffe, F.N. (1965). *Myxomatosis*. Cambridge University Press, Cambridge.

Fernandez, J., and López-Fanjul, C. (1996). Spontaneous mutational variances and covariances for fitness-related traits in *Drosophila melanogaster*. *Genetics* **143:** 829–837.

Ferris, P.J., Pavlovic, C., Fabry, S., and Goodenough, U.W. (1997). Rapid evolution of sex-related genes in *Chlamydomonas*. *Proceedings of the National Academy of Sciences of the USA* **94:** 8634–8639.

Ferris, P.J., Armbrust, E.V., and Goodenough, U.W. (2002). Genetic structure of the mating-type locus of *Chlamydomonas reinhardtii*. *Genetics* **160:** 181–200.

Fetcher, N., Cordero, R.A., and Voltzow, J. (2000). Lack of ecotypic differentiation: plant response to elevation, population origin, and wind in the Luquillo Mountains, Puerto Rico. *Biotropica* **32:** 225–234.

Field, D., Magnasco, M., Moxon, E.R., and Metzgar, D. (1999). Contingency loci, mutator alleles and their interactions: synergistic strategies for microbial evolution. *Annals of the New York Academy of Science* **870:** 378–381.

Figueroa, F., Gunther,E., and Klein, J. (1988). MHC polymorphism pre-dating speciation. *Nature* **335:** 265–267.

Filosa, M.F. (1962). Heterocystosis in cellular slime molds. *American Naturalist* **96:** 79–92

Finlay, B.J. (2002). Global dispersal of free-living microbial eukaryote species. *Science* **296:** 1061–1063.

Finlay, B.J., and Clarke, K.J. (1999). Apparent global ubiquity of species in the protist genus *Paraphysomonas*. *Protist* **150:** 419–430.

Finlay, B.J., and Fenchel, T. (2004). Cosmopolitan metapopulations of free-living microbial eukaryotes. *Protist* **155:** 237–244.

Finlay, B.J., Esteban, G.F., and Fenchel, T. (2004). Protist diversity is different? *Protist* **155:** 15–22.

Fisher, R.A. (1930). *The genetical theory of natural selection*. Oxford University Press, Oxford.

Fisher, R.A., Corbet, A.S., and Williams, C.B. (1943). The relation between the number of species and the number of individuals in a random sample of an animal population. *Journal of Animal Ecology* **12:** 42–58.

Flexon, P.B., and Rodell, C.F. (1982). Genetic recombination and directional selection for DDT resistance in *Drosophila melanogaster*. *Nature* **298:** 672–674.

Fontana, W., Stadler, P.F.; Bornbergbauer, E. G. *et al.* (1993). RNA folding and combinatory landscapes. *Physical Reviews* E **47:** 2083–2099.

Ford, E.B. (1964). *Ecological genetics*. Methuen, London.

Foster, P.L. (2005). Stress responses and genetic variation in bacteria. *Mutation Research* **569:** 3–11.

Foster, P.L., and Cairns, J. (1992). Mechanisms of directed mutation. *Genetics* **131:** 783–789.

Fox, S.F. (1975). Natural selection on morphological phenotypes of the lizard *Uta stanburiana*. *Evolution* **29:** 95–107.

Francis, J.C., and Hansche, P.E. (1972). Directed evolution of metabolic pathways in microbial populations. I. Modifications of the acid phosphatase pH optimum in *Saccharomyces cerevisiae*. *Genetics* **70:** 59–73.

Francis, J.C., and Hansche, P.E. (1973). Directed evolution of metabolic pathways in microbial populations. II. A repeatable adaptation in *Saccharomyces cerevisiae*. *Genetics* **74:** 259–265.

Frank, S.A. (1998). *Foundations of social evolution*. Princeton University Press, Princeton, NJ.

Frank, S.A., and Nowak, M.A. (2004). Problems of somatic mutation and cancer. *BioEssays* **26:** 291–299.

Frankham, R. (1990). Are responses to artificial selection for reproductive fitness characters consistently asymmetrical? *Genetical Research* **50:** 35–42.

Frankham, R. (1995). Effective population size/adult population size ratios in wildlife: a review. *Genetical Research* **66:** 95–107.

Frankham, R., Jones, L.P., and Barker, J.S.F. (1968a). The effects of population size and selection intensity for a quantitative characters in *Drosophila*. I. Short-term response to selection. *Genetical Research* **12:** 237–248.

Frankham, R., Jones, L.P., and Barker, J.S.F. (1968b). The effects of population size and selection intensity for a quantitative characters in *Drosophila*. III. Analyses of the lines. *Genetical Research* **12:** 267–283.

Frankham, R., Yoo, B.H., and Sheldon, B.L. (1988). Reproductive fitness and artificial selection in animal breeding: culling on fitness prevents a decline in reproductive fitness in lines of *Drosophila melanogaster* selected for increased inebriation time. *Theoretical and Applied Genetics* **76:** 909–914.

Fretwell, S.D., and Lucas, H.L. (1970). On territorial behaviour and other factors influencing habitat distribution in birds. *Acta Biotheoretica* **19:** 16–36.

Fry, J.D. (1990). Trade-offs in fitness on different hosts—evidence from a selection experiment with a phytophagous mite. *American Naturalist* **136:**569–580.

Fry, J.D. (2004). On the rate and linearity of viability declines in *Drosophila* mutation-accumulation experiments: genomic mutation rates and synergistic epistasis revisited. *Genetics* **166:** 797–806.

Fry, J.D., Keightley P.D., Heinsohn S.L., and Nuzhdin S.V. (1999). New estimates of the rates and effects of mildly deleterious mutations in *Drosophila melanogaster*. *Proceedings of the National Academy of Sciences of the USA* **96:** 574–579.

Funchain, P., Yeung, A., Stewart, J.L., Lin, R., Slupska, M.M, and Miller, J.H. (2000). The consequences of growth of a mutator strain of *Escherichia coli* as measured by loss of function among multiple gene targets and loss of fitness. *Genetics* **154:** 959–970.

Funk, D.J., Nosil, P., and Etges, W.G. (2006). Ecological divergence is consistently positively associated with reproductive isolation across disparate taxa. *Proceedings of the National Academy of Sciences of the USA* **103:** 3209–3213.

Futcher, B., Reid, E., and Hickey, D.A. (1988). Maintenance of the 2 μm circle plasmid of *Saccharomyces cerevisiae* by sexual transmission: an example of a selfish DNA. *Genetics* **118:** 411–415.

Futuyma, D.J. (1970). Variation in genetic response to intraspecific competition in laboratory populations of *Drosophila*. *American Naturalist* **104:** 239–252.

Gaffney, B., and Cunningham, E.P. (1988). Estimation of genetic trend in racing performance of thoroughbred horses. *Nature* **332:** 722–724.

Gage, M., and Morrow, E. (2003). Experimental evidence for the evolution of numerous, tiny sperm via sperm competition. *Current Biology* **13:** 754–757.

Galen, C. (1996). Rates of floral evolution: adaptation to bumblebee pollination in alpine wildflower, *Polemonium viscosum*. *Evolution* **50:** 120–125.

Galiana, A., Moya, A., and Ayala, F.J. (1993). Founder-flush speciation in *Drosophila pseudoobscura*: a large-scale experiment. *Evolution* **47:** 432–444.

Galloway, L.F., and Fenster, C.B. (2000). Population differentiation in an annual legume: local adaptation. *Evolution* **54:** 1173–1181.

Ganz, T. (2003). Defensins: antimicrobial peptides of innate immunity. *Nature Reviews Immunology* **3:** 710–720.

Garcia-Dorado, A., and López-Fanjul, C. (1983). Accumulation of lethals in highly selected lines of *Drosophila melanogaster*. *Theoretical and Applied Genetics* **66:** 221–223.

Garland, T.J. (2003). Selection experiments: an underutilized tool in biomechanics and organismal biology. In V.L. Bels, J.-P. Gasc, and A. Castnos (eds.), *Vertebrate biomechanics and evolution*, pp. 23–56. BIOS Scientific Publications, Oxford.

Garrigan, D., and Hedrick, P.W. (2003). Detecting adaptive molecular polymorphism: lessons from the MHC. *Evolution* **57:** 1707–1722.

Garwood, V.A., Lowe, P.C., and Bohren, B.B. (1980). An experimental test of the efficiency of family selection in chickens. *Theoretical and Applied Genetics* **56:** 5–9.

Gaston, K.J., Blackburn, T.M., and Lawton, J.H. (1997). Interspecific range-abundance relationships: an appraisal of mechanisms. *Journal of Animal Ecology* **66:** 579–601.

Gaston, K.J., Blackburn, T.M., Gregory, R.D., and Greenwood, J.J.D. (1998). The anatomy of the interspecific abundance-range size relationship for the British avifauna. I. Spatial patterns. *Ecology Letters* **1:** 38–46.

Gause, G.F. (1934). *The struggle for existence*. Williams & Wilkins, New York. (Reprinted 1971 by Dover, New York).

Genter, C.F. (1976). Mass selection in a composite of intercrosses of Mexican races of maize. *Crop Science* **16:** 556–558.

Gerdes, K. (1988). The *parB* (*hok/sok*) locus of plasmid R1: a general purpose plasmid stabilization system. *Nature Biotechnology* **6:** 1402–1405.

Gerdes, K., Gultyaev, A.P., Franch, T., Pederson, K., and Mikkelsen, N.D. (1997). Antisense RNA-regulated programmed cell death. *Annual Review of Genetics* **31:** 1–31.

Gerrish, P. (2001). The rhythm of microbial adaptation. *Nature* **413:** 299–302.

Gerrish, P.J., and Lenski, R.E. (1998). The fate of competing beneficial mutations in an asexual population. *Genetica* **102/103:** 127–144.

Getty, T. (1999). What do experimental studies tell us about group selection in nature? *American Naturalist*. **154:** 596–598.

Ghiselin, M.T. (1974). *The economy of nature and the evolution of sex*. University of California Press, Berkeley, CA.

Giaever, G., Chu, A.M., Ni, L. *et al.* (2002). Functional profiling of the *Saccharomyces cerevisiae* genome. *Nature* **418:** 387–391.

Gibbs, H.L., and Grant, P.R. (1987a). Ecological consequences of an exceptionally strong El Nino event on Darwin's finches. *Evolution* **68:** 1735–1746.

Gibbs, H.L., and Grant, P.R. (1987b). Oscillating selection in Darwin's finches. *Nature* 327: 511–513.

Gibson, J.B., and Bradley, B.P. (1974). Stabilizing selection in constant and fluctuating environments. *Heredity* **33:** 293–302.

Gibson, T.C., Scheppe, M.L., and Cox, E.C. (1970). Fitness of an *Escherichia coli* mutator gene. *Science* **169:** 686–688.

Gillespie, J.H. (1977). Natural selection for variance in offspring numbers: a new evolutionary principle. *American Naturalist* **111:** 1010–1014.

Gillespie, J.H. (1984). Molecular evolution over the mutational landscape. *Evolution* **38:** 1116–1129.

Gillespie, J.H. (1991). *The causes of molecular evolution*. Oxford University Press, Oxford.

Gillespie, J.H., and Langley, C.H. (1974). A general model to account for enzyyme variation in natural populations. *Genetics* **76:** 837–884.

Gillis, D.M., Kramer, D.L., and Bell, G. (1986). Taylor's Power Law as a consequence of Fretwell's Ideal Free Distribution. *Journal of Theoretical Biology* **123:** 281–287.

Gillooly, J.F., Brown, J.H., West, G.B., Savage, V.M., and Charnov, E.L. (2002). Effects of size and temperature on metabolic rate. *Science* **293:** 2248–2251.

Gingerich, P.D. (1983). Rates of evolution: effects of time and temporal scaling. *Science* **222:** 159–161.

Gingerich, P.D. (2001). Rates of evolution over the time scale of the evolutionary process. *Genetica* **112/113:** 127–144.

Giraud, A., Matic, I., Tenallion, O. *et al.* (2001a). Costs and benefits of high mutation rates: Adaptive evolution of bacteria in the mouse gut. *Science* **291:** 2606–2608.

Giraud, A., Radman, M., Matic, I., and Taddei, F. (2001b). The rise and fall of mutator bacteria. *Current Opinion in Microbiology* **4:** 582–585.

Glass, N.L., Grotelueschen, J., and Metzenberg, R.L. (1990). *Neurospora crassa* A mating-type region. *Proceedings of the National Academy of Sciences of the USA* **87:** 4912–4916.

Goddard, M.R., Greig, D., and Burt, A. (2001). Outcrossed sex allows a selfish gene to invade yeast populations. *Proceedings of the Royal Society of London Series B* **268:** 2537–2542.

Goddard, M.R., Godfray, H.C., and Burt, A. (2005). Sex increases the efficacy of natural selection in experimental yeast populations. *Nature* **434:** 571–573.

Godelle, B. (1997). Role of mutator alleles in adaptive evolution. *Nature* **387:** 700–702.

Godin, J.-G.J., and McDonough, H.E. (2003). Predator preference for brightly colored males in the guppy: a viability cost for a sexually selected trait. *Behavioral Ecology* **14:** 194–200.

Godoy, V.G., Gizatullin, F.S., and Fox, M.S. (2000). Some features of the mutability of bacteria during non-lethal selection. *Genetics* **154:** 49–59.

Goho, S., and Bell, G. (2000). Mild environmental stress elicits mutations affecting fitness in *Chlamydomonas*. *Proceedings of the Royal Society of London Series B* **267:** 123–129.

Gomendio, M., Martin-Coello, J., Crespo, C., Magaña C., and Roldan, E.R.S. (2006). Sperm competition enhances functional capacity of mammalian spermatozoa. *Proceedings of the National Academy of Sciences of the USA* **103:** 15113–15117

Gomulkiewicz, R. and Holt, R.D. (1995). When does evolution by natural selection prevent extinction? *Evolution* **49:** 201–207.

Good, A.G., Meister, G.A., Brock, H.W., Grigliatti, T.A., and Hickey, D.A. (1989). Rapid spread of transposable P elements in experimental populations of *Drosophila melanogaster*. *Genetics* **122:** 387–396.

Goodman, D. (1979). Competitive hierarchies in laboratory *Drosophila*. *Evolution* **33:** 207–219.

Goodnight, C.J. (1990a). Experimental studies of community evolution. I. The response to selection at the community level. *Evolution* **44:** 1614–1624.

Goodnight, C.J. (1990b). Experimental studies of community evolution. II. The ecological basis of the response to community selection. *Evolution* **44:** 1625–1636.

Goodnight, C.J., and Stevens, L. (1997). Experimental studies of group selection: What do they tell us about group selection in nature? *American Naturalist* **150:** S59–S79.

Goodwill, R. (1974). Comparison of three selection programs using *Tribolium castaneum*. *Journal of Heredity* **65:** 8–14.

Gordo, I., and Charlesworth, B. (2000). The degeneration of asexual haploid populations and the speed of Muller's Ratchet. *Genetics* **154:** 1379–1387.

Gordon, D.R., and Rice, K.J. (1998). Patterns of differentiation in wiregrass (*Aristida beyrichiana*): implications for restoration efforts. *Restoration Ecology* **6:** 166–174.

Gorodetskii, V.P., Zhuchenko, A.A., and Korol, A.B. (1990). Efficiency of feedback selection for recombination in *Drosophila*. *Genetika* **26:** 1942–1952.

Gottschal, J.C. (1986). Mixed substrate utilization by mixed cultures. In Poindexter, J.S. and Leadbetter, E.R.

(eds.), *Bacteria in nature*, vol. 2, pp. 261–296. Plenum, New York.

Gould, F. (1979). Rapid host range evolution in a population of the phytophagous mite *Tetranychus urticae* Koch. *Evolution* **33:** 791–802.

Gould, S.J., and Lewontin, R.C. (1979). The spandrels of San Marco and the Panglossian paradigm: a critique of the adaptationist programme. *Proceedings of the Royal Society of London Series B* **205:** 581–598.

Gowe, R.S., and Fairfull, R.W. (1985). The direct response to long-term selection for multiple traits in egg stocks and changes in genetic parameters with selection. In Hill, W.G., Manson, J.M. and Hewitt, D. (eds.) *Poultry genetics and breeding*, pp. 125–146. Longman, Harlow, Essex.

Grafen, A. (1985). A geometric view of relatedness. *Oxford Surveys on Evolutionary Biology* **2:** 28–89.

Grant, B., and Mettler, L.E. (1969). Disruptive and stabilizing selection on the 'escape' behavior of *Drosophila melanogaster. Genetics* **62:** 625–637.

Grant, B.R. (1985). Selection on bill characters in a population of Darwin's finches, *Geospiza conirostris*, on Isla Genovesa, Galapagos. *Evolution* **39:** 523–532.

Grant, P.R., and Grant, B.R. (1995). Predicting microevolutionary responses to directional selection on heritable variation. *Evolution* **49:** 241–251.

Grant, P.R., Grant, B.R., Smith, J.M.N., Abbott, I.J., and Abbott, I.K. (1976). Darwin's finches: population variation and natural selection. *Proceedings of the National Academy of Sciences of the USA* **13:** 257–261.

Grant, P.R., Grant, B.R., and Abzhanov, A. (2006). A developing paradigm for the development of bird beaks. *Biological Journal of the Linnean Society* **88:** 17–22.

Graves, J.L., Luckinbill, L.S., and Nichols, A. (1988). Flight duration and wing beat frequency in long- and short-lived *Drosophila melanogaster. Journal of Insect Physiology* **34:** 1021–1026.

Gray, A.J., Marshall, D.F., and Raybould, A.F. (1991). A century of evolution in *Spartina anglica. Advances in Ecological Research* **21:** 1–62.

Greig, D., and Travisano, M. (2004). The Prisoner's Dilemma and polymorphism in yeast *SUC* genes. *Proceedings of the Royal Society of London Series B* **271:** S25–S26.

Greig, D., Borts, R.H., and Louis, E.J. (1998). The effect of sex on adaptation to high temperature in heterozygous and homozygous yeast. *Proceedings of the Royal Society of London Series B* **265:** 1017–1023.

Greig, D., Louis, E.J., Borts, R.H., and Travisano, M. (2002). Hybrid speciation in experimental populations of yeast. *Science* **298:** 1773–1775.

Griffin, A.S., West, S.A., and Buckling, A. (2004). Cooperation and competition in pathogenic bacteria. *Nature* **430:** 1024–1027.

Griffiths, A.D., and Tawfik, D.S. (2003). Directed evolution of an extremely fast phosphotriesterase by in vitro compartmentalization. *EMBO Journal* **22:** 24–35.

Gromko, M.H., Briot, A. Jensen, S.C., and Fukui, H.H. (1991). Selection on copulation duration in *Drosophila melanogaster*: predictability of direct response versus unpredictability of correlated response. *Evolution* **45:** 69–81.

Grosberg, R.K. (1988). The evolution of allorecognition specificity in clonal invertebrates. *Quarterly Review of Biology* **63:** 377–412.

Gross, H.P. (1978). Natural selection by predators on the defensive apparatus of the three-spined stickleback, *Gasterosteus aculeatus* L. *Canadian Journal of Zoology* **56:** 398–413.

Gross, M.D., and Siegel, E.C. (1981). The incidence of mutator strains in *Escherichia coli* and coliforms in nature. *Mutation Research* **91:** 107–110.

Grundmann, G.L. (2004). Spatial scales of soil bacterial diversity—the size of a clone. *FEMS Microbiology Ecology* **48:** 119–127.

Grundmann, G.L., and Normand, P. (2000). Microscale diversity of the genus *Nitrobacter* in soil on the basis of the analysis of genes encoding rRNA. *Applied Environmental Microbiology* **66:** 4543–4546.

Grundmann, G.L., Dechesne, N., Bartoli, F., and Flandrois, J.P. (2001). Spatial modelling of nitrifier microhabitats in soil. *Soil Science Society of America Journal* **65:** 1709–1716.

Haddrill, P.R., Halligan, D.L., Tomaras, D., and Charlesworth, B. (2007). Reduced efficacy of selection in regions of the *Drosophila* genome that lack crossing over. *Genome Biology* **8:** R18.

Hagen, D.W., and Gilbertson, L.G. (1973). Selective predation and the intensity of selection acting on the lateral plates of three-spine sticklebacks. *Heredity* **30:** 273–287.

Haigh, J. (1978). The accumulation of deleterious genes in a population—Muller's Ratchet. *Theoretical Population Biology* **14:** 251–267.

Haldane, J.B.S. (1924). *Transactions of the Cambridge Philosophical Society* **23:** 19–41.

Haldane, J.B.S. (1927). *Transactions of the Cambridge Philosophical Society* **23:** 838–844.

Haldane, J.B.S. (1932). *The causes of evolution*. Longman, Green and Co., London.

Haldane, J.B.S. (1937). The effect of variation on fitness. *American Naturalist* **71:** 337–349.

Haldane, J.B.S. (1949). Suggestions as to quantitative measurement of rates of evolution. *Evolution* **3:** 51–56.

Haldane, J.B.S. (1957). The cost of natural selection. *Journal of Genetics* **55:** 511–524.

Haldane, J.B.S., and Jayakar, S.D. (1963). Polymorphism due to selection of varying direction. *Journal of Genetics* **58:** 237–242.

Haley, C.S., and Birley, A.J. (1983). The genetical response to natural selection by varied environments. II. Observations on replicate populations in spatially varied environments. *Heredity* **51:** 581–606.

Halkka, O., Halkka, L., and Raatikainen, M. (1975). Transfer of individuals as a means of investigating natural selection in operation. *Hereditas* **80:** 27–34.

Hall, B.G. (1990). Spontaneous point mutations that occur more often when advantageous than when neutral. *Genetics* **126:** 5–16.

Hall, B.G. (2003). The EBG system of *E. coli*: origin and evolution of a novel β-galactosidase for the metabolism of lactose. *Genetica* **118:** 143–156.

Hall, M., Lindholm, A.K., and Brooks, R. (2004). Direct selection on male attractiveness and female preference fails to produce a response. *BMC Evolutionary Biology* 4: 1. doi: 10.(1186)/1471-2148-4-1.

Hallauer, A.R., Ross, A.J., and Lee, M. (2004). Long-term divergent selection for ear length in maize. *Plant Breeding Reviews* **24**(2): 153–168.

Halley, J.M. (1996). Ecology, evolution and 1/*f* noise. *Trends in Ecology and Evolution* **11:** 33–37.

Halligan, D.L., and Keightley, P.D. (2006) Ubiquitous genetic constraints in the *Prosophila* genome revealed by a genome-wide interspecies comparison. *Genome Research* **16:** 875–884.

Hamilton, W.D. (1964a). The genetical evolution of social behaviour. I. *Journal of Theoretical Biology* **7:** 1–16.

Hamilton, W.D. (1964b). The genetical evolution of social behaviour. II. *Journal of Theoretical Biology* **7:** 17–52.

Hamilton, W.D. (1967). Extraordinary sex ratios. *Science* **156:** 477–488.

Hamilton, W.D. (1980). Sex versus non-sex versus parasite. *Oikos* **35:** 282–290.

Hamilton, W.D., and Zuk, M. (1982). Heritable true fitness and bright birds: a role for parasites? *Science* **218:** 384–387.

Hämmerli, H., and Reusch, T.B.A. (2002). Local adaptation and transplant dominance in genets of the marine clonal plant *Zostera marina*. *Marine Ecology Progress Series* **242:** 111–118.

Hamon, T.R., Foote, C.J., Hilborn, R., and Rogers, D.E. (2000). Selection on morphology of spawning wild sockeye salmon by a gill-net fishery. *Transactions of the American Fisheries Society* **129:** 1300–1315.

Hamrick, J.L., and Godt, M.J.W. (1996). Effects of life history traits on genetic diversity in plant species. *Philosophical Transactions of the Royal Society of London* B **351:** 1291–1298.

Hancock, J.F. (1992). *Plant evolution and the origin of crop species.* Prentice-Hall, Englewood Cliffs, NJ.

Hancock, R.E. (2001). Cationic peptides: effectors in innate immunity and novel antimicrobials. *Lancet Infectious Disease* **1:** 156–164.

Hanczyc, M.M., and Dorit, R.L. (2000). Replicability and recurrence in the experimental evolution of a group I ribozyme. *Molecular Biology and Evolution* **17**(7): 1050–1060.

Handford, P., Bell, G., and Reimchen, T. (1977). A gillnet fishery considered as an experiment in artificial selection. *Journal of the Fisheries Research Board of Canada* **34:** 954–961.

Hanel, E. (1908). Verebung bei ungeschlechtlicher Fortplanzung von *Hydra grisea*. *Jenaische Zeitschrift* **43:** 321–372. (Not seen.).

Hanes, J., and Plückthon, A. (1997). In vitro selection methods for screening of peptide and protein libraries. *Current Topics in Microbiology and Immunology* **243:** 107–122.

Hanes, J., Schaffitzel, C., Knappik, A., and Plückthon, A. (2000). Picomolar affinity antibodies from a fully synthetic naive library selected and evolved by ribosome display. *Nature Biotechnology* **18:** 1287–1292.

Hansche, P.E. (1975). Gene duplication as a mechanism of genetic adaptation in *Saccharomyces cerevisiae*. *Genetics* **79:** 661–674.

Hansen, S.R., and Hubbell, S.P. (1980). Single-nutrient microbial competition: qualitative agreement between experimental and theoretically forecast outcomes. *Science* **207:** 1491–1493.

Harder, W., and Dijkhuizen, L. (1982). Strategies of mixed substrate utilization in microorganisms. *Philosophical Transactions of the Royal Society of London* B **297:** 459–480.

Hardin, G. (1968). The tragedy of the commons. *Science* **162:** 1243–1248.

Harding, J., Allard, R.W., and Smeltzer, D.G. (1966). Population studies in predominantly self-pollinating species. IX. Frequency-dependent selection in *Phaseolus lunatus*. *Proceedings of the National Academy of Sciences of the USA* **56:** 99–104.

Harlan, H.V., and Martini, M.I. (1938). The effect of natural selection in a mixture of barley varieties. *Journal of Agricultural Research* **57:** 189–199.

Harlan, J.R. (1992). *Crops and man.* American Society of Agronomy, Madison, Wisconsin.

Harrison, P., Kumar, A., Lan, N., Echols, N., Snyder, M., and Gerstein, M. (2002). A small reservoir of disabled ORFs in the yeast genome and its implications for the dynamics of proteome evolution. *Journal of Molecular Biology* **316:** 409–419.

Hartl, D.L., Dykhuizen, D.E., Miller, R.D., Green, L., and de Framond, J. (1983). Transposable element IS50 improves growth rate of *E. coli* cells without transposition. *Cell* **35:** 503–510.

Hartley, B.S. (1984). Experimental evolution of ribitol dehydrogenase. In R.P. Mortlock (ed.), *Microorganisms as model systems for studying evolution*, pp. 23–54. Plenum, New York.

Hartley, B.S., Altosaar, I., Dothie, J.M., and Neuberger, M.S. (1976). Experimental evolution of a xylitol dehydrogenase. In R. Markham and R.W. Horne (eds.), *Structure-function relationship of proteins*, pp. 191–200. North-Holland, Amsterdam.

Hatfield, T., and Schluter, D. (1999). Ecological speciation in sticklebacks: environment-dependent hybrid fitness. *Evolution* **53:** 866–873.

Hayes, B., and Goddard, M.E. (2001). The distribution of the effects of genes affecting quantitative traits in livestock. *Genetics, Selection and Evolution* **33:** 209–229.

Hayes, F. (2003). Toxins-antitoxins: plasmid maintenance, programmed cell death, and cell cycle arrest. *Science* **301:** 1496–1499.

Hayflick, L., and Moorhead, P.S. (1961). The serial cultivation of human diploid cell strains. *Experimental Cell Research* **25:** 585–621.

Hector, A., Schmid, B., Beierkuhnlein, C. *et al.* (1999). Plant diversity and productivity experiments in European grasslands. *Science* **286:** 1123–1127.

Hedrick, P.W. (2000). *Genetics of Populations*, 2nd edn. Jones & Bartlett, Sudbury, MA.

Hegeman, G.D. (1966). Synthesis of the enzymes of the mandelate pathway by *Pseudomonas putida* II. Isolation and properties of blocked mutants. *Journal of Bacteriology* **91:** 1155–1160.

Hekimi, S., Burgess, J., Bussiere, F., Meng, Y., and Benard, C. (2001). Genetics of lifespan in *C. elegans*: molecular diversity, physiological complexity, mechanistic simplicity. *Trends in Genetics* **17:** 712–718.

Helenurm, K. (1998). Outplanting and differential source population success in *Lupinus guadalupensis*. *Conservation Biology* **12:** 118–127.

Helling, R.B., Vargas, C.N., and Adams, J. (1987). Evolution of *Escherichia coli* during growth in a constant environment. *Genetics* **116:** 349–358.

Hendry, A.P. (2005). The power of natural selection. *Nature* **433:** 694–695.

Hendry, A.P., and Kinnison, M.T. (1999). The pace of modern life: measuring rates of contemporary microevolution. *Evolution* **53:** 1637–1653.

Hendry, A.P., Taylor, E.B., and McPhail, J.D. (2002). Adaptive divergence and the balance between selection and gene flow: lake and stream stickleback in the Misty system. *Evolution* **56:** 1199–1216.

Henikoff, S., and Malik, H.S. (2002). Selfish drivers. *Nature* **417:** 227.

Henikoff, S., Ahmad, K., and Malik, H.S. (2001). The centromere paradox: stable inheritance with rapidly evolving DNA. *Science* **293:** 1098–1102.

Hereford, J., Hansen, T.F., and Houle, D. (2004). Comparing strengths of directional selection: how strong is strong? *Evolution* **58:** 2133–2143.

Hersch, E., and Phillips, P.C. (2004). Power and potential bias in the detection of selection in natural populations. *Evolution* **58:** 479–485.

Hetzer, H.O. (1954). Effectiveness of selection for extension of black spotting in Beltsville no. 1 swine. *Journal of Heredity* **45:** 215–223.

Hewlett, B.S., and Cavalli-Sforza, L.L. (1986). Cultural transmission among Aka pygmies. *American Anthropologist* **88:** 922–934.

Hickey, D. (1982). Selfish DNA: a sexually-transmitted nuclear parasite. *Genetics* **101:** 519–531.

Hickey, D.A., and McNeilly, T. (1975). Competition between metal tolerant and normal plant populations: a field experiment on normal soil. *Evolution* **29:** 458–464.

Higgie, M., Chenoweth, S., and Blows, M.W. (2000). Natural selection and the reinforcement of mate recognition. *Science* **290:** 519–521.

Hill, D., and Blumenthal, T. (1983). Does Qβ replicase synthesize RNA in the absence of template? *Nature* **301:** 350–352.

Hill, W.G., and Bunger, L. (2004). Inferences on the genetics of quantitative traits from long-term selection in laboratory and domestic animals. *Plant Breeding Reviews* **24:** 169–210.

Hill, W.G., and Mackay, T.F.C. (eds.) (1988). *Evolution and animal breeding*. CAB International, Wallingford.

Hill, W.G., and Mbaga, S.H. (1998). Mutation and conflicts between artificial selection and natural selection for quantitative traits. *Genetica* **102/103:** 171–181.

Hill, W.G., and Robertson, A. (1966) The effect of linkage on limits to artificial selection. *Genetical Research* 8: 269–294.

Hill, W.G., and Robertson, A. (1968) Linkage disequilibrium in finite populations. *Theoretical and Applied Genetics* 38: 226–231.

Hillesheim, E., and Stearns, S.C. (1991). The responses of *Drosophila melanogaster* to artificial selection on body weight and its phenotypic plasticity in two larval food environments. *Evolution* **45:** 1909–1923.

Hinson, K., and Hanson, W.D. (1962). Competition studies in soybeans. *Crop Science* **2:** 117–123.

Hiscock, S.J., and McInnis, S.M. (2003). The diversity of self-incompatibility systems in flowering plants. *Plant Biology (Stuttgart)* **5:** 23–32.

Hoekstra, H.E., Hoekstra, J.M., Berrigan, D. *et al.* (2001). Strength and tempo of natural selection in the wild. *Proceedings of the National Academy of Sciences of the USA* **98:** 9157–9160.

Hoffman, A.A., and Parsons, P.A. (1989). Selection for increased desiccation resistance in *Drosophila melanogaster*: additive genetic control and correlated responses for other stresses. *Genetics* **122:** 837–845.

Hoffmann, J.A., and Reichart, J.-M. (2002). *Drosophila* innate immunity: an evolutionary perspective. *Nature Immunology* **3:** 121–126.

Holder, K.K., and Bull, J.J. (2001). Profiles of adaptation in two similar viruses. *Genetics* **159:** 1393–1404.

Holman, E.W. (1987). Recognizability of sexual and asexual species of rotifers. *Systematic Zoology* **36:** 381–386.

Holt, M., Vangen, O., and Meuwissen, T. (2005) Long-term responses, changes in genetic variances and inbreeding depression from 122 generations of selection on increased litter size in mice. *Journal of Animal Breeding Genetics* **122:** 199–209.

Hooper, J. (2002). *Of moths and men.* Fourth Estate, London.

Hori, M., Tomikawa, I., Przybo, E., and Fujishima, M. (2006). Comparison of the evolutionary distances among syngens and sibling species of *Paramecium. Molecular Phylogenetics and Evolution* **38:** 697–704.

Houde, A.E. (1994). Effect of artificial selection on male color patterns on mating preference of female guppies. *Proceedings of the Royal Society of London Series B* **256:** 125–130.

Houle, D., and Kondrashov, A.S. (2001). Coevolution of costly male choice and condition-dependent display of good genes. *Proceedings of the Royal Society of London Series B* **269:** 97–104.

Houle, D., Mezey, J., and Galpern, P. (2002). Interpretation of the results of common principal component analyses. *Evolution* **56:** 433–440.

Huang, H.W. (2000). Action of antimicrobial peptides: two-state model. *Biochemistry* **39:** 8347–8352.

Huang, S.L., Singh, M., and Kojima, K. (1971). A study of frequency-dependent selection observed in the Esterase-6 locus of *Drosophila melanogaster* using a conditioned media method. *Genetics* **68:** 97–104.

Hubbell, S.P. (1995) Towards a theory of biodiversity and biogeography on continuous landscapes. In Carmichael, G.R., Folk, G.E., and Schnoor, J.L. (eds) *Preparing for global change: a Midwestern perspective,* pp 173–201. SPB Academic Publishing, Amsterdam.

Hubbell, S.P. (2001). *The unified neutral theory of biodiversity and biogeography.* Princeton University Press, Princeton, NJ.

Hudak, M.J., and Gromko, M.H. (1989). Responses to selection for early and late development of sexual maturity in *Drosophila melanogaster. Animal Behaviour* **38:** 344–351.

Hudson, R.R., Kreitman, M., and Aguade, M. (1987). A test of neutral molecular evolution based on nucleotide data. *Genetics* **116:** 153–159.

Huffaker, C.B. (1958). Experimental studies on predation: dispersion factors and predator-prey oscillations. *Hilgardia* **27:** 343–383.

Hughes, J.S., and Otto, S.P. (1999). Ecology and the evolution of biphasic life cycles. *American Naturalist* **154:** 306–320.

Hughes, K.A., Alipaz, J.A., Drnevich, J.M., and Reynolds, R.M. (2002). A test of evolutionary theories of aging. *Proceedings of the National Academy of Sciences of the USA* 99(22): 14286–14291.

Hulbert, R.C. (1993). Taxonomic evolution in North American Neogene horses (subfamily Equinae): the rise and fall of an adaptive radiation. *Paleobiology* **19:** 216–234.

Hunt, G.R. (2003). Diversification and cumulative evolution in New Caledonian crow tool manufacture. *Proceedings of the Royal Society of London Series B* **270:** 867–874.

Hunter, P.E. (1959). Selection of *Drosophila melanogaster* for length of larval period. *Zeitschrift für Vererbungslehre* **90:** 7–28.

Hurd, L.E., and Eisenberg, R.M. (1975). Divergent selection for geotactic response and evolution of reproductive isolation in sympatric and allopatric populations of houseflies. *American Naturalist* **109:** 353–358.

Hurst, L.D., and Hamilton, W.D. (1992). Cytoplasmic fusion and the nature of sexes. *Proceedings of the Royal Society of London Series B* **247:** 189–194.

Hurst, L.D., and Randerson, J.P. (2000). Transitions in the evolution of meiosis. *Journal of Evolutionary Biology* **13:** 466–479.

Hutchinson, E.W., and Rose, M.R. (1991). Quantitative genetics of postponed aging in *Drosophila melanogaster.* I. Analysis of outbred populations. *Genetics* **127:** 719–727.

Hutchinson, E.W., Shaw, A.J., and Rose, M.R. (1991). Quantitative genetics of postponed aging in *Drosophila melanogaster.* II. Analysis of selected lines. *Genetics* **127:** 729–737.

Hutson, V., and Law., R. (1981). Evolution of recombination in populations experiencing frequency-dependent selection with time delay. *Proceedings of the Royal Society of London Series B* **213:** 345–359.

Ichimura, T. (1985). Geographical distribution and isolating mechanisms in the *Closterium ehrenbergii* species complex (Chlorophyceae, Closteriaceae). In Ham, H. (ed.) *Origin and evolution of diversity in plants and plant communities*, pp. 295–303. Academia Scientific Book, Tokyo.

Ichimura, T. (1996). Genome rearrangement and speciation in freshwater algae. *Hydrobiologia* **336:** 1–17.

Imhof, M., and Schötterer, C. (2001). Fitness effects of advantageous mutations in evolving *Escherichia coli* populations. *Proceedings of the National Academy of Science of the USA* **98**(3): 1113–1117.

Inchausti, P., and Halley, J. (2002). The long-term temporal variability and spectral colour of animal populations. *Evolutionary Ecology Research* **4:** 1033–1048.

Inderlied, C.B., and Mortlock, R.P. (1977). Growth of *Klebsiella aerogenes* on xylitol: Implications for bacterial enzyme evolution. *Journal of Molecular Evolution* **9:** 181–190.

Inger, R.F. (1942). Differential selection of variant juvenile snakes. *American Naturalist* **76:** 104–109.

Iriberri, J., Unanue, M., Barcina, I., and Egea, L. (1987). Seasonal variation in population density and heterotrophic activity of attached and free-living bacteria in coastal waters. *Applied and Environmental Miocrobiology* **53:** 2308–2314.

Istock, C.A., Duncan, K.E., Ferguson, N., and Zhou, X. (1992). Sexuality in a natural population of bacteria: *Bacillus subtilis* challenges the clonal paradigm. *Molecular Ecology* **1:** 95–103.

Iwasa, Y., and Sasaki, A. (1983). Evolution of the number of sexes. *Evolution* **41:** 49–65.

Jablonski, D. (1986). Larval ecology and macroevolution in marine invertebrates. *Bulletin of Marine Science* **39:** 565–587.

Jaenike, J. (1978). An hypothesis to account for the maintenance of sex within populations. *Evolutionary Theory* **3:** 191–194.

Jaenike, J., and Dombeck, I. (1998). General-purpose genotypes for host species utilization in a nematode parasite of *Drosophila*. *Evolution* **52:** 832–840.

Jain, S.K., and Bradshaw, A.D. (1966). Evolution in closely adjacent plant populations. I. The evidence and its theoretical analysis. *Heredity* **21:** 407–441.

Jakobsson, A., and Dinnetz, P. (2005). Local adaptation and the effects of isolation and population size—the semelparous perennial *Carlina vulgaris* as a study case. *Evolutionary Ecology* **19:** 449–466.

Jamieson, I.G., Blouw, D.M., and Colgan, P.W. (1992). Field observations on the reproductive biology of a newly discovered stickleback (*Gasterosteus*). *Canadian Journal of Zoology* **70:** 1057–1063.

Jarne, P., and Charlesworth, D. (1993). The evolution of the selfing rate in functionally hermaphroditic plants and animals. *Annual Reviews of Ecology and Systematics* **24:** 441–466.

Jarosz, A.M., and Burdon, J.J. (1991). Host-pathogen interactions in natural populations of *Linum marginale* and *Melampsora lini*: II. Local and regional variation in patterns of resistance and racial structure. *Evolution* **45:** 1618–1627.

Jenkins, C.D. (1993). Selection and the evolution of genetic life cycles. *Genetics* **133:** 401–410.

Jennings, H.S. (1908). Heredity, variation and evolution in protozoa. II. Heredity, and variation in size and form in *Paramecium*, with studies of growth, environmental action and selection. *Proceedings of the American Philosophical Society* **47:** 393–546.

Jennings, H.S. (1910). Experimental evidence on the effectiveness of selection. *American Naturalist* **44:** 136–145.

Jennings, H.S. (1916). Heredity, variation and the results of selection in the uniparental reproduction of *Difflugia corona*. *Genetics* **1:** 407–534.

Jennings, P.R., and de Jesus, J. (1968). Studies on competition in rice. I. Competition in mixtures of varieties. *Evolution* **22:** 119–124.

Jeon, K.W., and Jeon, M.S. (1976). Endosymbiosis in amoebae: recently established endosymbionts have become required cytoplasmic components. *Journal of Cellular Physiology* **89:** 337–344.

Jinks, J.L., and Connolly, V. (1973). Selection for specific and general response to environmental differences. *Heredity* **30:** 33–40.

Jinks, J.L., and Connolly, V. (1975). Determination of the environmental sensitivity of selection lines by the selection environment. *Heredity* **34:** 401–406.

Joakimsen, O., and Baker, R.L. (1977). Selection for litter size in mice. *Acta Agricultura Scandinavica* **27:** 301–318.

Johannsen, W. (1903). *über Erblichkeit in Populationen und in reinen Linien*. Gustav Fischer, Jena.

Johannsen, W. (1911). The genotype conception of heredity. *American Naturalist* **45:** 129–159.

Johnston, R.F., and Selander, R.K. (1964). House sparrows: rapid evolution of races in North America. *Science* **144:** 548–550.

Johnson, T., and Barton, N.H. (2002). The effect of deleterious alleles on adaptation in asexual populations. *Genetics* **162:** 395–411.

Johnstone, M.O., and Schoen, D.J. (1995). Mutation rates and dominance levels of genes affecting total fitness in two angiosperm species. *Science* **267:** 226–229.

Jones, D.R., Anderson, K.E., and Davis, G.S. (2001). The effects of genetic selection on production parameters of single comb white leghorn hens. *Poultry Science* **80:** 1139–1143.

Jones, J.S., and Parkin, D.T. (1977). Attempts to measure selection by altering gene frequencies in natural populations. In F.B. Christiansen and T.M. Fenchel (eds.), *Measuring selection in natural populations. Lecture Notes in Biomathematics* 19, pp. 83–96. Springer-Verlag, Heidelberg.

Jones, J.S., Leith, B., and Rawlings, P. (1977). Polymorphism in *Cepaea*: a problem with too many solutions? *Annual Review of Ecology and Systematics* **8:** 109–143.

Jones, R.N. (1991). B-chromosome drive. *American Naturalist* **137:** 430–442.

Jones, R.N. (1995). B chromosomes in plants. *New Phytologist* **131:**411–434.

Jones, R.N., and Rees, H. (1982). *B chromosomes*. Academic Press, London.

Jordan, I.K., Rogozin, I.B., Wolf, Y.I., and Koonin, E.V. (2002). Essential genes are more evolutionarily conserved than are nonessential genes in bacteria. *Genome Research* **12:** 962–968.

Joseph, S.B., and Hall, D.W. (2004). Spontaneous mutations in diploid *Saccharomyces cerevisiae*: more beneficial than expected. *Genetics* **168:** 1817–1825.

Joshi, A., and Mueller, L.D. (1988). Evolution of higher feeding rate in *Drosophila* due to density-dependent natural selection. *Evolution* **42:** 1090–1093.

Joshi, J., Schmid, B., Caldeira, M.C. *et al.* (2001). Local adaptation enhances performance of common plant species. *Ecology Letters* **4:** 536–544.

Joyce, G.F. (1992). Directed molecular evolution. *Scientific American* **267:** 90–97.

Juchault, P., and Legrand, J.J. (1989). Sex determination and monogeny in terrestrial isopods *Armadillidium vulgare* (Latreille, (1804) and *Armadillidium nasutum* (Budde-Lund, (1885). *Monitore Zoologico Italiana Monographico* **4:** 359–375.

Jyssum, K. (1960). Observations on two types of genetic instability in *Escherichia coli*. *Acta Pathologica Microbiologica Scandinavica* **48:** 113–120.

Kaar, P., Jokela, J., Helle, T., and Kojola, I. (1996). Direct and correlative phenotypic selection on life-history traits in three pre-industrial human populations. *Proceedings of the Royal Society of London Series B* **263:** 1475–1480.

Kacian, D.L., Mills, D.R., Kramer, F.R., and Spiegelman, S. (1972). A replicating RNA molecule suitable for a detailed analysis of extracellular evolution and replication. *Proceedings of the National Academy of Sciences of the USA* **69:** 3038–3042.

Kacser, H. (1988). Quantitative variation and the control analysis of enzyme systems. In W.G. Hill and T.F.C. Mackay (eds.), *Evolution and animal breeding*, pp. 219–226. CAB International, Wallingford.

Kacser, H., and Burns, J.A. (1973) The control of flux. *Symposia of the Society for Experimental Biology* 27: 65–104.

Kaltz, O., and Bell, G. (2002). The ecology and genetics of fitness in *Chlamydomonas*. XII. Repeated sexual episodes increase rates of adaptation to novel environments. *Evolution* **56:** 1743–1753.

Kaltz, O., and Shykoff, J. (1998). Local adaptation in host-parasite systems. *Heredity* **81:** 361–370.

Kannenberg, L.W., and Hunter, R.B. (1972). Yielding ability and competitive influence in hybrid mixtures of maize. *Crop Science* **12:** 274–277.

Karlin, S., and Lessard, S. (1986). *Theoretical studies on sex ratio evolution*. Princeton Univ Press, Princeton, NJ.

Kassen, R., and Bataillon, T. (2006). Distribution of fitness effects among beneficial mutations before selection in experimental populations of *Pseudomonas fluorescens*. *Nature Genetics* **38:** 484–488.

Kassen, R., and Bell, G. (1998). Experimental evolution in *Chlamydomonas*. IV. Selection in environments that vary through time at different scales. *Heredity* **80:** 732–741.

Kassen, R., and Bell, G. (2000). The ecology and genetics of fitness in *Chlamydomonas*. X. The relationship between genetic correlation and genetic distance. *Evolution* **54:** 425–432.

Katz, A.J., and Enfield, F.D. (1977). Response to selection for increased pupa weight in *Tribolium castaneum* as related to population structure. *Genetical Research* **30:** 237–246.

Katz, A.J., and Young, S.Y.Y. (1975). Selection for high adult body weight in *Drosophila* populations with different structures. *Genetics* **81:** 163–175.

Kaufman, P.K., Enfield, F.D., and Comstock, R.E. (1977). Stabilizing selection for pupa weight in *Tribolium castaneum*. *Genetics* **87:** 327–341.

Kauffman, S.A., and Levin, S. (1987). Towards a general theory of adaptive walks on rugged landscapes. *Journal of Theoretical Biology* **128:** 11–45.

Kawai, Y., and Otsuka, J. (2004). The deep phylogeny of land plants inferred from a full analysis of nucleotide base changes in terms of mutation and selection. *Journal of Molecular Evolution* **58:** 479–489.

Kawecki, T.J., and Ebert, D. (2004). Conceptual issues in local adaptation. *Ecology Letters* **7:** 1225–1241.

Kayasthar, B.N., and Heyne, E.G. (1978). Interaction of near-isogenic populations of wheat in infested and uninfested environments of wheat soilborne mosaic virus. *Crop Science* **18:** 840–844.

Kearsey, M.J., and Farquhar, A.G.L. (1998). QTL analysis in plants: where are we now? *Heredity* **80:** 137–142.

Keefe, A.D., and Szostak, J.W. (2001). Functional proteins from a random-sequence library. *Nature* **410:** 715–718.

Keeling, P.J., Burger, G., Durnford, D.G. *et al.* (2005). The tree of eukaryotes. *Trends in Ecology and Evolution* **20:** 670- 676.

Keightley, P.D. (1998). Inference of genome-wide mutation rates and distribution of mutation effects for fitness traits: a simulation study. *Genetics* **150:** 1283–1293.

Keightley, P.D. (2004). Mutational variation and long-term selection response. *Plant Breeding Reviews* **24:** 227–248.

Keightley, P.D., and Caballero, A. (1997). Genomic mutation rates for lifetime reproductive output and lifespan in *Caenorhabditis elegans. Proceedings of the National Academy of Sciences of the USA* **94**(8): 3823–3827.

Keightley P.D., and Lynch M. (2003). Toward a realistic model of mutations affecting fitness. *Evolution* **57:** 683–685.

Kelley, J., Walter, L., and Trowsdale, J. (2005). Comparative genomics of major histocompatibility complexes. *Immunogenetics* **56:** 683–695.

Kennedy, B.M., Thompson, B.W., and Luecke, C. (2006). Ecological differences between two closely related morphologically similar benthic whitefish (*Prosopium spilonotus* and *Prosopium abyssicola*) in an endemic whitefish complex. *Canadian Journal of Fisheries and Aquatic Science* **63:** 1700–1709.

Kerr, B., Riley, M.A., Feldman, M.W., and Bohannan, B.J.M. (2002). Local dispersal promotes biodiversity in a real-life game of rock-paper-scissors. *Nature* **418:** 171–174.

Kessler, S. (1966). Selection for and against ethological isolation between *Drosophila pseudoobscura* and *Drosophila persimilis. Evolution* **20:** 634–645.

Kessler, S. (1969). The genetics of *Drosophila* mating behaviour. II. The genetic architecture of mating speed in *Drosophila pseudoobscura. Genetics* **62:** 421–433.

Kettlewell, H.B.D. (1955). Selection experiments on industrial melanism in the Lepidoptera. *Heredity* **9:** 323–342.

Kettlewell, H.B.D. (1973). *The evolution of melanism: the study of a recurring necessity.* Oxford University Press, Oxford.

Khalifa, M.A., and Qualset, C.O. (1974). Intergenotypic competition between tall and dwarf wheats. I. In mechanical mixtures. *Crop Science* **14:** 795–799.

Kibota, T.T., and Lynch, M. (1996). Estimate of the genomic mutation rate deleterious to overall fitness in *E. coli. Nature* **381:** 694–696.

Kidwell, J.F., Clegg, M.T., Stewart, F.M., and Prout, T. (1977). Regions of stable equilibria for models of differential selection in the two sexes under random mating. *Genetics* **85:** 171–183.

Kidwell, M.G. (1972a). Genetic change of recombination value in *Drosophila melanogaster.* I. Artificial selection for high and low recombination and some properties of recombination-modifying genes. *Genetics* **70:** 419–432.

Kidwell, M.G. (1972b). Genetic change of recombination value in *Drosophila melanogaster.* II. Simulated natural selection. *Genetics* **70:** 433–443.

Kilias, G., Alahiotis, S.N., and Pelecanos, M. (1980). A multifactorial genetic investigation of speciation theory using *Drosophila melanogaster. Evolution* **34:** 730–737.

Kimura, M. (1957). Some problems of stochastic processes in genetics. *Annals of Mathematical Statistics* **28:** 882–901.

Kimura, M. (1983). *The neutral theory of molecular evolution.* Cambridge University Press, Cambridge.

Kingman, J.F.C. (1982). The coalescent. *Stochastic Process and their Applications* **13:** 235–248.

Kingsolver, J.G., Hoekstra, H.E., Hoekstra, J.M. *et al.* (2001). The strength of phenotypic selection in natural populations. *American Naturalist* **157:** 245–261.

Kinney, T.B., Bohren, B.B., Craig, J.V., and Lowe, P.C. (1970). Responses to individual, family or index selection for short term rate of egg production in chickens. *Poultry Science* **49:** 1052–1064.

Kinnison, M.T., and Hendry, A.P. (2001). The pace of modern life II. From rates of contemporary microevolution to pattern and process. *Genetica* **112/113:** 145–164.

Kirk, D.L. (1998). *Volvox.* Cambridge University Press, Cambridge.

Kirk, D.L. (2003). Seeking the ultimate and proximate causes of *Volvox* multicellularity and cellular differentiation. *Integrative and Comparative Biology* **43:** 247–253.

Kirk, K.M., Blomberg, S.P., Duffy, D.L., Heath, A.C., Owens, I.P.F., and Martin, N.G. (2001). Natural selection and quantitative genetics of life-history traits in Western women: a twin study. *Evolution* **55:** 423–435.

Kirkpatrick, M. (1982). Sexual selection and the evolution of female choice. *Evolution* **36:** 1–12.

Kirkpatrick, M., and Jenkins, C.D. (1989). Genetic segregation and the maintenance of sexual reproduction. *Nature* **339:** 300–301.

Kirkpatrick, M., and Ravigné, V. (2002). Speciation by natural and sexual selection: models and experiments. *American Naturalist* **159:** S22–S35.

Kirkwood, T.B.L. (1977). Evolution of aging. *Nature* **270:** 301–304.

Kittelson, P.M., and Maron, J.L. (2001). Fine-scale genetically based variation of life history traits in the perennial shrub *Lupinus arboreus. Evolution* **55:** 2429–2438.

Knight, T.F., and Miller, T.E. (2004). Local adaptation within a population of *Hydrocotyle bonariensis. Evolutionary Ecology Research* **6:** 103–114

Knop, M. (2006). Evolution of the hemiascomycete yeasts: on life styles and the importance of inbreeding. *Bioessays* **28:** 696–708.

Kojima, K. (1971). Is there a constant fitness value for a given phenotype? No! *Evolution* **25:** 281–285.

Kojima, K., and Kelleher, T.H. (1963). A comparison of purebred and crossbred selection schemes with two populations of *Drosophila pseudo-obscura*. *Genetics* **48:** 57–72.

Kojima, K., and Tobari, Y.N. (1969). The pattern of viability changes associated with genotype frequency at the alcohol dehydrogenase locus in a population of *Drosophila melanogaster*. *Genetics* **61:** 201–209.

Kojima, K., and Yarborough, K.M. (1967). Frequency-dependent selection at the Esterase-6 locus in *Drosophila melanogaster*. *Proceedings of the National Academy of Sciences of the USA* **57:** 645–649.

Kondrashov, A.S. (1988). Deleterious mutations and the evolution of sexual reproduction. *Nature* **336:** 435–440.

Kondrashov, A.S. (1995). Contamination of the genome by very slightly deleterious mutations: why have we not died 100 times over? *Journal of Theoretical Biology* **175:** 583–594.

Kondrashov, A.S., and Crow, J.F. (1993). A molecular approach to estimating the human deleterious mutation rate. *Human Mutation* **2:** 229–235.

Kondrashov, A.S., and Houle, D. (1994). Genotype-environment interactions and the estimation of the genomic mutation rate in *Drosophila melanogaster*. *Proceedings of the Royal Society of London Series B* **258:** 221–227.

Korol, A.B., and Iliadi, K.G. (1994). Increased recombination frequencies resulting from directional selection for geotaxis in *Drosophila*. *Heredity* **72:** 64–68.

Korona, R. (1996a). Genetic divergence and fitness convergence under uniform selection in experimental populations of bacteria. *Genetics* **143:** 637–644.

Korona, R. (1996b). Adaptation to structurally different environments. *Proceedings of the Royal Society of London Series B* **263:** 1665–1669.

Korona, R. (1999). Genetic load of the yeast *Saccharomyces cerevisiae* under diverse environmental conditions. *Evolution* **53**(6): 1966–1971.

Korona, R., and Levin, B.R. (1993). Phage-mediated selection and the evolution and maintenance of restriction-modification. *Evolution* **47:** 556–575.

Korona, R., Nakatsu, C.H., Forney, L.J., and Lenski, R.E. (1994). Evidence for multiple adaptive peaks from populations of bacteria evolving in a structure habitat. *Proceedings of the National Academy of Sciences of the USA* **91:** 9037–9041.

Kosaka, T. (1991). Life cycle of *Paramecium bursaria* syngen I in nature. *Journal of Protozoology* **38:** 140–148.

Koscielny-Bunde, E., Bunde, A., Havlin, S., Roman, H.E., Goldreich, Y., and Schellnhuber, H.-J. (1998). Indication of a universal persistence law governing atmospheric variability. *Physical Review Letters* **81:** 729–732.

Koufopanou, V., and Bell, G. (1993). Soma and germ: an experimental approach using *Volvox*. *Proceedings of the Royal Society of London Series B* **254:** 107–113.

Koufopanou, V., Hughes, J., Bell, G., and Burt, A. (2006). The spatial scale of differentiation in a model organism: the wild yeast *Saccharomyces paradoxus*. *Proceedings of the Royal Society of London Series B* **361:** 1941–1946.

Kownacki, M. (1979). Effect of reciprocal crossing of selected lines of mice. *Theoretical and Applied Genetics* **54:** 169–175.

Kozlowski, J., and Stearns, S.C. (1989). Hypotheses for the production of excess zygotes: models of bet-hedging and selective abortion. *Evolution* **43:** 1369–1377.

Kramer, F.R., Mills, D.R., Cole, P.E., Nishihara, T., and Spiegelman, S. (1974). Evolution in vitro: Sequence and phenotype of a mutant RNA resistant to ethidium bromide. *Journal of Molecular Biology* **89:** 719–736.

Krause, A.E., Frank, K.A., Mason, D.M., Ulanowicz, R.E., and Taylor, W.W. (2003). Compartments revealed in food-web structure. *Nature* **426:** 282–285.

Kreitman, M. (2000). Methods to detect selection in populations with applications to the human. *Annual Review of Genomics and Human Genetics* **1:** 539–559.

Kruckeberg, A.R. (1954). The ecology of serpentine soils. III. Plant species in relation to serpentine soils. *Ecology* **35:** 267–274.

Kucera, M., and Darling, K.F. (2002). Cryptic species of oceanic foraminifera: their effect on palaeoceanographic reconstructions. *Philosophical Transactions of the Royal Society* **360:** 695–718.

Kvist, L., Ruokonen, M., Lumme, J., and Orell, M. (1999). The colonization history and present-day population structure of the European Great Tit (*Parus major major*). *Heredity* **82:** 495–502.

Lack, D. (1966). *Population studies of birds*. Clarendon Press, Oxford.

Lambio, A.L. (1981). Response to divergent selection for 4-week body weight, egg production and total plasma phosphorus in Japanese quail. *Dissertation Abstracts International* B **42:** (2694). (Not seen.).

LaMunyon, C.W., and Ward, S. (1998). Larger sperm outcompete smaller sperm in the nematode *Caenorhabditis elegans*. *Proceedings of the Royal Society of London Series B* **265:** 1997–2002.

LaMunyon, C.W., and Ward, S. (2002). Evolution of larger sperm in response to experimentally increased sperm competition in *Caenorhabditis elegans*. *Proceedings of the Royal Society of London Series B* **269:** 1125–1128.

Land, R.B., and Falconer, D.S. (1968). Genetic studies of ovulation rate in the mouse. *Genetical Research* **13:** 25–46.

Lande, R. (1979). Quantitative genetic analysis of multivariate evolution, applied to brain:body size allometry. *Evolution* **33:** 402–416.

Lande, R. (1981). Models of speciation by sexual selection on polygenic traits. *Proceedings of the National Academy of Sciences of the USA* **78:** 3721–3725.

Lande, R. (1985). Expected time for random genetic drift of a population between stable phenotypic states. *Proceedings of the National Academy of Sciences of the USA* **82:** 7641–7645.

Lande, R., and Arnold, S.J. (1983). The measurement of selection on correlated characters. *Evolution* **37:** 1210–1226.

Lande, R., and Schemske, D. (1984). The evolution of self-fertilization and inbreeding depression in plants. I. Genetic models. *Evolution* **39:** 24–40.

Lande, R., and Thompson, R. (1990). Efficiency of marker-assisted selection in the improvement of quantitative traits. *Genetics* **124:** 743–756.

Lander, E.S., Linton, L.M., Birren, B. *et al.* (2001). Initial sequencing and analysis of the human genome. *Nature* **409:** 860–921.

Lang, A.L., Pendleton, J.W., and Dungan, G.H. (1956). Influence of population and nitrogen levels on yield and protein and oil contents of nine corn hybrids. *Agronomy Journal* **48:** 284–289.

Lang, A.S., and Beatty, J.T. (2000). Genetic analysis of a bacterial genetic exchange element: the gene transfer agent of *Rhodobacter capsulatus*. *Proceedings of the National Academy of Sciences of the USA* **97:** 859–864.

Langefors, Å., Lohm, J., and von Schantz, T. (2001). Allelic polymorphism in MHC Class II B in four populations of Atlantic salmon (*Salmo salar*). *Immunogenetics* **53:** 329–336.

Lashley, K.S. (1916). Results of continued selection in *Hydra*. *Journal of Experimental Zoology* **20:** 19–26.

Latter, B.D.H. (1964). Selection for a threshold character in *Drosophila*. I. An analysis of the phenotypic variance on the underlying scale. *Genetical Research* **5:** 198–210.

Latter, B.D.H., and Robertson, A. (1962). The effects of inbreeding and artificial selection on reproductive fitness. *Genetical Research* **3:** 110–138.

Lavin, P.A., and McPhail, J.D. (1986). Adaptive divergence of trophic phenotype among freshwater populations of the threespine stickleback (*Gasterosteus aculeatus*). *Canadian Journal of Fisheries and Aquatic Sciences* **43:** 2455–2463.

Lawrence, M.J. (2000). Population genetics of homomorphic self-incompatibility polymorphisms in flowering plants. *Annals of Botany* **85**(Suppl A): 221–226.

Lawton, J. H. (1989) What is the relationship between population density and body size in animals? Oikos **55:** 429–434.

Lazzaro, B.P., and Clark, A.G. (2003). Molecular population genetics of inducible antibacterial peptide genes in *Drosophila melanogaster*. *Molecular Biology and Evolution* **20:** 914–923.

Lefebvre, L., and Palameta, B.(1988). Mechanisms, ecology and population diffusion of socially-learned food-finding behavior in feral pigeons. In Zentall, T.R. and Galef, B.G. (eds) *Social learning*. Lawrence Erlbaum Associates, Hillsdale, NJ.

Lehman, N., and Joyce, G.F. (1993a). Evolution in vitro: analysis of a lineage of ribozymes. *Current Biology* **3:** 723–734.

Lehman, N., and Joyce, G.F. (1993b). Evolution in vitro of an RNA enzyme. *Nature* **361:** 182–185.

Lehmann, L., and Keller, L. (2006). The evolution of cooperation and altruism—a general framework and a classification of models. *Journal of Evolutionary Biology* **19:** 1365–1376.

Leigh, E.G. (1971) Adaptation and Diversity. Cooper, San Francisco.

Leigh, E.G. (1983). When does the good of the group override the advantage of the individual? *Proceedings of the National Academy of Sciences of the USA* **80:** 2985–2989.

Leiss, K.A. and Müller-Schärer, H. (2001). Performance of reciprocally sown populations of I from ruderal and agricultural habitats. *Oecologia* **128:** 210–216.

Lengeler, K.B., Fox, D.S., Fraser, J.A. *et al.* (2002). Mating-type locus of *Cryptococcus neoformans*: a step in the evolution of sex chromosomes. *Eukaryotic Cell* **1:** 704–718.

Lenormand, T., and Otto, S.P. (2000). The evolution of recombination in a heterogeneous environment. *Genetics* **156:** 423–438.

Lenski, R.E. (1988a). Experimental studies of pleiotropy and epistasis in *Escherichia coli*. I. Variation in competitive fitness among mutants resistant to virus T4. *Evolution* **42:** 425–432.

Lenski, R.E. (1988b). Experimental studies of pleiotropy and epistasis in *Escherichia coli*. II. Compensation for maladaptive effects associated with resistance to virus T4. *Evolution* **42:** 433–440.

Lenski, R.E., and Hattingh, S.E. (1986). Coexistence of two competitors on one resource and one inhibitor: a chemostat model based on bacteria and antibiotics. *Journal of Theoretical Biology* **122:** 83–93.

Lenski, R.E., and Travisano, M. (1994). Dynamics of adaptation and diversification: a 10,000-generation experiment with bacterial populations. *Proceedings of the National Academy of Sciences of the USA* **91:** 6808–6814.

Lenski, R.E., Rose, M.R., Simpson, S.C., and Tadler, S.C. (1991). Long-term experimental evolution in *Escherichi coli*. 1. Adaptation and divergence during 2,000 generations. *American Naturalist* **138:**1315–1341.

Lenski, R.E., Winkworth, C.L., and Riley, M.A. (2003). Rates of DNA sequence evolution in experimental populations of *Escherichia coli* during 2000 generations. *Journal of Molecular Evolution* **56:** 498–508.

Leonard, K.J. (1969). Selection in heterogeneous populations of *Puccinia graminis* f. sp. *avenae*. *Phytopathology* **59:** 1851–1857.

Lerner, I.M., and Hazel, L.N. (1947). Population genetics of a poultry flock under artificial selection. *Genetics* **32:** 325–339.

Lerner, S.A., Wu, T.T., and Lin, E.C.C. (1964). Evolution of a catabolic pathway in bacteria. *Science* **146:** 1313–1314.

Leroi, A.M., Bennett, A.F., and Lenski, R.E. (1994). Temperature-acclimation and competitive fitness – an experimental test of the beneficial acclimation assumption. *Proceedings of the National Academy of Sciences of the USA* **91:** 1917–1921.

Levene, H. (1953). Genetic equilibrium when more than one ecological niche is available. *American Naturalist* **87:** 331–333.

Levene, H., Pavlovsky, O., and Dobhansky, T. (1954). Interaction of the adaptive values in polymorphic experimental populations of *Drosophila pseudoobscura*. *Evolution* **8:** 335–349.

Levin, B.R. (1981). Periodic selection, infectious gene exchange and the genetic structure of *E. coli* populations. *Genetics* **99:** 1–23.

Levin, B.R. (1988). Frequency-dependent selection in bacterial populations. *Philosophical Transactions of the Royal Society of London* **319:** 459–472.

Levin, B.R. (1996). The evolution and maintenance of virulence in microparasites. *Emerging Infectious Diseases* **2:** 93–102.

Levin, B.R., and Kilmer, W.L. (1975). Interdemic selection and the evolution of altruism: a computer simulation study. *Evolution* **28:** 527–545.

Levin, B.R., Stewart, F.M., and Rice, V.A. (1979). The kinetics of conjugative plasmid transmission: fit of a simple mass-action model. *Plasmid* **2:** 247–260.

Levin, B.R., Lipsitch, M., Perrot, V. *et al.* (1997). The population genetics of antibiotic resistance. *Clinical and Infectious Disease* **24:** S9–16.

Levin, D.A. (1975). Pest pressure and recombination systems in plants. *American Naturalist* **109:** 437–451.

Levin, D.A., and Wilson, A.C. (1976). Rates of evolution in seed plants: net increase in diversity in chromosome numbers and species numbers through time. *Proceedings of the National Academy of Sciences of the USA* **73:** 2086–2090.

Levin, S., and Pimentel, D. (1981). Selection of intermediate rates of increase in parasite-host systems. *American Naturalist* **11:** 308–315.

Levitan, D.R. (2000). Optimal egg size in marine invertebrates: theory and phylogenetic analysis of the critical relationship between egg size and development time in echinoids. *American Naturalist* **156:** 175–192.

Lewis, D. (1942). The evolution of sex in flowering plants. *Biological Reviews* **17:** 46–67.

Lewis, W.M. (1985). Nutrient scarcity as an evolutionary cause of haploidy. *American Naturalist* **125:** 692–701.

Lewontin, R.C. (1955). The effects of population density and composition on viability in *Drosophila melanogaster*. *Evolution* **9:** 27–41.

Lewontin, R.C. (1974). *The genetic basis of evolutionary change*. Columbia University Press, New York.

L'Héritier, P., and Tessier, G. (1934). Une expérience de sélection naturelle. Courbe d'élimination du gène 'Bar' dans une population de Drosophiles en équilibre. *Comptes Rendues de la Société de Biologie de Paris* **117:** 1049–1051.

Liberman, U., and Feldman, M.W. (1986). Modifiers of mutation rate: a general reduction principle. *Theoretical Population Biology* **30:** 125–142.

Licht, T.R., Christiansen, B.B., Krogfelt, K.A., and Molin, S. (1999). Plasmid transfer in the animal intestine and other dynamic bacterial populations: the role of community structure and environment. *Microbiology* **145:** 2615–2622.

Lin, E.C.C., and Wu, T.T. (1984). Functional divergence of the L-fucose system in mutants of *Escherichia coli*. In R.P. Mortlock (ed.), *Microorganisms as model systems for studying evolution*, pp. 135–164. Plenum, New York.

Lin, E.C.C., Hacking, A.J., and Aguilar, J. (1976). Experimental models of acquisitive evolution. *BioScience* **26:** 548–555.

Lindholm, A.K., and Breden, F. (2002). Sex chromosomes and sexual selection in poeciliid fishes. *American Naturalist* **160:** S214-S224.

Lindholm, A.K., Brooks, R., and Breden, F. (2004). Extreme polymorphism in a Y-linked sexually selected trait. *Heredity* **92:** 156–162.

Lindström, L., Alatalo, R.V., Lyytinen, A., and Mappes, J. (2001). Strong anti-apostatic selection against novel rare aposematic prey. *Proceedings of the National Academy of Sciences of the USA* 98: 9181–9184.

Liu, X. Jiang, N., Hughes, B., Bigras, E., Shoubridge, E, and Hekimi, S. (2005). Evolutionary conservation of the *clk-1*-dependent mechanism of longevity: loss of *mclk1* increases cellular fitness and life span in mice. *Genes and Development* **19:** 2424–2434.

Little, T.J. (2002). The evolutionary significance of parasitism: do parasite-driven genetic dynamics occur ex silico? *Journal of Evolutionary Biology* **15:** 1–9.

Liu, H., and Kubli, E. (2003). Sex-peptide is the molecular basis of the sperm effect in *Drosophila melanogaster. Proceedings of the National Academy of Sciences of the USA* **100:** 9929–9933.

Lloyd, D.G. (1979). Some reproductive factors affecting the selection of self-fertilization in plants. *American Naturalist* **113:** 67–79.

Loeuille, N., and Loreau, M. (2005). Evolutionary emergence of size-structured food webs. *Proceedings of the National Academy of Sciences of the USA* **102:** 5761–5766.

Loewe, L., Textor, V., and Scherer, S. (2003). High deleterious mutation rate in stationary phase of *Escherichia coli. Science* **302:** 1558–1560.

Lohe, A.R., Moriyama, E.N., Lidholm, D.A., and Hartl, D.L. (1995). Horizontal transmission, vertical inactivation, and stochastic loss of mariner-like transposable elements. *Molecular Biology and Evolution* **12:** 62–72.

Lohm, J., Grahn, M., Langefors, Å., Andersen, O., Storset, A., and von Schantz, T. (2002). Experimental evidence for major histocompatibility complex-allele-specific resistance to a bacterial infection. *Proceedings of the Royal Society of London Series B* **269:** 2029–2033.

Long, T.A.F., Montgomerie, R., and Chippindale, A.K. (2006). Quantifying the gender load: can population crosses reveal interlocus sexual conflict? *Proceedings of the Royal Society of London Series B* **361:** 363–374.

Longhurst, A.R. (1955). Evolution in the Notostraca. *Evolution* **9:** 84–86.

López, M.A., and López-Fanjul, C. (1993). Spontaneous mutation for a quantitative trait in *Drosophila melanogaster.* II. Distribution of mutant effects on the trait and fitness. *Genetical Research, Cambridge* **61:** 117–126.

Lopez-Reynoso, J.J., and Hallauer, A.R. (1998). Twenty-seven cycles of divergent mass selection for ear length in maize. *Crop Science* **38:** 1099–1107.

Loreau, M., and Hector, A. (2001) Partitioning selection and complementarity in biodiversity experiments. *Nature* **412:** 72–76 [erratum published in *Nature* 2001; **413:** 548].

Losos, J.B., Warheitt, K.I., and Schoener, T.W. (1997). Adaptive differentiation following experimental island colonization in *Anolis* lizards. *Nature* **387:** 70–73.

Losos, J.B., Jackman, T.R., Larson, A., de Queiroz, K., and Schettino, L.R. (1998). Contingency and determinism in replicated adaptive radiations of island lizards. *Science* **279:** 2115–2118.

Losos, J.B., Schoener, T.W., Warheitt, W.I., and Creer, D. (2001). Experimental studies of adaptive radiation in Bahamian *Anolis* lizards. *Genetica* **112/113:** 399–415.

Lotz, L.A.P. (1990). The relation between age and size at first flowering *of Plantago major* in various habitats. *Journal of Ecology* **78:** 757–771.

Louette, G., and de Meester, L. (2003). High dispersal capacity of cladoceran zooplankton in newly founded communities. *Ecology* **86:** 353–359.

Lovelock, J. (1979). *Gaia.* Oxford University Press, Oxford.

Lovelock, J. (2003). The living Earth. *Nature* **426:** 769–770.

Lovett Doust, L. (1981). Population dynamics and local specialization in a clonal perennial (*Ranunculus repens*). II. The dynamics of leaves, and a reciprocal transplant-replant experiment. *Journal of Ecology* **69:** 757–768.

Lu, G., and Bernatchez, L. (1998). Experimental evidence for reduced hybrid viability between dwarf and normal ecotypes of lake whitefish (*Coregonus clupeaformis* Mitchill). *Proceedings of the Royal Society of London Series B* **265:** 1025–1030.

Lu, G., and Bernatchez, L. (1999). Correlated trophic specialization and genetic divergence in sympatric lake whitefish ecotypes (*Coregonus clupeaformis*): support for the ecological speciation hypothesis. *Evolution* **53:** 1491–1505.

Luckinbill, L.S. (1984). An experimental analysis of a life-history theory. *Ecology* **65:** 1170–1184.

Luckinbill, L.S, Arking R., and Clare, M.J. (1984). Selection for delayed senescence in *Drosophila melanogaster. Evolution* **38:** 996-(1003).

Luckinbill, L.S., Clare, M.J., Krell, W.L., Cirocco, W.C., and Richards, P.A. (1987). Estimating the number of genetic elements that defer senescence in *Drosophila. Evolutionary Ecology* **1:** 37–46.

Luckinbill, L.S., Graves, J.L., Reed, A.H., and Koetsawang, S. (1988). Localizing genes that defer senescence in *Drosophila melanogaster. Heredity* **60:** 367–374.

Lunzer, M., Natarajan, A., Dykhuizen, D.E., and Dean, A.M. (2002). Enzyme kinetics, substitutable resources and competition: from biochemistry to frequency-dependent selection in *lac. Genetics* **162:** 485–499.

Lush, J.L. (1947a). Family merit and individual merit as bases for selection. Part I. *American Naturalist* **81:** 241–261.

Lush, J.L. (1947b). Family merit and individual merit as bases for selection. Part II. *American Naturalist* **81:** 362–379.

Lynch, C.B. (1980). Response to divergent selection for nesting behavior in *Mus musculus*. *Genetics* **96:** 757–765.

Lynch, M. (1987). The consequences of fluctuating selection for isozyme polymorphisms in *Daphnia*. *Genetics* **115:** 657–669.

Lynch, M., Latta, L., Hicks, J., and Giorgianni, M. (1998). Mutation, selection and the maintenance of life-history variation in a natural population. *Evolution* **52:** 727–733.

Lyttle, T.W. (1977). Experimental population genetics of meiotic drive systems. I. Pseudo-Y chromosomal drive as a means of eliminating cage populations of *Drosophila melanogaster*. *Genetics* **86:** 413–445.

Lyttle, T.W. (1979). Experimental population genetics of meiotic drive systems. II. Accumulation of genetic modifiers of Segregation Distorter (SD) in laboratory populations. *Genetics* **91:** 339–357.

Mable, B.K. (2001). Ploidy evolution in the yeast *Saccharomyces cerevisiae*: a test of the nutrient limitation hypothesis. *Journal of Evolutionary Biology* **14:** 157–170.

Mable, B.K., and Otto, S.P. (2001). Masking and purging mutations following EMS treatment in haploid, diploid and tetraploid yeast (*Saccharomyces cerevisiae*). *Genetical Research* **77:** 9–26.

MacAlpine, D.M., Kolesar, J., Okamoto, K., Butow, R.A., and Perlman, P.S. (2001). Replication and preferential inheritance of hypersuppressive petite mitochondrial DNA. *EMBO Journal* **20:** 1807–1817.

MacArthur, J.W. (1949). Selection for small and large body size in the house mouse. *Genetics* **34:** 194–209.

Macarthur, R.H. (1969). Patterns of communities in the tropics. *Biological Journal of the Linnean Society* **1:** 19–30.

Macarthur, R.H., and Levins, R. (1964). Competition, habitat selection and character displacement in a patchy environment. *Proceedings of the National Academy of Sciences of the USA* **51:** 1207–1210.

MacDowell, E.C. (1919). Bristle inheritance in *Drosophila*. II. Selection. *Journal of Experimental Zoology* **23:** 109–146.

Mackay, T.F.C. (1985). Transposable element-induced response to artificial selection in *Drosophila melanogaster*. *Genetics* **111:** 351–374.

Mackay, T.F.C. (1996). The nature of quantitative genetic variation revisited: lessons from *Drosophila* bristles. *BioEssays* **18:** 113–121.

Mackay, T.F.C. (2001). The genetic architecture of quantitative traits. *Annual Review of Genetics* **35:** 303–339.

Mackay, T.F.C., and Lyman, R.F. (2005). *Drosophila* bristles and the nature of quantitative genetic variation. *Proceedings of the Royal Society of London Series B* **360:** 1513–1527.

Mackay, T.F.C., Fry, J.D., Lyman, R., and Nuzhdin, S.V. (1994). Polygenic mutation in *Drosophila melanogaster*: estimates from response to selection of inbred strains. *Genetics* **136:** 937–951.

Mackinnon, M.J., and Read, A.F. (1999a). Genetic relationships between parasite virulence and transmission in the rodent malaria *Plasmodium chabaudi*. *Evolution* **53:** 689–703.

Mackinnon, M.J., and Read, A.F. (1999b). Selection for high and low virulence in the malaria parasite *Plasmodium chabaudi*. *Proceedings of the Royal Society of London Series B* **266:** 741–748.

MacLean, R.C., and Bell, G. (2002). Experimental adaptive radiation in *Pseudomonas*. *American Naturalist* **160:** 569–581.

MacLean, R.C., and Bell, G. (2003). Divergent evolution during an experimental adaptive radiation. *Proceedings of the Royal Society of London Series B* **270:** 1645–1650.

MacLean, R.C., and Bell, G. (2004). The evolution of a phenotypic fitness trade-off in *Pseudomonas fluorescens*. *Proceedings of the National Academy of Sciences of the USA* **101:** 8072–8077.

MacLean, R.C., Dickson, A., and Bell, G. (2005). Resource competition and adaptive radiation in a microbial microcosm. *Ecology Letters* **8:** 38–46.

Macnair, M.R. (1987). Heavy metal tolerance in plants: a model evolutionary system. *Trends in Ecology and Evolution* **2:** 354–359.

Macnair, M.R. (1991). Why the evolution of resistance to anthropogenic toxins normally involves major gene changes: the limits to natural selection. *Genetica* **84:** 213–219.

Madalena, F.E., and Robertson, A. (1975). Population structure in artificial selection: studies with *Drosophila melanogaster*. *Genetical Research* **24:** 113–126.

Maguire, B. (1963). The passive dispersal of small aquatic organisms and their colonization of isolated bodies of water. *Ecological Monographs* **33:** 161–185.

Majerus, M. (1998). *Melanism: evolution in action*. Oxford University Press, Oxford.

Mallet, J., and Barton, N.H. (1989). Strong natural selection in a warning color hybrid zone. *Evolution* **43:** 421–431.

Malmberg, R.L. (1977). The evolution of epistasis and the advantage of recombination in populations of bacteriophage. *Genetics* **86:** 607–621.

Manning, A. (1961). The effects of artificial selection for mating speed in *Drosophila melanogaster*. *Animal Behaviour* **9:** 82–92.

Manning, A. (1963). Selection for mating speed in *Drosophila melanogaster* based on the behaviour of one sex. *Animal Behaviour* **11:** 116–120.

Manning, A. (1968). The effects of artificial selection for slow mating in *Drosophila* simulans. *Animal Behaviour* **16:** 108–113.

Manning, A., and Hirsch, J. (1971). The effects of artificial selection for slow mating in *Drosophila* simulans. II. Genetic analysis of the slow mating line. *Animal Behaviour* **19:** 448–453.

Mao, E.F., Lane, L., Lee, J., and Miller, J.H. (1997). Proliferation of mutators in a cell population. *Journal of Bacteriology* **179:** 417–422.

Mareck, J.H., and Gardner, C.O. (1979). Responses to mass selection in maize and stability of resulting populations. *Crop Science* **19:** 779–783.

Marien, D. (1958). Selection for developmental rate in *Drosophila pseudoobscura*. *Genetics* **50:** 3–15.

Marin, I., and Baker, B.S. (1998). The evolutionary genetics of sex determination. *Science* **281:** 1990–1994.

Markow, T.A. (1975). A genetic analysis of phototactic behavior in *Drosophila melanogaster*. I. Selection in the presence of inversions. *Genetics* **79:** 527–534.

Marks, C.O., and Lechowicz, M.J. (2006). Alternative designs and the evolution of functional diversity. *American Naturalist* **167:** 55–67

Marks, H.L. (1996). Long-term selection for body weight in Japanese quail under different environments. *Poultry Science* **75:** 1198–1203.

Martin, G.A., and Bell, A.E. (1960). An experimental check on the accuracy of prediction of response during selection. In O. Kempthorne (ed.), *Biometrical genetics*, pp. 178–187. Pergamon Press, Oxford.

Martin, G., Otto, S.P., and Lenormand, T. (2005). Selection for recombination in structured populations. *Genetics* **172:** 593–609.

Martorell, C., Toro, M.A., and Gallego, C. (1998). Spontaneous mutation for life-history traits in *Drosophila melanogaster*. *Genetica* **103:** 315–324

Mason, L.G. (1964). Stabilizing selection for mating fitness in natural populations of *Tetraopes*. *Evolution* **18:** 492–497.

Mastrobattista, E., Taly, V., Chadunet, E., Treacy, P., Kelly, B.T., and Griffiths, A.D. (2005). High-throughput screening of enzyme libraries: in vitro evolution of a β-galactosidase fluorescence-activated sorting of double emulsions. *Chemistry and Biology* **12:** 1291–1300.

Mather, K. (1941). Variation and selection of polygenic characters. *Journal of Genetics* **41:** 159–193.

Mather, K. (1983). Response to selection. In M. Ashburner, H.L. Carson, and J.N. Thompson (eds.), *The genetics and biology of* Drosophila, Vol. 3c, pp. 155–221. Academic Press, New York.

Mather, K., and Cooke, P. (1962). Differences in competitive ability between genotypes of *Drosophila*. *Heredity* **17:** 381–407.

Mather, K., and Harrison, B.J. (1949). The manifold effects of selection. *Heredity* **3:** 1–52 and 131–162.

Mather, K., and Wigan,, L.G. (1942). The selection of invisible mutants. *Proceedings of the Royal Society of London Series B* **131:** 50–64.

Mattheakis, L.C., Bhjatt, R.R., and Dower, W.J. (1994). An in vitro polysome display system for identifying ligands from very large peptide libraries. *Proceedings of the National Academy of Sciences of the USA* **91:** 9022–9026.

Maxwell, A.I., Morrison, G.M., and Dorin, J.R. (2003). Rapid sequence divergence in mammalian beta-defensins by adaptive evolution. *Molecular Immunology* **40:** 413–421.

May, H.G. (1917). Selection for higher and lower facet numbers in the bar-eyed race of *Drosophila* and the appearance of reverse mutations. *Biological Bulletin* **33:** 361–395.

Maynard Smith, J. (1956). Fertility, mating behaviour and sexual selection in *Drosophila subobscura*. *Journal of Genetics* **54:** 261–279.

Maynard Smith, J. (1962). Disruptive selection, polymorphism and sympatric speciation. *Nature* **195:** 60–62.

Maynard Smith, J. (1966). Sympatric speciation. *American Naturalist* **100:** 637–650.

Maynard Smith, J. (1970). Time in the evolutionary process. *Studium Generale* **23:** 266–272.

Maynard Smith, J. (1976). Group selection. *Quarterly Review of Biology* **51:** 277–283.

Maynard Smith, J. (1978). *The evolution of sex*. Cambridge University Press, Cambridge.

Maynard Smith, J. (1982). *Evolution and the theory of games*. Cambridge University Press, Cambridge.

Maynard Smith, J. (1991). The population genetics of bacteria. *Proceedings of the Royal Society of London, Series B* **245:** 37–41.

Maynard Smith, J., and Haigh, J. (1974). The hitch-hiking effect of a favourable gene. *Genetical Research* **23:** 23–35.

Maynard Smith, J., and Hoekstra, R. (1980). Polymorphism in a varied environment: how robust are the models? *Genetical Research* **35:** 45–57.

Maynard Smith, J., and Price, G.R. (1973). The logic of animal conflict. *Nature* **246:** 15–18.

Mayr, E. (1940). Speciation phenomena in birds. *American Naturalist* **74:** 249–278.

Mayr, E. (1963). *Animal species and evolution*. Oxford University Press, Oxford.

Mazel, D. (2006). Integrons: agents of bacterial evolution. *Nature Reviews Microbiology* **4:** 608–620.

McCann, K. (2000). The diversity-stability debate. *Nature* **405:** 228–233.

McCarthy, M.A., and Parris, K.M. (2004). Clarifying the effect of toe clipping on frogs with Bayesian statistics. Journal of *Applied Ecology* **41:**780–786.

McCaskill, J.S., and Bauer, G.J. (1993). Images of evolution: the origin of spontaneous RNA replication waves. *Proceedings of the National Academy of Sciences of the USA* **90:** 4191–4195.

McCauley, D.E., and Wade, M.J. (1980). Group selection: the genetic and demographic basis for the phenotypic differentiation of small populations of *Tribolium castaneum*. *Evolution* **34:** 813–821.

McCleery, R.H., Pettifor, R.A., Armbruster, P., Meyer, K. Sheldon, B.C., and Perrins, C.M. (2004). Components of variance underlying fitness in a natural population of the great tit *Parus major*. *American Naturalist* **164:** E62–E72.

McDonald, J.H., and Kreitman, M. (1991). Adaptive evolution at the *Adh* locus in *Drosophila*. *Nature* 351: 652–654.

McGill, A., and Mather, K. (1972). Competition in *Drosophila*. I. A case of stabilizing selection. *Heredity* **27:** 473–478.

McGraw, J.B. (1987). Experimental ecology of *Dryas octopetala* ecotypes. *Oecologia* **73:** 465–468.

McKinnon, J.S., and Rundle, H.D. (2002). Speciation in nature: the threespine stickleback model system *Trends in Ecology and Evolution* **17:** 480–488.

McLennan, D.A. (1996). Integrating phylogenetic and experimental analyses: the evolution of male and female nuptial coloration in the stickleback fishes (Gasterosteidae). *Systematic Biology* **45:** 261–277.

McNeal, F.H., Qualset, C.O., Baldridge, D.E., and Stewart, V.R. (1978). Selection for yield and yield components in wheat. *Crop Science* **18:** 795–801.

McNeilly, T. (1968). Evolution in closely adjacent plant populations. III. *Agrostis tenuis* on a small copper mine. *Heredity* **23:** 99–108.

McNeilly, T., and Bradshaw, A.D. (1968). Evolutionary processes in populations of copper tolerant *Agrostis tenuis* Sibth. *Evolution* **22:** 108–118.

McPhail, J.D. (1993). Speciation, the evolution of reproductive isolation in the sticklebacks (*Gasterosteus*) of southwestern British Columbia. In Bell, M.A. and Foster, S.A. (eds), *Evolutionary biology of the threespine stickleback*, pp. 399–437. Oxford University Press, Oxford.

McPhee, C.P., and Robertson, A. (1970). The effect of suppressing crossing-over on the response to selection in *Drosophila melanogaster*. *Genetical Research* **16:** 1–16.

Mead, L.S., and Arnold, S.J. (2004). Quantitative genetic models of sexual selection. *Trends in Ecology and Evolution* **19:** 264–271.

Mead-Briggs, A.R., and Vaughan, J.A. (1975). The differential transmissibility of Myxoma virus strains of differing virulence grades by the rabbit flea *Spilopsyllus cuniculi* (Dale). *Journal of Hygiene* **75:** 237–247.

Medawar, P.B. (1952). *An unsolved problem in biology*. H.K. Lewis, London.

Memmott, J., Fowler, S.V., Paynter, Q., Sheppard, A.P., and Syrett, P. (2000). The invertebrate fauna on broom, *Cytisus scoparius*, in two native and two exotic habitats. *Acta Oecologica* **21:** 213–222.

Mendiola, N.B. (1919). Variation and selection within clonal lines of *Lemna minor*. *Genetics* **4:** 151–182.

Merilä, J., and Sheldon, B.C. (2000). Lifetime reproductive success and heritability in nature. *American Naturalist* **155:** 301–310.

Mertz, D.R. (1975). Senescent decline in flour beetle strains selected for early adult fitness. *Physiological Zoology* **48:** 1–23.

Mesoudi, A., Whiten, A., and Laland, K.N. (2004). Is human cultural evolution Darwinian? Evidence reviewed from the perspective of the *Origin of Species*. *Evolution* **58:** 1–11.

Messenger, S.L., Molineux, I.J., and Bull, J.J. (1999). Virulence evolution in a virus obeys a trade-off. *Proceedings of the Royal Society of London Series B*, **266:** 397–404.

Metz, E.C., Robles-Sikisaka, R., and Vacquier, V.D. (1998). Nonsynonymous substitution in abalone sperm fertilization genes exceeds substitution in introns and mitochondrial DNA. *Proceedings of the National Academy of Sciences of the USA* **95:** 10676–10681.

Michael, C.A., Gillings, M.R., Holmes, A.J. *et al.* (2004). Mobile gene cassettes: a fundamental resource for bacterial evolution. *American Naturalist* **164:** 1–12.

Middleton, A.R. (1915). Heritable variations and the results of selection in the fission rate of *Stylonichia pustulata*. *Journal of Experimental Zoology* **19:** 451–503.

Miller, J.H. (1996). Spontaneous mutators in bacteria: insights into pathways of mutagenesis and repair. *Annual Review of Microbiology* **50:** 625–643.

Miller, R.E., and Fowler, N.L. (1993). Variation in reaction norms among populations of the grass *Bouteloa rigidiseta*. *Evolution* **47:** 1446–1455.

Mills, D.R., Peterson, R.L., and Spiegelman, S. (1967). An extracellular Darwinian experiment with a self-duplicating nucleic acid molecule. *Proceedings of the National Academy of Sciences of the USA* **58:** 217–224.

Miralles, R., Moya, A., and Elena, S.F. (1997). Is group selection a factor modulating the virulence of RNA viruses? *Genetical Research* **69:** 165–172.

Miralles, R., Gerrish, P.J., Moya, A., and Elena, S.F. (1999). Clonal interference and the evolution of RNA viruses. *Science* **285:**1745–1747.

Miralles, R., Moya, A., and Elena, S.F. (2000). Diminishing returns of population size in the rate of RNA virus adaptation. *Journal of Virology* **74:**3566–3571.

Modi, R.I., and Adams, J. (1991). Coevolution in bacterial-plasmid populations. *Evolution* **45:** 656–667.

Modi, R.I., Castilla, L.H., Puskas-Rozsa, S., Helling, R.B., and Adams, J. (1992). Genetic changes accompanying increased fitness in evolving populations of *Escherichia coli*. *Genetics* **130:** 241–249.

Montoya, J.M., and Solé, R.V. (2003). Topological properties of food webs: from real data to community assembly models. *Oikos* **102:** 614–622.

Moodie, G.E.E. (1982). Why asymmetric mating preferences may not show the direction of evolution. *Evolution* **36:** 1096–1097.

Moore, F.B., Rozen, D.E., and Lenski, R.E. (2000) Pervasive compensatory adaptation in *Escherichia coli*. *Proceedings of the Royal Society of London, Series B* **267:** 515–522.

Moose, S.P., Dudley, J.W., and Rocheford, T.R. (2004). Maize selection passes the century mark: a unique resource for 21st century genomics. *Trends in Plant Sciences* **9:** 358–364.

Moran, P.A.P. (1962). *The statistical processes of evolutionary theory*. Clarendon Press, Oxford.

Moreira, D., and Lopez-Garcia, P. (2002). The molecular ecology of microbial eukaryotes unveils a hidden world. *Trends in Microbiology* **10:** 31–38.

Morgan, P. (1976). Frequency-dependent selection at two enzyme loci in *Drosophila melanogaster*. *Nature* **263:** 765–767.

Moriwaki, D., and Fuyama, Y. (1963). Responses to selection for rate of development in *Drosophila melanogaster*. *Drosophila Information Service* **38:** 74.

Morris, J.A. (1963). Continuous selection for egg production using short-term records. Australian *Journal of Agricultural Research* **14:** 909–925.

Mortlock, R.P. (1984a). The utilization of pentitols in studies of the evolution of enzyme pathways. In. R.P. Mortlock (ed.), *Microorganisms as model systems for studying evolution*, pp. 1–22. Plenum, New York.

Mortlock, R.P. (ed.) (1984b). *Microorganisms as model systems for studying evolution*. Plenum, New York.

Mortlock, R.P., and Wood, W.A. (1964). Metabolism of pentoses and pentitols by *Aerobacter aerogenes*. I. Demonstration of pentose isomerase, pentulokinase, and pentitol dehydrogenase enzyme families. *Journal of Bacteriology* **88:** 835–844.

Mortlock, R.P., Fossitt, D.D., Petering, D.H., and Wood, W.A. (1965). A basis for utilization of unnatural pentoses and pentitols by *Aerobacter aerogenes*. *Proceedings of the National Academy of Sciences of the USA* **54:** 572–579.

Morton, R.A, and Hall, S.C. (1985). Response of dysgenic and non-dysgenic populations to mutation exposure. *Drosophila Information Service* **61:** 126–128.

Mousseau, T.A., and Roff, D.A. (1987). Natural selection and the heritability of fitness components. *Heredity* **59:** 181–197.

Moxon, R., Rainey, P.B., Nowak, M.A., and Lenski, R.E. (1994). Adaptive evolution of highly mutable loci in pathogenic bacteria. *Current Biology* **4:** 24–33.

Moxon, R., Bayliss, C., and Hood, D. (2006). Bacterial contingency loci: the role of simple sequence DNA repeats in bacterial adaptation. *Annual Review of Genetics* **40:** 307–333.

Mueller, L.D. (1987). Evolution of accelerated senescence in laboratory populations of *Drosophila*. *Proceedings of the National Academy of Sciences of the USA* **84:** 1974–1977.

Mueller, L.D. (1991). Ecological determinants of life-history evolution. *Philosophical Transactions of the Royal Society of London Series B* **332:** 25–30.

Mueller, L.D., and Ayala, F.J. (1981). Trade-off between r-selection and K-selection in *Drosophila* populations. *Proceedings of the National Academy of Sciences of the USA* **78:** 1303–1305.

Mueller, L.D., and Sweet, V.F. (1986). Density-dependent natural selection in *Drosophila*: evolution of pupation height. *Evolution* **40:** 1354–1356.

Mueller, L.D., Guo, P., and Ayala, F.J. (1991). Density-dependent natural selection and trade-offs in life history traits. *Science* **253:** 433–435.

Muir, W.M., and Craig, J.V. (1998). Improving animal well-being through genetic selection. *Poultry Science* **77:** 1781–1788.

Muir, W.M., Miles, D., and Bell, A.E. (2004). Long-term selection for pupal weight in *Tribolium castaneum*. *Plant Breeding Reviews* **24**(2): 211–224.

Mukai, T. (1964). The genetic structure of natural populations of *Drosophila melanogaster*. I. Spontaneous mutation rate of polygenes controlling viability. *Genetics* **50:** 1–19.

Mukai, T., Chigusa, S.T., Mettler, L.E., and Crow, J.F. (1972). Mutation rate and dominance of genes affecting viability in *Drosophila melanogaster*. *Genetics* **72:** 335–355.

Mukherjee, A.S. (1961). Effect of selection on crossing-over in the males of *Drosophila* ananassae. *American Naturalist* **95:** 57–59.

Mulcahy, D.L, Mulcahy, G.B., and Ottaviano, E. (1978). Further evidences that gametophytic selection modifies the genetic quality of the sporophyte. *Société Botanique de France Actualités Botaniques* **1:** 57–60.

Muller, H.J. (1932). Some genetic aspects of sex. *Nature* **66**: 118–138.

Nachman, M.W., and Crowell, S.L. (1997). Estimation of the mutation rate per nucleotide in humans. *Genetics* **156:** 297–304.

Nagel, L., and Schluter, D. (1998). Body size, natural selection, and speciation in sticklebacks. *Evolution* **52:** 209–218.

Nagylaki, T. (1976). A model for the evolution of self-fertilization and vegetative reproduction. *Journal of Theoretical Biology* **58:** 55–58.

Nanney, D.L., and McCoy, J.W. (1976). Characterization of the species of the *Tetrahymena pyriformis* complex. *Transactions of the American Microscopical Society* **95:** 664–682.

Narain, P. Joshi, C., and Prabhu, S.S. (1960). Response to selection for fecundity in *Drosophila melanogaster. Drosophila Information Service* **36:** 96–99.

Nassar, R., Muhs, H.J., and Cook, R.D. (1973). Frequency-dependent selection at the Payne inversion in *Drosophila melanogaster. Evolution* **27:** 558–564.

Nee, S. (1989). Antagonistic coevolution and the evolution of genotypic randomization. *Journal of Theoretical Biology* **140:** 499–518.

Nei, M. (2005). Selectionism and neutralism in molecular evolution. *Molecular Biology and Evolution* **22:** 2318–2342.

Nei, M., and Graur, D. (1984). Extent of protein polymorphism and the neutral mutation theory. *Evolutionary Biology* **17:** 73–118.

Nei, M., and Rooney, A.P. (2005). Concerted and birth-and-death evolution of multigene families. *Annual Review of* Genetics **39:** 121–152.

Neigel, J.E. (2002). Is F_{ST} obsolete? *Conservation Genetics* **3:** 167–173.

Nestmann, E.R., and Hill, R.F. (1973). Population changes in continuously growing mutator cultures of *Escherichia coli. Genetics* (Suppl.) **73:** 41–44.

Nestor, K.E., Noble, D.O., Zhu, J., and Moritsu, Y. (1996). Direct and correlated responses to long-term selection for increased body weight and egg production in turkeys. *Poultry Science* **75:**1180–1191.

Neuhauser, C., and Krone, S.M. (1997). The genealogy of samples in models with selection. *Genetics* **145:** 519–534.

Nevo, E. *et al.* (1984). The evolutionary significance of genetic diversity ecological, demographic and life history correlates, in *Evolutionary dynamics of genetic diversity*, pp. 13–213. Lecture Notes in Biomathematics 53, ed. G.S. Mani. Springer-Verlag, Heidelberg.

Newman, E.I., and Rovira, A.D. (1975). Allelopathy among some British grassland species. *Journal of Ecology* **63:** 727–737.

Nicolas, P., Vanhoye, D., and Amiche, M. (2003). Molecular strategies in biological evolution of antimicrobial peptides. *Peptides* **24:** 1669–1680.

Nielsen, R. (2001). Statistical tests of neutrality in the age of genomics. *Heredity* **86:** 641–647.

Notley-McRobb, L., and Ferenci, T. (1999a). Adaptive *mgl*-regulatory mutations and genetic diversity evolving in glucose-limited *Escherichia coli* populations. *Environmental Microbiology* **1**(1):33–43.

Notley-McRobb, L., and Ferenci, T. (1999b). The generation of multiple co-existing *mal*-regulatory mutations through polygenic evolution in glucose-limited populations of *Escherichia coli. Environmental Microbiology* **1**(1):45–52.

Notley-McRobb, L., and Ferenci, T. (2000). Experimental analysis of molecular events during mutational periodic selections in bacterial evolution. *Genetics* **156:** 1493–1501.

Novick, A. (1958). Genetic and physiological studies with the chemostat. In: Málek, A (ed.) *Continuous cultivation of microorganisms*. Czechoslovak Academy of Sciences, Prague.

Novick, A., and Szilard, L. (1951). Genetic mechanisms in bacteria and bacterial viruses. I. Experiments on spontaneous and chemically induced mutations of bacteria growing in the chemostat. *Cold Spring Harbor Symposia on Quantitative Biology* **16:** 337–343.

Nowak, M.A., and May, R.M. (1994). Superinfection and the evolution of parasite virulence. *Proceedings of the Royal Society of London Series B* **255:** 81–89.

Nowak, M., and Sigmund, K. (1992). Tit for tat in heterogeneous populations. *Nature* **355:** 250–252.

Nowak, M.A., and Sigmund, K. (2004). The evolutionary dynamics of biological games. *Science* **303:** 793–799.

Nowell, P.C. (1975). The clonal evolution of tumor cell populations. *Science* **194:** 23–28.

Nunan, N., Wu, K., Young, I.M., Crawford, J.W., and Ritz, K. (2002). In situ spatial patterns of soil bacterial populations, mapped at multiple scales, in an arable soil. *Microbial Ecology* **44:** 296–305.

Nunney, L. (1999). Lineage selection and the evolution of multistage carcinogenesis. *Proceedings of the Royal Society of London, Series B* **266:** 493–498.

Nunney, L., and Elam, D.R. (1994). Estimating the effective population size of conserved populations. *Conservation Biology* **8:** 175–184.

Nuzhdin, S.V., and Mackay, T.F.C. (1995). The genomic rate of transposable element movement in *Drosophila melanogaster. Molecular Biology and Evolution* **12:** 180–191.

O'Donald, P. (1973). A further analysis of Bumpus's data: the intensity of natural selection. *Evolution* **27:** 398–404.

O'Hara, R.B. (2005). Comparing the effects of genetic drift and fluctuating selection on genotype frequency changes in the scarlet tiger moth. *Proceedings of the Royal Society of London Series B* **272:** 211–217.

Orgel, L.E. (1979). Selection in vitro. *Proceedings of the Royal Society of London Series B* **205:** 435–442.

Orgel, L.E., and Crick, F.H.C. (1980). Selfish DNA: the ultimate parasite. *Nature* **284:** 604–607.

Orme, C.D.L., Quicke, D.L.J., Cook, J.M., and Purvis, A. (2002). Body size does not predict species richness among the metazoan phyla. *Journal of Evolutionary Biology* **15:** 235–247.

Orr, H.A. (1998). The population genetics of adaptation: the distribution of factors fixed during adaptive evolution. *Evolution* **52:** 935–949.

Orr, H.A. (2001). The genetics of species differences. *Trends in Ecology and Evolution* **16:** 343–350.

Orr, H.A. (2002). The population genetics of adaptation: the adaptation of DNA sequences. *Evolution* **56:** 1317–1330.

Orr, H.A. (2003). The distribution of fitness effects among beneficial mutations. *Genetics* **163:** 1519–1526.

Orr, H.A., and Otto, S.P. (1994). Does diploidy increase the rate of adaptation? *Genetics* **136:** 1475–1480.

Østbye, K., Naesje, T.F., Bernatchez, L., Sandlund, O.T., and Hindar, K. (2005). Morphological divergence and origin of sympatric populations of European whitefish (*Coregonus lavaretus* L.) in Lake Femund, Norway. *Journal of Evolutionary Biology* **18:** 683–702.

Ostergren, G. (1945). Parasitic nature of extra fragment chromosomes. *Botaniska Notiser* **2:** 157–163. (Not seen).

Otto, S.P., and Barton, N.H. (2001). Selection for recombination in small populations. *Evolution* **55:** 1921–1931.

Otto, S.P., and Jones, C.D. (2000). Detecting the undetected: estimating the total number of loci underlying a quantitative trait. *Genetics* **156:** 2093–2017.

Otto, S.P., and Lenormand, T. (2002). Resolving the paradox of sex and recombination. *Nature Reviews Genetics* 3: 252–261.

Paine, R.T. (1992). Food-web analysis through field measurement of per capita interaction strength. *Nature* **355:** 73–75.

Palenzona, D.L., Fini, C., and Scossiroli, R.E. (1971). Comparative study of natural selection in *Cardium edule* L. and *Drosophila melanogaster* Meig. *Monitori Zoologica Italiana* **5:** 165–172.

Palumbi, S.R. (2002). *The evolution explosion*. Norton, New York.

Pancer, Z., and Cooper, M.D. (2006). The evolution of adaptive immunity. *Annual Review of Immunology* **24:** 497–518.

Pancer, Z., Gershon, H., and Rinkevich, B. (1995). Coexistence and possible parasitism of somatic and germ cell lines in chimeras of the colonial urochordate *Botryllus schlosseri. Biological Bulletin* **189:** 106–112.

Panhuis, T.M., Butlin, R., Zuk, M., and Tregenza, T. (2001). Sexual selection and speciation. *Trends in Ecology and Evolution* **16:** 364–371.

Pannell, J.R., and Charlesworth, B. (2000). Neutral genetic diversity in a metapopulation with recurrent local extinction and recolonization. *Evolution* **53:** 664–676.

Paquin, C., and Adams, J. (1983). Frequency of fixation of adaptive mutations is higher in evolving diploid than haploid yeast populations. *Nature* **302:** 495–500.

Parker, G.A. (1979). Sexual selection and sexual conflict. In Blum, M.S. & Blum, N.A. (eds) *Sexual selection and reproductive competition in insects*, pp. 123–166. Academic Press, New York.

Parker, G.A. (1982). Why are there so many tiny sperm? Sperm competition and the maintenance of two sexes. *Journal of Theoretical Biology* **96:** 281–294.

Parker, G.A. (2006). Sexual conflict over mating and fertilization: an overview. *Philosophical Transactions of the Royal Society of London* **361:** 235–259.

Parker, G.A., Baker, R.R., and Smith, V.G.F. (1972). The origin and evolution of gamete dimorphism and the male–female phenomenon. *Journal of Theoretical Biology* **36:** 181–198.

Parker, H.G., Kim, L.V., Sutter, N.B. *et al.* (2004). Genetic structure of the pure-bred domestic dog. *Science* **304:** 1160–1164.

Parker, M.A. (1985). Local population differentiation for compatibility in an annual legume and its host-specific fungal pathogen. *Evolution* **39:** 713–723.

Parker, M.A. (1994). Pathogens and sex in plants. *Evolutionary Ecology* **8:** 560–584.

Parker, M.A. (1996). The nature of plant-parasite specificity. *Evolutionary Ecology* **10:** 319–322

Parsons, P.A. (1958). Selection for increased recombination in *Drosophila melanogaster. American Naturalist* **88:** 255–256.

Partridge, L. (1980). Mate choice increases a component of offspring fitness in fruit flies. *Nature* **283:** 290.

Partridge, L., and Barton, N.H. (1993). Optimality, mutation and the evolution of aging. *Nature* **362:** 305–311.

Pasyukova, E.G., Belyaeva, E.S., Kogan, G.L., Kaidanov, L.Z., and Gvozdev, V.A. (1986). Concerted transpositions of mobile genetic elements coupled with fitness changes in *Drosophila melanogaster. Molecular Biology and Evolution* **34:** 299–312.

Paterson, S., Wilson, K., and Pemberton, J.M. (1998). Major histocompatibility complex variation associated with juvenile survival and parasite resistance in a large unmanaged ungulate population (*Ovis aries* L.). *Proceedings of the National Academy of Sciences of the USA* **95:** 3714–3719.

Payne, F. (1912). *Drosophila ampelophila* Loew bred in the dark for sixty-nine generations. *Biological Bulletin* **28:** 297–301.

Pearl, R. (1917). The selection problem. *American Naturalist* **51:** 65–91.

Peck, J.R. (1994). A ruby in the rubbish: beneficial mutations, deleterious mutations and the evolution of sex. *Genetics* **137:** 597–606.

Peichel, C.L., Nereng, K.S., Ohgi, K.A. *et al.* (2001). The genetic architecture of divergence between threespine stickleback species. *Nature* **414:** 901–905.

Pelletier, J.D. (1997) Analysis and modeling of the natural variability of climate. *Journal of Climate* **10:** 1331–1342.

Penn, D.J., and Potts, W.K. (1999). The evolution of mating preferences and major histocompatibility complex genes. *American Naturalist* **153:** 145–164.

Perron, G.G., Zasloff, M., and Bell, G. (2006). Experimental evolution of resistance to an antimicrobial peptide. *Proceedings of the Royal Society of London Series B* **273:** 251–256.

Peters, A.D., and Keightley, P.D. (2000). A test for epistasis among induced mutations in *Caenorhabditis elegans. Genetics* **156:** 1635–1647.

Peters, R.H. (1983). *The ecological implications of body size.* Cambridge University Press, Cambridge.

Peters, R.H. (1986). Seasonal and trophic effects on the-size structure of the planktonic communities in four lakes in northern Italy. *Memorie dell'Istituto Italiano di Idrobiologia* **43:** 91–103.

Peters, R.H., and Wassenberg, K. (1983). The effect of body size on animal abundance. *Oecologia* **60:** 89–96.

Pfeiffer, T., and Schuster, S. (2005). Game-theoretical approaches to studying the evolution of biochemical systems. *Trends in Biochemical Sciences* **30:** 20–25.

Pfeiffer, T., Schuster, S., and Bonhoeffer, S. (2001). Cooperation and competition in the evolution of ATP-producing pathways. *Science* **292:** 504–507.Erratum in: *Science* 2001;**293:** 1436.

Phillips, P.C., and Johnson, N.A. (1998). The population genetics of synthetic lethals. *Genetics* **150:** 449–458.

Pielou, E.C., and Robson, D.S. (1972). 2*k* contingency tables in ecology. *Journal of Theoretical Biology* **34:** 337–352. (Appendix by Robson.).

Piertney, S.B., and Oliver, M.K. (2006). The evolutionary ecology of the major histocompatibility complex. *Heredity* **96:** 7–21.

Pignatelli, P.M., and Mackay, T.F.C. (1989). Hybrid dysgenesis-induced response to selection in *Drosophila melanogaster. Genetical Research* **54:** 183–195.

Pimentel, D., Levin, S.A., and Soans, A.B. (1975). On the evolution of energy balance in some exploiter-victim systems. *Ecology* **56:** 381–390.

Pimm, S.L., and Redfearn, A. (1988). The variability of population densities. *Nature* **334:** 613–614.

Platenkamp, G.A.J. (1991). Phenotypic plasticity and population differentiation in seeds and seedlings of the grass *Anthoxanthum odoratum. Oecologia* **88:** 515–520.

Potts, D.C. (1984). Natural selection in experimental populations of reef-building corals (*Scleractinia*). *Evolution* **38:** 1059–1078.

Powell, J.R. (1978). The founder-flush speciation theory: an experimental approach. *Evolution* **32:** 465–474.

Preston, F.W. (1962). The canonical distribution of abundance and rarity. *Ecology* **43:** 185–215.

Price, G.R. (1972). Fisher's 'fundamental theorem' made clear. *Annals of Human Genetics* **36:** 129–140.

Price, T.D., and Schluter, D. (1991). On the low heritability of life-history traits. *Evolution* **45:** 833–861.

Price, M.V., and Waser, N.M. (1979). Pollen dispersal and optimal outcrossing in *Delphinium nelsoni. Nature* **277:** 294–297.

Primack, R.B. (1978). Regulation of seed yield in *Plantago. Journal of Ecology* **66:** 835–847.

Primack, R.B. (1979). Reproductive effort in annual and perennial species of *Plantago* (Plantinaginaceae). *American Naturalist* **114:** 51–62.

Primack, R.B., and Antonovics, J. (1981). Experimental ecological genetics in *Plantago.* V. Components of seed yield in the ribwort plantain *Plantago lanceolata* L. *Evolution* **35:** 1069–1079.

Primack, R.B., and Antonovics, J. (1982). Experimental ecological genetics in *Plantago.* VII. Reproductive effort in populations of *P. lanceolata* L. *Evolution* **36:** 742–752.

Proctor, J. (1971). The plant ecology of serpentine. III. The influence of a high magnesium/calcium ratio and high nickel and chromium levels in some British and Swedish serpentine soils. *Journal of Ecology* **59:** 827–842.

Promislow, D.E.L. (1991). Senescence in natural populations of mammals: a comparative study. *Evolution* **45:** 1869–1887.

Pröschold, T., Harris, E.H.& Coleman, A.W. (2005). Portrait of a species: *Chlamydomonas reinhardtii. Genetics* **170:** 1601–1610.

Prout, T. (1962). The effects of stabilizing selection on the time of development in *Drosophila melanogaster*. *Genetical Research* **3:** 364–382.

Prugnolle, F., Manica, A., Charpentier, M., Guégan, J.F., Guernier, V., and Balloux, F. (2005). Pathogen-driven selection and worldwide HLA class I diversity. *Current Biology* **15**(11): 1022–1027.

Pruzan-Hotchkiss, A., Perelle, I.B., Hotchkiss, F.H.C., and Ehrman, L. (1980). Altered competition between two reproductively isolated strains of *Drosophila melanogaster*. *Evolution* **34:** 445–452.

Queller, D.C. (1985). Kinship, reciprocity and synergism in the evolution of social behaviour. *Nature* **318:** 366–367.

Queller, D.C. (1992). A general model for kin selection. *Evolution* **46:** 376–380.

Quinones, R.A., Platt, T., and Rodriguez, J. (2003). Patterns of biomass-size spectra from oligotrophic waters of the Northwest Atlantic. *Progress in Oceanography* **57:** 405–427.

Radman, M. (1999). Mutation: enzymes of evolutionary change. *Nature* **401:** 866–869.

Rainey, P.B., and Rainey, K. (2003). Evolution of cooperation and conflict in experimental bacterial populations. *Nature* **425:** 72–74.

Rainey, P.B., and Travisano, M. (1998). Adaptive radiation in a heterogeneous environment. *Nature* **394:** 69–72.

Rapson, G.L., and Wilson, J.B. (1988). Non-adaptation in *Agrostis capillaris* L. *Functional Ecology* **2:** 479–490.

Rasmuson, M. (1956). Reciprocal recurrent selection. Results of three model experiments on *Drosophila* for improvement of quantitative characters. *Hereditas* **42:** 397–414.

Raup, D.M. (1991). A kill curve for Phanerozoic marine species. *Paleobiology* **17:** 37–48.

Rausher, M.D. (1992). The measurement of selection on quantitative traits: biases due to environmental covariances between traits and fitness. *Evolution* **46:** 616–626.

Rauw, W. M., Kanis, E., Noordhuizen-Stassen E.N., and Grommers, F.J. (1998). Undesirable side effects of selection for high production efficiency in farm animals: a review. Livestock Production Science 56: 15–33.

Ray, T.S. (1991). An approach to the synthesis of life. In C. Langton, C. Taylor, D. Farmer, and S. Rasmussen, (eds.), *Artificial life II*, pp. 371–408. Santa Fe Institute Studies in the Science of Complexity Vol. X. Addison-Wesley, Redwood City, CA.

Reboud, X., and Bell, G. (1997). Experimental evolution in *Chlamydomonas*. 3. Evolution of specialist and generalist types in environments that vary in space and time. *Heredity* **78:** 507–514.

Reimchen, T.E. (1989). Loss of nuptial color in threespine sticklebacks (*Gasterosteus aculeatus*). *Evolution* **43:** 450–460.

Reimchen, T.E. (2000). Predator handling failures of lateral plate morphs in *Gasterosteus aculeatus*: functional implications for the ancestral plate condition. *Behaviour* **137:** 1081–1096.

Reimchen, T.E., Stinson, E.M. & Nelson, J.S. (1985). Multivariate differentiation of parapatric and allopatric populations of threespine stickleback in the Sangan River watershed, Queen Charlotte Islands. *Canadian Journal of Zoology* **63:** 2944–2951.

Remold, S.K., and Lenski, R.E. (2001). Contribution of individual random mutations to genotype-by-environment interactions in *E. coli*. *Proceedings of the National Academy of Sciences of the USA* **98:** 11388–11393.

Renaut, S., Replansky, T., Heppleston, A., and Bell, G. (2006). The ecology and genetics of fitness in *Chlamydomonas*. XIII. Fitness of long-term sexual and asexual populations in benign environments. *Evolution* **60:** 2272–2279.

Rendel, J.M. (1943). Variation in the weights of hatched and unhatched duck's eggs. *Biometrika* **33:** 48–56.

Rendón, B., and Núñez-Farfán, J. (2001). Population differentiation and phenotypic plasticity of wild and agrestal populations of the annual *Anoda cristata* (Malvaceae) growing in two contrasting habitats. *Plant Ecology* **156:** 205–213.

Renne, U., Langhammer, M., Wytrwat, E., Dietl, G., and Bunger, L. (2003). Analysis of growth in selected and unselected mouse lines. *Journal of Experimental Animal Science* **42:** 218–232.

Reuter, M., Bell, G., and Greig, D. (2007). Increased outbreeding in yeast in response to dispersal by an insect vector. *Current Biology* **17:** R81–R83.

Reznick, D., and Endler, J.A. (1982). The impact of predation on life history evolution in Trinidadian guppies (*Poecilia reticulata*). *Evolution* **36:** 160–177.

Reznick, D.N., Shaw, F.H., Rodd, F.H., and Shaw, R.G. (1997). Evaluation of the rate of evolution in natural populations of guppies (*Poecilia reticulata*). *Science* **275:** 1934–1937.

Rhodes, G., Parkhill, J., Bird, C. *et al.* (2004). Complete nucleotide sequence of the conjugative tetracycline resistance plasmid pFBAOT6, a member of a group of IncU plasmids with global ubiquity. *Applied and Environmental Microbiology* **70:** 7497–7510.

Rice, K.J., and Mack, R.N. (1991). Ecological genetics of *Bromus tinctorum*. *Oecologia* **88:** 91–101.

Rice, W.R. (1992). Sexually antagonistic genes: experimental evidence. *Science* **256:** 1436–1439.

Rice, W.R. (1994). Degeneration of a nonrecombining chromosome. *Science* **263:** 230–232.

Rice, W.R. (1996). Sexually antagonistic male adaptation triggered by experimental arrest of female evolution. *Nature* **381:** 232–234.

Rice, W.R. (1998). Male fitness increases when females are eliminated from gene pool: implications for the Y chromosome. *Proceedings of the National Academy of Sciences of the USA* **95:** 6217–6221.

Rice, W.R. (2002). Experimental tests of the adaptive significance of sexual recombination. *Nature Reviews Genetics* **3:** 241–251.

Rice, W.R., and Chippindale, A.K. (2001). Sexual recombination and the power of natural selection. *Science* **294:** 555–559.

Rice, W.R., and Hostert, E.E. (1993). Laboratory experiments on speciation: What have we learned in forty years? *Evolution* **47:** 1637–1653.

Rice, W.R., and Salt, G.W. (1990). The evolution of reproductive isolation as a correlated character under sympatric conditions: experimental evidence. *Evolution* **44:** 1140–1152.

Rich, S.S., Bell, A.E., and Wilson, S.P. (1979). Genetic drift in small populations of *Tribolium*. *Evolution* **33:** 579–584.

Richard, M., Bernhardt, T., and Bell, G. (2000). Environmental heterogeneity and the spatial structure of fern species diversity in one hectare of old-growth forest. *Ecography* **23:** 2311–2345.

Richardson, R.H., Kojima, K., and Lucas, H.L. (1968). An analysis of short-term selection experiments. *Heredity* **23:** 493–506.

Richman, A.D., and Kohn, J.R. (2000). Evolutionary genetics of self-incompatibility in the Solanaceae. *Plant Molecular Biology* **42:** 169–179.

Richmond, R.C., Gilpin, M.E., Perez Salas, S., and Ayala, F.J. (1975). A search for emergent competitive phenomena: the dynamics of multispecies *Drosophila* systems. *Ecology* **56:** 709–714.

Ricker, J.P., and Hirsch, J. (1988). Genetic changes occurring over 500 generations in lines of *Drosophila melanogaster* selected divergently for geotaxis. *Behavior Genetics* **18:** 13–25.

Ricker, W.E. (1981). Changes in the average size and average age of Pacific salmon. *Canadian Journal of Fisheries and Aquatic Science* **38:** 1636–1656.

Ricklefs, R.E. (1998). Evolutionary theories of aging: confirmation of a fundamental prediction, with implications for the genetic basis and evolution of life span. *American Naturalist* **152:** 24–44.

Ridgway, M.S., and McPhail, J.D. (1984). Ecology and evolution of sympatric sticklebacks (*Gasterosteus*): mate choice and reproductive isolation in the Enos Lake species pair. *Canadian Journal of Zoology* **62:** 1813–1818.

Riehle, M.M., Bennett, A.F., and Long, A.D. (2001). Genetic architecture of thermal adaptation in *Escherichia coli*. *Proceedings of the National Academy of Sciences of the USA* **98:** 525–530.

Rieseberg, L.H. (2001). Chromosomal rearrangements and speciation. *Trends in Ecology and Evolution* **16:** 351–358.

Rigby, P.W.J., Burleigh, B.D., and Hartley, B.S. (1974). Gene duplication in experimental enzyme evolution. *Nature* **251:** 200–204.

Ripley, P.O. (1941). The influence of crops upon those which follow. *Scientific Agriculture* **21:** 522–583.

Rivero, A., Balloux, G.F., and West, S.A. (2003). Testing for epistasis between deleterious mutations in a parasitoid wasp. *Evolution* **57:** 1698–1703.

Roberts, R.C. (1966a). The limits to artificial selection for body weight in the mouse. I. The limits attained in earlier experiments. *Genetical Research* **8:** 347–360.

Roberts, R.C. (1966b). The limits to artificial selection for body weight in the mouse. II. The genetic nature of the limits. *Genetical Research* **9:** 73–85.

Roberts, R.C. (1967). The limits to artificial selection for body weight in the mouse. III. Selection from crosses between previously selected lines. *Genetical Research* **9:** 73–85.

Roberts, R.W., and Szostak, J.W. (1997). RNA-peptide fusions for the in vitro selection of peptides and proteins. *Proceedings of the National Academy of Sciences of the USA* **94:** 12297–12302.

Robertson, A. (1959a). Experimental design in the measurement of heritabilities and genetic correlations. In Kempthorne, O.(ed.) *Biometrical genetics*, pp 79–91. Pergamon, New York.

Robertson, A. (1959b). The sampling variance of the genetic correlation coefficient. *Biometrics* **15:** 469–485.

Robertson, A. (1960). A theory of limits in artificial selection. *Proceedings of the Royal Society of London Series B* **153:** 234–249.

Robertson, F.W. (1966). A test of sexual isolation in *Drosophila*. *Genetical Research* **8:** 181–187.

Robertson, F.W., and Reeve, E. (1952). Studies in quantitative inheritance. I. Effects of selection of wing and thorax length in *Drosophila melanogaster*. *Journal of Genetics* **50:** 414–448.

Rodd, H., Hughes, K.A., Grether, G.F. & Baril, C.T. (2002). A possible non-sexual origin of mate preference: are male guppies mimicking fruit? *Proceedings of the Royal Society of London Series B* **269:** 475–481.

Rodriguez, J., and Mullin, M.M. (1986). Relation between biomass and body weight of plankton in a steady state oceanic ecosystem. *Limnology and Oceanography* **31:** 361–370.

Roff, D.A. (1992). *The evolution of life histories.* Springer, New York.

Roff, D.A. (1997). *Evolutionary quantitative genetics.* Chapman & Hall, London.

Rogers, S.M., and Bernatchez, L. (2005). Integrating QTL mapping and genome scans towards the characterization of candidate loci under parallelel selection in the lake whitefish (*Coregonus clupeaformis*). *Molecular Ecology* **14:** 351–361.

Rogers, S.M., and Bernatchez, L. (2006). The genetic basis of intrinsic and extrinsic post-zygotic reproductive isolation jointly promoting speciation in the lake whitefish species complex (*Coregonus clupeaformis*). *Journal of Evolutionary Biology* **19.** 1979–1994.

Rogers, S.M., Isabel, N., and Bernatchez, L. (2006). Linkage maps of the dwarf and normal lake whitefish (*Coregonus clupeaformis*) species complex and their hybrids reveal the genetic architecture of population divergence. *Genetics* (in press)

Rokyta, D., Badgett, M.R., Molineux, I.J., and Bull, J.J. (2002). Experimental genomic evolution: Extensive compensation for loss of DNA ligase activity in a virus. *Molecular Biology and Evolution* **19:** 230–238.

Rokyta, D.R., Joyce, P., Caudle, S.B., and Wichman, H.A. (2005). An empirical test of the mutational landscape model of adaptation using a single-stranded DNA virus. *Nature Genetics* **37:** 441–444.

Roper, C., Pignatelli, P., and Partridge, L. (1993). Evolutionary effects of selection on age at reproduction in larval and adult *Drosophila melanogaster. Evolution* **47:** 445–455.

Rose, M.R. (1984). Laboratory evolution of postponed senescence in *Drosophila melanogaster. Evolution* **38:** 1004–1010.

Rose, M.R. (1991). *Evolutionary biology of ageing.* Oxford University Press, Oxford.

Rose, M.R., and Charlesworth, B. (1980). A test of evolutionary theories of senescence. *Nature* **287:** 141–142.

Rose, M.R., and Charlesworth, B. (1981). Genetics of life history in *Drosophila melanogaster.* II. Exploratory selection experiments. *Genetics* **9:** 187–196.

Rose, M.R., Dorey, M.L., Coyle, A.M., and Serivce, P.M. (1984). The morphology of postponed senescence in *Drosophila melanogaster. Canadian Journal of Zoology* **62:** 1576–1580.

Rose, M.R., Passananti, H.B., Chippindale, A.K. *et al.* (2005). The effects of evolution are local: evidence from experimental evolution in *Drosophila. Integrative and Comparative Biology* **45:** 486–491.

Rosenberg, S.M., Longerich, S., Gee, P., and Harris, R.S. (1994). Adaptive mutation by deletions in small mononucleotide repeats. *Science* **265:** 405–407.

Rosenberg, N.A., Pritchard, J.K., Weber, J.L. *et al.* (2002). Genetic structure of human populations. *Science* **298:** 2381–2385.

Rosenweig, R.F., Sharp, R.R., Treves, D.S., and Adams, J. (1994). Microbial evolution in a simple unstructured environment—genetic differentiation in *Escherichia coli. Genetics* **137:** 903–917.

Rosenzweig, M.L. (1995). *Species diversity in space and time.* Cambridge University Press, Cambridge.

Ross, J. (1982). Myxomatosis: the natural evolution of the disease. *Symposia of the Zoological Society of London* **50:** 77–92.

Rowe Magnus, D.A., and Mazel, D. (2001). Integrons: natural tools for bacterial genome evolution. *Current Opinion in Microbiology* **4:** 565–569.

Roy, K.R. (1961). Interaction between rice varieties. *Journal of Genetics* **57:** 137–152.

Rozen, D., and Lenski, R.E. (2000). Long-term experimental evolution in *Escherichia coli.* VIII. Dynamics of a balanced polymorphism. *American Naturalist* **105:**24–35.

Rozen, D., de Visser, J.A.G.M., and Gerrish, P. (2002). Fitness effects of fixed beneficial mutations in microbial populations. *Current Biology* **12:** 1040–1045.

Ruano, R.G., Orozco, F., and López-Fanjul, C. (1975). The effect of different selection intensities on selection response in egg-laying of *Tribolium castaneum. Genetical Research* **25:** 17–27.

Rundle, H.D., and Nosil, P. (2005). Ecological speciation. *Ecology Letters* **8:** 336–352.

Ruzzante, D.E., and Doyle, R.W. (1991). Rapid behavioral changes in medaka (*Oryzias latipes*) caused by selection for competitive and noncompetitive growth. *Evolution* **45:** 1936–1946.

Saadeh, H.K., Craig, J.V., Smith, L.T., and Wearden, S. (1968). Effectiveness of alternative breeding systems for increasing rate of egg production in chicken. *Poultry Science,* 47, 1057–1072.

Sachs, J.L., Mueller, U.G., Wilcox, T.P, and Bull, J.J. (2004). The evolution of cooperation. *Quarterly Review of Biology* **79:** 135–160.

Sacks, M. (1964). Life history of an aquatic gastrotrich. *Transactions of the American Microscopical Society* **83:** 358–362.

Saffhill, R., Schneider-Bernloehr, H., Orgel, L.E., and Spiegelman, S. (1970). In vitro selection of bacteriophage Qβ RNA variants resistant to ethidium bromide. *Journal of Molecular Biology* **51:** 531–539.

Sang, J.H., and Clayton, G.A. (1957). Selection for larval development time in *Drosophila*. *Journal of Heredity* **48:** 265–270.

Sanjuan, R., Moya, A., and Elena, S.F. (2004). The distribution of fitness effects caused by single-nucleotide substitutions in an RNA virus. *Proceedings of the National Academy of Sciences of the USA* **101:** 8396–8401.

Santamaria, L., Figuerola, J., Pilon, J.J. *et al.* 2003. Plant performance across latitude: the role of plasticity and local adaptation in an aquatic plant. *Ecology* **84:** 2454–2461.

Sargent, T.D., Millar, C.D., and Lambert, D.M. (1998). The 'classical' explanation of industrial melanism. *Evolutionary Biology* **30:** 299–322.

Scharloo, W., den Boer, M., and Hoogmoed, M.S. (1967a). Disruptive selection on sternopleural chaeta number. *Genetical Research* **9:** 115–118.

Scharloo, W., Hoogmoed, M.S., and Ter Kuile, A. (1967b). Stabilizing and disruptive selection on a mutant character in *Drosophila*. I. The phenotypic variance and its components. *Genetics* **56:** 709–726.

Schat, H., Vooijs, R., and Kuiper, E. (1996). Identical major gene loci for heavy metal tolerances have independently evolved in different local populations and subspecies of *Silene vulgaris*. *Evolution* **50:** 1888–1895.

Scheiring, J.F. (1977). Stabilizing selection for size as related to mating fitness in *Tetraopes*. *Evolution* **31:** 447–449.

Schemske, D.W. (1984). Population structure and local selection in *Impatiens pallida* (Balsaminaceae), a selfing annual. *Evolution* **38:** 817–832.

Schluter, D. (1994). Experimental evidence that competition promotes divergence in adaptive radiation. *Science* **266:** 798–801.

Schluter, D. (2000). *The ecology of adaptive radiation*. Oxford University Press, Oxford.

Schluter, D. (2001). Ecology and the origin of species. *Trends in Ecology and Evolution* **16:** 372–380.

Schluter, D. &. McPhail, J.D. (1992). Ecological character displacement and speciation in sticklebacks. *American Naturalist* **140:** 85–108.

Schmidt, A.L., and Anderson, L.M. (2006). Repetitive DNA elements as mediators of genomic change in response to environmental cues. *Biological Reviews*. To appear.

Schmidt, K.P., and Levin, D.A. (1985). The comparative demography of reciprocally sown populations of *Phlox drummondi* Hood. I. Survivorships, fecundities, and finite rates of increase. *Evolution* **39:** 396–404.

Schmitt, J., and Antonovics, J. (1986). Experimental studies of the evolutionary significance of sexual reproduction. IV. Effect of neighbor relatedness and aphid infestation on seedling performance. *Evolution* **40:** 830–836.

Schmitt, J., and Gamble, S.E. (1990). The effect of distance from the parental site on offspring performance and inbreeding depression in *Impatiens capensis*: a test of the local adaptation hypothesis. *Evolution* **44:** 2022–2030.

Schoen, D.J., Stewart, S.C., Lechowicz, M.J., and Bell, G. (1986). Partitioning the transplant site effect in reciprocal transplant experiments with *Impatiens capensis* and *Impatiens pallida*. *Oecologia* **70:** 149–154.

Schoen, D.J., Bell, G., and Lechowicz, M.J. (1994). The ecology and genetics of fitness in forest plants. IV. Quantitative genetics of fitness components in *Impatiens pallida*. *American Journal of Botany* **81:** 232–239.

Schoustra, S.E., Slakhorst, M., Debets, A.J.M., and Hoekstra, R.F. (2005). Comparing artificial and natural selection in rate of adaptation to genetic stress in *Aspergillus nidulans*. *Journal of Evolutionary Biology* **18:** 771–778.

Scowcroft, W.R. (1968). Variation of scutellar bristles in *Drosophila*. XI. Selection for scutellar microchaetae and the correlated responses of scutellar bristles. *Genetical Research* **11:** 125–134.

Seaton, A.P.C., and Antonovics, J. (1967). Population interrelationships. I. Evolution in mixtures of *Drosophila* mutants. *Heredity* **22:** 19–33.

Seeley, R.H. (1986). Intense natural selection caused a rapid morphological transition in a living marine snail. *Proceedings of the National Academy of Sciences of the USA* **83:** 6897–6901.

Senior, E., Bull, A.T., and Slater, J.H. (1976). Enzyme evolution in a microbial community growing on the herbicide Dalapon. *Nature* **263:** 476–479.

Sepkoski, J.J. (1998). Rates of speciation in the fossil record. *Philosophical Transactions of the Royal Society of London* **353:** 315–326.

Sepkoski, J.J. (1999). Rates of speciation in the fossil record. In Magurran, A. and May, R.M. (eds.) *The evolution of biological diversity*. Oxford University Press, Oxford.

Service, P.M. (1987). Physiological mechanisms of increased stress resistance in *Drosophila melanogaster* selected for postponed senescence. Physiological Zoology **60:** 321–326.

Service, P.M. (1993). Laboratory evolution of longevity and reproductive fitness components in male fruit flies: mating ability. *Evolution* **47:** 387–399.

Service, P.M., Hutchinson, E.W., MacKinley, M.D., and Rose, M.R. (1985). Resistance to environmental stress in *Drosophila melanogaster* selected for postponed senescence. *Physiological Zoology* **58:** 380–389.

Sevilla, N., Ruiz-Jarabo, C.M., Gomez-Mariano, G., Baranowski, E., and Domingo, E. (1998). An RNA virus can adapt to the multiplicity of infection. *Journal of General Virology* **79:** 2971–2980.

Shackney, S.E., and Shankey, T.V. (1998). Common patterns of genetic evolution in human solid tumors. *Cytometry* **29:** 1–27.

Shapiro, M.D., Marks, M.E., Peichel, C.L. *et al.* (2004). Genetic and developmental basis of evolutionary pelvic reduction in threespine sticklebacks. *Nature* **428:** 717–723.

Sharp, G.L., Hill, W.G., and Robertson, A. (1984). Effects of selection on growth, body composition and food intake in mice. 1. Responses in selected traits. *Genetical Research* **43:** 75–92.

Shaw, F.H., Geyer, C.J., and Shaw, R.G. (2002). A comprehensive model of mutations affecting fitness and inferences for *Arabidopsis thaliana. Evolution* **56:** 453–463.

Shaw, R.F., and Mohler, J.D. (1953). The selective significance of the sex ratio. *American Naturalist* **87:** 337–342.

Shay, J.W., and Wright, W.E. (2000). Hayflick, his limit, and cellular aging. *Nature Reviews Molecular and Cell Biology* **1:** 72–76.

Sheldon, B.L. (1963a). Studies in artificial selection on quantitative characters. I. Selection for abdominal bristles in *Drosophila melanogaster. Australian Journal of Biological Science* **16:** 490–515.

Sheldon, B.L. (1963b). Studies in artificial selection on quantitative characters. II. Selection for body weight in *Drosophila melanogaster. Australian Journal of Biological Science* **16:** 516–541.

Sheldon, R.W., Prakash, A., and Sutcliffe, W.H. (1972). The size distribution of particles in the ocean. *Limnology and Oceanography* **17:** 327–340.

Sheridan, A.K. (1988). Agreement between estimated and realized genetic parameters. *Animal Breeding Abstracts* **56:** 877–889.

Sheridan, A.K., and Barker, J.S.F. (1974a). Two-trait selection and the genetic correlation. I. Prediction of responses in single-trait and two-trait selection. *Australian Journal of Biological Sciences* **27:** 75–88.

Sheridan, A.K., and Barker, J.S.F. (1974b). Two-trait selection and the genetic correlation. II. Changes in the genetic correlation during two-trait selection. *Australian Journal of Biological Sciences* **27:** 89–101.

Sherwin, R.N. (1975). Selection for mating ability in two chromosomal arrangements of *Drosophila pseudoobscura. Evolution* **29:** 519–530.

Siegel, P.B. (1965). Genetics of behavior: selection for mating ability in chickens. *Genetics* **52:** 1269–1277.

Siemann, E., Tilman, D., and Haarstad, J. (1996). Insect species diversity, abundance and body size relationships. *Nature* **380:** 704–706.

Simons, K.J., Fellers, J.P., Trick, H.N. *et al.* (2006). Molecular characterization of the major wheat domestication gene *Q. Genetics* **172:** 547–555.

Simonsen L. (1991). The existence conditions for bacterial plasmids: theory and reality. *Microbial Ecology* **22:** 187–205.

Simpson, G.G., Roe, A., and Lewontin, R.C. (1960). *Quantitative zoology.* Harcourt, Brace and Co., New York.

Skÿlason, S., Snorrason, S.S., Noakes, D.L.G., and Ferguson, M.M. (1996). Genetic basis of life history variations among sympatric morphs of Arctic char, *Salvelinus alpinus. Canadian Journal of Fisheries and Aquatic Science* 53: 1807–1813.

Slapeta, J., Moreira, D., and Lopez-Garcia, P. (2005). The extent of protist diversity: insights from molecular ecology of freshwater eukaryotes. *Proceedings of the Royal Society of London Series B* **272:** 2073–2081.

Slapeta, J., Lopez-Garcia, P., and Moreira, D. (2006). Global dispersal and ancient cryptic species in the smallest marine eukaryotes. *Molecular Biology and Evolution* **23:** 23–29.

Sliwa, P., Kluz, J., and Korona, R. (2004). Mutational load and the transition between diploidy and haploidy in experimental populations of the yeast *Saccharomyces cerevisiae. Genetica* **121:** 285–293.

Smith, C., Barber, I, Wootton, R.J., and Chittka, L. (2004b). A receiver bias in the origin of three-spined stickleback mate choice. *Proceedings of the Royal Society of London Series B* **271:** 949–955.

Smith, K.P., and Bohren, B.B. (1974). Direct and correlated responses to selection for hatching time in the fowl. *British Poutry Science* **15:** 597–604.

Smith, N.G., and Eyre-Walker, A. (2002). Adaptive protein evolution in *Drosophila. Nature* **415:** 1022–1024.

Smith, O.D., Kleese, R.A., and Stuthman, D.D. (1970). Competition among oat varieties grown in hill plots. *Crop Science* **10:** 381–384.

Smith, S.A., Bell, G., and Bermingham, E. (2004a). Cross-Cordillera exchange mediated by the Panama Canal increased the species richness of local freshwater fish assemblages. *Proceedings of the Royal Society of London Series B* **271:** 1889–1896.

Smouse, P., and Kosuda, K. (1977). The effects of genotypic frequency and population density on fitness differentials in *Escherichia coli. Genetics* **64:** 399–411.

Snaydon, R.W., and Davies, M.S. (1976). Rapid population differentiation in a mosaic environment. IV. Populations of *Anthoxanthum odoratum* at sharp boundaries. *Heredity* **37:** 9–25.

Snaydon, R.W., and Davies, M.S. (1982). Rapid divergence of plant populations in response to recent changes in soil conditions. *Evolution* **36:** 289–297.

Sniegowski, P.D., Gerrish, P.J., and Lenski, R.E. (1997). Evolution of high mutation rates in experiemental populations of *E. coli. Nature* **387**(6634): 703–705.

Snijders, A.M., Fridlyand, J., Mans, D.A. *et al.* (2003). Shaping of tumor and drug-resistant genomes by instability and selection. *Oncogene* **22:** 4370–4379.

Snow, A.A., and Mazer, S.J. (1988). Gametophytic selection in *Raphanus raphinastrum*: A test for heritable variation in pollen competitive ability. *Evolution* **42:** 1065–1075.

Soans, A.B., Pimentel, D., and Soans, J.S. (1974). Evolution of reproductive isolation in allopatric and sympatric populations. *American Naturalist* **101:** 493–504.

Sohal, R.S., and Weindruch, R. (1996). Oxidative stress, caloric restriction, and aging. *Science* **273:** 59–63.

Sokal, R.R., and Sullivan, R.L. (1963). Competition between mutant and wild-type house-fly strains at varying densities. *Ecology* **44:** 314–322.

Soliman, M.H. (1982). Directional and stabilizing selection for developmental time and correlated response in reproductive fitness in *Tribolium castaneum. Theoretical and Applied Genetics* **63:** 111–116.

Sonneborn, T.M. (1957). Breeding systems, reproductive methods and species problems in Protozoa. In Mayr, E. (ed.) *The species problem*, pp 155–324. AAAS, Washington, DC.

Souza, V., Nguyen, T.T., Hudson, R.R., Pinero, D., and Lenski, R.E. (2002). Hierarchical analysis of linkage disequilibrium in *Rhizobium* populations: Evidence for sex? *Proceedings of the National Academy of Sciences of the USA* **89:** 8389–8393.

Speakman, J.R. (2005). Body size, energy metabolism and lifespan. *Journal of Experimental Biology* **208:** 1717–1730.

Specht, J.E., Hume, D.J., and Kumudini, S.V. (1999). Soybean yield potential: a genetic and physiological perspective. *Crop Science* **39:** 993–998.

Spiegelman, S. (1971). An approach to the experimental analysis of precellular evolution. *Quarterly Reviews of Biophysics* **4:** 213–253.

Spiess, E.B., and Stankevych, A.J. (1973). Mating speed selection and egg chamber correlation in *Drosophila persimilis. Egyptian Journal of Genetics and Cytology* **2:** 177–194.

Spiess, E.B., and Wilke, C.M. (1984). Still another attempt to achieve assortative mating by disruptive selection in *Drosophila. Evolution* **38:** 505–515.

Spuhler, K.P., Crumpacker, D.W., Williams, J.S., and Bradley, B.P. (1978). Response to selection for mating speed and changes in gene arrangement frequencies in descendants from a single population of *Drosophila pseudoobscura. Genetics* **89:** 729–749.

Stanton, M., and Young, H.J. (1994). Selecting for floral character associations in wild radish, *Raphanus sativus* L. *Journal of Evolutionary Biology* **7:** 271–285.

Stauffer, D., Kunwar, A., and Chowdhury, D. (2005). Evolutionary ecology in silico: evolving food webs, migrating population and speciation. *Physica A* **352:** 202–215.

Stearns, S.C. (1983). The genetic basics of differences in life-history traits among six populations of mosquitofish (*Gambusia affinis*) that shared ancestors in 1905. *Evolution* **37:** 618–627.

Stearns, S.C. (1992). *The evolution of life histories*. Oxford University Press, Oxford.

Stebbins, G.L., and Hill, G.J.C. (1980). Did multicellular plants invade the land? *American Naturalist* **115:** 342–353.

Steele, J.H. (1985). A comparison of terrestrial and marine ecological systems. *Nature* **313:** 355–358.

Stein, J.R., and McCauley, M.J. (1976). Sexual compatibility in *Gonium pectorale* (Volvocales: Chlorophyceae) from soil of a single pond. *Canadian Journal of Botany* **54:** 1126–1130.

Stephan, W., Chao, L., and Smale, J.G. (1993). The advance of Muller's Ratchet in a haploid asexual population: approximate solutions based on diffusion theory. *Genetical Research* **61:** 225–231.

Stewart E., Madden R., Paul G., and Taddei F. (2005). Aging and death in an organism that reproduces by morphologically symmetric division. *PloS Biology* **3:** e45.

Stewart, F.M., and Levin, B.R. (1973). Partitioning of resources and the outcome of interspecific competition: a model and some general considerations. *American Naturalist* **107:** 171–198.

Stewart, F.M., and Levin, B.R. (1977). The population biology of bacterial plasmids: a priori conditions for the existence of conjugationally transmitted factors. *Genetics* **87:** 209–228.

Stewart, S.C., and Schoen, D.J. (1987). Pattern of phenotypic viability and fecundity selection in a natural population of *Impatiens pallida. Evolution* **41:** 1290–1301.

Stoner, D.S., Rinkevich, B., and Weissman, I.L. (1999). Heritable germ and somatic cell lineage competitions in chimeric colonial protochordates. *Proceedings of the National Academy of Sciences of the USA* **96:** 9148–9153.

Stouthamer, R., Luck, R.F., and Hamilton, W.D. (1990). Antibiotics cause parthenogenetic *Trichogramma* (Hymenoptera/Trichogrammatidae) to revert to sex. *Proceedings of the National Academy of Sciences of the USA* **87:** 2424–2427.

Strauss, S.Y., Rudgers, J.A., Lau, J.A., and Irwin, R.E. (2002). Direct and ecological costs of resistance to herbivory. *Trends in Ecology and Evolution* **17:** 278–285.

Strobeck, C. (1975). Haploid selection with n alleles in m niches. *American Naturalist* **113:** 439–444.

Sudarshana, P., and Knudsen, G.R. (1995). Effect of parental growth on dynamics of conjugative plasmid transfer in the pea spermosphere. *Applied and Environmental Microbiology* **61:** 3136–3141.

Sulzbach, D.S. (1980). Selection for competitive ability: Negative results in *Drosophila*. *Evolution* **34:** 431–436.

Sulzbach, D.S., and Emlen, J.M. (1979). Evolution of competitive ability in mixtures of *Drosophila melanogaster*: Populations with an initial asymmetry. *Evolution* **33:** 1138–1149.

Sumper, M., and Luce, R. (1975). Evidence for de novo production of self-replicating and environmentally adapted RNA structures by bacteriophage Qβ replicase. *Proceedings of the National Academy of Sciences of the USA* **72:** 162–166.

Suneson, C.A. (1949). Survival of four barley varieties in a mixture. *Agronomy Journal* **41:** 459–461.

Surface, F.M., and Pearl, R. (1915). Selection in oats. *Maine Agricultural Experimental Station Annual Report 1915:* 1–40.

Sutherland, G.D., Harestad, A.S., Price, K., and Lertzman, K.P. (2000). Scaling of natal dispersal distances in terrestrial birds and mammals. *Conservation Ecology* **4:** 16. URL: *http://www.consecol.org/vol4/iss1/art16.*

Swanson, W.J., and Vacquier, V.D. (2002). The rapid evolution of reproductive proteins. *Nature Reviews Genetics* **3:** 127–144.

Swenson, W., Arendt, J., and Wilson, D.S. (2000a). Artificial selection of microbial ecosystems for 3- chloroaniline biodegradation. *Environmental Microbiology* **2:** 564–571.

Swenson, W., Wilson, D.S., and Elias R. (2000b). Artificial ecosystem selection. *Proceedings of the National Academy of Sciences of the USA* **97:** 9110–9114.

Taddei, F., Radman, M., Maynard Smith, J., Toupance, B., Gouyon, P.H., and Godelle, B. (1997). Role of mutator alleles in adaptive evolution. *Nature* **387**(6634): 700–702.

Tajima, F. (1989). Statistical method for testing the neutral mutation hypothesis by DNA polymorphism. *Genetics* **123:** 585–595.

Talbert, P.B., Bryson, T.D., and Henikoff, S. (2004). Adaptive evolution of centromere proteins in plants and animals. *Journal of Biology* **3:** 18.

Tantawy, A.O., and Tayel, A.A. (1970). Studies on natural populations of *Drosophila*. X. Effects of disruptive and stabilizing selection on wing length and the correlated response in *Drosophila melanogaster*. *Genetics* **65:** 121–132.

Tawfik, D.S., and Griffiths, A.D. (1998). Man-made cell-like compartments for molecular evolution. *Nature Biotechnology* **16:** 652–656.

Taylor, C.E., and Condra, C. (1980). r- and K-selection in *Drosophila pseudoobscura*. *Evolution* **34:** 1183–1193.

Taylor, C.R., and Weibel, E.R. (1981). Design of the mammalian respiratory system. I. Problem and strategy. *Respiratory Physiology* **44:** 1–10.

Taylor, E.B., and Bentzen, P. (1993). Evidence for multiple origins and sympatric divergence of trophic ecotypes of smelt (*Osmerus*) in northeastern North America. *Evolution* **47:** 813–832.

Taylor, E.B., Boughman, J. W., Groenenboom, M., Schluter, D., Sniatynski, M., and Gow, J.L. (2006). Speciation in reverse: morphological and genetic evidence of the collapse of a three-spined stickleback (*Gasterosteus aculeatus*) species pair. *Molecular Ecology* **15:** 343–355.

Taylor, E.D., and McPhail, J.D. (1999). Evolutionary history of an adaptive radiation in species pairs of threespine sticklebacks (*Gasterosteus*): insights from mitochondrial DNA. *Biological Journal of the Linnean Society* **66:** 271–291.

Taylor, L.R. (1961). Aggregation, variance and the mean. *Nature* **189:** 732–735.

Taylor, L.R., Woiwod, I.P., and Perry, J.N. (1978). The density dependence of spatial behaviour and the rarity of randomness. *Journal of Animal Ecology* **47:** 383–406.

Taylor, L.R., Taylor R.A.J., Woiewood, I.P., and Perry, J.N. (1983). Behavioural dynamics. *Nature* **303:** 801–804.

Taylor, P.A., and Williams, P.J. (1975). Theoretical studies on the coexistence of competing species under continuous-flow conditions. *Canadian Journal of Microbiology* **21:** 90–98.

Tenaillon, O., Toupance, B., Nagard, L.H., Taddei, F., and Godelle, B. (1999). Mutators, population size, adaptive landscape and the adaptation of asexual populations of bacteria. *Genetics* **152:** 485–493.

Tenaillon, O., Taddei, F., Radman, M., and Matic, I. (2001). Second-order selection in bacterial evolution: selection acting on mutation and recombination rates in the course of adaptation. *Research in Microbiology* **152:** 11–16.

Tennessen, J.A. (2005). Molecular evolution of animal antimicrobial peptides: widespread moderate positive selection. *Journal of Evolutionary Biology* **18:** 1387–1394.

Thatcher, J.W., Shaw, J.M., and Dickinson, W.J. (1998). Marginal fitness contributions of nonessential genes in yeast. *Proceedings of the National Academy of Sciences of the USA* **95:** 253–257.

Thoday, J.M. (1959). Effects of disruptive selection. I. Genetic flexibility. *Heredity* **14:** 406–409.

Thoday, J.M., and Gibson, J.B. (1962). Isolation by disruptive selection. *Nature* **193:** 1164–1166.

Thomas, C.M., and Nielsen, K.M. (2005). Mechanisms of, and barriers to, horizontal gene transfer between bacteria. *Nature Reviews Microbiology* **3:** 711–721.

Thomas, H. (2002). Ageing in plants. *Mechanisms of Ageing and Development* **123:** 747–753.

Thompson, C.E., Taylor, E.B., and McPhail, J.D. (1997). Parallel evolution of lake-stream pairs of threespine sticklebacks (*Gasterosteus*) inferred from mitochondrial DNA variation. *Evolution* **51:** 1955–1965.

Thompson, J.N., and Burdon, J.J. (1992). Gene-for-gene coevolution between plants and parasites. *Nature* **360:** 121–125.

Thompson, L.W., and Krawiec, S. (1983). Acquisitive evolution of ribitol dehydrogenase in *Klebsiella pneumoniae*. *Journal of Bacteriology* **154:** 1027–1031.

Thompson, V. (1977). Recombination and response to selection in *Drosophila melanogaster*. *Genetics* **85:** 125–140.

Thornber, C.S., and Gaines, S.D. (2004). Population demographics in species with biphasic life cycles. *Ecology* **85:** 1661–1674.

Tian, D., Araki, H., Stahl, E., Bergelson, J., and Kreitman, M. (2002). Signature of balancing selection in *Arabidopsis*. *Proceedings of the National Academy of Sciences of the USA* **99:** 11525–11530.

Tilman, D. (1977). Resource competition between planktonic algae: An experimental and theoretical approach. *Ecology* **58:** 338–348.

Tilman, D. (1981). Experimental tests of resource competition theory using four species of Lake Michigan algae. *Ecology* **62:** 802–815.

Tindell, D., and Arze, C.G. (1965). Sexual maturity of male chickens selected for mating ability. *Poultry Science* **44:** 70–72.

Tishkoff, S., and Verrelli, B.C. (2003). Patterns of human genetic diversity: implications for human evolutionary history and disease. *Annual Review of Genomics and Human Genetics* **4:** 293–340.

Torkamanzehi, A., Moran, C., and Nicholas, F.W. (1988). P-element-induced mutation and quantitative variation in *Drosophila melanogaster*: Lack of enhanced response to selection in lines derived from dysgenic crosses. *Genetical Research* **51:** 231–238.

Torkamanzehi, A., Moran, C., and Nicholas, F.W. (1992). P element transposition contributes substantial new variation for a quantitative trait in *Drosophila melanogaster*. *Genetics* **131:** 73–78.

Torkelson, J., Harris, R.S., Lombardo, S.-J., Nagendran, J., Thulin, C., and Rosenberg, S.M. (1997). Genome-wide hypermutation in a subpopulation of stationary-phase cells underlies recombination-dependent adaptive mutation. *EMBO Journal* **16:** 3303–3311.

Torsvik, V., Goksoyr, J., and Daae, F.L. (1990). High diversity in DNA of soil bacteria. *Applied and Environmental Microbiology* **56:** 782–787.

Tracy, W.F., Goldman, I.L., Tiefenthaler, A.E., and Schaber, M.A. (2004). Trends in productivity of US crops and long-term selection. *Plant Breeding Reviews* **24:** 89–108.

Travisano, M., Mongold, J.A., Bennett, A.F., and Lenski, R.E. (1995). Experimental tests of the roles of adaptation, chance, and history in evolution. *Science* **267:** 87–90.

Trenbath, B.R. (1974). Biomass productivity of mixtures. *Advances in Agronomy* **26:** 177–210.

Treves, D.S., Manning, S., and Adams, J. (1998). Repeated evolution of an acetate-crossfeeding polymorphism in long-term populations of *Escherichia coli*. *Molecular Biology and Evolution* **15:** 789–797.

Trivers, R.L. (1971). The evolution of reciprocal altruism. *Quarterly Review of Biology* **46:** 35–57.

Trivers, R.L. (1974). Parent-offspring conflict. *American Zoologist* **14:** 249–264.

Turgeon, B.G. (1998). Application of mating type gene technology to problems in fungal biology. *Annual Review of Phytopathology* **36:** 115–137.

Turner, J.R.G. (1979). Genetic control of recombination in the silkworm. I. Multigenetic control of chromosome 2. *Heredity* **43:** 273–293.

Turner, P.E., and Chao, L. (1998). Sex and the evolution of intrahost competition in RNA virus Π6. *Genetics* **150:** 523–532.

Turner, P.E., and Chao, L. (1999). Prisoner's dilemma in an RNA virus. *Nature*. **398:** 441–443.

Turner, P.E., Souza, V., and Lenski, R.E.. (1996). Tests of ecological mechanisms promoting the stable coexistence of two bacterial genotypes. *Ecology* **77:** 2119–2129.

Vacquier, V.D. (1998). Evolution of gamete recognition proteins. *Science* **281:** 1995–1998.

Valero, M., Richerd, S., Perrot, V., and Destombe, C. (1992). Evolution of alternation of haploid and diploid phases in life cycles. *Trends in Ecology and Evolution* **7:** 25–29.

van Delden, W. (1970). Selection for competitive ability. *Drosophila Information Service* **45:** 169.

van Dijken, F.R., and Scharloo, W. (1979a). Divergent selection on locomotor activity in *Drosophila melanogaster*. I. Selection response. *Behavior Genetics* **9:** 543–553.

van Dijken, F.R., and Scharloo, W. (1979b). Divergent selection on locomotor activity in *Drosophila melanogaster*.

II. Test for reproductive isolation between selected lines. *Behavior Genetics* **9:** 555–561.

van Tienderen, P.H., and van der Toorn, J. (1991). Genetic differentiation between populations of *Plantago lanceolata*. II. Phenotypic selection in a transplant experiment in three contrasting habitats. *Journal of Ecology* **79:** 43–59.

Vandel, A. (1938). Recherches sur la sexualité des isopods. III. Le determinisme du sexe et de la monogénie chez *Trichoniscus (Spiloniscus) provisorius* Racovitza. *Bulletin de Biologie* **72:** 147–186.

Vassilieva, L.L., and Lynch, M. (1999). The rate of spontaneous mutation for life-history traits in *Caenorhabditis elegans*. *Genetics* **151:** 119–129.

Vassilieva, L.L., Hook, A.M., and Lynch ,M. (2000). The fitness effects of spontaneous mutations in *Caenorhabditis elegans*. *Evolution* **54**(4): 1234–1246.

Velicer, G .J., & Lenski, R .E . (1999) . Evolutionary trade-offs under conditions of resource abundance and scarcity: experiments with bacteria . *Ecology* **80:** 1168–1179.

Velicer, G.J., Kroos, L., and Lenski, R.E. (1998). Loss of social behaviors by *Myxococcus xanthus* during evolution in an unstructured habitat. *Proceedings of the National Academy of Sciences of the USA* **95:** 12376–12380.

Velicer, G.J., Kroos, L., and Lenski, R.E. (2000). Developmental cheating in the social bacterium *Myxococcus xanthus*. *Nature*. **404:** 598.

Verhulst, S., Perrins, C.M., and Riddington, R. (1997). Natal dispersal of great tits in a patchy environment. *Ecology* **78:** 864–872.

Via, S. (1984). The quantitative genetics of polyphagy in an insect herbivore. I. Genotype-environment interaction in larval performance on different host plant species. *Evolution* **38:** 881–895.

Via, S. (2001). Sympatric speciation in animals: the ugly duckling grows up. *Trends in Ecology and Evolution* **16:** 381–390.

Via, S., and Lande, R. (1985). Genotype-environment interaction and the evolution of phenotypic plasticity. *Evolution* **39:** 505–522.

Vickery, W.L., Giraldeau, L.A., Templeton, J..J, Kramer, D.L., and Chapman, C.A. (1991). Producers, scroungers and group foraging. *American Naturalist* **137:** 847–865.

Vidondo, B., Prairie, Y., Blanco, J.M., and Duarte, C.M. (1997). Some aspects of the analysis of size spectra in aquatic ecology. *Limnology and Oceanography* **42:** 184–192.

Vogel, J., Normand, P., Thioulouse, J., Nesme, X., and Grundmann, G.L. (2003). Relationship between spatial and genetic distance in *Agrobacterium* spp in 1 cubic centimeter of soil. *Applied and Environmental Microbiology* **69:** 1482–1487.

Vogler, D.W., and Kalisz, S. (2001). Sex among the flowers: the distribution of plant mating systems. *Evolution* **55:** 202–204.

Vogul, F., and Motulsky, A.G. (1997). *Human genetics: problems and approaches*. Springer-Verlag, Berlin.

Wade, M.J. (1977). An experimental study of group selection. *Evolution* **31:** 134–153.

Wade, M.J. (1980). Group selection, population growth rate, and competitive ability in the flour beetles, *Tribolium* spp. *Ecology* **61:** 1056–1064.

Wade, M.J. (1982). Group selection: migration and the differentiation of small populations. *Evolution* **36:** 949–961.

Wade, M.J. (1984). Changes in group-related traits that occur when group selection is relaxed. *Evolution* **38:** 1039–1046.

Wade, M.J., and Beeman, R.W. (1994). The population dynamics of maternal-effect selfish genes. *Genetics* **138:** 1309–1314.

Wade, M.J., and Goodnight, C.J. (1991). Wright shifting balance theory—an experimental study. *Science* **253:** 1015–1018.

Wade, M.J., and McCauley, D.E. (1984). Group selection: the interaction of local deme size and migration in the differentiation of small populations. *Evolution* **38:** 1047–1058.

Wahl, L.M., Gerrish, P.J., and Saika-Voivod, I. (2002). Evaluating the impact of population bottlenecks in experimental evolution. *Genetics* **162:** 961–971.

Wallace, B. (1958). The average effect of radiation-induced mutations on viability in *Drosophila melanogaster*. *Evolution* **12:** 532–556.

Wallace, B. (1982). *Drosophila melanogaster* populations selected for resistance to NaCl and $CuSO_4$ in both allopatry and sympatry. *Journal of Heredity* **73:** 35–42.

Wallinga, J.H., and Bakker, H. (1978). Effect of long-term selection for litter size in mice on lifetime reproduction. *Journal of Animal Science* **46:** 1563–1571.

Waloff, N. (1968). Studies on the insect fauna on Scotch Broom *Sarothamnus scoparius* (L.) Wimmer. *Advances in Ecological Research* **5:** 87–208.

Waloff, N., and Richards, O.W. (1977). The effect of insect fauna on the growth mortality and natality of broom, *Sarothamnus scoparius*. *Journal of Applied Ecology* **14:** 787–798.

Walsh, B. (2004). Population- and quantitative-genetic models of selection limits. *Plant Breeding Reviews* **24**(1): 177–225.

Wang, X.Y., and Redmann, R.E. (2005). Adaptation to salinity in *Hordeum jubatum* L. populations studied using reciprocal transplants. *Plant Ecology* **123:** 65–71.

Waples, R.S., and Gaggiotti, O. (2006). What is a population? An empirical evaluation of some genetic methods for identifying the number of gene pools and their degree of connectivity. *Molecular Ecology* **15:** 1419–1439.

Waser, N.M., and Price, M.V. (1985). Reciprocal transplant experiments with *Delphinium nelsonii* (Ranunculaceae): evidence for local adaptation. *American Journal of Botany* **72:** 1726–1732.

Waser, N.M., and Price, M.V. (1989). Optimal outcrossing in *Ipomopsis aggregata*: seed set and offspring fitness. *Evolution* **43:** 1097–1109.

Waser, N.M., and Price, M.V. (1991). Outcrossing distance effects in *Delphinium nelsoni*: Pollen loads, pollen tubes and seed set. *Ecology* **72:** 171–179.

Waterston, R.H., Lindblad-Toh, K., Birney, E. *et al.* (2002). Initial sequencing and comparative analysis of the mouse genome. *Nature* **420:** 520–562.

Waterway, M.J., & Starr, J.R. (2006) Phylogenetic relationships in Tribe Cariceae (Cyperareae) based on nested analyses of four molecular data sets. *Aliso* **23**.

Watterson, G.A. (1978). The homozygosity test of neutrality. *Genetics* **88:** 405–416.

Waxman, D., and Peck, J.R. (1999). Sex and adaptation in a changing environment. *Genetics* **153:** 1041–1053.

Weber, K.E. (1990). Increased selection response in larger populations. I. Selection for wing-tip height in *Drosophila melanogaster* at three population sizes. *Genetics* **125:** 579–584.

Weber, K.E., and Diggins, L.T. (1990). Increased selection response in larger populations. II. Selection for ethanol vapor resistance in *Drosophila melanogaster* at two population sizes. *Genetics* **125:** 585–597.

Webster, J.P., and Woolhouse, M.E.J. (1998). Selection and strain specificity of compatibility between snail intermediate hosts and their parasitic schistosomes. *Evolution* **52:** 1627–1634.

Webster, R.K., Saghai-Maroof, M.A., and Allard, R.W. (1986). Evolutionary response of barley Composite Cross II to *Rhynchosporium secalis* analyzed by pathogenic complexity and by gene-by-race relationships. *Phytopathology* **76:** 661–668.

Wegner, K.M., Kalbe, M., Schaschl, H., and Reusch, T.B.H. (2004). Parasites and individual major histocompatibility complex diversity—an optimal choice? *Microbes and Infection* **6:** 1110–1116.

Weigensberg, I., and Roff, D.A. (1996). Natural heritabilities: can they be reliably estimated in the laboratory? *Evolution* **50:** 2149–2157.

Weisbrot, D.R. (1966). Genotypic interactions among competing strains and species of *Drosophila*. *Genetics* **53:** 427–435.

Weismann, A. (1889). The significance of sexual reproduction in the theory of natural selection. In Poulton, E.B., Schauonland, S. and Shipley, A.E. (eds) *Essays upon heredity*, pp 251–332. Clarendon Press, Oxford.

Weldon, W.F.R. (1901). A first study of natural selection in *Clausilia laminata* (Montagu). *Biometrika* **1:** 109–124.

Weldon, W.F.R. (1904). Note on a race of *Clausilia italica* (von Martens). *Biometrika* **3:** 299–307.

Wells, H. (1979). Self-fertilization: advantageous or deleterious? *Evolution* **33:** 252–255.

Werren, J.H. (1980). Sex ratio adaptations to local mate competition in a parasitic wasp. *Science* **208:** 1157–1159.

Werren, J.H., Nur, U., and Eickbush, D. (1987). An extrachromosomal factor causing loss of paternal chromosomes. *Nature* **327:** 75–76.

West, S.A., Peters, A., and Barton, N.H. (1998). Testing for epistasis between deleterious mutations. *Genetics* **149:** 435–444.

West, S.A., Griffin, A.S., Gardner, A., and Diggle, S.P. (2006). Social evolution theory for microbes. *Nature Reviews Microbiology* **4:** 597–607.

Wettberg, E.J. von, Huber, H., and Schmitt, J. (2005) Interacting effects of microsite quality, plasticity, and dispersal distance from the parental site on fitness in a natural population of *Impatiens capensis*. *Evolutionary Ecology Research* **7:** 531–548.

Whitkus, R. (1988). Experimental hybridizations among chromosome races of *Carex pachystachya* and the related species *C. macloviana* and *C. preslii* (Cyperaceae). *Systematic Botany* **13:** 146–153.

Whitlock, M.C. (2002). Selection, load and inbreeding depression in a large metapopulation. *Genetics* **160:** 1191–1202.

Whitlock, M.C., and McCauley, D.E. (1999). Indirect measures of gene flow and migration: $F_{ST} \neq 1/(4Nm + 1)$. *Heredity* **82:** 117–125.

Whitman, W.B., Coleman, D.C., and Wiebe, W.J. (1998). Prokaryotes: the unseen majority. *Proceedings of the National Academy of Sciences of the USA* **95:** 6578–6583.

Whittam, T.S., Ochman, H., and Selander, R.K. (1983). Multilocus genetic structure of natural populations of *Escherichia coli*. *Proceedings of the National Academy of Sciences of the USA* **80:** 1751–1755.

Whoriskey, F.G., and Fitzgerald, G.J. (1985). The effects of bird predation on an estuarine stickleback (Pisces: Gasterosteidae) community. *Canadian Journal of Zoology* **63:** 301–307.

Wichman, H.A., Badgett, M.R., Scott, L.A., Boulianne, C.M., and Bull, J.J. (1999). Different trajectories of parallel evolution during viral adaptation. *Science* **285:** 422–424.

Wigby, S., and Chapman, T. (2004). Female resistance to male harm evolves in response to manipulation of sexual conflict. *Evolution* **58:** 1028–1037.

Wilke, C.M., and Adams, J. (1992). Fitness effects of Ty transposition in *Saccharomyces cerevisiae*. *Genetics* **131:** 31–42.

Williams, C.K., and Moore, R.J. (1989). Phenotypic adaptation and natural selection in the wild rabbit, *Oryctolagus cuniculus*, in Australia. *Journal of Animal Ecology* **58:** 495–507.

Williams, E.E. (1972). The origin of faunas. Evolution of lizard congeners in a complex island fauna: a trial analysis. *Evolutionary Biology* **6:** 47–89.

Williams, G.C. (1975). *Sex and evolution.* Princeton University Press, Princeton, NJ.

Williams, G.C. (1966). *Adaptation and natural selection: a critique of some current evolutionary thought.* Princeton University Press, Princeton, NJ.

Williams, G.C. (1992a). Gaia, nature worship and biocentric fallacies. *Quarterly Review of Biology* **67:** 479–486.

Williams, G.C. (1992b). *Natural selection. domains, levels, and challenges.* Oxford Series in Ecology and Evolution, Vol. 4. Oxford University Press, Oxford.

Williamson, M. (1988). Relationship of species number to area, distance and other variables. In Myers, A.A. and Giller, P.S. (eds.) *Analytical biogeography*, pp 91–115. Chapman & Hall, London.

Willson, M. (1993). Dispersal mode, seed shadows, and colonization patterns. *Plant Ecology* **107/108:** 261–280.

Wilson, D.S. (1975). A theory of group selection. *Proceedings of the National Academy of Sciences of the USA* **72:** 143–146.

Wilson, D.S. (1980). *The natural selection of populations and communities.* Benjamin/Cummings, Menlo Park, CA.

Wilson, D.S. (1983). The group selection controversy: history and current status. *Annual Reviews of Ecology and Systematics* **14:** 159–187.

Wilson, D.S., & Colwell, R.K. (1981). Evolution of sex ratio in structured demes. *Evolution* **35:** 882–897.

Wilson, D.S., and Szostak, J.W. (1999). In vitro selection of functional nucleic acids. *Annual Review of Biochemistry* **68:** 611–647.

Wilson, S.P. (1974). An experimental comparison of individual, family and combination selection. *Genetics* **76:** 823–836.

Wilson, S.P., Goodale, H.P., Kyle, W.H., and Godfrey, E.F. (1971). Long-term selection for body weight in mice. *Journal of Heredity* **62:** 228–234.

Winzeler, E.A, Shoemaker, D.D., Astromoff, A. *et al.* (1999). Functional characterization of the *Saccharomyces cerevisiae* genome by precise deletion and parallel analysis. *Science* **285:** 901–906.

Wloch, D.M., Szafraniec, K., Borts, R.H., and Korona, R. (2001). Direct estimate of the mutation rate and the distribution of fitness effects in the yeast *Saccharomyces cerevisiae*. *Genetics* **159:** 441–452.

Wolfe, M.S. (1987). Trying to understand and control powdery mildew. In M.S. Wolfe and C.E. Caten (eds.), *Populations of Plant Pathogens,* pp. 253–273. Blackwell, Oxford.

Wolfe, M.S., and Barrett, J.A. (1981). The agricultural value of variety mixtures. *Barley Genetics* **4:** 435–440.

Wooding, F.J. (1981). Performance of three barley cultivars grown in different cropping sequences in central Alaska. *Barley Genetics* **4:** 147–152.

Woodruff, L.L. (1926). Eleven thousand generations of *Paramecium*. *Quarterly Review of Biology* **1:** 436–438.

Wright, S. (1931). Evolution in Mendelian populations. *Genetics* **16:** 97–159.

Wright, S. (1932). The roles of mutation, inbreeding, crossbreeding and selection in evolution. *Proceedings of the 6th International Congress of Genetics* **1:** 356–366.

Wright, S. (1943). Isolation by distance. *Genetics* **28:** 114–138.

Wright, S. (1945). Tempo and mode in evolution: A critical review. *Ecology* **26:** 415–419.

Wright, S. (1946). Isolation by distance under diverse systems of mating. *Genetics* **31:** 39–59.

Wright, S. (1951). The genetical structure of populations. *Annals of Eugenics* **15:** 323–354.

Wright, S.I., and Gaut, B.S. (2004). Molecular population genetics and the search for adaptive evolution in plants. *Molecular Biology and Evolution* **22:** 506–519.

Wu, T.T. (1976). Growth of a mutant of *Escherichia coli* K-12 on xylitol by recruiting enzymes for D-xylose and L-1,2-propanediol metabolism. *Biochimica et Biophysica Acta* **428:** 656–663.

Wu, T.T., Lin, E.C.C., and Tanaka, S. (1968). Mutants of *Aerobacter aerogenes* capable of utilizing xylitol as a novel carbon source. *Journal of Bacteriology* **96:** 447–456.

Wych, R.D., and Rasmusson, D.C. (1983). Genetic improvement in malting barley cultivars since 1920. *Crop Science* **23:** 1037–1040.

Wynne-Edwards, V.C. (1962). *Animal dispersion in relation to social behaviour.* Oliver & Boyd, Edinburgh.

Wynne-Edwards, V.C. (1986). *Evolution through group selection.* Blackwell Scientific Publications, Oxford.

Xu, S.Z. (2003). Estimating polygenic effects using markers of the entire genome. *Genetics* **163:** 789–801.

Yamazaki, T. (1971). Measurement of fitness at the Esterase-5 locus in *Drosophila pseudoobscura*. *Genetics* **67:** 579–603.

Yarborough, K., and Kojima, K. (1967). The model of selection at the polymorphic esterase-6 locus in cage populations of *Drosophila melanogaster*. *Genetics* **57:** 677–686.

Yeaman, M.R., and Yount, N.Y. (2003). Mechanisms of antimicrobial peptide action and resistance. *Pharmacological Reviews* **55:** 27–56.

Yedid, G., and Bell, G. (2001). Microevolution in an electronic microcosm. *American Naturalist*. **157:** 465–487.

Yedid, G., and Bell, G. (2002). Macroevolution simulated with autonomously replicating computer programs. *Nature* **420:** 810–812.

Yoo, B.H. (1980a). Long-term selection for a quantitative character in large replicate populations of *Drosophila melanogaster*. I. Response to selection. *Genetical Research* **35:** 1–17.

Yoo, B.H. (1980b). Long-term selection for a quantitative character in large replicate populations of *Drosophila melanogaster*. II. Lethals and visible mutants with large effects. *Genetical Research* **35:** 19–31.

Yoo, B.H. (1980c). Long-term selection for a quantitative character in large replicate populations of *Drosophila melanogaster*. III. The nature of residual genetic variability. *Theoretical and Applied Genetics* **57:** 25–32.

Yoshida, T., Jones, L.E., Ellner, S.P., Fussmann, G.F., and Hairston, N.G. (2003). Rapid evolution drives ecological dynamics in a predator–prey system. *Nature* **424:** 303–326.

Zahavi, A. (1975). Mate selection—a selection for a handicap. *Journal of Theoretical Biology* **53:** 205–214.

Zambrano, M.M., Siegele, D.A., Almirón, M., Tormo, A., and Kolter, R. (1993). Microbial competition: *Escherichia coli* mutants that take over stationary phase cultures. *Science* **259:** 1757–1760.

Zasloff, M. (2002). Antimicrobial peptides of multicellular organisms. *Nature* **415:** 389–395.

Zeleny, C. (1921). Decrease in sexual dimorphism of bar-eye *Drosophila* during the course of selection for low and high facet number. *American Naturalist* **55:** 404–411.

Zeleny, C. (1922). The effect of selection for eye facet number in the white bar-eye race of *Drosophila melanogaster. Genetics* **7:** 1–115.

Zeleny, C., and Mattoon, E.W. (1915). The effect of selection up on the 'bar-eye' mutant of *Drosophila. Journal of Experimental Zoology* **19:** 515–529.

Zeng, Z.-B., and Hill, W.G. (1986). The selection limit due to the conflict between truncation and stabilizing selection with mutation. *Genetics* **114:** 1313–1328.

Zeyl, C., and Bell, G. (1995). Symbiotic DNA in eukaryote genomes. *Trends in Ecology and Evolution* **11:** 10–15.

Zeyl, C., and Bell, G. (1997). The advantage of sex in evolving yeast populations. *Nature* **388:** 465–468.

Zeyl, C., Bell, G., and da Silva, J. (1994). Transposon abundance in sexual and asexual populations of *Chlamydomonas reinhardtii. Evolution* **48:** 1406–1409.

Zeyl, C., Bell, G., and Green, D.M. (1996). Sex and the spread of retrotransposon Ty3 in experimental populations of *Saccharomyces cerevisiae. Genetics* **143:** 1567–1577.

Zeyl, C., Mizesko, M., and de Visser, J.A.G.M. (2001). Mutational meltdown in laboratory yeast populations. *Evolution* **55**(5): 909–917.

Zeyl, C., Vanderford, T., and Carter, M. (2003). An evolutionary advantage of haploidy in large yeast populations. *Science* **299:** 555–558.

Zuker, M. (2003). *www.bioinfo.rpi.edu/~zukerm/lectures/RNAfold-html.*

Index